Geophysical Monograph Series

Including

IUGG Volumes

Maurice Ewing Volumes

Mineral Physics Volumes

GEOPHYSICAL MONOGRAPH SERIES

Geophysical Monograph Volumes

44 Modeling Magnetospheric Plasma *T. E. Moore and J. H. Waite, Jr. (Eds.)*

45 Perovskite: A Structure of Great Interest to Geophysics and Materials Science *Alexandra Navrotsky and Donald J. Weidner (Eds.)*

46 Structure and Dynamics of Earth's Deep Interior (IUGG Volume 1) *D. E. Smylie and Raymond Hide (Eds.)*

47 Hydrological Regimes and Their Subsurface Thermal Effects (IUGG Volume 2) *Alan E. Beck, Grant Garven, and Lajos Stegena (Eds.)*

48 Origin and Evolution of Sedimentary Basins and Their Energy and Mineral Resources (IUGG Volume 3) *Raymond A. Price (Ed.)*

49 Slow Deformation and Transmission of Stress in the Earth (IUGG Volume 4) *Steven C. Cohen and Petr Vaníček (Eds.)*

50 Deep Structure and Past Kinematics of Accreted Terranes (IUGG Volume 5) *John W. Hillhouse (Ed.)*

51 Properties and Processes of Earth's Lower Crust (IUGG Volume 6) *Robert F. Mereu, Stephan Mueller, and David M. Fountain (Eds.)*

52 Understanding Climate Change (IUGG Volume 7) *Andre L. Berger, Robert E. Dickinson, and J. Kidson (Eds.)*

53 Plasma Waves and Instabilities at Comets and in Magnetospheres *Bruce T. Tsurutani and Hiroshi Oya (Eds.)*

54 Solar System Plasma Physics *J. H. Waite, Jr., J. L. Burch, and R. L. Moore (Eds.)*

55 Aspects of Climate Variability in the Pacific and Western Americas *David H. Peterson (Ed.)*

56 The Brittle-Ductile Transition in Rocks *A. G. Duba, W. B. Durham, J. W. Handin, and H. F. Wang (Eds.)*

57 Evolution of Mid Ocean Ridges (IUGG Volume 8) *John M. Sinton (Ed.)*

58 Physics of Magnetic Flux Ropes *C. T. Russell, E. R. Priest, and L. C. Lee (Eds.)*

59 Variations in Earth Rotation (IUGG Volume 9) *Dennis D. McCarthy and Williams E. Carter (Eds.)*

60 Quo Vadimus Geophysics for the Next Generation (IUGG Volume 10) *George D. Garland and John R. Apel (Eds.)*

61 Cometary Plasma Processes *Alan D. Johnstone (Ed.)*

62 Modeling Magnetospheric Plasma Processes *Gordon R. Wilson (Ed.)*

63 Marine Particles: Analysis and Characterization *David C. Hurd and Derek W. Spencer (Eds.)*

64 Magnetospheric Substorms *Joseph R. Kan, Thomas A. Potemra, Susumu Kokubun, and Takesi Iijima (Eds.)*

65 Explosion Source Phenomenology *Steven R. Taylor, Howard J. Patton, and Paul G. Richards (Eds.)*

66 Venus and Mars: Atmospheres, Ionospheres, and Solar Wind Interactions *Janet G. Luhmann, Mariella Tatrallyay, and Robert O. Pepin (Eds.)*

67 High-Pressure Research: Application to Earth and Planetary Sciences (Mineral Physics Volume 3) *Yasuhiko Syono and Murli H. Manghnani (Eds.)*

68 Microwave Remote Sensing of Sea Ice *Frank Carsey, Roger Barry, Josefino Comiso, D. Andrew Rothrock, Robert Shuchman, W. Terry Tucker, Wilford Weeks, and Dale Winebrenner*

69 Sea Level Changes: Determination and Effects (IUGG Volume 11) *P. L. Woodworth, D. T. Pugh, J. G. DeRonde, R. G. Warrick, and J. Hannah*

70 Synthesis of Results from Scientific Drilling in the Indian Ocean *Robert A. Duncan, David K. Rea, Robert B. Kidd, Ulrich von Rad, and Jeffrey K. Weissel (Eds.)*

71 Mantle Flow and Melt Generation at Mid-Ocean Ridges *Jason Phipps Morgan, Donna K. Blackman, and John M. Sinton (Eds.)*

72 Dynamics of Earth's Deep Interior and Earth Rotation (IUGG Volume 12) *Jean-Louis Le Mouël, D.E. Smylie, and Thomas Herring (Eds.)*

73 Environmental Effects on Spacecraft Positioning and Trajectories (IUGG Volume 13) *A. Vallance Jones (Ed.)*

74 Evolution of the Earth and Planets (IUGG Volume 14) *E. Takahashi, Raymond Jeanloz, and David Rubie (Eds.)*

75 Interactions Between Global Climate Subsystems: The Legacy of Hann (IUGG Volume 15) *G. A. McBean and M. Hantel (Eds.)*

76 Relating Geophysical Structures and Processes: The Jeffreys Volume (IUGG Volume 16) *K. Aki and R. Dmowska (Eds.)*

77 The Mesozoic Pacific: Geology, Tectonics, and Volcanism—A Volume in Memory of Sy Schlanger *Malcolm S. Pringle, William W. Sager, William V. Sliter, and Seth Stein (Eds.)*

78 Climate Change in Continental Isotopic Records *P. K. Swart, K. C. Lohmann, J. McKenzie, and S. Savin (Eds.)*

79 The Tornado: Its Structure, Dynamics, Prediction, and Hazards *C. Church, D. Burgess, C. Doswell, R. Davies-Jones (Eds.)*

80 Auroral Plasma Dynamics *R. L. Lysak (Ed.)*

81 Solar Wind Sources of Magnetospheric Ultra-Low Frequency Waves *M. J. Engebretson, K. Takahashi, and M. Scholer (Eds.)*

82 **Gravimetry and Space Techniques Applied to Geodynamics and Ocean Dynamics** *Bob E. Schutz, Allen Anderson, Claude Froidevaux, and Michael Parke (Eds.)*

83 **Nonlinear Dynamics and Predictability of Geophysical Phenomena** *William I. Newman, Andrei Gabrielov, and Donald L. Turcotte (Eds.)*

84 **Solar System Plasmas in Space and Time** *J. Burch, J. H. Waite, Jr. (Eds.)*

85 **The Polar Oceans and Their Role in Shaping the Global Environment** *O. M. Johannessen, R. D. Muench, and J. E. Overland (Eds.)*

Maurice Ewing Volumes

1 **Island Arcs, Deep Sea Trenches, and Back-Arc Basins** *Manik Talwani and Walter C. Pitman III (Eds.)*

2 **Deep Drilling Results in the Atlantic Ocean: Ocean Crust** *Manik Talwani, Christopher G. Harrison, and Dennis E. Hayes (Eds.)*

3 **Deep Drilling Results in the Atlantic Ocean: Continental Margins and Paleoenvironment** *Manik Talwani, William Hay, and William B. F. Ryan (Eds.)*

4 **Earthquake Prediction—An International Review** *David W. Simpson and Paul G. Richards (Eds.)*

5 **Climate Processes and Climate Sensitivity** *James E. Hansen and Taro Takahashi (Eds.)*

6 **Earthquake Source Mechanics** *Shamita Das, John Boatwright, and Christopher H. Scholz (Eds.)*

IUGG Volumes

1 **Structure and Dynamics of Earth's Deep Interior** *D. E. Smylie and Raymond Hide (Eds.)*

2 **Hydrological Regimes and Their Subsurface Thermal Effects** *Alan E. Beck, Grant Garven, and Lajos Stegena (Eds.)*

3 **Origin and Evolution of Sedimentary Basins and Their Energy and Mineral Resources** *Raymond A. Price (Ed.)*

4 **Slow Deformation and Transmission of Stress in the Earth** *Steven C. Cohen and Petr Vaníček (Eds.)*

5 **Deep Structure and Past Kinematics of Accreted Terrances** *John W. Hillhouse (Ed.)*

6 **Properties and Processes of Earth's Lower Crust** *Robert F. Mereu, Stephan Mueller, and David M. Fountain (Eds.)*

7 **Understanding Climate Change** *Andre L. Berger, Robert E. Dickinson, and J. Kidson (Eds.)*

8 **Evolution of Mid Ocean Ridges** *John M. Sinton (Ed.)*

9 **Variations in Earth Rotation** *Dennis D. McCarthy and William E. Carter (Eds.)*

10 **Quo Vadimus Geophysics for the Next Generation** *George D. Garland and John R. Apel (Eds.)*

11 **Sea Level Changes: Determinations and Effects** *Philip L. Woodworth, David T. Pugh, John G. DeRonde, Richard G. Warrick, and John Hannah (Eds.)*

12 **Dynamics of Earth's Deep Interior and Earth Rotation** *Jean-Louis Le Mouël, D.E. Smylie, and Thomas Herring (Eds.)*

13 **Environmental Effects on Spacecraft Positioning and Trajectories** *A. Vallance Jones (Ed.)*

14 **Evolution of the Earth and Planets** *E. Takahashi, Raymond Jeanloz, and David Rubie (Eds.)*

15 **Interactions Between Global Climate Subsystems: The Legacy of Hann** *G. A. McBean and M. Hantel (Eds.)*

16 **Relating Geophysical Structures and Processes: The Jeffreys Volume** *K. Aki and R. Dmowska (Eds.)*

17 **Gravimetry and Space Techniques Applied to Geodynamics and Ocean Dynamics** *Bob E. Schutz, Allen Anderson, Claude Froidevaux, and Michael Parke (Eds.)*

18 **Nonlinear Dynamics and Predictability of Geophysical Phenomena** *William I. Newman, Andrei Gabrielov, and Donald L. Turcotte (Eds.)*

Mineral Physics Volumes

1 **Point Defects in Minerals** *Robert N. Schock (Ed.)*

2 **High Pressure Research in Mineral Physics** *Murli H. Manghnani and Yasuhiko Syona (Eds.)*

3 **High Pressure Research: Application to Earth and Planetary Sciences** *Yasuhiko Syono and Murli H. Manghnani (Eds.)*

Geophysical Monograph 86

Space Plasmas: Coupling Between Small and Medium Scale Processes

Maha Ashour-Abdalla
Tom Chang
Paul Dusenbery
Editors

American Geophysical Union

Published under the aegis of the AGU Books Board.

Library of Congress Cataloging-in-Publication Data

Space plasmas : coupling between small and medium scale processes / Maha Ashour-Abdalla, Tom Chang, Paul Dusenbery, editors.
p. cm. — (Geophysical monograph, ISSN 0065-8448 ; 86)
Includes bibliographical references (p.).
ISBN 0-87590-043-7 (alk. paper)
1. Space plasmas. 2. Geophysics. 3. Astrophysics.
I. Series.
QC809.P5S68 1995
538'.36—dc20 95-3752
CIP

ISSN 0065-8448

ISBN 0-87590-043-7

This book is printed on acid-free paper. ♾

Printed in the United States of America.

CONTENTS

CONTENTS

CONTENTS

PREFACE

This volume addresses the issue of achieving agreement between space observations and that body of plasma theory that attempts to account for the interrelations of plasma behavior and electromagnetic inhomogeneities and fluctuations. It is difficult to assign a single name to this area of research, because most of the terms that might be used have specific and limiting technical connotations. However, though a single name has not yet been found, we can say that the space plasma phenomena considered in this volume may all be characterized as fluctuating, nonequilibrium, or turbulent, and that the processes which play important roles in determining the plasma behavior are stochastic, quasi-linear, nonlinear, inhomogeneous, or nonlocal.

An understanding of these processes is of vital importance to this research discipline. One of the central themes articulated by the NASA Theory Panel (1990) was the necessity that future theory efforts bridge the gap between medium- and small-scale phenomena. Bridging this gap is also of paramount importance to the NSF GEM (Geospace Environmental Modeling), cross-scale, and global coherence programs. It is important that we move toward achieving a consensus between observations and theory: general theories must be tested with observations, and observational data must be analyzed to evaluate specific theoretical predictions.

In recognition of the importance of this area of space research, an AGU Chapman Conference "Micro and Meso Scale Phenomena in Space Plasmas" was held to bring together the international community of experimentalists and theoreticians engaged in such studies in order to bring their collective efforts to bear on addressing theory-data closure. The program committee that worked towards this end consisted of Maha Ashour-Abdalla, Jean Berchem, Jim Burch, Robert Carovillano, Tom Chang, Geoffrey Crew, Paul Dusenbery, and Robert Strangeway. This volume presents a cross-section of the papers presented at the conference and addresses such diverse subjects as solar processes, shocks, the magnetopause, the low-latitude boundary layer, magnetosphere-ionosphere coupling, waves, stochasticity, magnetotail dynamics, reconnection, auroral phenomena, transport theory, cusp phenomena, particle acceleration, current sheet, and wave-particle interactions.

Bill Lewis contributed an enormous amount of his valuable time serving as associate editor of this volume. We are indebted to Mary Terhune and Amy Schubert for their assistance with the conference and the volume. We thank the panel members, Jim Burch, Robert Carovillano, Tom Chang, Paul Dusenbery, Don Gurnett, Jack Jasperse, Goetz Paschmann, Bengt Sonnerup, and Lev Zelenyi, whose efforts contributed significantly to the success of the conference. We are grateful to the U.S. National Science Foundation and the Geophysics Directorate of the Phillips Laboratory for their financial support.

Maha Ashour-Abdalla
University of California at Los Angeles
Tom Chang
Massachusetts Institute of Technology, Cambridge
Paul Dusenbery
University of Colorado, Boulder
Editors

The Evolution of Magnetic Helicity in Compressible Magnetohydrodynamics with a Mean Magnetic Field

Sanjoy Ghosh

Applied Research Corporation, Landover, Maryland

W. Troy Stribling

National Research Council, NASA, Goddard Space Flight Center, Greenbelt, Maryland

Melvyn L. Goldstein

Laboratory for Extraterrestrial Physics, NASA, Goddard Space Flight Center, Greenbelt, Maryland

William H. Matthaeus

Bartol Research Institute, University of Delaware, Newark, Delaware

The solar wind is often modeled as an incompressible magnetofluid. Such a description has been substantiated for two-dimensional (2-D) geometries through compressible magnetohydrodynamic (MHD) simulations which show that many features of incompressible 2-D MHD turbulent relaxation are preserved at subsonic, but non-negligible Mach numbers. However, analogous studies do not exist for three-dimensional (3-D) dynamics, and in particular, it is not known whether significant changes occur in the evolution of the magnetic helicity in going from the purely incompressible model to a low Mach number compressible model. Here we use a two and half dimensional (velocity and magnetic components in all three spatial directions, but functions of only two spatial variables) compressible MHD code to investigate the evolution of magnetic helicity in the presence of a mean magnetic field. The system is closed by assuming a polytropic equation of state. We consider the global and spectral features of magnetic helicity evolution, and discuss our results relative to incompressible 3-D decay simulations as well as to observations.

1. INTRODUCTION

Many features of the solar wind appear to be those of a turbulent magnetohydrodynamic (MHD) medium. This has prompted space physicists to study MHD turbulence using numerical simulations in an effort to better understand the inherent physics. Influenced largely by the limitations of computational resources as well as a reflection of increasing theoretical understanding, MHD turbulence simulations have ranged from incompressible to ideal gas models, two-dimensions (2-D) to three-dimensions (3-D) models, and homogeneous to models containing an ambient magnetic field. Surprisingly, many solar wind features seem to be explainable using one of the simplest MHD models: the incompressible homogeneous 2-D system. The reasons why this is true is a matter of ongoing research [*Zank and Matthaeus,* 1992] .

An important aspect of MHD turbulent relaxation is the preferential decay of certain ideal invariants. For incompressible homogeneous 3-D geometries, these global invariants of the ideal (dissipation-free) system include the total energy (E), the cross helicity (H_c), and the

Space Plasmas: Coupling Between Small
and Medium Scale Processes
Geophysical Monograph 86

magnetic helicity (H_m). H_c is also an invariant of the 2-D system and has been shown through numerous 2-D simulations to have a dynamical behavior that may explain certain highly Alfvénic states in the solar wind. The purely 3-D quantity, $H_m = \frac{1}{2} < \mathbf{A} \cdot \mathbf{B} >$, where the brackets denote a spatial average, is a correlation measure between the vector potential and the fluctuating magnetic field. When an ambient magnetic field is present a new invariant,

$$\hat{H}_m = H_m + \mathbf{A}_0 \cdot \mathbf{B}_0, \qquad (1)$$

emerges in place of H_m [*Matthaeus and Goldstein,* 1982]. $\hat{H}_m$ is a less well studied quantity. In fact, the work of *Stribling and Matthaeus* [1992] is perhaps the only study of its turbulence characteristics to date.

In this paper we compare the dynamics of H_m and $\hat{H}_m$ as obtained by *Stribling and Matthaeus* [1992] for an incompressible 3-D system to those of a compressible system. In addition, we discuss the wavenumber distribution of normalized magnetic helicity from our simulations relative to observations. We employ a restricted 3-D formulation, the so-called two and half dimensional ($2\frac{1}{2}$-D) model. Here all spatial vectors have three components, but each component is a function of only two spatial directions, (x, y). This model greatly reduces computational times over a full 3-D compressible model while preserving nonlinear couplings in the parallel direction and at least one perpendicular direction relative to the ambient magnetic field. Compressibility is incorporated in the form of a polytropic equation of state.

Our strategy is to introduce a random initial spectrum of fixed magnetic helicity and study the evolution of H_m during the ensuing turbulent dynamics under a variety of conditions which include the presence of dissipation or lack thereof and the presence or absence of an ambient magnetic field. Our goals are two-fold. First, we seek to determine whether weak compressibility (low sonic Mach number) causes significant changes in magnetic helicity dynamics over the incompressible 3-D system. Second, we want to know how well H_m measurements in the solar wind can be characterized by the behavior of H_m in our polytropic $2\frac{1}{2}$-D MHD model.

2. NUMERICAL METHOD AND INITIAL CONDITIONS

Our simulations are based on a standard dimensionless representation of the compressible MHD system [*Ghosh and Matthaeus,* 1990]. The polytropic $2\frac{1}{2}$-D MHD system of continuity, momentum, and magnetic induction equations is as follows.

$$\frac{\partial}{\partial t}\rho = -\nabla \cdot (\rho \mathbf{u}), \qquad (2a)$$

$$\frac{\partial}{\partial t}\mathbf{u} = \mathbf{u} \times \tilde{\omega} - \frac{1}{2}\nabla[\mathbf{u} \cdot \mathbf{u} + \frac{2}{M_{s0}^2(\gamma - 1)}\rho^{\gamma-1}] + \frac{\mathbf{J} \times \mathbf{B}}{M_{a0}^2 \rho}$$
$$+\frac{1}{\rho}\nu_c \nabla^2 \mathbf{u} + \frac{1}{\rho}(\zeta_c + \frac{1}{3}\nu_c)\nabla(\nabla \cdot \mathbf{u}), \qquad (2b)$$

$$\frac{\partial}{\partial t}\mathbf{A} = \mathbf{u} \times \mathbf{B} + \mu_c \mathbf{J} + \nabla F. \qquad (2c)$$

Here the vorticity, $\tilde{\omega}$, is derived from the velocity field by $\tilde{\omega} = \nabla \times \mathbf{u}$. The magnetic field, $\mathbf{B}$, is related to the current, $\mathbf{J}$, and the vector potential, $\mathbf{A}$, through $\mathbf{J} = \nabla \times \mathbf{B}$ and $\nabla \times \mathbf{A} = \mathbf{B}$, respectively. The system of equations is closed assuming a polytropic relation between pressure and density, $P = \rho^\gamma$, where we have chosen $\gamma = 5/3$. The function F in (2c) is chosen to preserve the Coulomb gauge condition, $\nabla \cdot \mathbf{A} = 0$.

Constants present in (2) include a characteristic sonic Mach number, M_{s0}, and a characteristic Alfvénic Mach number, M_{a0}. The last two terms in (2b) represent a standard model of viscous dissipation, involving viscosity coefficients ζ_c and ν_c. The dimensionless dynamic viscosity, ν_c, and resistivity, μ_c, are related to the mechanical (R) and magnetic (R_m) Reynolds numbers, respectively, by $R = u_{rms}/\nu_c$ and $R_m = u_{rms}/\mu_c$ where $u_{rms} = [< |\mathbf{u}|^2 >]^{1/2}$ is the root-mean-square value of the velocity.

The simulations were performed on the Cray Y-MP at NASA-Goddard using a pseudospectral algorithm. The grid resolution is 64^2. This implies a wavenumber range of $0 \leq |k_i| \leq 32$ for each component, $i = x, y$.

The initial condition is a flat spectrum of random kinetic and magnetic fluctuations between $1 \leq |\mathbf{k}| \leq 5$. The total energy is unity and is equipartitioned between kinetic and magnetic fluctuations. The cross helicity is low, $H_c = 0.001$, signifying that the velocity and magnetic fields are essentially uncorrelated. The magnetic helicity is fixed at $H_m = 0.1$. The sonic Mach number is set to $M_{s0} = 0.3$ while the Alfvénic Mach number is $M_{a0} = 1.0$. The density is $\rho = 1.0 + \delta\rho$ where $\delta\rho$ is a pseudosound correction [*Ghosh and Matthaeus,* 1990] that minimizes the influence of acoustic waves in the initial state and maintains a nearly incompressible ordering. We set the Reynolds numbers to $R = R_m = 250$ for the dissipative runs, and the ambient magnetic field is $\mathbf{B}_0 = 1.0\hat{x}$ when present. These are basically the same initial conditions used by *Stribling and Matthaeus* [1992] for their incompressible 3-D simulations after neglecting fluctuations along the $\hat{z}$ direction and including the pseudosound correction in the density.

3. SIMULATION RESULTS

Four simulations were run using the initial condition described above. All the simulations were run to approximately ten eddy turnover times, $T_E \approx 10$. The first run, Run (a), contains an ambient magnetic field and dissipation. The second run, Run (b), contain no ambient magnetic field, but retains dissipation. The third run, Run (c), contains an ambient magnetic field, but is ideal (non-dissipative). Finally, the last run, Run (d), is both ideal and lacks an ambient magnetic field.

As expected, H_m is constant in Run (d), the ideal, homogeneous case. However, the mere introduction of dissipation as in Run (b) does not lead to a significant decrease in H_m. We find that H_m decays monotonically to about sixty percent of the initial value after $T_E \approx 7$ in Run (b). On the other hand, the presence of an ambient magnetic field, as in Runs (a) and (c), causes substantial decreases in H_m. In both these cases, H_m decays abruptly to about twenty percent of the initial value by $T_E \approx 0.5$. Eventually, less than ten percent of the initial H_m is remains after $T_E \approx 10$. The time history of H_m from Runs (a), (b) and (c) are shown in Figure 1. It is clear that for the adopted values of R_m and $\mathbf{B}_0$, the presence of an ambient magnetic field is a much stronger influence on the decay of H_m than the presence of dissipation. This result is consistent with Fig. 4 of *Stribling and Matthaeus* [1992] and implies that weak compressibility ($M_{s0} = 0.3$) does not substantially alter the dynamics of H_m from the incompressible 3-D model for the selected initial state.

Although H_m decays rapidly from its initial value, the change in $\hat{H}_m$ is much more gradual in our simulations. $\hat{H}_m$ retains more than eighty percent of its original value at $T_E \approx 10$ in Run (a). As is evident from (1), this results in a rapid buildup of $\mathbf{A}_0$. The growth of $\mathbf{A}_0$ is associated with a mean electric field pulse, $\vec{E}$, which can be derived by taking the spatial average of (2c):

$$\vec{E} = -<\mathbf{u} \times \mathbf{B}> = -\frac{\partial}{\partial t}\mathbf{A}_0.$$

The time histories of H_m, $\hat{H}_m$, and $\vec{E}$ from Run (2) are shown in Figure 2. Again, the similarity of our compressible $2\frac{1}{2}$-D result to the incompressible 3-D model [*Stribling and Matthaeus,* 1992] is clear.

The spectral distribution of magnetic helicity in the solar wind has been considered by *Goldstein, Roberts and Fitch* [1991, 1992] in studying the wavenumber distribution of the normalized reduced magnetic helicity. The normalized magnetic helicity distribution can be studied from our compressible $2\frac{1}{2}$-D simulations as well. We define the normalized magnetic helicity as

$$\sigma_m(k) = kH_m(k)/E_B(k)$$

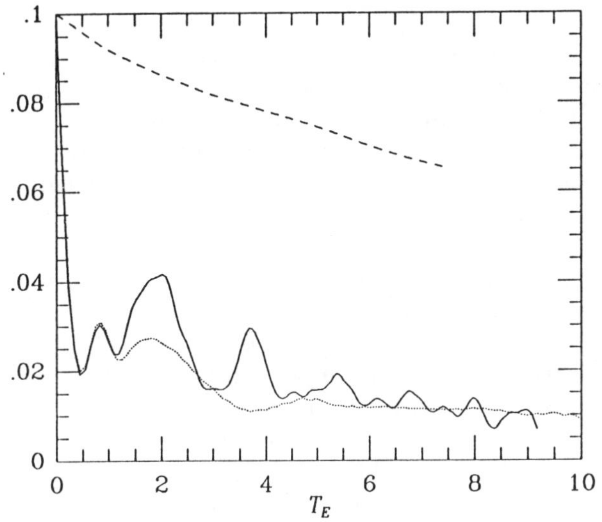

Fig. 1. Time history of H_m from a simulation with dissipation and an ambient magnetic field, Run (a) (solid line), a simulation with dissipation and no ambient magnetic field, Run (b) (dashed line), and a simulation with a ambient magnetic field but no dissipation, Run (c) (dotted line).

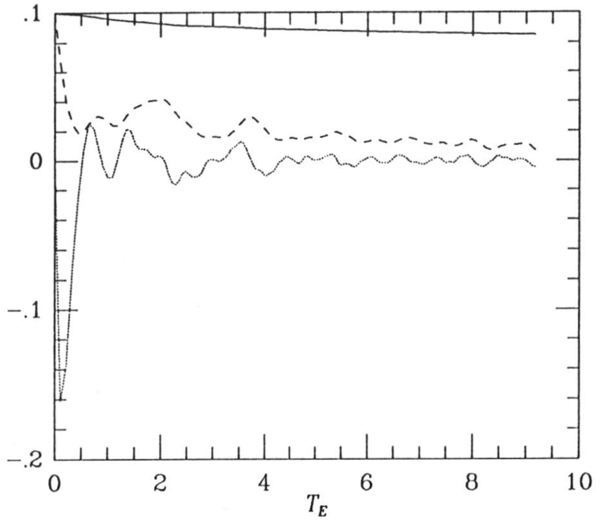

Fig. 2. Time histories of $\hat{H}_m$ (solid line), H_m (dashed line), and $\vec{E}$ (dotted line) from Run (a), the simulation with dissipation and an ambient magnetic field.

where $E_B(k)$ is the magnetic power spectrum. The distribution of $\sigma_m(k)$ over three time intervals from Run (a) is shown in Figure 3. The time intervals are $0 \leq T_E \leq 3$ (solid line), $3 \leq T_E \leq 6$ (dashed line), and $6 \leq T_E \leq 9$ (dotted line). For these simulations the initial state would appear as a delta function at $\sigma_m(k) = 0.1$. Hence, it is interesting that the distribution covers the entire range of $\sigma_m(k)$ after only a few eddy turnover times. The $\sigma_m(k)$ distribution is biased towards positive values during the early time interval. The biase changes and lists towards negative values during the intermediate times. Finally, a peaked distribution, centered on zero, emerges during the late time interval. A similar change from positive to negative biase with increasing heliocentric distance, followed by a distribution centered on zero, is seen in the normalized reduced helicity distribution of certain solar wind data [*Goldstein et al.,* 1992].

This result is not necessarily in contradiction with other studies which indicate the persistence of H_m at large heliocentric distances [*Bieber et al.,* 1987; *Bieber and Smith,* 1992]. In the presence of a Kolmogorov-like magnetic energy spectrum as in the solar wind, H_m will reflect the magnetic helicity of the large-scale energy-containing eddies while $\sigma_m(k)$ will tend to emphasize the magnetic helicity of the turbulent inertial- and dissipation-range eddies. It is possible for H_m to retain a non-zero value due to the same sign of magnetic helicity among the large-scale eddies while the $\sigma_m(k)$ distribution is still centered approximately on zero.

In contrast to the above, the normalized helicity distribution from a simulation where no ambient magnetic field is present shows a somewhat different evolution. The $\sigma_m(k)$ distribution between the time intervals $0 \leq T_E \leq 3$ (solid line) and $3 \leq T_E \leq 8$ (dotted line) from Run (b) is shown in Figure 4. Here the distribution remains biased toward positive values at late times during the simulation.

We do not attempt a more rigorous comparison with observations here. Nevertheless, a similarity between Run (a) and the outwardly expanding solar wind does appear to exist. One difficulty with establishing a closer comparison is that our $\sigma_m(k)$ has not been derived from reduced spectra as in the observations. The lack of multi-spacecraft observations means that spectra derived from observations are typically reduced one-dimensional functions in the direction of the outwardly expanding solar wind. A careful study would require establishing an angle, θ, between the direction of the solar wind flow and the direction of the mean magnetic field. One could then return to the simulations, select this direction θ relative to the $\mathbf{B}_0$-direction, and calculate the equivalent one-dimensional spectra of $H_m(k_\theta)$ and $E_B(k_\theta)$ for comparison with the observations. We defer this detailed study to a future paper.

4. DISCUSSION

We have studied the time evolution of H_m and $\hat{H}_m$

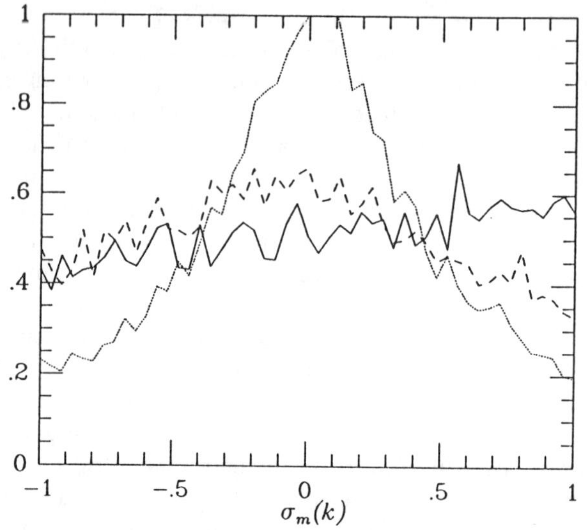

Fig. 3. The distribution of $\sigma_m(k)$ in the time intervals: $0 \leq T_E \leq 3$ (solid line), $3 \leq T_E \leq 6$ (dashed line), and $6 \leq T_E \leq 9$ (dotted line), from Run (a), the simulation with dissipation and an ambient magnetic field.

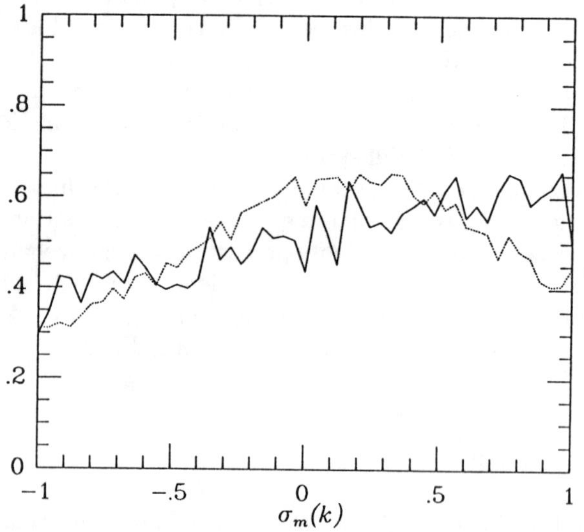

Fig. 4. The distribution of $\sigma_m(k)$ in the time intervals: $0 \leq T_E \leq 3$ (solid line) and $3 \leq T_E \leq 8$ (dotted line) for Run (b), the simulation with dissipation and no ambient magnetic field.

and have found virtually no difference in their behavior between an incompressible 3-D model presented elsewhere [*Stribling and Matthaeus,* 1992] and the compressible $2\frac{1}{2}$-D model presented here. For the chosen set of parameters, H_m decays rapidly relative to $\hat{H}_m$ whenever an ambient magnetic field is present. In addition, a strong electric field pulse is associated with the initial decay of H_m.

Our study of the $\sigma_m(k)$ distribution indicates that normalized helicity distributions in the solar wind evolve more in accord with turbulent MHD where an ambient magnetic field is present than with turbulent dynamics where it is absent.

Acknowledgments. This research has been supported by NASA through the Solar Terrestrial Theory Programs at the Goddard Space Flight Center and at the Bartol Research Institute.

REFERENCES

Bieber, J.W., P. Evenson and W.H. Matthaeus, Magnetic Helicity of the IMF and the Solar Modulation of Cosmic Rays, *Geophys. Res. Lett., 14,* 864, 1987.

Bieber, J.W., and C.W. Smith, Magnetic Helicity in the Solar Wind: Radial Variation and Cosmic Ray Effects, *EOS, 73,* 247, 1992.

Ghosh, S., and W.H. Matthaeus, Relaxation processes in a turbulent compressible magnetofluid, *Phys. Fluids B, 2,* 1520, 1990.

Goldstein, M.L., D.A. Roberts, and C.A. Fitch, The structure of helical interplanetary magnetic fields, *Geophys. Res. Lettr., 18,* 1505, 1991.

Goldstein, M.L., D.A. Roberts, and C.A. Fitch, Observed properties of helical interplanetary magnetic fields, (this publication).

Matthaeus, W.H., and M.L. Goldstein, Measurement of the rugged invariants of magnetohydrodynamic turbulence, *J. Geophys. Res., 87,* 6011, 1982.

Stribling, T., and W.H. Matthaeus, Decay of magnetic helicity in ideal magnetohydrodynamics with a DC magnetic field, (this publication).

Zank, G.P., and W.H. Matthaeus, Waves and turbulence in the solar wind, *J. Geophys. Res., 97,* 17189, 1992.

S. Ghosh, W.T. Stribling and M.L. Goldstein, Code 692, NASA - Goddard, Greenbelt, MD 20771.

W.H. Matthaeus, Bartol Research Institute, Univ. of Delaware, DE 19716.

Magnetohydrodynamic Turbulence and its Relationship to Interplanetary Magnetic Fluctuations

Melvyn L. Goldstein

Laboratory for Extraterrestrial Physics, NASA Goddard Space Flight Center, Greenbelt, Maryland

During the past decade, we have initiated a synergy of data analysis, analytic theory, and numerical simulations to investigate the extent to which the solar wind is a turbulent magnetofluid. This paper reviews the present status of that work. From Helios and Voyager data a clear evolution in the properties of the low frequency magnetic and velocity fluctuations is evident. In the inner heliosphere and in undisturbed flows, the interplanetary fluctuations have preserved strong signatures of coronal processes, while in the outer heliosphere and near regions of strong shear in velocity, the interplanetary medium appears to have been stirred with the concomitant *in situ* generation of turbulence. Using solutions of the equations describing magnetohydrodynamics (MHD), we have demonstrated how shear-driven turbulence produces many aspects of the evolution of the cross-helicity at small scales. Large-scale shear can nonlinearly produce a cascade to smaller scale fluctuations even when the linear Kelvin-Helmholtz mode is stable. A (roughly) power-law inertial range is established by this process. Simulations with Alfvénic fluctuations at high wave numbers, both with and without velocity shear layers, indicate that it is the low cross-helicity at low wave numbers that is critical to the cross-helicity evolution, rather than the geometry of the flow or the dominance of kinetic energy at large scales. Fluctuations produced by velocity shear effects evolve more slowly in the presence of a larger mean field and are anisotropic with a preferred direction of spectral transfer perpendicular to the mean field. MHD simulations with velocity shear on either side of a central current sheet show that even in the absence of compressibility, the absence of a mean magnetic field parallel to the current sheet leads to rapid nonlinear evolution and the observed characteristics of "Elsässer spectra" of the fields in the inner heliosphere.

PHILOSOPHY OF MODELS

When trying to model a complex inhomogeneous system such as the solar wind, the questions one is trying to answer determine, to a large extent, the approach taken. For example, to describe the overall global aspects of the solar wind flow, perhaps including its interaction with planets and comets, one might construct a model in which the complete set of equations describing the physical system and its boundaries are solved. Available computer resources will necessarily limit the resolution and accuracy of the solutions. One may also need to parametrize subgridscale phenomena and forfeit any realistic description of small scale fluctuations.

An alternative approach is to approximate the complete set of equations with a simplified set of model equations that contain essential elements of the physics, but which are still sufficiently simple to permit accurate solution. Particle-in-cell simulations that describe various plasma phenomena are one example. This approach is the one we have taken, but we use a fluid viewpoint to describe phenomena smaller than the global scale of Astronomical Units (AU), but larger than the kinetic scale of wave-particle interactions. One can thereby gain an accurate description of turbulence and its temporal evolution.

Because our goal has been to examine the nature of MHD turbulence and its relationship to solar wind evolu-

Space Plasmas: Coupling Between Small and Medium Scale Processes
Geophysical Monograph 86

tion and structure, our numerical models have had to describe small-scales accurately, *i.e.* we wish to model fluctuations in the inertial range of the turbulence, which at 1 AU means length scales between 10^6 km (~1/100th of an AU), the correlation length of the fluctuations [*Matthaeus and Goldstein,* 1982] and about 400 km, where kinetic effects control the dissipation of the fluctuations. At present, accurate modeling of these nearly three decades of scale sizes is not possible in three dimensions (a little more than two decades is the present practical limit). Because the power spectra of turbulent fluid fluctuations fall rapidly with frequency or wave number (in three dimensions the spectral index is approximately –5/3), very accurate algorithms are required to simulate the inertial range of the fluctuation power spectrum. Furthermore, because the primary diagnostics of (incompressible and homogeneous) MHD are the global invariants of the equations, including, in three dimensions, energy, cross helicity and magnetic helicity, very accurate numerical techniques, with no numerical dissipation, are required. (An invariant is global, or rugged, if it is preserved after truncation of the spectral representation of the equations.) The excitation of turbulence can produce both direct and inverse cascades of these invariants and accurate description of the resulting power spectra also requires using numerical methods with minimal numerical viscosity and resistivity. Spectral techniques are a natural choice for achieving the required numerical accuracy albeit with the disadvantage that the boundary conditions are usually idealized.

Additional idealizations and approximations are often necessary, including restriction to less than three dimensions and the use of an incompressible rather than a compressible model of the magnetofluid. Within these restrictions, the models we have developed have shown that the solar wind is a dynamically evolving turbulent magnetofluid that can be surprisingly well characterized in terms of nearly incompressible MHD processes [*Roberts and Goldstein,* 1991]. The principal sources of the evolution of interplanetary turbulence are the radial expansion of the wind, which reduces the amplitudes of all quantities, and velocity shears between and within solar wind streams. This view of the nature of solar wind fluctuations provides a framework for understanding other aspects of the interplanetary medium [*Roberts et al.*, 1991,1992]. Simulations have also been useful for understanding better the observations that the magnetic helicity of the interplanetary field has a strong tendency toward randomness [*Ghosh et al.*, 1994; *Goldstein et al.*, 1991,1993,1994; *Stribling and Matthaeus,* 1994; *Stribling et al.*, 1992a,b]. Related work supports the idea [*Burlaga*, 1990] that the solar wind contains a vortex street in the outer heliosphere at solar minimum [*Siregar et al.,* 1992a,b,c,d].

Analytic theory and simulations have also be used to understand the stability properties of Alfvén waves [*Ghosh et al.*, 1992; *Viñas and Goldstein*, 1991a,b]. These and related studies have led us to conclude that the Alfvén wave flux near the Sun is insufficient to accelerate the solar wind [*Roberts*, 1989], contrary to a commonly held belief. In the next section, we review briefly the interplanetary data that motivates this project. A description of the numerical models developed and a comparison of those simulations with the observations follows.

BACKGROUND

Coleman [1968] first suggested that the solar wind was a turbulent evolving medium in which turbulence was driven by stream-shear instabilities. The primary motivation for this idea was the observation that the slope of the power spectrum of magnetic and velocity fluctuations was usually close to the values expected for fluid turbulence [*Kolmogoroff*, 1941] or magnetofluid turbulence [*Kraichnan*, 1965]. Coleman also noted the possibility that dissipation of the turbulence at high wave number could account for the anomalously high proton temperatures observed in the solar wind at 1 AU. This viewpoint initially suffered from two major shortcomings: First, as argued by *Belcher and Davis* [1971] (also see, *Unti and Neugebauer* [1968]), it is difficult to see how stream-shear instabilities can produce the preponderance of waves observed traveling outward from the Sun. Second, the Kelvin-Helmholtz instability in the solar wind was thought to be nearly or completely damped [*Parker*, 1964]. In addition, *Bavassano et al.* [1978] had argued that even if excited, the instability could not produce fluctuations on sufficiently large scales.

Other mechanisms for generating turbulence by *in situ* mechanisms involving, for example, kinetic instabilities, have generally been discarded because of their failure to generate outward traveling waves and because of their tendency to produce waves at too high frequencies and wave numbers than is required to account for the power-law inertial range of the fluctuations that extends from about $k = 10^{-12}$ cm^{-1} to 10^{-7} cm^{-1} [*Barnes*, 1979; *Bavassano et al.*, 1978; *Hollweg*, 1975].

In light of these difficulties, the major source of interplanetary fluctuations had been assumed to lie below the Alfvénic critical point [*Belcher and Davis*, 1971]. This assumption has the advantage that it naturally accounts for outward traveling waves, and the Alfvénic character of the fluctuations is ensured by the rapid damping of magnetoacoustic modes [*Barnes*, 1966].

The relative importance of these two generation mechanisms was clarified substantially in analyses of Helios and Voyager data by *Roberts et al.* [1987a,b], who showed that the solar wind intervals became less Alfvénic in an average sense with increasing distance (*cf.* Figure 1). No obvious tendency was found for the fluctuations to acquire the velocity-magnetic field correlation characteristic of inward propagating Alfvénic waves, as would be expected if sufficient time had elapsed in the minority species picture [*Dobrowolny et al.*, 1980a,b]. The highly Alfvénic character of many intervals near 0.3 AU down to scales longer than the transit time to the spacecraft also indicated that no reservoir of inward correlation was apparent as was suggested by *Matthaeus et al.* [1984]. Thus, the solar atmosphere appeared to be the main source of the outward wave flux at 0.3 AU. In addition, *Roberts et al.* [1987a,b] noted that regions of high Alfvénicity were associated with *small* velocity gradients, not necessarily with the trailing edges of high speed streams; this was especially true for highly correlated regions observed in the outer heliosphere [*Whang and Burlaga*, 1985].

Roberts et al. [1987a,b] also provided evidence that regions of strong shear accompanied a rapid evolution from a purely Alfvénic to a more mixed state in which the degree of correlation between $\delta\mathbf{v}$ and $\delta\mathbf{b}$ decreased rapidly with heliocentric distance with the largest scales decreasing the most rapidly. Near and beyond 1 AU less than 15% of the fluctuations in **b** and **v** were purely Alfvénic and outward propagating. As the outward propagating fluctuations decreased, the fraction of fluctuations propagating *inward* increased. A general property of the cross helicity analyzed in these Voyager and Helios data is that the degree of Alfvénicity was nearly independent of the degree of compression in the wind—being only slightly less in compression regions than in stream rarefaction regions. Furthermore, interaction of fast and slow streams to form compression regions was not associated with an increase of inward propagating fluctuations.

These analyses indicated that Belcher and Davis were correct in concluding that the high degree of Alfvénicity implied a source for the fluctuations that lay below the super-Alfvénic point in the solar corona. However, the strong evolution of the fluctuations with heliocentric distance and in the vicinity of large velocity shears showed that Coleman had correctly inferred that the free energy in stream-stream interactions would drive an *in situ* turbulent stirring of the medium that determined the subsequent evolution of the fluctuations.

At scales of less than a day, the energy in the magnetic fluctuations exceeded that in the velocity (the opposite situation prevailed at longer scales), indicating that the compressibility of the wind was not very important in the inertial range of the turbulence. Density fluctuations are also generally small ($\langle\delta n/n\rangle\sim$ 0.1),.except at shocks and in the vicinity of stream compression regions. The density and magnitude of magnetic field fluctuations are generally negatively correlated, indicating that the fluctuations represent neither slow modes nor tangential discontinuities. At scales less than a day, the energy in the magnetic field fluctuations exceeded that in the velocity fluctuations (*i.e.,* the Alfvén ratio $r_A(k) = E_V(k)/E_B(k) < 1$).

COMPARISON OF SOLAR WIND OBSERVATIONS WITH SIMULATIONS

An indication of the importance of velocity shear instabilities in the solar wind is available from linear instability analysis. *Korzhov et al.* [1984] have shown that the estimate made by *Parker* [1964] of the linear growth rate of the Kelvin-Helmholtz instability was overly pessimistic when wave vectors oblique to the shear direction were included. Earlier, *Southwood* [1968] had conjectured that unstable modes might always exist in a compressible system. However, linear theory generally predicts very small growth rates for waves with wave lengths small compared to the shear layer thickness, so the question of how high wave numbers were excited remained. *Miura* [1987] and others have carried out simulations of the fully nonlinear system using finite-difference techniques, but the algorithms used had significant dissipation and were thus not suited to studying spectral transfer. To avoid that problem, we have used spectral method codes capable of resolving wave numbers near the grid scale [*Ghosh et al.*, 1994; *Stribling and Matthaeus*, 1994; *Stribling et al.*, 1992a,b].

Evidence for Stream-Shear Instabilities

Coleman [1968] realized that the energy available in stream shear is typically about four times the energy of the magnetic fluctuations at the same large scales—often comparable with the energy in the mean field [*Roberts et al.*, 1990] in the region beyond 0.3 AU where direct observations can be made. While it seems plausible that these speed differences are the cause of much of the observed evolution, it is not altogether clear whether

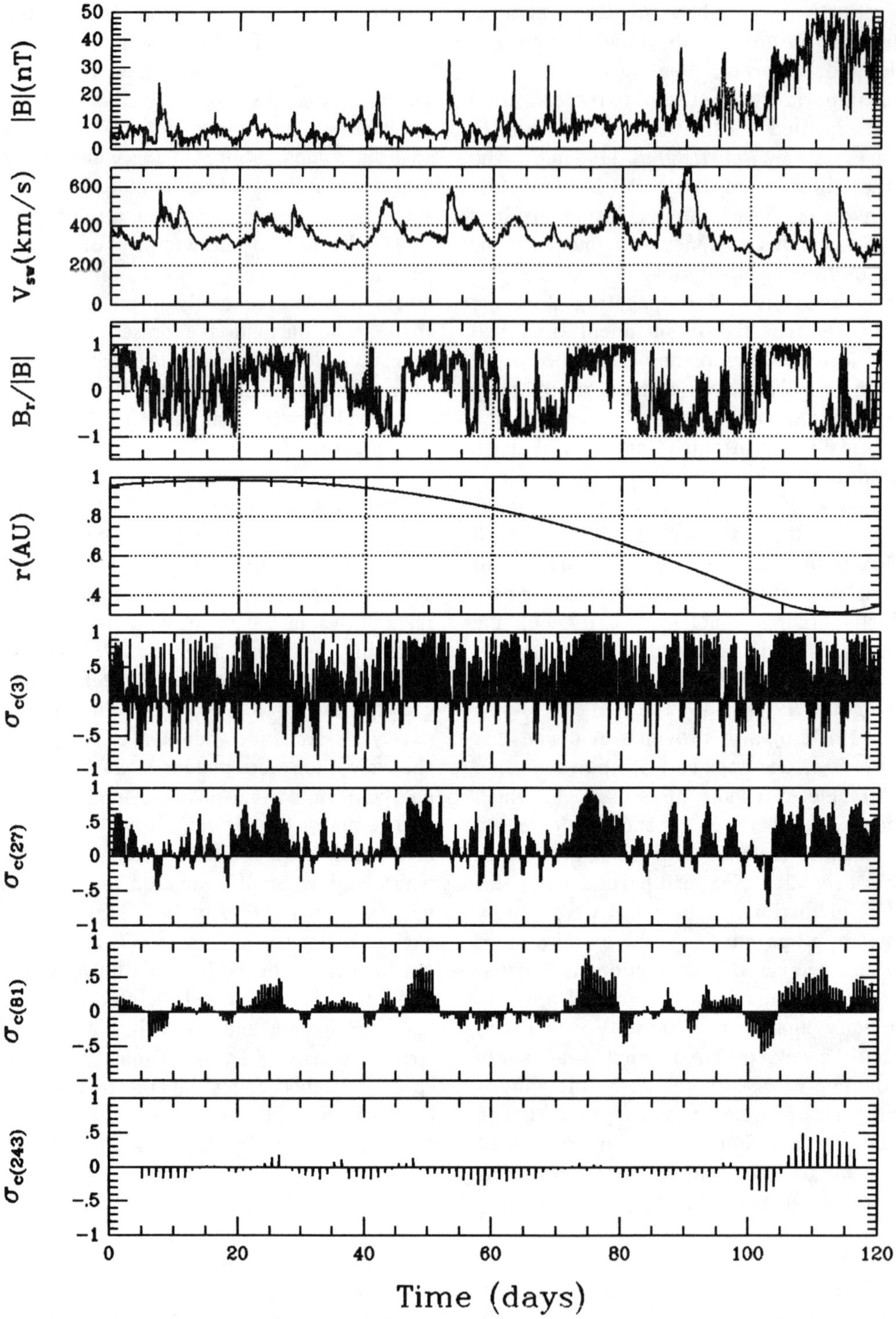

Fig. 1. Panels of (top to bottom) |**B**|, V_{SW}, B_r, heliocentric distance, and 3-, 27-, 81-, and 243-hour running values of σ_c for Helios data from early 1980 (from *Roberts et al.*, 1987b]).

the streams drive a turbulent cascade by compressing the plasma or by shearing it. Generally speaking shear will be more important in the inner heliosphere and compression in the outer, but a reliable quantitative estimate of their relative importance is not yet available. The issue is complicated by the fact that in a spiral field geometry regions of enhanced shear tend to be accompanied by enhanced compression [*Marsch and Tu*, 1990; *Roberts et al.*, 1987a,b].

A place where strong shears clearly exist is near the heliospheric current sheet at solar minimum; and we have concentrated much of our recent modeling efforts on this region. We know from both spacecraft observations and radio scintillation experiments [*Kojima and Kakinuma*, 1987] that at solar minimum the speed of the solar wind is slow along the equatorial current sheet and fast in the region outside about 10° above (and below) the current sheet. The transition between these regions is often quite sharp [*Burlaga et al.*, 1978]. Because the current sheet is not perfectly flat, the plasma is compressed where the slow and fast streams interact, but this is delayed by the nearly latitudinal stratification of the flow. Observations near the equatorial current sheet at solar minimum near and inside 1 AU are thus nearly ideal for determining the influence of velocity-shear on the flow.

Figure 1 illustrates many of these properties. Plotted are the magnitude $B = |\mathbf{B}|$ of the magnetic field, the solar wind speed, V_{SW}, radial component B_R, the radial distance of the spacecraft r, and the normalized cross helicity

$$\sigma_c = \frac{2H_c}{E} = \frac{2\langle \delta\mathbf{v} \cdot \delta\mathbf{B} \rangle}{\langle \delta\mathbf{v}^2 + \delta\mathbf{B}^2 \rangle} \qquad (1)$$

where the energy and the cross helicity (per unit mass) are $E = 1/2 \int d^3x (\mathbf{v}^2 + \mathbf{b}^2)$ and $H_c = 1/2 \int d^3x\, \mathbf{v} \cdot \mathbf{b}$, respectively. The data is from Helios 1 taken near 0.3 AU on days 93–110 of 1976. The sign of σ_c has been normalized so that positive values always correspond to outward propagating Alfvén waves. The fluctuations are taken about 3–. 27–, 81–, and 243–hour running averages. The magnetic field is measured in Alfvén speed units by dividing $\mathbf{B}$ by $\sqrt{(4\pi\rho)}$ with ρ equal to the local hour-averaged density.

Sector structure is indicated by the radial magnetic field. Note that low and negative values of σ_c are found preferentially near 1 AU. During the first 20 days of data the spacecraft is probably near the heliospheric current sheet and the low values of σ_c may reflect increased generation of turbulence in that region. The most

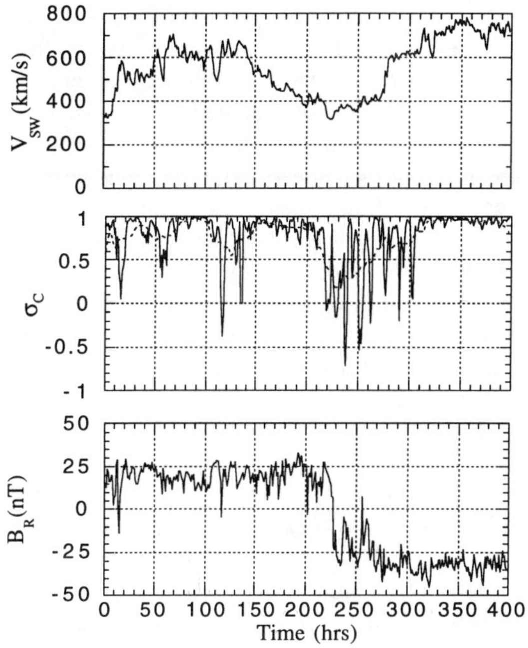

Fig. 2. Hourly-averaged data from Helios 2 for days 93-110 of 1976, showing (top) the speed of the wind, (middle) σ_c at the scale of about 3 hours.(solid) along with 25-hour running means (dashed), and (bottom) the radial component ("B_x") of the magnetic field, Note that the minimum values of σ_c occur in the low-speed region at the current sheet where the radial magnetic field is ≅ 0.

Alfvénic fluctuations are not necessarily associated with neither fast wind nor with the trailing edges of high speed flows. Near day 100 and day 50 there are a regions of slow wind with speed less than 400 km/s that contains fluctuations with very high cross helicity.

In Figure 2 we show a shorter interval of data near the current sheet, this time from Helios 2 (days 93–110 of 1976). Here the fluctuations are taken about 3–hour (3-point) running means. The values of σ_c are plotted as the solid line in Figure 2 (middle panel) along with a daily running average (dashed line). The fluctuations are the least Alfvénic in the center of the low speed region, very near the current sheet. The whole region of slow wind appears to be sheared, and thus it is quite plausible that the low cross helicity is caused by the shear. However, the strongest shear is at the edges of the low-speed region, while the strongest effect occurs closer to the current sheet. While this asymmetry might arise because the low speed region was smaller closer to the Sun and the strongest shear would then have arisen earlier in a way that affected the entire region, the fact that the mean field is transverse to the flow where σ_c is smallest implies that shear instabilities will develop faster there because they are not inhibited by the magnetic field. *Tu et al.* [1990] showed that the energy associated with compressive fluctuations is usually much less than even the "inward propagating" wave energy throughout the stream-shear region and thus compression does not seem to be the generator of the low cross helicity. These arguments provide the most direct evidence that we have that velocity shear must play an important role in the evolution of solar wind fluctuations. Wc will return to these observations below and compare them to both incompressible and compressible simulations designed to study how velocity shear in the vicinity of a current sheet produces the observed decrease in the cross helicity.

Spectral Simulations—General Considerations

To date our simulations of shear-generated turbulence assume that the background plasma has a uniform density and magnetic field. Because the solar wind plasma is three-dimensional, inhomogeneous, and compressible, some aspects of these simulations are not realistic in important respects. For example, the Fourier method spectral codes all use periodic boundary conditions, while the Chebyshev-Fourier algorithm can be non-periodic in only one direction [*Siregar et al.*, 1992a,b,c].

The assumption of periodic boundaries of course limits the domain of solar wind to which the simulation results can be applied. The assumption of incompressibility, however, appears to be rather reasonable. A theoretical justification for treating the solar wind as a nearly incompressible fluid was first developed by *Montgomery et al.* [1987] and later extended to an asymptotic expansion in which incompressible theory is the leading order approximation to compressible MHD in the limit of low Mach number [*Matthaeus and Brown*, 1988; *Matthaeus et al.*, 1991; *Zank and Matthaeus*, 1990,1992a,b; *Zank et al.*, 1990]. Between shocks and stream interfaces, the Mach number of the solar wind flow based on a root-mean-square (rms) speed is, in fact, usually $\ll 1$, thus justifying the nearly incompressible approximation. The simulations reviewed here have the advantage of isolating incompressive shear effects, and thus provide a basis for comparison with both the results of compressive simulations [*Roberts et al.*, 1991] and incompressive multi-length-scale MHD turbulence modeling [*Marsch and Tu*, 1989; *Zhou and Matthaeus*, 1989,1990a,b]. The latter models can also include the effects of the expansion of the medium with increasing heliocentric distance [*Grappin et al.*, 1993].

Two-and-a-half-dimensional simulations (in which all three components of the fluctuating magnetic and velocity fields are included), or three-dimensional simulations, are necessary to study the evolution of the magnetic helicity (in two dimensions, the analogous invariant is $A = 1/2\int a^2 d^2x = \langle a^2 \rangle$, the mean-squared vector potential). Preliminary results from studies of the evolution of the magnetic helicity are reported in these proceedings and elsewhere [*Ghosh et al.*, 1994; *Stribling and Matthaeus*, 1994; *Stribling et al.*, 1992a,b]. In three-dimensions, because of the increased number of degrees of freedom available for nonlinear coupling, perturbations orthogonal to the two-dimensional plane may excite new nonlinear couplings that exhibit both fast growth and new physics—preliminary results suggest that this indeed happens with velocity shears. Nevertheless, we shall assume that spectral transfer in three dimensions will, in general, occur in a manner analogous to, and at least as quickly as in two dimensions. In a sense, two-dimensional simulations provide a lower limit on the rate of evolution of turbulence. Because they are so much less expensive to run, two-dimensional simulations also allow us to explore a wide range of well-resolved spatial scales. The relevance of a two-dimensional description is also enhanced because the strong mean interplanetary magnetic field tends to force the turbulence into a quasi-two-dimensional state (see [*Matthaeus et al.*, 1990], and references therein).

The equations of incompressible MHD that we solve (in dimensionless units) are

$$\begin{aligned}
&\frac{\partial \mathbf{v}}{\partial t} + \mathbf{v}\cdot\nabla\mathbf{v} = \mathbf{B}\cdot\nabla\mathbf{B} - (1/2)\nabla B^2 + \upsilon\nabla^2\mathbf{v} \\
&\frac{\partial \mathbf{B}}{\partial t} + \mathbf{v}\cdot\nabla\mathbf{B} = \mathbf{B}\cdot\nabla\mathbf{v} + \eta\nabla^2\mathbf{B} \qquad (2) \\
&\nabla\cdot\mathbf{v} = \nabla\cdot\mathbf{B} = 0
\end{aligned}$$

where **v** is the velocity field, **B** is the magnetic field, ν is the reciprocal of the mechanical Reynolds number R, η is the reciprocal of the magnetic Reynolds number R_m, and the mass density is assumed uniform and constant. The simulation box is 2π length units on a side, and the dimensionless unit of time is the "eddy-turnover-time" given by the characteristic length divided by the characteristic large scale speed (both of which are order 1).

When solved in two dimensions, these equations simplify considerably and $\mathbf{B} = (B_x, B_y, 0)$, $\mathbf{v} = (v_x, v_y, 0)$, and $\partial/\partial z = 0$ for all variables. When **v** and **B** are written in terms of a stream function $\psi(x,y,t)$ and a vector potential $\mathbf{a} = a(x,y,t)\hat{\mathbf{e}}_\mathbf{z}$, eqs (2) are equivalent to

$$\frac{\partial \omega}{\partial t} + \mathbf{v} \cdot \nabla \omega = \mathbf{B} \cdot \nabla j + \upsilon \nabla^2 \omega$$
$$\frac{\partial a}{\partial t} + \mathbf{v} \cdot \nabla a = \eta \nabla^2 a \quad (3)$$

where the dimensionless vorticity and current densities are defined by $\nabla \times \mathbf{v} = \omega \hat{\mathbf{e}}_\mathbf{z}$ and $\nabla \times \mathbf{B} = j \hat{\mathbf{e}}_\mathbf{z}$, respectively, with $j = -\nabla^2 a$, and $\omega = -\nabla^2 \psi$.

The two-dimensional simulations discussed here had spatial resolutions ranging from 32×32 to 256×256 points, which allowed maximum magnetic and kinetic Reynolds numbers of 2000. The Reynolds number is generally chosen so that the dissipation wave number, *i.e.* the wave number at which the spectrum becomes exponential, is approximately that of the largest wave number in the box. A 256×256 code cannot simulate the observed spectral shapes (one needs resolution of order 1024×1024 together with hyperresistivity and hyperviscosity [*Biskamp and Welter*, 1989] to achieve a good power law inertial range). For the phenomena we are interested in describing, this is not a serious limitation. The geometry of the simulation domain used for studies of velocity shear in MHD is a square 2π on a side containing two velocity shear layers roughly at $y = \pi/2$ and $3\pi/2$ in the center of momentum frame of reference. The shear layer is defined by truncating the Fourier series for a step function—the width of the layer is determined by the highest mode included.

Velocity-Shear in Two Dimensions with a Mean Magnetic Field

We describe a few results here to indicate general turbulent properties of the solutions; for more examples and details see *Roberts et al.* [1991,1992]. We begin with a 256×256 mode simulation in which initially there is no energy in modes above $|\mathbf{k}| = 16$. The magnetic and kinetic Reynolds numbers were both 1000 and the first 16 modes in $|\mathbf{k}|$ defined the stream structure. The kinetic energy in the shear represented about 85% of the total energy of the fluctuations. We included a slight shear in the magnetic field (defined for modes $|\mathbf{k}| \leq 8$) which represented all but 1% of the rest of the energy; the final 1% was random noise necessary to trigger the nonlinear evolution. The strength of the mean field in dimensionless units was 1.244, equal to the Alfvén speed. Consequently, the energy in the mean field was nearly equal to the fluctuating energy, and the normalized cross helicity σ_c was only 0.12. This situation suppresses dynamical development in a number of ways. The strong mean field tends to stabilize the shear layer against the (linear) Kelvin-Helmholtz instability because the perturbation wave vector is along the magnetic field so that the velocity difference across the layer must be twice the Alfvén speed [*Chandrasekhar*, 1961], which was only marginally true in this simulation because the kinetic energy per unit area $((1/2)(2\pi^{-2})\ \delta v^2$ summed over the box) was 1, and, since $|v|$ was nearly constant at the shear layer, Δv was about 2.8, just greater than $2V_A$ = 2.48. The only modes with significant (although small) growth rates, however, were those with $|\mathbf{k}| \leq 16$, since the shear layer was relatively wide. The 1% perturbation, while relatively large by the standards of simulations of linear instabilities, represented only a fraction of the energy that would actually be available to initiate a cascade of energy to smaller wave numbers in the solar wind.

Figure 3 shows the rapid evolution of the wave number spectra of magnetic energy (solid) and kinetic energy (dashed). The spectra presented here are modal (energy per Fourier mode) meaning that the energies are summed in annular rings in the wave vector space, and then each sum is divided by the number of modes in the ring. Note the appearance of energy in the high-k modes at early times. The time unit in the simulation is an eddy-turn-over time. In the solar wind the eddy-turn-over time changes with radial distance, but at 1 AU the correlation length of the fluctuations is about 5×10^6 km [*Matthaeus and Goldstein*, 1982] and a typical rms speed fluctuation is ~100 km/s for scales that include streams—this means that the characteristic time is about a day and there are only a few eddy-turn-over times between the Sun and Earth.

The significance, then, of the spectrum in Figure 3 for T = 1 is that even with the various inhibitions to spectral transfer imposed by the conditions of the simulation, there is still substantial nonlinear evolution producing high-k fluctuations on a time-scale small enough to be important in the inner heliosphere. The spectrum at T = 9 (bottom panel of Figure 3) suggests that a power law spectrum can arise from velocity-shear-driven turbulence. Figure 4 (top) shows the cross helicity evolution accompanying the energy evolution. In the initial condition σ_c is small at all scales and is undefined outside the initially excited band of k's specifying the shear. At later times $\sigma_c(k)$ fluctuates considerably, but at any given k its random oscillations about zero indicate that the cascade of energy from large scales favors neither "outward" nor "inward" propagation (see, for example, *Bavassano et al.* [1978]). The low values of σ_c are not due to a lack of equipartition between kinetic and magnetic energy as is shown by Figure 4 (bottom) in which

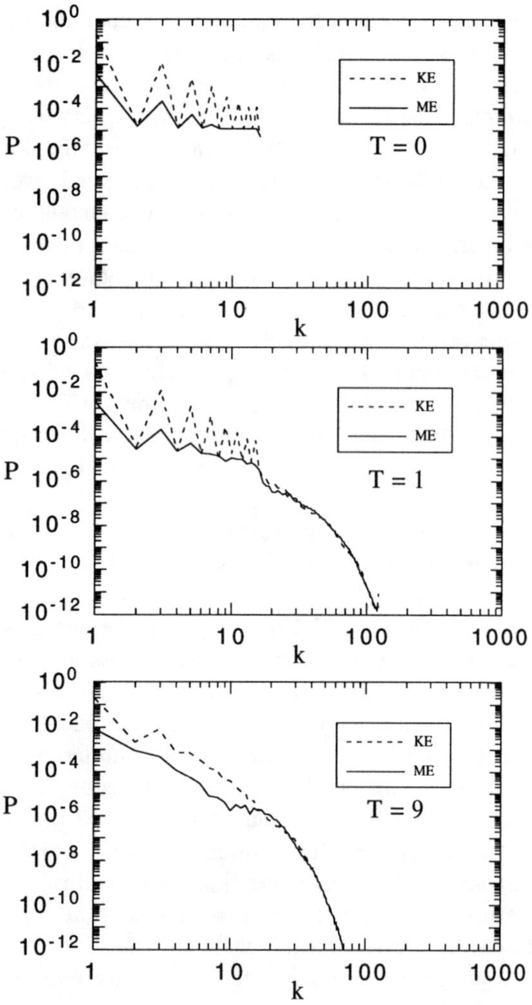

Fig. 3. Kinetic (dashed) and magnetic (solid) energy spectra for various times in the simulation with no energy in small scale modes initially. The flow quickly develops small scale fluctuations.

the values of $r_A = E_k/E_m$ are plotted. Thus the velocity shear instability, as represented by this simulation, will generate high-k fluctuations that are nearly equipartitioned in magnetic and kinetic energy, but it will not produce the outward traveling Alfvén waves observed in the inner heliosphere.

Velocity Shear in the Presence of Alfvén Waves

The second simulation, also 256×256 modes, was initialized with an isotropic distribution of Alfvén waves with a power law spectrum. This situation more closely mimics the solar wind, where the high-k cross helicity is frequently nearly maximal close to the Sun. Each mode

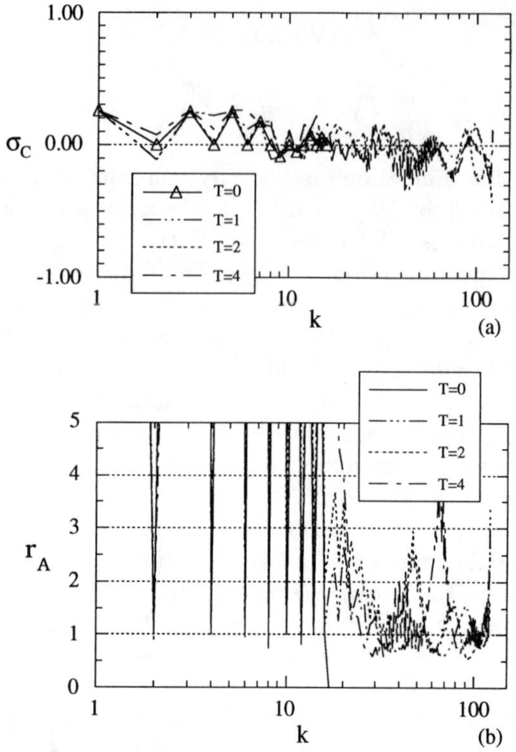

Fig. 4. Plots of $\sigma_c(k)$ and $r_A(k)$f or the run with no initial high k energy.

had maximal cross helicity so that r_A was unity. The spectrum at wave numbers above those that represent the shear layer (here defined using modes out to $|\mathbf{k}| = 4$) had a modal spectral index of –4.

The overall level of the high-k spectrum for this run increases with time as the shear layer energy cascades down to the dissipation range. There is also a related (relative) weakening of the shear layer peaks. On the other hand, Figure 5a shows that σ_c systematically decreases at high wave numbers, reminiscent of the decrease in $\sigma_c(k)$ reported by *Roberts et al.* [1987a] in narrowly peaked histograms of $\sigma_c(k)$. The magnetic and velocity fields remain nearly equipartitioned throughout the run, as shown in Figure 5b where r_A is plotted at various times. Thus the numerically simulated shear layer mechanism decreases the cross helicity of the small scale fluctuations, much as is observed in the solar wind.

A clearer picture of the role played by the velocity shear in destroying correlations can be gained from Figure 6 where $\sigma_c(x)$ is plotted as a function of the distance through the simulation box for the $k = 10$ to $k = 30$ band of fluctuations. The plot was constructed by filtering the fluctuations to include only the indicated k band, and

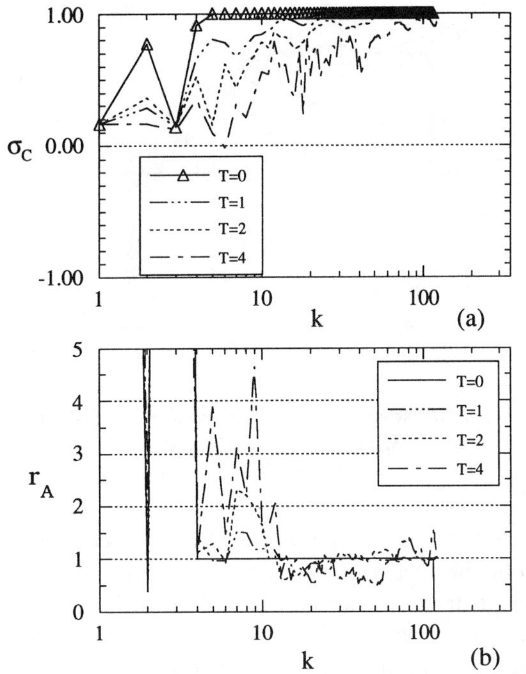

Fig. 5. a) The normalized cross helicity spectra as a function of time for the shear layer with high k Alfvénic fluctuations showing a relatively uniform decay of σ_c at small scales. b) The ratio of kinetic to magnetic energies for the same case, indicating near equipartition throughout the run.

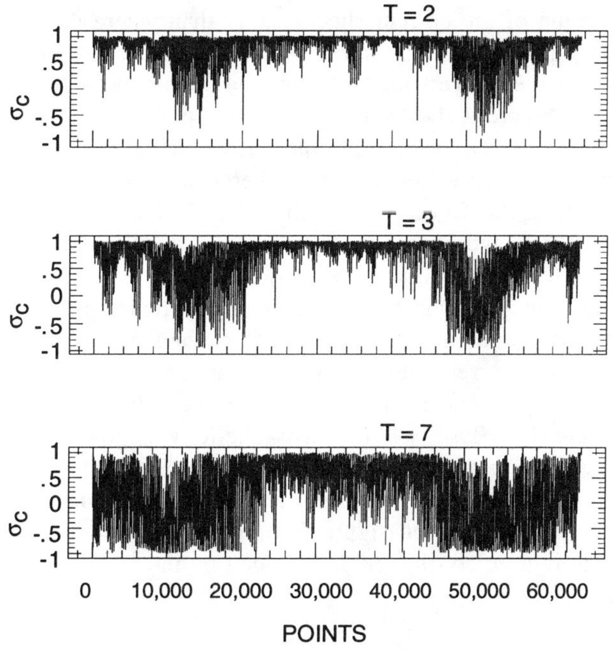

Fig. 6. A "spacecraft" view of σ_c moving slowly upward through the (periodic) simulation box. As time increases (downward), the fluctuations become less Alfvénic first in the shear layers and later nearly everywhere.

then forming one-dimensional 9-point correlations for successive rows of the simulation. Initially all values of $b(x)$ and $v(x)$ in this spatial band were completely aligned, but by $T = 2$ the values of σ_c are significantly reduced at positions close to and in the velocity shear layers. The alignment is systematically destroyed by the end of the run ($T = 7$). *Bavassano and Bruno* [1989] found at least one case where a large shear was associated with a low cross helicity region, consistent with this simulation. In general, however, there is the problem in this example because the lowest values of σ_c are not associated with the largest shears. Below we argue that the behavior of $\sigma_c(k)$ in the solar wind is due to the rapid evolution of the fluctuations near the heliospheric current sheet in the inner heliosphere where the magnetic field in the direction of the shear layer is small, thereby allowing a rapid nonlinear evolution.

Velocity Shear Near the Heliospheric Current Sheet

We now consider the effect of the heliospheric current sheet on the evolution of cross helicity and discuss some effects introduced by compressibility. The initial conditions of the simulations are essentially similar to those above with the addition of a current sheet embedded in the middle of the "slow" wind region. Initially, the simulation box contained a population of purely Alfvénic fluctuations with a fairly flat spectrum (modal spectral index −1) at numbers above those defining the shear layers. Random fluctuations (about 1% of the total energy) were added isotropically to modes with $|\mathbf{k}| \le 6$. The Alfvénic population models the observed fluctuations—we assume that both slow and fast wind start out Alfvénic and outward propagating at the critical point near $20\,R_S$ (≈ 0.1 AU) [*Bavassano and Bruno*, 1989; *Belcher and Davis*, 1971; *Bruno and Bavassano*, 1991; *Tu and Marsch*, 1990].

Behavior of Elsässer Variables.

Figure 7 shows one-dimensional energy spectra in terms of Elsässer variables in the incompressible case. (The Elsässer variables are defined as $\mathbf{z}^{\pm} = \mathbf{v} \pm \mathbf{b}$ and are useful for distinguishing "outward" ($\mathbf{z}^{+}$) and "inward" ($\mathbf{z}^{-}$) propagating waves.) In the top panel the dominant "outward propagating" fluctuations relax rapidly toward a steep spectrum, and then evolve slowly [*Tu*, 1988]. Recall that $T = 3$ is roughly equivalent to 1 AU. Initially the low-k fluctuations are nonAlfvénic, but the high-k fluctuations are purely Alfvénic. While the small scale z^{+} variations are rapidly dissipated, the z^{-} fluctuations

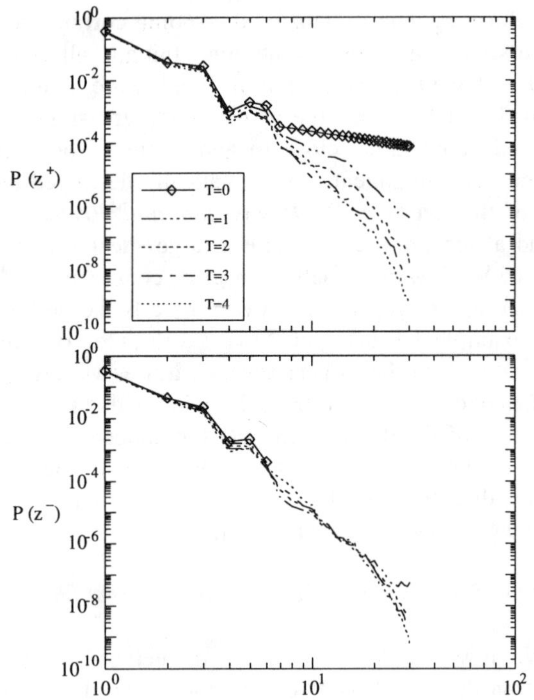

Fig. 7. Time evolution of the power spectra of the Elsässer variables z^+ and z^- for the incompressible run containing current sheets.

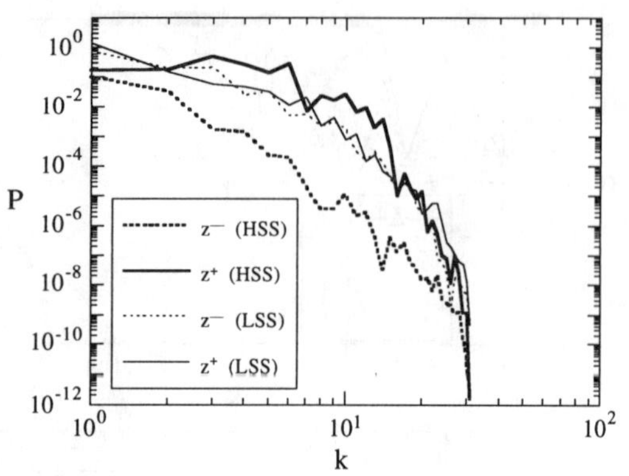

Fig. 8. Elsässer spectra for high speed streams (HSS) and low-speed streams (LSS) taken from cuts along x in the simulation box.

establish and maintain a steady state. As shown above, shear driven turbulence has a nearly equal large-scale input to the cascade of both z^+ and z^- energies. The z^- energy attains a steady state matched by dissipation, whereas the z^+ dissipation is greatly enhanced by the high power levels at high k. This behavior is observed in the inner heliosphere [*Tu and Marsch*, 1990] with a "background" spectrum of z^- that is approached in time by z^+.

Subsequently, in the outer heliosphere the spectra gradually evolve toward each other, presumably with a slow cascade matched by slow dissipation [*Roberts et al.*, 1987a,b], and this quasi-steady state leads to an evolution of the amplitudes that is nearly that of dissipationless Alfvénic turbulence [*Roberts et al.*, 1990].

The "background" spectrum of z^- is observed to vary, and, in particular, there are increases in its level associated (at solar minimum) with low speed wind. Figure 8 shows Elsässer spectra for a typical cut in x in the middle of the "high-speed stream" (HSS) of the simulation, and for a cut near the current sheet—the "low-speed stream" (LSS). We see the characteristic "breathing" that was pointed out originally by *Grappin et al.* [1990], although the z^- increases somewhat more here than is observed. Spectra for a cut near the current sheet at the edge of the box, in the middle of the HSS look like the LSS spectra, indicating that proximity to the current sheet is the critical factor and not the speed of the wind or even the proximity to shear layers.

The spatial structure of the fluctuations can be analyzed by filtering out all power with $|k_x|$ or $|k_y| < 8$, and then inverse transforming the resulting coefficients. At $T = 1$ for the case with current sheets, σ_c derived from such a procedure deviates significantly from +1 at the location of the current sheets rather than where the vorticity is large. By contrast, without current sheets the strong shear layers are the dominant source of the evolution. Figure 9 shows the spatial evolution of B_x, V_x, σ_c, and ω for the incompressible case with current sheets using plots as a function of y averaged over x. The top panel of the figure shows that the field and vorticity structure have been maintained on average, but that σ_c at high k (originally +1 everywhere) has been significantly decreased in the region of the current sheets near the edge and the middle of the box and especially so when the current sheet is near the strong velocity shear layers. This effect is observed in the solar wind (*cf.* Figure 2). By $T = 4$ the cross helicity is low everywhere, on average, although a two-dimensional spatial plot shows that both positive and negative regions contribute. There is no significant spreading of the current sheet region, considered as bounded by the vorticity layers.

Compressible Simulations

A compressible simulation with current sheet, an initial turbulent sonic Mach number of 0.3 (typical of solar wind values), and $\gamma = 5/3$, yields nearly identical evolu-

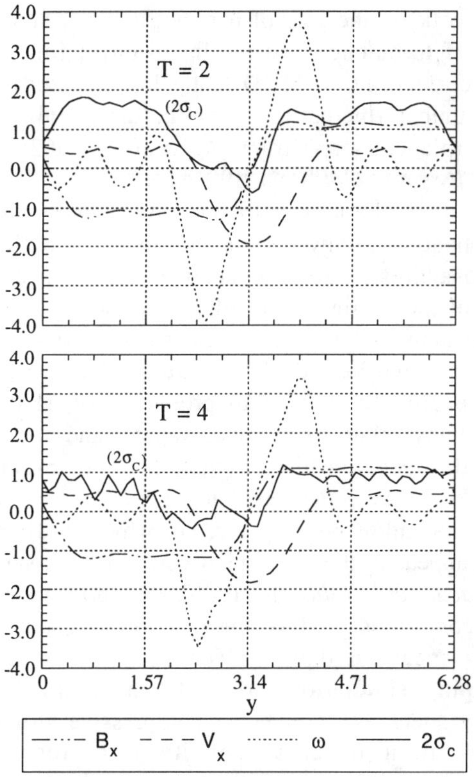

Fig. 9. Quantities averaged over x at T = 2 and T = 4 for the incompressible run with current sheets.

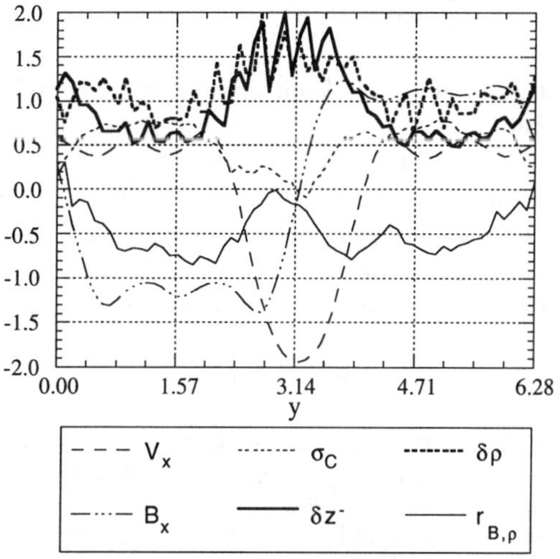

Fig. 10. Quantities averaged over x at T = 2 for the compressible run with current sheets. The density and z^- fluctuations are normalized to twice their maximum values (0.042 and 0.075, respectively).

tion of the "incompressive" quantities σ_c, v_x, and b_x (see Figure 10). In this case, the equations being integrated are

$$\frac{\partial}{\partial t}\rho = -\nabla \cdot (\rho \mathbf{u})$$

$$\frac{\partial}{\partial t}\mathbf{v} = \mathbf{v} \times (\omega \hat{e}_z) - \left(\frac{1}{2}\right)\nabla\left(\mathbf{v}\cdot\mathbf{v} + \frac{2}{M_1^2(\gamma-1)}\rho^{(\gamma-1)}\right) +$$

$$\frac{j}{M_2^2\rho}\nabla a + \frac{\nu}{\rho}\nabla^2\mathbf{v} + \frac{1}{\rho}\left(\zeta + \frac{\nu}{3}\right)\nabla(\nabla\cdot\mathbf{v})$$

$$\frac{\partial a}{\partial t} + \mathbf{v}\cdot\nabla a = \eta\nabla^2 a \tag{4}$$

where the thermal pressure p is assumed to obey $p = \rho^\gamma$; ζ is the bulk viscosity; $M_1^2 = (U_o/C_s)^2$ with U_o the characteristic speed of the medium; C_s the typical sound speed, $M_2^2 = (U_o/V_A)^2$ and V_A is the characteristic Alfvén speed. The new features are a persistent anti-correlation between the magnetic field magnitude and the density ($r_{B,\rho}$), suggestive of nearly pressure balanced "pseudo-sound," and a density fluctuation that correlates well with the fluctuations in z^-. This correlation has been pointed out [*Bavassano and Bruno*, 1989; *Bruno and Bavassano*, 1991; *Grappin et al.*, 1990] and is associated with the appearance of higher compression. The present simulations indicate that these density fluctuations arise naturally from nearly incompressive evolution [*Matthaeus and Brown*, 1988; *Matthaeus et al.*, 1991; *Montgomery et al.*, 1987].

In addition to the characteristics presented above, the simulations show essentially the correct behavior of the Alfvén ratio r_A. The values of r_A remain near 1, and although the mean value is in fact very close to 1 at T = 2 (somewhat higher than the observed 0.5), the distribution of r_A values peaks strongly near 0.5. There are regions of lower r_A associated with low cross helicity regions such as were recently reported by *Tu and Marsch* [1991]. While there is no simple association between r_A and σ_c in either the simulations or observations, both quantities tend to have their lowest values in regions that are more strongly developed, and both tend to be near +1 in more purely Alfvénic regions, as expected.

We conclude that much of the evolution of interplanetary field fluctuations is accounted for by nonlinear nearly incompressive MHD processes. The shear in the velocity field produced near the Sun leads to approximately equal injection of power in z^+ and z^- at large scales. Because z^- is initially small at high wave numbers, this injection leads to a spectrum that grows until it is balanced by dissipation; both injection and dissipation then remain at a relatively low, nearly constant, level. The z^- spectrum thus is well developed, or "old," very near the Sun. The injection of z^+ is inadequate to balance the rapid dissipation of the large high-k fluctuations, and thus the z^+ spectrum initially decreases rapidly. When the z^+ level is nearly equal to the z^- level the two spectra continue to evolve slowly toward each other with their overall amplitudes decreasing at nearly the rate they would if no spectral transfer were occurring.

This process is accelerated in regions where the local mean field has only a small component parallel to and in proximity of strong shear layers, as occurs especially at the heliospheric current sheet at solar minimum. There, the flat z^+ spectrum has already dissipated by 0.3 AU and thus the turbulence is "older" and the observed temperature evolution is nearly adiabatic [*Freeman*, 1988]. By contrast, regions experiencing the rapid z^+ decay outside 0.3 AU have a slower than adiabatic temperature decrease because the dissipation heats the plasma. Broad slow speed regions that do not contain a current sheet have properties similar to high-speed streams at solar minimum. Compressive effects are largely a result of the incompressive dynamics, as is to be expected in flows with low turbulent sonic Mach numbers.

SUMMARY

In this review, I was asked to discuss the status of the work of our group at Goddard which has been studying the properties and evolution of solar wind turbulence. In so doing, it has not been possible to discuss, except for anecdotal references, the many important contributions of colleagues working in other institutions. For a more comprehensive picture of those aspects of solar wind research that could not be covered here, I recommend reviews by *Roberts and Goldstein* [1991]; *Marsch* [1991]; and *Isenberg* [1991]. Space has also not allowed for a complete review of work by my colleagues at Goddard, but details of that work can be found in the many references cited above. In particular, I have not discussed in any depth the origin of anisotropic spectra in MHD turbulence nor such evidence as exists in the data for such anisotropies. Similarly, I have only briefly mentioned lack of support in the data for the hypothesis that Alfvén waves heat and accelerate the solar wind [*Roberts*, 1989,1990].

Space also has not permitted a discussion of more recent work on the role of parametric instabilities in initiation of turbulence and the inclusion of finite Larmor radius corrections to MHD that allow for study of high wave number fluctuations near the dissipation range of the interplanetary spectrum [*Agim et al.*, 1993; *Roberts*, 1993; *Viñas and Goldstein*, 1991a,b,c].

Important challenges to our understanding of solar wind phenomenology remain that theory and simulation have to address. Some concern generic aspects of the theory of the origin of the solar wind, such as the physical mechanism(s) by which fast solar wind streams are accelerated and heated. There are outstanding problems more specifically related to properties of MHD turbulence; many relate to the photospheric and coronal processes that generate the fluctuations. For example, we do not know the processes that generate Alfvén waves in the corona and/or photosphere, nor why such fluctuations should appear in the solar wind with a k^{-1} power spectra [*Matthaeus and Goldstein*, 1986]. Nor do we know what initial conditions produce the randomly signed magnetic helicity spectrum that is observed. We are currently developing simulations to study the relative role of velocity shear and convected compressive structures in the evolution of cross helicity (see, for example, [*Bavassano and Bruno*, 1989, *Roberts et al.*, 1991, *Roberts et al.*, 1987a,b,1992]). In addition, we are attempting to determine whether dissipation of turbulence can explain the observed heating of solar wind protons with heliospheric distance [*Freeman*, 1988, *Gazis and Lazarus*, 1982, *Tu et al.*, 1989].

Acknowledgments. I am grateful for discussions and collaborations with W. H. Matthaeus, D. A. Roberts, L. F. Burlaga, S. Ghosh, T. Stribling, E. Siregar, L. Klein, and Z. Agim, among others. The research was supported by NASA's Space Physics Theory Program and by a Space Research and Technology grant to the Goddard Space Flight Center, L. F. Burlaga, Principal Investigator.

REFERENCES

Agim, Y. Z., A. F. Viñas, and M. L. Goldstein, Magnetic and density fluctuations by parametric instabilities near interplanetary shocks, in: *Proc. of 7th Scientific Assembly,* Buenos Aires, edited by Local Organizing Committee, *Bulletin No. 55,* p. 382, IAGA, 1993.

Barnes, A., Collisionless damping of hydromagnetic waves, *Phys. Fluids, 9*, 1483, 1966.

Barnes, A., Hydromagnetic waves and turbulence in the solar wind, in *Solar System Plasma Physics*, edited by E. N. Parker, C. F. Kennel and L. J. Lanzerotti, p. 249, North-Holland,, 1979.

Bavassano, B., M. Dobrowolny, and G. Moreno, Local insta-

bilities of Alfvén waves in high-speed streams, *Solar Phys.*, *57*, 445, 1978.
Bavassano, B., and R. Bruno, Evidence of local generation of Alfvénic turbulence in the solar wind, *J. Geophys. Res.*, *94*, 11977, 1989.
Belcher, J. W., and L. Davis, Large-amplitude Alfvén waves in the interplanetary medium, 2, *J. Geophys. Res.*, *76*, 3534, 1971.
Biskamp, D., and H. Welter, Dynamics of decaying two-dimensional magnetohydrodynamic turbulence, *Phys. Fluids B*, *1*, 1964, 1989.
Bruno, R., and B. Bavassano, Solar wind fluctuations at large scale: A comparison between low and high solar activity conditions, *J. Geophys. Res.*, *96*, 7841, 1991.
Burlaga, L. F., N. F. Ness, F. Mariani, B. Bavassano, U. Villante, H. Rosenbauer, R. Schwenn, and J. Harvey, Magnetic fields and flows between 1 and 0.3 AU during the primary mission of Helios 1, *J. Geophys. Res.*, *83*, 5167, 1978.
Burlaga, L. F., A heliospheric vortex street?, *J. Geophys. Res.*, *95*, 4333, 1990.
Chandrasekhar, S., *Hydrodynamic and Hydromagnetic Stability*, Clarendon, 1961.
Coleman, P. J., Turbulence, viscosity, and dissipation in the solar wind plasma, *Astrophys. J.*, *153*, 371, 1968.
Dobrowolny, M., A. Mangeney, and P. Veltri, Fully developed anisotropic turbulence in interplanetary space, *Phys. Rev. Lett.*, *45*, 144, 1980a.
Dobrowolny, M., A. Mangeney, and P. Veltri, Properties of mhd turbulence in the solar wind, *Astrophys. J.*, *83*, 2632, 1980b.
Freeman, J. W., Estimates of solar wind heating inside 0.3 AU, *Geophys. Res. Lett.*, *15*, 88, 1988.
Gazis, P. R., and A. J. Lazarus, Voyager observations of solar wind proton temperature: 1–10 AU, *Geophys. Res. Lett.*, *9*, 431, 1982.
Ghosh, S., A. F. Viñas, and M. L. Goldstein, Parametric instabilities of a large amplitude circularly polarized Alfvén wave: Linear growth in 2–D geometries, *J. Geophys. Res.*, *submitted*, 1992.
Ghosh, S., T. Stribling, M. L. Goldstein, and W. H. Matthaeus, Evolution of magnetic helicity in compressible magnetohydrodynamics with a mean magnetic field, in *Micro- and Mesoscale Phenomena in Space Plasma Physics*, edited by M. Ashour-Abdalla and T. Chang, *Geophys. Monogr. Ser.*, AGU, 1994.
Goldstein, M. L., D. A. Roberts, and C. A. Fitch, The structure of helical interplanetary magnetic fields, *Geophys. Res. Lett.*, *18*, 1505, 1991.
Goldstein, M. L., D. A. Roberts, and C. A. Fitch, Properties of the fluctuating magnetic helicity in the inertial and dissipation ranges of solar wind turbulence, *J. Geophys. Res.*, *submitted*, 1993.
Goldstein, M. L., D. A. Roberts, and C. A. Fitch, Observed properties of helical interplanetary magnetic fields, in *Micro- and Mesoscale Phenomena in Space Plasma Physics*, edited by M. Ashour-Abdalla and T. Chang, *Geophys. Monogr. Ser.*, AGU, 1994.
Grappin, R., A. Mangeney, and E. Marsch, On the origin of solar wind MHD turbulence: Helios data revisited, *J. Geophys. Res.*, *95*, 8197, 1990.
Grappin, R., M. Velli, and A. Mangeney, MHD simulations of solar wind turbulence in comobile coordinates, in: *Proc. of Spatio-Temporal Analysis for Resolving Plasma Turbulence (START)*, Aussois, France, *ESA WPP–325*, p. 325, ESA, 1993.
Hollweg, J. V., Waves and instabilities in the solar wind, *Rev. Geophys.*, *13*, 263, 1975.
Isenberg, P. A., The solar wind, in *Geomagnetism*, edited by J. A. Jacobs, p. Academic Press, 1991.
Kojima, M., and T. Kakinuma, Solar cycle evolution of solar wind speed structure between 1973 and 1985 observed with the interplanetary scintillation method, *J. Geophys. Res.*, *92*, 1169, 1987.
Kolmogoroff, A. N., The local structure of turbulence in incompressible viscous fluid for very large Reynolds numbers, *C. R. Acad. Sci. URSS*, *30*, 301, 1941.
Korzhov, N. P., V. V. Mishin, and V. M. Tomozov, On the role of plasma parameters and the Kelvin-Helmholtz instability in a viscous interaction of solar wind streams, *Planet. Space. Sci.*, *32*, 1169, 1984.
Kraichnan, R. H., Inertial range of hydromagnetic turbulence, *J. Geophys. Res.*, *87*, 6011, 1965.
Marsch, E., and C.-Y. Tu, Dynamics of correlation functions with Elsässer variables for inhomogeneous MHD turbulence, *J. Plasma Phys*, *41*, 479, 1989.
Marsch, E., and C.-Y. Tu, On the radial evolution of MHD turbulence in the inner heliosphere, *J. Geophys. Res.*, *95*, 8211, 1990.
Marsch, E., MHD turbulence in the solar wind, in *Physics of the Inner Heliosphere*, edited by R. Schwenn and E. Marsch, p. 159, Springer-Verlag, 1991.
Matthaeus, W. H., and M. L. Goldstein, Measurement of the rugged invariants of magnetohydrodynamic turbulence, *J. Geophys. Res.*, *87*, 6011, 1982.
Matthaeus, W. H., M. L. Goldstein, and D. C. Montgomery, Turbulent generation of outward traveling interplanetary Alfvénic fluctuations, *Phys. Rev. Lett.*, *51*, 1484, 1984.
Matthaeus, W. H., and M. L. Goldstein, Low-frequency 1/f noise in the interplanetary magnetic field, *Phys. Rev. Lett.*, *57*, 495, 1986.
Matthaeus, W. H., and M. R. Brown, Nearly incompressible magnetohydrodynamics at low Mach number, *Phys. Fluids*, *31*, 3634, 1988.
Matthaeus, W. H., M. L. Goldstein, and D. A. Roberts, Evidence for the presence of quasi-two-dimensional, nearly incompressible fluctuations in the solar wind, *J. Geophys. Res.*, *95*, 20673, 1990.
Matthaeus, W. H., L. W. Klein, S. Ghosh, and M. R. Brown, Nearly incompressible magnetohydrodynamics, pseudosound, and solar wind fluctuations, *J. Geophys. Res.*, *96*, 1991.
Miura, A., Simulation of Kelvin-Helmholtz instability at the magnetopause boundary, *J. Geophys. Res.*, *92*, 3195, 1987.
Montgomery, D., M. Brown, and W. H. Matthaeus, Density fluctuation spectra in magnetohydrodynamic turbulence, *J. Geophys. Res.*, *92*, 282, 1987.
Parker, E. N., Dynamical properties of solar and stellar winds. III, *Astrophys. J.*, *139*, 690, 1964.
Roberts, D. A., L. W. Klein, M. L. Goldstein, and W. H. Matthaeus, The nature and evolution of magnetohydrodynamic fluctuations in the solar wind: Voyager observations, *J. Geophys. Res.*, *92*, 11021, 1987a.

Roberts, D. A., M. L. Goldstein, L. W. Klein, and W. H. Matthaeus, Origin and evolution of fluctuations in the solar wind: Helios observations and Helios-Voyager comparisons, *J. Geophys. Res., 92*, 12023, 1987b.

Roberts, D. A., Interplanetary observational constraints on Alfvén wave acceleration of the solar wind, *J. Geophys. Res., 94*, 6899, 1989.

Roberts, D. A., Turbulent polar heliospheric fields, *Geophys. Res. Lett., 17*, 567, 1990.

Roberts, D. A., M. L. Goldstein, and L. W. Klein, The amplitudes of interplanetary fluctuations: Stream structure, heliocentric distance, and frequency dependence, *J. Geophys. Res., 95*, 4203, 1990.

Roberts, D. A., S. Ghosh, M. L. Goldstein, and W. H. Matthaeus, MHD simulation of the radial evolution and stream structure of solar wind turbulence, *Phys. Rev. Lett., 67*, 3741, 1991.

Roberts, D. A., and M. L. Goldstein, Turbulence and waves in the solar wind, in *U. S. National Report to International Union of Geodesy and Geophysics*, edited by M. A. Shea, p. 932, American Geophysical Union, 1991.

Roberts, D. A., M. L. Goldstein, W. H. Matthaeus, and S. Ghosh, Velocity shear generation of solar wind turbulence, *J. Geophys. Res., 97*, 17115, 1992.

Roberts, D. A., Nonequilibrium, large-amplitude MHD fluctuations in the solar wind, *J. Geophys. Res., in press*, 1993.

Siregar, E., D. A. Roberts, and M. L. Goldstein, The formation of vortex streets through nonlinear interactions of two vortex sheets in a two dimensional compressible flow, *Phys. Fluids, in preparation*, 1992a.

Siregar, E., D. A. Roberts, and M. L. Goldstein, The formation of vortex streets through the nonlinear interactions of vortex sheets in a two-dimensional compressible flow, *Geophys. Res. Lett., 19*, 1427, 1992b.

Siregar, E., D. A. Roberts, and M. L. Goldstein, Quasi-periodic transverse plasma flow associated with an evolving MHD vortex street in the outer heliosphere, *J. Geophys. Res., submitted*, 1992c.

Siregar, E., D. A. Roberts, and M. L. Goldstein, Magnetohydrodynamics of interacting vortex sheets, in *Micro- and Mesoscale Phenomena in Space Plasma Physics*, edited by M. Ashour-Abdalla and T. Chang, *Geophys. Monogr. Ser.*, AGU, 1994.

Southwood, D. J., The hydromagnetic stability of the magnetospheric boundary, *Planet. Space. Sci., 16*, 587, 1968.

Stribling, T., W. H. Matthaeus, and S. Ghosh, Nonlinear decay of magnetic helicity in magnetohydrodynamic turbulence with a mean magnetic field, *J. Geophys. Res.*, in press, 1992a.

Stribling, T., W. H. Matthaeus, and S. Oughton, Magnetic helicity in magnetohydrodynamic turbulence with a mean magnetic field, *Phys. Fluids B, submitted*, 1992b.

Stribling, T., and W. H. Matthaeus, Decay of magnetic helicity in ideal magnetohydrodynamics with a DC magnetic field, in *Micro- and Mesoscale Phenomena in Space Plasma Physics*, edited by M. Ashour-Abdalla and T. Chang, *Geophys. Monogr. Ser.*, AGU, 1994.

Tu, C.-Y., The damping of interplanetary Alfvénic fluctuations and the heating of the solar wind, *J. Geophys. Res., 93*, 7, 1988.

Tu, C.-Y., J. W. Freeman, and R. E. Lopez, The proton temperature and the total hourly variance of the magnetic field components in different solar wind regions, *Solar Phys., 119*, 197, 1989.

Tu, C.-Y., and E. Marsch, Evidence for a "background" spectrum of solar wind turbulence in the inner heliosphere, *J. Geophys. Res., 95*, 4337, 1990.

Tu, C.-Y., E. Marsch, and H. Rosenbauer, The dependence of MHD turbulence spectra on the inner solar wind stream structure near solar minimum, *Geophys. Res. Lett., 17*, 283, 1990.

Tu, C. Y., and E. Marsch, A case study of very low cross-helicity fluctuations in the solar wind, *Ann. Geophys., 9*, 319, 1991.

Unti, T. W., and M. Neugebauer, Alfvén waves in the solar wind, *Phys. Fluids, 11*, 563, 1968.

Viñas, A. F., and M. L. Goldstein, Parametric instabilities of circularly polarized, large amplitude, dispersive Alfvén waves: Excitation of obliquely propagating daughter and sideband waves, *J. Plasma Phys., 46*, 129, 1991a.

Viñas, A. F., and M. L. Goldstein, Parametric instabilities of circularly polarized, large amplitude, dispersive Alfvén waves: Excitation of parallel propagating electromagnetic daughter waves, *J. Plasma Phys., 46*, 107, 1991b.

Viñas, A. F., and M. L. Goldstein, Parametric instabilities of large amplitude Alfvén waves with obliquely propagating sidebands, in: *Proc. of 3rd COSPAR Colloquium,* Goslar, Germany, edited by E. Marsch and R. Schwenn, *Vol.* p. 577, Pergamon Press, Oxford, 1991c.

Whang, Y. C., and L. F. Burlaga, Evolution and interaction of interplanetary shocks, *J. Geophys. Res., 90*, 10765, 1985.

Zank, G. P., and W. H. Matthaeus, Nearly incompressible hydrodynamics and heat conduction, *Phys. Rev. Lett., 64*, 1243, 1990.

Zank, G. P., W. H. Matthaeus, and L. W. Klein, Temperature and density anti-correlations in solar wind fluctuations, *Geophys. Res. Lett., 17*, 1239, 1990.

Zank, G. P., and W. H. Matthaeus, Nearly incompressible fluids II: Magnetohydrodynamics, turbulence, and waves, *Phys. Fluids A, in press*, 1992a.

Zank, G. P., and W. H. Matthaeus, Waves and turbulence in the solar wind, *J. Geophys. Res., 97*, 17189, 1992b.

Zhou, Y., and W. H. Matthaeus, Non-WKB evolution of solar wind fluctuations: A turbulence modeling approach, *Geophys. Res. Lett., 16*, 755, 1989.

Zhou, Y., and W. H. Matthaeus, Models of inertial range spectra of interplanetary magnetohydrodynamic turbulence, *J. Geophys. Res., 95*, 14881, 1990a.

Zhou, Y., and W. H. Matthaeus, Transport and turbulence modeling of solar wind fluctuations, *J. Geophys. Res., 95*, 10291, 1990b.

M. L. Goldstein, Laboratory for Extraterrestrial Physics, Code 692, NASA—Goddard Space Flight Center, Greenbelt, MD 20771

Observed Properties of Helical Interplanetary Magnetic Fields

M. L. Goldstein, D. A. Roberts, and C. A. Fitch

Laboratory For Extraterrestrial Physics, NASA-Goddard Space Flight Center

We report the results of an analysis of magnetic helicity in the solar wind over the heliocentric distance range of 0.5 to 10 AU. Using data that ranged in resolution from 25 samples per second to 9.6 second averages, we computed occurrence probability distributions of the normalized magnetic helicity. In the 9.6 second data, these were centered about zero. At high resolution, we found that the normalized magnetic helicity had a clear skewness in sign. We also constructed a temporal measurement of the normalized magnetic helicity that sometimes indicated an association between particular field directions and highly helical structures.

1. BACKGROUND AND MOTIVATION

The three integral (or "rugged") invariants of ideal incompressible magnetohydrodynamics (MHD) were first discussed by *Woltjer* [1958a,b] and have since become an important aspect of MHD turbulence studies. The magnetic helicity is one of those rugged invariants and provides a description of the helical structure of the magnetic field. The other two invariants are the energy and the cross helicity [*Frisch et al.*, 1975].

While the properties of the energy and cross helicity have been studied extensively in the past ten years (see *Roberts and Goldstein* [1991]), the properties of the magnetic helicity in the inertial range (10^{-7}–10^{-4} km^{-1}) are, however, largely unknown. In 1982, Matthaeus and collaborators first showed how to determine the magnetic helicity from single point measurements in a super-Alfvénic magnetofluid [*Matthaeus and Goldstein*, 1982; *Matthaeus et al.*, 1982]. Subsequently, *Bieber et al.* [1987a,b] studied the large scale magnetic helicity and its role in the 22–year cycle of the solar modulation of cosmic rays (also see *Bieber and Smith* [1992] and *Smith and Bieber* [1993]). In this paper we seek to obtain greater insight into the nature and dynamical importance of the helical structure of the interplanetary magnetic field through extension of the heliocentric distance and frequency range of the observations and through various analytical tools. In this paper, we will not concern ourselves with determinations of the *net* magnetic helicity because the result would depend too sensitively on the frequency range of the analysis. We begin with a brief discussion of magnetic helicity and its determination from observations followed by a review of some of the currently known characteristics of the fluctuations. Section 3 contains the extended analysis and Section 4 concludes with a summary and discussion.

2. MAGNETIC HELICITY

Magnetic helicity can be thought of as a combination of the internal twist and the external linkages of magnetic flux tubes. In the absence of a mean component to the magnetic field, the magnetic helicity is defined as [*Moffatt*, 1978]

$$H_m \equiv \int d^3x\, \mathbf{A} \cdot \mathbf{B} \tag{2}$$

where $\mathbf{A}$ is the fluctuating magnetic vector potential, usually defined in the Coulomb gauge, and $\mathbf{B}$ is the fluctuating magnetic field defined by $\mathbf{B} = \nabla \times \mathbf{A}$. H_m is the total magnetic helicity, and the integral in eq. (2) extends over all field containing regions. When a mean magnetic field is present, H_m is no longer invariant because

$$\frac{dH'_m}{dt} = -2\frac{d}{dt}\left(\mathbf{B}_0 \cdot \langle \mathbf{A} \rangle\right) \tag{3}$$

where $H'_m \equiv \left(1/L^3\right)\int d^3x\, \delta\mathbf{A} \cdot \delta\mathbf{B}$, $d\left(\mathbf{B}_0 \cdot \langle \mathbf{A} \rangle\right)/dt \equiv$

Space Plasmas: Coupling Between Small
and Medium Scale Processes
Geophysical Monograph 86

$-\mathbf{B}_0 \cdot \langle \mathbf{E} \rangle$ with $\langle \mathbf{E} \rangle$ the volume average of the fluctuating electric field. The new invariant is $\hat{H}_m \equiv H_m' + 2\mathbf{B}_0 \cdot \langle \mathbf{A} \rangle$ (see, for example, *Matthaeus et al.* [1982], *Moffatt* [1978], and *Stribling et al.* [1992a]). The term $2\,\mathbf{B}_0 \cdot \langle \mathbf{A} \rangle$ can be interpreted as accounting for flux linkages of structures larger than the size of the domain, for example, larger than the periodic box of the simulations [*Matthaeus et al.*, 1982; *Moffatt*, 1978]. In our analysis of solar wind data, we follow *Matthaeus and Goldstein* [1982] in assuming that the quantity of theoretical interest is the fluctuating magnetic helicity density, the volume average of (2), denoted by

$$H_m = \langle \mathbf{A} \cdot \mathbf{B} \rangle \tag{4}$$

even though it is not a conserved quantity in the solar wind due to the presence of the Parker field. Simulations of two-and-a-half and three-dimensional turbulence by *Stribling and Matthaeus* [1994], *Stribling et al.* [1991a,b], and *Ghosh et al.* [1994], suggest that in the presence of a mean magnetic field an initially large value of H_m' will decay rapidly a few eddy-turn-over times.

The quantity that is measurable from observations is the "reduced" magnetic helicity [*Matthaeus and Goldstein*, 1982; *Matthaeus et al.*, 1982] given by

$$H_m(k_1) = \frac{2\,\mathrm{Im}\, S_{23}(k_1)}{(k_1)} \tag{5}$$

where $S_{23}(k_1)$ is the skew-symmetric component of the reduced magnetic energy spectrum tensor [*Batchelor*, 1970] obtained from the two-point correlation matrix under the assumption that the MHD analog of G. I. Taylor's "frozen-in-flow" approximation is valid [*Taylor*, 1938]. The 1-direction is the direction about which the spectrum is reduced; for interplanetary data this is the flow direction of the solar wind.

The previous work of *Matthaeus and Goldstein* [1982] determined several characteristics of the solar wind magnetic helicity in the inertial range. All the intervals they analyzed exhibited similar magnetic helicity spectra in that the values were highly helical but alternated rapidly in sign across the entire wave number range. They found the spectrum for each sign of the helicity varied as $k^{8/3}$ in all of the intervals.

A convenient dimensionless measure of the magnetic helicity is σ_m defined by

$$\sigma_m(k) \equiv \frac{kH_m(k)}{S(k)} \tag{6}$$

where $S(k)$ is the energy spectrum in mode k. This quantity measures the degree of twist in the field and the handedness of the fluctuations. The upper bound of +1 on $\sigma_m(k)$ indicates a pure right-handed circular twist in the field, while the lower bound (–1) indicates pure left-handed circular twist (see eq. 1).

3. OBSERVATIONS

The present analysis includes 9 intervals of data from Mariner 10 and Voyagers 1 and 2 and covers radial distances from 0.5 to 10 AU. The time resolution of the data sets range from 0.04 to 9.6 seconds and the frequency range covered is $3\times10^{-6} - 4.2$ Hz. Both these ranges extend the investigation by *Matthaeus and Goldstein* [1982] whose analysis was restricted to radial distances of 1–5 AU and frequencies of $6.9\times10^{-6} - 1.7\times10^{-2}$ Hz. Table 1 lists the spacecraft, the starting time of the specific intervals, the time between points Δt, the heliocentric distance R, the length in minutes of the interval, and $\overline{P}_3(\sigma)$ which is a measure of the degree of asymmetry in $\sigma_m(k)$. The value of $\overline{P}_3(\sigma)$ is obtained by first calculating $P(\sigma)$, the percentage of occurrence of $\sigma_m(k)$. The weighted average is then defined as $\overline{P}_3(\sigma) \equiv \sum_i \sigma_i^3 P_i(\sigma_i)$, where σ_i is the median value of σ in the ith bin. Our analysis procedure included editing the data to remove spurious points and using linear interpolation to fill data gaps. The highest-resolution data were filtered and decimated to eliminate aliasing. All intervals were linearly detrended and windowed with a 10% cosine taper.

The spectra of interval (1) are plotted in Figure 1 as an example of $H_m(k)$. The data consist of 21 minutes of Mariner 10 observations on day 59 of year 1974. The solid line indicates the trace of the power spectral matrix (here denoted E_B), and the top (bottom) panel contains the positive (negative) values of $kH_m(k)$.

Table 1

	Interval	Δt sec	R AU	Interval Length	$\overline{P}_3(\sigma)$
1	M10 1974:59:0:0	0.12	0.5	21.6 min	0.1979
2	M10 1974:59:0:23	0.12	0.5	21.6 min	0.1058
3	V1 1977:250:8:38	0.18	11	32.4 min	-0.3701
4	V1 1977:250:9:56	0.18	11	32.4 min	-0.2985
5	V1 1977:250:8:12	9.6	1	28.8 hr	0.0584
6	V2 1979:177:1:27	0.18	5	32.4 min	-0.0377
7	V2 1979:177:8:26	0.18	5	32.4 min	-0.0161
8	V2 1979:177:0:0	9.6	5	28.8 hr	-0.0586
9	V2 1981:186:0:0	9.6	10	28.8 hr	-0.0530

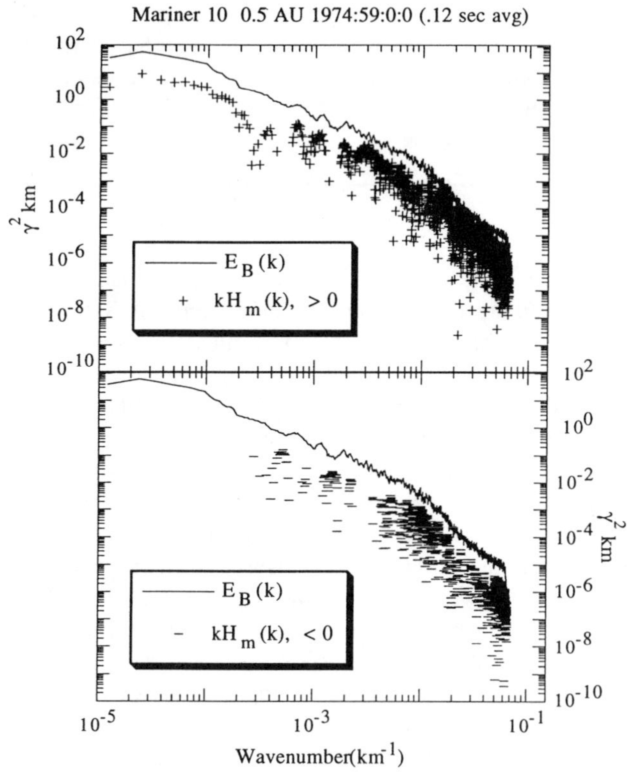

Fig. 1. (Top panel) The positive values of $kH_m(k)$ (+'s) bounded by the trace of the power spectral density matrix (solid line). The spectrum, in units of γ^2 *km*, is plotted versus wave number. Data is from interval 1 of Table 1. (Bottom panel) The negative values of $kH_m(k)$ are plotted as –'s for the same interval—the trace is repeated for clarity.

To compare the magnetic helicity of the different intervals and to ascertain the degree of randomness of the helicity values, we first computed $P(\sigma)$ and plotted the distributions. The top panel of Figure 2 displays the occurrence probability plots for both of the Mariner 10 intervals. There is a clear tendency toward positive helicity in these distributions which is confirmed in the values of $\overline{P}_3(\sigma)$, both are positive in magnitude (see Table 1). These results are also evident in the spectra of the first interval plotted in Figure 1. The right-hand modes of $\sigma_m(k)$ are consistently higher in magnitude than the corresponding left-hand modes within a small frequency band. In addition $\sigma_m(k)$ at the lowest wave numbers is virtually all right-hand.

Matthaeus and Goldstein's analyses at lower time resolutions showed no indication of a preference in the sign of $\sigma_m(k)$. For comparison to their study, the lower panel in Figure 2 displays the occurrence probability for 9.6 second data obtained by Voyagers 1 and 2 at 1 AU and 5 AU (solid and dashed lines, respectively). The distributions of $\sigma_m(k)$ for these intervals are both centered about zero. Also plotted in this panel, as a dotted line, is a distribution constructed from the 10 AU data set (cf. Table 1). No apparent change occurs in the magnetic helicity even out to 10 AU at these lower frequencies. Again this is confirmed in the values of $\overline{P}_3(\sigma)$ which have an average value significantly closer to zero than the first two high resolution intervals (see Table 1).

Figure 2 indicates that at high wave numbers and small heliocentric distances (0.5 AU), the distributions of $\sigma_m(k)$ are asymmetric and that at lower wave numbers and larger heliocentric distances (1–10 AU) this skewness in the distributions disappears. To investigate this as a function of heliocentric distance we have analyzed high-resolution data at 1 AU (intervals 3 and 4) and 5 AU (intervals 6 and 7). In Figure 3, top panel, we find that at 1 AU a skewed helicity still persists since both of the 0.18 second Voyager 1 intervals display a dominance of negative helicity (solid and dashed lines). These two intervals have the greatest degree of skewness as quantified in $\overline{P}_3(\sigma)$ (-0.3701 and -0.2985 respectively). However, the high-time-resolution from 5 AU (intervals 6, 7, and 8) appears to be symmetric—$\overline{P}_3(\sigma)$ is smaller by an order of magnitude. In the bottom panel we show the variation with resolution at 5 AU. The high resolution data is repeated (solid and dashed lines) with the addition of a lower resolution Voyager 2 example at the

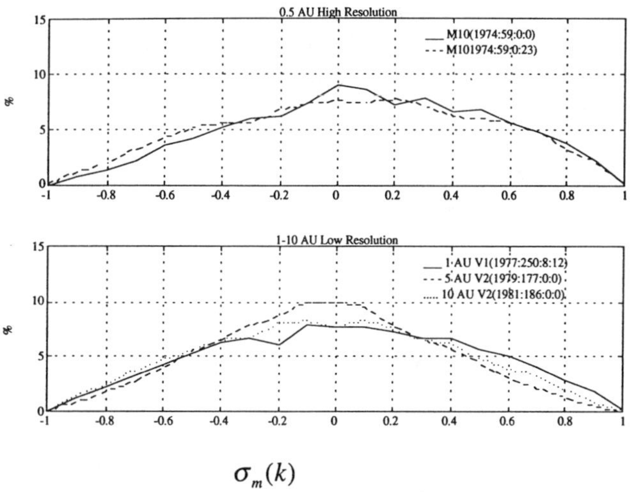

Fig. 2. (Top panel) The occurrence probability of $\sigma_m(k)$ computed using 0.12 sec averages from two Mariner 10 data sets (*cf.* Table 1) obtained near 0.5 AU. (Bottom panel) The percentage of occurrence of values of $\sigma_m(k)$ from 9.6 second averages at 1, 5, and 10 AU from Voyagers 1 and 2.

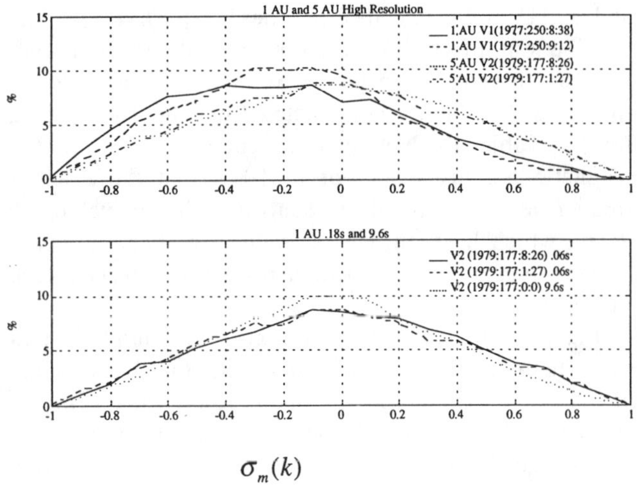

Fig. 3. (Top panel) The percentage of occurrence of values of $\sigma_m(k)$ from 0.18 second averages at 1, and 5 AU. (Bottom panel) The percentage of occurrence of values of $\sigma_m(k)$ from 9.6 and 0.18 second averages at 5 AU.

same radial distance (dotted line). All curves are now centered about zero, which is not always the case. *Goldstein et al.* [1993] included in their analysis a time period just one hour earlier than interval (6) and concluded that the distribution of $\sigma_m(k)$ was still highly asymmetric.

Another approach to the study of global invariants in the interplanetary medium that has proven very informative in the past [*Roberts et al.,* 1987a,b] is a correlation technique that shows the temporal behavior of a normalized quantity at various scales. Roberts used this procedure to study the evolution of the cross helicity in the heliosphere. However, the technique is somewhat more complicated for the magnetic helicity than it is for the normalized cross helicity σ_c, since in the latter case σ_c can be computed from the time series alone, while σ_m is usually computed from the Fourier transform of the magnetic field. Nonetheless, it is possible to form a configuration-space measure of the normalized magnetic helicity..

The procedure is to choose one of the transverse components of **B**, *e.g.* the *T* component in RTN coordinates, transform into Fourier space, and then shift the phase by 90° by multiplying the first half of the Fourier terms by $-i$ and the complex conjugate half by $+i$. Inverse transforming the rotated component then produces a new *T* component time series whose correlation with the original *N* component of the field is then a measure of the spatial twist in the field and, effectively, gives the space dependence of the normalized measure of the magnetic helicity. Before calculating the correlation, we first subtract a running mean over a time interval *n* from all components of **B**, including the phase-shifted *T* component. This measure of the magnetic helicity is defined by

$$\langle \sigma_m(t) \rangle_n \equiv 2 \frac{\langle b_T^r(t) b_N(t) \rangle_n}{\langle b^2 \rangle_n} \tag{6}$$

where the superscript *r* indicates the phase shift of 90° and the b_i are the fluctuating components of **B** resulting from the subtraction. The $\langle \, \rangle_n$ indicates an additional averaging by *n* which yields $\langle \sigma_m(t) \rangle_n$ on various temporal scales. The quantity $\langle \sigma_m(t) \rangle_n$ is essentially a horizontal cut across a dynamic spectrum.

Figure 4 displays the results of one such analysis using 9.6 second data from Voyager 1 on day 250 of year 1977. The first three panels are the field values in units of nT, δ is the polar angle measured with respect to the ecliptic, and λ is the azimuthal angle (0° is radially outward). The lower three panels display the temporal evolution of σ_m at three different scales. As the number of points included in the averaging interval increases from 3 (28.8 sec) in panel *d* to 27 (259.2 sec) in panel *f*, the magnitude of $\langle \sigma_m(t) \rangle_n$ systematically decreases. We have superposed (in white) on $\langle \sigma_m(t) \rangle_{28.8}$ a 50 point smoothing of the highly fluctuating values to give a clearer sense of the behavior of the function.

Within this interval, there are three periods when λ approaches 90° (near hours 250.4, 250.7, and 251.0). Decreases in the magnitude of σ_m correspond to these values of λ. This is especially evident at the longest wave length scales (panel *f*). The correlation of $\lambda \approx 90°$ with near zero values of σ_m is also present in the 5 AU 9.6 second data set and in the 10 AU data set (neither is shown here). This phenomena can result, at least in part, from a geometrical projection effect because when $\lambda \approx 90°$ any parallel propagating plane circularly polarized wave will be observed "edge-on" so that the reduced magnetic helicity computed from eq. (5) would be zero. The observed decrease in the magnitude of σ_m suggests that the fluctuations may be planar. This point is discussed and analyzed in more detail in *Goldstein et al.* [1993].

The higher resolution data also exhibits interesting correlations with λ, but the relationship is more complex and cannot be explained simply in terms of a geometrical effect. For example, in Figure 5, we have plotted an interval of high time resolution data obtained from Voyager 1 at 1 AU (0.18 second averages). At hour 8.7, λ makes a sharp transition from ~135° to ~80° and δ rotates from positive to negative values. At that time,

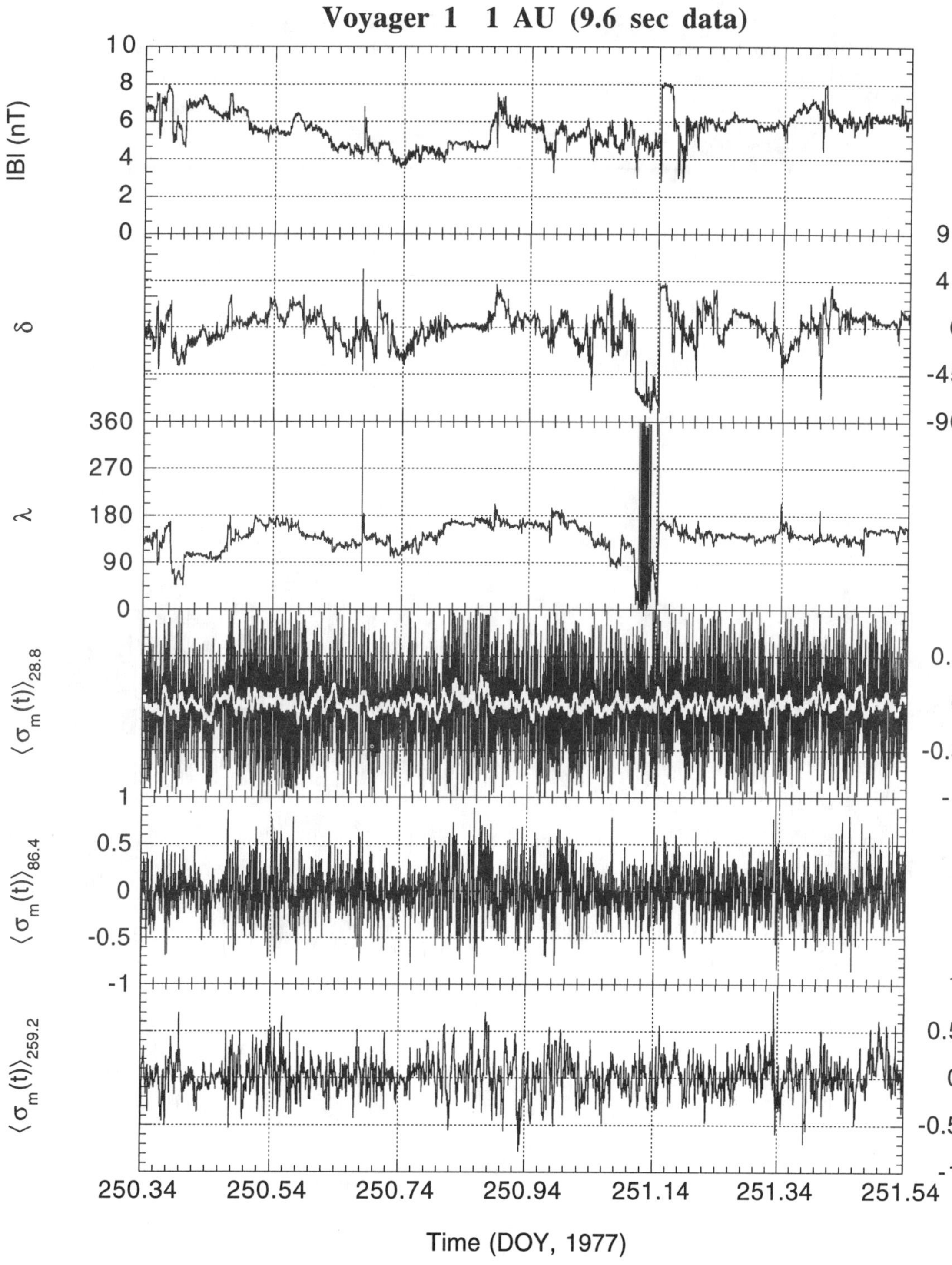

Fig. 4. Voyager 1 low resolution magnetic field data obtained on day 250 of year 1977 when Voyager 1 was near 1 AU (interval 5). Panels *a–c* are the magnetic field components B (in nT), δ, and λ. Panels *d–f* show $\langle\sigma_m(t)\rangle_n$ at 28.8 sec, 86.4 sec, and 259.2 sec scales computed from the correlation technique described in the text. A 50 point smoothing of the highly fluctuating values of $\langle\sigma_m(t)\rangle_{28.8}$ are superposed in white.

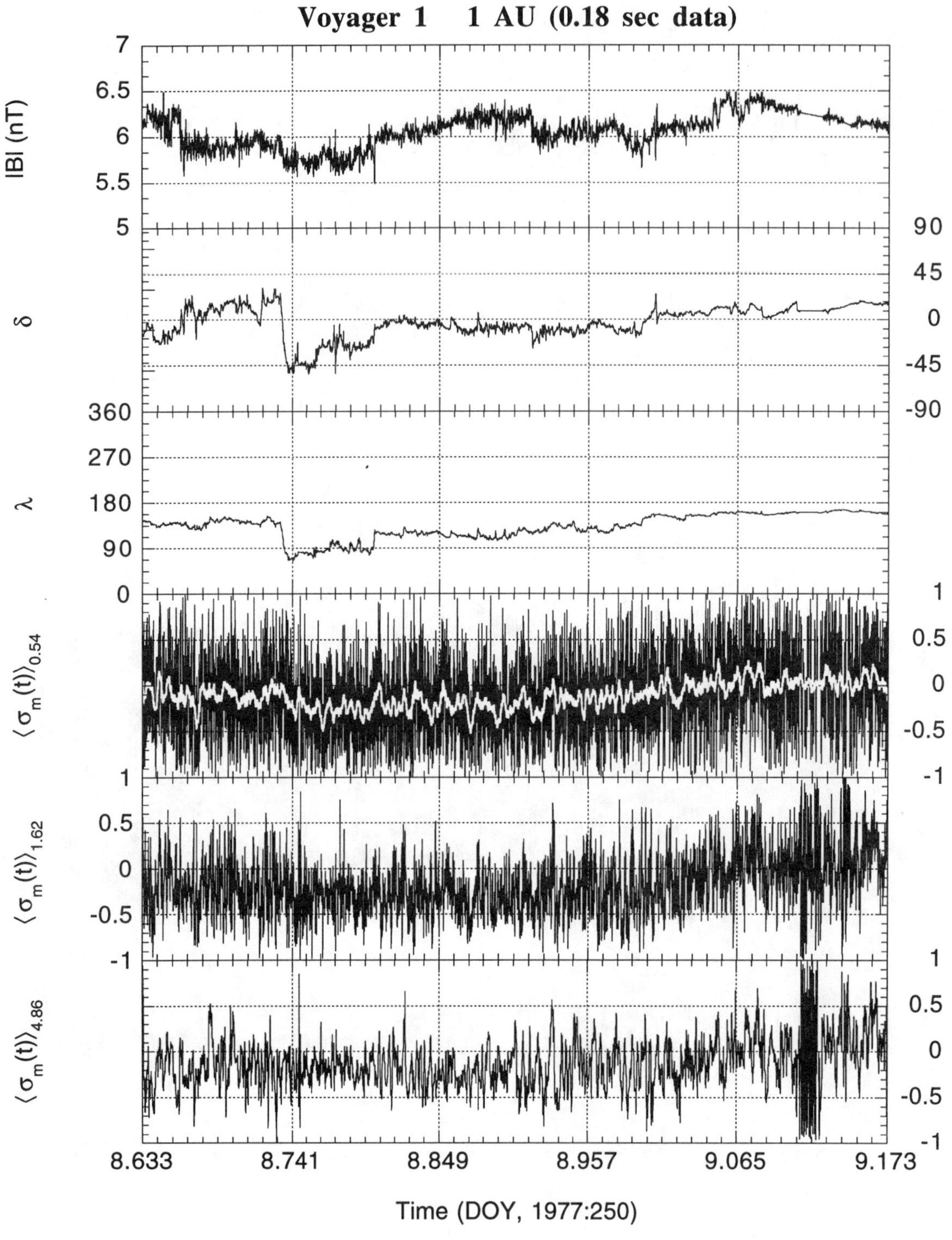

Fig. 5. Voyager 1 high resolution magnetic field data obtained on day 250 of year 1977 when Voyager 1 was near 1 AU (interval 3). Panels *a–c* are the magnetic field components B (in nT), δ, and λ. Panels *d–f* show $\langle \sigma_m(t) \rangle_n$ at 0.54 sec, 1.62 sec, and 4.86 sec scales computed from the correlation technique described in the text. A 50 point smoothing of the highly fluctuating values of $\langle \sigma_m(t) \rangle_{0.54}$ are superposed in white.

$\sigma_m(k)$ becomes even more negative (see panel *e*). As λ approaches 180° toward the end of the interval, $\sigma_m(k)$ becomes more symmetric about 0. We also performed similar analyses on Mariner 10 data sets and the reader is referred to *Goldstein et al.* [1993] for a discussion of those results.

4. SUMMARY AND DISCUSSION

In this preliminary study, we have found a tendency for the magnetic helicity to have a dominant sign at small spatial scales (*i.e.* high time resolutions). In *Goldstein et al.* [1993] the asymmetries are investigated in more detail and it is shown there that they are correlated with sector structure and that the sign of the correlation is consistent with the interpretation that the dissipation range of the spectrum is populated by whistler waves.

The time series of values of σ_m at varying wave number resolutions that we constructed provides a new diagnostic technique for studying magnetic helicity in the solar wind. The fact that the magnetic helicity tends toward zero when the interplanetary magnetic field lies in the ecliptic at 90° to the radial direction is consistent with interpretation that the interplanetary medium consists in part of plane polarized Alfvénic fluctuations that are propagating approximately parallel to the direction of the average direction of the magnetic field [cf. *Matthaeus et al.*, 1990].

Constructing a comprehensive picture of the variation of the fluctuating magnetic helicity with heliocentric distance requires more analyses, including the use of longer time averages of magnetic field data in the inner heliosphere, and additional examples in the outer heliosphere. A comparison with other solar wind flow parameters will also broaden the scope of understanding the magnetic helicity. The results of such a study have been reported by *Goldstein et al.* [1993].

REFERENCES

Frisch, U., A. Pouquet, J. Léorat, and A. Mazure, Possibility of an inverse cascade of magnetic helicity in magnetohydrodynamic turbulence, *J. Fluid Mech.*, *68*, 769, 1975.

Batchelor, G. K., *Theory of Homogeneous Turbulence*, Cambridge University Press, 1970.

Bieber, J. W., P. Evenson, and W. H. Matthaeus, Magnetic helicity of the IMF and the solar modulation of cosmic rays, *Geophys. Res. Lett.*, *14*, 864, 1987a.

Bieber, J. W., P. A. Evenson, and W. H. Matthaeus, Magnetic helicity of the Parker field, *Astrophys. J.*, *315*, 700, 1987b.

Bieber, J. W., and C. W. Smith, Magnetic helicity in the solar wind: Radial variation and cosmic ray effects, *EOS*, 247, 1992.

Ghosh, S., T. Stribling, M. L. Goldstein, and W. H. Matthaeus, Evolution of magnetic helicity in compressible magnetohydrodynamics with a mean magnetic field, in *Micro- and Mesoscale Phenomena in Space Plasma Physics*, edited by M. Ashour-Abdalla and T. Chang, *Geophys. Monogr. Ser.*, AGU, 1994.

Goldstein, M. L., D. A. Roberts, and C. A. Fitch, Properties of the fluctuating magnetic helicity in the inertial and dissipation ranges of solar wind turbulence, *J. Geophys. Res., submitted*, 1993.

Matthaeus, W. H., and M. L. Goldstein, Measurement of the rugged invariants of magnetohydrodynamic turbulence, *J. Geophys. Res.*, *87*, 6011, 1982.

Matthaeus, W. H., M. L. Goldstein, and C. Smith, Evaluation of magnetic helicity in homogeneous turbulence, *Phys. Rev. Lett.*, *58*, 1256, 1982.

Matthaeus, W. H., M. L. Goldstein, and D. A. Roberts, Evidence for the presence of quasi-two-dimensional, nearly incompressible fluctuations in the solar wind, *J. Geophys. Res.*, *95*, 20673, 1990.

Moffatt, H. K., *Magnetic Field Generation in Electrically Conducting Fluids*, Cambridge University Press, 1978.

Roberts, D. A., L. W. Klein, M. L. Goldstein, and W. H. Matthaeus, The nature and evolution of magnetohydrodynamic fluctuations in the solar wind: Voyager observations, *J. Geophys. Res.*, *92*, 11,021, 1987a.

Roberts, D. A., M. L. Goldstein, L. W. Klein, and W. H. Matthaeus, Origin and evolution of fluctuations in the solar wind: Helios observations and Helios-Voyager comparisons, *J. Geophys. Res.*, *92*, 12,023, 1987b.

Roberts, D. A., and M. L. Goldstein, Turbulence and waves in the solar wind, in *U. S. National Report to International Union of Geodesy and Geophysics*, edited by M. A. Shea, p. 932, AGU, 1991.

Smith, C. W., and J. W. Bieber, Detection of steady magnetic helicity in low-frequency IMF turbulence, in: *Proc. of Proc. 23rd International Cosmic Ray Conference,* Calgary, Alberta, Canada, 1993.

Stribling, T., W. H. Matthaeus, and S. Ghosh, Nonlinear decay of magnetic helicity in magnetohydrodynamic turbulence with a mean magnetic field, *J. Geophys. Res., in press*, 1992a.

Stribling, T., W. H. Matthaeus, and S. Oughton, Magnetic helicity in magnetohydrodynamic turbulence with a mean magnetic field, *Phys. Fluids B, submitted*, 1992b.

Stribling, T., and W. H. Matthaeus, Decay of magnetic helicity in ideal magnetohydrodynamics with a DC magnetic field, in *Micro- and Mesoscale Phenomena in Space Plasma Physics*, edited by M. Ashour-Abdalla and T. Chang, *Geophys. Monogr. Ser.*, AGU, 1994.

Taylor, G. I., The spectrum of turbulence, *Proc. R. Soc. London Ser. A*, *164*, 476, 1938.

Woltjer, L., A theorem on force free fields, *Proc. Natl. Acad. Sci. USA*, *44*, 489, 1958a

Woltjer, L., On hydromagnetic equilibrium, *Proc. Natl. Acad. Sci. USA*, *44*, 833, 1958b.

M. L. Goldstein, D. A. Roberts, and C. A. Fitch, Laboratory for Extraterrestrial Physics, Code 692, NASA-Goddard Space Flight Center, Greenbelt, MD 20771

A Simulation Study of the Formation of Solar Prominences

L. C. Lee, G. S. Choe, and S.-I. Akasofu

Geophysical Institute, University of Alaska, Fairbanks, Alaska

The formation of solar prominences is investigated in magnetic arcades under photospheric footpoint motions by numerical simulations, in which the gravity, radiative cooling, heat conduction and a simplified form of coronal heating are taken into account. Three different cases are presented. In the first case, a footpoint-shear induces the expansion of the arcade and the accompanying adiabatic cooling. The enhanced radiative cooling leads to plasma condensation by the thermal instability. The successive condensation of plasma in the vertical direction results in a sheet-like structure and the high density material presses down field lines to form dips creating a Kippenhahn-Schlüter type prominence. Above the prominence, the cooled material slides down along the field lines and leaves a low density cavity. The minimum value of plasma β for the dip formation is estimated. In the second case, it is found that photospheric converging motions imposed to a sheared magnetic arcade give rise to development of a current layer and drive magnetic reconnection. The compressed material in the magnetic island cools down through a thermal instability and slides along the field lines to the lower part of the island, leading to the formation of a Kuperus-Raadu type prominence. In the third case, a current sheet is found to form between two bipolar regions by photospheric shearing or converging motions. The magnetic reconnection in the current sheet leads to condensation of coronal material, resulting in a Kippenhahn-Schlüter type prominence.

1. INTRODUCTION

Solar prominences consist of cool (5000–8000 K) and dense ($\sim 10^{11}\,\mathrm{cm}^{-3}$) material embedded in the hot (10^6 K) and tenuous ($\sim 10^9\,\mathrm{cm}^{-3}$) solar corona. The location of prominences above the magnetic neutral lines [*Babcock and Babcock*, 1955] indicates the support of dense plasma by magnetic fields. Prominences can be grouped into two categories according to the direction of the magnetic field component perpendicular to the prominence plane and that of the plausible potential field deduced from the photospheric magnetic polarities across the neutral line. In the Kippenhahn-Schlüter type [*Kippenhahn and Schlüter*, 1957] or the normal polarity (hereafter NP) prominences, the two field directions are parallel, while in the Kuperus-Raadu type [*Kuperus and Raadu*, 1974] or the inverse polarity (hereafter IP) prominences, the two field directions are antiparallel. The ratio of the population of the two types observed by coronagraphs is about 3 (IP) to 1 (NP) [*Leroy*, 1989]. However, NP prominences are located lower than IPs so that the former are more difficult to detect than the latter. According to studies on the prominence heights, low-lying prominences are more abundant than high prominences [*Billings and Kober*, 1957].

There have been good reviews on the observations of solar prominences [*e.g.*, *Martin*, 1973, 1990; *Hirayama*, 1985; *Priest*, 1989]. Here we briefly summarize the observations relevant to the formation of prominences. The magnetic field vector in prominences is highly aligned along the prominence axis with an angle of about 25° [*Leroy*, 1989]. The photospheric fields also tend to align along the magnetic neutral line before a prominence is formed, and this magnetic shear is considered as one of the necessary conditions for promi-

Space Plasmas: Coupling Between Small and Medium Scale Processes
Geophysical Monograph 86

nence formation [*Martin*, 1990]. Sometimes shearing mass motions are also observed during prominence formation in two different patterns. In one type, the velocity vectors are oppositely directed across the neutral line, while in the other the velocity vectors are parallel on both sides of the neutral line but increase in magnitude towards the neutral line [*Rompolt and Bogdan*, 1986]. Another remarkable feature associated with prominence formation is the flux cancellation, in which two flux elements of opposite polarities come close to each other and disappear in the magnetogram. The observation of flux cancellation indicates the existence of converging flows towards the neutral line and magnetic reconnection either above or below the neutral line. In addition, arcades of closed field lines are observed to lie over prominences and a dark cavity with a low electron density is often observed between the prominence and the overlying arcade [*Waldmeier*, 1970]. It was naturally speculated that cavities are consequences of the depletion caused by condensation of material into the prominence [*Engvold*, 1989]. *Saito and Tandberg-Hanssen* [1973], however, found that the plasma mass in a cavity before depletion is not sufficient to supply the mass of the prominence. The presence of the overlying arcades may suggest that prominences are formed above neutral lines within a single bipolar arcade. However, a survey of quiescent prominences by *Tang* [1987] revealed that more prominences are formed between bipolar regions than inside a bipolar region.

Investigations of prominence formation are usually performed in two directions. The first direction is related to the cooling and condensation of plasma, which is attributed to the thermal instability. In the coronal plasma, the loss of energy per volume by radiation is proportional to the density squared and a function of temperature, which increases with decreasing temperature in some temperature range [*e.g., Cox and Tucker*, 1969]. In a perturbed plasma, which is dense and cool enough to overcome adiabatic and nonadiabatic heat flows, the perturbation can grow until a new equilibrium is reached where the radiative loss decreases with decreasing temperature. The condensation by this thermal instability has been studied analytically [*Field*, 1965; *Chiuderi and Van Hoven*, 1979; *Van Hoven and Mok*, 1984; *Steinolfson and Van Hoven*, 1984; *Van Hoven et al.*, 1986; *An*, 1985; 1986] and numerically [*Hildner*, 1974; *Sparks et al.*, 1990]. The second direction of prominence studies is related to dynamic mechanisms, which can perturb plasma strongly enough for the onset of thermal instability. By a numerical simulation, *Forbes and Malherbe* [1991] found that a loop prominence, which appears after a flare, can be formed by fast shocks generated by reconnection jets. *Wu et al.* [1990] performed a numerical simulation of the formation of a quiescent prominence by injecting high density material ballistically from the lower atmosphere. This assumption, however, can hardly be regarded as a general condition for prominence formation. Recently *Choe and Lee* [1992, hereafter CL] showed that a photospheric footpoint shear in a magnetic arcade can create a coronal condition where a prominence can be formed.

In this paper, we present a brief summary of numerical simulations on the prominence formation, where only the observationally established photospheric conditions are imposed as boundary conditions. Case 1 (Section 3) deals with the prominence formation by a shearing motion, Case 2 (Section 4) is concerning the formation of an IP prominence by a shearing and converging motion, and Case 3 (Section 5) treats the prominence formation between two bipolar regions. Mechanisms of prominence formation under various photospheric motions and field geometries are discussed.

2. DESCRIPTION OF MODELLING

In our simulation, a Cartesian geometry is adopted ignoring the curvature of the solar surface. Since the prominence-field configurations look more or less alike along the prominence axis, all the variables are assumed to be invariant in this direction (the z-coordinate axis) although the z-component of vector quantities are taken into account. The vertical coordinate is designated by y, and the horizontal coordinate perpendicular to the neutral line by x. Assuming the symmetry of the initial condition and perturbations across the y axis, our computation is performed only in one quadrant of the x-y plane.

The governing equations for the simulation are a full set of MHD equations including gravity, radiative cooling, thermal conduction and heating as follows:

$$\frac{\partial \rho}{\partial t} + \nabla \cdot (\rho \mathbf{v}) = 0 \tag{1}$$

$$\rho(\frac{\partial \mathbf{v}}{\partial t} + \mathbf{v} \cdot \nabla \mathbf{v}) = -\nabla p + \mathbf{J} \times \mathbf{B} + \rho \mathbf{g} + \rho \nu \nabla^2 \mathbf{v} \tag{2}$$

$$\begin{gathered} \frac{\partial \psi}{\partial t} + \mathbf{v} \cdot \nabla \psi = \eta_n \nabla^2 \psi, \\ B_x = -\frac{\partial \psi}{\partial y}, \qquad B_y = \frac{\partial \psi}{\partial x} \end{gathered} \tag{3}$$

$$\frac{\partial B_z}{\partial t} = \frac{\partial}{\partial x}(v_z B_x - v_x B_z)$$

$$+\frac{\partial}{\partial y}(v_z B_y - v_y B_z) + \eta_n \nabla^2 B_z \tag{4}$$

$$\frac{\partial T}{\partial t} = -\mathbf{v}\cdot\nabla T - (\gamma - 1)T\nabla\cdot\mathbf{v} + (\gamma-1)\frac{T}{p}(\nabla\cdot\kappa_{\parallel}\nabla_{\parallel}T + \eta J^2 + H - C) \tag{5}$$

$$\mathbf{J} = \nabla\times\mathbf{B} \tag{6}$$

$$p = \rho T \tag{7}$$

where ρ is the density, $\mathbf{v}$ the velocity vector (v_x, v_y, v_z), p the pressure, $\mathbf{B}$ the magnetic field vector (B_x, B_y, B_z), ψ the flux function, $\mathbf{J}$ the current density, T the temperature and γ the ratio of specific heats. All the quantities are expressed in nondimensionalized forms by a proper normalization, which is detailed in CL. The length scale unit L_0 is the initial pressure scale height at $y = 0$. The gravity $\mathbf{g}$ is a function of y such that $\mathbf{g} = -g\hat{y}$ and g is given by

$$g = \frac{4\pi G M_{\odot}}{(R_{\odot} + y)^2} = g_0 \frac{R_{\odot}{}^2}{(R_{\odot} + y)^2} \tag{8}$$

where $g_0 = 2.74 \times 10^4\,\mathrm{cm\,s^{-2}}$ is the surface gravity. The inclusion of the resistivity η_n and the viscosity ν is solely for the purpose of numerical smoothing. The adopted values are $\eta_n \approx 10^{-4} - 10^{-3}$ and $\nu \approx 10^{-3}$ while η in Equation (5) takes a realistic value $\sim 2\times10^{-14}$ [*Spitzer*, 1962]. The heat conduction is considered only along the field lines because the perpendicular conductivity is more than ten orders of magnitude smaller than the parallel conductivity whose value is taken from *Spitzer* [1962] as

$$\kappa_{\parallel} = k_0 T^{5/2} = 9\times10^{-7} T^{5/2}\,\mathrm{erg\,s^{-1}\,cm^{-1}\,K^{-1}}. \tag{9}$$

For the optically thin atmosphere, the radiative cooling term C in Equation (5) is of the form

$$C = n_e n_H Q(T) = n^2 Q(T), \tag{10}$$

where n_e is the electron number density and $Q(T)$ the cooling function [*Cox and Tucker*, 1969; *Tucker and Koren*, 1971; *McWhirter et al.*, 1975; *Raymond and Smith*, 1977]. An analytical fitting of $Q(T)$ to a piecewise continuous function $Q(T) = \chi_i T^{\alpha_i}$ has been attempted by *Hildner* [1974] and *Rosner et al.* [1978]. The forms of the two functions are shown in Figure 1. There are two major differences in the two functions, which can affect numerical results. First, the difference $Q(10^5\,\mathrm{K}) - Q(10^6\,\mathrm{K})$ is larger in *Hildner*'s function than in *Rosner et al.*'s. The cooling time scale is thus shorter with the former than with the latter. Secondly the exponent α at $T = 10^6\,\mathrm{K}$ is -1 by *Hildner* while 0 by *Rosner et al.*. Thus the cooling term with the latter function is sensitive only to the density perturbation until the temperature reaches the value where the exponent α changes. We have tested both functions and found that the difference in cooling time scales is enhanced (reduced) in cases dominated by the temperature (density) perturbation. For the coronal heating term H, we assume that

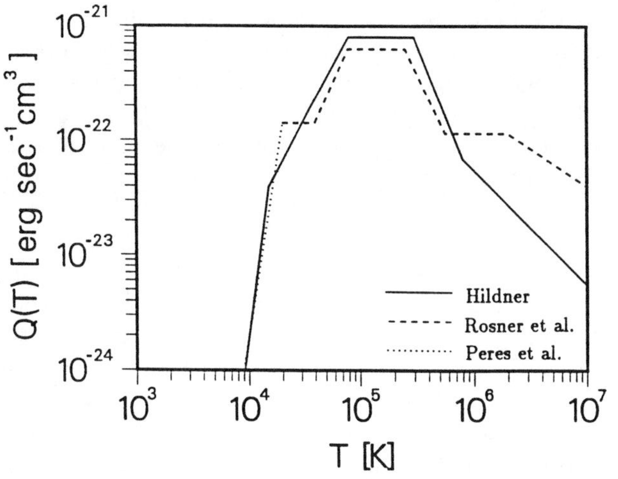

Fig. 1. Analytic fits of the radiation loss function $Q(T)$. The solid line is from *Hildner* [1974] and the dashed from *Rosner et al.* [1978]. *Peres et al.*'s [1982] curve (dotted line) is attached to *Rosner et al.*'s below 2×10^4 K [from *Choe and Lee*, 1992].

$$H = nh(y), \tag{11}$$

where $h(y)$ is a time-independent function. The initial thermal equilibrium is achieved by setting $C = H$ initially, whereby we obtain

$$h(y) = n_0(y)\chi_0 T_0{}^{\alpha_0}. \tag{12}$$

In this study, the solar corona is assumed to consist of fully ionized hydrogen gas and be initially isothermal with a temperature of 10^6 K and in hydrostatic equilibrium. This temperature gives the scale height $L_0 = 6\times10^4$ km at $y = 0$. The initial magnetic field is assumed to be potential and obtained by numerically solving the Laplace equation $\nabla^2\psi = 0$ with a given B_y profile at the bottom boundary. With no initial motions, the corona in the computational domain is in a mechanical and thermal equilibrium.

The boundary conditions at $x = 0$ are determined by the symmetry property. The other lateral boundary and the upper boundary are assumed to be open. A detailed description is given in CL. At the bottom boundary, the horizontal velocities v_x and v_z are given and no flow across the boundary is allowed, *i.e.*, $v_y = 0$, considering the far higher density of the photosphere. This condition for v_y allows us to determine the time evolution of $\psi(x, y = 0)$ by simply integrating the governing equation (Equation (3)). The density and temperature are set to be constant so that the bottom boundary may behave like a pressure and heat reservoir. The toroidal field B_z is obtained by integrating the governing equation (Equation (4)) with one side spatial differencing with respect to y.

A uniform 127 × 257 grid is used for Case 1 and non-uniform 101 × 101 grids are employed for Cases 2 and 3. Our test simulations show that a finer resolution yields a higher maximum density in the prominence. Therefore, the maximum density reported in this paper should be interpreted as a low end of the possible maximum density in reality.

The governing equations are finite differenced and integrated in time using the semi-implicit scheme [*Harned and Schnack*, 1986]. Although this scheme is free from the time-step restriction by the Alfvén and fast modes, the employed time-step size is 4-10 times the CFL time-step due to the nonlinearity of the problems. The heat conduction term is treated by the alternating direction implicit method as in the simulation of *Sparks et al.* [1990]. The readers are referred to CL for a detailed description of the numerical procedure.

3. FORMATION OF NORMAL POLARITY PROMINENCES BY FOOTPOINT SHEAR (CASE 1)

In this section, we present the formation of NP prominences by the simplest footpoint motion, the shear, following CL. The initial B_y profile at $y = 0$ is given by

$$B_y(x, y=0, t=0) = B_{y0}\left\{\exp\left(-\frac{(x+X_0)^2}{2{X_w}^2}\right) -\exp\left(-\frac{(x-X_0)^2}{2{X_w}^2}\right)\right\} \quad (13)$$

where $B_{y0} = 12\,\mathrm{G}$, $X_0 = 0.4\,L_0$, and $X_w = 0.27\,L_0$. The shear velocity pattern imposed at the bottom boundary is

$$v_z(x) = \frac{1}{2}V_{z0}\left\{\tanh\left(\frac{x-X_a}{X_c}\right) - \tanh\left(\frac{x-X_b}{X_c}\right) +\tanh\left(\frac{x+X_a}{X_c}\right) - \tanh\left(\frac{x+X_b}{X_c}\right)\right\}, \quad (14)$$

where $V_{z0} = 4\,\mathrm{km\,s^{-1}}$, $X_a = 0.21\,L_0$, $X_b = 0.59\,L_0$, $X_c = 0.04\,L_0$. The shear is thus localized around $x = \pm(X_a + X_b)/2$ with an approximate halfwidth of $(X_b - X_a)/2$ (see CL). Staring from zero, the shear velocity is increased to the above value in 470 s. The electron number density at the bottom is taken to be $5 \times 10^9\,\mathrm{cm^{-3}}$ and the plasma β at the origin is initially 0.2. The cooling function by *Hildner* [1974] is employed for the reported case.

The shear velocity v_z creates the field component B_z, whose pressure force expands the field lines with material attached to them. The first stage of evolution is almost adiabatic; effects of the heating, radiative cooling, and thermal conduction are small compared to the adiabatic cooling. As shown in Figure 2, the expansion of the arcade decreases the temperature over the whole domain. The local density is decreased in the lower part of the arcade where the expansion is most active, while the local density above that part is increased because higher density material from below is transported upward by the expanding field lines. The adiabatic cooling causes the enhancement of radiative cooling, which becomes more and more dominant in the temperature variation of plasma as the temperature drops. A runaway cooling—thermal instability—is thus effected. At $t \approx 7500\,\mathrm{s}$ the temperature drops quickly to about $10^4\,\mathrm{K}$ in an oval-shaped region at a height of $0.5\,L_0 \approx 3 \times 10^4\,\mathrm{km}$ as shown in Figure 2. Due to the sudden drop of pressure the material in the vicinity is sucked into the cool region with a speed of several $10\,\mathrm{km\,s^{-1}}$. Although the material flows into the oval-shaped region from all directions, the dominant flow is along the field lines from the lower atmosphere. The oval-shaped patch contracts along the field lines increasing the plasma density. Above and below this region, thermal instability takes place successively upward and downward. This is due to the variation of the onset time for thermal instability for different field lines. The condensed material thus forms a sheet-like structure as shown in the third column of Figure 2 ($t \approx 9400\,\mathrm{s}$). The density of the condensed material reaches at the peak about 70 times its initial value, which corresponds to the number density of $n = 2 \times 10^{11}\,\mathrm{cm^{-3}}$. It should be mentioned that we observed a higher value of maximum density by employing a finer grid. This high density material presses down the field lines to form a configuration with a dip, which is the essential signature of the Kippenhahn-Schlüter prominence model.

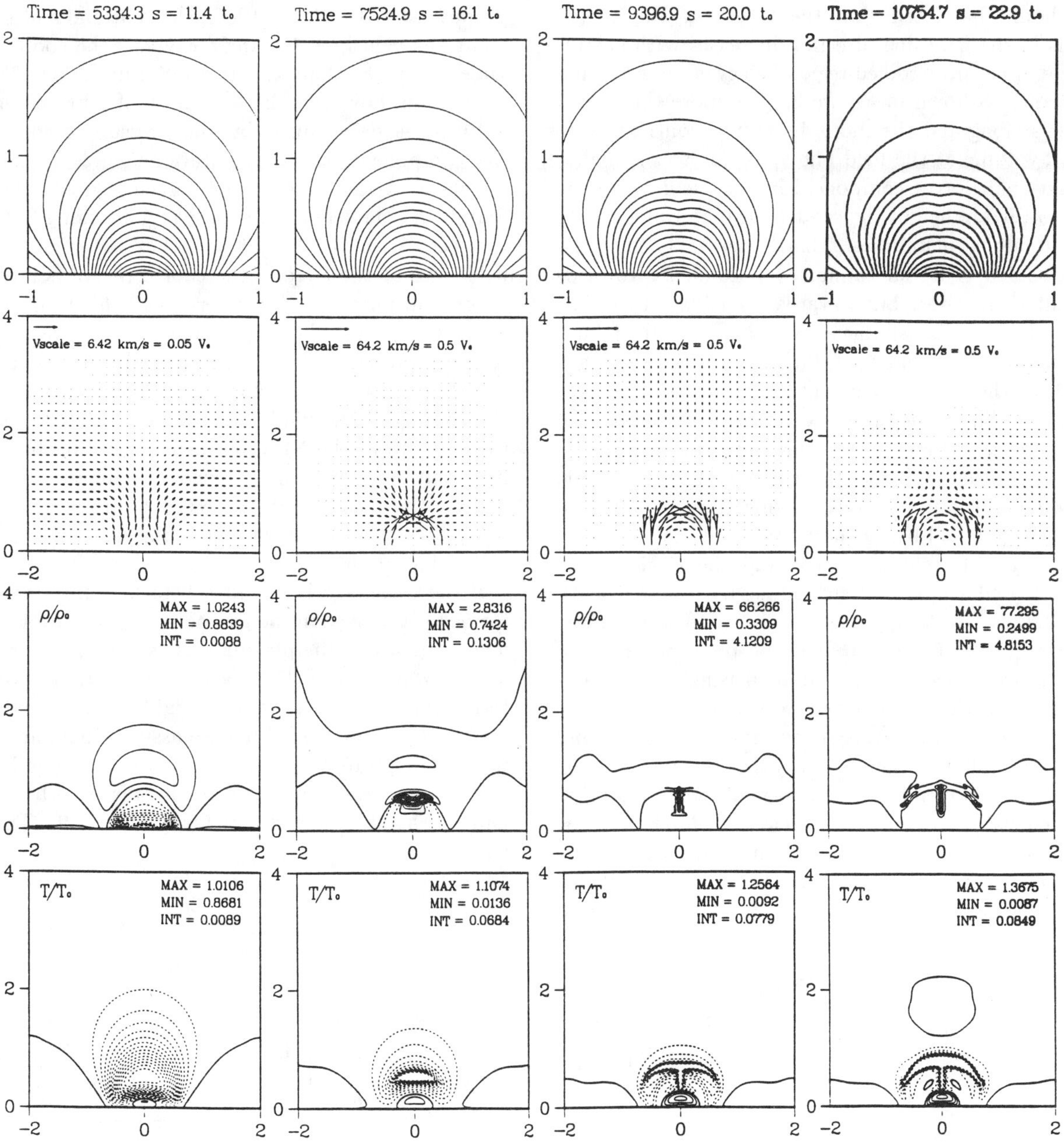

Fig. 2. Time evolution of the magnetic arcade in Case 1. From the top, the four rows show respectively the magnetic field lines, velocity fields, ratio of density to the initial local value, and ratio of temperature to the initial value. The time increases from left to right. In the contour plots, the solid lines represent values greater than 1, the dashed lines values less than 1, and the thick solid lines the value of 1. In the upper right corner of each contour plot are shown the maximum and minimum values and the interval between adjacent contours [from *Choe and Lee*, 1992].

It should be noted that the length of the cooled regions in the field line direction increases with height, and the uppermost cooled region has even an umbrella-like shape as shown in Figure 2. The more the cooled region is elongated the more difficult it contracts into a small volume at the field line apex due to the higher gravitational force component in the field direction. Above a certain height the condensed material is partly shedded along field lines. In a greater height, the newly cooled material cannot even condense towards the center any more but towards the edges of the low temperature region sliding down along the field lines (see the fourth column of Figure 2). The upward growth of the vertically elongated prominence is thus saturated. Since the sliding condensed material is of low temperature and of low pressure, material from below is still sucked into the condensed region while the whole structure moves downward. As seen in Figure 3, at $t \approx 14500\,\mathrm{s}$ the falling material is finally merged into the chromosphere and the depletion of material results in an arc-shaped cavity of low density (0.01–$0.2\,\rho_0$), low temperature (~ 1–$2 \times 10^5\,\mathrm{K}$) and low pressure ($10^{-3}$–$10^{-2}\,p_0$) around the prominence. The temperature, however, is still increasing due to the dominance of the heating and expected to recover the initial coronal temperature as observed by *Bessey and Liebenberg* [1984]. It is noteworthy that the cavity in our simulation is formed by the drainage of cooled material. The pre-existing material in the cavity volume does not contribute to the prominence mass and most of the prominence mass is supplied by the siphon-type field-aligned flows. In the mechanism of cavity formation proposed by *An et al.* [1985], the material slides down along the field lines leaving a cavity if the cooling time scale is longer than the free-fall time scale. In our simulation, however, the formation of a prominence or a cavity is determined by the pressure gradient and the component of the gravitational force in the field direction at the edge of the cooled region. We have found that the length of cooled regions along field lines increases with the height and that the inclination of field lines at the edge of the cooled region is larger for outer field lines than for inner ones. This result can be expected for a self-similar expansion of arcades. The cooled material in the lower part of the arcade can thus readily contract towards the field line apex whereas in the upper part the condensed material slides down along the field lines. The NP prominences can thus be formed favorably on a rather flattened field lines. This may explain the reason why the NP prominences lie in the low corona because newly emerging flux tubes may be rather flat at their apex.

If the plasma β is too low, field line dips cannot be formed. In our prominence model, dips are formed on field lines where the plasma β is as low as 0.1, which is quite low in view of the "conventional" criterion for dip formation, $\beta \geq 1$. Here we roughly derive a criterion for the dip formation. First we assume that the corona without a prominence is in a hydrostatic equilibrium embedding a magnetic arcade which is force-free in the neighborhood of the symmetry axis $x = 0$. Then we have

$$\rho_o g = -\frac{dp}{dy}, \tag{15}$$

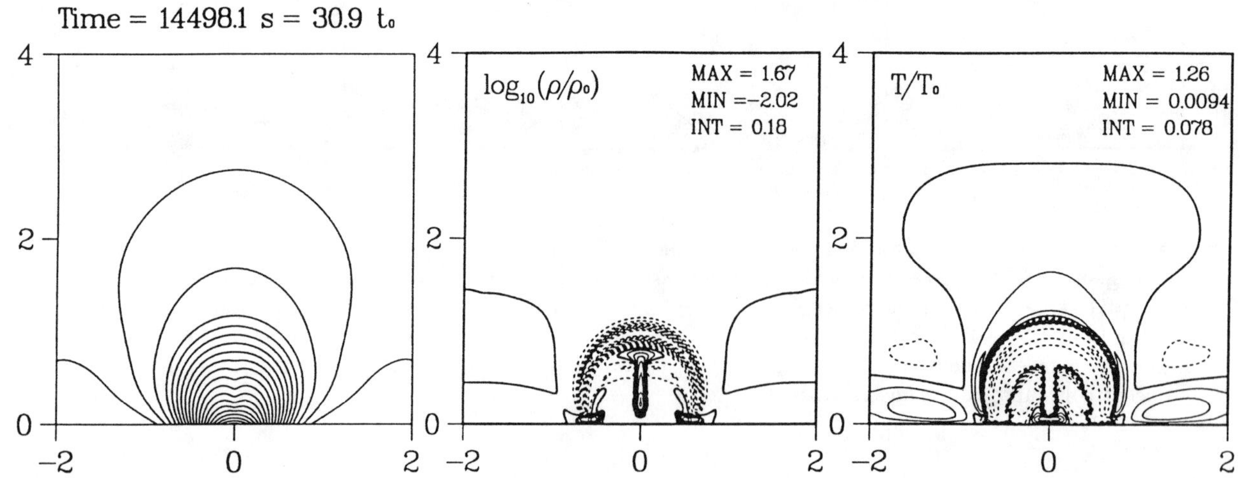

Fig. 3. Field lines and contours of $\log_{10}(\rho/\rho_0)$ and T/T_0 at $t = 1.45 \times 10^4$ s in Case 1 showing a coronal cavity surrounding the prominence [from *Choe and Lee*, 1992].

$$\frac{B^2}{4\pi R_o} = -\frac{d}{dy}\left(\frac{B^2}{8\pi}\right), \tag{16}$$

where R is the radius of curvature of field lines. Throughout our derivation, all the variables are evaluated at $x = 0$ and thus functions of y only. The subscript "o" denotes values just before the prominence formation while the subscript "p" will indicate values just after the prominence formation. Secondly we also assume that the pressure $p(x = 0, y)$ and the magnitude of the magnetic field $B(x = 0, y)$ change little after the formation of a prominence in a region inside the prominence. Although this assumption cannot be generally applied, it approximately holds if the depth of a dip is small compared to the radius of curvature of the original field lines as in our simulation. By this assumption, $p_o \approx p_p$ and $B_o \approx B_p$, the subscripts will be deleted for these two variables. After a prominence is formed, the plasma is in an MHD equilibrium with dips on field lines. Then

$$\rho_p g = -\frac{dp}{dy} + \frac{B^2}{4\pi R_p} - \frac{d}{dy}\left(\frac{B^2}{8\pi}\right). \tag{17}$$

Note that the curvature normal of the field lines is now upward. Setting $\Lambda_o = (k_B T_o)/(m_H g)$, $\Lambda_p = (k_B T_p)/(m_H g)$, $\beta = 8\pi p/B^2$, we can rewrite the above equation as

$$\frac{p}{\Lambda_o} = \frac{p}{\Lambda_p} + \frac{2p}{\beta R_p} + \frac{2p}{\beta R_o}. \tag{18}$$

When the field line just starts to be depressed, $R_p = \infty$ and

$$\beta = \frac{2\Lambda_o}{R_o(1 - \frac{\Lambda_p}{\Lambda_o})}. \tag{19}$$

Since $(\Lambda_p/\Lambda_o) \approx 10^2 \gg 1$, the criterion for dip formation is approximately

$$\beta > \frac{2\Lambda_p}{R_o} \equiv \beta_{min}. \tag{20}$$

Considering the width of the weak field corridor ($\sim 2 \times 10^4$ km), the smallest radius of curvature of field lines over this corridor may be about 10^4 km. With $R_o \approx 10^4$ km and $\Lambda_p \approx 500$ km, the minimum value of plasma β is $\beta_{min} \approx 0.1$. Since outer field lines or sheared field lines have a larger radius of curvature the lowest allowable β can be even smaller.

4. FORMATION OF INVERSE POLARITY PROMINENCES BY SHEARING AND CONVERGING MOTIONS (CASE2)

The field configuration of the inverse polarity prominence has a magnetic island within an arcade, which can be obtained from a potential-like field geometry through magnetic reconnection. It is known that a shear imposed on a single magnetic arcade cannot develop a current sheet and reconnection of field lines [*Biskamp and Welter*, 1989; *Klimchuk and Sturrock*, 1989; *Finn and Chen*, 1990; *Steinolfson*, 1991]. The observation of flux cancellation [*Martin*, 1990] suggests that converging motions imposed on a sheared field may drive magnetic reconnection. A plausible scenario has been presented by *van Ballegooijen and Martens* [1989] based on the observation. The simulation studies by *Lee* [1990] and *Inhester et al.* [1992] have shown that a converging motion superposed to a shear can drive magnetic reconnection and result in a field configuration with a magnetic island. The present study proceeds one step further with all the thermodynamic effects taken into account.

In this case, the initial B_y profile at $y = 0$ is given by

$$B_y(x, y = 0, t = 0) = B_{y0} e^{\frac{1}{2}} \frac{x}{X_0} \exp\left(-\frac{x^2}{2X_0^2}\right), \tag{21}$$

where $B_{y0} = 50\,\mathrm{G}$ and $X_0 = 0.32\,L_0$. The resulting field configuration has a large curvature compared to the footpoint distance, which is in contrast with the field in the former NP case. The shear pattern employed is

$$v_z(x) = V_{z0} \tanh\left(\frac{x}{X_0}\right) \exp\left(-\frac{x^2}{4X_0^2}\right), \tag{22}$$

where $V_{z0} = 4\,\mathrm{km\,s^{-1}}$. The converging velocity profile is given by

$$v_x(x) = -\frac{1}{2} V_{x0} \left(\tanh\left(\frac{x + X_a}{X_b}\right) + \tanh\left(\frac{x - X_a}{X_b}\right) \right) \times \tanh^2\left(\frac{30x}{X_L}\right) \exp\left(\frac{3x^2}{X_L^2}\right), \tag{23}$$

where $V_{x0} = 2\,\mathrm{km\,s^{-1}}$, $X_a = 0.23\,L_0$, $X_b = 0.45\,L_0$ and $X_L = 4\,L_0$. With this converging velocity, the magnetic flux through $y = 0$ plane always remains finite. The shear velocity is increased from zero at $t = 0$ to the full value in 430 s, kept constant up to $t = 1.7 \times 10^4$ s and decreased to zero at $t = 2.6 \times 10^4$ s. The converging velocity is increased from zero to its full value between $t = 6.0 \times 10^3$ s and $t = 6.9 \times 10^3$ s, kept constant until

$t = 2.6 \times 10^4$ s and decreased to zero at $t = 3.0 \times 10^4$ s. The density at the bottom is taken to be 2×10^9 cm^{-3}, which gives a longer cooling time than in Case 1.

It was shown in CL that a mild converging motion, when added to a shear, promotes the cooling of material by increasing the density. The initial evolution of the arcade in the present case is similar to that in Case 1, *i.e.*, as the plasma is cooled down adiabatically by the arcade expansion, the radiative cooling becomes more and more dominant over the heating. The central part of the arcade is first drastically cooled down to the prominence temperature, but the cooled material bifurcates and condenses downwards along the field lines (see the first row of Figure 4), which is attributed to the high field-aligned gravity due to a high curvature of field lines. Above and below this region, plasma cools successively and condenses downwards along the field lines. A cavity is thus formed without any prominence at the field line apices.

As the footpoints get closer a current layer develops along the lower y-axis and fields lines are reconnected (the second row of Figure 4). The enhanced density in the magnetic island triggers thermal instability and the swift cooling to the prominence temperature takes place at the upper part of field lines where material is most compressed. Due to the high curvature of the looped field lines, the cooled material cannot condense on the top of the field lines, but moves downwards along the field lines. The condensed material is finally accumulated at the bottom of field lines in the island

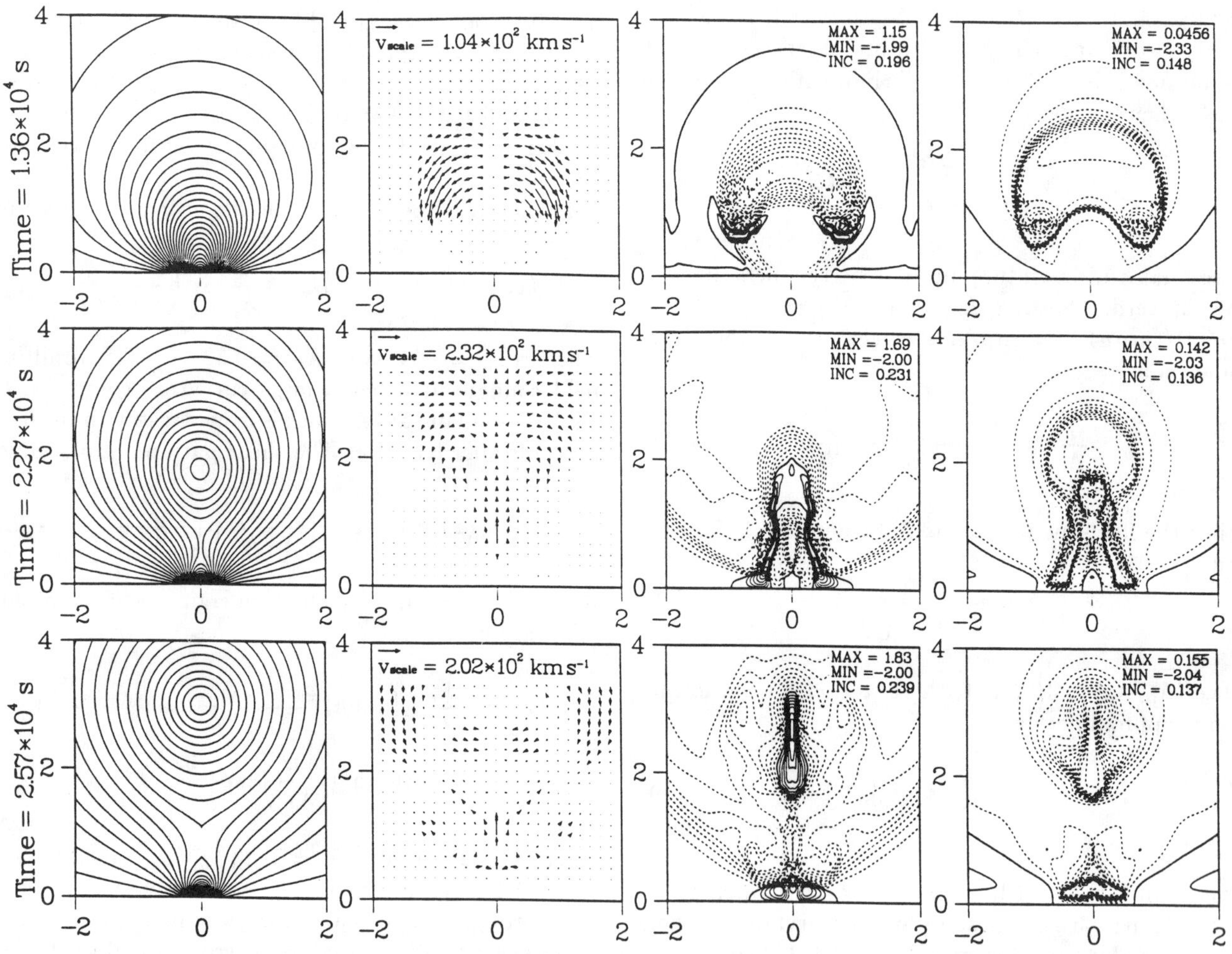

Fig. 4. Time evolution of the magnetic arcade in Case 2. Time increases from the top to the bottom. From left to right, the four columns show respectively the magnetic field lines, velocity fields, contours of $\log_{10}(\rho/\rho_0)$ and contours of $\log_{10}(T/T_0)$. In the contour plots, the solid lines represent values greater than 0, dashed lines values less than 0, and the thick solid lines the value of 0. The maximum and minimum values and the increment between adjacent contour levels are shown in the upper right corner of each contour plot.

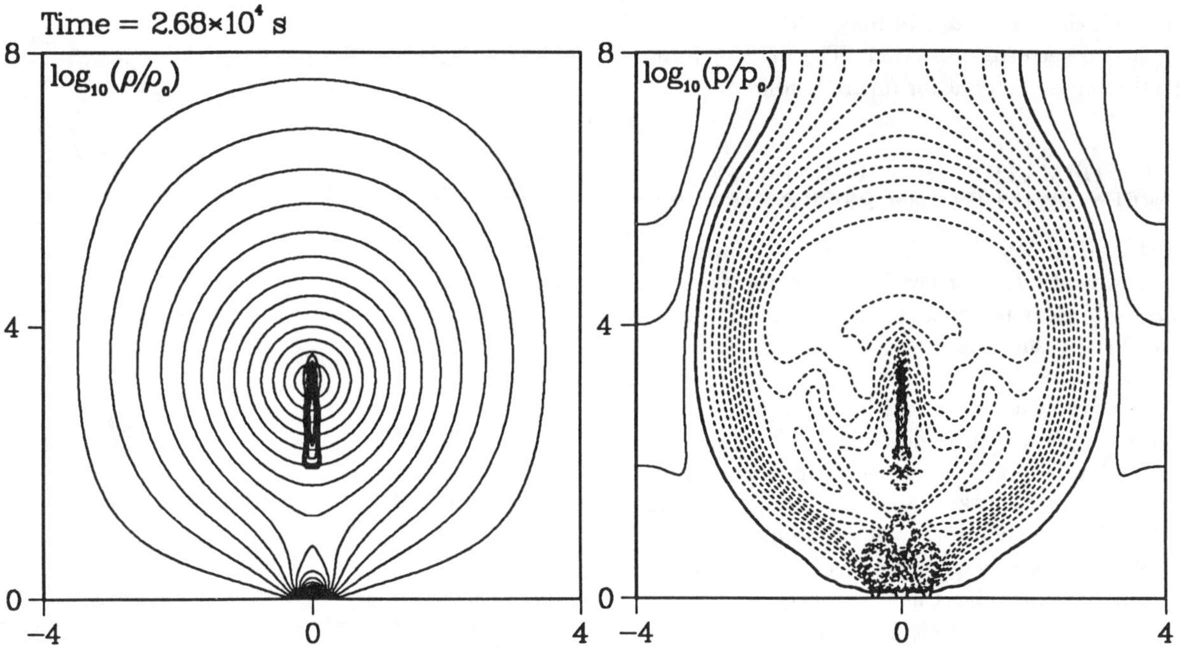

Fig. 5. On the left, contours of $\log_{10}(\rho/\rho_0) \geq 0$ at $t = 2.68 \times 10^4$ s in Case 2 are superimposed on the magnetic field lines. The corresponding contours of $\log_{10}(p/p_0)$ are shown on the right.

to form a sheet-like structure which extends from the center to the lower boundary of the island. A minimum temperature of 8500 K and a density increase of 110 times are attained at $t = 26800$ s (Figure 5). The prominence formed in Case 2 is located far higher ($\sim$ 2-4 L_0) than the NP prominence in Case 1, which is consistent with the statistics that IP prominences have a higher population than NP prominences above some height [*Leroy*, 1989]. It is also noted that this mechanism is not affected by the plasma β. The plasma β in our prominence is about 0.01. As shown in Figure 5 the region in the island where material is depleted to the prominence has formed a cavity which has a lower pressure ($\sim 4 \times 10^3 p_0$) than the surrounding cavity region formed before magnetic reconnection.

At the end of our simulation, the island contains about 62% of the total mass in the volume surrounded by the flux surface crossing the X-line, while the rest 38% is held in the arcade volume below the X-line. The prominence possesses 92% of the mass in the island. Since our simulation domain consists of the corona only, the mass of a prominence is limited to a portion of the coronal mass. The magnetic reconnection responsible for flux cancellation is, however, thought to take place just near or in the photosphere, for otherwise the reconnected flux tube would have to make a journey too long and too hard to reach the photospheric level where the magnetic flux is measured. If the magnetic reconnection takes place in the photosphere or chromosphere, material of far higher density can be transported into the corona than in our simulation and even such a big prominence may be able to form that its mass is comparable to the whole coronal mass. The siphon flow can also contribute to the mass supply in a three dimensional geometry. Observations show that prominences are anchored in the photosphere by feet which are located with a 3×10^4 km separation [*Schmieder*, 1989]. The feet can thus act as a mass transport channel between the prominence and the photosphere both for a siphon-type mass inflow and for the leaking of condensed material.

The stability of IP prominences is quite robust against lateral perturbations. However, one may not exclude the possibility of disintegration of the prominence structure if a reverse magnetic reconnection takes place. In one simulation, we turned off the footpoint motion after the prominence formation and let the system relax by itself for 10^4 s. The magnetic island and the prominence are found to rise to the end of simulation though slow. Observations show that most prominences have a tendency to rise rather than to descend [*Zirin*, 1988]. Thus a reverse magnetic reconnection is thought to be relatively improbable in reality. However, if there exists a photospheric flow which takes energy away from the coronal field, the line-tied field may collapse downward leading to a reverse reconnection. The investigation of this possibility is reserved for future studies.

5. FORMATION OF PROMINENCES BETWEEN TWO BIPOLAR REGIONS (CASE 3)

The statistical study by *Tang* [1987] showed that more prominences are formed between bipolar regions than inside bipolar regions. This is not surprising because a current sheet is expected to form inevitably when two bipolar regions are brought into contact. By magnetic reconnection, field lines will be connected from two originally unconnected arcades to form a field geometry favorable to supporting a prominence. If the reconnection takes place at multiple X-lines a Kuperus-Raadu geometry will be obtained whereas a Kippenhahn-Schlüter geometry is expected to form with a single X-line. It is not necessary to bring the footpoints of two arcade close together to develop a current sheet. The expansion of arcades by shearing motions will also cause a current sheet formation. Furthermore the shear velocity pattern need not be antiparallel across the neutral line. No matter whether we have an antiparallel shear or a parallel shear like the one observed by *Rompolt and Bogdan* [1986], each arcade will expand separately and come into contact to form a current sheet. With an antiparallel shear, the photospheric magnetic field component along the neutral line, B_z, has the same direction on both sides of the neutral line as shown in Figure 6a. Observations sometimes show that the magnetic field component along the neutral line, B_z, is oppositely directed across the neutral line [*Martin*, 1990]. This is definitely suggestive of a parallel footpoint motion (see Figure 6b). However, for a single arcade overlying the neutral line, an antiparallel B_z distribution cannot be achieved even if a parallel shear is imposed. Therefore, we may assert that where an antiparallel B_z distribution across the neutral line is observed, field lines anchored on the two sides are probably not interconnected.

The initial field profile in our simulation is adopted from *Low* [1992] taking on a quadrupolar topology:

$$\psi(x,y) = \frac{B_0}{k}\left[\exp(-ky)\cos(kx) - a\exp(-3ky)\cos(3kx)\right] \tag{24}$$

where $B_0 = 50\,\mathrm{G}$, $k = \pi/(2\,L_x)$ with $L_x = 2\,L_0$ and a is taken to be 0.32. Note that, in the region $y > 0$, the field given above is bipolar for $a \le 1/9$, quadrupolar

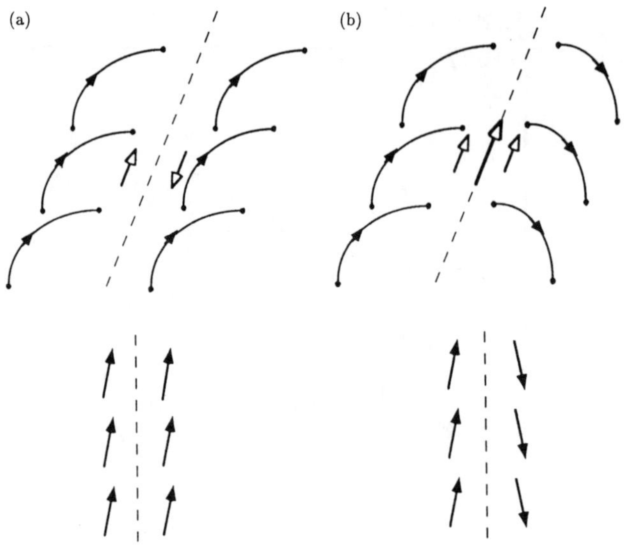

Fig. 6. Schematic sketch of the shear patterns (light arrows) and the associated photospheric field directions (dark arrows) between two bipolar regions. (a) With an antiparallel shear, the photospheric fields are parallel on both sides of the neutral line. (b) With a parallel shear, the photospheric field components along the neutral line are antiparallel across the neutral line.

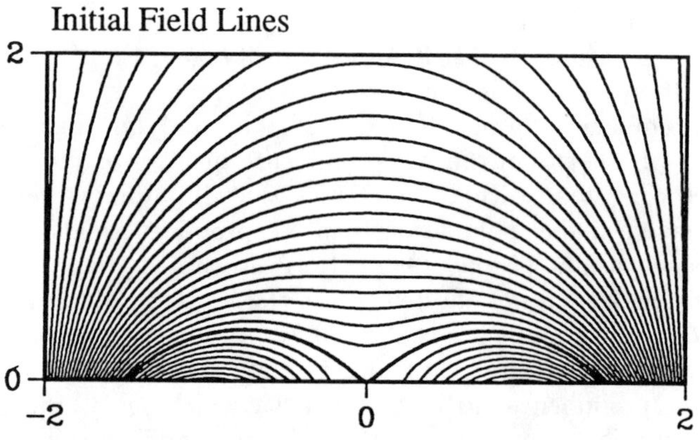

Fig. 7. Initial magnetic field configuration in Case 3. The height of the computational domain is three times as large as the vertical size of the panel. The thick solid line represents the field line passing through the origin, which initially separates three flux systems.

with no neutral point for $1/9 < a \le 1/3$, and has an X-type neutral point for $a > 1/3$. The shape of the initial field lines is shown in Figure 7, in which three neutral lines are shown at $x = \pm X_n = \pm 0.99\,L_0$ as well as at $x = 0$ in $y = 0$ plane. An antiparallel shear profile is employed as follows:

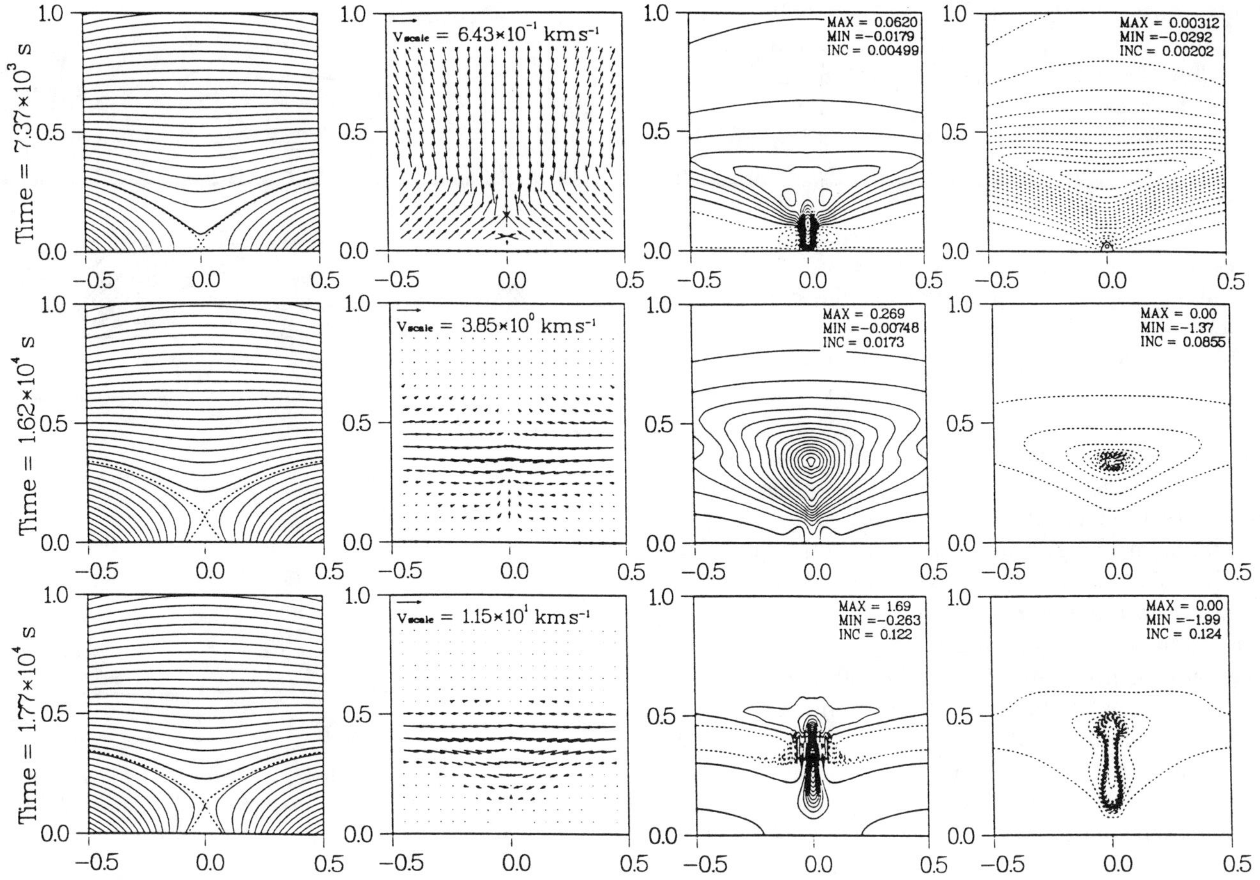

Fig. 8. Time evolution of the magnetic arcade in Case 3. Time increases from the top to the bottom. From left to right, the four columns show respectively the magnetic field lines, velocity fields, contours of $\log_{10}(\rho/\rho_0)$ and contours of $\log_{10}(T/T_0)$. In the plot of field lines, the thick solid line represents the field line passing through the origin at $t = 0$ and the dashed line the field lines passing through the X-point. In the contour plots the solid lines represent values greater than 0, dashed lines values less than 0, and the thick solid lines the value of 0.

$$v_z(x, y=0) = \begin{cases} -V_{z0}\sin(\dfrac{\pi x}{X_n}), & \text{if } |x| \le X_n; \\ 0, & \text{if } |x| > X_n. \end{cases} \tag{25}$$

To accelerate the current sheet formation, a converging motion is added to the above shears.

$$v_x(x, y=0) = \begin{cases} -V_{x0}\sin(\dfrac{\pi x}{X_n}), & \text{if } |x| \le X_n; \\ 0, & \text{if } |x| > X_n. \end{cases} \tag{26}$$

The maximum shear and converging velocities are taken respectively as $V_{z0} = 2\,\mathrm{km\,s^{-1}}$ and $V_{x0} = 0.5\,\mathrm{km\,s^{-1}}$. Both v_z and v_x are increased starting from zero at $t = 0$ to their full values at $t \approx 860\,\mathrm{s}$. An electron number density of $2 \times 10^9\,\mathrm{cm^{-1}}$ is assumed along $y = 0$. With the above field and density, a minimum plasma β of 1.4×10^{-3} is obtained, but the β at the origin is rather high (~ 2) since a virtual X-point is located just below the origin. In the present case the cooling function by *Rosner et al.* [1978] is employed. As mentioned in Section 2, we set $v_y\,(y = 0) = 0$ in our calculation. In this respect, it is noted that there has been a controversy about the effect of $v_y(y = 0)$ on the current sheet formation in the field configuration with two arcades [*Karpen et al.*, 1990; *Low*, 1991].

Our field configuration consists of three separate flux systems. As shown in Figure 8, the expansion of both arcades expels magnetic flux towards the neutral line at $x = 0$, and the overlying arcade is lifted up. The current layer first forms at the origin, where the first magnetic reconnection takes place. As time grows, the X-point slowly moves up with successively reconnecting field lines. By the expansion of all three arcades, the temperature drops over the whole domain except just below the X-point. The temperature minimum is located in the lower part of the overlying arcade. The density is decreased inside the two expanding arcades, but increased in the overlying arcade. The highest density enhancement is, however, achieved in the current layer region, especially just above the X-point. The density enhancement maximum moves up slowly while the temperature minimum remains at the same site.

The cooling mechanism in this case is very similar to that in Case 1. In the first evolutionary stage, the adiabatic cooling leads the nonadiabatic processes. In Case 3, however, the density near the temperature minimum is increased not only by the shear-induced expansion but also by magnetic reconnection. After $t \approx 15000$ s, the temperature decreasing rate is abruptly accelerated at the site of temperature minimum. As material is sucked into the cool region the density enhancement maximum also moves from the current layer region to the temperature minimum. The cooling to the prominence temperature takes place successively above and below the first cooled spot forming a vertical sheet-like structure. By the end of the simulation, the prominence extends to a height of $1.2\,L_0$ (Figure 9). Different from Case 1, however, neither shedding of condensed material nor cavity formation can be seen due to the very flat field geometry. The small field line curvature also enables high density material to stay at the apex of upper field lines without dips although this equilibrium seems to be unstable to a strong perturbation. The plasma β in the prominence is less than 0.01, which makes it impossible for a dip to be formed by the gravity of the material. The dips on field lines in Case 3 are due to the intrinsic field topology associated with the presence of the X-point.

Judging from our results, prominences between two bipolar regions are more likely to have a Kippenhahn-Schlüter geometry than a Kuperus-Raadu geometry. However, we cannot exclude the possibility of Kuperus-Raadu type prominence formation. The numerical resistivity or the strong overlying field may have prevented the current layer from growing long enough to allow the multiple X-line reconnection [*Lee and Fu*, 1986]. More observations are necessary to model the prominences between bipolar regions.

A case with a parallel shear is also simulated and found to yield a result similar to that of Case 3. A detailed comparison will be presented in a later paper.

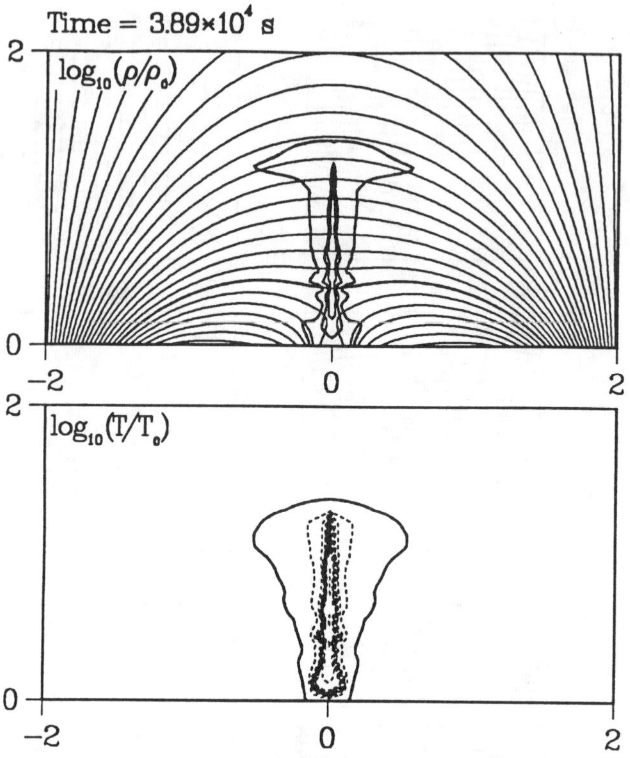

Fig. 9. In the top panel, contours of $\log_{10}(\rho/\rho_0) \geq 0$ are superimposed on the magnetic field lines. At the bottom, contours of $\log_{10}(T/T_0)$ are shown for Case 3 at $t = 3.89 \times 10^4$ s.

6. SUMMARY

In summary, we have demonstrated prominence formation by numerical simulation in three different cases. In Case 1, the shear-induced expansion of a magnetic arcade is shown to lead to a thermal instability through adiabatic cooling and upward material transport by magnetic fields. In Case 2, a converging motion combined with a shear is found to create a current layer and drive magnetic reconnection in a single arcade. The compression of material in the upper part of the island generates thermal instability. In Case 3, a shearing or a converging motion can bring two arcades into contact to form a current sheet. The magnetic reconnection and plasma expansion give a perturbation in temperature and density above the X-point, which leads to thermal instability. Formation of prominence cavity is also observed in Case 1 and Case 2.

In addition, we have found that the minimum value of plasma β for the formation of NP prominences within one magnetic arcade is approximately $\beta_{min} = 2\Lambda_p/R_o$.

Acknowledgments. This work was supported by DOE grant DE-FG06-91ER 13530 to the University of Alaska. Supercomputing resources were provided by the San Diego Supercomputer Center.

REFERENCES

An, C.-H., Formation of prominences by condensation modes in magnetized cylindrical plasma, *Astrophys. J., 298,* 409, 1985.

An, C.-H., Condensation modes in sheared magnetic fields, *Astrophys. J., 304,* 532, 1986.

An, C.-H., S. T. Suess, E. Tandberg-Hanssen, and R. S. Steinolfson, On the formation of coronal cavities, *Solar Phys., 102,* 165, 1985.

Babcock, H. W., and H. D. Babcock, The Sun's magnetic field, 1952–1954, *Astrophys. J., 121,* 349, 1955.

Bessey, R. J., and D. H. Liebenberg, Coronal temperature measurements near a helmet structure base at the 1973 solar eclipse, *Solar Phys., 94,* 239, 1984.

Billings, D. E., and C. Kober, Distribution of prominence heights, *Astron. J., 62,* 242, 1957.

Biskamp, D., and H. Welter, Magnetic arcade evolution and instability, *Solar Phys., 120,* 49, 1989.

Chiuderi, C., and G. Van Hoven, The dynamics of filament formation: the thermal instability in a sheared magnetic field, *Astrophys. J., 232,* L69, 1979.

Choe, G. S., and L. C. Lee, Formation of solar prominences by photospheric shearing motions, *Solar Phys., 138,* 291, 1992.

Cox, D. P., and W. H. Tucker, Ionization equilibrium and radiative cooling of a low density plasma, *Astrophys. J., 157,* 1157, 1969.

Engvold, O., Prominence environment, in *Dynamics and Structure of Quiescent Solar Prominences,* edited by E. R. Priest, pp. 47-76, Kluwer Academic Publishers, Dordrecht, Holland, 1989.

Field, G. B., Thermal instability, *Astrophys. J., 142,* 531, 1965.

Finn, J. M., and J. Chen, Equilibrium of solar coronal arcades, *Astrophys. J., 349,* 345, 1990.

Forbes, T. G., and J. M. Malherbe, A numerical simulation of magnetic reconnection and radiative cooling in line-tied current sheets, *Solar Phys., 135,* 361, 1991.

Harned, D. S., and D. D. Schnack, Semi-implicit method for long time scale magnetohydrodynamic computations in three dimensions, *J. Comput. Phys., 65,* 57, 1986.

Hildner, E., The formation of solar quiescent prominences by condensations, *Solar Phys., 35,* 123, 1974.

Hirayama, T., Modern observations of solar prominences, *Solar Phys.,100,* 415, 1985.

Inhester, B., J. Birn, and M. Hesse, The evolution of line-tied coronal arcades including a converging footpoint motion, *Solar Phys., 138,* 257, 1992.

Karpen, J. T., S. K. Antiochos, and C. R. DeVore, On the formation of current sheets in the solar corona, *Astrophys. J., 356,* L67, 1990.

Kippenhahn, R., and A. Schlüter, Eine theorie der solaren filamente, *Z. Astrophys., 43,* 36, 1957.

Klimchuk, J. A., and P. A. Sturrock, Force-free magnetic fields: Is there a "loss of equilibrium"?, *Astrophys. J., 345,* 1034, 1989.

Kuperus, M., and M. A. Raadu, The support of prominences formed in neutral sheets, *Astron. Astrophys., 31,* 185, 1974.

Lee, L. C., a presentation given in the Workshop on Solar Flares and Magnetospheric Substorms, Hawaii, 1990.

Lee, L. C., and Z. F. Fu, Multiple X line reconnection, 1. A criterion for the transition from a single X line to a multiple X line reconnection, *J. Geophys. Res., 91,* 6807, 1986.

Leroy, J. L., Observation of prominence magnetic fields, in *Dynamics and Structure of Quiescent Solar Prominences,* edited by E. R. Priest, pp. 77-114, Kluwer Academic Publishers, Dordrecht, Holland, 1989.

Low, B. C., On the spontaneous formation of electric current sheets above a flexible solar photosphere, *Astrophys. J., 381,* 295, 1991.

Low, B. C., Formation of electric-current sheets in the magnetostatic atmosphere, *Astron. Astrophys., 253,* 311, 1992.

McWhirter, R. W. P., P. C. Thonemann, and R. Wilson, The heating of the solar corona. II. A model based on energy balance, *Astron. Astrophys., 40,* 63, 1975.

Martin, S. F., The evolution of prominences and their relationship to active centers, *Solar Phys., 31,* 3, 1973.

Martin, S. F., Conditions for the formation of prominences as inferred from optical observations, in *Dynamics of Quiescent Prominences, Proceedings of the No. 117 Colloquium of the IAU, Hvar, Yugoslavia,* edited by V. Ruždjak and E. Tandberg-Hanssen, *Lecture Notes in Physics,363,* pp. 1-48, Springer Verlag, Berlin, 1990.

Peres, G., R. Rosner, S. Serio, and G. S. Vaiana, Coronal closed structures. IV. Hydrodynamical stability and response to heating perturbations, *Astrophys. J., 252,* 791, 1982.

Priest, E. R., *Dynamics and Structures of Quiescent Solar Prominences,* Kluwer Academic Publishers, Dordrecht, Holland, 1989.

Raymond, J. C., and B. W. Smith, Soft X-ray spectrum of a hot plasma, *Astrophys. J. Suppl., 35,* 419, 1977.

Rompolt, B., and T. Bogdan, On the formation of active region prominences (H_α filaments), in *Coronal and Prominence Plasmas,* edited by A. Poland, pp. 81-87, NASA Conference Publication 2442, Washington, DC, 1986.

Rosner, R., W. H. Tucker, and G. S. Vaiana, Dynamics of the quiescent solar corona, *Astrophys. J., 220,* 643, 1978.

Saito, K., and E. Tandberg-Hanssen, The arch systems, cavities and prominences in the helmet streamer observed at the solar eclipse, November 12, 1966, *Solar Phys., 31,* 105, 1973.

Schmieder, B., Overall properties and steady flows, in *Dy-*

namics and Structure of Quiescent Solar Prominences, edited by E. R. Priest, pp. 15-46, Kluwer Academic Publishers, Dordrecht, Holland, 1989.

Sparks, L., G. Van Hoven, and D. D. Schnack, The nonlinear evolution of magnetized solar filaments, *Astrophys. J., 353,* 297, 1990.

Spitzer, L., *Physics of Fully Ionized Gases,* Interscience, New York, 1962.

Steinolfson, R. S., Coronal evolution due to shear motion, *Astrophys. J., 382,* 677, 1991.

Steinolfson, R. S., and G. Van Hoven, Radiative tearing: magnetic reconnection on a fast thermal-instability time scale, *Astrophys. J., 276,* 391, 1984.

Tang, F., Quiescent prominences - Where are they formed?, *Solar Phys., 107,* 233, 1987.

Tucker, W. H., and M. Koren, Radiation from a high temperature, low-density plasma: the X-ray spectrum of the solar corona, *Astrophys. J., 168,* 283, 1971.

van Ballegooijen, A. A., and P. C. H. Martens, Formation and eruption of solar prominences, *Astrophys. J., 343,* 971, 1989.

Van Hoven, G., and Y. Mok, The thermal instability in a sheared magnetic field: filament condensation with anisotropic heat conduction, *Astrophys. J., 282,* 267, 1984.

Van Hoven, G., L. Sparks, and T. Tachi, Ideal condensations due to perpendicular thermal conduction in a sheared magnetic field, *Astrophys. J., 300,* 249, 1986.

Waldmeier, M., The structure of monochromatic corona in the surroundings of prominences, *Solar Phys., 15,* 167, 1970.

Wu, S. T., J. J. Bao, C. H. An, and E. Tandberg-Hanssen, The role of condensation and heat conduction in the formation of prominences: an MHD simulation, *Solar Phys., 125,* 277, 1990.

Zirin, H., *Astrophysics of the Sun,* Cambridge University Press, Cambridge, 1988.

S.-I. Akasofu, G. S. Choe, and L. C. Lee, Geophysical Institute, University of Alaska, Fairbanks, AK 99775-0800.

Laval Nozzle Effects in Solar Wind - Exosphere Interaction

K. Sauer[1], K. Baumgärtel[1], Th. Roatsch[2] and J.F. McKenzie[3]

[1]*Max-Planck-Institut für extraterrestrische Physik, Außenstelle Berlin, 12489 Berlin, Germany*
[2]*Institut für Planetare Erkundung, DLR, 12489 Berlin, Germany*
[3]*Max-Planck-Institut für Aeronomie, 37189 Katlenburg-Lindau, Germany*

Bi-ion fluid models describing the electrostatic interaction between two ion species in the presence of neutralizing (massless) hot electrons are used to describe the penetration of the solar wind into unmagnetized ionospheres of planets (as Venus and Mars) and comets. In a first step, an unmagnetized bi-ion plasma is considered in which the 'heavy' ion component is assumed to be fixed spatially. It is shown that the shocked plasma flow behaves like an ideal gas through a Laval nozzle permitting a subsonic-supersonic transition. A flow boundary is formed at the "sonic layer" where the light ions (protons) are deflected. The so-called planetopause found at Mars is proposed to be a flow boundary of this new type.

Introduction

A problem of great importance concerning the plasma flow signatures around comets or ionospheres of unmagnetized planets is the behaviour of the solar wind if it penetrates the region occupied by heavy ions of cometary or ionospheric origin. In this context, a question of special interest is whether the interaction of a light-ion plasma flow with a "heavy" plasma which is steadily reproduced by background neutral gas ionization is characterized by a continuous transition between the inflowing plasma and the "target" plasma overtaking the momentum or a discontinuity is formed separating two plasmas of different constitution (mass, density, temperature etc.).

After analyzing the plasma and magnetic field data form the Phobos 2 mission around Mars in 1989, there are controversal opinions on the nature of the boundary which was detected inside the bow shock. The term "planetopause (PP)" was introduced by the magnetometer team (Riedler et al., 1989). The difficulty in understanding the physics behind this boundary is due to the PP's unusual property of forming an "obstacle" for the proton flow but not for the solar wind magnetic field, unlike the magnetopause at Earth.

Such unusual behaviour of both the proton flow and the magnetic field is not contained in MHD theories. Obviously, it requires plasma models in which the individual dynamics of light and heavy ions is included and a momentum transfer between both ion populations is possible. This is especially important for Mars where the interaction scale is small compared with the heavy ion gyroradius. In recent studies by Moore et al. (1991) and Brecht and Ferrante (1991) hybrid codes have been used in order to describe finite gyroradius effects at Venus and Mars. However, these kinetic simulations are very computer intensive and owing to necessary numerical limitations no answer was obtained on the possible formation of a plasma boundary which acts as an obstacle for the proton flow, as suggested by the Phob os measurements.

An alternative way for studying multiple-ion effects is the use of multifluid models which have proved to be very practicable. In this paper a simplified bi-ion model is presented. It describes the flow of light ions in the presence of an inhomogeneous distribution of immobile ions (called heavy ion cloud = HIC) representing the

Space Plasmas: Coupling Between Small
and Medium Scale Processes
Geophysical Monograph 86

heavy ion component. It is shown that a subsonic flow is accelerated if it penetrates the HIC. Thus, the plasma flow behaves like an ideal gas stream through a Laval nozzle permitting a subsonic-supersonic transition. A flow boundary is formed at the "sonic layer" where the protons are deflected. The relevance of the results for understanding the planetopause features is discussed.

Plasma flow through a heavy ion cloud

The plasma flow through a heavy ion cloud is formulated in terms of a simple bi-ion fluid model (Sauer et al., 1992; Baumgärtel and Sauer, 1992) which represents a reduced version of the more general multi-fluid concept used in earlier papers (Sauer et al., 1990a,b). The analysis is simplified by the assumption of immobile "heavy" ions and the neglect of Lorentz force terms. Thus, the proton dynamics is only influenced by self-consistent electrostatic fields which result from the presence of fixed (heavy ion) density gradients.

The heavy ions are described by a prescribed density distribution for which we take a spherical Gaussian profile,

$$n_c = n_{c,max} \exp(-r^2/h^2)$$

where r and h are respectively the radial distance and the half-width. (A Gaussian profile is not a good fitting of real ionospheric density profiles; it was mainly used for mathematical convience.)

The protons are described by cold fluid equations for the density n and the velocity $\mathbf{u}$. The electrons are massless and hot, so that the driving electrostatic field is given by

$$\boldsymbol{E} = -\frac{1}{en_e}\nabla p_e$$

where n_e and p_e are the electron density and the electron thermal pressure, resp.. Throughout, quasi-neutrality ($n_e=n+n_c$) is imposed. Then, under the assumption of axial-symmetry, the continuity, momentum and electron energy equations take the form

$$\frac{\partial n}{\partial t} + \frac{1}{r}\frac{\partial rnu_r}{\partial r} + \frac{\partial nu_z}{\partial z} = 0$$

$$\frac{\partial nu_z}{\partial t} + \frac{1}{r}\frac{\partial rnu_r u_z}{\partial r} + \frac{\partial nu_z^2}{\partial z} = -\frac{1}{m_p}\frac{n}{n_e}\frac{\partial p_e}{\partial z}$$

$$\frac{\partial nu_r}{\partial t} + \frac{1}{r}\frac{\partial rnu_r^2}{\partial r} + \frac{\partial nu_r u_z}{\partial z} = -\frac{1}{m_p}\frac{n}{n_e}\frac{\partial p_e}{\partial r}$$

$$\frac{\partial p_e}{\partial t} + \frac{1}{r}\frac{\partial ru_{er}p_e}{\partial r} + \frac{\partial u_{ez}p_e}{\partial z} =$$

$$-(\gamma-1)\, p_e\left[\frac{1}{r}\frac{\partial ru_{er}}{\partial r} + \frac{\partial u_{ez}}{\partial z}\right]$$

The electron velocity $\mathbf{u}_e$ follows from the zero-current condition $\mathbf{u}_e=n\mathbf{u}/n_e$.

As an initial state for the numerical integration we take a homogeneous flow with given density (n_∞) and velocity ($\mathbf{u}_\infty$). These initial values are used as boundary conditions at z = - 5 where the plasma flow enters the integration box. For the other boundaries ($r=r_0$, $z=z_0$) free boundary conditions in form of linear extrapolation are introduced.

Generally, subsonic flows ($M_{s,\infty} < 1$) were studied. Keeping the other parameters fixed, the (normalized) cloud peak density $n_{c,max}/n_\infty$ was found to be a crucial parameter which determines the flow pattern in the vicinity of the heavy ion cloud (HIC).

For $n_{c,max} \ll n_\infty$ the HIC is penetrable and it acts as a converging-diverging nozzle. That means that the flow is accelerated up to the centre of the cloud (nozzle throat) and then it is decelerated down to its initial velocity. The flow signatures change drastically when the peak density $n_{c,max}$ exceeds a certain threshold which is near to the undisturbed proton density. (For this case the term "overcritical cloud" is used.) The flow pattern for an "overcritical" cloud with a maximum density of $n_{c,max}$ = 1.5 n_∞ which is put into a subsonic flow ($M_{s,\infty}$ = 0.4) is shown in Fig. 1. The "heavy" ion density contours (n_c/n_∞ = 0.2, 0.5, 1.2) are plotted as dashed lines in the vector plot of proton velocities. As clearly seen, the resulting steady-state configuration is that of a flow around an "active obstacle": The flow is stopped in front of the cloud and diverges along the flanks where an acceleration to supersonic velocities takes place. As a result, a proton cavity in the wake of the cloud is formed.

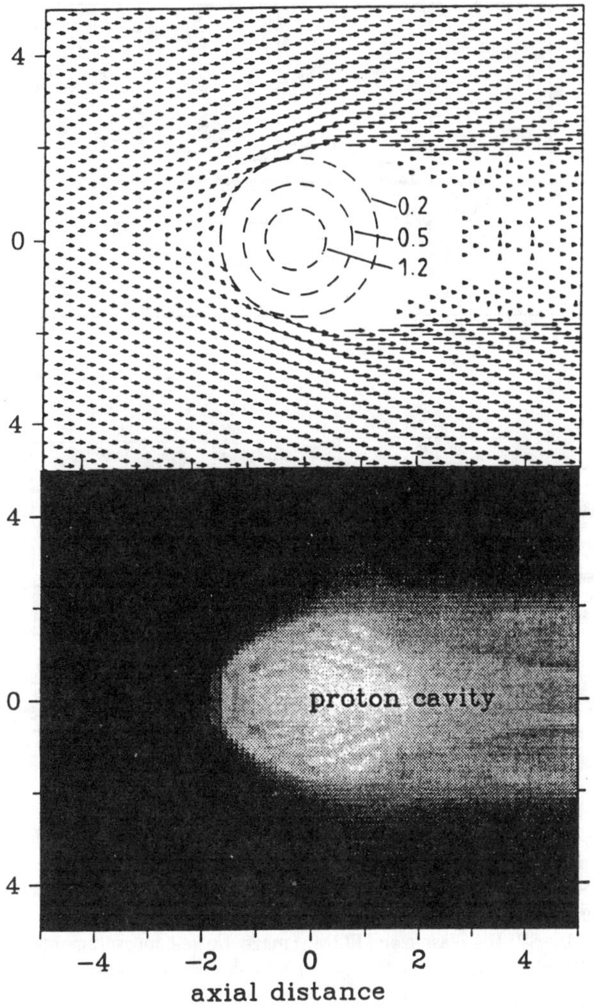

Fig. 1 Axially-symmetric, unsteady flow through an "overcritical" ion cloud; formation of a plasma flow boundary at the "sonic layer". Top panel: Vector plot of proton velocity; the HIC is located at z=0, its peak density is $n_{c,max} = 1.5n_\infty$. The density contours are plotted as dashed circles. The incident flow is subsonic, $M_{s,\infty} = 0.4$. The acceleration associated with formation of a flow boundary is clearly seen. Lower panel: Colour plot of proton density. The bright region marks the proton cavity.

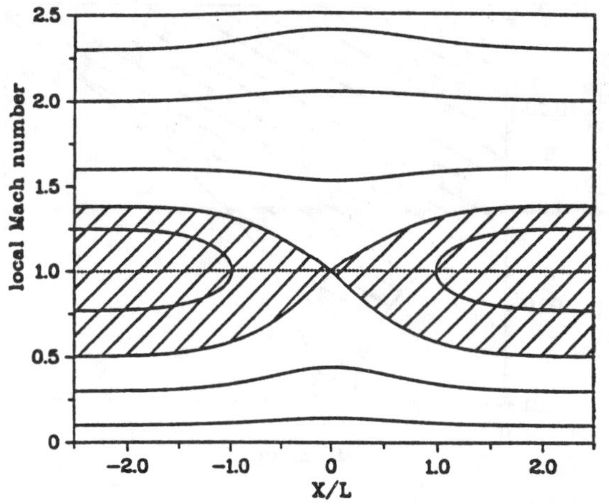

Fig. 2. Trajectories in the u-x-plane providing a comprehensive look at the flow feature. Cloud parameters are: $n_{c,max}/n_\infty=0.2$, h/L=1.0. Separatrices enclose a region where no steady-state solution exists. u is normalized by the "local" sound speed v_{ph}.

The flow pattern shown in Fig. 1 is exclusively caused by the electrostatic field which changes its sign if the flow penetrates the HIC. In the low-density front region the field is positive and accelerates the protons. At a certain HIC density, however, the field becomes negative and prevents the protons from penetrating the denser part of the cloud.

Laval nozzle flow and critical densities

The flow behaviour described above can easily be interpreted by means of a simple steady-state one-dimensional model. It is shown that the interaction between the flow and the fixed ion cloud is similar to an ideal gas flow through a Laval nozzle. Further, the critical (fixed ion) density above which the HIC acts as an "obstacle" can be determined. Particularly simple analytical considerations are possible if isothermal electrons are assumed. A "wind equation" for the protons velocity u (the velocity in x direction) is then readily obtained in the form

$$(u^2-v_{ph}^2)\ \frac{1}{u}\frac{du}{dx} = -\frac{v_{ph}^2}{n}\frac{dn_c}{dx}$$

where

$$v_{ph}^2 = c_s^2\ \frac{n}{(n+n_c)}\ ,\qquad c_s^2 = \frac{kT_e}{m}\ .$$

c_s is the ion-sound velocity of the undisturbed flow. Eq. (3) governing the acceleration (deceleration) of the streaming protons, is analogous to a Laval nozzle flow. The phase trajectories in the ($M^* = u/v_{ph}$, x) plane are depicted in Fig. 2 and illustrate incoming supersonic (subsonic) flows being decelerated (accelerated) up to the throat and then being accelerated (decelerated) after the throat. As further seen, in the upstream region there

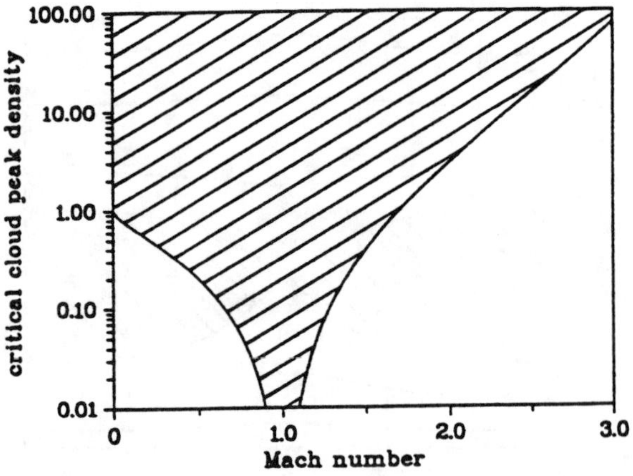

Fig. 3 Constraints an HIC peak density for a steady-state response. The plasma stream can penetrate the cloud for peak densities below the critical.

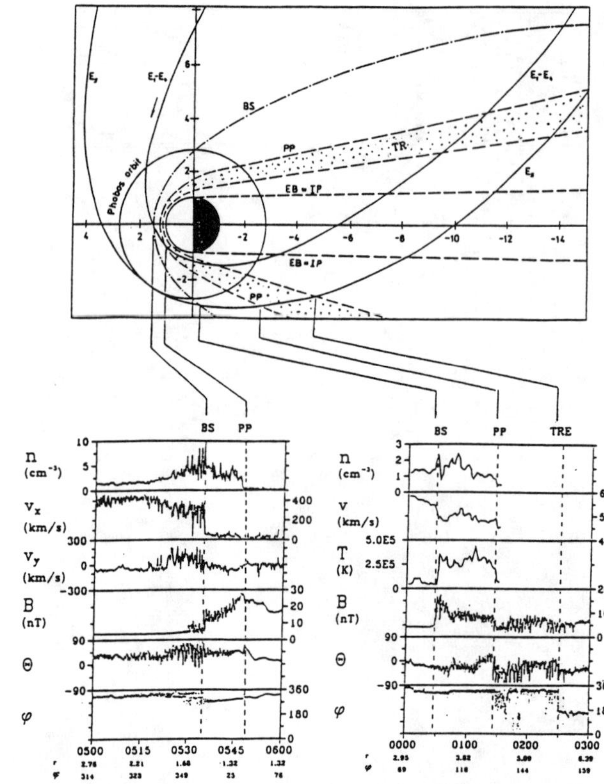

Fig.4 Top panel: Phobos 2 spacecraft trajectories during five elliptical orbits near closest approach. The plasma boundaries and layers, bow shock (BS), planetopause (PP), transition region (TR), eclipse boundary (EB) and orbit of the Phobos moon are shown. Lower left panel: ASPERA plasma and FGMM magnetic field measurements from the third elliptical orbit (Febr. 8, 1989) showing the subsolar BS and PP. Lower right panel: TAUS plasma and FGMM magnetic field measurements of the fifth elliptical orbit (Febr. 14/15, 1989) showing BS and PP in the tail (after Sauer et al., 1992).

are two critical Mach numbers, one subsonic and the other supersonic, between which no steady solution exists. These critical Mach numbers may be obtained from the flow integral

$$\frac{1}{2} M^2 + \log\left(\frac{M_\infty}{M} + \frac{n_c}{n_\infty}\right) = \frac{1}{2} M_\infty^2$$

where $M = u/c_s$. Fig. 3 shows the critical ion density ratio ($n_{c,crit}/n_\infty$) versus the upstream Mach number, where the hatched region belongs to continuous solutions. In the subsonic regime ($M_\infty \leq 1$) the critical density decreases to zero as the Mach number approaches one. For supersonic flows the critical density increases with increasing Mach number. (For the relevance of the above studies to the planetopause problem at Mars it is important to point out that for magnetized flows with $\mathbf{B}_\infty \perp \mathbf{v}_\infty$ a similar dependence with respect to the magneto-sonic Mach number is obtained; Baumgärtel and Sauer (1992) and Motschmann et al. (1992)).

Planetopause at Mars – a new type of flow boundary

As result of the plasma and magnetic field measurements by the ASPERA/TAUS (Lundin et al., 1989; Rosenbauer et al., 1989) and FGMM/MAGMA instruments (Riedler et al., 1989) aboard Phobos 2 around Mars a plasma flow boundary inside the bow shock, called planetopause (Riedler et al., 1989) was detected. The location of both boundaries is shown in Fig. 3 together with observations from the third (lower left panel) and fifth elliptical orbit (lower right panel). The characteristic features of the planetopause are: (1) a sharp decrease of proton density, (2) a drastic change in the spectrum of magnetic fluctuations and (3) the appearance of exospheric ions (O^+), whereby its density becomes comparable to the proton density.

It was proved by Schwingenschuh et al. (1990) that the magnetic field inside the PP is a draped solar wind magnetic field. Near the PP no indication was found for a contribution from an intrinsic field. Thus, the PP represents a new type of plasma boundary, which stops the proton flow without forming a boundary for the magnetic field. As originally pointed out by Sauer et al. (1990), such a property can only be explained in terms of multi-ions models the simplest form of which was presented above.

Summary and discussion

A simple bi-ion fluid model has been used in order to simulate day-side features of the interaction of the solar wind with planetary exospheres, such as those at Venus and Mars. Only the outer regions where the heavy ion density reaches that of the protons have been modelled. It has been shown that the shocked proton flow is essentially influenced by the localized electrostatic field at the "sonic layer" where the protons are deflected and accelerated to supersonic velocities. Although the acceleration of "subsonic" flows as consequence of massloading processes was discussed already in earlier papers (Biermann et al., 1967; Cloutier et al., 1987), the importance of the "sonic layer" as an obstacle boundary which simultaneously acts as ion composition boundary was not pointed out before, since the effect cannot be described by one-fluid models.

We suggest that the disputed planetopause at Mars can be interpreted as "sonic layer". The observational facts presented above seem to be in favour of such explanation. Especially, the exospheric ion density near the PP was measured to be comparable with the proton density (Lundin et al., 1991; Breus et al., 1991). Moreover, a critical densitiy of $n_{crit} \leq 0.2\ n_{\infty}$ in a subsolar height of the obstacle surface of about 500 km (Slavin et al., 1991) is consistent with the exospheric ion production which follows from calculated neutral gas profiles (Ip, 1988).

At a first glance, however, problems within this model arise with respect to the magnetic field. How is it possible to understand the observations that the planetopause is a border for the proton flow, but not a boundary for the solar wind magnetic field? This contradiction can easily be overcome if the motion of the heavy ions (which actually exists owing to momentum coupling) is taken into consideration. Then, a proton flow boundary is not equivalent with a magnetic field boundary, since the heavy ion motion allows charge-neutralizing electrons to penetrate the boundary which means a magnetic field transport across the boundary. The observed magnetic field changes at the PP (see Fig. 4) may reflect jumps in the electron velocity, which is essentially the proton velocity upstream and should be the heavy ion velocity downstream the PP, see also Fig. 11 of Sauer et al. (1990a).

Note added in proof: In a recent bi-ion fluid simulation of a magnetized plasma with mobile heavy ions from a distributed source, the formation of an ion composition boundary with the expected features has been seen.

References

Baumgärtel, K., and K. Sauer, Interaction of a magnetized plasma stream with an immobile ion cloud,*Ann. Geophys.*,*10*, 763-771, 1992.

Biermann et al., The interaction of the solar wind with a comet, *Solar Physics*, *1*, 254, 1967.

Brecht, S.H., and J.R. Ferrante, *J. Geophys. Res.*, *96*, 11,209-11,220, 1991.

Breus, T.K. al., The solar wind interaction with Mars: Consideration of Phobos-2 mission observation of an ion composition boundary on the dayside, *J. Geophys. Res.*, *96*, 11,165-11,174, 1991.

Cloutier,P.A. et al., Steady State Flow/Field Model of Solar Wind Interaction With Venus: Global Implications of Local Effects, *J. Geophys. Res.*, *92*, 7289-7307, 1987.

Ip, W.-H., *Geophys. Res. Letters*, *17*, 2289-2292, 1990.

Lundin, R. et al., First measurements of the ionospheric plasma escape from Mars, *Nature*, *341*, 609-612, 1989.

Moore, K.R. et al., Global Hybrid Simulations of the Solar Wind Interaction with the Dayside of Venus,

Motschmann, U. et al., Simulation of ion acceleration in a charged dust cloud, *Geophys. Res. Letters*, *19*,1992.

Riedler, W. et al., Magnetic fields near Mars: First results of the Phobos mission, *Nature*, *341*, 604-607, 1989.

Rosenbauer et al., Ions of martian origin and plasma sheet in the martian magnetosphere: Initial results of the TAUS experiment, *Nature*, *341*, 612-614, 1989.

Sauer, K. et al., Plasma boundaries at Mars discovered by the Phobos 2 magnetometers, *Ann. Geophys.*, *8*, 661-670, 1990.

Sauer, K. et al., Observation of Plasma Boundaries and Phenomena around Mars with Phobos-2, *J. Geophys. Res.*, *97*, 6227-6233, 1992a.

Sauer, K. et al., Critical density layer as obstacle at solar wind - exospheric ion interaction, *Geophys. Res. Letters*, *19*, 645-648, 1992b.

Schwingenschuh et al., Martian bow shock: Phobos observations, *Geophys. Res. Letters*, *17*, 889-892, 1990.

Slavin et al., The solar wind interaction with Mars: Mariner 4, Mars 2, Mars 3, Mars 5 and Phobos observations of bow shock position and shape, *J. Geophys. Res.*, *47*, 11,235-11,241, 1991.

On Interacting Plasma Vortex Sheets

EDOUARD SIREGAR, D. AARON ROBERTS, AND MELVYN L. GOLDSTEIN

Laboratory for Extraterrestrial Physics, NASA-Goddard Space Flight Center, Greenbelt, Maryland

Abstract. We study the evolution of a plasma vortex street formed through the nonlinear interaction of two vortex sheets which are initially in equilibrium and perturbed by four modes. This leads to a staggered vorticity distribution along the plasma flow direction. The combined effects of vortex repulsion and inertial ejection of plasma from vortices lead to an expansion of the central region along the symmetry axis of the flow, and to transport of plasma and magnetic energy away from this axis. This evolution is preserved in a large area of parameter space and in the presence of noise. This MHD model reproduces the signatures of the quasi-periodic plasma fluctuations observed by Voyager 2 in the outer heliosphere.

1. INTRODUCTION AND MODEL

The Vortex streets are frequently observed in the earth's atmosphere and generally in wakes, but they are also relevant to astrophysical phenomena. In recent papers [*Burlaga*,1990; *Siregar et al.*, 1992, 1993, 1994] this model was used to explain the surprising quasi-periodic solar wind fluctuations [*Lazarus et al.*, 1988] observed by Voyager 2 in the outer heliosphere. We study various physical processes underlying the evolution of a magnetized vortex street formed through the interaction of two neighboring vortex sheets.

Interest in vortex streets began after Von Kármán [*Von Kármán*, 1911] published the first paper on the theory of vortex streets, where the emphasis was on the stability analysis of rows of alternate point vortices. Today, it is still an active subject of research (e.g., [*Ohle et al.*, 1990; *Hayot and Rajlakshmi*, 1989]). Experiments show that at sufficient Reynolds numbers, vortices are shed periodically from a cylinder. The no-slip boundary condition near the cylinder creates vorticity and a boundary layer and instabilities in the boundary layer lead to a staggered vorticity distribution in the wake. In this paper we generate a vortex street through the nonlinear interaction of two sheets of vorticity.

We start with the two-dimensional MHD equations in the following dimensionless form

$$\partial_t \rho = -\bar{\nabla}\cdot(\rho\bar{u}) \tag{1}$$

$$\partial_t \bar{u} = -\bar{u}\cdot\bar{\nabla}\bar{u} - \frac{1}{\gamma\rho M_a^2}\bar{\nabla}\rho^\gamma + \frac{R_e}{\rho}\left[\nabla^2\bar{u} + \frac{1}{3}\bar{\nabla}(\bar{\nabla}\cdot\bar{u})\right] - \frac{1}{\rho M_m^2}\bar{\nabla}a(\nabla^2 a) \tag{2}$$

$$\partial_t a = -\bar{u}\cdot\bar{\nabla}a + \frac{1}{R_m}\nabla^2 a \tag{3}$$

where $\bar{u}$, ρ, a, R_e, R_m are the velocity, density, vector potential, fluid and magnetic Reynolds numbers respectively, M_a and M_m scale as acoustic and Alfvénic Mach numbers, and g=5/3 is the polytropic index. The algorithm used is a Chebyshev-Fourier spectral collocation method where a dependent variable is decomposed in each spatial direction in orthogonal Chebyshev $T_n(y)$ and Fourier exp[-ikmx] functions. The Chebyshev y-direction is either periodic of length scale 2 or infinite through the use of an algebraic map on the Chebyshev collocation points. The computation in the Fourier x-direction is periodic of length 2p. Time advancing is done in Chebyshev-Fourier spectral space using a second order Adams-Bashforth Crank-Nicholson (ABCN) semi-implicit scheme. The CN implicit part allows reasonable time steps sizes on the Chebyshev dissipation operator which contains very small spatial scales near the outer Chebyshev collocation points $y=\pm1$.

We initialize the velocity field as two equilibrium vortex sheets (very thin hyperbolic tangent shear layers compared to the perturbation wavelength) and perturbed them with random small-scale noise plus deterministic

Space Plasmas: Coupling Between Small
and Medium Scale Processes
Geophysical Monograph 86

perturbations. The coherent part of the perturbation is composed of the four most unstable modes that can fit in the computation domain. there is a slow plasma flow region surrounded by two fast flow regions. This perturbation consists of two modes for each sheet, so we initialize four independent velocity functions to induce an antisymmetric displacement of the sheets. This class of perturbations leads to the staggered vorticity distribution.

A current sheet is placed along the center of the slow region with a tiny passive magnetic field used as a tracer. We start with a constant unit density throughout the computational domain. The only spatial scales initially in the flow are the initial distance d between the vortex sheets and the initial perturbation wavelength l=2p/m. The dimensionless ratio S=d/l (S=0.075 is used here) and the relative phase between the wave perturbations along each of the two sheets are important bifurcation parameters for the time asymptotic solution taken by the flow, but these aspects are not discussed here. The Reynolds number Re=uL/n scales as the ratio of the long momentum diffusion time scale $T_d = l_{min}^2/n$ (where l_{min} is the smallest computational scale of the order of a grid spacing and n the kinematic viscosity) to the short inertial nonlinear scale $T_{nl}=1/|w|$. The disparity between T_d and T_{nl} ($Re=10^3$ is used here) means that in between these two scales there is a quasi-steady inertial state during which the vortex street evolves slowly.

The street evolves in a quasi-static pressure equilibrium where acoustic waves play no significant role. This is confirmed by doing the simulation in a semi-infinite domain. In this case, the acoustic waves generated by the initial conditions simply propagate out to infinity in a short acoustic time. The flow is polytropic so there is no temperature, but we can still define a plasma beta through the ratio of sound to Alfven speeds $(c_s/c_a)^2$ or through the ratio of gas to magnetic pressures p/p_{mag}. In our computations, b is infinite using the first definition (since the Alfven speed is defined in terms of the mean magnetic field which is zero), or b>>1 using the pressure ratio definition (which is in a sense more robust in nonlinear situations where fluctuations can be large compared to mean quantities). The relevance to the outer heliosphere comes from the assumption (verified in our 3D study [*Siregar et al.*, 1994]) that the main nonlinear couplings and fastest growing modes occur in the velocity shear (x,y)-plane perpendicular to mean Parker field in the outer heliosphere. So this beta describes only the magnetic field component in the (x,y)-plane, which is used only as a passive tracer to study the evolution of magnetic flux in the current sheet region. We find that in the central current sheet region, the magnetic flux decreases rapidly during the nonlinear inertial time scales. This flux decrease is related to the mean plasma expansion around the current sheet region. This is discussed further in Section II. Our 3D study shows that a magnetic field in the z-direction stabilizes the vortex tubes along the street by preventing kinking and the growth of three-dimensional modes. The Mach numbers used lie in the range 0.2-0.6 (M_m=1 was always used) do not influence significantly the large-scale features of the street, but higher filamentary vorticity is observed at higher Mach numbers. A flow aligned magnetic field in the (x,y) velocity shear plane such as would be found in the inner heliosphere leads to a quasi-periodic exchange of energy between magnetic and kinetic forms. When the ratio of magnetic to fluid (kinetic + internal) energy E_m/E_f increases to about 10^{-2} the vortices are significantly squeezed through their centers. For higher such ratios E_m/E_f (but still less than unity), the magnetic field becomes strongly stabilizing and prevents the formation of the street altogether. For further details on the discussion in this paragraph, see [*Siregar et al.*, 1993].

Figure 1 illustrates the evolution of the street. White vorticity contours are superimposed to the gray density background. The alternating vortices and their associated density/pressure cavities (in black) are formed during the inertial time scale through the nonlinear interaction of the initial vortex sheets. Then, the isolated vortices decay on the long time dissipative scales. We have shown [*Siregar et. al* 1992] that the characteristic signatures of the quasi-periodic plasma fluctuations (which involve variables and correlations between them) observed by Voyager 2 are well reproduced by the vortex street.

2 DYNAMICS OF VORTICITY AND MAGNETIC VARIABLES

During the evolution of the street, there is a complex interplay between the vorticity, the density/pressure and magnetic quantities in the plasma. We proceed to some MHD analysis in order to clarify some of the underlying physical processes in the flow. Since the magnetic field is only used as a tracer in this paper, we start by investigating fluid quantities. The effects of a magnetic field perpendicular to the shear plane are studied elsewhere [*Siregar et al.*,1992, 1993].

There are simple free vortex solutions to the Navier-Stokes compressible flow. Axisymmetric (i.e., non-elliptical) free vortex solutions satisfy a pressure centrifugal equilibrium $\partial p/\partial r = ru^2/r$ (where r is the radial distance to the center of the vortex). Elliptical free vortex steady solutions also exist. Additional information is needed to completely specify a compressible vortex solution. For example, a line vortex u(r)=a/r for a polytropic flow has a density profile (for r>0) $r(r)=[(g/(g-1)][-a^2/2r^2 + C]^{1/g-1}$ which shows that vortices are also density cavities. Inertia ejects plasma out of the free spinning vortex until a pressure equilibrium is reached. More complex and realistic free vortex solutions are available [*Colonius et al.,* 1991] and more generally, the ideal dynamics of vorticity is governed by the global flow invariants; mass and momentum (eq. (1) and (2)).

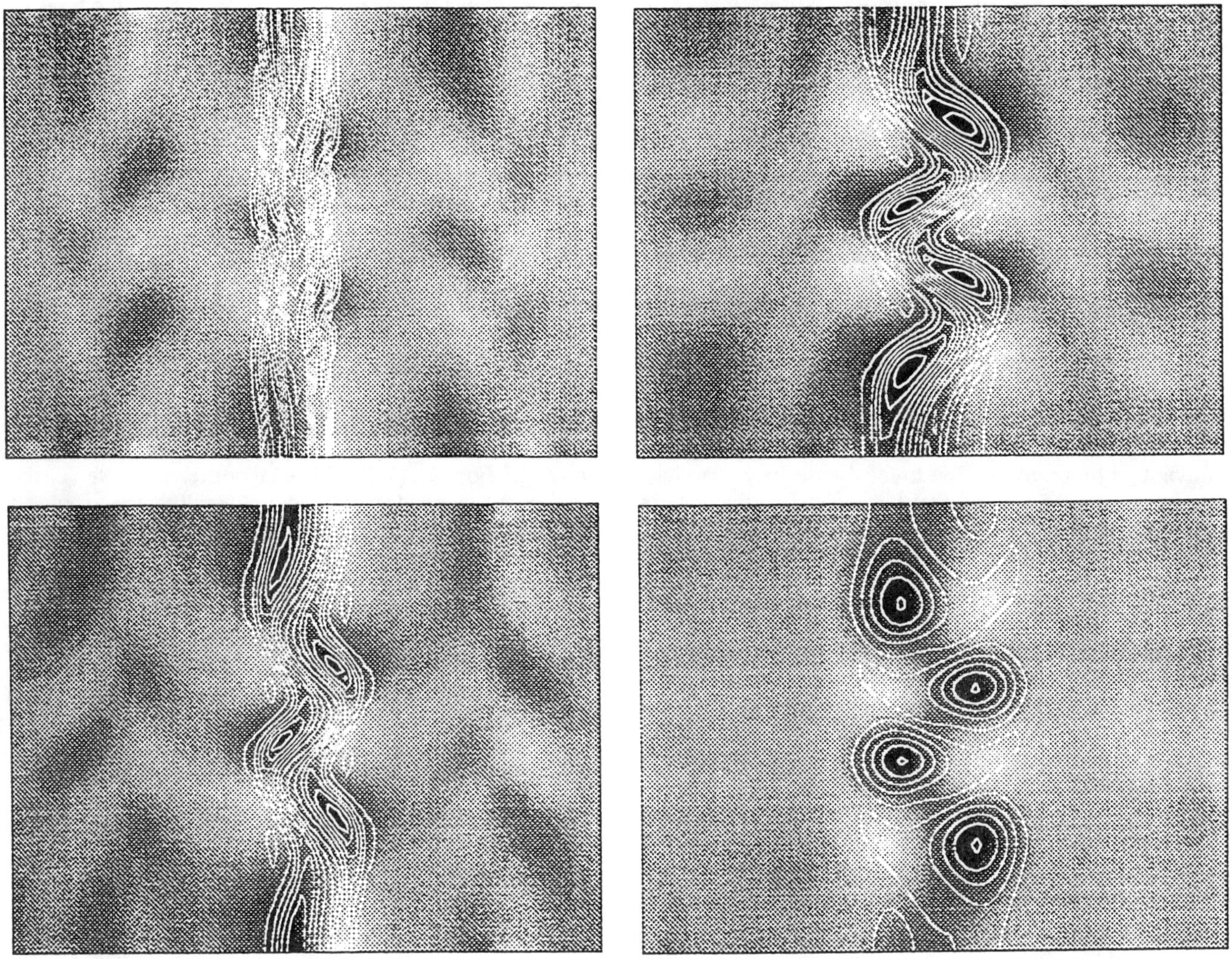

Fig. 1. Time evolution of the vortex street. White vorticity contours are superimposed on the density gray background. Density depletion areas are in black and compression regions are in white.

Starting from (1) and (2) we can obtain the following form of the vorticity equation:

$$\partial_t \vec{\omega} = -\vec{u}\cdot\vec{\nabla}\vec{\omega} + (\vec{\omega}\cdot\vec{\nabla})\vec{u} - \vec{\omega}(\vec{\nabla}\cdot\vec{u}) + \vec{\nabla}\times\vec{f} + \frac{1}{\rho^2}(\vec{\nabla}\rho\times\vec{\nabla}p) \qquad (4)$$

We can extract information on the dynamical evolution of a given fluid element using a local invariant, which can be found from (4). In a three-dimensional polytropic flow and neglecting the Lorentz force f, the two last terms on the r.h.s. of equation (4) vanish. By introducing the Lagrangian derivative D_t and continuity in the form $(1/\rho)D_t(\rho) = -\vec{\nabla}\cdot\vec{u}$ we get $D_t(\vec{\omega}/\rho) = [(\vec{\omega}/\rho)\cdot\vec{\nabla}]\vec{u}$. In the two-dimensional case, the vortex stretching term on the right hand side of this equation vanishes since vorticity remains perpendicular to the velocity, and by integration we get $w/r=w_0/r_0$ a local invariant as we follow a parcel of fluid. The fluid elements in the density cavities-vortex structures ($r(t)<r_0=1$) are those parts of the fluid that started with the highest vorticity ($w(t)<w_0$). Regions where vorticity increases also see their density increase by the same amount, so there is plasma transport through mixing of vorticity. This seems to be the case in figure 1, where there are density enhancements (lighter regions) bordering the outer boundaries of the vortex street (dark regions), where vorticity is spreading.

In the two-dimensional non polytropic case, the departure from the local invariance of w/r measures the departure from the polytropic assumption. This can be seen by setting to zero the vortex stretching, external force terms in equation (4) and applying a procedure similar to the above one obtains $D_t(\vec{\omega}/\rho) = (1/\rho^3)[\vec{\nabla}\rho \times \vec{\nabla}p]$, so the ideal model predicts that low density regions such as the cavities inside the vortices, are the ones that will depart most from this constant specific heat model.

We are interested in following macroscopic quantities which always refer to a given set of particles. Indeed, on the inertial vortex formation time scales, there is little mixing of fluid from the central vortex street region with the outer regions. The fundamental laws of motion themselves (eqs. 1, 2 and 3) actually refer to a collection of fluid/plasma particles of fixed identity. The quantities associated to a given set of plasma particles are represented by material integrals. Since enstrophy $z=(1/2)w^2$ is a practical measure of the evolution of vorticity, we look at the time evolution of the total enstrophy in a volume V(t) made up of a given set of fluid particles. By applying Reynold's transport relation to local enstrophy z we have the time evolution of the total enstrophy Z of the material plasma volume V

$$\frac{d}{dt}Z = \int_{V(t)} \vec{\omega}\cdot D_t\, \vec{\omega}\; d^3x + \int_{V(t)} \frac{1}{2}\,\omega^2(\vec{\nabla}\cdot\vec{u})\; d^3x$$

By using this relation and (4), we can put the general compressible evolution equation for the total enstrophy associated to V(t) in the form

$$\frac{d}{dt}Z = \int_{V(t)} [\vec{\omega}\cdot(\vec{\omega}\cdot\vec{\nabla})\,\vec{u} - \frac{1}{2}\omega^2(\vec{\nabla}\cdot\vec{u})$$

$$+\vec{\omega}\cdot(\vec{\nabla}\times\vec{f}) + \frac{\vec{\omega}}{\rho^2}\cdot(\vec{\nabla}\rho\times\vec{\nabla}p)]\; d^3x \qquad (5)$$

The first term on the right hand side integral is the 3D vortex stretching term that we do not consider here. In the 2D polytropic approximation and neglecting body forces, only the second term on the right hand side of (5) is non-vanishing so that compressibility is, in this case, the unique source of variation for Z. More specifically, the total enstrophy of a given parcel of fluid tends to increase as it contracts and compressibility effects are more important for highly vortical elements. As we shall see, magnetic energy density obeys a similar evolution equation, which is important for the evolution of magnetic quantities in the central region. Non polytropic effects (which are more important inside the low pressure/density vortices) would also contribute to changes in Z.

Now we turn our attention to magnetic quantities. For application to the ecliptic solar wind, a current sheet is initially placed in the central region between the two vortex sheets. This allows us to locate the position of the "magnetic sector" boundaries. Again, during the inertial nonlinear interaction time scale the mixing of quantities between the vortex street region with the outer regions is small, so that we can concentrate on given set plasma particles in the inner region. To study the evolution of magnetic quantities in this region we look at the transport of magnetic quantities from the Lagrangian viewpoint. Different magnetic quantities obey different transport equations and one must be careful about the terminology. We focus on the magnetic field itself and magnetic energy density. Starting from the induction equation and introducing the Lagrangian derivative, one can derive the standard magnetic transport equation similar to the vorticity equation, namely (in the case of zero resistivity)

$$D_t(\vec{b}/\rho) = \frac{1}{\rho}(\vec{b}\cdot\vec{\nabla})\vec{u} \qquad (6)$$

so that as in the case of vorticity, the b/r lines are always composed of the same fluid particles, but they can rotate, stretch or be compressed. Note that in equations (6) b is not constrained to be the curl of $\vec{u}$ so that (6) admits a far broader class of initial conditions than the vorticity analog. For the 2D case, the main difference with vorticity is that the magnetic field is not generally perpendicular to the velocity. Hence, the field line stretching term does not vanish in the magnetic case, and b/r is not conserved as we follow a given piece of plasma. Note that the stretching of field lines means that even in this "frozen in" case, the relative change in the magnetic field is not simply proportional to the change in plasma density.

To compare the vortex street model with the solar wind, we need to work with a variable closer to |b| which is the measured quantity by spacecraft. Magnetic energy $b^2/2$ is such quantity. To obtain a transport equation for magnetic energy density, we form the dot product of b with (6) and get $D_t(b^2/2) = \rho D_t\,(b^2/\rho) - \vec{b}\cdot(\vec{b}\cdot\vec{\nabla})\,\vec{u}$, which shows the two components responsible for the local change of energy density $b^2/2$ associated with a given piece of plasma: magnetic line stretching itself and the transport of b^2/r. The similarity in the evolution of the magnetic field lines and vortex lines in a compressible plasma, extends to that of magnetic energy and enstrophy. To show this we apply the Reynolds transport theorem to $b^2/2$ and using the ideal version of the induction equation, we obtain the local evolution of the magnetic energy E_m of a given parcel of plasma

$$\frac{d}{dt}E_m = \int_{V(t)} [\vec{b}\cdot(\vec{b}\cdot\vec{\nabla})\,\vec{u} - \frac{\vec{b}^2}{2}(\vec{\nabla}\cdot\vec{u})]\; d^3x \qquad (7)$$

This equation is similar to the evolution of enstrophy for polytropic flows (eq. (5)). The stretching of field lines can occur in two dimensions and magnetic energy of a given parcel of plasma tends to decrease when its volume increases. Expansion is clearly active during the evolution of the vortex street, in particular in the region along the central current sheet. This can be understood as follows: the two layers of opposite vorticity tend to repel each other, broadening the street as time goes by. Since the highly vortical areas are also density cavities, the combined processes of vortex repulsion and inertial ejection of

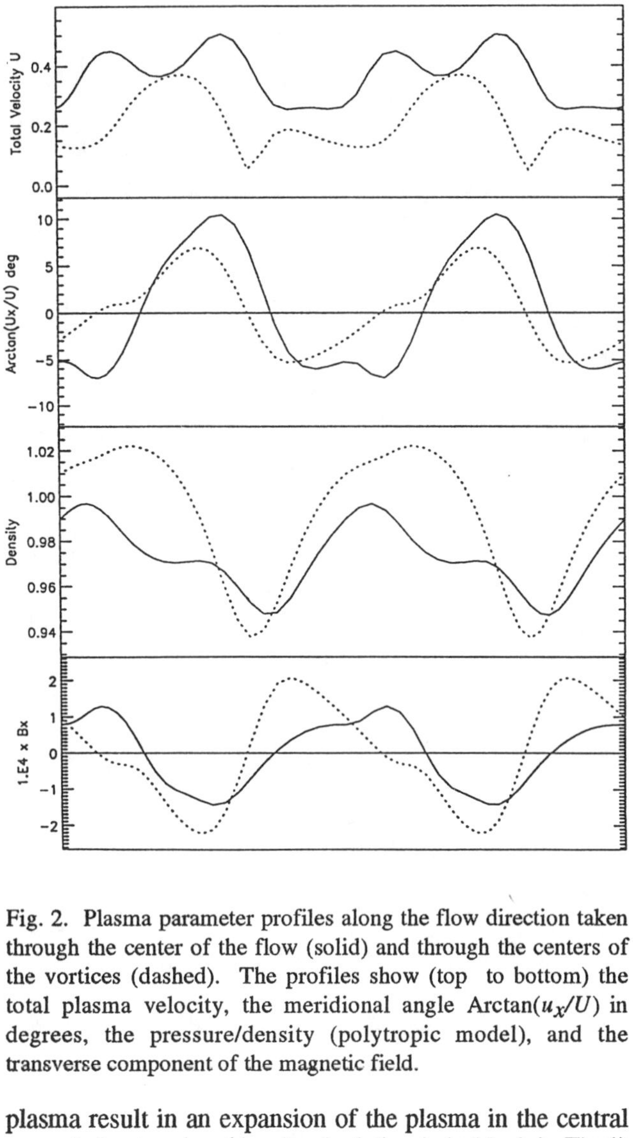

Fig. 2. Plasma parameter profiles along the flow direction taken through the center of the flow (solid) and through the centers of the vortices (dashed). The profiles show (top to bottom) the total plasma velocity, the meridional angle Arctan(u_x/U) in degrees, the pressure/density (polytropic model), and the transverse component of the magnetic field.

plasma result in an expansion of the plasma in the central current sheet region (density depletion is in black in Fig.1) towards the outer boundaries of the street where vorticity is close to zero (white compression regions). This process could explain the "flux deficit" problem in the solar wind (see [*Siregar et al.*, 1993] for more details). Deficit from the Parker field value seems to be present in the outer heliosphere when the quasi-periodic flow is observed (*L. Burlaga*, private communication, 1992).

A close comparison can be made between the variations of plasma quantities in the computed evolving vortex street model and the observations of quasi-periodic fluctuations [*Lazarus et. al* 1988] made by Voyager 2 in the outer heliosphere. In Figure 2, we take profiles of plasma parameters along the streamwise direction which are similar to spacecraft samplings as the solar wind plasma flows by. The solid lines are profiles taken along the symmetry axis of the flow, while the dotted lines are taken along a path joining the center of the vortices. We see that the signatures of an underlying vortex street are (Fig. 2 from top to bottom) double peaked maxima for the total plasma velocity, a single peaked meridional angle and correlations between pressure fluctuations maxima and the rise of total plasma velocity. Figure 2 (bottom) also shows that the cross-stream (x) component of the magnetic field undergoes two reversals of its sign.

These basic features are robust under reasonable changes in Mach numbers and the initial geometry of the flow (that is, within a reasonable range of the two bifurcation parameters mentioned above). They are also in good agreement with the observations (see [*Siregar et al.*, 1992]). We found that a strong flow-aligned magnetic field, such as would exist in the inner heliospheric solar wind, inhibits the development of the vortex street. This can explain the lack of observations of these signatures in the inner heliosphere.

Acknowledgments. This research was supported by the Space Physics Theory Program at the Goddard Space Flight Center and the National Academy of Sciences. Computations were made at the NASA Center for Computational Sciences.

REFERENCES

Burlaga, L. F., A heliospheric Vortex Street?, *J. Geophys. Res.*, 95, 4333-4336, 1990.

Siregar E., D.A. Roberts and M.L. Goldstein, An Evolving MHD Vortex Street Model for Quasi-Periodic Solar Wind Fluctuations, *Geophys. Res. Lett.*, 19, 1427, 1992.

Siregar E., D.A. Roberts and M.L. Goldstein, Quasi-Periodic Transverse Plasma Flow Associated With an Evolving MHD Vortex Street in the Outer Heliosphere, J. *Geophys. Res.*, 98, 13,233, 1993.

Siregar E., W.T. Stribling and M.L. Goldstein, On the Dynamics of a Plasma Vortex Street and its Topological Signatures, *Physics of Plasmas*, Vol. 1, 7, 2125-2134, 1994.

Lazarus, A.J., Yedidia, B., Villanueva, L., McNutt, R. L. and Belcher, Jr. J. W., Meridional Plasma Flow in the Outer Heliosphere, *Geophys. Res. Lett.*, 15, 1519-1522, 1988.

Kármán, T. V., *Gottinger Nachrichten*, *Math-Phys. KL.*, 509-17,1911.

Ohle F. et al., Description of transient states of Von Karman vortex streets by low dimensional differential equations, *Phys. Fluids A*, 1990, Vol 2, Iss 4, p 479-481.

Hayot F. and Rajlakshmi M., Cylinder wake in lattice gas hydrodynamics, *Physica D*, 1989, Vol 40, Iss 3, p 415-420.

Colonius, T., Lele, S. K. and Moin, P., *J. Fluid Mech.*, Vol 230, 45-73, 1991.

Batchelor, G. K., An introduction to fluid dynamics, Cambridge Univ. Press, Cambridge, 1967.

Lazarus, A. J., Yedidia, B., Villanueva, L., McNutt, R. L. and Belcher, Jr. J. W., Meridional plasma flow in the outer heliosphere, *Geophys. Res. Lett.*, *15*, 1519-1522, 1988.

M. L. Goldstein, D. Aaron Roberts, and E. Siregar, Laboratory for Extraterrestrial Physics, NASA Goddard Space Flight Center, Greenbelt, MD 20771

Decay of Magnetic Helicity in Ideal Magnetohydrodynamics with a DC Magnetic Field

Troy Stribling

National Research Council, NASA Goddard Space Flight Center, Greenbelt, Maryland

William H. Matthaeus

Bartol Research Institute, University of Delaware Newark, Delaware

We show that the magnetic helicity associated with fluctuations in homogeneous three-dimensional incompressible ideal magnetohydrodynamics (MHD) with a mean magnetic field decays in time due to nonlinear processes. Evidence is obtained numerically by use of nondissipative spectral method simulations. The process of nonlinear decay is described in terms of a generalized ideal helicity invariant, and is characterized by the transient production of a mean induced electric field aligned with the applied magnetic field. Dissipative simulations are also performed which suggest that the nonlinear decay process causes the magnetic helicity associated with fluctuations to decay at a much faster rate than it would by resistivity alone. The described effect stands in contrast to expectations based on studies of MHD turbulence without an applied mean field, in which magnetic helicity is transferred nonlinearly to long wavelengths, and is preserved in time due to selective decay when dissipation is present.

1. BACKGROUND

The magnetic helicity of a turbulent magnetofluid may be defined as,

$$H_m = \frac{1}{2}\langle \mathbf{a} \cdot \mathbf{b} \rangle, \tag{1}$$

where $\mathbf{b}$ is the magnetic field, $\mathbf{b} = \nabla \times \mathbf{a}$ defines the vector potential $\mathbf{a}$ and $\langle \cdots \rangle$ denotes a spatial average. A popular interpretation of H_m, suggested by Moffatt [1978], is that H_m measures the linkage of magnetic flux tubes. Another interpretation, commonly used in turbulence applications, is obtained by considering the Fourier transform of $\mathbf{b}$, $\mathbf{b}(\mathbf{k})$. We can think of H_m at wavenumber $\mathbf{k}$ as a measure of polarization of the $\mathbf{b}(\mathbf{k})$. Plane polarized $\mathbf{b}(\mathbf{k})$ have zero H_m and circularly polarized $\mathbf{b}(\mathbf{k})$ have a maximal value of H_m whose magnitude is equal to $|\mathbf{b}(\mathbf{k})|^2/k$. Thus H_m defined by Eq. (1) gives a measure of the mean polarization of the $\mathbf{b}(\mathbf{k})$. It is this interpretation that is most convenient for this work. The principal result presented here is that when a DC magnetic field is present and $\mathbf{b}$ is defined so that $\langle \mathbf{b} \rangle = 0$, H_m defined by Eq. (1) decays from its initial value and fluctuates about zero in the absence of resistivity.

Magnetic helicity is an important measure in magnetohydrodynamics (MHD) because it is conserved in the absence of both dissipation and a DC magnetic field. This conservation law is believed responsible for back transfer of H_m in turbulent MHD [Frisch et al., 1975]. Back transfer of H_m has been invoked as a mechanism leading to efficient turbulent magnetic dynamo activity in forced [Pouquet et al., 1976; Meneguzzi et al., 1981] and freely decaying MHD [Pouquet and Patterson, 1978; Stribling and Matthaeus, 1991]. Selective decay relaxation, which describes some properties of RFP plasma devices, is also a consequence of H_m back transfer in freely decaying three dimensional (3D) MHD [Taylor, 1974; Montgomery et al., 1978; Matthaeus and

Space Plasmas: Coupling Between Small and Medium Scale Processes
Geophysical Monograph 86

Montgomery, 1980; Riyopoulos et al., 1982; Horiuchi and Sato, 1986; Stribling and Matthaeus, 1991]. The presence of magnetic helicity has been shown to slow down the energy decay rate in simulations 3D MHD decay as well [Stribling and Matthaeus, 1991].

On the other hand, particularly in space physics and astrophysics applications, homogeneous MHD turbulence concepts are applied to local samples of the fluid plasma, while externally supported magnetic fields can thread the region without closing in the region of interest. This alternative situation may eventually prove to be more relevant than the case of "isolated" homogeneous turbulence which is often modeled using periodic boundary conditions with zero volume averaged magnetic field. The basic underpinning of helicity dynamics in MHD turbulence theory, i.e., the ideal conservation of H_m, needs to be reconsidered in this case. In this paper we examine the conservation properties of H_m for a simple model, in which the MHD turbulence is periodic but evolves in the presence of a specified, externally supported uniform magnetic field. We show, by considering the helicity conservation law, and by use of direct numerical solution of the 3D MHD equations, that helicity behaves dynamically in a somewhat different way than in the case of magnetically isolated MHD turbulence.

Investigation of magnetic helicity in a homogeneous MHD turbulence model with a mean magnetic field might prove relevant for interplanetary applications. In particular the homogeneous MHD turbulence model has been used to interpret MHD scale magnetic field field fluctuations in the solar wind [Matthaeus and Goldstein, 1982], motivating models using periodic boundaries. However, interplanetary space also contains the large scale solar magnetic field, which appears locally (on the scale of a correlation length) as a uniform field. For example at 1 AU the correlation length of the magnetic field fluctuations is of the order of .01 AU while the length scale for gradiants in the solar magnetic field is of the order of 1 AU [Matthaeus and Goldstein, 1982]. Fluctuations in H_m have been observed in the solar wind [Matthaeus and Goldstein, 1982; Bieber et al., 1987a; Goldstein et al., 1991]. The inertial range fluctuations in these observations have been shown to be randomly distributed with zero mean, although there is evidence for a net magnetic helicity at the energy containing fluctuation scale [Bieber et al., 1987b; Bieber and Smith, 1992]. It is also believed that H_m plays a role in cosmic ray transport in the heliosphere [Bieber et al., 1987a; Bieber et al., 1987b].

We investigate the time behavior of H_m for constant density unforced 3D MHD which written in Alfvén speed units has the form,

$$\frac{\partial \mathbf{v}}{\partial t} + \mathbf{v}\cdot\nabla\mathbf{v} = -\nabla p^* + \mathbf{b}\cdot\nabla\mathbf{b} + \mathbf{B}_0\cdot\nabla\mathbf{b} \quad (2)$$

$$+\nu\nabla^2\mathbf{v}, \quad (3)$$

$$\frac{\partial \mathbf{b}}{\partial t} + \mathbf{v}\cdot\nabla\mathbf{b} = \mathbf{b}\cdot\nabla\mathbf{v} + \mathbf{B}_0\cdot\nabla\mathbf{v} + \eta\nabla^2\mathbf{b}, \quad (4)$$

$$\nabla\cdot\mathbf{v} = 0, \quad (5)$$

$$\nabla\cdot\mathbf{b} = 0. \quad (6)$$

The fluctuating velocity and magnetic fields, $\mathbf{v}$ and $\mathbf{b}$, each have zero mean, and $\mathbf{B}_0$ is the mean magnetic field which is assumed to be constant. The total pressure, p^*, is determined by a Poisson equation when use is made of the incompressibility condition, Eq. (4). The viscosity and resistivity are ν and η, respectively. The time unit is taken to be an eddy turnover time for unit velocity field at unit length scale. A dealiased spectral method is used in the numerical solution of Eqs. (2-5) which assumes periodic boundary conditions in a cube with sides of length 2π.

Of particular importance in the following discussion are quantities known as "rugged" invariants. These invariants are conserved by arbitrary truncations of the Galerkin representations of Eqs. (2-5) when $\eta = \nu = 0$, and are usually considered to be the relevant isolating invariants in a statistical mechanics treatment [Kraichnan, 1975]. When $\mathbf{B}_0 = 0$ there are believed to be only three rugged invariants [Frisch et al., 1975; Stribling and Matthaeus, 1990]. They are the total energy, cross helicity and magnetic helicity. For periodic 3D MHD the rugged invariants are explicitly given by,

$$E = \frac{1}{2V}\int |\mathbf{v}|^2 + |\mathbf{b}|^2\, d^3x = E_v + E_b, \quad (7)$$

$$H_c = \frac{1}{2V}\int \mathbf{v}\cdot\mathbf{b}\, d^3x, \quad (8)$$

$$H_m = \frac{1}{2V}\int \mathbf{a}\cdot\mathbf{b}\, d^3x, \quad (9)$$

where $V = 8\pi^3$. Each rugged invariant is expressed as a per unit mass quantity, and the integrals are evaluated over a cube with sides of length 2π. The cross helicity can be viewed as a measure of correlation between the velocity and magnetic fields, and possible interpretations of H_m were previously mentioned. If $\mathbf{B}_0 \neq 0$ the total energy and cross helicity remain rugged invariants. However, with $\mathbf{B}_0 \neq 0$, magnetic helicity is not an ideal invariant; instead one now has an invariant which we call the total magnetic helicity [Mattheaus and Goldstein, 1982], defined by

$$\hat{H}_m = H_m + \mathbf{B}_0 \cdot \mathbf{A}_0. \tag{10}$$

Here $d\mathbf{A}_0/dt = \frac{1}{V}\int \mathbf{v} \times \mathbf{b}\ d^3x = -\vec{\mathcal{E}}$ represents the time derivative of the $\mathbf{k} = 0$ component of the vector potential. The mean induced electric field is given by $\vec{\mathcal{E}}$ and H_m is the fluctuating magnetic helicity given by Eq. (8). It is unclear at present whether the ideal invariant $\hat{H}_m$ should be considered to be "rugged", since, although it is quadratic, it is nonlocal in time. A suggested interpretation of the $\mathbf{B}_0 \cdot \mathbf{A}_0$ term of $\hat{H}_m$ is that it represents magnetic flux linkages associated with structures having scales much larger than the periodic box [Moffatt, 1978; Matthaeus and Goldstein, 1982]. Thus, $\hat{H}_m$ can be viewed as a superpositioning of the magnetic helicity of the fluctuating magnetic fields and the linkage of the fluctuating magnetic field with the uniform field. A proper accounting of both of these magnetic helicity contributions is necessary in the construction of the magnetic helicity invariant for homogeneous MHD turbulence in the presence of a DC magnetic field. It should also be noted that $\hat{H}_m$ is not the magnetic helicity one would compute using Eq. (1) and also including the DC contributions $\mathbf{A}_0$ and $\mathbf{B}_0$. The result so obtained would differ from $\hat{H}_m$ by an amount $\frac{1}{2}\mathbf{B}_0 \cdot \mathbf{A}_0$.

In the ideal simulations discussed in the following paragraphs we show that $\hat{H}_m$ is conserved and that whatever initial H_m is in the system decays and proceeds to fluctuate about zero. The decay of H_m is concurrent with a growth in $\mathbf{B}_0 \cdot \mathbf{A}_0$ in accordance with the conservation law given by Eq. (9). Thus the nonlinear decay process acting on H_m is, in fact, a conversion of H_m into $\mathbf{B}_0 \cdot \mathbf{A}_0$.

2. NUMERICAL RESULTS

Each of the contributions to $\hat{H}_m$ from an 32^3 ideal simulation is given in Fig. (1). For this particular run $\mathbf{B}_0 = 1\hat{\mathbf{z}}$, $\sqrt{\langle|\mathbf{b}|^2\rangle}/B_0 = 1$, $E = 1$, $H_c = 10^{-3}$ and at $t = 0$, $H_m = 0.1$. The initial energy spectrum was flat with the kinetic and magnetic energy equipartitioned for wavenumbers in the band $1 \leq k \leq 5$. The ratio $\sqrt{\langle|\mathbf{b}|^2\rangle}/B_0 \sim 1$ is characteristic of the solar wind. The time history of H_m is given in Fig. (1a). It is seen that H_m decays very quickly to zero while $\hat{H}_m$ remains constant, see Fig. (1d). In Fig. (1b) the time history of the $\hat{\mathbf{z}}$ component of spatially averaged electric field $\mathcal{E}_z = -\frac{1}{V}\int(\mathbf{v}\times\mathbf{b})_z\ d^3x$ is given. Only the $\hat{\mathbf{z}}$ component of $\vec{\mathcal{E}}$,

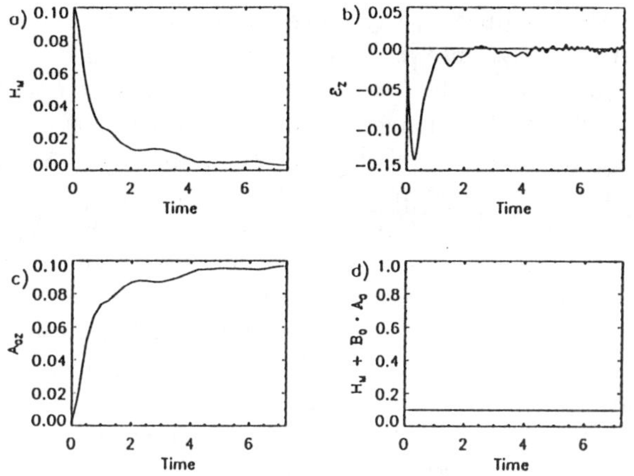

Fig. 1. Time histories of the quantities which form $\hat{H}_m$ for an ideal 32^3 simulation with $E = 1$, $H_c = 10^{-3}$, $\sqrt{\langle|\mathbf{b}|^2\rangle}/B_0 = 1$ and $\hat{H}_m = .1$. The quantities depicted in each plot are: (a) magnetic helicity of fluctuating $\mathbf{b}$, (b) component of mean electric field parallel to $\mathbf{B}_0$, (c) component of mean vector potential parallel to $\mathbf{B}_0$, which is the time integral of the curve in Fig. (1b) and (b) the total magnetic helicity $\hat{H}_m$.

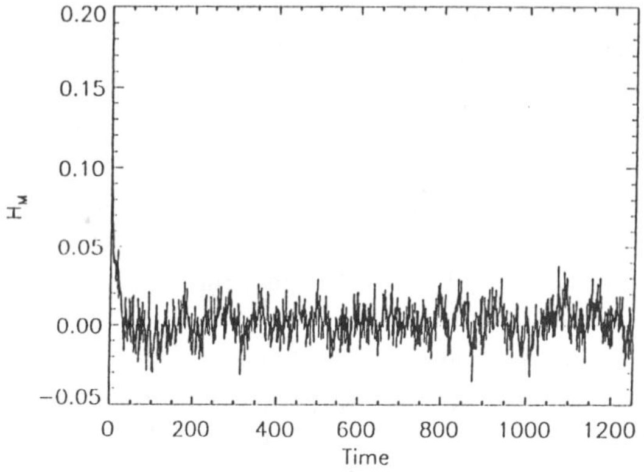

Fig. 2. Time history of H_m for an ideal 8^3 simulation with $E = 1$, $H_c = 0.2$, $\sqrt{\langle|\mathbf{b}|^2\rangle}/B_0 = 1$ and $\hat{H}_m = 0.2$ which was integrated out to 1200 eddy turnover times.

which in the simulations lies along $\mathbf{B}_0$, enters into $\hat{H}_m$. Most noticeable is the pulse in $\mathcal{E}_z$ which coincides with the period of most rapid decay in H_m. The time integral of $\mathcal{E}_z$, the $\hat{\mathbf{z}}$ component of $\mathbf{A}_0$, is depicted in Fig. (1c). The pulse in $\mathcal{E}_z$ gives rise to rapid growth in A_{0z} which signifies the conversion of H_m into $\mathbf{B}_0 \cdot \mathbf{A}_0$. Because $\hat{H}_m$ is conserved during this process, the fluctuations in $\mathcal{E}_z$ are perfectly correlated with the dH_m/dt fluctuations.

In Fig. (2) the time history of H_m is given for an 8^3 simulation integrated for 1200 eddy turn over times. This is much longer than the time history given in Fig. (1a) and illustrates the fluctuating character of H_m after the nonlinear decay process has converted the majority of H_m into $\mathbf{B}_0 \cdot \mathbf{A}_0$. For this run $\mathbf{B}_0 = 1\hat{\mathbf{z}}$, $\sqrt{\langle|\mathbf{b}|^2\rangle}/B_0 = 1$, $E = 1$, $H_c = 0.2$ and at $t = 0$, $H_m = 0.2$. This longer simulation is used to compute long time averages of the H_m spectrum. A convenient measure of the polarization of the $\mathbf{b}$ fluctuations at a wavenumber $\mathbf{k}$ is given by $\sigma_m(\mathbf{k}) = kH_m(\mathbf{k})/E_b(\mathbf{k})$, where $E_b(\mathbf{k})$ is the magnetic energy in wavenumber $\mathbf{k}$. A value of ± 1 for $\sigma_m(\mathbf{k})$ indicates that the fluctuations are circularly polarized with handedness determined by the sign. A value of zero indicates plane polarization. Fig. (3) depicts the omnidirectional spectrum of $k\langle H_m(\mathbf{k})\rangle/\langle E_b(\mathbf{k})\rangle$. Here $\langle\cdots\rangle$ represents a time average. This quantity gives a measure of the mean polarization of the $\mathbf{b}$ fluctuations at wavenumbers with magnitude k. We see that the mean polarization is near zero across the spectrum which is consistent with the behavior of the time history of H_m depicted in Fig. (3). This differs significantly from the $B_0 = 0$ case in which H_m is conserved and thus a nonzero mean polarization at each wavenumber is obtained in nondissipative simulations [Stribling and Matthaeus, 1990]. In the analysis of interplanetary magnetic field fluctuations investigators have generally found that the high wavenumber magnetic field polarizations average to zero Matthaeus and Goldstein, 1982; Goldstein et al., 1991]. The present result is in qualitative agreement with the observations, suggesting that the MHD scales of the interplanetary environment may be consistent with a model which in which $\hat{H}_m$ is the relevant ideal invariant instead of H_m. However, this feature may not be a sensitive discriminator, since higher wavenumber fluctuations in dissipative MHD without an applied magnetic field may share this characteristic.

An interesting consequence of the conservation of $\hat{H}_m$ and the subsequent decay of H_m is the previously mentioned transient $\mathcal{E}_z$ which persists while H_m is decaying. If $\hat{H}_m$ is examined and it is assumed that H_m goes to zero some properties of the $\mathcal{E}_z$ pulse can be determined. First, assume that $H_m(t=0) > 0$. Since H_m decays as time progresses, in order to conserve $\hat{H}_m$ the $\mathbf{B}_0 \cdot \mathbf{A}_0$ term must be positive and increasing while H_m is decreasing. This in turn requires that the component of the electric field along $\mathbf{B}_0$, $\mathcal{E}_\parallel$, have the sign opposite of $\mathbf{B}_0$ as H_m decreases (recall that $d\mathbf{A}_0/dt = -\vec{\mathcal{E}}$). In a similar vein if $H_m(t=0) < 0$ then a transient $\mathcal{E}_\parallel$ is generated in the same direction of $\mathbf{B}_0$. The transient $\mathcal{E}_\parallel$ for the 32^3 ideal simulation can be seen in Fig. (1b).

To verify that ideal magnetic helicity decay process is not significantly modified in dissipative MHD, we performed two 32^3 dissipative simulations with $\eta = \nu =$

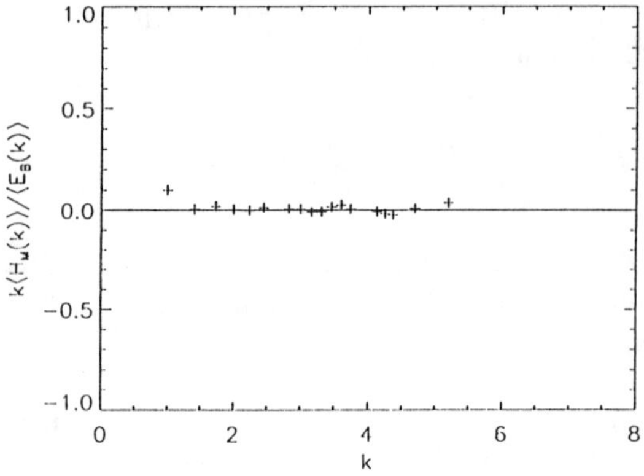

Fig. 3. Time averaged normalized magnetic helicity spectrum for the 8^3 simulation.

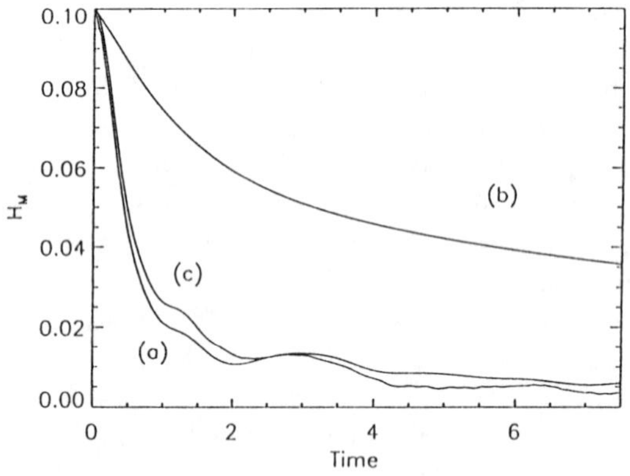

Fig. 4. Comparison of H_m time histories for Runs (a), (b) and (c).

0.01. In particular we are interested in examining the dissipative and nondissipative contributions to the time rate of change of H_m, namely,

$$\frac{dH_m}{dt} = -\mathbf{B}_0 \cdot \langle \mathbf{v} \times \mathbf{b} \rangle - \eta \langle \mathbf{j} \cdot \mathbf{b} \rangle, \tag{11}$$

as $\mathbf{B}_0$ and η are varied. Here $\mathbf{j} = \nabla \times \mathbf{b}$ is the current and $\langle\cdots\rangle$ represents a spatial average. The $-\mathbf{B}_0 \cdot \langle \mathbf{v} \times \mathbf{b} \rangle$ term represents the ideal contribution and the $-\eta\langle \mathbf{j} \cdot \mathbf{b}\rangle$ term the resistive contribution of the H_m time derivative. The initial conditions for two dissipative runs are identical to the initial condition of the 32^3 ideal simulation previously discussed. The values given to $\mathbf{B}_0$ and η for this parameter space scan are: Run (a), $\mathbf{B}_0 = 1\hat{\mathbf{z}}$, $\eta = 0.01$; Run (b) $\mathbf{B}_0 = 0$, $\eta = 0.01$; Run (c) $\mathbf{B}_0 = 1\hat{\mathbf{z}}$, $\eta = 0$. The time histories of H_m are compared for the three runs in Fig. (4). The slowest decay rate occurs in the simulation where there is only resistive decay of H_m, Run (b). A much faster decay rate for H_m is real-

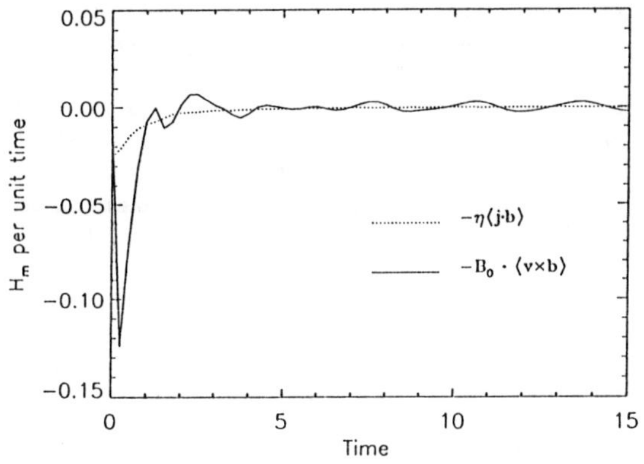

Fig. 5. Comparison of time histories of nonlinear and resistive decay terms (see Eq. 10) for Run (a).

ized in the presence of a mean magnetic field, Runs (a) and (c). It is also interesting to note that the contribution of the resistive term in the time evolution of H_m is quite small for nonzero $\mathbf{B}_0$, since there is little difference in the H_m time histories for Runs (a) and (c). In Fig. (5) the time histories of the nonlinear and resistive contributions to dH_m/dt are shown for Run (a). Upon examination of Fig. (5) it can be seen that the suspicion that resistive decay contributes little compared to the ideal decay term in the time evolution of H_m was correct. It should be kept in mind that here only the strong $\mathbf{B}_0$, (i.e. $\mathbf{B}_0 \approx \sqrt{\langle|\mathbf{b}|^2\rangle}$ the parameter regime most relevant to interplanetary plasmas) limit has been investigated. If $\mathbf{B}_0$ is small relative to $\sqrt{\langle|\mathbf{b}|^2\rangle}$ it is possible that resistive decay will dominate.

3. CONCLUSIONS

The principal results of this paper can be summarized as follows. Simulations of ideal and dissipative 3D MHD with a mean magnetic field have been presented. We have shown the magnetic helicity defined by Eq. (8) decays from its initial value and fluctuates about zero in the ideal simulations. The decay of H_m is concurrent with a pulse in the component of the mean electric field parallel to $\mathbf{B}_0$. The time average of the magnetic helicity spectrum also has properties qualitatively similar to the average helicity spectrum measured for $\mathbf{b}$ fluctuations in the solar wind. This result raises the possibility that the MHD scales of the solar wind may be described by a model in which $\hat{H}_m$ is the relevant ideal invariant as opposed to H_m.

The dissipative simulations suggest that when $B_0 \sim \sqrt{\langle|\mathbf{b}|^2\rangle}$, H_m decays at a faster rate that it would in the absence of $\mathbf{B}_0$. This could lead to quite significant differences in the structure of high Reynolds number turbulence for systems with and without a mean magnetic field. Most notable is that the presence of $\mathbf{B}_0$ could cause a diminution of H_m back transfer effects, such as selective decay [Matthaeus and Montgomery, 1980], upon the long wavelength fluctuations. It is possible that for this reason no conclusive evidence of H_m back transfer has been found in interplanetary magnetic field fluctuations [Goldstein et al., 1984].

Our work thus far, though not discussed here, indicates that the rate of H_m decay by the nonlinear decay process varies greatly with H_c and $\sqrt{\langle|\mathbf{b}|^2\rangle}/B_0$. Hence, it is likely that for some values of H_c and $\sqrt{\langle|\mathbf{b}|^2\rangle}/B_0$, different from those chosen here, the strength of the nonlinear decay process in dissipative MHD could lessen.

Acknowledgments. The authors would like to thank S. Ghosh, M. L. Goldstein, S. Oughton and D. A. Roberts for helpful comments, the NASA Center for Computational Sciences for computer time and the National Academy of Sciences for financial support.

REFERENCES

Bieber, J. W., P. Evenson and W. H. Matthaeus, *Magnetic Helicity of the Parker Field*, Astrophysical J., 315, 700 (1987a).

Bieber, J. W., P. Evenson and W. H. Matthaeus, *Magnetic Helicity of the IMF and the Solar Modulation of Cosmic Rays*, Geophysical Res. Lett., 14, 864 (1987b).

Bieber, J. W. and C. W. Smith *Magnetic Helicity in the Solar Wind: Radial Variation and Cosmic Ray Effects*, EOS, 73, 247 (1992).

Frisch, U., A. Pouquet, J. Léorat and A. Mazure, *Possibility of an Inverse Cascade of Magnetic Helicity in Magnetohydrodynamic Turbulence*, J. Fluid Mech., 68, 769 (1975).

Goldstein, M. L., L. F. Burlaga and W. H. Matthaeus, *Power Spectra of Interplanetary Corotating and Transient Flows*, J. Geophys. Res., 89, 3747 (1984).

Goldstein, M. L., D. A. Roberts and C. A. Fitch, *The Structure of Helical Interplanetary Magnetic Fields*, GRL, 18, 1505 (1991).

Horiuchi, R. and T, Sato, *Self-Organization and Energy Relaxation in a Three-Dimensional Magnetohydrodynamic Plasma*, Phys. Fluids, 29, 1161 (1986).

Kraichnan, R. H., *Statistical Dynamics of Two-Dimensional Flow*, J. Fluid Mech., 67, 155 (1975).

Matthaeus, W. H. and D. Montgomery, *Selective Decay Hypotheses at High Mechanical and Magnetic Reynolds Numbers*, Ann. N.Y. Acad. Sci., 357, 202 (1980).

Matthaeus, W. H., and M. L. Goldstein, *Measurement of the Rugged Invariants of Magnetohydrodynamic Turbulence in the Solar Wind*, J. Geophys. Res., 87, 6011 (1982).

Meneguzzi, M., U. Frisch and A. Pouquet, *Helical and Nonhelical Turbulent Dynamos*, Phys. Rev. Lett., 47, 1069

(1981).

Moffatt, H. K., *Magnetic Field Generation in Electrically Conducting Fluids*, Cambridge University Press, Cambridge (1978).

Montgomery, D., L. Turner and G. Vahala, *Three-Dimensional Magnetohydrodynamic Turbulence in Cylindrical Geometry*, Phys. Fluids, 21 , 757 (1978).

Pouquet, A., U. Frisch and J. Léorat, *Strong MHD Helical Turbulence and the Nonlinear Dynamo Effect*, J. Fluid Mech., 77, 321 (1976).

Pouquet, A. and G. S. Patterson, *Numerical Simulation of Helical Magnetohydrodynamic Turbulence*, J. Fluid Mech., 85, 305 (1978).

Riyopoulos, S., A. Bondeson and D. Montgomery, *Relaxation toward States of Minimum Energy in a Compact Torus*, Phys. Fluids, 25, 107 (1982).

Stribling, T. and W. H. Matthaeus, *Relaxation Processes in a Low-Order Three-Dimensional Magnetohydrodynamics Model*, Phys. Fluids B, 8, 1848 (1991).

Stribling, T. and W. H. Matthaeus, *Statistical Properties of Ideal Three-Dimensional Magnetohydrodynamics*, Phys. Fluids B, 2, 1979 (1990).

Taylor, J. B., *Relaxation of Toroidal Plasma and Generation of Reverse Magnetic Fields*, Phys. Rev. Lett., 33, 1139 (1974).

T. Stribling, NASA Goddard Space Flight Center,Code 692, Greenbelt Rd., Greenbelt MD 20771

W. H. Matthaeus Bartol Research Institute, University of Delaware , Newark, Delaware 19716

Some Electron Conic Generation Mechanisms

Mats André

Swedish Institute of Space Physics, Umeå University, Umeå, Sweden

Lars Eliasson

Swedish Institute of Space Physics, Kiruna, Sweden

Electron conics can be generated both by monochromatic and by rather broadband fluctuations at roughly one Hz in a parallel electric field accelerating auroral electrons at altitudes of several thousand kilometers. Electron conical distributions may also be generated by parallel or perpendicular diffusion in velocity space caused by some waves. Our simulations show that it often may be hard to distinguish between generation mechanisms just by using the observed electron distributions. However, we investigate a model based on an observed broadband parallel electric field emission. The results from our simulations including such time-varying electric fields are in good agreement with observed electron distributions. The simulation parameters are consistent with the observed ion beam and electron conic energies. Our calculations are used to predict for example oscillations in the low altitude electron fluxes. These results can be used to compare different possible mechanisms for the generation of electron conics.

1. INTRODUCTION

There are many satellite observations of electron distributions with enhanced fluxes at keV energies near the edge of the upward loss cone. These distributions usually occur at altitudes of several thousand kilometers on auroral field lines. These so-called "electron conics" were first noted in data from the DE 1 satellite by *Menietti and Burch* [1985]. They have also been observed by the Viking satellite [*André et al.*, 1987; *Lundin et al.*, 1987; *Lundin and Eliasson*, 1991; *André and Eliasson*, 1992], by the S3-3 spacecraft [*Roth et al.*, 1989] and by the Akebono satellite. The acceleration of upgoing electrons to keV energies is interesting in its own right. Furthermore, understanding of this energization mechanism should help to clarify the formation, stability and dynamics of the mechanism accelerating downgoing auroral electrons to keV energies.

Electron conics are observed in regions of upgoing ion beams and expanded electron loss cones, indicating acceleration parallel to the geomagnetic field below the satellite. Auroral electrons are usually thought to be accelerated by a "static" potential structure parallel to the geomagnetic field [*Lundin and Eliasson*, 1991, and references therein]. However, since the electron conics we discuss consist of upgoing electrons with higher energies than the downgoing electrons, magnetic mirroring above the atmosphere alone can not explain these distributions. Rather, some time-varying electric field must be involved. One possibility is perpendicular and oblique heating at low and mid-altitudes by e.g. upper hybrid waves as discussed by *Menietti and Burch* [1985], *Wong et al.* [1988], *Roth et al.* [1989], *Lin et al.* [1990], and *Menietti et al.* [1992]. Adiabatic motion up the inhomogeneous geomagnetic field would then cause the observed shape of the electron distributions. Another possibility is that parallel heating by e.g. Langmuir, acoustic-like or lower hybrid modes together with loss of electrons in the atmosphere can cause electron conics [*Roth et al.*, 1989; *Bryant et al.*, 1990]. Using a Monte Carlo technique to simulate electron heating over an extended region in space, *Temerin and Cravens* [1990] argued that parallel heating is a likely mechanism for the generation of electron conics.

Space Plasmas: Coupling Between Small and Medium Scale Processes
Geophysical Monograph 86

Previous detailed models of electron conic generation include particle interaction with some wave mode(s). Wave modes over a wide range of frequencies may contribute to the formation of electron conics and indeed we show that both perpendicular and parallel wave heating might, if intense enough waves exist, generate electron conical distributions in good agreement with observations. Some previous models include also a parallel potential drop. This electric field is usually assumed to be static on all interesting timescales, although the details of the formation of such a static structure are not well understood. Since a parallel auroral electric field can accelerate charged particles to high energies, it is natural to investigate the consequences of a non-static parallel electric field $E_{||}$ on the electron distributions. It has been suggested by *Lundin et al.* [1987], *Lundin and Eliasson* [1991] and *André and Eliasson* [1992] that fluctuations in $E_{||}$ on times of the order of one second can generate electron conics. To understand this mechanism we note that the travel time for an auroral electron from the acceleration region at an altitude of several thousand kilometer, via magnetic mirroring above the ionosphere, and back to the acceleration region, is about 1 second. Thus several electrons may be accelerated by one $E_{||}$ on their way down, and then slowed down by a smaller $E_{||}$ on their way up, in this way gaining energy. This mechanism is attractive e.g. since large amounts of energy are associated with the parallel electric field and since variations of electric fields on auroral field lines at roughly 1 Hz often are observed. In the following we show that variations of $E_{||}$ at frequencies of about 1 Hz can generate electron conical distributions in good agreement with observations.

2. OBSERVATIONS

Examples of electron conics have been observed by the Viking satellite at various magnetic local times, at altitudes from a few thousand kilometers to 13 500 km (apogee). An example of an electron conic observed by the ESP 1 electron detector on Viking [*Lundin et al.*, 1987] is shown in Figure 1. This figure shows countrate (which here is proportional to differential energy flux) as a function of pitch angle and energy. The narrow angular peaks near the atmospheric loss cone at energies of about 1 keV are the electron conic. The same data displayed as a distribution function in velocity space are shown in Figure 2. Electron conics are observed in regions of upgoing ion beams and expanded electron loss cones, indicating parallel acceleration below the spacecraft [*Menietti and Burch*, 1985; *André et al.*, 1987; *Lundin et al.*, 1987; *Lundin and Eliasson*, 1991; *Menietti et al.*, 1992]. The ion beam energy is larger than the electron conic energy for all Viking events we have studied. As an example of Viking data, Figure 3 shows the ratio of electron conic and ion beam energies for several 20 second spins of the satellite. These energies are determined from peaks in the countrate, but the same tendency is obtained by using e.g. peaks in flux. Our particle observations are consistent with an $E_{||}$ with a nearly static part and a smaller component varying on timescales of about 1 second. Electron conics may be generated by the time-varying part, while the ions would spend several wave periods inside the acceleration region, and the net effect on these would come mainly from the mean (static) electric field.

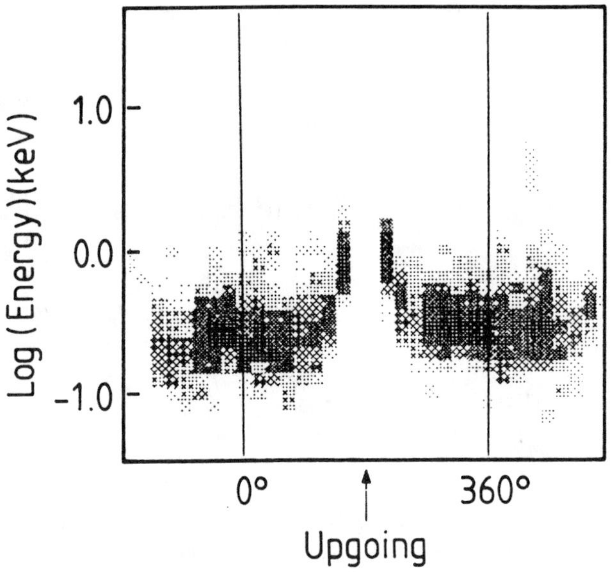

Fig. 1. Countrate of electrons showing an electron conic near the edge of the upgoing (180°) loss cone (Viking, May 8, 1986, 2024:04 to 2024:44) UT. Viking is at an altitude of about 10 000 km (16 400 km geocentric) with magnetic local time ≈ 18.25 and invariant latitude ≈ 79.5°. Logarithmic scale of counts from 3 to 80 counts.

3. SIMULATIONS

Using the observation that the time-varying part of the electric field seems to be smaller than the static part, and guided by the idea that oscillations of $E_{||}$ at roughly 1 Hz can generate electron conics, we create a simple model of the particle acceleration region on an auroral field line. The variation with geocentric distance r of the magnetic field strength is taken to be r^{-3} (dipole) and an absorbing atmosphere is placed at

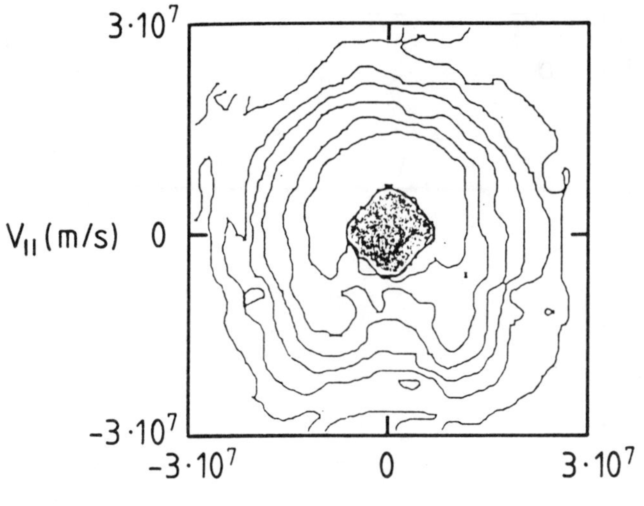

Fig. 2. a) Viking electron distribution, May 8, 1986, 2024:14 to 2024:34 UT. There are two contours per decade of the distribution function. See Figure 1 for satellite position.

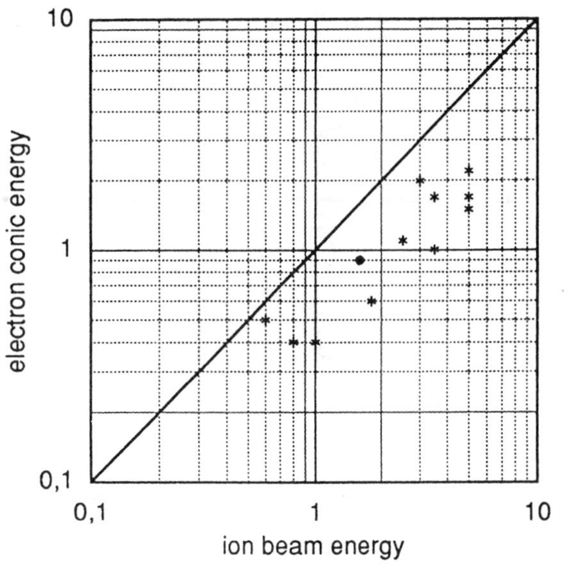

Fig. 3. Ratio of electron conic and ion beam energy determined from individual Viking spins on orbits 343 and 418.

$r = 6500$ km. A time-varying $E_{||}(t) = E_s + E_t(t)$ is assumed to exist between $r = 14500$ and 16500 km. Here $E_s = 1$ mV/m is static, directed upward, and corresponds to a total potential drop of 2 kV. Furthermore, $E_t(t) = E_v \sin(2\pi t)$ and $E_v = 0.35$ mV/m gives a component varying with a frequency of 1 Hz. Downgoing electrons are injected at the upper boundary of the acceleration region by randomly selecting 5×10^4 particles from a 100 eV Maxwellian distribution. This is a reasonable approximation of the observed downgoing electron population, see Figures 1 and 2. The particle start times are spread out over one wave period. The electrons are then followed numerically in small timesteps until they are either lost in the atmosphere or reach $r = 16500$ km again. The result is displayed in Figures 4a and 5a, which show an obvious electron conic.

It is sometimes useful to plot the simulation results as differential energy flux (Figure 4), e.g. since this is the format used when routinely searching satellite data for electron conics. Figure 4 has been obtained by sorting the simulation particles in 21 logarithmically spaced energy bins between 40 eV and 10 keV, using a pitch angle resolution of 18°. This is the pitch angle resolution used when routinely plotting Viking electron data, although better resolution is available with the detectors on this satellite.

The model electron conic is not sensitive to reasonable changes in the oscillation frequency, and wave frequencies between 0.3 and 6 Hz give distributions with clear indications of electron conics. High frequencies (above 10 Hz in this model) mean that the electrons spend at least roughly one wave period inside the acceleration region (where $E_{||} \neq 0$), and the effect of the time-varying electric field will essentially be averaged out. Low frequencies (less than 0.1 Hz) mean that $E_{||}$ has not changed much when the particles come back after mirroring above the atmosphere. The most "efficient" frequency for our specific model with a single sinewave, as measured by the ratio between upgoing and downgoing energy flux above the acceleration region, is about 2 Hz. For frequencies of one or a few Hz, the energy flux is roughly the same in both directions along the field line. Changing the model will of course change the most efficient frequency, but one or a few Hz should be a reasonable approximation for many realistic situations.

Assuming the acceleration region to be much thinner than the 2000 kilometers used in our model will increase the importance of higher frequencies. The electrons now spend only a short time inside the region where $E_{||} \neq 0$, and the effect of the oscillations may not average out. Furthermore, the electron travel time of roughly one second below the acceleration region is much longer than the period of high frequency oscillations of $E_{||}$. A reflected, upgoing electron can hit the acceleration region during any phase of the oscillation. As in our previous model, some electrons may gain energy, and an electron conic may form. This version of electron

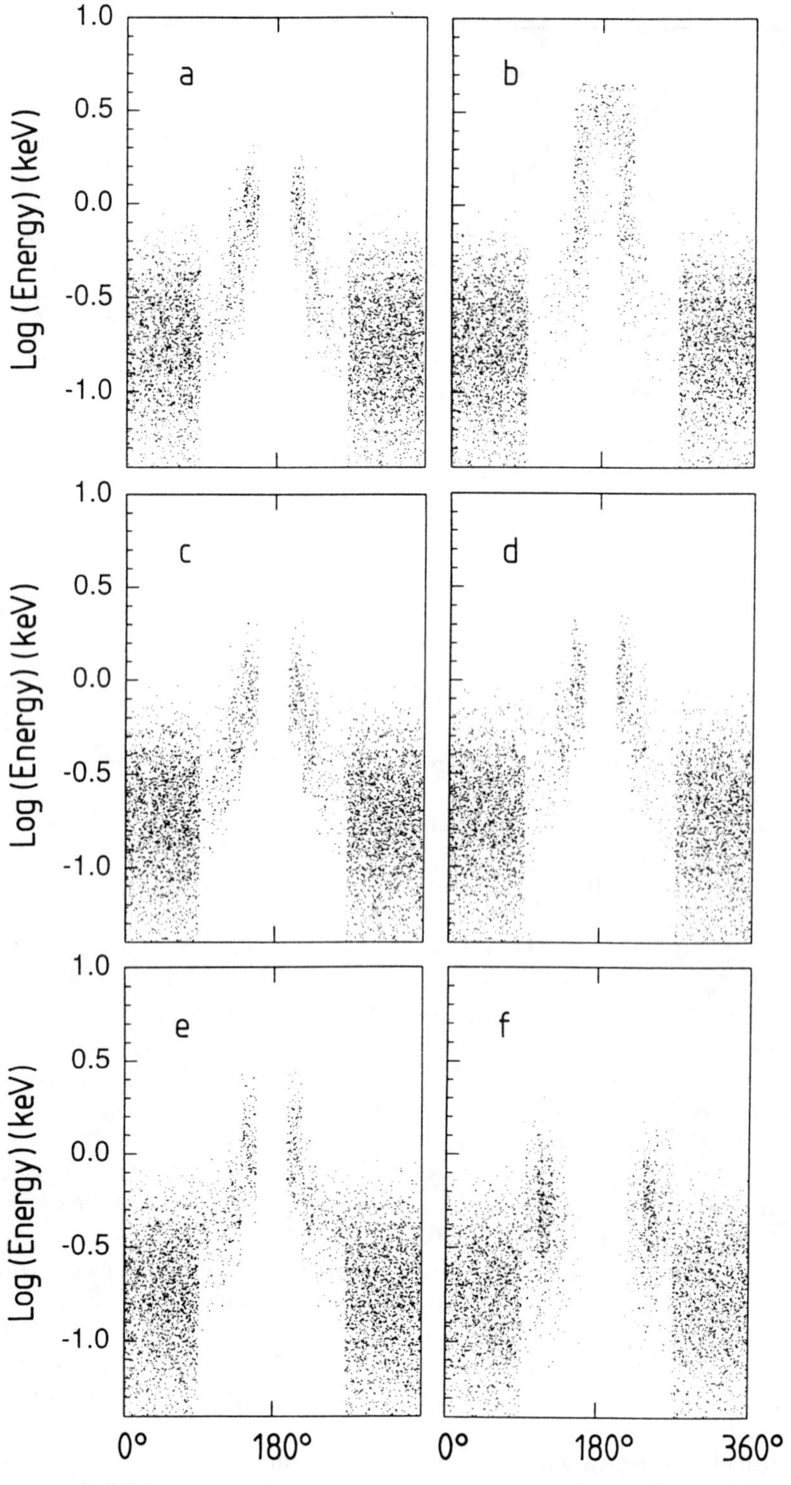

Fig. 4. Differential energy flux at a geocentric distance of 16500 km obtained from model calculations. The corresponding distribution functions are given in Figure 5. a) 1 Hz fluctuations in $E_{\|}$, b) high amplitude 1 Hz fluctuations, c) broadband fluctuations in $E_{\|}$, d) parallel wave diffusion at high altitude, e) perpendicular wave diffusion at low altitude, and f) perpendicular wave diffusion at high altitude. The number of points in each energy–pitch angle bin is proportional to the differential energy flux.

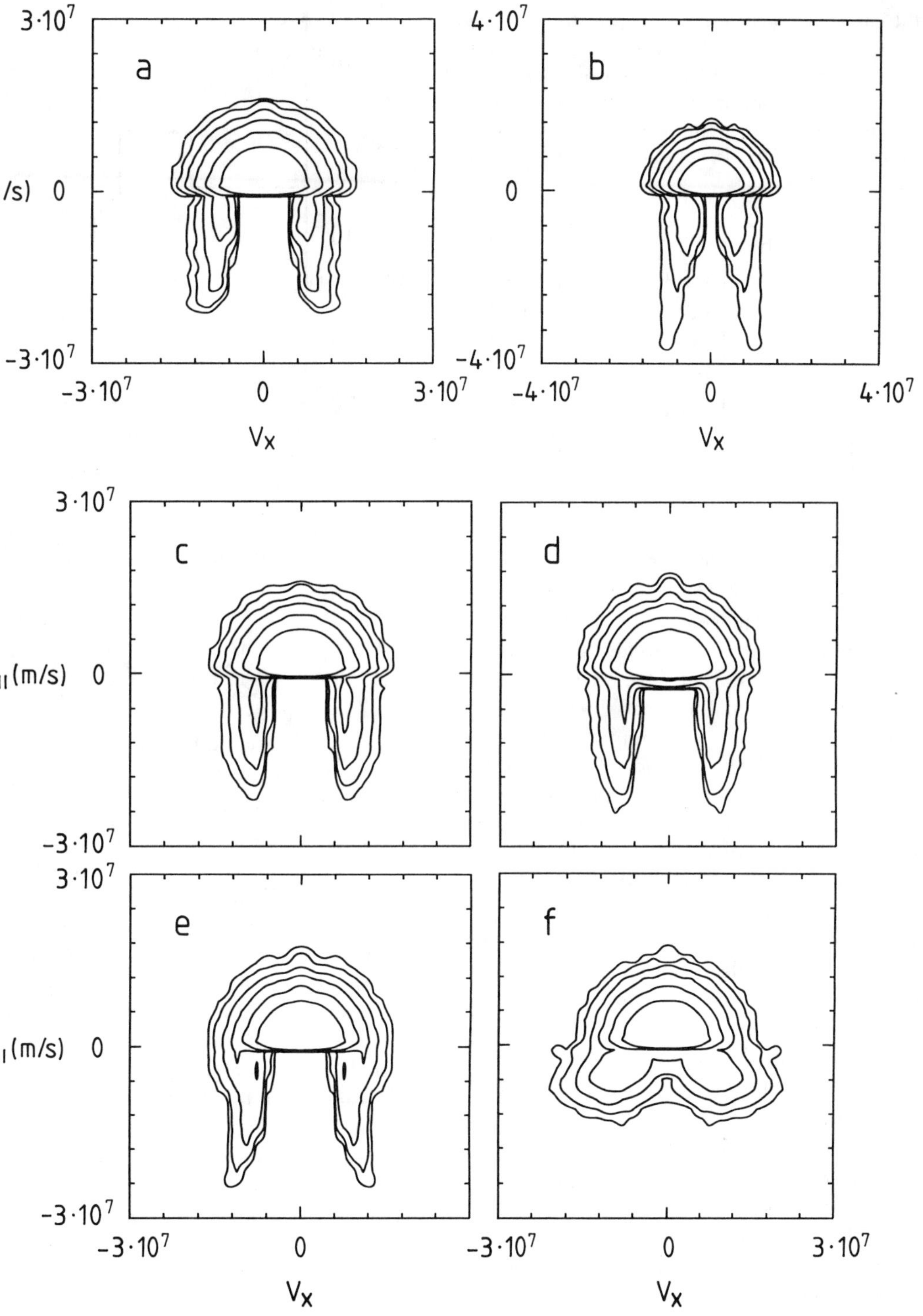

Fig. 5. Distribution functions at a geocentric distance of 16500 km obtained from model calculations. The corresponding differential energy flux is given in Figure 4. a) 1 Hz fluctuations in $E_{\|}$, b) high amplitude 1 Hz fluctuations, c) broadband fluctuations in $E_{\|}$, d) parallel wave diffusion at high altitude, e) perpendicular wave diffusion at low altitude, and f) perpendicular wave diffusion at high altitude. There are two contours per decade of the distribution function in all plots.

Table 1 Center frequencies $\widehat{f^m}$, bandwidths $\widehat{\delta f^m}$, wave intensities E_v^m, and number of waves N_m per frequency region in a model of broadband fluctuations in $E_{||}$ discussed in the text. More than one significant digit is used for E_v^m at higher frequencies to keep the spectral density nearly constant above 2 Hz. This is not meant to indicate that the model is very exact.

$\widehat{f^m}$ [Hz]	$\widehat{\delta f^m}$ [Hz]	E_v^m [mV/m]	N_m
0.3	0.4	0.09	5
0.95	0.9	0.2	10
1.7	0.6	0.1	3
3.5	3.0	0.116	3
7.5	5.0	0.15	5
20.	20.	0.3	10

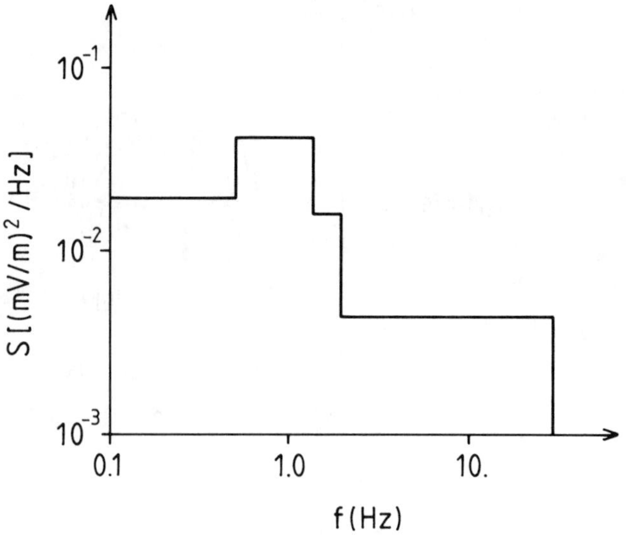

Fig. 6. Sketch of the spectral density S corresponding to a model of broadband fluctuations on $E_{||}$, defined in Table 1.

conic generation requires rather strong parallel electric fields, and is not further discussed in this report.

We now turn back to the generation of electron conics by a time-varying $E_{||}$ at geocentric distances between 14500 and 16500 kilometers. It is interesting to investigate the consequencies of an increased fluctuation level on the model electron conic. Keeping the static part of the parallel electric field to be $E_s = 1$ mV/m, but increasing E_v to 1 mV/m gives the result in Figures 4b and 5b. Here the electrons are accelerated to high parallel velocities and thus to pitch angles near 180°. Noting that the electrons spend rather short time inside the acceleration region, we find that the maximum electron energization now is larger than the acceleration given by the static E_s. This model is not consistent with any of the observations presented in Figure 3, and the resulting distribution in Figure 4b has too many particles close to 180° to be classified as an electron conic.

Using a single monochromatic sinewave to model the fluctuations in $E_{||}$ is instructive but not very realistic. A more realistic model is to spread the wave energy over M frequency regions centered at frequencies $\widehat{f^m}$, including N_m waves with frequencies f_n^m in each region [e.g. *Ball and André*, 1991]. In a model used by *André and Eliasson* [1992] waves around 1 Hz were included (which gave the electron conic), and also waves at higher and lower frequencies (to show that these waves did not interfere much with the conic generation). Here we investigate a more realistic broadband emission. In each frequency region m we randomly choose $2N_m$ numbers $f_1^m, \ldots, f_{N_m}^m$ and $\phi_1^m, \ldots, \phi_{N_m}^m$ satisfying $|f_n^m - \widehat{f^m}| \leq \widehat{\delta f^m}/2$ and $0 \leq \phi_n^m < 2\pi$, and use

$$E_{||}(t) = E_s + \sum_{m=1}^{M} ((E_v^m/\sqrt{N_m}) \sum_{n=1}^{N_m} \sin(2\pi f_n^m\, t + \phi_n^m)\ )$$

As a specific example, we choose $M = 6$ frequency regions. The center frequencies $\widehat{f^m}$ together with the corresponding bandwidths $\widehat{\delta f^m}$, wave intensities E_v^m, and number of waves N_m per frequency region are given in Table 1. The corresponding wave spectrum is sketched in Figure 6. Using the same position of the acceleration region as before and $E_s = 1$ mV/m, and spreading the particle start times over 20 s gives the result in Figures 4c and 5c. Selecting new frequencies and phases for each individual particle gives similar results.

The shape of the spectrum in Figure 6 may be compared with the estimate of a broadband parallel wave emission given by *Block and Fälthammar* [1990] in their Figure 6. Broadband emissions of this kind are common in the auroral zone [*Gurnett et al.*, 1984; *Block and Fälthammar*, 1990]. These emissions usually have a large perpendicular component which can generate ion conics [e.g. *Chang et al.*, 1986; *André et al.*, 1990; *Crew et al.*, 1990; *Ball and André*, 1991], but which is not in resonance with the electrons.

The result from a broadband model emission is similar to that in Figures 4a and 5a. These figures correspond to a single monochromatic wave at 1 Hz with an amplitude of 0.35 mV/m. The broadband emission used to obtain Figures 4c and 5c is similar in the sense that the wave intensity around 1 Hz is roughly the same, see column 3 in Table 1. However, the broadband emission includes rather strong waves at lower and higher frequencies, but these do not interfere much with the conic generation in this model.

4. COMPARISONS WITH OTHER MODELS

It is interesting to compare the model presented above with other models of electron conic generation. One possibility is parallel wave diffusion described by a diffusion coefficient $D_{||}$ [*Temerin and Cravens*, 1990]. To understand the difference between this diffusion model and our model, we assume that the diffusion occurs inside the acceleration region. We then note that for the diffusion model, the electron travel time below the acceleration region is of no importance. In this model, the electron heating is simply proportional to the time spent in the heating region. As an example of a diffusion model we keep the position of the acceleration region as before, $E_s = 1$ mV/m, and take $E_v = 0$, and $D_{||} = 1.5 \times 10^{14}$ m^2/s^3 (corresponding to a heating rate of ≈ 2 keV/s). The result in Figures 4d and 5d is very similar to our results obtained from oscillations in $E_{||}$. As noted by *Temerin and Cravens* [1990], parallel wave diffusion extending all the way down to the atmosphere will also give distributions similar to electron conics.

Perpendicular electron heating at low altitudes together with adiabatic folding by the inhomogeneous geomagnetic field, can generate distributions similar to those obtained from parallel electron heating. In our next example we follow the original suggestion by *Menietti and Burch* [1985] and include electron energization only at rather low altitudes. Taking the position of the acceleration region ($E_{||} \neq 0$) to be the same as before, $E_s = 1$ mV/m, $E_v = 0$, and including perpendicular wave diffusion ($D_\perp = 1.5 \times 10^{14}$ m^2/s^3) at geocentric distances between 7500 and 8500 km, gives the results in Figures 4e and 5e. Again, the results are similar to those obtained with our model with a time-varying $E_{||}$. Thus all our models discussed above generate electron conics with rather similar shape.

Perpendicular electron energization at high altitudes may give distributions which are significantly different from the distributions discussed above. This is simply because the magnetic mirror force has been changing the direction of the velocity vectors of upgoing electrons over a smaller altitude range. In an example by *Temerin and Cravens* [1990] perpendicular wave diffusion was included all the way from the atmosphere up to the observation altitude. This gave an electron distribution with an obvious "V-shape" in velocity space (Figure 3 of *Temerin and Cravens*). Such electron conics are similar to many ion conics. This is not surprising since the assumed electron energization is similar to the mechanism believed to generate many ion conics. Perpendicular electron heating in some limited altitude region can also give V-shaped distributions, again similar to ion conics. Since not many electron conics with a clear V-shape have been identified, *Temerin and Cravens* [1990] argued that perpendicular heating is not a likely candidate for electron conic generation.

However, detailed studies of some DE 1 data indicate that V-shaped electron conics may occur [*Menietti and Burch*, 1985; *Menietti et al.*, 1992]. Quantitatively the V-shape of a distribution may be found by investigating the density of upgoing (n_{up}) and downgoing (n_{down}) particles for some interval of perpendicular velocities [*Temerin and Cravens*, 1990]. *Menietti and Burch* [1992] showed that for at least one electron conic observed by DE 1, the ratio n_{up}/n_{down} was larger than unity for perpendicular velocities corresponding to the conic. This is a strong indication that perpendicular energization contributed to the formation of this distribution. Furthermore, *Menietti et al.* [1992] found that electron conics sometimes, but certainly not always, are correlated with upper hybrid emissions observed at the same altitude. Thus it is of interest to study perpendicular electron heating at and just below the observation altitude. As an example we take the position of the acceleration region to be the same as before, $E_s = 1$ mV/m, $E_v = 0$, and include perpendicular diffusion ($D_\perp = 0.8 \times 10^{14}$ m^2/s^3) at geocentric distances between 14500 and 16500 km (i.e. inside the region where $E_{||} \neq 0$). The result is given in Figures 4f and 5f. This distribution can possibly be called an electron conic, although it is significantly different from the conic caused by parallel diffusion (Figures 4d and 5d). A particle with a pitch angle of 90° at 14500 km will by adiabatic folding be moved to about 125° at 16500 km. This angle can be compared with the peak angle of the differential energy flux in Figure 4f. Many observed electron distributions (e.g. Figures 1 and 2) can not be explained by simulated distributions with peaks so far off from the loss cone. Thus, perpendicular energization may play an important role in the formation of some electron conics, but then the heating might take place significantly below the observation altitude.

It is likely that diffusion caused by e.g. upper hybrid waves is not strictly perpendicular but rather peaks at some oblique angle [*Wong et al.*, 1988]. However, models similar to the one used to obtain Figures 4f and 5f,

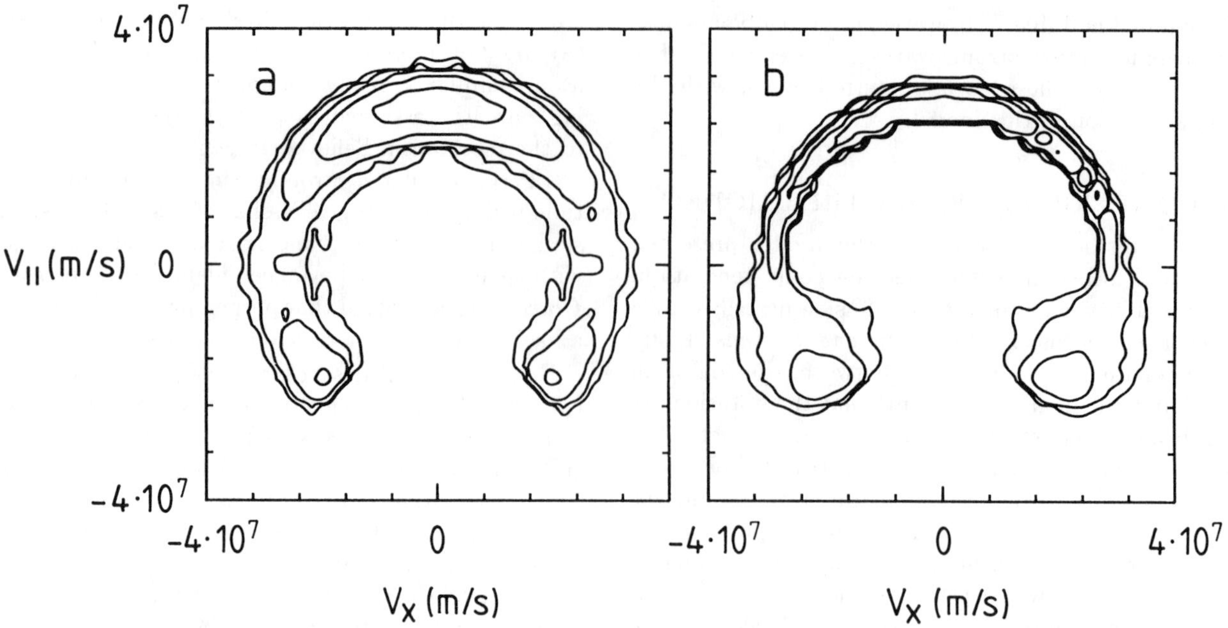

Fig. 7. Distribution functions at a geocentric distance of 8500 km obtained from model calculations. a) broadband fluctuations in $E_{\|}$, and b) perpendicular wave diffusion at low altitude.

but including both parallel and perpendicular diffusion with equal strengths, give distributions that are more spread out in pitch angle, but still show no peak near the loss cone. Although many observed electron distributions can not be explained by perpendicular and oblique heating near the observation altitude, detailed studies may reveal that such heating can explain some electron observations.

One problem with perpendicular and oblique diffusion is that it might require the presense of e.g. upper hybrid waves with rather large amplitudes [*Wong et al.*, 1988], although it is not yet fully clear how strong emissions that are needed [*Lin et al.*, 1990]. There does not seem to be any satellite data showing that upper hybrid emissions with amplitudes of several mV/m are common at or below altitudes of a few thousand km in the auroral zone [*Beghin et al.*, 1989]. However, *Menietti et al.* [1992] have recently presented some upper hybrid emissions associated with electron conics at altitudes around 11 000 km. These waves have amplitudes from 0.015 to 2 mV/m.

Some electron conics can possibly be generated by mechanisms involving perpendicular or oblique heating. Parallel diffusion in velocity space can also give electron conics. Furthermore, we find that oscillations in $E_{\|}$ at roughly 1 Hz can generate many electron conics. However, in many cases the shape of the electron distribution alone can not be used to determine which generation mechanism dominates.

5. LOW ALTITUDES

Magnetically conjugate satellites on an auroral field line would in principle be able to observe both the time-varying $E_{\|}$ in the acceleration region suggested in some of our models, and electron conics at high altitude. Such observations are usually not available. However, one test of models of electron conic generation could be to investigate electron distributions below the acceleration region.

Figure 7a corresponds to parallel energization by a broadband wave (the electron conic in Figures 4c and 5c), and is obtained at a geocentric distance of 8500 km. Downgoing electrons are first sampled when they reach 8500 km. Upgoing electrons that were reflected above the atmosphere are again sampled at the same altitude, and are then removed from the simulation. All models including a time-varying $E_{\|}$ or parallel wave diffusion which give similar conics at high altitude (Figures 4a, c and d) correspond to distributions like the one in Figure 7a at low altitude. All these distributions have been broadened in energy by the time-varying $E_{\|}$ or some wave. Increasing the wave amplitude will of course increase the energy broadening. These distributions have also been broadened in pitch angle, due to adiabatic particle motion in an inhomogeneous magnetic field.

Figure 7b corresponds to perpendicular diffusion at low altitude (the conic in Figures 4e and 5e). Figure 7b is obtained at 8500 km, just above the region of perpendicular diffusion. The downgoing part of this distribution has not been broadened. However, note that no broadening due to waves generated by anisotropies in the distribution are included. Furthermore, neither cool ionospheric electrons, nor backscattered electrons, are included. Keeping all these approximations in mind, we again find that it might be difficult to determine which electron conic generation mechanism is important by using only observed electron distributions.

It may be noted that perpendicular energization (Figure 7b) does not simply give a Maxwellian distribution where the perpendicular temperature is higher than the parallel temperature ("a 90°-conic"). This is partly due to adiabatic folding inside the heating region. The fact that all particles in the model have been accelerated by a 2 kV potential drop before the heating also contributes. Adding backscattered and cool ionospheric particles to the models may change the distributions at low altitudes. However, the more energetic electrons in the conic at high altitudes should still be those with reasonably high energies at low altitudes, i.e. those which originated from the acceleration region.

Including perpendicular diffusion only at high altitude (Figures 4f and 5f) will give a distribution which is narrow in energy at low altitudes (like the downgoing part of Figure 7b). However, again such narrow distributions may be broadened by waves generated by the highly anisotropic particle distribution.

6. TIME-VARYING ELECTRON FLUXES

All the results in Figures 4, 5, and 7 have been averaged over time. Since some models involve a time-varying $E_{\|}$, we expect the corresponding electron fluxes to vary with time. Figures 8a–c show downgoing field-aligned electron fluxes from the model with a broadband $E_{\|}$ (Figure 7a) as a function of time. The results are from a geocentric distance of 8500 km. Fluxes at 1-2 keV and 2-3 keV are clearly varying with time, with a frequency of very approximately 1 Hz. Examination of panels 8a and 8b reveal that the fluxes are roughly anticorrelated. This agrees with the flux sampled between 1.5 and 2.5 keV, which vary much less with time (panel 8c).

In Figure 7a, the fixed part of $E_{\|}$ (E_v = 1 mV/m) corresponds to a potential drop of 2 kV, giving a mean

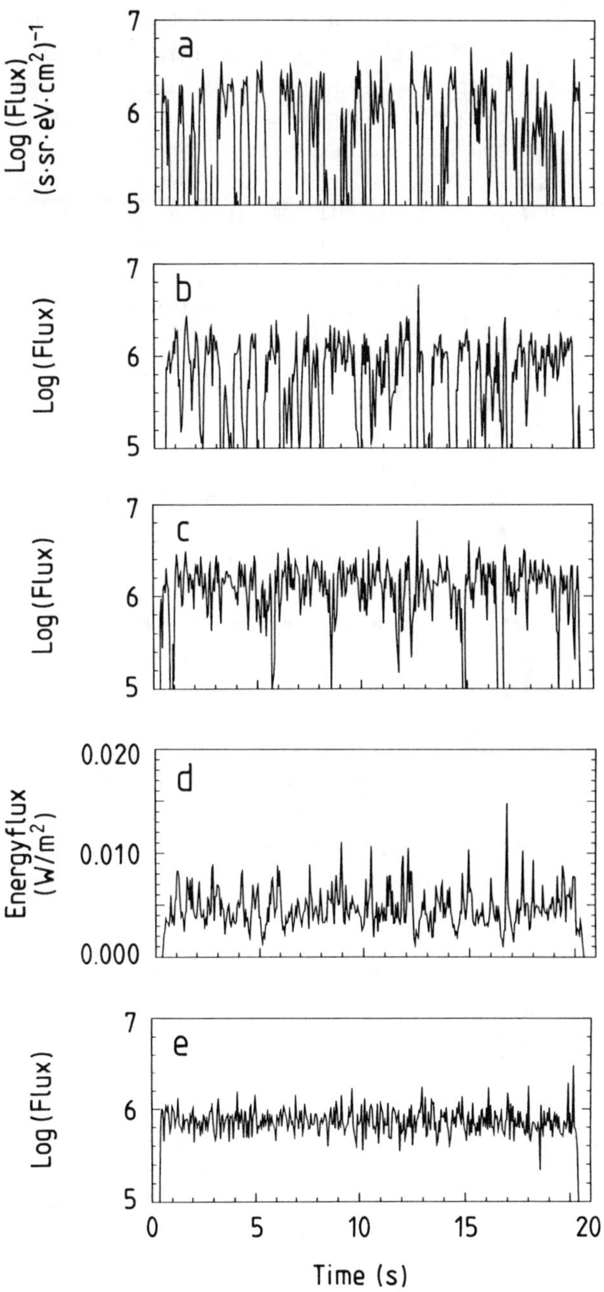

Fig. 8. a) Electron flux between 2 and 3 keV, b) electron flux between 1 and 2 keV, and c) electron flux between 1.5 and 2.5 keV, in each case for pitch angles less than $\approx 20°$ (downgoing field aligned), obtained at a geocentric distance of 8500 km from broadband fluctuations in $E_{\|}$ at higher altitudes, assuming that the initial 100 eV downgoing electron distribution has a density of 1 cm^{-3}. d) The energy flux of this distribution that will hit the atmosphere. e) Electron flux between 1 and 2 keV for pitch angles less than $\approx 20°$ (downgoing field aligned) for a model with parallel diffusion at high altitude, to be compared with panel b.

energy of the distribution of about 2 keV. The time-varying part of $E_{||}$ gives a spread in energy in the time-averaged distribution, approximately from 1 to 3 keV. Investigating electron fluxes as a function of time, we note that when $E_{||} > 1$ mV/m, the electron energies are high. Similarly, a low $E_{||}$ gives low electron energies. For an $E_{||}$ oscillating with high frequency, downgoing high energy electrons could overtake slower electrons started half an $E_{||}$ cycle later before the electrons arrive at low altitudes. However, at frequencies around 1 Hz, the time of flight from the acceleration region to the observation altitude (a distance of 6000 km) does not give any significant aliasing. The difference in travel time of electrons with energies of 1 and 3 keV is about $0.32 - 0.18 = 0.14$ seconds, significantly less than one second. Thus, in this model electrons with high and low energies will occur out of phase at low altitudes. An electron detector with high enough time resolution should be able to observe these fluctuations. Another way to detect these oscillations would be to investigate visible auroras. Panel 8d shows the energy flux which will hit the atmosphere. As expected, this énergy flux is roughly in phase with the flux of high energy electrons in panel 8a. For comparison with panel 8b, panel 8e shows field-aligned downgoing electron flux between 1 and 2 kev for the model with parallel wave diffusion at high altitude (Figures 4d and 5d). There are about 13000 simulation particles in both these panels. The time resolution of 50 ms gives an average of more than 30 particles per timestep. No variation as a function of time is expected for the simple diffusion model, and the fluctuations in panel 8d represent the "noise-level" of the simulation. The oscillations in panel 8b are significantly larger, corresponding to oscillations that possibly can be detected by spacecraft.

Auroral emissions are often fluctuating and oscillations of the order of 1 Hz are not uncommon [e.g. *Berkey et al.*, 1980]. However, not many spacecraft have carried particle detectors with high enough resolution to resolve 1 Hz oscillations of auroral electron fluxes [*McFadden et al.*, 1987, and references therein]. Some published rocket observations show fluctuations at several Hz in the downgoing, nearly field aligned electron flux. These electrons, which carry a small fraction of the total energy flux, will mainly be lost in the atmosphere and may cause flickering aurora [*McFadden et al.*, 1987]. These fluctuations may be generated by downgoing Alfvén waves, maybe by acceleration of electrons of ionospheric origin [*Temerin et al.*, 1986]. However, a mechanism that is used to explain electron conics must correspond to electron distributions which can return to high altitudes, and which carry a significant energy flux. Both observations [*Menietti and Burch*, 1985] and our simulations show that the upgoing energy flux may be larger than the downgoing above the acceleration region. Considering that a significant part of the electrons are lost in the atmosphere, this suggests that effects on the main auroral peak and not only on some minor part of the distribution are important for the generation of electron conics.

Oscillations in $E_{||}$ can cause fluctuations also in upgoing electron fluxes at high altitude. Figure 9 shows upgoing electron flux at energies of roughly one keV as a function of time, using the same model with a broadband $E_{||}$ oscillation as in Figures 4c, 5c and 7a. Particles are started during the first 20 seconds. During this time there are fluctuations in the flux, with a frequency of very approximately one Hz. After this time the flux decreases rapidly. However, a few electrons return to the start altitude only after bouncing several times between a mirror point above the atmosphere and the acceleration region. Oscillations of the type in Figure 8 can not easily be observed by existing electron detectors on satellites such as Viking, DE 1, S3-3 and Akebono in near polar orbit at altitudes of several thousand kilometers. Many detectors sweep in energy over a range of several keV on time-scales of at least several tenths of a second. A satellite spin rate of 6–20 seconds and a velocity along the orbit of several kilometers per second further complicates the observations. However, some electron instruments with high time resolution such as the ESP4/5 magnetic electron spectrometers on Viking may possibly be used to observe this type of oscillations.

7. DISCUSSION AND CONCLUSSIONS

By using a realistic broadband fluctuation in $E_{||}$ at an altitude of about 10 000 km we show that such a time-varying electric field can generate electron conics. This mechanism has been suggested by *Lundin et al.* [1987] and *Lundin and Eliasson* [1991], and investigated by *André and Eliasson* [1992]. Our result that this mechanism works also when tested with a broadband electric field spectrum with a realistic shape is one major conclusion of the present investigation.

It has previously been shown that perpendicular waves near the ion gyrofrequency in a broadband spectrum may cause significant resonant ion heating [e.g. *Chang et al.*, 1986; *André et al.*, 1990; *Crew et al.*, 1990; *Ball and André*, 1991] and that waves around 1 Hz in the same spectrum also may cause some ion energization [e.g. *Lundin and Eliasson*, 1991]. We now show that similar broadband oscillations in $E_{||}$ around

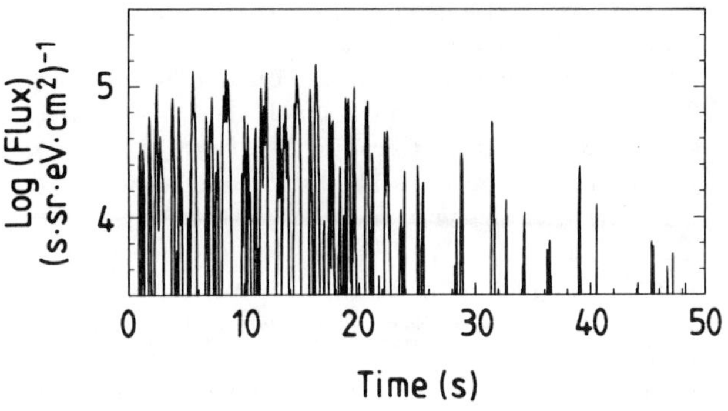

Fig. 9. Electron flux at parallel velocities between -20×10^6 and -10×10^6, and perpendicular velocities between 5×10^6 and 15×10^6 m/s, obtained at a geocentric distance of 16500 km from broadband fluctuations in $E_{||}$. This corresponds to the high energy part of the upgoing electron conic in Figure 5c.

1 Hz can generate electron conics. This mechanism requires another type of resonance, involving the travel time of electrons below the acceleration region and the frequency of the time-varying $E_{||}$.

In addition to fluctuations in $E_{||}$, we have investigated some other electron generation mechanisms. These include diffusion in velocity space parallel to the geomagnetic field, and perpendicular diffusion at various altitudes. We conclude that fluctuations in $E_{||}$, and also parallel and perpendicular diffusion, in principle can generate electron conics. In the case of parallel fluctuations large enough amplitudes seem to exist [e.g. *Block and Fälthammar*, 1990], while for the other mechanisms this should be further investigated.

Fluctuations in $E_{||}$ in a limited region in altitude are not likely to be the only time variations in the electric field on an auroral field line. There are probably e.g. Alfveń waves. Such waves with frequencies around 1 Hz may be trapped between the ionosphere and a density gradient at several thousand kilometers altitude [e.g. *Lysak*, 1991]. Alfvén waves with frequencies of several Hz can also cause parallel acceleration of electrons [*Temerin et al.*, 1986]. As noted by *Temerin and Cravens* [1990] even coherent processes covering a substantial altitude range with different plasma parameters may imitate the effects of stochastic processes such as diffusion. For parallel acceleration, we should not look for a single electron conic generation mechanism. Rather the relative importance of e.g. coherent versus stochastic processes, and of fluctuations in $E_{||}$ (in a limited region of altitude) versus traveling Alfvén waves, should be investigated.

The need for any perpendicular energization to produce electron conics is not as obvious as the necessity of perpendicular heating to generate ion conics. However, a statistical study of the shape of electron conical distributions should clarify whether perpendicular energization is involved in the acceleration process. The possibility that only some electron conics have experienced significant perpendicular energization should be kept in mind. Data from the Viking, DE 1, Akebono and S3-3 satellites should be useful in a statistical study. Data from these spacecraft can also be used to investigate the correlation between low frequency fluctuations in $E_{||}$ and electron conics. This should give an indication of the importance of the generation mechanism involving low frequency oscillations in $E_{||}$. This mechanism can give fluctuations in electron fluxes which possibly can be observed by high resolution detectors on the Freja (launched 1992) and FAST (to be launched 1994) satellites, and by rockets. Travelling Alfvén waves may give similar fluctuations at various frequencies. Wave observations by spacecraft at rather low altitude (a few thousand kilometer) are also of interest for investigations of electron conic generation. Intense upper hybrid waves may indicate the importance of perpendicular electron energization. Low frequency (roughly one to one hundred Hz) Alfvén waves observed together with oscillations in the electron flux may indicate electron acceleration by such electromagnetic waves. For comparison of waves and particles to be meaningful, theoretical studies are needed of the wave amplitudes required to give significant electron energization.

Fluctuations in $E_{||}$ at frequencies around 1 Hz can generate electron conics. However, the relative importance of this and other possible processes is not yet clear. Further studies of these processes should give understanding not only of electron conic generation, but of particle acceleration on auroral field lines in general.

Acknowledgments. Useful discussions with R. Lundin, D. Menietti, and A. Eriksson are gratefully acknowledged. This research was partially supported by the Swedish National Science Research Council. The Viking project is managed by the Swedish Space Corporation under contract from the Swedish National Space Board. Part of this study was carried out during a visit by one of the authors (M. A.) to MIT, where he was supported by Contract No. AFOSR-90-0085, NASA-NAGW-1532, and AFGL Contract No. F19628-91-K-0043.

REFERENCES

André, M. and L. Eliasson, Electron acceleration by low frequency electric field fluctuations: electron conics, *Geophys. Res. Lett., 19*, 1073, 1992.

André, M., H. Koskinen, G. Gustafsson, and R. Lundin, Ion waves and upgoing ion beams observed by the Viking satellite, *Geophys. Res. Lett., 14*, 463, 1987.

André, M., G. B. Crew, W. K. Peterson, A. M. Persoon, C. J. Pollock and M. J. Engebretson, Ion heating by broadband low-frequency waves in the cusp/cleft, *J. Geophys. Res., 95*, 20 809, 1990.

Ball, L., and M. André, What parts of broadband spectra are responsible for ion conic production?, *Geophys. Res. Lett., 18*, 1683, 1991.

Beghin, C., J. L. Rauch, and J. M. Bosqued, Electrostatic plasma waves and HF auroral hiss generated at low altitude, *J. Geophys. Res., 94*, 1359, 1989.

Berkey, F. T., M. B. Silevitch, and N. R. Parsons, Time sequence analysis of flickering auroras 1. Application of fourier analysis, *J. Geophys. Res., 85*, 6827, 1980.

Block, L. P., and C.-G. Fälthammar, The role of magnetic-field-aligned electric fields in auroral acceleration, *J. Geophys. Res., 95*, 5877, 1990.

Bryant, D. A., Auroral electron acceleration, *Physica Scripta, T30*, 215, 1990.

Chang, T., G. B. Crew, N. Hershkowitz, J. R. Jasperse, J. M. Rettere, and J. D. Winningham, Transverse acceleration of oxygen ions by electromagnetic ion cyclotron resonance with broadband left-hand-polarized waves *Geophys. Res. Lett., 13* 636, 1986.

Crew, G. B., T. Chang, J. M. Retterer, W. K. Peterson, D. A. Gurnett and R. L. Huff, Ion cyclotron resonance heated conics: Theory and observations, *J. Geophys. Res., 95*, 3959, 1990.

Gurnett, D. A., R. L. Huff, J. D. Menietti, J. L. Burch, J. D. Winningham, and S. D. Shawan, Correlated low-frequency electric and magnetic noise along the auroral field lines, *J. Geophys. Res., 89*, 8971, 1984.

Lin, C. S., J. D. Menietti, and H. K. Wong, Perpendicular heating of electrons by upper hybrid waves generated by a ring distribution, *J. Geophys. Res., 95*, 12295, 1990.

Lundin, R., and L. Eliasson, Auroral energization processes, *Ann. Geophysicae, 9*, 202, 1991.

Lundin, R., L. Eliasson, B. Hultqvist, and K. Stasiewicz, Plasma energization on auroral field lines as observed by the Viking spacecraft, *Geophys. Res. Lett., 14*, 443, 1987.

Lysak, R. L., Feedback instability of the ionospheric resonant cavity, *J. Geophys. Res., 96*, 1553, 1991

McFadden, J. P., C. W. Carlson, M. H. Boehm, and T. J. Hallinan, Field-aligned electron flux oscillations that produce flickering aurora, *J. Geophys. Res., 92*, 11133, 1987.

Menietti, J. D., and J. L. Burch, "Electron Conic" signatures observed in the nightside auroral zone and over the polar cap, *J. Geophys. Res., 90*, 5345, 1985.

Menietti, J. D., C. S. Lin, H. K. Wong, A. Bahnsen, and D. A. Gurnett, Association of electron conical distributions with upper hybrid waves, *J. Geophys. Res., 97*, 1353, 1992.

Roth, I., M. K. Hudson and M. Temerin, Generation models of electron conics, *J. Geophys. Res., 94*, 10095, 1989.

Temerin, M., J. McFadden, M. Boehm, C. W. Carlson, and W. Lotko, Production of flickering aurora and field-aligned electron flux by electromagnetic ion cyclotron waves, *J. Geophys. Res., 91*, 5769, 1986.

Temerin, M. A., and D. Cravens, Production of electron conics by stochastic acceleration parallel to the magnetic field, *J. Geophys. Res., 95*, 4285, 1990.

Wong, H. K., J. D. Menietti, C. S. Lin, and J. L. Burch, Generation of electron conical distributions by upper hybrid waves in the Earth's polar region, *J. Geophys. Res., 93*, 10025, 1988.

Mats André, Swedish Institute of Space Physics, Umeå University, S-901 87 Umeå, Sweden.

Lars Eliasson, Swedish Institute of Space Physics, P. O. Box 812, S-981 28 Kiruna, Sweden.

Kinetic Alfven Wave Instability and Wave-Particle Interaction at the Magnetopause

S. Y. Fu and Z. Y. Pu

Department of Geophysics, Peking University, Beijing

S. C. Guo

Institute of Physics, Academia Sinica, Beijing

Z. X. Liu

Institute of Space Physics, Academia Sinica, Beijing

Low-frequency drift kinetic instabilities at the magnetopause are investigated by using the gyrokinetic formalism. Two unstable branches of the drift kinetic Alfven wave (DKAW) are found, which have magnetic field perturbations in the direction normal to the magnetopause. Once the amplitude of DKAWs grows and perturbations become turbulent, magnetic percolation-like phenomena associated with wave-particle interaction takes place, leading to substantial energy, momentum and mass transfer from the solar wind to the magnetosphere. Quasilinear anomalous diffusivity is estimated to be on the order of 10^{13} cm^2/s when $\delta B_x \sim 1$nT.

1. INTRODUCTION

Anomalous transport resulting from wave-particle interaction has widely been accepted as a possible mechanism responsible for the transfer of energy, momentum and mass from the solar wind to the magnetosphere. In order to account for the observed boundary layer properties, $D_\perp$, the coefficient of diffusion normal to the magnetopause, should reach ~10^{13} cm^2/s (Sonnerup, 1980; Haerendel and Paschmann, 1982). Much effort has been devoted to finding micro- instabilities which are able to produce such a viscous-like interaction (Huba, 1980; Pu et al.,1986; Labelle and Treumann, 1988). By taking into account all inhomogeneities existing in the magnetopause transition region, Pu et al. (1986) found two drift kinetic unstable modes which may lead to $D_\perp$~10^{13} cm^2/s at the magnetopause interface. In this work, we extend their results by using the gyrokinetic formalism (Chen and Tsai, 1983). A self-consistent theory for the drift kinetic Alfven wave (DKAW) instability is obtained. It is found that two growing modes may appear at the magnetopause, which are shown to coincide with those previously studied by Pu et al. (1986). The properties of DKAWs are discussed. A striking feature of DKAWs lies in the fact that they have magnetic field perturbations in the direction normal to the magnetopause. Once the amplitude of waves grows and perturbations become turbulent, magnetic percolation-like phenomena associated with wave-particle interactions may take place, which can lead to substantial energy, momentum and mass transfer from the solar wind to the magnetosphere. The quasilinear theory is developed which shows that $D_\perp$ can reach ~10^{13} cm^2/s for both branches. Thus, the DKAW turbulence may be considered to be one of the mechanisms responsible for micro-anomalous transport phenomena at the

Space Plasmas: Coupling Between Small
and Medium Scale Processes
Geophysical Monograph 86

magnetopause during the period of northward interplanetary magnetic field.

2. THE INITIAL STATE AND DISPERSION EQUATION

It is assumed that in the initial state $\mathbf{B}_0=B_0(X)\mathbf{e}_z$, $V_0=V_0(X)\mathbf{e}_z$, $T_{\alpha\|}=T_{\alpha\|}(X)$, $T_{\alpha\perp}=T_{\alpha\perp}(X)$ $(\alpha=i,e)$, and $N_{i0}=N_{e0}=N_0(X)$. The unperturbed distribution function of particles takes the form of

$$F_{\alpha 0}(v,X)=\frac{N_0(X)}{\pi^{3/2}v_{\alpha\|}(X)v_{\alpha\perp}^2(X)}\times \exp\left(-\frac{(v_\| - V_0(X))^2}{v_{\alpha\|}^2(X)}-\frac{v_\perp^2}{v_{\alpha\perp}^2(X)}\right) \tag{1}$$

where $v_{\alpha\|}^2=2T_{\alpha\|}/m_\alpha$, $v_{\alpha\perp}^2=2T_{\alpha\perp}/m_\alpha$, X represents the guiding center coordinate. Starting from the WKB formalism of gyrokinetic equations (Chen and Tsai, 1983), after a long tedious procedure, we obtain the following dispersion equation of DKAWs:

$$Det\left(\left(k^2-\frac{\omega'^2}{c^2}\right)I+\epsilon+\epsilon'\right)=0 \tag{2}$$

where $\omega'=\omega-k_\| v_0$, the dielectric tensor $\boldsymbol{\varepsilon}$ is expressed as (Pu et al., 1991)

$$\begin{aligned}\epsilon=&\sum_\alpha\left(-2\frac{\omega_{p\alpha}^2}{v_{\|\alpha}^2}\right)e_z e_z\\&+\sum_\alpha 2\frac{\omega_{p\alpha}^2}{c^2}\sum_l\Bigg(\Bigg(\Bigg(\frac{-\omega'\,\Omega_\alpha}{k_\| v_{\|\alpha}^2}\Bigg)+\Bigg(\frac{k_\perp\nabla_x N_o}{2k_\| N_o}\Bigg)\\&+\Bigg(\frac{l\Omega_\alpha^2}{k_\|}\Bigg)\Bigg(\frac{1}{v_{\|\alpha}^2}-\frac{1}{v_{\perp\alpha}^2}\Bigg)\Bigg)M_1\Bigg)+\Bigg(\frac{k_\perp\nabla_x v_{\|\alpha}}{k_\| v_{\|\alpha}}\Bigg)\\&\times\Bigg(M_3-\frac{M_1}{2}\Bigg)+\Bigg(\frac{k_\perp\nabla_x v_{\perp\alpha}}{k_\| v_{\perp\alpha}}\Bigg)(M_2-M_1)\\&+\Bigg(\frac{v_{\perp\alpha}^2}{v_{\|\alpha}^2}-1\Bigg)\Bigg(\frac{\nabla_x B_0}{2B_0}\Bigg)M_2+\Bigg(\frac{k_\perp\nabla_x V_0}{k_\| v_{\|\alpha}}\Bigg)M_4\Bigg)\end{aligned} \tag{3}$$

In obtaining (2) and (3), we have assumed that the scale length of the magnetopause current sheet, $L>r_{i\perp}$, the gyroradius of ions, and the local approximation has been used. The first two terms on right hand side of Equation (3) correspond to the general terms which give the conventional kinetic Alfven wave. The fourth term comes from temperature anisotropy. The remaining terms contain the free energy (i.e., the shear flow, the gradients of the temperature, number density, and magnitude of the magnetic field) which cause the drift kinetic Alfven waves to grow. The detailed derivation of Equation (3) and the expressions for ε' and M_j (j=1, 2, 3, and 4) were given by Pu et al. (1991).

3. PROPERTIES OF THE LINEAR UNSTABLE MODES

In inhomogeneous plasmas with $\beta\sim 1$, Equation (2) requires a numerical analysis. Typical parameters used in calculations are listed in Table 1, where subscript 1 (or 2) indicates the magnetosphere (or magnetosheath). For simplicity we set $T_{\alpha\|}=T_{\alpha\perp}=T$ and L=600 km in the present calculation. Following are a part of our main results:

(1) Two growing modes, A and B, and two damping modes, C and D, are present. The former ones are shown to essentially coincide with those previously studied by Pu et al. (1986). Figure 1 plots how γ/Ω_i, the growth rate normalized by the ion gyrofrequency, varies as a function of $\lambda_i=(k_y r_i)^2$. Figure 2 shows how the normalized real frequency, ω'_r/Ω_i, varies with λ_i.

(2) Detailed analyses reveal (Pu et al., 1991) that A and B are evolved from two branches of the kinetic Alfven wave in low β homogeneous plasmas (Hasegawa and Mima, 1978) propagating in opposite directions, hence we refer them as DKAWs; while C and D are originated from the slow ion acoustic wave.

(3) Figure 3 shows how the maximum growth rate γ_{max} varies with β. We see γ_{max} increases with increasing β, a result in contrast with that of Huba (1980), who dealt with cases where $\nabla T=0$. This can be understood as the following. The presence of non-zero $\nabla T \| -\nabla N$ in finite β plasmas reduces drastically the ∇B drift. Thus, for example, the A mode can be amplified owing to the weakness of the resonance between the ∇B drift of electrons and the wave. The presence of $\nabla T \| -\nabla N$ is shown to be crucial for exciting kinetic drift modes in finite β plasmas (Pu et al., 1991).

(4) The velocity shear provides significant free energy for driving DKAWs. Figure 4 plots how γ_{max} varies with λ_i for different magnetosheath flow V_0, or equally, the intensity of the shear flow V_0/L. The larger V_0/L is, the stronger the growth rate becomes. When V_0 vanishes, the growth rate can still be positive, provided non-zero ∇T exists which is anti-parallel to N_0.

(5) Figures 5a and 5b depict the variations of $\delta B_x/\delta B_\|$ with β for mode A and B, respectively, where $\delta B_\|$ and δB_x represent the magnetic field perturbations parallel to $\mathbf{B}_0$ and normal to the magnetopause, respectively. It is seen that near $\beta=1$, both $\delta B_\|$ and δB_x are non-negligible. We will see later that the presence of δB_x is of particular importance in

TABLE 1. Parameters Used in Calculation

N_1	N_2	V_0	B_1	B_2	T_{i1}	T_{i2}	T_{e1}	T_{e2}
(cm^{-3})		(10^2 km/s)	(nT)		(10^6 °K)			
2.0	10.0	2.5	30	24	7.0	2.0	1.0	0.4

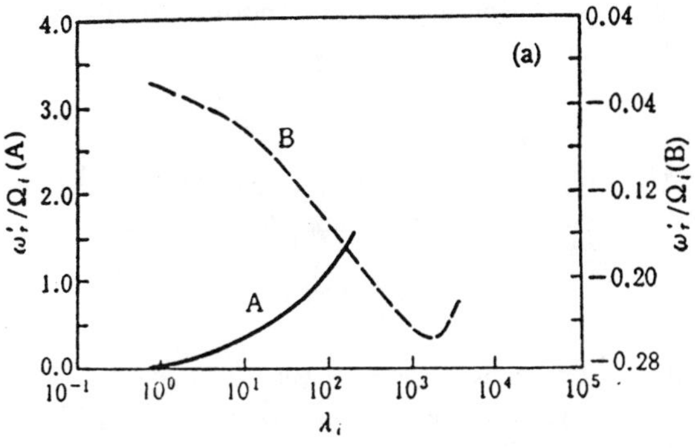

Fig. 1. The effective time growth rate as a function of λ_i.

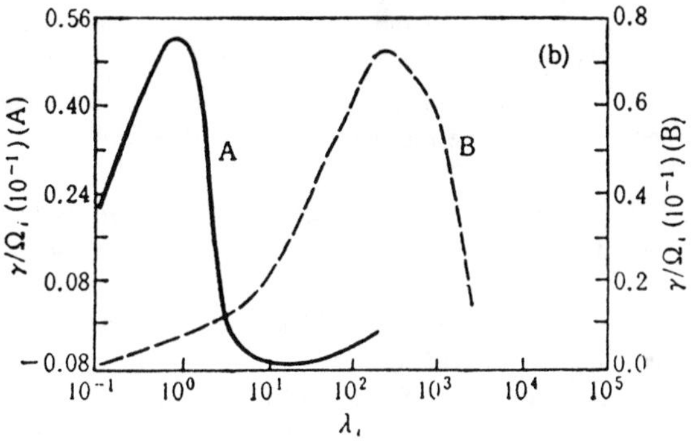

Fig. 2. The normalized real frequency as a function of λ_i.

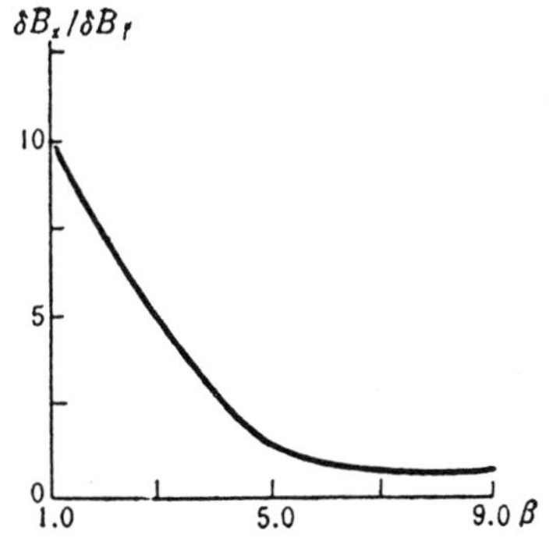

Fig. 3. The maximum growth rate varies as a function of β.

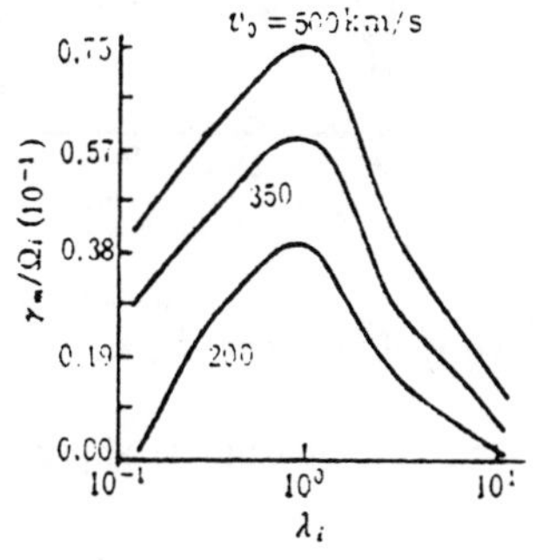

Fig. 4. The maximum growth rates of mode A for different V_0.

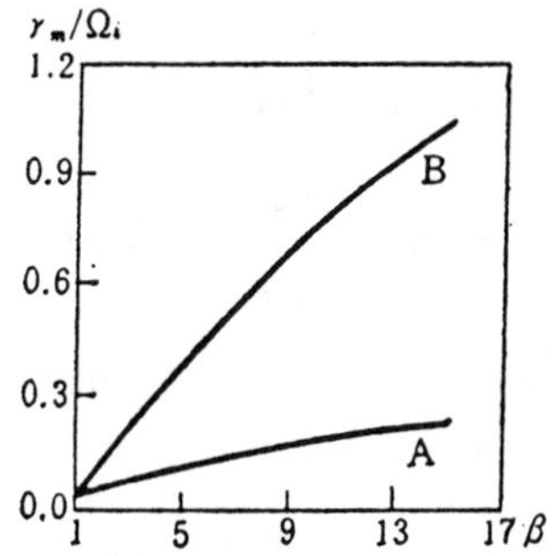

Fig. 5. $\delta B_X/\delta B_\parallel$ varies as a function of β for (a) mode and (b) mode B.

leading to the magnetic percolation-like phenomena at the magnetopause when the amplitude of DKAWs grows and perturbation becomes turbulent.

4. QUASILINEAR TRANSPORT EFFECTS

Since $|\omega_r'| << \Omega_\alpha$, for the slab model we used, the quasilinear equation can be expressed as

$$\frac{\partial\langle F_\alpha\rangle}{\partial t} = -\langle\nabla(\delta v_D \delta f_\alpha)\rangle - \frac{q_\alpha}{m_\alpha}\langle\delta E_\parallel \frac{\partial \delta f_\alpha}{\partial v_\parallel}\rangle \tag{4}$$

where < > means average, δf_α and δv_D represent the disturbed distribution and drift velocity. Following the

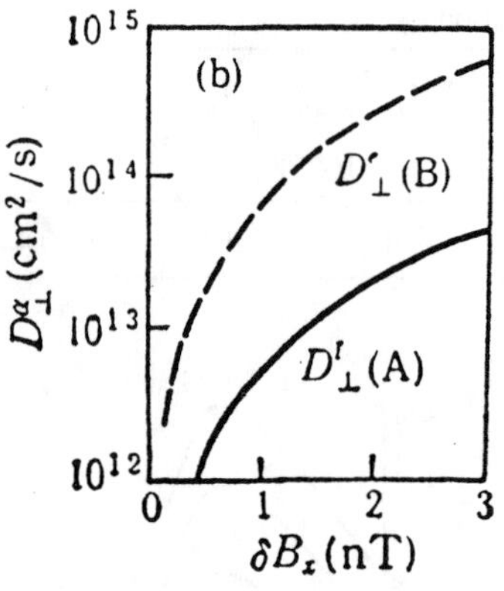

Fig. 6. The dependence of $D_{\perp\perp}{}^{\alpha}$ on δB_X.

common procedure in the quasilinear approach, we find that (Pu et al., 1991):

$$D^{\alpha}_{\perp} = \frac{\pi}{8} \sum_k \frac{|\delta B_{kx}|^2}{B_0^2} \frac{\bar{\omega}^2_{\alpha}}{k^3_{\parallel} v_{\alpha}} \tag{5}$$

$$\bar{\omega}^2_{\alpha} = \int_0^{\infty} (\omega_r - \omega_{\alpha\beta})^2 J_0(\xi_{\alpha}) \exp\left(-\left(\frac{\omega_r - \omega_{\alpha\beta}}{k_{\parallel} v_{\alpha}}\right)^2 - S^2\right) ds \tag{6}$$

where $\xi_{\alpha}=k_y v_{\perp}/\Omega_{\alpha}$, $\omega_{\alpha B}=(\xi_{\alpha} v_{\perp}{}^2/2\Omega_{\alpha})(\partial \ln B/\partial x)$, $S=v_{\perp}/v_{\alpha}$. We calculated $D_{\perp}{}^{\alpha}$ for modes A and B by assuming the spectrum of δB_{kx} to be a δ-function at k_m. The results are plotted in Figure 6. It is seen that both $D_{\perp}{}^{i}$(A) and $D_{\perp}{}^{e}$(B) reach 10^{13} cm^2 /s for $\delta B_x \sim 1$nT. Furthermore, we estimated the anomalous viscosity $\mu^{i}_{\perp} = N_0 m_i D_{\perp}{}^{i} \sim 0.5 \times 10^{-6}$ $gcm^{-1}s^{-1}$, which is close to the value required by the Axford-Hines model (Axford, 1964).

Hasegawa and Mima (1978) first noted that kinetic Alfven wave turbulence can yield substantial anomalous transport. This can be understood as the following. In the presence of δB_x, the field-aligned and cross-field random walk of particles both contribute to the transport across the boundary. The former is known to be much 'faster' than the latter. Thus, the resulting effective diffusivity must be greatly larger than that of 'purely' cross-field diffusion. It is this percolation-like phenomena of DKAW turbulence that may lead to substantial anomalous transport across the magnetopause during the period of northward interplanetary magnetic field.

Acknowledgements. The authors wish to thank Dr. F. Li for her helpful assistance. This work is supported by CNSF.

REFERENCES

Axford, W.I., Viscous interaction between the solar wind and earth's magnetosphere, *Planet.Space Sci., 12,* 45, 1964.

Chen, L., and S.D. Tsai, Linear oscillation in general magnetically confined plasmas, *Plasma Phys., 25,* 349, 1984.

Haerendel, G., and G. Paschmann, Interaction of the solar wind with the dayside magnetopause, in *Magnetospheric Plasma Physics,* ed. A. Nishida, p.49, Center for Academic Publications, Tokyo, Japan, 1982.

Hasegawa, A., and K. Mima, Anomalous transport produced by kinetic Alfven wave turbulence, *J. Geophys. Res., 83,* 1117, 1978.

Huba, J. D., Stabilization of the electrostatic Kelvin-Helmholtz instability in high β plasmas, *J. Geophys. Res., 86,* 8991, 1981.

Labelle, J., and R.A. Treumann, Plasma waves at the dayside magnetopause, Space Sci. Rev., 47, 175, 1988.

Pu, Z. Y., C.Q. Wei, and Z.X. Liu, Drift kinetic instabilities at the magnetopause, in *Proceedings of International Symposium on Space Physics in Beijing,* 4-066, 1986.

Pu, Z.Y., S.Y. Fu, S.C. Guo, and Z.X. Liu, Unstable drift kinetic Alfven modes at the magnetopause, *ACTA GEOPHYSICA SINICA, 34,* 404, 1991.

Sonnerup, B. U. Ö., Theory of the low-latitude boundary layer, *J. Geophys. Res., 85,* 2017, 1980.

High Frequency Electrostatic Plasma Instabilities and Turbulence Layers in the Lower Ionosphere

J. R. Jasperse[1]

Centre de Recherche en Physique de l'Environnement, Saint-Maur-des-Fossés, France

B. Basu and J. M. Retterer

Phillips Laboratory, Hanscom Air Force Base, Bedford, Massachusetts

D. T. Decker

Institute for Space Research, Boston College, Chestnut Hill, Massachusetts

T. Chang

Center for Space Research, Massachusetts Institute of Technology, Cambridge, Massachusetts

It has been known for some time that calculations of the collisional electron distribution function in the 2 to 5 eV energy range which neglect wave-particle interactions disagree with rocket measurements in both the auroral and sunlit, midlatitude ionospheres. The latter show a filling-in of the 2 to 5 eV valley in the isotropic distribution function predicted by the zero-order calculations. Several narrow layers (each about 4 to 10 km thick) of intense electron heating from 98 to 130 km and a broad layer of weaker electron heating from about 135 to 210 km in the sunlit, midlatitude ionosphere have also been observed. More recently, a rocket experiment in the auroral zone recorded intense plasma waves near the upper hybrid frequency in the altitude range of 130 to 170 km. We can tie these observations together by invoking the existence of the electron cyclotron and upper hybrid plasma instabilities described by *Basu et al.* [1981, 1982]. The free energy source for these instabilities is the suprathermal, shell-like bump in the electron energy distribution near 5 eV. In this paper, we review and extend the linear collisional stability analysis. We also present a new collisional quasilinear analysis and particle-in-cell plasma simulations which we have carried out in our study of this problem.

1. INTRODUCTION

In this paper, we present a theoretical analysis and some experimental evidence in support of the existence of high frequency, electrostatic, instability and turbulent layers in the E- and lower F-region ionosphere. The free energy source for these instability layers is the suprathermal, isotropic, shell-like bump in the secondary electron energy distribution near 5 eV. We feel that our analysis applies to three ionospheric subregions: (1) the daytime mid- to high-latitude ionosphere; (2) the diffuse aurora; and (3) possibly to a stable auroral arc. Two kinds of instability layers will be discussed. In the E region from about 100 to 130 km, one or several narrow (~ 2 km thick) layers of electron cyclotron waves can be excited by the nonresonant interaction of the suprathermal bump with the thermal plasma ($\underline{k} = \underline{k}_\perp$, $\underline{k}_\parallel = 0$). This occurs for the so-called "double resonance condition" that one of the cyclotron harmonic frequencies is sufficiently close to the local upper hybrid frequency. In the E and lower F region from about 125 to 170 km and possibly higher, a single broad layer of upper hybrid waves can be excited by the resonant interaction of the suprathermal bump with the thermal plasma ($\underline{k} = \underline{k}_\perp + \underline{k}_\parallel$, $k_\perp >> k_\parallel \neq 0$). Here, $k_\perp$ ($k_\parallel$) is the

Space Plasmas: Coupling Between Small
and Medium Scale Processes
Geophysical Monograph 86

component of the wavevector perpendicular (parallel) to the geomagnetic field.

Before the stability of the plasma can be studied, the zero-order plasma state must be determined. Detailed calculations of the zero-order, photoelectron distribution function using similar methods have been carried out by *Ashihara and Takayanagi* [1974], *Jasperse* [1976, 1977], and *Khazanov et al.* [1978, 1979] and for the steady-state auroral ionosphere by *Strickland et al.* [1976] and others. They show a near Maxwellian, thermal distribution of electrons smoothly joined to a suprathermal, secondary distribution which contains a pronounced isotropic, shell-like bump near 5 eV.

Linear stability studies of the suprathermal bump have been carried out by a number of authors for the daytime and auroral ionospheres. The nonresonant ($k_{||} = 0$) excitation of electron cyclotron waves by the 5 eV bump has been studied by *Basu et al.* [1981, 1982] and *Jasperse et al.* [1982]. The resonant ($k_{||} \neq 0$) excitation of upper hybrid waves by the 5 eV bump has been studied by *Bloomberg* [1975], *Khazanov* [1979], *Ivanov et al.* [1980], *Ivanov et al.* [1982], *Basu et al.* [1981, 1982], *Jasperse et al.* [1982], and *Gefan et al.* [1985].

In their work on the nonresonant excitation of electron cyclotron waves by the 5 eV bump, the authors solved for the collisional dielectric function, ε^{ν}, where the effect of electron-neutral particle collisions on the linear stability of the waves as a function of ω and k was found. In the work by *Basu et al.* [1981, 1982], the linearized, complete Boltzmann collision integrals were approximated by the BGK collision integral [*Bhatnagar et al.*, 1954] and, in the work of *Jasperse et al.* [1982], by an approximate Boltzmann-type collision integral. The work predicted that one and sometimes two narrow altitude layers (~ 2 km thick) of electron cyclotron waves could be excited by the 5 eV bump in the daytime, midlatitude E-region ionosphere from about 100 to 130 km. This work is discussed and extended in Sections 3-6 of this paper.

The work on the resonant excitation of upper hybrid waves by the 5 eV bump may be summarized as follows. *Bloomberg* [1975] solved a collisionless dispersion relation for the excitation of upper hybrid waves and compared the collisionless growth rate to the collisional damping rate, $-\nu_s$, where ν_s is the electron-neutral particle collision frequency for the suprathermal electrons. He concluded that there was an upper hybrid instability in the daytime ionosphere above about 160 km. In the work by *Khazanov* [1979], *Ivanov et al.* [1980], *Ivanov et al.* [1982], and *Gefan et al.* [1985], a dielectric function for the problem was derived using the relaxation approximation for treating collisional effects. In this approximation, the complete Boltzmann collision integrals which appear in the linearized kinetic-Poisson equations are replaced by $-\nu_s f_e$, where f_e is the perturbed distribution function. This is tantamount to solving for a collisionless dielectric function, ε, and then comparing the collisionless growth rate to the collisional damping rate which, in this case, is given by $-\nu_s/2$. The work of *Bloomberg* [1975] and *Khazanov* [1979] was criticized by *Ivanov et al.* [1982] for its neglect of the finite $b(= k_{\perp}^2 r_L^2)$ effect, and the work of *Ivanov et al.* [1980] for not including more terms in the Bessel function sum which appears in the expression for ε. When the earlier work was extended by *Ivanov et al.* [1982], the authors concluded that upper hybrid waves were stable at all altitudes in the daytime ionosphere. In the work of *Basu et al.* [1981, 1982] and *Jasperse et al.* [1982], a different approach was taken. The linearized Boltzmann collision integrals were retained in the analysis and a collisional dielectric function, ε^{ν}, was found which gives a more general treatment of the effect of collisions on the stability of plasma waves. For example, their collisional analysis showed that it is the thermal (Maxwellian) electrons, not the suprathermal electrons, that damp the waves and that the collisional damping rate depends on the ratio of the thermal electron plasma frequency to the electron cyclotron frequency. Since $\nu_t \cong \nu_s/10$, where ν_t is the thermal (Maxwellian) collision frequency, collisional damping is significantly overestimated unless the collisional theory is used. These issues were discussed in the work of *Basu et al.* [1981, 1982] who concluded that the upper hybrid instability may be excited in a broad altitude layer from about 125 to at least 170 km. The correct upper altitude of the layer depends on the effect of charged particle collisions as they dominate electron-neutral particle collisions above about 170 km. This work is also discussed and extended in Sections 3-6 of this paper.

In the auroral F-region ionosphere, several groups have studied the stability of the electron distribution function at primary (a few KeV) and secondary (a few 10's eV) energies. The distribution function is sometimes stable and sometimes unstable according to the specific auroral situation. For example, in the work by Pottelette's group [*Pottelette and Illiano*, 1982], they find that a ring-like electron distribution function exists in the auroral F region but the ring is sufficiently weak so that it is stable to high frequency, electrostatic perturbations ($\gamma < 0$, where γ is the imaginary part of the wave frequency). Even though the distribution was stable, it produced electrostatic fluctuation spectra that were observed to be enhanced over those for thermal fluctuations. In their study of such electron distributions in the ionosphere, they used the double dipole probe method which they developed for the diagnosis of laboratory plasmas. See *Pottelette* [1979], *Pottelette et al.* [1981], and *Pottelette and Storey* [1981]. In the work of the New Hampshire group [*Kaufmann et al.*, 1978; and *Kaufmann*, 1980] and later the Berkeley group [*McFadden et al.* 1986], auroral F-region primary electron distributions were reported that were more beam-like than ring-like. These distributions were found to be unstable to high frequency, electrostatic perturbations with growth rates on the order of $\gamma \sim 10^3\ s^{-1}$ ($\gamma/\omega_r \sim 10^{-3}$, where ω_r is the real part of the wave frequency).

We now turn our attention to the E- and lower F-region ionosphere. We discuss some experimental evidence for the existence of several narrow layers (a few kilometers thick) of plasma turbulence in the lower E region and for the existence of a broad layer of high frequency plasma turbulence and wave-particle interactions near the upper hybrid frequency in the upper E and lower F regions. The experimental evidence for the plasma instability layers comes from three rocket experiments which are discussed below.

First, we present the rocket data of *McMahon and Heroux* [1978, 1983] for the midlatitude, photoelectron flux in the 2 to 5 eV energy range. This is shown in the left panel of Figure 1. A valley-like structure in the 125-173 km altitude region is evident from these data. This structure reaches a maximum in prominence, measured by the peak-to-valley ratio of the electron fluxes, in the 140-150 km region. Above 150 km, the valley diminishes with increasing altitude until it disappears in the 200-250 km region (not shown in the figure). Below 140 km, the valley also diminishes with decreasing altitude and ultimately disappears at about 110 km. In order to realize what these observations indicate, let us compare them with the zero-order results as calculated by *Jasperse* [1976, 1977] which are shown in the right panel of Figure 1. As can be seen, the valley-like structure in the 2 to 5 eV energy range is also predicted by the zero-order calculations. Above 170 km, the theoretical peak-to-valley ratios of the electron fluxes compare well with the observed values, and, in particular, the theory predicts the disappearance of the valley-like structure at or above 210 km in complete agreement with the observations. However, the theory predicts that the structural prominence should increase with decreasing altitude. This is in clear disagreement with the observations and indicates that the photoelectron energy distribution in the lower ionosphere (below 170 km) cannot be adequately described by particle-particle collisions alone. The diminished valley-like structure at high altitudes and the flattening of the structure at lower altitudes could result from anomalous diffusion in energy space due to wave-particle interactions. Moreover, the observed trend of the structural prominence, namely, that it increases with altitude above 110 km and then reverses its altitude dependence above the 140-150 km region, indicates that two types of plasma instabilities may be operative in the two altitude regions. It is also apparent that the suprathermal, shell-like bump near 5 eV in the calculated electron energy distribution could be the source of free energy for the excitation of the plasma instabilities, as the peak-to-valley ratios in the zero-order calculations are much more pronounced than those observed in the rocket measurements. We mention here that similar discrepancies between the zero-order calculations and observations of the electron energy spectra in the 2 to 5 eV energy region also exist at auroral latitudes [*Sharp et al.*, 1979]. Here again,

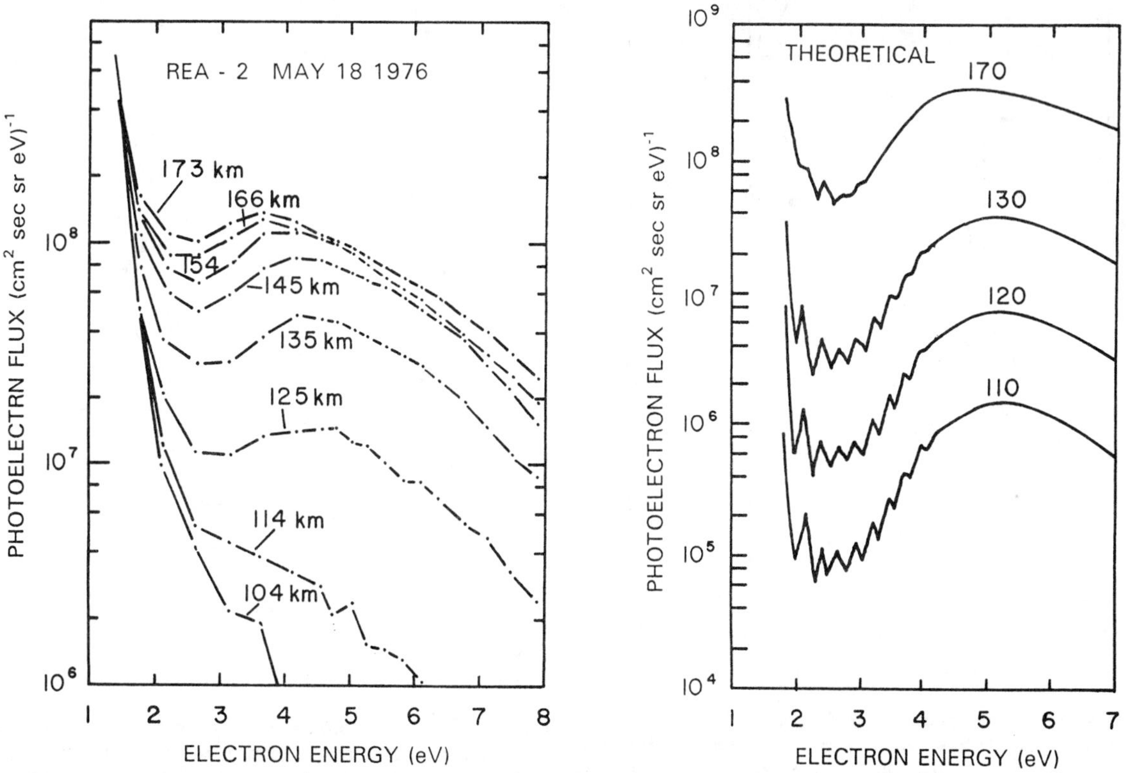

Fig. 1. Measured [*McMahon and Heroux*, 1983] and calculated [*Jasperse*, 1977] midlatitude photoelectron flux as a function of energy at various altitudes.

wave-particle effects, in addition to collisional effects, need to be included in order to explain the data, and the suprathermal bump which is also present at auroral latitudes may provide the free energy for excitation of the plasma waves.

The next set of rocket data was also taken in the daytime, midlatitude ionosphere and was reported by *Oyama and Hirao* [1979] and by *Oyama et al.* [1983]. These data are shown in Figures 2 and 3, respectively. In Figure 2, seven electron temperature altitude profiles measured during different rocket flights are presented. The leftmost curve is the neutral temperature profile from CIRA 72 (T_{ex} = 600 K). As is seen in Figure 2, three profiles exhibit greatly enhanced electron temperatures in the 95 to 130 km altitude region, and these enhancements occur within narrow altitude layers of a few kilometers in thickness. Figure 3 shows enhanced tails in the thermal electron energy distribution associated with the localized hot electron layers shown in Figure 2. In their rocket experiments, the authors also found a broad layer of weakly enhanced thermal electron energy tails which started appearing at 135 km, became most intense around 150 km, and gradually vanished at an altitude of about 210 km. These observations once again indicate the presence of plasma instabilities and wave-particle interactions in the E- and lower F-region ionosphere. Like the rocket measurements of McMahon and Heroux, these particle measurements did not include any simultaneous high frequency plasma wave measurements.

Finally, we present the rocket data from a direct measurement of plasma waves in the upper E and lower F regions of the morningside auroral zone (Figures 4 and 5). These data were reported by *Kelley and Earle* [1988]. The data show intense plasma wave emissions in a narrow band of frequencies close to the local upper hybrid frequency when the dipole antenna was nearly perpendicular to the geomagnetic field $\underline{B}_0$, i.e., when the wave propagation vector $\underline{k}$ is nearly perpendicular to $\underline{B}_0$ (Figure 4).

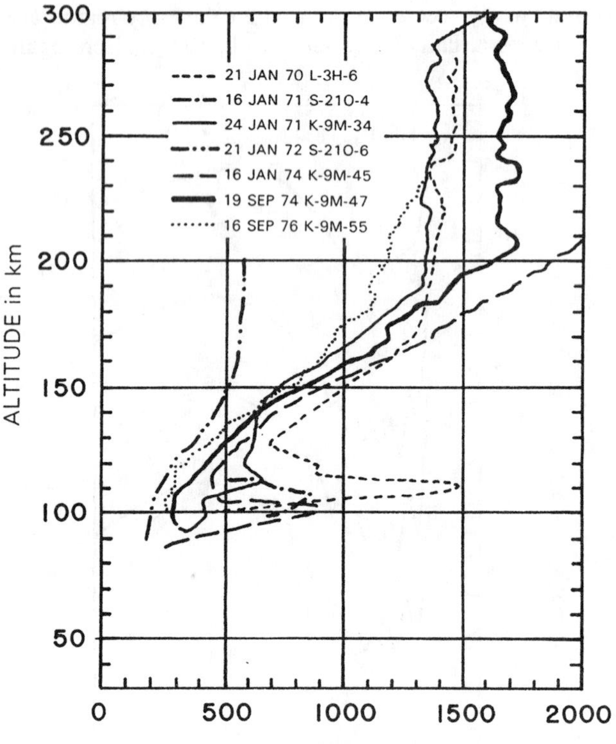

Fig. 2. Seven electron temperature altitude profiles obtained at approximately 1100 hrs local time for the midlatitude ionosphere. The standard deviations are ± 100 K. The leftmost curve is the neutral temperature profile from CIRA 72 (T_{ex} = 600K). Taken from *Oyama and Hirao* [1979].

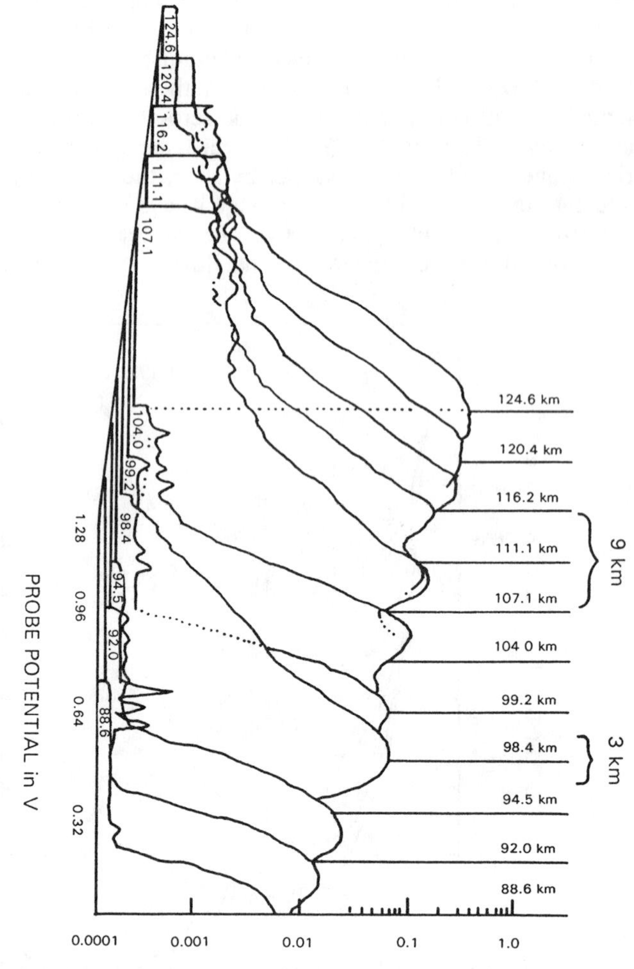

Fig. 3. Nested semilogarithmic curves of the second harmonic current component as a function of the probe potential for the daytime midlatitude ionosphere. The data show enhanced thermal electron tails in two narrow altitude layers as indicated in the figure. Taken from *Oyama et al.* [1983].

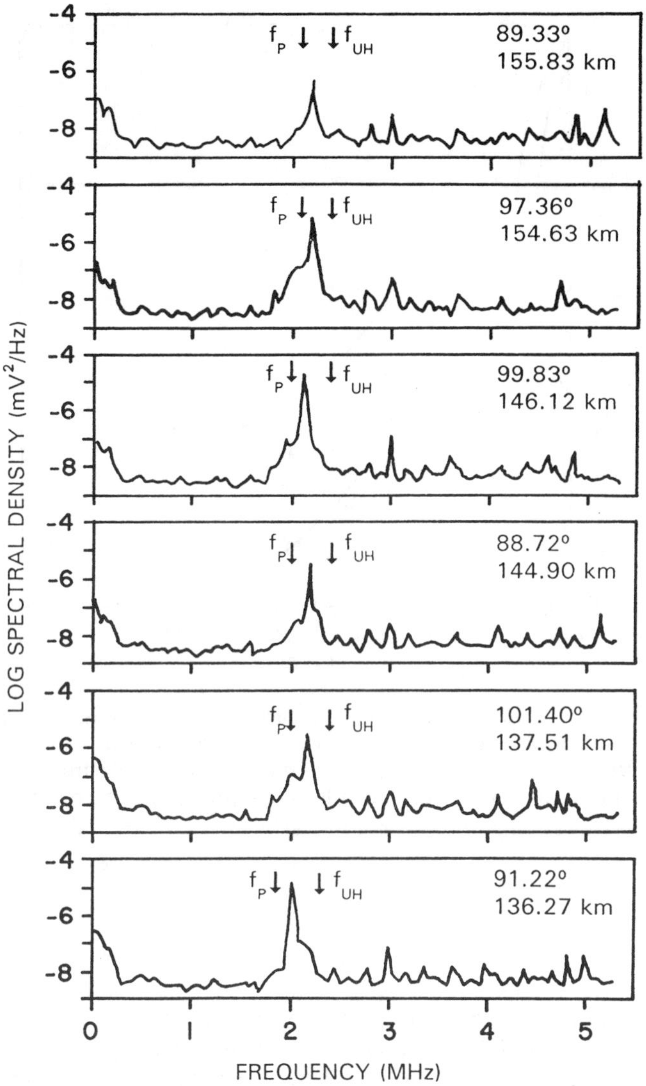

Fig. 4. Swept Frequency Analyzer (SFA) data for the spectral wave power density as a function of frequency. The local plasma and upper hybrid frequencies are indicated by arrows. The numbers in the upper right corner of each panel are the altitiude and the angle between the detector and $\underline{B}_0$. Taken from *Kelley and Earle* [1988].

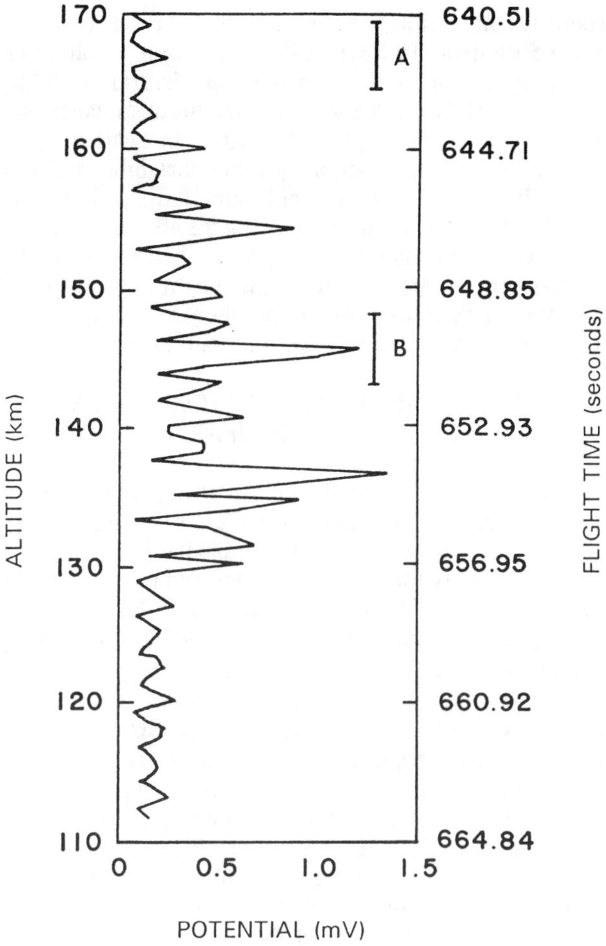

Fig. 5. The potential detected by the Swept Frequency Analyzer (SFA) integrated over the wave emission band (i.e., the total signal strength detected in the emissions) as a function of altitude. Taken from *Kelly and Earle* [1988].

Furthermore, this high frequency plasma wave emission is confined within an altitude region of 130-170 km (Figure 5). Simultaneous measurements of the spectra of the precipitating, high energy (a few keV) electrons showed a diffuse auroral-like spectrum in the altitude region of intense wave emissions (region B) and a weak auroral arc-like spectrum above the cut-off altitude of the waves (region A). In view of this, it seems unlikely that the waves were caused by a precipitating electron beam, and a different source of free energy for the excitation of the observed high frequency plasma waves needs to be identified.

We note that the high-frequency instabilities at upper hybrid and harmonics of electron cyclotron frequencies described by *Basu et al.* [1981, 1982] have characteristics that are very similar to the observations described above and we propose that these instabilities play a role in the theoretical explanation of the observations. As mentioned before, the free energy source for the excitation of the instabilities is the shell-like bump near 5 eV in the E- and lower F-region secondary electron energy distribution. The physical origin of this suprathermal bump is discussed in Section 2. In Section 3, we derive a general formula for the collisional dielectric function and discuss two approximate forms which can be solved exactly. In Sections 4-6, we carry out a linear stability analysis of the problem and discuss: (1) the narrow, electron cyclotron instability altitude layers in the lower E region; and (2) the broad, upper hybrid instability altitude layer in the upper E and lower F region. In Section 7, we focus on the resonant upper hybrid instability and discuss some ways that the wave growth could be limited. In Section 8, we perform a collisional quasilinear analysis of the resonant upper hybrid instability and, from an H-theorem analysis of the problem, determine the form of the time-asymptotic state. In Section 9, we study the nonresonant electron cyclotron instability and give the results of a particle-in-cell simulation of a two-dimensional ring distribution including the effect of electron-neutral particle collisions. Although the shell-like and ring-like instabilities are different, the qualitative effect of collisions on their time evolution should be similar. In Section 10, we discuss and summarize our results.

2. PHYSICAL ORIGIN OF THE SUPRATHERMAL SHELL-LIKE BUMP NEAR 5 eV

The theoretical explanation for the structure in the electron energy distributions in the 2-5 eV energy range, which are obtained by including particle-particle collisional effects only, is as follows. The cross section for electron energy loss due to the impact excitation of the vibrational states of the ground electronic state of N_2 has a sharp maximum at 2.3 eV. This explains the dip in the electron energy distribution at 2.3 eV, as the electrons near 2.3 eV rapidly lose energy by colliding with N_2. Beyond 2.3 eV, the electron flux rises sharply to a maximum at about 5 eV, as the vibrational excitation cross section decreases sharply. Above 5 eV, the electron flux decreases again as the cross section for the excitation of the metastable states of atomic oxygen increases. This explains the formation of the shell-like bump in the energy distribution near 5 eV. For a more quantitative discussion, see *Jasperse* [1981]. It should be noted here that this suprathermal bump will always be produced as long as there is an ionizing source creating secondary electrons in the E layer. For example, it will occur for the daytime mid- to high-latitude ionosphere, the diffuse aurora, and a stable auroral arc. In Figure 6, we show the E-region electron energy spectra resulting from precipitating keV Maxwellian electrons, which are typical of the diffuse aurora. The shell-like bump in the secondary electron distribution around 5 eV and its altitude dependence are quite similar to those found in the midlatitude, photoelectron energy spectra. At high altitudes (>170 km), the peak-to-valley ratio decreases until the structure disappears at or above 210 km. This can be attributed to the exponential decrease in the N_2 density as well as to the increase in the temperature of the thermal part of the electron distribution which starts to fill in the 2.3 eV valley. See Figure 4 of *Jasperse* [1977] for details.

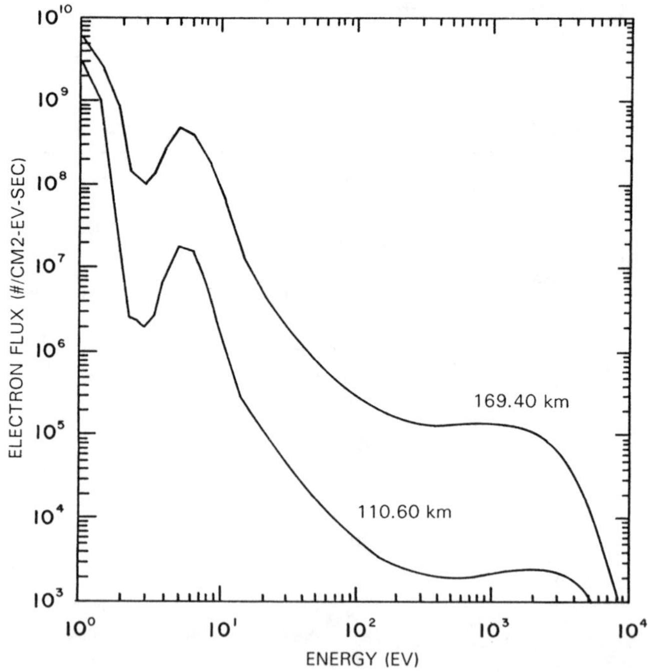

Fig. 6. Calculated zero-order auroral electron flux integrated over the angles as a function of energy at two selected altitudes produced by a precipitating primary Maxwellian electron flux with a low energy tail (LET) having characteristic energy of 1 keV and energy flux of 1 erg $cm^{-2}s^{-1}$.

3. COLLISIONAL DIELECTRIC FUNCTION

3.1. *Assumptions*

In the analysis presented in this paper, we neglect the presence of a steady-state, zero-order E field. In the daytime lower ionosphere, the steady E fields are on the order of the ambipolar E field and can be neglected. The same is also true for the diffuse aurora except where it overlaps with the auroral electrojet. In a stable auroral arc, there is a parallel E field associated with the strong electron precipitation. If its value in the lower ionosphere approaches that of the Dreicer field [*Dreicer*, 1960], then thermal electrons will tend to run away and the zero-order distribution function could be quite different from that considered in our analysis. A perpendicular E field associated with the electrojet is present in auroral arcs. If the turbulence generated, which is low frequency and long wavelength compared to the high frequency and short wavelength turbulence discussed in this paper, does not significantly alter the near isotropic Maxwellian form of the thermal electrons or the shape of the 5 eV bump, then our analysis may also apply to stable auroral arcs.

3.2. *Collisional Dielectric Function*

We start with the Boltzmann-Fokker-Planck and Poisson equations for the thermal and low-energy secondary electrons.

$$\left\{\frac{\partial}{\partial t}+\underline{v}\cdot\frac{\partial}{\partial \underline{r}}+\left[\frac{q}{m}\underline{E}(\underline{r},t)+\underline{a}_0(\underline{v})\right]\cdot\frac{\partial}{\partial \underline{v}}\right\}f(\underline{r},\underline{v},t)=$$

$$S-\mathcal{R}(f)+\mathcal{L}_{en}(f)+\mathcal{L}_{ei}(f)+\mathcal{L}_{ee}(f), \quad (1)$$

$$\frac{\partial}{\partial \underline{r}}\cdot\underline{E}(\underline{r},t)=4\pi\left\{q_i n_i+q\int d\underline{v}\, f(\underline{r},\underline{v},t)\right\}, \quad (2)$$

$$\int d\underline{v}\, f(\underline{r},\underline{v},t)=n(\underline{r},t). \quad (3)$$

Here, $\underline{a}_0(\underline{v}) = q\, \underline{v} \times \underline{B}_0/mc$, S is the photoelectron source function and/or the secondary electron source function due to primary auroral electrons, $\mathcal{R}$ is the electron-ion recombination operator, and the $\mathcal{L}_{en}$, $\mathcal{L}_{ei}$ and $\mathcal{L}_{ee}$ operators describe electron-neutral, electron-ion, and electron-electron collisions, respectively. Also, $\underline{B}_0$ is the geomagnetic field, the ions are assumed to provide a fixed neutralizing background with charge density $q_i n_i$, and the other quantities have their usual meanings. The $\mathcal{R}$ and $\mathcal{L}$ operators act only on the velocity space coordinates of f and are discussed in detail in *Jasperse* [1976, 1977].

In order to carry out a stability analysis, we first must determine the steady-state, zero-order electron distribution function. Equation (1) has been solved for the photoelectron flux as a function of altitude and energy by *Jasperse* [1976, 1977] and for the auroral electron flux as a function of altitude, energy and pitch angle by *Strickland* [1976]. Typical results for each are shown in the right panel of Figure 1 and in Figure 6.

We now wish to study the stability of the kinetic-Poisson equations to small electrostatic perturbations about the above, zero-order solutions. We seek a solution of the form $f = f_0(\underline{v}) + f_1(\underline{r},\underline{v},t)$ and $\underline{E} = \underline{E}_1(\underline{r},t)$ where $|f_1| \ll f_0$. For the moment, we leave the velocity dependence of f_0 arbitrary but later specialize. Linearizing the system using standard procedures, we obtain

$$\left[\frac{\partial}{\partial t}+\underline{v}\cdot\frac{\partial}{\partial \underline{r}}+\underline{a}_0(\underline{v})\cdot\frac{\partial}{\partial \underline{v}}-L\right]f_1(\underline{r},\underline{v},t)$$

$$=\frac{q}{m}\frac{\partial}{\partial \underline{r}}\phi(\underline{r},t)\cdot\frac{\partial}{\partial \underline{v}}f_0(\underline{v}), \quad (4)$$

$$\nabla^2\phi(\underline{r},t)=-4\pi q\int d\underline{v}\, f_1(\underline{r},\underline{v},t), \quad (5)$$

$$\int d\underline{v}\, f_1(\underline{r},\underline{v},t)=n_1(\underline{r},t), \quad (6)$$

where we have introduced the electrostatic potential $\underline{E}_1 = -\partial\phi/\partial\underline{r}$, and $L = -R+L_{en}+L_{ei}+L_{ee}$. Here, the linearized collision operators are defined by

$$\mathcal{L}(f_0+f_1)-\mathcal{L}(f_0)=Lf_1,$$

with a similar expression for R. Equations (4) and (5) may be solved formally for the collisional dielectric function using the method of *Jasperse* [1984]. In this method, the Green function (or propagation operator) for (4) is found in transform space $(\underline{k},\underline{v},\omega)$ and used to find the collisional dielectric function, ε^V. The result is

$$\varepsilon^V(\underline{k},\omega)=1-i\left(\frac{4\pi q^2}{mk^2}\right)\int d\underline{v}\,\tilde{U}^V(\underline{k},\underline{v},\omega)\underline{k}\cdot\frac{\partial}{\partial \underline{v}}f_0(\underline{v}), \quad (7)$$

$$\tilde{U}^V(\underline{k},\underline{v},\omega)=\int d\underline{v}'\,\tilde{G}^V(\underline{k},\underline{v},\underline{v}',\omega). \quad (8)$$

Here, $\tilde{U}^V$ is the collisional propagation operator and $\tilde{G}^V$ is the Fourier transform of the Green function for (4). In our analysis, we have considered perturbations of the form $\exp(i\underline{k}\cdot\underline{r} - i\omega t)$. If $\tilde{G}^V$ can be found, then the effect of collisions on electrostatic plasma waves as a function of ω and $\underline{k}$ can be determined.

3.3. *Solution for an Approximate Boltzmann Collision Operator*

At altitudes below about 170 km, electron-neutral particle collisions dominate charged particle collisions, and recombination time scales are much slower than wave-particle time scales. Therefore, at altitudes below about 170 km, we may neglect L_{ei}, L_{ee} and R compared to L_{en}. At these altitudes, the N_2 density exceeds the O and O_2 densities. Therefore, the two collisional processes which are dominant for electron energies from 0 to about 10 eV are: (1) electron - N_2 momentum transfer collisions; and (2) electron - N_2 vibrational excitation collisions. See *Jasperse* [1976, 1977].

It can be shown that, since the electron distribution function is nearly isotropic, a good approximation to the complete Boltzmann collision integrals discussed in *Jasperse* [1976, 1977] in the 0 to 10 eV energy range is

$$L_{en}{}^{B}f_1(\underline{r},\underline{v},t)=-\nu(|\underline{v}|)f_1(\underline{r},\underline{v},t)$$
$$+\nu(|\underline{v}|)\int\frac{d\Omega_v}{4\pi}f_1(\underline{r},\underline{v},t), \quad (9)$$

$$\nu(|\underline{v}|)=N|\underline{v}|2\pi\int_0^\pi d\theta\sin\theta\frac{\partial 6}{\partial\Omega_T}(|\underline{v}|,\cos\theta), \quad (10)$$

$$\frac{\partial 6}{\partial\Omega_T}=\frac{\partial 6}{\partial\Omega_{MOM}}+\frac{\partial 6}{\partial\Omega_{VIB\,EX}}. \quad (11)$$

Here, $\partial\sigma/\partial\Omega_T$ is the sum of the momentum transfer and the vibrational excitation differential cross sections, and N is the N_2 density. In this approximation, (4) becomes

$$\left[\frac{\partial}{\partial t}+\underline{v}\cdot\frac{\partial}{\partial \underline{r}}+\underline{a}_0(\underline{v})\cdot\frac{\partial}{\partial \underline{v}}\right]f_1(\underline{r},\underline{v},t)+\nu(|\underline{v}|)f_1(\underline{r},\underline{v},t)$$

$$=\frac{q}{m}\frac{\partial}{\partial \underline{r}}\phi(\underline{r},t)\cdot\frac{\partial}{\partial \underline{v}}f_0(\underline{v})$$

$$+\nu(|\underline{v}|)\int\frac{d\Omega_v}{4\pi}f_1(\underline{r},\underline{v},t). \tag{12}$$

This is a linear, partial differential-integral equation for f_1, in terms of ϕ and f_0. The collisional propagation operator for the problem, $\tilde{U}^V$, may be found by the method of characteristics. See *Krall and Trivelpiece* [1973] for a review of the method. The reason that this method yields the solution is because ν and

$$\int d\Omega_v f_1(\underline{r},\underline{v},t) \tag{13}$$

depend only on $|\underline{v}|$ and are therefore independent of the orbit integration in a uniform B field. Also, the transform of (13) may be solved explicitly in terms of ϕ and f_0 and eliminated from the final expression for the transform of f_1. In operator notation, the solution for $\tilde{U}^V$ is

$$\tilde{U}^V(\underline{k},\underline{v},\omega)=\tilde{K}^V(\underline{k},\underline{v},\omega)+\nu(v)\tilde{K}^V(\underline{k},\underline{v},\omega)$$

$$\times\left[1-\nu(v)\int\frac{d\Omega_v}{4\pi}\tilde{K}^V(\underline{k},\underline{v},\omega)\right]^{-1}$$

$$\times\int\frac{d\Omega_v}{4\pi}\tilde{K}^V(\underline{k},\underline{v},\omega), \tag{14}$$

$$\tilde{K}^V(\underline{k},\underline{v},\omega)=\int_{-\infty}^{0}d\tau\exp\{i\underline{k}\cdot\underline{R}-i[\omega+i\nu(v)]\tau\}. \tag{15}$$

Here, $\underline{R}=\underline{r}'-\underline{r}$, $\tau=t'-t$ and $|\underline{v}'|=|\underline{v}|$ where $\underline{r}'$, $\underline{v}'$ and t' are the orbit variables. See *Krall and Trivelpiece* [1973], page 403. Also, we have used the notation $\nu(v)=\nu(|\underline{v}|)$. The collisional dielectric function is given by (7) where (14) is substituted for $\tilde{U}^V$. When we let the $\tilde{K}^V$ operator act, we obtain the following expression

$$\varepsilon^V(\underline{k},\omega)_B=1+\frac{\omega_p^2}{k^2}\int d\underline{v}\left[1-\frac{i}{2}\nu(v)\sum_{n=-\infty}^{\infty}H_n(v)\right]^{-1}$$

$$\times\sum_{n=-\infty}^{\infty}\frac{J_n^2(k_\perp v_\perp/\Omega)}{(\omega+i\nu(v)-n\Omega-k_\parallel v_\parallel)}$$

$$\times\left[\frac{n\Omega}{v_\perp}\frac{\partial}{\partial v_\perp}+k_\parallel\frac{\partial}{\partial v_\parallel}\right]\frac{f_0(v_\perp^2,v_\parallel)}{n_0}, \tag{16}$$

$H_n(v)$

$$=\int_0^\pi d\theta\sin\theta\frac{J_n^2(k_\perp v\sin\theta/\Omega)}{(\omega+i\nu(v)-n\Omega-k_\parallel v\cos\theta)}, \tag{17}$$

where $\Omega=|q|B_0/mc$ is the cyclotron frequency, $\omega_p^2=4\pi q^2 n_0/m$, and $k^2=k_\perp^2+k_\parallel^2$. Here, we have considered f_0 as independent of the azimuthal angle ϕ and the other quantities have their usual meanings. This is the collisional dielectric function for the problem using the approximate Boltzmann collision operator. It gives the effect of electron - N_2 collisions at energies from about 0 to 10 eV for all values of $\underline{k}$, ω and $\nu(v)$.

3.4. *Solution for the BGK Collision Operator*

The collisional dielectric function for the problem using the BGK collision operator to approximate the electron - N_2 collisions has been found by *Basu et al.* [1981, 1982]. In solving the problem, they approximately decoupled the solutions into two energy ranges, thermal (j=t) and suprathermal (j=s). Thus, $f_0(\underline{v})\rightarrow f_{0t}(\underline{v})+f_{0s}(\underline{v})$ and $f_1(\underline{r},\underline{v},t)\rightarrow f_{1t}(\underline{r},\underline{v},t)+f_{1s}(\underline{r},\underline{v},t)$. The BGK collision operators for the problem are

$$L_{enj}^{BGK}f_{1j}(\underline{r},\underline{v},t)=-\nu_j f_{1j}(\underline{r},\underline{v},t)$$

$$+\nu_j\left[f_{0j}(\underline{v})/n_{0j}\right]n_{1j}(\underline{r},t), \tag{18}$$

$$n_{1j}(\underline{r},t)=\int d\underline{v}f_{1j}(\underline{r},\underline{v},t). \tag{19}$$

Here, ν_j is the average collision frequency for the thermal (j=t) and the suprathermal (j=s) electrons. Using procedures identical to the ones described above, Basu et al. obtained the following expressions for the thermal and suprathermal collisional propagation operators

$$\tilde{U}^{V}{}_j(\underline{k},\underline{v},\omega)=\tilde{K}^{V}{}_j(\underline{k},\underline{v},\omega)+\left(\frac{v_j}{n_{0j}}\right)$$

$$\times\tilde{K}^{V}{}_j(\underline{k},\underline{v},\omega)f_{0j}(\underline{v})\left[1-\frac{v_j}{n_{0j}}\int d\underline{v}\tilde{K}^{V}{}_j(\underline{k},\underline{v},\omega)f_{0j}(\underline{v})\right]^{-1}$$

$$\times\int d\underline{v}\tilde{K}^{V}{}_j(\underline{k},\underline{v},\omega). \tag{20}$$

Using these expressions in (7) and summing over the thermal and suprathermal electrons, they obtained

$$\varepsilon^{V}(\underline{k},\omega)_{BGK}=1+\frac{\omega_p^{\,2}}{k^2}\sum_{j=t,s}\frac{N_j}{D_j}, \tag{21}$$

$$N_j=\sum_{n=-\infty}^{\infty}\int d\underline{v}\frac{J_n^{\,2}(k_\perp v_\perp/\Omega)}{\left(\omega+i\nu_j-n\Omega-k_\parallel v_\parallel\right)}$$

$$\times\left[\frac{n\Omega}{v_\perp}\frac{\partial}{\partial v_\perp}+k_\parallel\frac{\partial}{\partial v_\parallel}\right]\frac{f_{0j}\left(v_\perp^{\,2},v_\parallel\right)}{n_0}, \tag{22}$$

$$D_j=1-i\nu_j\sum_{n=-\infty}^{+\infty}\int d\underline{v}\frac{J_n^{\,2}(k_\perp v_\perp/\Omega)}{\left(\omega+i\nu_j-n\Omega-k_\parallel v_\parallel\right)}$$

$$\times\frac{f_{0j}\left(v_\perp^{\,2},v_\parallel\right)}{n_{0j}}. \tag{23}$$

This is the collisional dielectric function for the problem using the BGK operator valid for all values of $\underline{k}$, ω and ν_j.

3.5. *Equivalence of the Boltzmann and BGK Dielectric Functions for ν/ω<<1.*

Now we wish to examine both dielectric functions for the limiting case when ν/ω<<1. Consider the expression for $\varepsilon^V{}_B$ first. In the cold plasma limit, $k_\perp v_\perp/\Omega$ <<1, and the H_n integrals will be dominated by the n=0 term. Also, at long wavelengths, Landau damping will be small compared to collisional damping ($k_\parallel v_\parallel$<<ω). Therefore, the inverse of the term in the square brackets in (16) $\cong 1+i\nu(v)/\omega$. If the analysis leading to (16) is repeated for two electron populations, thermal and suprathermal, then an index j appears on $\nu(v)$ and on $f_0(\underline{v})$. Since $\nu_j(v)/\omega$<<1, $\nu_j(v)$ may be replaced under the integral sign by its average value at thermal and suprathermal energies, $\nu_j(v)\rightarrow\bar{\nu}_j$. The expression for $\varepsilon^V{}_B$ reduces to

$$\varepsilon^{V}(\underline{k},\omega)_B\cong1+\frac{\omega_p^{\,2}}{k^2}\sum_{j=t,s}\left(1+\frac{i\bar{\nu}_j}{\omega}\right)$$

$$\times\int d\underline{v}\sum_{n=-\infty}^{\infty}\frac{J_n^{\,2}(k_\perp v_\perp/\Omega)}{\left(\omega+i\bar{\nu}_j-n\Omega-k_\parallel v_\parallel\right)}$$

$$\times\left[\frac{n\Omega}{v_\perp}\frac{\partial}{\partial v_\perp}+k_\parallel\frac{\partial}{\partial v_\parallel}\right]\frac{f_{0j}\left(v_\perp^{\,2},v_\parallel\right)}{n_0}, \tag{24}$$

which is valid to the first order in $\bar{\nu}_j/\omega$.

Now consider $\varepsilon^V{}_{BGK}$. Using a similar argument, we see that $D_j^{-1}\cong 1+i\nu_j/\omega$ to the first order. Thus (21) becomes identical to (24) where $\bar{\nu}_j$ is replaced by ν_j. We see that for the above conditions, the two dispersion relations are equivalent to the first order in $\bar{\nu}_j/\omega$.

3.6. *Generalizations*

It is trivial to include the effect of electron-ion recombination in the above analysis. According to *Jasperse* [1976], the linearized collision operator has the form $Rf_1 = -\nu_r(|\underline{v}|)\,f_1$ and may be included on the left-hand side of (12) without altering the analysis.

At altitudes above about 170 km for the daytime ionosphere, charged particle collisions become important. The correct collision operators to use are the Balescu-Lenard operators. The effect of Coulomb collisions on high frequency electrostatic modes as a function of $\underline{k}$ and ω has been studied by *Jasperse and Basu* [1986] in the absence of an applied B field. The effect of Coulomb collisions on the dispersion relation in the presence of an applied B field is currently under study and will not be discussed here.

4. LINEAR STABILITY ANALYSIS OF THE COLLISIONAL DISPERSION RELATION

In this section, we carry out a stability analysis of the collisional dielectric function. Since the thermal population is Maxwellian with temperature T_e, we may evaluate the velocity space integrations, and either (24) or (21) for small $\bar{\nu}_j/\omega$ becomes

$$\varepsilon^V(\underline{k},\omega) = 1 - \frac{\omega_{pt}^2}{\omega(\omega + i\bar{\nu}_t)}\left(\frac{k_{\parallel}^2}{k^2}\right) I_0(b)\exp(-b)$$

$$- \sum_{n=1}^{\infty} \frac{2n^2\omega_{pt}^2(1 + i\bar{\nu}_t/\omega)}{\left[(\omega + i\bar{\nu}_t)^2 - n^2\Omega^2\right]}\left(\frac{k_{\perp}^2}{k^2}\right)\frac{I_n(b)\exp(-b)}{b}$$

$$+ \frac{\omega_p^2}{k^2}\left(1 + \frac{i\bar{\nu}_s}{\omega}\right)N_s, \qquad (25)$$

where the subscripts t and s refer to the Maxwellian and suprathermal electrons, respectively. Here $I_n(b)$ is the modified Bessel function of the first kind, $b = k_{\perp}^2 T_e/m\Omega^2$, and $\omega_{pt}^2 = 4\pi q^2 n_{0t}/m$. In deriving (25), we have assumed that $k_{\perp}^2 \gg k_{\parallel}^2$ and $(\omega - n\Omega)/[|k_{\parallel}|(2T_e/m)^{1/2}] \gg 1$ for all n, so that both Landau and cyclotron damping can be neglected.

Equation (25) can be solved without the suprathermal term in order to find the waves that can be supported by the Maxwellian electrons in the presence of collisional damping. We may write $\omega \cong \omega_k + i\gamma_k^V$, where $|\gamma_k^V| \ll \omega_k$, which is consistent with $\bar{\nu}_t \ll \omega_k$. In the lower ionosphere, we note that $\omega_{pt}^2 > \Omega^2$. Using standard methods, we obtain the following results for the real and imaginary parts of the wave frequency. The lowest frequency mode is given by

$$\omega_k = \omega_2 \cong 2\Omega\left[1 - \left(\frac{3}{8}\right)\frac{\omega_{pt}^2 b}{\left(\omega_{pt}^2 - 3\Omega^2\right)}\right],$$

$$\omega_{pt}^2 > 3\Omega^2. \qquad (26)$$

The next higher frequency mode is the so-called upper hybrid mode and is given by

$$\omega_k \equiv \omega_{uh} \cong (\omega^2_{pt} + \Omega^2)^{1/2} \qquad (27)$$

The next higher frequency modes (just below or above the harmonics) depend on ω_{pt}^2/Ω^2 and are given by (n = 3, 4, ...)

$$\omega_k \equiv \omega_n \cong \begin{cases} n\Omega\left[1 - o\left(b^{n-1}\right)\right], \omega_{pt}^2 > \left(n^2 - 1\right)\Omega^2 \\ n\Omega\left[1 + o\left(b^{n-1}\right)\right], \omega_{pt}^2 < \left(n^2 - 1\right)\Omega^2 \end{cases}. \qquad (28)$$

These are the *Bernstein* [1958] modes for the problem and are separated by frequency intervals for which there is no propagation. In (26) - (28) there are small corrections of the order of $k_{\parallel}^2/k_{\perp}^2 \ll 1$, which have been omitted for the sake of convenience. The corresponding collisional damping rates are:

$$\gamma_k^V \cong \begin{cases} -4\bar{\nu}_t\frac{\Omega^2}{\omega_2^2}\left[1 - \frac{3}{8}\frac{\omega_{pt}^4 b}{\left(\omega_{pt}^2 - 3\Omega^2\right)^2}\right], & (29) \\ -\bar{\nu}_t\left(\frac{\omega_{uh}^2 + \Omega^2}{2\omega_{uh}^2}\right), & (30) \\ -\bar{\nu}_t\left(\frac{\omega_n^2 + n^2\Omega^2}{2\omega_n^2}\right), & (31) \end{cases}$$

respectively, for the modes ω_2, ω_{uh} and ω_n.

We note that at long wavelengths when Landau and cyclotron damping are negligible, the Maxwellian thermal plasma in the lower ionosphere will support Bernstein modes which are weakly damped by electron - N_2 collisions. We also see that the collisional damping rate depends on $\bar{\nu}_t$ (the collision frequency for the thermal electrons) and on the electron cyclotron frequencies.

Now that we have described the collisional Bernstein modes that can be supported by the thermal electrons, we wish to solve the full dispersion relation given by (25) and examine the destabilizing effect of the suprathermal electrons on these modes. The destabilizing effect is contained in the last term on the right-hand side of (25). In particular, we are interested in finding growth rates that are larger than the collisional damping rates given by (29) - (31). We consider two types of instabilities: (1) nonresonant ($\underline{k} = \underline{k}_{\perp}$, $k_{\parallel} = 0$); and (2) resonant ($\underline{k} = \underline{k}_{\perp} + \underline{k}_{\parallel}$, $k_{\parallel} \ll k_{\perp}$). In each case, we write $\omega = \omega_k + i\gamma_k^V + i\gamma_k$, where γ_k is the growth rate to be determined. This problem has been solved by *Basu et al.* [1981, 1982] using standard procedures. In the following two sections, we briefly review their results.

5. NARROW ELECTRON CYCLOTRON INSTABILITY LAYERS IN THE LOWER E REGION (~ 100-130 KM)

In this altitude region, Bernstein modes with $\underline{k} \perp \underline{B}_0$ and $\omega_k \cong n\Omega$, where $n \geq 2$, are potentially unstable. This is a nonresonant type of instability which derives its energy from the suprathermal bump, and, because of the plasma parameters involved, the potentially unstable harmonics are excited within narrow altitude layers about 2 km thick. The physical reason for the narrow altitude layers is that as the plasma density increases with altitude in the lower E region, the upper hybrid frequency increases and becomes coincident with one of the cyclotron harmonics. This "double resonance condition" has a narrow width associated with it, and, as a result, there is a narrow altitude range over which the instability can be excited. We find that for typical daytime midlatitude ionospheric conditions, one or several layers can be excited from about 100 to 130 km. The low altitude cut-off is due to an amount of damping by electron-neutral collisions that is larger than the collisionless growth rate, and the high altitude cut-off is due to stabilization by the finite b ($\equiv k^2T_e/m\Omega^2$) effect as the electron temperature increases with altitude. This is shown in Figure 7.

6. A BROAD UPPER HYBRID INSTABILITY LAYER IN THE UPPER E AND LOWER F REGIONS (~ 125 TO GREATER THAN 170 KM)

In this region, obliquely propagating upper hybrid modes with $\omega_k \cong (\omega_{pt}^2 + \Omega^2)^{1/2}$ and $k_\perp >> k_\parallel$ are excited due to the resonant interaction between the waves and the suprathermal electrons in the shell-like bump near 5 eV. Since our analysis shows that the upper hybrid mode has the largest growth rate, we will focus on it. The low altitude cut-off is due to the increased damping by electron - N_2 collisions and the high altitude cut-off is due to two effects. One is the increased damping due to electron-ion and electron-electron collisions and the other is the flattening out of the zero-order electron distribution function, both of which occur somewhere above 170 km. This is shown in Figure 8 where the high altitude cut-off is not calculated explicitly.

The maximum growth rates of both the nonresonant and resonant instabilities discussed in Sections 5 and 6 are 10^3-10^4-s^{-1} and the corresponding transverse wavelengths are 11-20 cm. In Figure 9, we show the spectral characteristics of both types of unstable modes. Here, K is the normalized wavevector and v_0 is the velocity associated with the peak in the suprathermal bump at 5 eV.

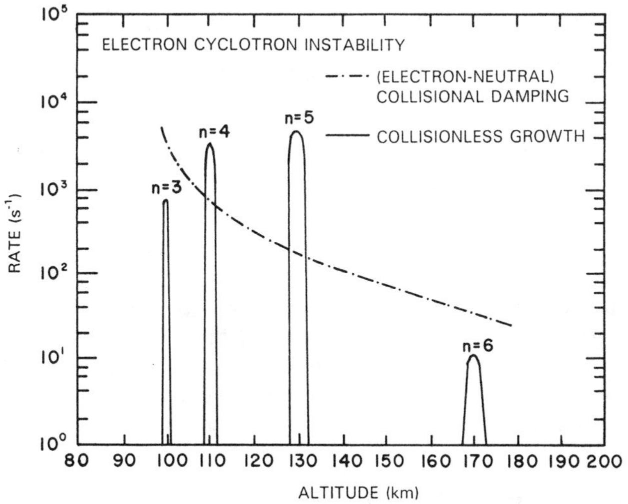

Fig. 7. Collisionless growth rates of several electron cyclotron harmonic modes and collisional damping rate versus altitude. The figure shows that the unstable Bernstein modes are excited only within narrow altitude layers a few kilometers thick in the lower E region.

7. LIMITATIONS ON WAVE GROWTH

The growth of the unstable waves can be limited by: (1) propagation out of a limited spatial region of wave growth; (2) wave refracton out of resonance by density variations; (3) elimination of the suprathermal bump by quasilinear wave-particle interactions, and (4) nonlinear wave-particle and wave-wave interactions. Considering the first two processes and following the approximate analysis of *McFadden et al.* [1986], we find that neither of them has any significant effect on the unstable modes described above. This is due to the fact that the source of free energy (suprathermal bump) is not spatially limited in the direction perpendicular to $\underline{B}_0$, and the wavelengths and group velocities of the modes are very small. However, if long wavelength and low frequency turbulence from another instability is present with sufficient

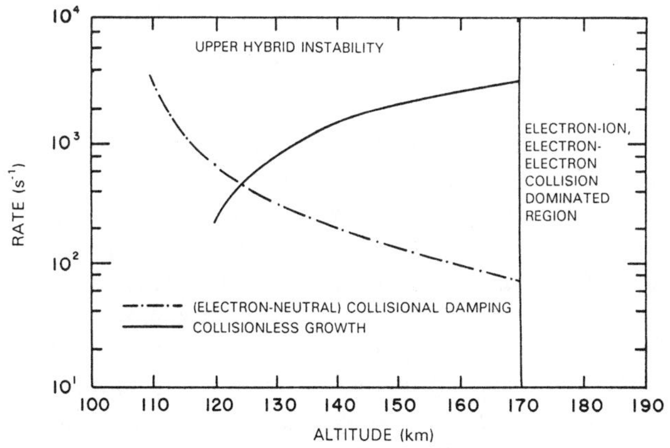

Fig. 8. Collisionless growth rate of the upper hybrid mode and collisional damping rate versus altitude. The mode is gradually stabilized somewhere above 170 km (not shown in the figure).

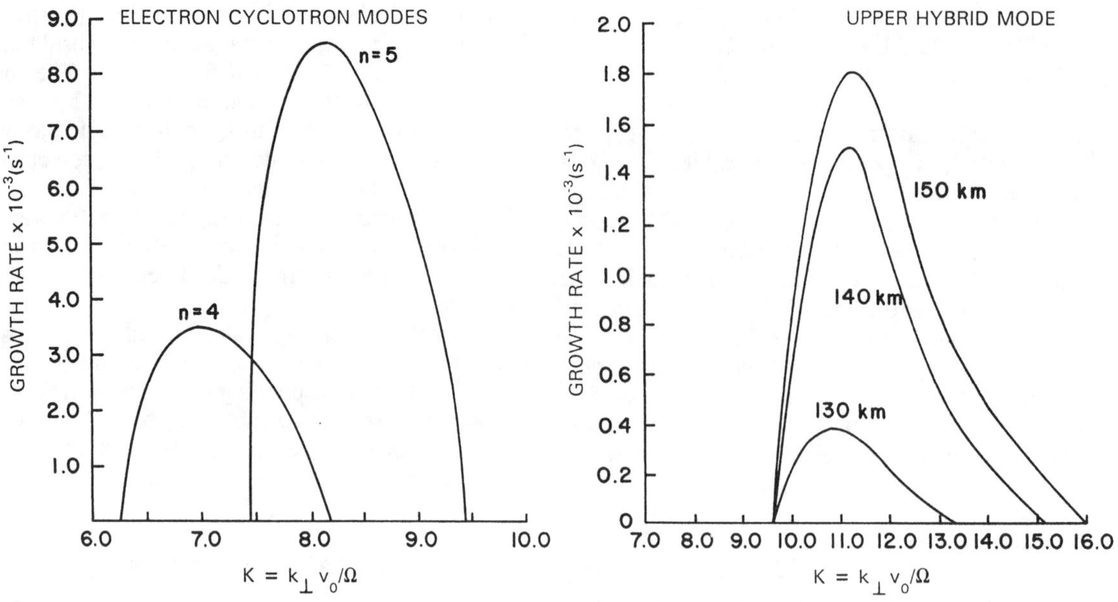

Fig. 9. Normalized wave number spectra of unstable Bernstein modes and the upper hybrid mode at different altitudes.

amplitude, then it could cause the short wavelength and high frequency waves to refract out of resonance and limit wave growth. Neglecting such a possibility, the important mechanisms for limiting wave growth are those described in (3) and (4)

In Section 8, we will consider the quasilinear evolution of the resonant upper hybrid instability. The first step is to study quasilinear evolution in order to see how large the waves become. If the quasilinear analysis reveals wave amplitudes at saturation that are too large, then nonlinear effects need to be considered.

8. COLLISIONAL QUASILINEAR ANALYSIS OF THE RESONANT UPPER HYBRID INSTABILITY

8.1. *Assumptions*

The quasilinear equations for a collisionless plasma have been derived from the Vlasov-Poisson equations by *Vedenov et al.* [1961, 1962] and by *Drummond and Pines* [1962]. In that work, the analysis includes the effect of linear wave-particle interactions and was carried out to the first order in η (= $\varepsilon_f/\varepsilon_{kin}$), where $\eta << 1$, where ε_f is the total field energy density, and where ε_{kin} is the total electron kinetic energy density. Weak turbulence equations have also been derived which, in addition to linear wave-particle processes, include nonlinear wave-particle and wave-wave processes, i.e., interactions to the second order in η. In addition, two-particle collisions among charged particles have been included where it was assumed that $g(=1/n\lambda_D^3)$ was second order in η. See *Frieman and Rutherford* [1964], *Rogister and Oberman* [1969], and *Davidson* [1972].

In this paper, we seek a collisional quasilinear theory where the electron-neutral particle collisional damping rate operates on the same time scale as the collisionless growth rate, $|\gamma_k^\nu|/\omega_k \sim \gamma_k/\omega_k$. Here, the electron-neutral particle collisional processes are first order in η. Since we wish to keep the analysis simple, we neglect second-order effects in η, i.e., nonlinear wave-particle and wave-wave interactions.

We begin with the Boltzmann-Fokker-Planck and Poisson equations and carry out a collisional quasilinear analysis to the first order in η in a way similar to what was done by *Vedenov et al.* [1961, 1962] and *Drummond and Pines* [1962]. We give the collisional quasilinear equations for the time evolution of the resonant upper hybrid instability discussed above, discuss some approximations which render the equations solvable, and give an H-theorem analysis of the form of the time-asymptotic state. Here, we neglect the response of the neutral particles so that total momentum and energy conservation laws between the electrons and neutral particles are not preserved. Preliminary results of this analysis were discussed in *Jasperse et al.* [1982].

8.2. *Collisional Quasilinear Equations*

We return to the Boltzmann-Fokker-Planck and Poisson equations for the problem which contain an electron source term. They are given by (1) and (2). We consider only one type of mode, the upper hybrid, and $\Gamma_k>0$, where $\Gamma_k/\omega_k<<1$

and $\Gamma_k = \gamma_k^v + \gamma_k$. Following standard procedures, we obtain

$$\left[\frac{\partial}{\partial t} + \underline{a}_0(\underline{v}) \cdot \frac{\partial}{\partial \underline{v}}\right] f_0(\underline{v},t) = S - \mathcal{R}(f_0) + \mathcal{L}_{en}(f_0)$$

$$+\mathcal{L}_{ei}(f_0) + \mathcal{L}_{ee}(f_0) + \frac{\partial}{\partial \underline{v}} \cdot \underline{\underline{T}}^V(\underline{v},t) \cdot \frac{\partial}{\partial \underline{v}} f_0(\underline{v},t), \quad (32)$$

$$\frac{\partial}{\partial t} \varepsilon_k(t) = 2\Gamma_k(t)\varepsilon_k(t), \quad (33)$$

$$\underline{\underline{T}}^V(\underline{v},t) = \left(\frac{8\pi q^2}{m^2}\right) \int d\underline{k}\, \varepsilon_k(t) \hat{\underline{k}} \hat{\underline{k}} \tilde{U}^V{}_{\underline{k},\omega_k + i\Gamma_k}(\underline{v}). \quad (34)$$

Here, $\tilde{U}^v$ is the collisional propagation operator given by (8) evaluated at $\omega = \omega_k + i\Gamma_k$, ε_k is the spectral energy density associated with the electric field fluctuations, $\underline{\underline{T}}^V$ is a second-order tensor operator acting on the velocity dependence of $\partial f_0/\partial \underline{v}$, and $\hat{\underline{k}}$ is a unit vector along $\underline{k}$. These equations are supplemented with the collisional dielectric function evaluated at $\omega = \omega_k + i\Gamma_k$

$$\varepsilon^V{}_{k,\omega_k + i\Gamma_k} = 1 - i\left(\frac{4\pi q^2}{mk^2}\right) \int d\underline{v}\, \tilde{U}^V{}_{k,\omega_k + i\Gamma_k}(\underline{v})$$

$$\times \underline{k} \cdot \frac{\partial}{\partial \underline{v}} f_0(\underline{v},t) = 0. \quad (35)$$

Note that if S and the collision operators are set to zero, these quasilinear equations reduce to the familiar form for the collisionless diffusion of electrons in a uniform B field for $f_0 \rightarrow f_0(v_\perp^2, v_\parallel, t)$. See *Kennel and Engelmann* [1966].

It should be noted that the term that describes the Cherenkov emission of waves by the electrons [*Davidson*, 1972] is missing from the right-hand side of (33). This term is second order in the wave field and does not appear in our analysis.

8.3. *Approximate Collisional Quasilinear Equations*

The formulas given in (32) through (35) describe the slow time evolution of the electron distribution function when electron-neutral particle and charged-particle collisions operate on the same time scale as the collisionless growth rate. The equations have the expected conservation properties. In Section 3, we discussed two approximate forms of the electron - N_2 collision integral that could be solved explicitly in order to analyze the effect of collisions on the stability of plasma waves in the ionosphere at low altitudes. In this section, we give approximate quasilinear equations which should be adequate for the analysis of the time evolution and saturation of the resonant, upper hybrid instability. We make the following approximations. In the equation for f_0, we retain the complete expressions for the collision integrals. In order to evaluate ε^V and $\underline{\underline{T}}^V{}_s$ at low altitudes, we neglect the effect of Coulomb collision and use the BGK approximation for electron - N_2 collisions.

Approximately decoupling (1) and (2) into a thermal and a suprathermal part as we did in the BGK analysis, we obtain the quasilinear equations f_{0t} and f_{0s}. Now we wish to examine the form of the wave-particle term in the quasilinear equation for f_{0s}. In the expression for the tensor operator for the suprathermal electrons, $\underline{\underline{T}}^V{}_s$, we may use the dispersion relation to eliminate the explicit appearance of N_s. The quasilinear equation for f_{0s} becomes

$$\frac{\partial}{\partial t} f_{0s}\left(v_\perp{}^2, v_\parallel, t\right) = S_s - \mathcal{R}_s\left(f_{0s}\right)$$

$$+\mathcal{L}_{ens}\left(f_{0s}\right) + \mathcal{L}_{eis}\left(f_{0s}\right) + \mathcal{L}_{ees}\left(f_{0s}\right)$$

$$+\frac{\partial}{\partial \underline{v}} \cdot \underline{\underline{D}}^V{}_s\left(v_\perp, v_\parallel, t\right) \cdot \frac{\partial}{\partial \underline{v}} f_{0s}\left(v_\perp^2, v_\parallel, t\right)$$

$$-\frac{\partial}{\partial \underline{v}} \cdot \underline{F}^V{}_s\left(v_\perp, v_\parallel, t\right) f_{0s}\left(v_\perp^2, v_\parallel, t\right). \quad (36)$$

The expressions for the last two terms in (36) are

$$\frac{\partial}{\partial \underline{v}} \cdot \underline{\underline{D}}^V{}_s \cdot \frac{\partial}{\partial \underline{v}} f_{0s} = \left(\frac{8\pi q^2}{m^2}\right)$$

$$\times \sum_{n=-\infty}^{\infty} \int d\underline{k} \frac{\varepsilon_k(t)}{k^2} \left(\frac{n\Omega}{v_\perp} \frac{\partial}{\partial v_\perp} + k_\parallel \frac{\partial}{\partial v_\parallel}\right)$$

$$\times \frac{iJ_n^2\left(k_\perp v_\perp / \Omega\right)}{\left(\omega_k + i\Gamma_k + i\nu_s - n\Omega - k_\parallel v_\parallel\right)}$$

$$\times\left(\frac{n\Omega}{v_\perp}\frac{\partial}{\partial v_\perp}+k_\parallel\frac{\partial}{\partial v_\parallel}\right)f_{0s}\left(v_\perp^2,v_\parallel,t\right), \quad (37)$$

$$\frac{\partial}{\partial \underline{v}}\cdot \underline{F}^V{}_s f_{0s}=\left(\frac{8\pi q^2}{m^2}\right)\sum_{n=-\infty}^{\infty}\int d\underline{k}\frac{\mathcal{E}_k(t)}{k^2}$$

$$\times v_s\left(\frac{n_o}{n_{os}}\right)\left(\frac{k^2}{\omega_p^2}+\frac{N_t}{D_t}\right)$$

$$\times\left(\frac{n\Omega}{v_\perp}\frac{\partial}{\partial v_\perp}+k_\parallel\frac{\partial}{\partial v_\parallel}\right)\frac{iJ_n^2\left(k_\perp v_\perp/\Omega\right)}{\omega_k+i\Gamma_k+iv_s-n\Omega-k_\parallel v_\parallel}$$

$$\times f_{0s}\left(v_\perp^2,v_\parallel,t\right). \quad (38)$$

We see that when the BGK operator is used to approximate electron - N_2 collisions, the wave-particle interaction term has the form of a diffusion part and a friction part. The diffusion part spreads out the electrons in the resonance region of velocity space and the friction part decelerates the electrons in the resonance region of velocity space. At this point, we caution the reader that a different form for the approximate collision operator may produce a different form for $\underline{\underline{T}}^V{}_s$.

8.4. *H-Theorem Analysis of the Quasilinear Equation for* f_{0s}

In this section, we obtain an approximate H-theorem for the collisional problem which may be used to determine the form of the saturated state as $t \to \infty$. In *Jasperse* [1976, 1977], we found a steady-state solution to (32) without the wave particle interaction term by making the isotropic approximation, $f_0 \to f_0(E)$, and by using numerical methods. This led to a thermal, near-Maxwellian part smoothly joined to a suprathermal, non-Maxwellian part, $f_0(E) = f_{0t}(E) + f_{0s}(E)$. In the quasilinear equation for f_{0s}, we make the isotropic approximation and average (36) over the angle θ in velocity space to obtain an equation for $f_{0s}(E,t)$. We obtain

$$\frac{\partial}{\partial t}f_{0s}(E,t)$$

$$-\left[S_s-\mathcal{R}_s\left(f_{0s}\right)+\mathcal{L}_{ens}\left(f_{0s}\right)+\mathcal{L}_{eis}\left(f_{0s}\right)+\mathcal{L}_{ees}\left(f_{0s}\right)\right]$$

$$=\frac{\partial}{\partial E}D^V{}_{sr}(E,t)\frac{\partial}{\partial E}f_{0s}(E,t)$$

$$-\frac{\partial}{\partial E}F^V{}_{sr}(E,t)f_{0s}(E,t), \quad (39)$$

where $D^V{}_{sr}$ and $F^V{}_{sr}$ are the averaged diffusion and friction coefficients in the resonant region.

Since $\mathcal{R}_s$, $\mathcal{L}_{eis}$ and $\mathcal{L}_{ees}$ are small in the lower ionosphere, we see that the source and electron - N_2 vibrational excitation collision terms are attempting to create the bump and the diffusion and friction terms are attempting to diffuse and decelerate the bump. Now assume that the wave-particle-induced rearrangements in f_{0s} produce an effect on the terms in the brackets on the left-hand side of (39) which is small compared to each of the other two terms on the right-hand side of (39). This assumption can be checked a posteriori when a numerical solution of the problem has been found. We now multiply (39) by f_{0s} and integrate with respect to E in the resonant energy region to obtain as $t \to \infty$.

$$0 \cong -\int dE\, D^V{}_{sr}(E,\infty)\left[\frac{\partial}{\partial E}f_{0s}(E,\infty)\right]^2$$

$$+\int dE\left[\frac{\partial}{\partial E}f_{0s}(E,\infty)\right]F^V{}_{sr}(E,\infty)f_{0s}(E,\infty). \quad (40)$$

We see that the meaning of the above assumption is that in the resonant energy region, the time-asymptotic solution is determined primarily by a balancing of the diffusion and friction processes. We also note that a steady-state solution where $\mathcal{E}_k\ (t \to \infty) \to 0$ can be ruled out because the source term would tend to recreate the bump and start the waves growing again. Note also that the volume in velocity space of a shell-like resonance region is finite. Therefore, since $D^V{}_{sr}$ and $F^V{}_{sr}$ are positive functions, the steady-state solution with wave-particle interactions must be one where $\partial f_{0s}/\partial E$ in the resonant region is positive and is reduced from its value found from a solution of the zero-order equation. Based on the H-theorem analysis of the approximate equation for f_{0s}, we propose the following

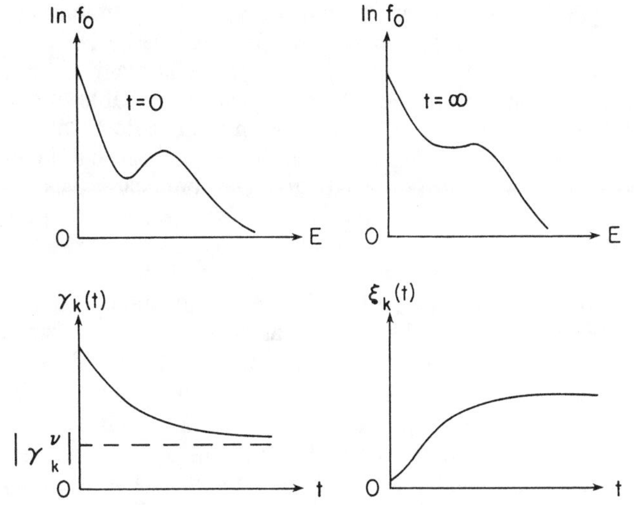

Fig. 10. Schematic representation of the time-asymptotic behavior of the electron distribution as a function of energy and the field fluctuation energy density as a function of time.

time-asymptotic solution. The resulting electron energy distribution and the spectral energy density of the field fluctuations in the time-asymptotic state are shown schematically in Figure 10. Due to the depletion of the suprathermal bump, the collisionless growth rate decreases with time from its initial value and this continues until it reaches the value equal to the magnitude of the collisional damping rate denoted by $|\gamma_k^\nu|$. The waves stop growing and the plasma reaches a marginally stable state. As the plasma approaches the quasilinear saturated state, nonlinear wave-particle and wave-wave interactions may become important and those processes may lead to additional changes in the electron energy distribution and the field fluctuation energy density.

9. PLASMA SIMULATIONS

In Section 8, we derived the general collisional quasilinear equations for the problem and used them to study the time evolution of the resonant upper hybrid instability. In this section, we wish to study the time evolution of the nonresonant electron cyclotron instability by performing particle-in-cell plasma simulations. The advantage of the particle simulation model for plasma phenomena is that it allows the entire range of plasma processes, from linear instability to quasilinear modification to fully nonlinear effects, to be described. To complete the simulation model, we included the appropriate collisional processes to allow the photoelectron instability in the lower ionosphere to be described. These processes include the photoelectron source, a loss process corresponding to recombination, a simple Fokker-Planck model to describe the effects of Coulomb collisions, and most importantly, an energy-dependent energy-loss term to describe the collisional excitation of molecular vibrational states, which carves out the unstable dip in the electron flux. The simulation was 1d-2v in the plane perpendicular to the magnetic field, so that the unstable feature is ring-like rather than shell-like in velocity space. This modifies the situation merely in a quantitative way, without altering the qualitative outcome, and makes the simulation practical to perform. We will present the results from two simulations here, using two different values of the cyclotron to plasma frequency ratio. In the first simulation, the plasma is stable, and the effects of the collisional processes are highlighted. In the second simulation, the upper hybrid frequency is near a multiple of the electron cyclotron frequency, and the plasma is unstable due to the ring-like velocity distribution created by collisional processes; the effects of the instability and the resulting electron acceleration will be demonstrated. Through the variation of the cyclotron to plasma frequency ratio, these two simulations together illustrate the strong altitude dependence of the plasma instability and the resulting electron heating.

The results of the simulation of the stable plasma are presented in the top panel of Figure 11. This figure is an overplot at regular time intervals of the electron velocity distribution function as a function of energy. (The energy and distribution function are in arbitrary units; the initial thermal energy in these energy units is 0.000288.) The simulation begins with only the thermal population, but as time progresses, photoionization produces energetic electrons and vibrational excitation causes the electrons to lose energy, carving out a velocity distribution with a deep dip around 0.01 energy units. The structure of the distribution function around the dip is solely dependent on the energy dependence of the collisional rates. A steady state is eventually reached in which the rates of photoproduction and collisional excitation balance at all energies. Because the instability cannot be excited, the dip in the electron velocity distribution remains deep and sharply defined, and the plasma waves remain at a low level.

The second simulation results show the linear instability (with a growth rate close to the predicted one) followed by an approach to a saturated state. The evolution of the electron distribution function in the simulation is shown in the lower part of Figure 11 and is created in the same way as the top panel. The simulation starts with the Maxwellian thermal population alone, but the photoionization process builds up the flux at higher energies at a steady rate. When the threshold of the strength of the energetic-electron flux is exceeded, the linear instability begins to act. As the plasma waves grow to significant amplitude, they begin to affect the velocities of the energetic electrons. The trend predicted by quasilinear theory is to reduce the growth rate of the instability by filling in the dip region in energy with electrons from the population produced by photoionization at higher energies. Because the vibrational-excitation cross section peaks in this region, however, this stabilization is

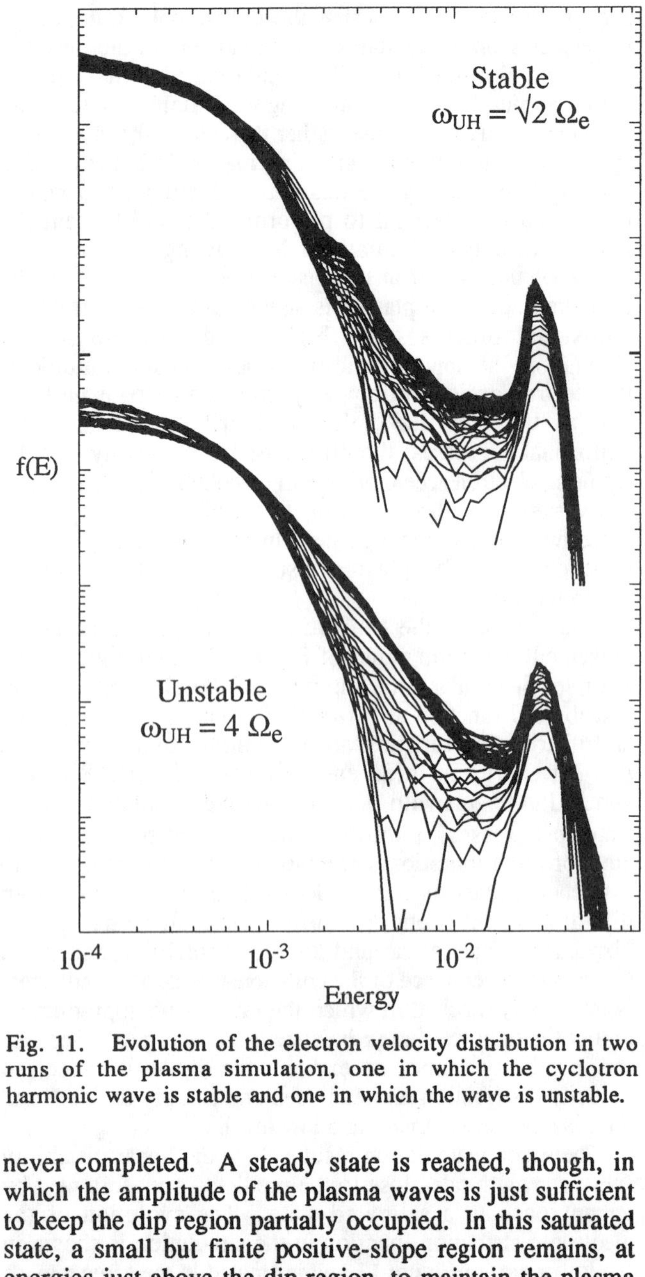

Fig. 11. Evolution of the electron velocity distribution in two runs of the plasma simulation, one in which the cyclotron harmonic wave is stable and one in which the wave is unstable.

never completed. A steady state is reached, though, in which the amplitude of the plasma waves is just sufficient to keep the dip region partially occupied. In this saturated state, a small but finite positive-slope region remains, at energies just above the dip region, to maintain the plasma waves at the required amplitude. (Note the overshoot of the photoelectron population at the beginning of the simulation, when the plasma waves have not yet reached their steady-state amplitude.) At lower energies in the thermal region, there is some evidence for bulk heating. The amount of bulk heating, however, is largely controlled by the strength of the vibrational excitation cooling of the electrons: any thermal electrons accelerated by the plasma waves find themselves in the dip region, where vibrational excitation reduces their energy. The steady state here represents a balance between the rates of these heating and cooling processes. As the result of nonlinear processes, both temporal (frequency) and spatial (wavenumber) harmonics of the linearly unstable waves are observed to be larger in amplitude than waves without such powerful relatives. (Of course, the amplitude of the linearly unstable mode dominates its harmonics.) Without more detailed analysis, however, we cannot report any effects on the particle fluxes that can be attributed especially to these nonlinear waves.

The general conclusion from the simulation is that the predictions of the quasilinear theory hold true in a qualitative sense. In the saturated state, the dip region in energy is filled in, but only partially filled in. To maintain a steady state in the face of the continuous generation of energetic photoelectrons, a small positive slope is established in the electron distribution function to support the plasma waves that control the energetic electron population. However, we must caution the reader that the collisional quasilinear equations used in Section 8 to study the resonant instability may only apply to the nonresonant instability in a qualitative sense, since at a later stage of the instability the adiabatic approximation may not be valid for the nonresonant situation.

10. DISCUSSION

We have shown that the suprathermal, shell-like bump in the electron energy distribution near 5 eV, which should be present at all latitudes, is a source of free energy for the excitation of Bernstein modes in the approximately 100-130 km altitude region and upper hybrid modes in the approximately 125-170 km region. The Bernstein modes are excited in narrow altitude layers of a few kilometers in thickness, while the upper hybrid modes are excited within a broad altitude layer with the high altitude cut-off above 170 km. Quasilinear estimates and plasma simulations suggest that a significant amount of flattening of the electron energy distribution in the 2-5 eV energy range occurs, which is consistent with both the auroral and photoelectron particle data.

We also recall that the rocket data of *McMahon and Heroux* [1978, 1983] indicate the possible existence of two different types of instabilities - one in the lower E region and the other in the upper E and lower F regions - much like our theoretical prediction. In addition, the rocket data of *Kelley and Earle* [1988] give direct evidence of the excitation of upper hybrid waves in the 130-170 km altitude region of the auroral zone with the waves disappearing above about 170 km. This is in complete agreement with our theoretical predictions. The observed disappearance of the upper hybrid waves above 170 km cannot be explained if the high energy (few keV) precipitating electrons are considered as the source of free energy for the excitation of the waves. Finally, the rocket data of *Oyama and Hirao* [1979] indicate the presence of narrow altitude layers of turbulence in the lower E-region, much like the Bernstein mode instabilities predicted by our theory. All these agreements between theory and the observations compel us to believe that the high frequency plasma instabilities excited by the suprathermal bump in the photo- and auroral electron energy distribution functions near 5 eV provide a theoretical explanation for the rocket observations. What we cannot explain yet is the observed [*Oyama and Hirao*, 1979] intense electron heating within the narrow altitude layers, although our model simulation indicates a modest amount of thermal electron heating. We also cannot explain the generation of the enhanced tail in the thermal electron energy distribution observed within the heated layers by *Oyama et al.* [1983]. Furthermore, the incoherent backscatter radar data of *Valladares et al.* [1988], which indicate excitation of plasma waves near upper hybrid frequency in the upper E region of the auroral zone, cannot be explained by our theory as these observed waves exist with $k_{||} >> k_{\perp}$. These data suggest that more theoretical work is necessary. At the same time, rocket experiments with simultaneous measurements of particle energy distribution and plasma wave spectral characteristics (ω-$\underline{k}$ spectra) are also necessary.

It has not escaped our attention that the one and sometimes two narrow altitude layers of the "enhanced aurora" could be related to our prediction that electron cyclotron harmonics are unstable in the lower auroral E region. These narrow altitude layers have been observed in the nighttime auroral region because of the enhanced optical emissions which they emit. For a discussion of the "enhanced aurora" (originally called the pulsating aurora and later renamed), see *Stenbaek-Nielsen and Hallinan* [1979], *Hallinan et al.* [1985], and *Whalund and Opgenoorth* [1989].

Acknowledgments. This work was primarily supported by the Air Force Office of Scientific Research Window on Europe Program. One of us (JRJ) wishes to acknowledge extensive discussions with Raymond Pottelette throughout the course of the work. One of us (JRJ) would also like to acknowledge Jean Jacque Berthelier and the Centre de Recherche en Physique de l'Environnement for supporting his activities while visiting the laboratory. The work was also supported by Phillips Laboratory contracts at Boston College and the Massachusetts Institute of Technology.

REFERENCES

Ashihara O., and K. Takayanagi, Velocity distribution of ionospheric low-energy electrons, *Planet. Space Sci.*, *22*, 1201, 1974.

Basu, B., T. Chang, and J. R. Jasperse, Electrostatic plasma instabilities in the daytime ionospheric E-region, in *Physics of Space Plasmas*, Ed. by T. Chang, B. Coppi, and J.R. Jasperse, SPI Conference Proceedings and Reprint Series (Scientific Publishers, Cambridge, MA, 1981), vol. 4, p. 141.

Basu, B., T. Chang, and J. R. Jasperse, Electrostatic plasma instabilities in the daytime lower ionosphere, *Geophys. Res. Lett.*, *9*, 68, 1982.

Bernstein, I. B., Waves in a plasma in a magnetic field, *Phys. Rev.*, *109*, 10, 1958.

Bhatnagar, P. L., E. P. Gross, and M. Krook, A model for collision processes in gases. I. Small amplitude processes in charged and neutral one-component systems, *Phys. Rev.*, *94*, 511, 1954.

Bloomberg, H. W., Effect of plasma instability on F-region photoelectron distributions, *J. Geophys. Res.*, *80*, 2851, 1975.

Davidson, R. C., *Methods in Nonlinear Plasma Theory*, Academic Press, New York, 1972.

Dreicer, H., Electron velocity distributions in a partially ionized gas, *Phys. Rev.*, *117*, 343, 1960.

Drummond, W. E. and D. Pines, Nonlinear stability of plasma oscillations, *Nucl. Fusion Suppl.*, *3*, 1049, 1962.

Frieman, E. A. and P. R. Rutherford, Kinetic theory of a weakly unstable plasma, *Ann. Phys.*, *28*, 134, 1964.

Gefan, G. D., A. A. Trukhan, and G. V. Khazanov, Ionospheric plasma instability effects in a diffuse auroral zone, *Ann. Geophys.*, *3*, 141, 1985.

Hallinan, T. J., H. C. Stenbaek-Nielsen, and C. S. Deehr, Enhanced aurora, *J. Geophys. Res.*, *90*, 8461, 1985.

Ivanov, V. B., A. A. Trukhan, and G. V. Khazanov, Stability of the midlatitude ionospheric plasma when photelectrons are present in it, *Radiofizika*, *23*, 143, 1980.

Ivanov, V. B., A. A. Trukhan, and G. V. Khazanov, Midlatitude F-region ionosphere plasma stability in the presence of photoelectrons, *Ann. Geophys.*, *38*, 33, 1982.

Jasperse, J. R., Boltzmann-Fokker-Planck model for the electron distribution function in the earth's ionosphere, *Planet. Space Sci.*, *24*, 33, 1976.

Jasperse, J. R., Electron distribution function and ion concentrations in the earth's lower ionosphere from Boltzmann-Fokker-Planck theory, *Planet. Space Sci.*, *25*, 743, 1977.

Jasperse, J. R., The photoelectron distribution function in the terrestrial ionosphere, in *Physics of Space Plasmas*, Ed. by T. Chang, B. Coppi, and J.R. Jasperse, SPI Conference Proceedings and Reprint Series (Scientific Publishers, Cambridge, MA, 1981), vol. 4, p. 53.

Jasperse, J. R., B. Basu, and T. S. Chang, Electrostatic instabilities and quasilinear effects in the terrestrial ionosphere, in *Proceeding of International* Conference on *Plasma Physics* (Goteborg, Sweden, 1982), vol. 1, p. 68.

Jasperse, J. R., A propagator expansion method for solving linearized plasma kinetic equations with collisions, *Phys. Lett.*, *106A*, 379, 1984.

Jasperse, J. R. and B. Basu, The dielectric function for the Balescu-Lenard-Poisson kinetic equations, *Phys. Fluids*, *29*, 110, 1986.

Kaufmann, R. L., P. B. Dusenbery, and B. J. Thomas, Stability of the auroral plasma: parallel and perpendicular propagation of electrostatic waves, *J. Geophys. Res.*, *83*, 5663, 1978.

Kaufmann, R. L., Electrostatic wave growth: secondary peaks in a measured auroral electron distribution function, *J. Geophys. Res.*, *85*, 1713, 1980.

Kelley, M. C., and G. D. Earle, Upper hybrid and Langmuir turbulence in the auroral E region, *J. Geophys. Res.*, *93*, 1993, 1988.

Kennel, C. F. and F. Engelmann, Velocity space diffusion from weak plasma turbulence in a magnetic field, *Phys. Fluids*, *9*, 2377, 1966.

Khazanov, G. V., M. A. Koen, and G. S. Kudryashev, Kinetics of electrons at altitudes of 100-260 km, *Radiofizika*, *21*, 646, 1978.
Khazanov, G. V., Kinetics of electron component of upper atmosphere plasma, *Nauka*, Moscow, 1979.
Krall, N. A. and A. W. Trivelpiece, *Principles of Plasma Physics*, San Francisco Press, San Francisco, 1986.
McFadden, J. P., C. W. Carlson, and M. H. Boehm, High-frequency waves generated by auroral electrons, *J. Geophys. Res.*, *91*, 12079, 1986.
McMahon, W. J., and L. Heroux, Rocket measurements of thermospheric photoelectron energy spectra, *J. Geophys. Res.*, *83*, 1390, 1978.
McMahon, W. J., and L. Heroux, The 2- to 5- eV energy spectra of thermospheric photoelectrons: measurements in apparent conflict with theory, *J. Geophys. Res.*, *88*, 9249, 1983.
Oyama, K. I., and K. Hirao, Anomalous heating of the thermal electrons near the focus of the Sq current vortex (Sq focus anomaly), *J. Geomag. Geoelectr.*, *31*, 11, 1979.
Oyama, K. I., K. Hirao, P. M. Banks, and P. R. Williamson, Nonthermal components of low energy electrons in the ionospheric E and F region, *J. Geomag. Geoelectr.*, *35*, 185, 1983.
Pottelette, R., Measurement of the cross spectrum of random linear electrostatic waves in an isotropic Maxwellian plasma, *Phys. Fluids*, *22*, 534, 1979.
Pottelette, R., M. Hamelin, J. M. Illiano, and B. Lembege, Interpretation of the fine structure of electrostatic waves excited in space, *Phys. Fluids*, *24*, 1517, 1981.
Pottelette, R., and L. R. O. Storey, Active and passive methods for the study of non-equilibrium plasmas using electrostatic waves, *J. Plasma Phys.*, *25*, 323, 1981.
Pottelette, R., and J. M. Illiano, Observation of weak HF electrostatic turbulence in the auroral ionosphere, *J. Geophys. Res.* *87*, 5151, 1982.
Rogister, A. and C. Oberman, On the kinetic theory of stable and weakly unstable plasma. Part I., *J. Plasma Phys.*, *2*, 33, 1968.
Sharp, W. E., M. H. Rees, and A. I. Stewart, Coordinated rocket and satellite measurements of an auroral event 2. The rocket observations and analysis, *J. Geophys. Res.*, *84*, 1977, 1979.
Stenbaek-Nielsen, H. C., and T. J. Hallinan, Pulsating auroras: Evidence for noncollisional thermalization of precipitating electrons, *J. Geophys. Res.*, *84*, 3257, 1979.
Strickland, D. J., D. L. Book, T. P. Coffey, and J. A. Fedder, Transport equations techniques for the deposition of auroral electrons, *J. Geophys. Res.*, *81*, 2755, 1976.
Valladares, C. E., M. C. Kelley, and J. F. Vickery, Plasma line observations in the auroral oval, *J. Geophys. Res.*, *93*, 1997, 1988.
Vedenov, A. A., E. P. Velikov, and R. Z. Sagdeev, Nonlinear oscillations of rarefied plasmas, *Nucl. Fusion*, *1*, 82, 1961.
Vedenov, A. A., E. P. Velikov, and R. Z. Sagdeev, Quasilinear theory of oscillating plasmas, *Nucl. Fusion Suppl.*, *2*, 465, 1962.
Wahlund, J. E. and H. J. Opgenoorth, Observations of thin auroral ionization layers by EISCAT in connection with pulsating aurora, *J. Geophys. Res.*, *94*, 17223, 1989.

J. R. Jasperse, Centre de Recherche en Physique de l'Environnement, 4 Avenue de Neptune, 94107 Saint-Maur, France.

B. Basu and J. M. Retterer, Phillips Laboratory, 29 Randolph Road, Hanscom AFB, Bedford, MA 01731.

D. T. Decker, Institute for Space Research, Boston College, Chestnut Hill, MA 02167.

T. S. Chang, Center for Space Research, Massachusettes Institute of Technology, Cambridge, MA 02139.

---

[1]Permanent address: Phillips Laboratory, 29 Randolph Road, Hanscom AFB, Bedford, MA 01731.

Laboratory Experiments on Particle Acceleration Processes Associated With Parallel Electric Fields

Z. Jin, J. Hamila, R. C. Allen, S. Meassick, and C. Chan

Department of Electrical and Computer Engineering, Northeastern University, Boston, Massachusetts

We have initiated a laboratory experiment to study the dynamics of charged particle acceleration across a quasi-static parallel electric field in a magnetized plasma column. The parallel electric field is set up by allowing two different sources of plasmas to interpenetrate each other. Since ions of different masses gain the same energy across the potential drop, the lighter ions will acquire higher velocities and the velocity difference between the two ion species of different masses can excite ion acoustic waves via the two-stream instability. In our experiment, two ion species (He^+ and Ar^+) are used. By actively launching waves near the unstable frequencies at the high potential side of the parallel electric field, the kinetics of the instabilities can be diagnosed. The results are quantitatively explained with a kinetic dispersion relation for the ion-ion two-stream instability.

1. INTRODUCTION

There are a number of phenomena that are often associated with the presence of parallel electric fields along the auroral geomagnetic field lines. They are the so-called "inverted" electron precipitation, the small scale ion acoustic double layers and solitary waves, and the heating of upflowing ion beams that are accelerated upward by the parallel electric field from the ionosphere.

Upflowing beams of ionospheric ions (e.g., H^+, He^+, and O^+) have been observed on evening auroral field lines above several thousand kilometers (4000-5000 km) altitude [*Bergmann*, 1989; *Ghielmetti et al.*, 1978; *Lundin et al.*, 1987]. This ion outflow is believed to be due to the presence of a quasi-static electric field parallel to the geomagnetic field. Satellite observations (DE, S3-3, etc.) show that these ion beams undergo significant heating in both parallel and perpendicular directions to the geomagnetic field as they traverse the acceleration region [*Kaufmann et al.*, 1986]. One proposed mechanism for the heating of the auroral ion beams suggests that it is due to the ion-ion two-stream instability [*Bergmann and Lotko*, 1986; *Dusenbery et al.*, 1988; *Kaufmann et al.*, 1986]. Ions of different masses gain the same energy across a potential drop, but they acquire a different velocity parallel to the magnetic field line i.e., lighter ions having the higher velocities. The velocity difference between two ion species of different masses (e.g., H^+ and O^+) can excite ion-acoustic waves via the two-stream instability mechanism.

Linear analysis of the ion-ion two-stream instability shows that the properties of excited waves as well as their growth rates depend on the relative drift speed of the ions (Δv), the density and mass ratios of the ion beams, and the temperature of the electrons (T_e) [*Bergmann and Lotko*, 1986; *Dusenbery and Martin*, 1987; *Dusenbery et al.*, 1988]. For the parallel propagation, there are two velocity limits on the H^+-O^+ instability: an upper limit determined by their fluid properties and a lower limit determined by their kinetic properties. When $T_e/T_H < 3.5$, where T_H is the temperature of H^+, the two-stream instability is kinetically suppressed for all Δv. When $T_e/T_H > 3.5$, the two-stream instability may be excited [*Bergmann and Lotko*, 1986].

A laboratory experiment has been initiated to study the properties of ion-ion two-stream instability in the presence of a parallel electric field. In this paper, we report results where the plasma is essentially unmagnetized. The experimental apparatus is introduced in section 2. Section 3

Space Plasmas: Coupling Between Small
and Medium Scale Processes
Geophysical Monograph 86

explains the stability properties of ion-ion interactions for He^+ and Ar^+ ions. The experimental results are contained in section 4 and the conclusion is given in section 5.

2. EXPERIMENTAL APPARATUS

The experiment was performed in a double-layer plasma chamber, as shown in Figure 1. The dimensions of the target chamber are 100cm in length and 50cm in diameter with an extension on the low potential side which is 30cm in length and 45cm in diameter.

The plasma was produced by the ionizing electrons thermionically emitted from tungsten filaments at both the high and low potential ends. A potential gradient can be set up by running two sources at different plasma potentials. The typical potential profile obtained in our experiment is shown in Figure 2. A steady state parallel electric field can be set up in the target chamber that is readily reproducible. Ions in the high potential side will be accelerated by this electric field and form the beams.

An emissive probe was used to measure the plasma potential. Both the electron temperature and density were measured with a Langmuir probe. The typical plasma parameters used in the experiments were as follows: electron temperature T_e=3~5eV, plasma ion temperature $T_i \approx T_e/10$, electron density $n_e=10^8\sim10^9 cm^{-3}$, total neutral gas (Helium and Argon) pressure $p\approx10^{-4}$Torr, and base pressure $p_b\approx10^{-6}$Torr.

External waves can be launched with a grid (5cm in diameter and 40cm away from the high potential end) and are detected by a movable Langmuir probe.

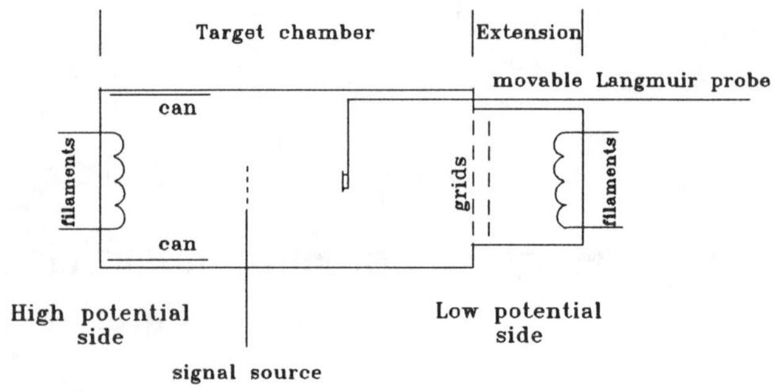

Fig. 1. Schematic of the double-layer plasma chamber.

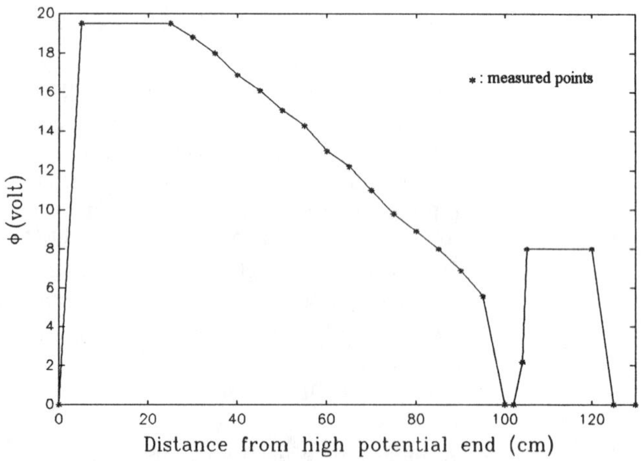

Fig. 2. The typical potential profile in the double-layer plasma chamber.

3. STABILITY PROPERTIES OF ION-ION INTERACTIONS

To elucidate the basic features of ion-ion interaction in parallel electric fields, a simple particle distribution model similar to that of *Bergmann and Lotko* [1986] is used. The particles are assumed to be Maxwellian distributions and are composed of non-drifting background electrons with mass m_e, density n_e and temperature T_e and two ion populations, the light (fast) and the heavy (slow) ion species (He^+ and Ar^+ in our experiment), with mass m_f and m_s, density n_f and n_s, and temperature T_f and T_s. The existing parallel electric field will accelerate both ion species to the same energy and the drift speeds, v_f and v_s, are related to one another with $v_s=(m_f/m_s)^{1/2}v_f$, corresponding to $\frac{1}{2}m_fv_f^2=\frac{1}{2}m_sv_s^2$. So the difference between the ion drift velocities is $\Delta v=v_f-v_s=[1-(m_f/m_s)^{1/2}]v_f$. Since the accelerated electrons are much fewer than the background electrons in our experiment, the effect of them will be neglected here.

The assumption of the model above leads to the following kinetic dispersion equation for the parallel propagating waves:

$$1-\frac{1}{2k^2\lambda_e^2}Z'(\frac{\omega}{ka_e})-\frac{1}{2k^2\lambda_f^2}Z'(\frac{\omega-kv_f}{ka_f})-\frac{1}{2k^2\lambda_s^2}Z'(\frac{\omega-kv_s}{ka_s})=0 \tag{1}$$

where perturbations are taken as $e^{i(kx-\omega t)}$ and λ_j, a_j and Z' denote the Debye length and the thermal velocity of the j-th particle (electrons, slow ions and fast ions), and the derivative of the plasma dispersion function [*Fried and Conte*, 1961] respectively. In this paper, we consider ω to be a real angular frequency and $k=k_r+ik_i$ to be complex.

A qualitative discussion of the stability properties of the ion-ion interaction can also be described by a simple dispersion relation derived from the fluid equations for cold plasmas:

$$1+\frac{1}{k^2\lambda_e^2}-\frac{\omega_f^2}{(\omega-kv_f)^2}-\frac{\omega_s^2}{(\omega-kv_s)^2}=0 \qquad (2)$$

where ω_j is the plasma angular frequency of the j-th ion species. Figure 3 shows the dispersion relation for He^+ and Ar^+ calculated from equation (2). With two ion beams present there exist four normal modes, the fast and slow light ion (He^+), and the fast and slow heavy ion (Ar^+) modes. If the beam velocities v_f and v_s are large enough, the phase velocities of the four modes are well separated. While at low beam velocities, the slow light ion mode and the fast heavy ion mode couple to form a pair of complex conjugate modes. One of the modes will have a negative k_i which will correspond to the wave growth. This ion-ion two-stream instability criterion for the parallel propagating waves is given by:

$$\frac{v_f}{c_f}\le\frac{(n_f/n_e)^{1/2}(1+\alpha^{1/3})^{3/2}}{[1-(m_f/m_s)^{1/2}](1+k^2\lambda_e^2)^{1/2}} \qquad (3)$$

where

$$\alpha\equiv\frac{n_s}{n_f}\frac{m_f}{m_s}$$

When the fast ion beam drifts faster than the upper limit speed given in equation (3), only obliquely propagating waves may be unstable.

When the kinetic dispersion equation (1) is computed, it is shown that a lower limit exists due to the kinetic properties while the upper limit can be still approximated by (3) [*Bergmann and Lotko*, 1986]. The results computed from equation (1) will be shown in a later section together with the experimental results.

4. EXPERIMENTAL RESULTS

Argon and Helium, with a mass ratio of 10, were used in our experiment. With a plasma potential gradient as shown in Figure 2, the He^+ and Ar^+ ions will be accelerated by the parallel electric field and have drift velocities $v_s=(m_f/m_s)^{1/2}v_f=(2e\Delta\phi/m_s)^{1/2}$. With an He/Ar density ratio of 9, wave growth in the frequency range of 400–500kHz is detected by a Langmuir probe connected to a spectrum analyzer. Since the wave is not easy to be followed by a Lock-in Amplifier, external signals are consequently launched into the plasma from a distance of 40cm from the high potential end to make the measurements. The propagation of these externally launched waves are shown in Figure 4. At frequencies 400kHz, 425kHz, 450kHz, and 500kHz we observe that the waves grow within a range of distance x≈10~20cm from the signal source (The obliquely propagating waves may still be unstable beyond x=20cm. However, since we measured the waves only along the ion beam flow, we did not observe this instability.). This observation can be qualitatively explained by the theories in section 3. The instability occurs within its lower and upper velocity limits; in our experiment these limits correspond to limits of the range of distance x. The reason for no wave growth in lower and higher frequencies is that, in lower frequencies, the damping due to collisions and impurities, such as N^+, O^+, etc., will overcome the growth provided by the coupling since k_i is approximately proportional to the signal frequency when $k\lambda_e<<1$, and in higher frequencies, the damping is mainly due to the impurities and is proportional to the signal frequency which will suppress the

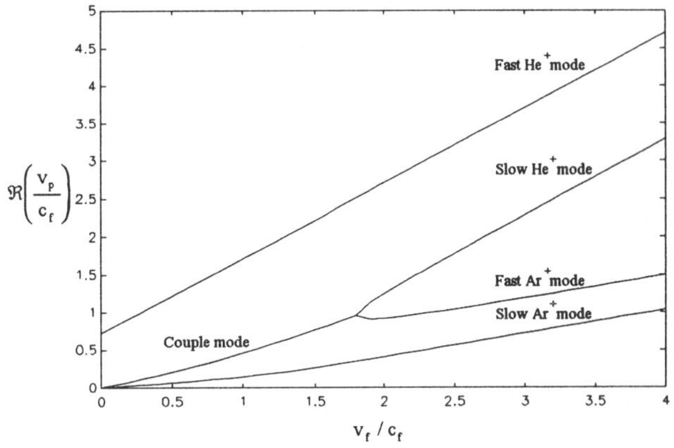

Fig. 3. Solutions to the fluid dispersion relation for He^+ and Ar^+, where $v_p=\omega/k$ and $c_f=(T_e/m_f)^{1/2}$. The parameters are: $n_f/n_s=1$, $T_e=3eV$, $n_e=5\times10^8cm^{-3}$, and f=400kHz.

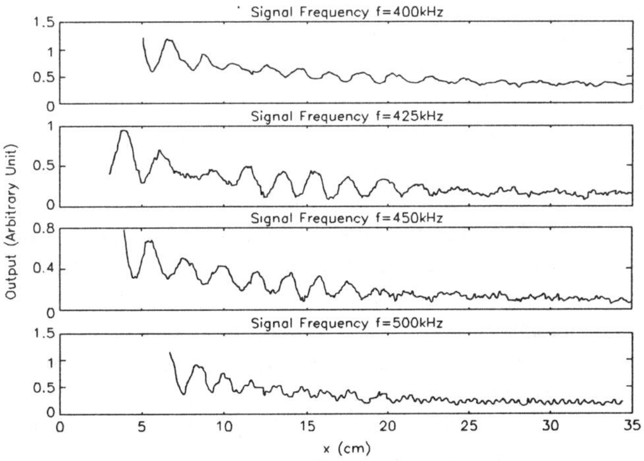

Fig. 4. Propagation of waves launched from a grid 40cm from the high potential end.

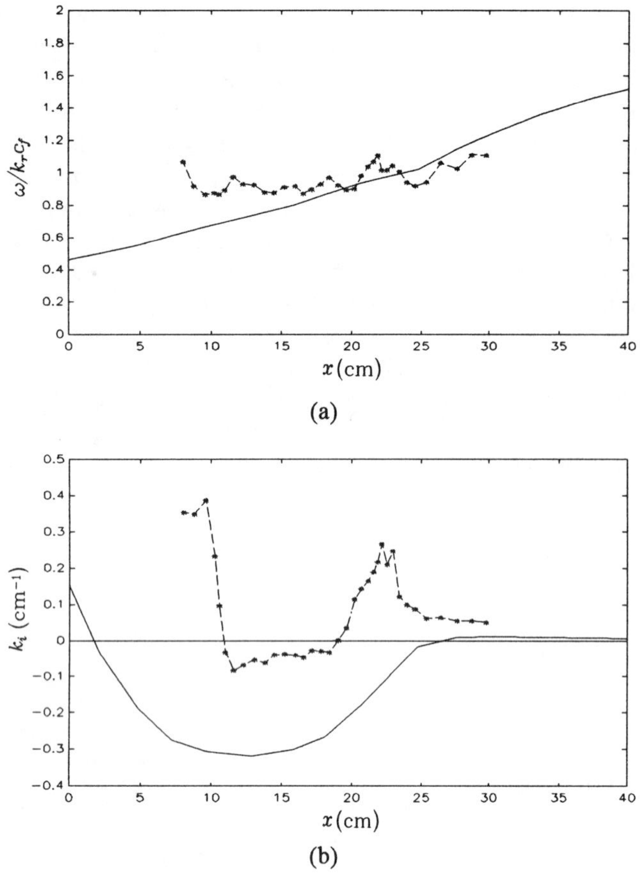

Fig. 5. The experimental and theoretical dispersion relation. (a). the normalized phase velocity $\omega/k_r c_f$ vs. x; (b). k_i vs. x. The parameter used for the calculation are: $n_f/n_s=9$, $T_e/T_f=T_e/T_s=10$, $T_e=3$eV, $n_e=5\times10^8$cm^{-3}, and f=400kHz.

growth. Figure 5 shows the plot of the phase velocities (ω/k_r) and k_i of the waves vs. distance x for signal frequency of f=400kHz (the accuracy for both quantities is about $\pm5\%$). The solid lines in Figure 5 are the theoretical values calculated from equation (1). The damping k_i due to collisions and impurities, which is not considered in equation (1), is usually about 0.2~0.3cm^{-1}. This is just the difference between the experimental and theoretical k_i in Figure 5b. This also explains why we did not observe wave growth at low frequencies in the experiment since k_i will be less than 0.3cm^{-1} when f<300kHz. The larger errors of the phase velocity and k_i values in small x values are caused by the direct coupled signal and the smaller k_i values in x>23cm are due to the computing errors caused by the very small detected signal. The figure shows that the experimental results coincide with the theories.

5. CONCLUSION

A laboratory experiment has been initiated to study the ion-ion two-stream instability in a quasi-static parallel electric field. The parallel electric field is set up by allowing two different sources of plasmas to interpenetrate each other. The velocity difference between the ion species of different masses can excite ion acoustic waves via the two-stream instability. Two ion species, He^+ and Ar^+, have been used in our experiment. The unstable waves generated by the co-streaming He^+ and Ar^+ ions are observed to exist only in a limited range of frequencies and distances. By actively launching waves in the plasmas, the growing waves are observed to agree with a kinetic dispersion relation for the ion-ion two-stream instability.

Acknowledgments. We would like to thank Joe Genevich for building much of the experimental apparatus. The work was supported by NSF grant number ATM-9013241.

REFERENCES

Bergmann, R., H^+-O^+ two-stream interaction on auroral field lines, in Physics of Space Plasma, *SPI Conf. Proc. and Reprint Ser.*, *9*, 229-243, 1989.

Bergmann, R., and W. Lotko, Transition to unstable ion flow in parallel electric fields, *J. Geophys. Res.*, *91*, 7033-7045, 1986.

Dusenbery, P. G., and R. F. Martin, Jr., Generation of broadband turbulence by accelerated auroral ions, *J. Geophys. Res.*, *92*, 3261-3272, 1987.

Dusenbery, P. G., R. F. Martin, Jr., and R. M. Winglee, Ion-ion waves in the auroral region: Wave excitation and ion heating, *J. Geophys. Res.*, *93*, 5655-5664, 1988.

Fried, B. D. and S. D. Conte, *The Plasma Dispersion Function*, Academic, San Diego, Calif., 1961.

Ghielmetti, A. G., R. G. Johnson, R. D. Sharp, and E. G. Shelly, The latitudinal, diurhal, and altitudinal distributions of upward flowing energetic ions of ionospheric origin, *Geophys. Res. Lett.*, *5*, 59, 1978.

Kaufmann, R. L., G. R. Ludlow, H. L. Collin, W. K. Peterson, and J. L. Burch, Interaction of auroral H^+ and O^+ beams, *J. Geophys. Res.*, *91*, 10080-10096, 1986.

Lundin, R., L. Eliasson, B. Hultqvist, and K. Stasiewicz, Plasma energization on auroral field lines as observed by the VIKING space craft, *Geophys. Res. Lett.*, *14*, 443, 1987.

Z. Jin, J. Hamila, R. C. Allen, S. Meassick, and C. Chan, Department of Electrical and Computer Engineering, Northeastern University, Boston, MA 02115.

Nonlinear Wave-Particle Interaction Leading to Chaotic Ion Motion in the Magnetosphere

S. P. KUO AND ANTONY Y. HO

Weber Research Institute, Polytechnic University, Farmingdale, New York

The interaction between a bouncing proton and a kinetic Alfvén wave in the magnetosphere is studied using a simplified model and a Hamiltonian averaged over the gyration period of the protons. Both transverse and parallel components of the wave electric fields are included in the formulation. Significant chaotic motion is observed for relatively small wave amplitudes, which may explain observed effects. The results also show that the chaotic level increases considerably when protons bounce resonantly with the wave.

INTRODUCTION

Geomagnetic micropulsations (magnetic field fluctuations) (Pc3-Pc5) are often observed in the magnetosphere. The magnetic activity as defined by Kp are monitored by the detectors on board of several satellites. For instance, SCATHA, GOES2, GOES3, and GEOS2 are those satellites used to examine any existence of wave structure along the ambient magnetic field line ($L \approx 6.6$) within a latitudinal region (-10°, 10°). Those geomagnetic fluctuations are usually explained by field line resonance with the externally excited Alfvén waves. If their perpendicular wavelengths become comparable to the thermal gyroradius of the background energetic protons (order of KeV kinetic energy), the kinetic effect modifies the nature of the Alfvén wave considerably and a sizable component of electric field parallel to the background magnetic field appears. This parallel field component of the kinetic Alfvén wave (KAW) can significantly affect the motion of energetic protons which are confined by the earth's dipolar magnetic field and execute bouncing motion about the equatorial plane. Similarly, it can also effectively perturb the motion of those energetic electrons trapped in the lower magnetosphere. Such a perturbation on the motion of charged particles provides a possible cause of charged particle precipitation observed in the polar region [*Hultqvist*, 1979] and provide a source mechanism of low-altitude trapped energetic protons in the equatorial region [*Guzik et al.*, 1989]. The KAW can be excited through mode conversion of surface MHD waves at magnetospheric boundaries [*Hasegawa*, 1976] or by bouncing electron beams [*Hasegawa*, 1979].

The nonlinear dynamics of a bouncing proton in the presence of a large amplitude KAW has been studied recently by *Prakash* [1989]. Considering one-dimensional oscillation motion, only the parallel component of the wave electric field was included in the equations of motion for analysis. The threshold field amplitude for the onset of chaos has been determined for a chosen set of initial conditions by surface of section method. It was shown that the threshold value can be marginally exceeded by the KAW in the magnetosphere.

It has been argued by *Liu* [1991] that the commencement of a proton chaos taking hundreds of KAW periods (~10 hr) as shown by the results of Prakash appeared to be a slow process in comparison with the convection time scale. Thus, the electron chaos could be more relevant to the magnetospheric processes. Electrons due to their lighter mass were believed to couple more strongly with the KAW in the region between L=2 and 3. He then examined the effect of a KAW on the motion of a test electron along a dipolar magnetic field. Again, the analysis was limited to one dimensional motion, thus, only the parallel component of the wave electric field was included in the equations of motion. It was shown that the electron motion can become strongly chaotic

Space Plasmas: Coupling Between Small and Medium Scale Processes
Geophysical Monograph 86

for a wide range of conditions. Moreover, the chaos led the kinetic energy of the test electron to exhibit a broad-band frequency spectrum.

Since the transverse field component is about an order of magnitude larger than the parallel field component [*Hasegawa,* 1979] and since the longitudinal and transverse motions of the proton are coupled together, the threshold fields for the onset of chaos determined by the previous analyses may be overestimated. The case of proton in bouncing resonance with the wave was also not included in Prakash's analysis.

In this work, the contributions from both the transverse and parallel components of the wave electric field are included in the total Hamiltonian of the proton [*Ho and Kuo,* 1993; *Ho and Kuo,* 1994]. This Hamiltonian is then averaged over the gyration period of the proton. The longitudinal equations of motion of the proton derived from such an average Hamiltonian keep the effect of transverse field component on the longitudinal motion of the proton.

Governing Equations

The KAW is accompanied by a sizable parallel component of the electric field. Its role in introducing anomalous diffusion of trapped protons into the loss cone and breaking down the adiabatic invariant of magnetic flux surface in order to permit cross field diffusion of trapped ions to low altitude is studied in the following. First, a coordinate system with the z-axis parallel to the geomagnetic field at the equator and with a reference x-axis on the equatorial plane so that the wave vector $\vec{k}$ of the KAW in the region around the equatorial plane is on the x-z plane is introduced, where the y-axis is in the direction of $\hat{z} \times \hat{x}$ and the origin of the coordinate system is fixed at the center of the local mirror field which coincides with the location of the equator. Secondly, the mirror field is modeled by a uniform magnetic field $B_0\hat{z}$ with an imposed restoring force $-m\omega_b^2 \hat{z}$ for those trapped energetic protons, where m and ω_b are the mass and bouncing frequency of protons, respectively. Thus, the feature of the bouncing motion of mirror trapped protons about the equatorial plane is retained and yet the analysis can be simplified.

The interaction of a proton with a large amplitude KAW in the modeled background fields is then studied. The Hamiltonian of the proton is given by

$$H = \frac{1}{2m}(\vec{P} - e\vec{A})^2 + e\phi \qquad (1)$$

where $\vec{P}$ is the canonical momentum of the proton, $\vec{A}$ and ϕ are the vector and scalar potentials of the total field in the system and are given by

$$\begin{aligned} \vec{A} &= [(k_z\hat{x} - k_x\hat{z})/k^2]B_1 \sin(k_x x + k_z z - \omega t) + \hat{y}B_0 x \\ \phi &= -[(\eta k_x + k_z)/k^2]E_1 \sin(k_x x + k_z z - \omega t) + \\ &\quad (1/2)(m/e)\omega_b^2 z^2 \end{aligned} \qquad (2)$$

The two first terms of $\vec{A}$ and ϕ are associated with the wave field, where the field amplitudes B_1 and E_1 are related by $B_1 = [(\eta k_z - k_x)/\omega]E_1$, with $\eta = (k_x/k_z)(k_x\rho_i)^{-2}$ and ρ_i is the thermal Larmour radius of the energetic protons. The second term of $\vec{A}$ defines a uniform background magnetic field $B_0\hat{z}$, while the second term of ϕ imposes a restoring force to the motion of the proton.

Since H is independent of y, $P_y = const. = 0$ is set without losing generality. This reduces the number of coupled equations of motion of the proton derived from the Hamiltonian from six to four. However, the degree of freedom of the resulting system $H(x, P_x, z, P_z, t)$ is still too high to use the straightforward surface of section method to examine the stochasticity of the system. Although there are special ways such as using the Lyapunov exponents to determine the stochasticity of a high dimensional system, the methods involve tedious numerical analysis leading to much less general results and physical pictures. On the other hand, it is realized that the transverse motion of the proton is dominated by the cyclotron motion since the cyclotron frequency $\Omega_i = eB_0/m$ is one order of magnitude larger than the bouncing frequency ω_b and the wave frequency ω. Hence, the stochasticity of the system is mainly determined by the behavior of the longitudinal motion of the proton, and an average of the Hamiltonian (1) over the cyclotron period is performed in the following to reduce the dimensionality of the considered Hamiltonian system to two.

Introducing the polar coordinates (I, θ) and relating (x, P_x) to them by

$$\begin{aligned} x &= (2I/m\Omega_i)^{1/2} \sin\theta = \rho \sin\theta \\ P_x &= (2m\Omega_i I)^{1/2} \cos\theta = m\Omega_i \rho \cos\theta \end{aligned} \qquad (3)$$

where $\rho = (2I/m\Omega_i)^{1/2}$.

Thus, the sinusoidal functions can be expanded in terms of the Bessel functions as $\sin(k_x x + k_z z - \omega t) = \sum_n J_n(k_x\rho)\sin(n\theta + k_z z - \omega t)$ and $\cos 2(k_x x + k_z z - \omega t) = \sum_n J_n(2k_x\rho)\cos[n\theta + 2(k_z z - \omega t)]$, and the average over the cyclotron period results $< \sin(k_x x + k_z z - \omega t) >= J_0(k_x\rho)\sin(k_z z - \omega t)$, $< \cos 2(k_x x + k_z z -$

$\omega t)>= J_0(2k_x\rho)\cos 2(k_z z-\omega t)$, and $< P_x \sin(k_x x + k_z z - \omega t) >= 0$. In terms of these average functions, the average Hamiltonian $\mathcal{H}(I,\theta,P_z,z,t)$ is found to be

$$\begin{aligned}\mathcal{H} &= \Omega_i I + P_z^2/2m + m\omega_b^2 z^2/2 + (m/4)(\Omega_1/k)^2 + \\ &\quad (k_x\Omega_1/k^2)(P_z - m\omega/k_x a)J_0(k_x\rho)\sin(k_z z-\omega t) \\ &\quad -(m/4)(\Omega_1/k)^2 J_0(2k_x\rho)\cos 2(k_z z-\omega t) \qquad (4)\end{aligned}$$

where $\Omega_1 = eB_1/m$, $a = (\eta\alpha - 1)/(\eta+\alpha)$, and $\alpha = k_z/k_x$.

Using (4), Hamilton's canonical equations are expressed explicitly as

$$\begin{aligned}\dot{I} &= 0 = \dot{\rho} \qquad (5)\\ \dot{\theta} &= \Omega_1 - (k_x/m\Omega_i\rho)(K_x\Omega_1/k^2)(P_z - m\omega/ak_x) \\ &\quad J_1(k_x\rho)\sin(k_z z-\omega t) + (k_x/2\Omega_i\rho)(\Omega_1/k)^2 \\ &\quad J_1(2k_x\rho)\cos 2(k_z z-\omega t) \qquad (6)\\ \dot{z} &= P_z/m + (k_x\Omega_1/k^2)J_0(k_x\rho)\sin(k_z z-\omega t) \qquad (7)\\ \dot{P_z} &= -m\omega_b^2 z - k_z(k_x\Omega_1/k^2)(P_z - m\omega/ak_x) \\ &\quad J_0(k_x\rho)\cos(k_z z-\omega t) - (m/2)k_z(\Omega_1/k)^2 \\ &\quad J_0(2k_x\rho)\sin 2(k_z z-\omega t) \qquad (8)\end{aligned}$$

Equation (5) leads to $\rho = const.$ and the RHS of (6)-(8) are functions of (ρ, z, P_z, t), hence, only (7) and (8) remain coupled to each other. Equation (6) can be integrated separately with the aid of the results of (7) and (8). Thus, only (7) and (8) are analyzed in the following for determining the stochasticity of the system.

It is convenient to introduce the dimensionless quantities, $\mathcal{P} = k_z P_z/m\omega$, $q = \omega/\omega_b$, $\epsilon = (k_z/k)^2(q\Omega_1/a\omega_b)$ (It is noted that ϵ in this work is different from that defined by *Prakash* [1989] which is smaller by a factor of 8.36), and $K = \epsilon(1 - a/\alpha)J_0(k_x\rho)$; and the dimensionless time $\omega t/2\pi \mapsto t$ and coordinates $(k_x\rho) \mapsto \rho$ and $k_z z/2\pi \mapsto z$. In addition we introduce the canonical transformation $Z = z$ and $P = \mathcal{P} + (\epsilon/\alpha q^2)aJ_0(\rho)sin2\pi(z-t)$ and a shift of the origin of time to the right by 1/4, i.e. $t' = t - 1/4 \mapsto t$. This converts (7) and (8) into

$$\begin{aligned}\dot{Z} &= P \qquad (9)\\ \dot{P} &= -(2\pi/q)^2\{Z - (K/2\pi)\sin 2\pi(Z-t) - (K^2/4\pi) \\ &\quad [a/(\alpha-a)q]^2[(1+\alpha^2)J_0(2\rho)/J_o^2(\rho) - 1] \\ &\quad \sin 4\pi(Z-t)\} \qquad (10)\end{aligned}$$

Results and Discussion

Equations (9) and (10) are to be integrated to find the motion of the proton. We are interested in resonant motion of the proton with the KAW, i.e. periodic motion with period one, the wave period on our normalized time scale. The periodic motion corresponds to one of the fixed points on the time one map where the proton starting at Z, P at $t = 0$ returns to the same Z, P point at $t = 1$. Consequently, we want to find such fixed points of the map and their dependence on the field intensity (i.e. ϵ) of the KAW.

This search is facilitated by the recognition that the map has a symmetry line. The transformation $Z \mapsto -Z$, $t \mapsto -t$ leaves Eqs. (9) and (10) invariant, revealing the P axis of the Z-P plane as the symmetry line. Therefore, we perform period one mapping starting at points along the P axis. By fixing the parameters ρ, a, α, q and K (i.e. ϵ) we integrate (9) and (10) from $t = 0$ to $t = 3000$ with initial conditions $Z = 0$, P arbitrary, and map the points at integer t on the Z-P plane. The fixed points must be those points on the P-axis and enclosed by concentric loops.

Figure 1 shows surface of sections plots for $q = 1/\sqrt{2}$, $\alpha = .4$, $a = .288$, $\rho = 1.6$ computed for $\epsilon = .8, 1.$ and 2., where ρ is the Larmour radius of the chosen proton which is different from the thermal Larmour radius ρ_i of the background energetic protons. The winding number q is chosen to be an irrational number so that bounce resonance with wave oscillation is ruled out. The transition from regular to chaotic trajectories is evident. We next examine the effect of bounce resonance. Consider the case $\omega = \omega_b/2$ or $q = 0.5$ with all the other parameters same as the previous case except now $\epsilon = .5, .6$ and 2., the corresponding set of surface of sections plots as shown in Figure 2 indicates that the bounce resonance enhances the perturbation on particle trajectory significantly. Nevertheless,one finds that significant chaos emerges generally for $1 < \epsilon < 2$, well below wave amplitudes observed in the magnetosphere.

Once the trajectories of the protons become chaotic, the geomagnetic field can no longer effectively confine them and consequently, many equilibrium processes in the magnetosphere are affected. For instance, an ion in the chaotic domain initially moving in the vicinity of the equatorial plane can penetrate into the polar region. Moreover, the turning points of the trapped bouncing ions vary significantly in time due to the interaction with the KAW. This phenomenon is shown in Figure 3 for Z as a function of time for a chaotic particle. It suggests that the interaction causes the breakdown of the second adiabatic invariant [*Schmidt,* 1966]. The third adiabatic invariant, implying the consistency of the flux surface, should also be violated [*Schmidt,* 1966], permitting energetic ions to cross flux surfaces and emerge as low altitude trapped protons at the geomagnetic equator.

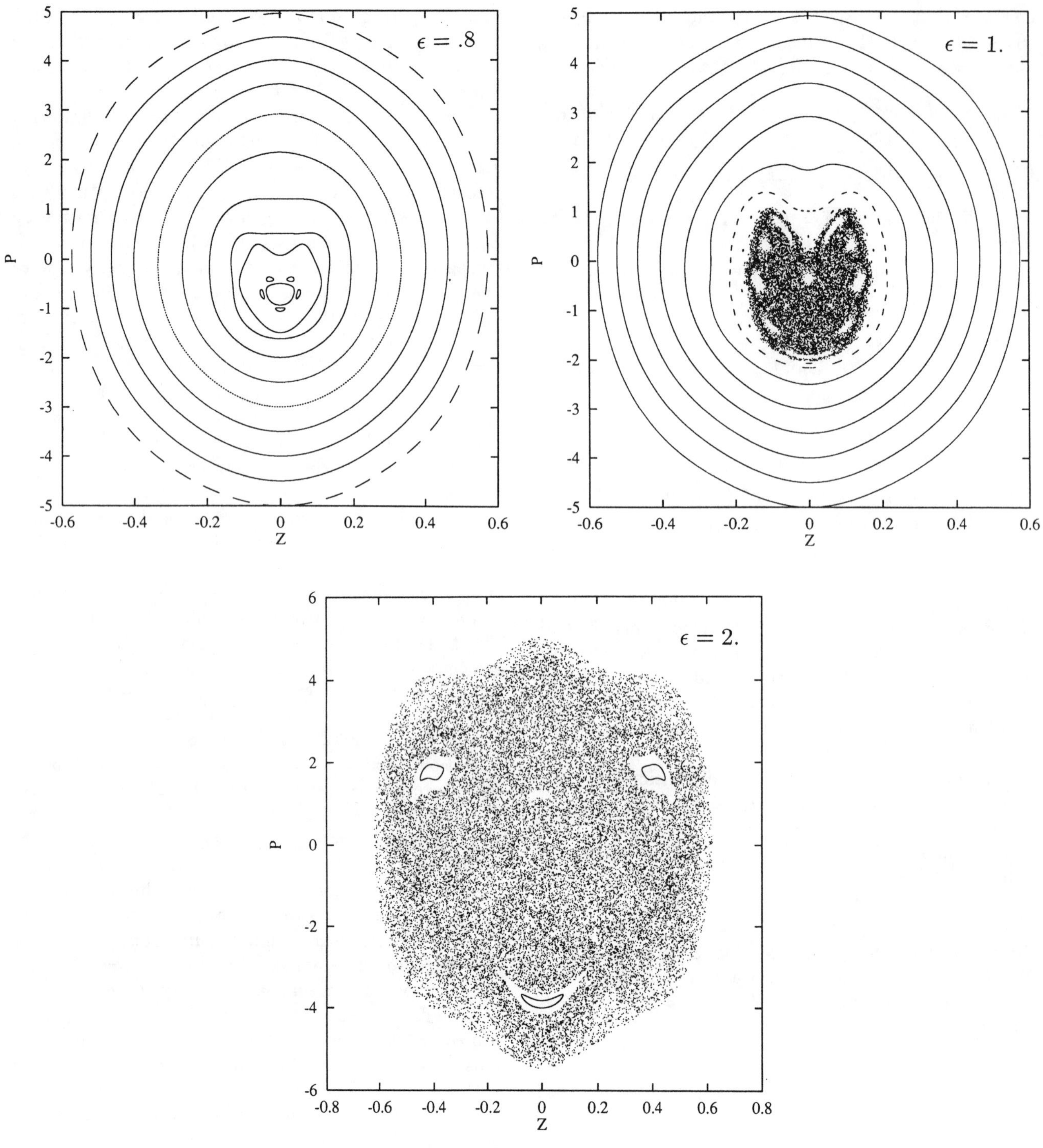

Fig. 1. Surface of section plot for proton's bouncing motion not in resonance with the KAW. $q = 1/\sqrt{2}$, $\alpha = .4$, $\rho = 1.6$, $a = .288$, $\epsilon = .8, 1., 2.$

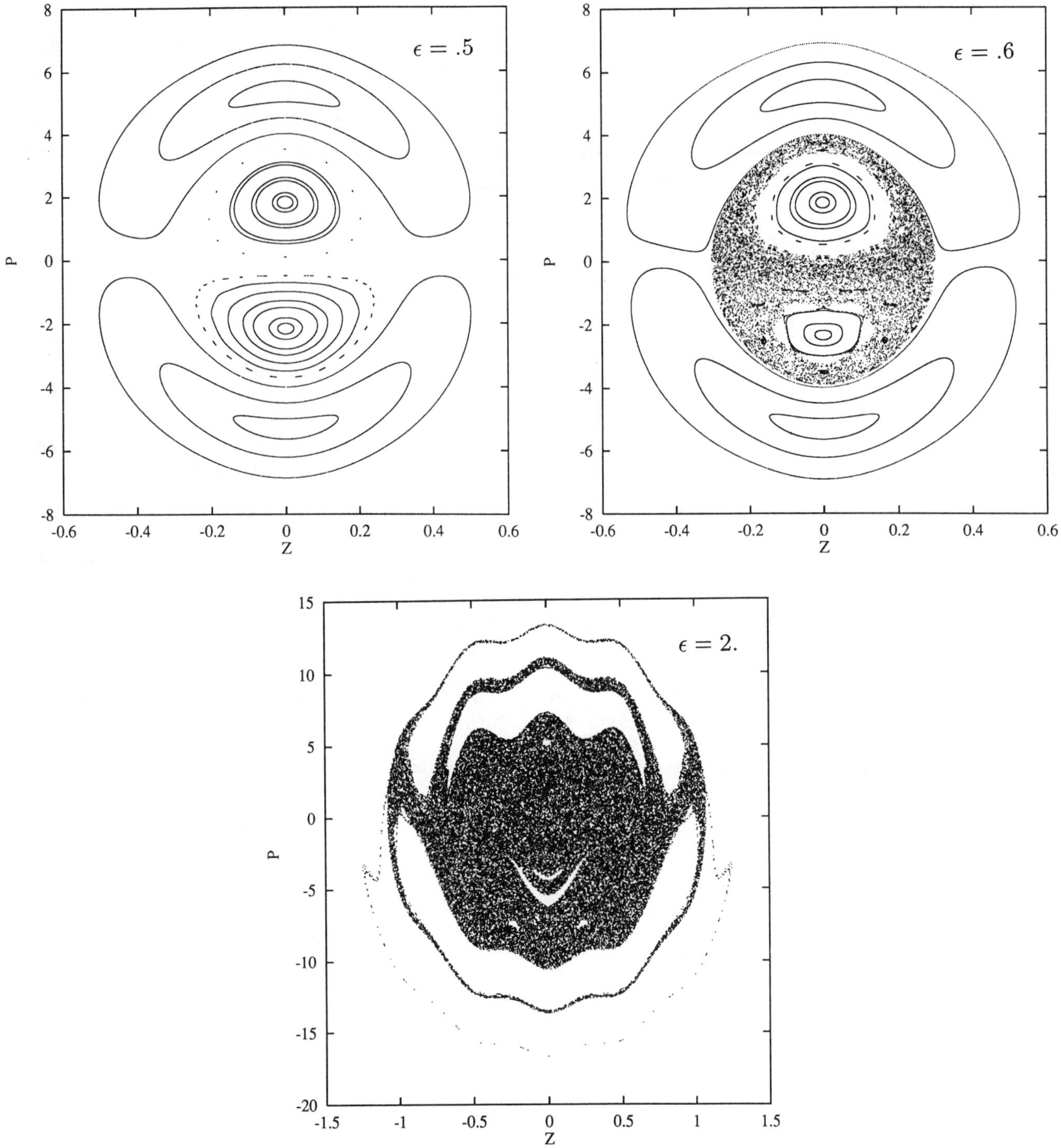

Fig. 2. Surface of section plot for proton's bouncing motion in resonance with the KAW. $q = .5$, $\alpha = .4$, $\rho = 1.6$, $a = .288$, $\epsilon = .5, .6, 2$.

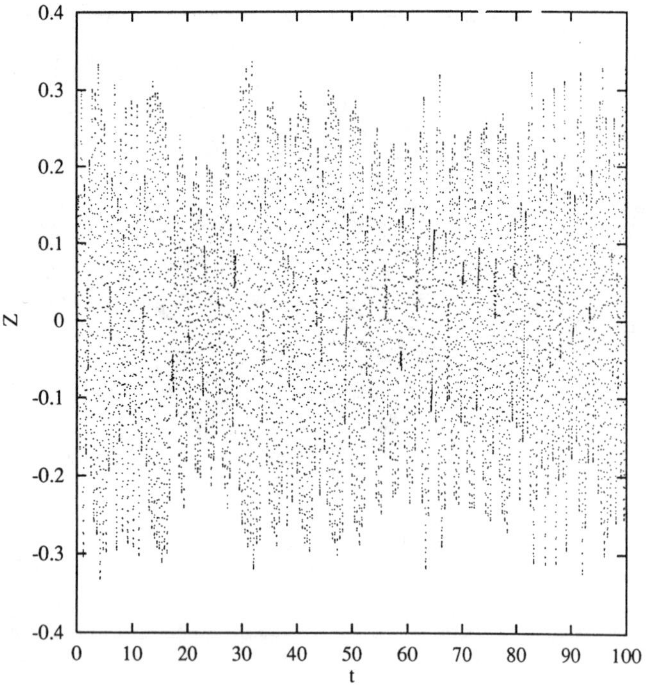

Fig. 3. Z vs. t for a chaotic proton with $q = .5$, $\alpha = .4$, $\rho = 1.6$, $a = .288$, $\epsilon = 2$, $Z(0) = 0$, and $P(0) = -1$.

Acknowledgments. The authors acknowledge useful discussions with Professor G. Schmidt and Dr. A. A. Chernikov of Steven's Institute of Technology. This work has been supported by the AFOSR grant AFOSR-91-0002 of the U.S. Air Force Command and by the NSF grant ATM-9024827.

References

Guzik, T.G., M.A. Miah, J.W. Mitchell, and J.P. Wefel, Low-altitude trapped protons at the geomagnetic equator, *J. Geophys. Res., 94*, 145, 1989.

Hasegawa, A., Particle acceleration by MHD surface wave and formulation of aurora, *J. Geophys. Res., 81*, 5083, 1976.

Hasegawa, A., Excitation of Alfvén waves by bouncing electron beams: An origin of nightside magnetic pulsations, *Geophys. Res. Lett., 6*, 664, 1979.

A. Ho, S. P. Kuo, A. A. Chernikov and G. Schmidt., Chaotic Ion Motion in the Magnetosphere, *Comments Plasma Phys. Controlled Fusion, 15*, 129, 1993.

Antony Y. Ho, S. P. Kuo and G. Schmidt, Chaotic Proton Motion Driven by Kinetic Alfvén waves in the Magnetosphere Leading to Polar and Equatorial Proton Precipitations, *J. Geophys. Res., in press.*

Hultqvist, B., The hot ion component of the magnetospheric plasma and some relations to the electron component, observations and physical implications , *Space Sci. Rev., 23*, 581, 1979.

Liu, W., Chaos driven by kinetic Alfvén waves, *Geophys. Res. Lett., 6*, 1611, 1991.

Prakash, M., Particle precipitation due to nonlinear wave-particle interactions in the magnetosphere, *J. Geophys. Res., 94*, 2487, 1989 and the references therein.

Schmidt, G., *Physics of High Temperature Plasmas*, Academic Press, New York, 1966.

Antony Y. Ho and S. P. Kuo, Weber Research Institute, Polytechnic University, Farmingdale, New York 11735.

Extended (Bi-Modal) Ion Conics at High Altitudes

W.K. Peterson, H.L. Collin, M.F. Doherty[1] and C.M Bjorklund

Lockheed Palo Alto Research Laboratory, Palo Alto, California

Ion energization both parallel and transverse to the local magnetic field was observed on every crossing of the auroral zone by the Dynamics Explorer -1 Satellite. Not infrequently, ion distributions characterized by both significant parallel and transverse energization were found. These distributions have flux maxima over extended energy and angular ranges and were initially called bi-modal in the belief that they are the result of quasi-independent ion energization processes acting transversely and parallel to the local magnetic field. Recently it has been shown that some extended (bi-modal) ion distributions can be formed by mesoscale processes acting over an extended altitude range. Knowledge of the occurrence frequency and characteristics of extended ion distributions can be used to improve our understanding of the physical process responsible for their formation. We have developed and used an automated procedure based on standard image processing techniques to identify and characterize ion distributions that have been significantly energized transverse to the local magnetic field. Our survey found approximately 10,000 events in over 2,500 hours of data. We found about half of the O^+ and He^+ ion conic distributions identified had, in addition, a significant upward field-aligned component. Most of these extended ion conic distributions were found above 15,000 km, while the ion conic distributions with a restricted angular distribution and without a field-aligned component had an approximately uniform occurrence frequency over the altitude range from 8,000 to 24,000 km.

INTRODUCTION

Ion conics are characterized by velocity space distributions with significant intensity maxima on a cone shaped surface centered on the magnetic field line direction. These strongly peaked angular distributions imply the existence of very intense and perhaps localized processes transferring energy to the ions. Investigation of the processes responsible for ion conic formation is required to understand the extraction and energization of low energy (~ 1 eV) ionospheric ions and their transport to the plasma sheet where they are quasi-isotropic and have energies on the order of 1 keV.

[1]Present address: 913 Laguna Ave., Burlingame, California 94010

Space Plasmas: Coupling Between Small
and Medium Scale Processes
Geophysical Monograph 86

Two types of ion conic distribution have been identified in the plasma flowing upwards above the auroral and polar cap ionosphere. Sharp et al. [1977] and other investigators using data from the S3-3 satellite assumed that the peak flux in ion conic distributions was characterized by one cone angle for all energies. Such distributions could be created by the localized transfer of energy to a cool plasma. The S3-3 measurements available to Sharp et al. were limited to energies above 500 eV. The high sensitivity and low energy range of the DE -1 EICS instrument, [Shelley et al., 1981], allowed Klumpar et al. [1984] to identify a new class of conics characterized by an extended range of pitch angles. Klumpar et al. noted that this new class of ion conic distribution was also characterized by a significant maximum in flux of energetic ions aligned with the magnetic field. Klumpar et al. called these distributions bi-modal conics and attributed them to the occurrence of both transverse and parallel (to the local magnetic field) acceleration processes. Temerin [1986], Chang et al. [1986], and Horwitz [1986] have suggested alternative mechanisms for forming extended ion conic distributions.

Figure 1 shows examples of the two types of ion conic distribution displayed in the flux image format.

Progress in understanding ion conic formation has been limited by a lack of detailed examples, including fluxes and temperatures, of ion conics, especially the extended type. There have been several reports on the average proprieties of a large number of ion conics, but the data for these reports (e.g. Gorney et al. 1981; Thelin et al. 1990; Miyake et al. 1993) has come from visual examination of high time resolution images. The notable exception was the series of papers by Yau et al. [1984], Sagawa et al. [1987] and Kondo et al. [1990]. Yau introduced a one dimensional algorithm to search for peaks in the angular distribution over three broad energy ranges (0.01-1, 1-4, and 4-17 keV) and applied it to the high altitude data to identify ion conic events in three broad energy ranges. This classification technique was extended by Sagawa el al. [1987] and Kondo et al. [1990] to include other types of ion distributions, but these studies were also limited by the three broad energy ranges and were not able to provide much information about extended ion conics. Recently Miyake et al. [1993] have been able to distinguish between the two basic types of ion conic distributions in a large number of events. To date Miyake et al. have limited their studies to conic angular distributions of the restricted type.

To progress further in our understanding of the processes responsible for ion conic formation we need to determine the relative importance of the types of conic distribution in various regions of the magnetosphere. Such information on conic type and related energy and angular characteristics can only come from detailed evaluation of the characteristics of thousands of ion conic distributions. To meet this need, we have developed a two-dimensional, model-based algorithm to identify and classify energetic ion conic events in the plasma flowing upward on auroral and polar cap field lines. Our model-based algorithm is based on casting observed ion distributions into two dimensional flux images and applying standard image processing techniques [Doherty et al., 1993]. Using our algorithm it is possible to classify and characterize a large number of ion conic events and to determine the relative importance of the bi-modal (extended) conics identified by Klumpar et al. [1984]. Our model and method are novel, and it is impractical to physically check the specific identification of each distribution. We will therefore not characterize ion distributions as bi-modal, rather we will characterize them as either restricted or extended.

The rest of this paper is organized as follows. We first discuss the large (over 100,000 samples) data base of flux images we have assembled and the model-based algorithm. We then present results of the application of our algorithm to

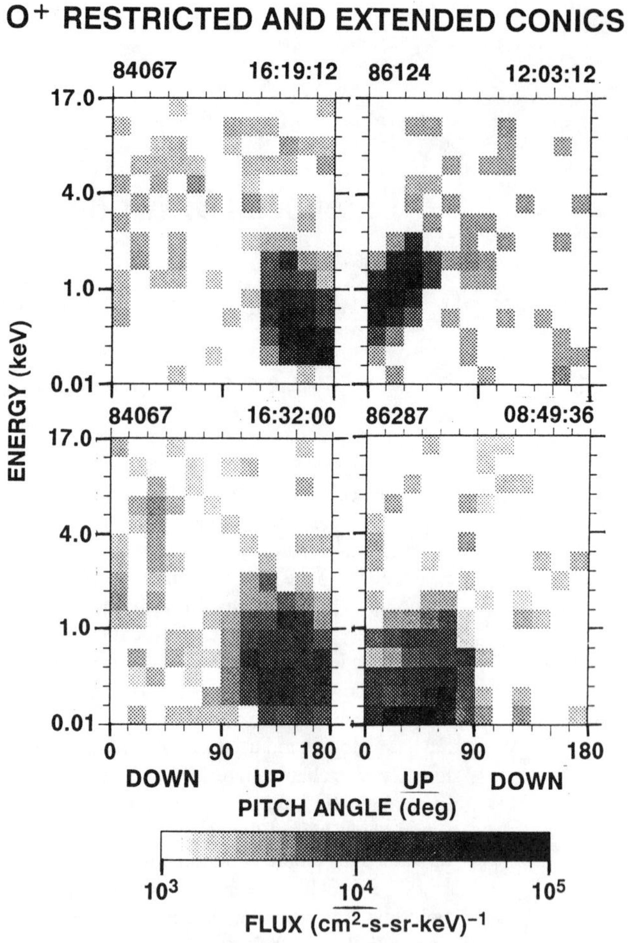

Fig. 1. Flux images of four oxygen ion conic distributions. The top two images are conic distributions with extended angular distributions. The bottom two images show distributions classified as having restricted angular distributions by the algorithm discussed in the text. Each image consists of a 15 × 12 pixel array where the oxygen ion flux intensity is encoded using the gray bar at the bottom. The energy/pitch angle sampling space is covered in 15 energy bands (ordinate) and 12 pitch angle bins (abscissa). The lowest energy band covers the range from 10 eV to ~ 100 eV. Successive energy bands are are quasi logarithmically spaced in energy over the range to 17 keV. The pitch angle range extends from 0° to 180°. Pitch angle is defined as the angle between the ion velocity and the local magnetic field. In the northern (southern) geomagnetic hemisphere 0° (180°) pitch angles correspond to ions flowing downward toward the ionosphere. The two images on left (right) were acquired in the northern (southern) geomagnetic hemisphere. The direction of ion flow is indicated on the respective pitch angle axes. Conic distributions have the most intense fluxes at pitch angles not aligned with the local magnetic field. The time that each image was acquired is indicated at the top of each image. For example the top right image was acquired at 16:19:12 universal time on the 67th day of 1984 (March 7).

the data base and make some observations about the distribution in space of transverse ion energization processes.

THE DATA BASE OF FLUX IMAGES

The data we are working with come from the Energetic Ion Composition Spectrometer (EICS) on the Dynamics Explorer (DE) -1 Satellite. The DE program has been described by Hoffman and Schmerling [1981], and the EICS instrument by Shelley et al. [1981]. Briefly DE -1 was launched in August 1981 and operated until February 1991 in a 23,500 x 550 km, 90° inclination orbit. The data used this report were restricted to those acquired above 8,000 km and at invariant latitudes greater than 58°. The EICS instrument returns measurements of the ambient, mass-resolved ion fluxes in the energy range from spacecraft potential to 17 keV/e. The complete range of ion pitch angles was sampled twice per spacecraft spin period, six seconds. Since the count rate from only one mass/energy/angle value was available during each measurement interval (1/32 s), the sequence of mass and energy steps scanned was programmable. The most commonly used modes covered the energy range from 10 eV/e to 17 keV/e in 15 logarithmically spaced steps. Mass information was obtained in three fundamentally different modes: Mass scan, drum and fast modes which are characterized by instrument cycle times of 128, 96 and 24 seconds respectively. Data transmitted to the ground was processed and inserted into two distinct data bases.

The high resolution data base contains all of the data acquired for all modes. Peterson et al. [1988] have presented examples of ion conic events found in the high resolution data base. The summary data base is assembled from the sub-set of the high resolution data obtained in the drum and fast modes for three of the major ion species (H^+, He^+ and O^+). The summary data base is composed of ion fluxes binned in energy and pitch angle and averaged over successive 96 second intervals. Approximately half of the ~ 40,000 hours of DE -1 EICS data acquired were captured in the high resolution data base before the NASA computer center processing them was shut down in 1991. Most (~90%) of these high resolution data were acquired in the drum or fast modes and inserted into the summary data base. Approximately half of the summary data have been transferred to a 12 inch worm optical disk system using NASA supplied hardware and software [Davis and Vaidta, 1989]. Because mounting tapes is so labor intensive we have elected to use the part of the summary data base resident on optical disk for this study. The ~10,000 hours of summary data on the optical platter represent approximately 1/4 of all DE -1 EICS data acquired.

The model based algorithm described in the next section uses ion data in the form of "flux images". A flux image is constructed for each ion species and each 96 s interval. It consists of a pitch angle/energy space with pitch angle divided into 12 bands each of 15° width, and 15 energy levels spaced quasi-logarithmically from 0.01 to 17 keV. Thus the space is partitioned into a matrix of 12 x 15 pixels. Each pixel contains the ion flux obtained from the summary data base for the corresponding energy and pitch angle. These fluxes form a surface in pitch angle/energy space which is an image of the ion distribution. Figures 1-10, described below, are examples of flux images. Associated with each flux image is an identically partitioned matrix of 12 x 15 pixels which contains the standard deviations (σ) of the corresponding fluxes and forms an "error image". The standard deviations are also obtained from the summary data base and are estimated on the basis of Poisson counting statistics only.

THE ALGORITHMIC MODEL OF CONICS

The algorithm examines data from each ion species in each 96 s interval in turn and searches for conics which are moving either parallel or antiparallel to the magnetic field direction. The field direction is downward towards the Earth in the northern geomagnetic hemisphere and upward away from the Earth in the southern hemisphere. Since the summary data base includes the geomagnetic latitude, the direction of the conic, upward or downward, can also be determined. If any conics are identified, a description of each is retained to from a data base of conics for further analysis.

Our first attempts to automatically identify conics in flux images were based on calculating one-dimensional properties of the image: the flux weighted pitch angle averaged over all energies, and the angular width of the angular peak in pitch angle at half the maximum flux intensity, averaged over all energies. This algorithm was not very sensitive to the differences between ion conic distributions extended in angle and trapped ion distributions because both distributions are characterized by a relatively large angular width. We found that extended and trapped ion distributions can be distinguished if the algorithm is extended to two dimensions (energy and angle).

The two-dimensional, model-based algorithm described here uses standard image processing techniques to process each flux image and the associated error image. In the flux image a conic forms a ridge running from low to high energy. A line along the length of the ridge is the conic path. Our algorithm identifies this conic path. In this respect it is identical to the way ion conics are visually identified on high resolution contour plots. The model requires that the following conditions hold true for a flux image to be identified as a conic:

1. Conics have a pitch angle specificity; thus, within any particular energy level, the distribution has a well-defined maximum;

2. Conics span at least three energy levels;

3. The conic path remains at a constant pitch angle or steadily moves away from the central (i.e. 90°) pitch angle as energy decreases;

4. Conic paths include pitch angles either completely less than 90° or completely greater than 90°. This requirement separates conic distributions into those flowing up and down magnetic field lines.

In addition to the above attributes or rules used to identify conics, it was discovered in the course of testing our algorithm that we needed to identify ion beams and trapped ion distributions, which were sometimes mistaken for conics, in order to minimize the number of false positive ion conic detections. Ion beams in the flux images input to our algorithm are characterized by a maximum flux level within a compact energy range and with a pitch angle range centered on the magnetic field direction. Figure 2 shows an example of an ion beam in the flux image format. Trapped ion distributions in flux image format have a high degree of symmetry about 90° pitch angle as shown in Figure 3.

Two characteristics of the data in the lowest energy channel complicate its use in the analysis and characterization of ion conic events. For oxygen ions the energy range of the lowest energy channel (10 eV to 100 eV) sometimes contains a contribution from a low (~ eV) energy cold rammed oxygen plasma which introduces an additional angular maximum. In addition, the broad energy width makes it extremely difficult to determine the average energy of the flux in the lowest energy channel. It is probable that the model-based algorithm we have developed to identify ion conic distributions can be extended to include data from the lowest energy channel. For the present study, however, we have restricted our interest to the highest 14 energy channels covering the energy range 150 eV to 17 keV.

Our algorithm has four stages: image formation, image compression, pixel qualification and conic path evaluation. The following paragraphs provide further detail on each of these stages.

Image formation.

A large number (roughly 50%) of the 96 second intervals included in the summary database were obtained in regions with little or no measurable ion flux. It is desirable to reliably identify empty flux images, reject them from further investi-

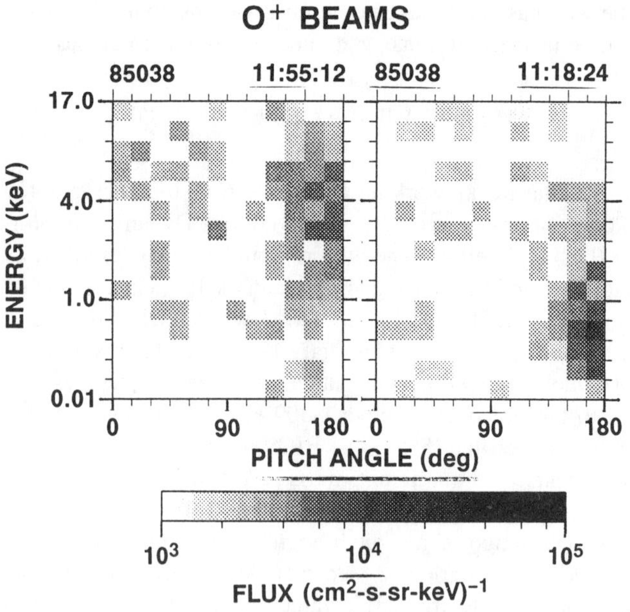

Fig. 2. Flux image of two oxygen ion beam distributions in the same format as Figure 1. The most intense fluxes in the ion beams occur along the magnetic field direction. The ion beam is further characterized by a more compact extent in energy and pitch angle. Note that the energy and particularly the angular resolution of the ion flux measurements place limits on the experimental differentiation between ion beams and ion conic distributions. In particular, restricted ion conics with a conic angle less than 15° will be characterized as ion beams. The examples in Figure two were both acquired in the northern geomagnetic hemisphere.

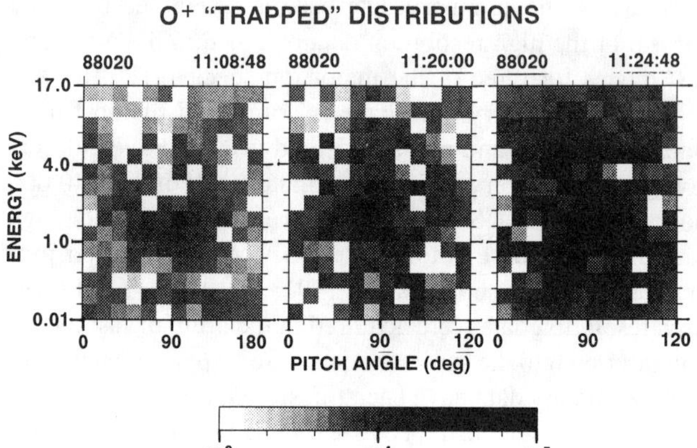

Fig. 3. Flux images of three oxygen ion trapped distributions in the same format as Figure 1. Note the symmetry about the 90° pitch angle axis.

gation, and to set a repeatable flux threshold for the detection of ion conic distributions. As noted above the data for each mass in each 96 second interval consist of a pair of a flux image and the corresponding error image containing ion fluxes and estimates of the standard deviations (σ) of the fluxes. Our algorithm requires that a minimum number of flux values in the flux image have magnitudes greater than the estimate of their standard deviation. The minimum number was set based on examination of the training results, and is specific to each mass type: 10 pixels for H^+, 7 for O^+, and 4 for He^+. This pre-selection considerably speeds up processing but is set low enough that it does not reject any conic distributions.

Image compression.

The flux images are next processed to remove certain energy bands from consideration, thus relieving the more computationally intensive qualification and evaluation stages of the algorithm. An energy band is declared valid if there exists some difference (i.e. angular dependence) in the scaled flux pixel values within that band. Homogeneous bands are assumed to indicate isotropic distributions or background noise and are ignored. (To discriminate against ion beams, flux values at pitch angles of 0° and 180° are not considered when performing this check.) A second test checks to see if the maximum flux value for a band exceeds a minimum threshold. (Again, flux values at the two outer pitch angles are ignored.) The minimum threshold value is set to the expected 1-σ value of the background noise of the flux measurements. This test ensures that a few pixels which exceed one standard deviation in their flux values do not adversely affect the algorithm performance. Finally, the largest flux value over all energy bands and pitch angles is compared with a pre-set minimum (chosen from a manual review of the training data set described below) for each possible ion type. If this value is not exceeded, the flux image is dropped from further processing.

Image compression is completed by eliminating the highest energy bands. Starting from the lowest energy, each energy band is checked for validity. If two adjacent energy bands are invalid, then those two bands and the remaining bands of higher energy levels are eliminated from further consideration. Subsequent processing uses this compressed image which can contain up to fourteen contiguous energy bands.

Pixel qualification.

For each flux pixel in the compressed image, a comparison is made between it and its standard deviation value (σ) from the error image. If the flux value exceeds the standard deviation, the pixel is marked as a candidate for inclusion in a conic structure. Figure 4 presents graphically the results of the pixel qualification test applied to the O^+ flux image obtained at 16:32 on March 7, 1984 shown in Figure 1

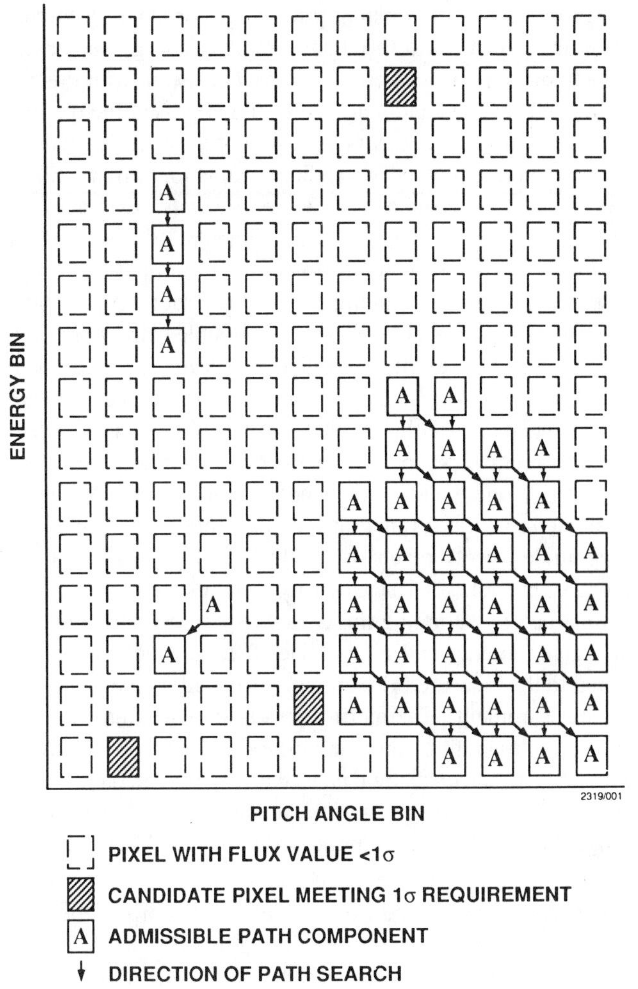

Fig. 4. Analysis of the conic path. See text for details.

.An additional qualification check eliminates trapped ion distributions. Our algorithm checks for trapped ion distributions at this stage by calculating the ratio of the number of symmetric (about the 90° pitch angle) pairs of candidate conic pixels within each energy band level. If the ratio of symmetric pairs of candidate pixels to total candidate pixels exceeds 70%, then the image is declared to be of a trapped ion distribution, and no further processing is done.

Conic path evaluation.

Our model requires that the conic path in energy/pitch angle space remains at a constant pitch angle or steadily moves away from the central (90°) pitch angle as energy decreases. The algorithm creates tree structures, consistent with this requirement, of the candidate pixels which passed the qualification step above. A recursive search of these tree structures is performed and a set of candidate paths identified. Each path must also extend at least 3 energy levels. Compare the diagram in Figure 4 with the conic image in the lower left of Figure 1; the pixels which qualify have been pictorially represented in Figure 4, with the possible path routes highlighted with arrows. Since there are several possible paths which can be selected through the candidate pixels for any one conic event, we have defined a merit function to select the best path based on quantifiable measurements. This function also yields a relative perspective as to how certain our algorithm is that any one event is, or is not, a conic event. We have evaluated two merit functions: M1) a simple sum of the flux magnitudes for all pixels in a path; and M2) a numerical integral of the velocity space density along the conic path. M2, is obtained by summing the velocity space density elements at a given energy and pitch angle $f(E,\theta)$ which is proportional to the measured flux intensity divided by the energy of the measurement. The merit function M2 takes into account the variable width of the energy bins. In sample runs, merit function M2 selected conic paths that were deemed more reasonable for conic distributions extending over a large energy range than did M1. We have accordingly used M2 to identify conic paths in the results presented below.

The merit function indicates the relative heights of the candidate paths running along the ridge formed by the conic in its flux image. By maximizing the merit function, the path running closest to the apex of the ridge is identified. All candidate paths are evaluated. The path which has the maximum merit function value is saved as part of the description of the conic and is reported to the end-user if a conic is found in the flux image. For every conic event, we also record and save the value of the merit function of the conic path with the highest figure of merit flowing in the opposite direction. Figure 5 shows an O^+ conic distribution with several distinct candidate paths on the left and the conic path with the largest figure of merit indicated on the right. At this point the algorithm has tentatively identified a candidate ion conic. In our training experiments, however, we observed that there were still large numbers of false positive detections due to confusion with another type of ion distribution ---- asymmetric counter streaming ion conics. Dual conic distributions or asymmetric counter streaming ion conics, are a common feature in the data as reported by Sagawa et al. [1987]. These appear to be a separate phenomenon from the restricted and extended types of conic and are rejected by the algorithm described here. Dual conic distributions exist when candidate

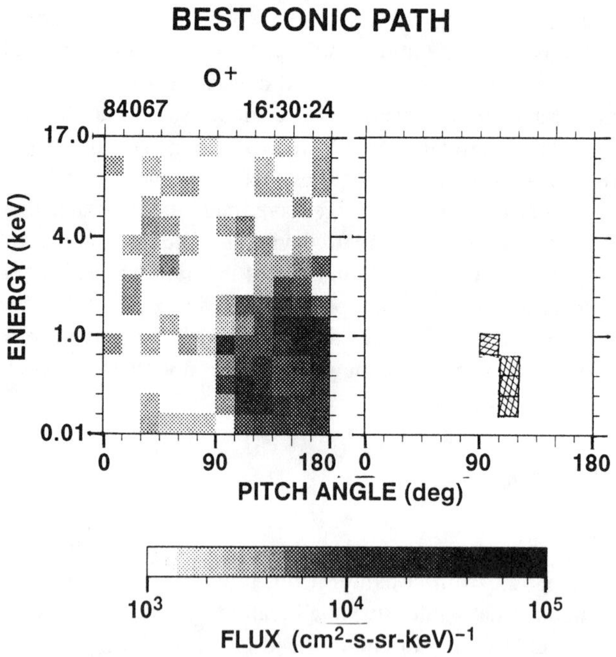

Fig. 5. Left: Oxygen ion flux image in the same format as Figure 1. Right: The conic path with the highest figure of merit identified by our algorithm is highlighted. This path identifies the flux image as that of a restricted conic distribution. The data in Figure 5 were acquired in the northern geomagnetic hemisphere.

conics are found flowing in both the upward and downward directions. These distributions resemble trapped distributions with the additional feature of a flux minimum at 90°. Additional checks are made to exclude counter streaming conic distributions from the results:

1) A comparison between the figures of merit (defined above) corresponding to both halves of counter streaming conics is made. Specifically, if the smaller figure of merit is greater than 10% of the average of the two figures of merit, we do not classify the distribution as conic.

2) If the lengths of the conic paths are small (i.e. less than 5 pixels), then a symmetry test is performed (as described in the qualification stage) over just those energy bands in included in the candidate conic paths. These highly symmetric small conics are also rejected.

Identification of restricted and extended ion conic distributions

Extended and restricted conic distributions are distinguished by their conic paths. The conic paths for extended distributions have their low energy pixel in the field aligned direction (pitch angles of 0° or 180°), and extend over at least 45° of pitch angle. The conic path for a restricted conic distribution terminates at the lowest energy used in the study and traverses at most 30° of pitch angle. Figures 6, 7, and 8, illustrate the way conic distributions are identified from the end points of the conic path. The conic path for the oxygen distribution in Figure 6 is characterized by a constant pitch angle and ends at the lowest energy included in the evaluation of conic paths (i.e. the second energy step). This is the type of conic distribution described by Sharp et al. [1977], which we term the restricted type. The conic path for the oxygen distribution in Figure 7 is characterized by a range of pitch angles terminating in a pixel in the magnetic field direction. This type of conic distribution was first described by Klumpar et al. [1984], and is termed the extended type. Figure 8 shows a candidate conic path that does not terminate at pixels either at the lowest energy or in the magnetic field direction. We do not consider images such as that shown in Figure 8 to be of an ion conic.

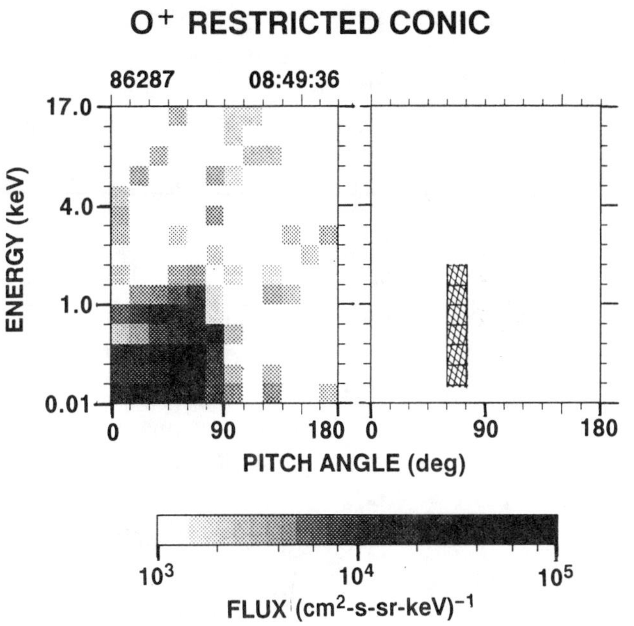

Fig. 6. Flux image of an oxygen restricted ion conic distribution in the format of Figure 5. The flux image was acquired in the southern geomagnetic hemisphere. Note that the conic path identified by our algorithm ends at the second energy channel, the lowest energy channel considered in this investigation.

TESTING THE ALGORITHM

The results of the supervised runs on the 400 training set events showed that we regularly identified all of the conic events and rejected all of the trapped ion events. The algorithm also identified as conics a small number of downflowing conic events, and was not capable of identifying ion conics in the presence of a large quasi-isotropic background ion flux. These anomalies are addressed below.

Figure 9 shows an interval where O^+ and He^+ conics were found, but where an intense, quasi-isotropic flux of H^+ ions obscured detection of an H^+ conic distribution. It would be possible to modify the algorithm to systematically subtract an isotropic flux from the observed distributions before searching for ion conic distributions. Because all of the O^+ and He^+ ion conic distributions found in the training set did not have a quasi-isotropic distribution obscuring the ion conic distribution, and because information on the energy and flux distributions of oxygen and helium ion conic distributions are of current interest, we elected to use the algorithm as described above, and restrict our initial investigations to O^+ and He^+ ion conic distributions. It is interesting to note that the general presence of the H^+ background isotropic flux with energetic ion conic events was not expected prior to commencement of this study.

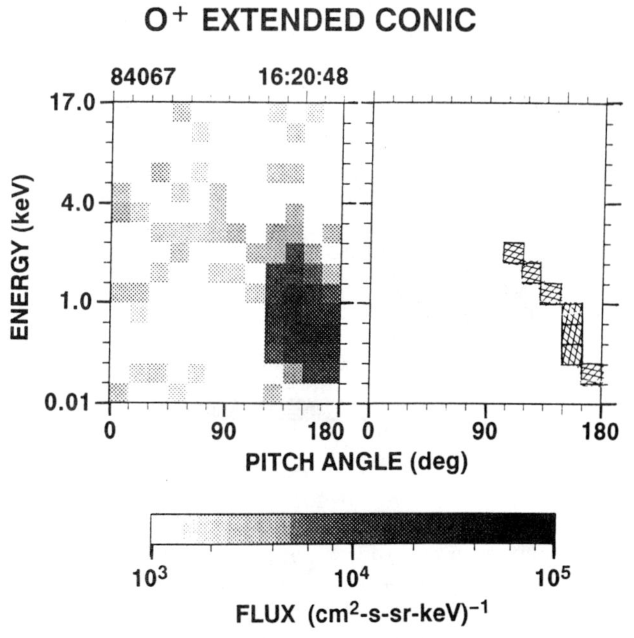

Fig. 7. Flux image of an oxygen extended ion conic distribution in the format of Figure 5. The data were acquired in the northern geomagnetic hemisphere. Note that the low energy end of the conic path is in the magnetic field direction (180° pitch angle).

Figure 10 shows examples of hydrogen, oxygen, and helium distributions that are identified as downflowing conics by our algorithm. Specifically our algorithm found that the can-

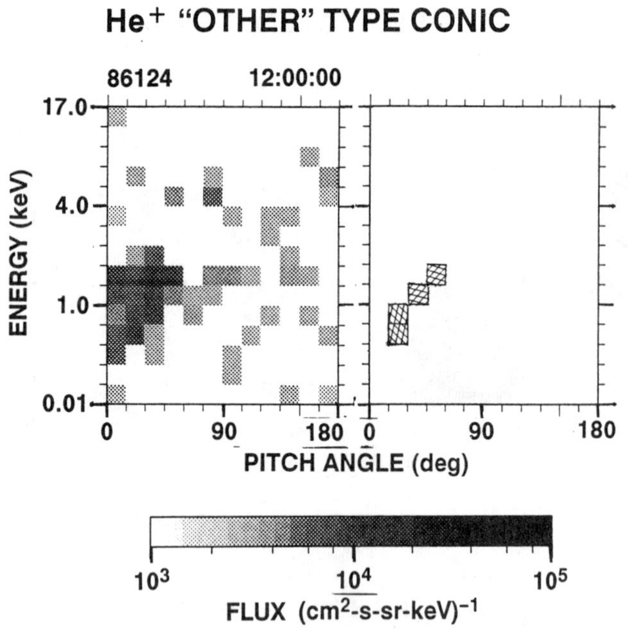

Fig. 8. Flux image of a helium ion distribution in the format of Figure 5. Here the identified conic path identified by our algorithm does not end in a pixel aligned with the magnetic field, or in the lowest energy channel considered. These distributions are not identified as conic distributions by our algorithm.

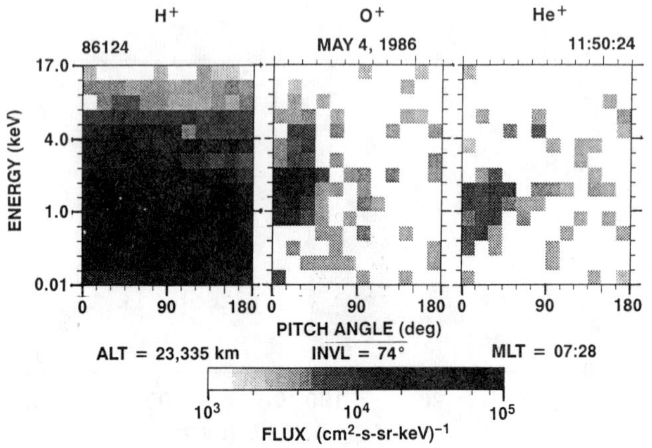

Fig. 9. Flux images of hydrogen, oxygen, and helium ion distributions acquired at 11:50:24 universal time on May 4, 1986 (i.e. day 124 of 1986) in the format of Figure 1. Our algorithm identified restricted oxygen and helium conic distributions. The intense, quasi-isotropic, flux of hydrogen ions obscured detection of any hydrogen conic distribution at this time.

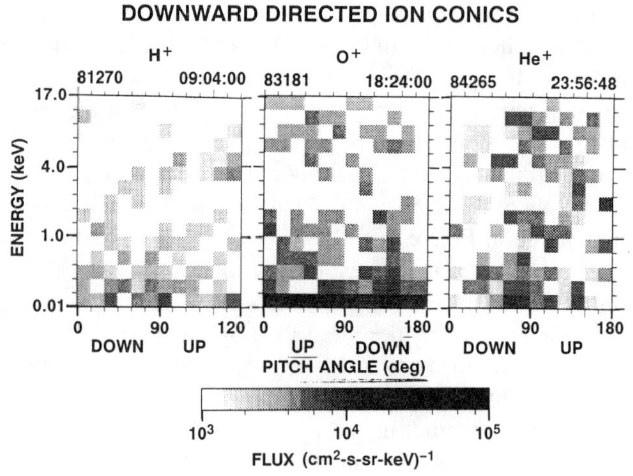

Fig. 10. Flux images of hydrogen, oxygen, and helium ion distributions where our algorithm identified downflowing, restricted, ion conics. The distributions were not acquired simultaneously. The hydrogen and helium distributions were acquired in the northern geomagnetic hemisphere and the oxygen was acquired in the southern. The direction of flow is indicated below the pitch angle axes (abscissa).

didate conic path with the highest figure of merit was characterized by pitch angles directing it downward into the ionosphere. Downward flowing ion conic distributions, if they exist at all, are rare and are restricted to pitch angles near 90°. See, for example, Gorney et al. [1985]. A large fraction of these downward flowing events identified by our algorithm were investigated. All downflowing events investigated were characterized by low flux levels and/or high background counting rates. We concluded that these were all false positive identifications. False positive identifications flowing up the field line were visually located in sample data runs after downward directed false positives were found. We could eliminate these upward false positive identifications by raising the threshold of detection but this would reduce the value of a survey because it would be restricted to only the most intense events. Instead we elected to make use of the downward false positives. The algorithm can be expected to make the same frequency of false identifications in both directions. We therefore can use the occurrence frequency of downward false positives as a measure of the frequency of occurrence of upward false positives and use this measure to correct the results. The occurrence frequencies reported below were calculated from the difference between the detected upward directed conics (including false positive) and downward directed false positives.

APPLICATION OF THE ALGORITHM TO THE DATA BASE

Table 1 summarizes the results from the application of the algorithm to 122,390 O^+ and 67,100 He^+ flux images. As noted above the data base is restricted to energies above ~150 eV, acquired between September 1981 and April 1988 at altitudes between 8,000 and 24,000 km, invariant latitudes above 56° and all local times. Table 1 presents the occurrence frequency of restricted and extended conics in the sample data set. It shows that about 3 times more O^+ conics than He^+ are found, and that the ratio of restricted to extended type conics is about the same for both species (~ 3:2). An examination of the ion conic events shows that they have a flux threshold, referenced to 1000 km altitude, of ~ 10^4 $(cm^2\text{-s-sr-keV})^{-1}$ and characteristic energy threshold of ~ 200 eV, a typical flux of > 10^6 $(cm^2\text{-s-sr-keV})^{-1}$ and a typical characteristic energy of ~ 700 eV. These average results are representative of conics observed at apogee where most of the data were acquired. Figure 11 shows the distribution in altitude, invariant latitude, and magnetic local time of the 122,390 O^+ and 67,100 He^+ flux images searched. Figure 11 shows that, in general, the sampling in invariant latitude and magnetic local time was uniform. No samples are shown for invariant latitudes above 85°, because data acquired at high latitudes was treated as if it was obtained at 84° invariant latitude. Note that there are relative maxima in the number of samples at 64° invariant latitude and 14:00 magnetic local time. The average conic in Table 1 was detected at ~ 20,000 km and an invariant latitude (INVL) of ~ 70°, not where the average data interval was acquired. To proceed further we need to look at the occurrence frequency of ion conic events by species and type in the altitude, invariant latitude, and magnetic local time range sampled by the DE -1 satellite.

Figure 12 presents the frequency of occurrence as a function of altitude, invariant latitude (INVL) and magnetic local time (MLT) of the conic events listed in Table 1. O^+ and He^+ restricted and extended ion conics have similar MLT and INVL occurrence frequencies; the altitude dependencies are different. He^+ restricted and extended conic distributions have approximately the same occurrence frequency over the full altitude range (8,000 to 24,000 km), but the occurrence frequency of He^+ restricted conics increases slightly above ~15,000 km. O^+ restricted conic distributions show a larger increase in occurrence frequency above ~15,000 km than do the He^+ restricted conics. The occurrence of O^+ extended conics has the largest variation with altitude, increasing from ≤ 1.5% at 15,000 km to ≥ 3.5% at the highest altitudes sampled.

Energetic ion conics are found most frequently on the morning side of the magnetosphere between 06:00 and 13:00 MLT. There are also slight but systematic differences in the MLT distribution of the two types of conic events for both O^+ and He^+ identified by our algorithm. The O^+ conics, both

TABLE 1. Conics Found by Species and Type

	O^+		He^+	
Type:	Restricted	Extended	Restricted	Extended
UP				
	4,149	2,070	1,469	506
	3.39%	1.69%	2.20%	0.75%
DOWN				
	818	131	849	129
	0.67%	0.11%	1.27%	0.19%
NET				
	2.72%	1.58%	0.92%	0.56%

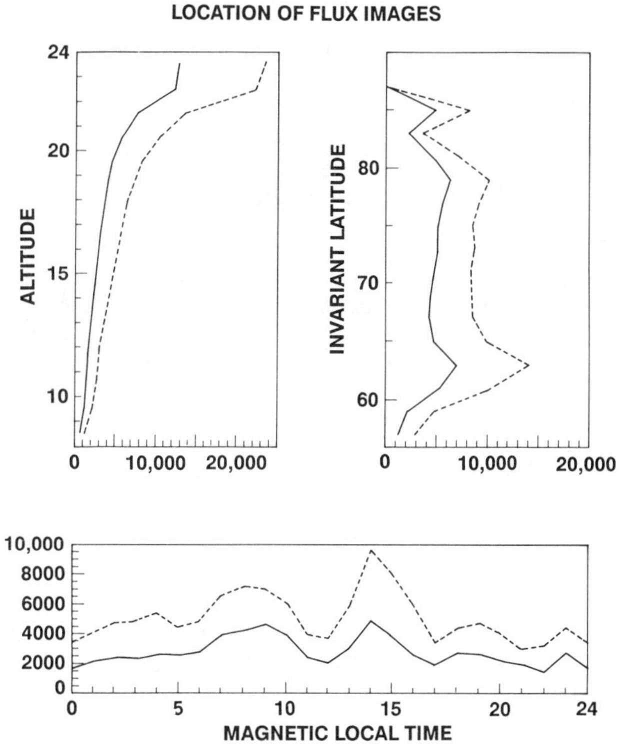

Fig. 11. Distribution of flux images examined in: 1,000 km altitude bins (top right), 2° invariant latitude bins (top left), and 1 hour magnetic local time bins (bottom) The distribution of oxygen flux images is indicated by the dotted lines and helium flux images by the solid lines. A total of 122,390 oxygen and 67,100 helium flux images are included in the data base used for this investigation.

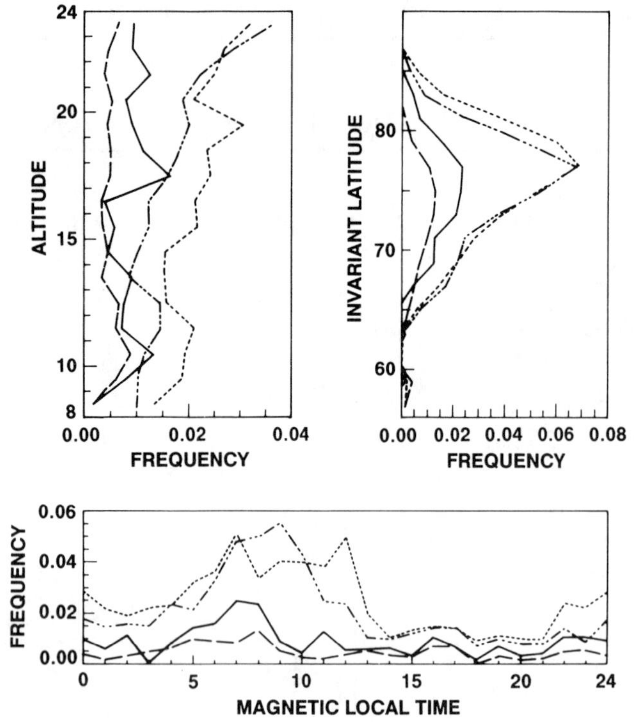

Fig. 12. Distribution in altitude (top left), invariant latitude (top right) and magnetic local time (bottom) of the corrected occurrence frequency of the four types of ion conic distribution identified by our algorithm. Key: ... O^+ restricted; _ .. _ O^+ extended; ____ He+ restricted; and _ _ _ He^+ extended.

restricted and extended, occur over a broader range of MLT than the He^+ conics. There is an increased occurrence frequency of both types of conics for both species starting at ~06:00 MLT and extending to ~13:00 MLT for O^+ restricted conics and earlier MLTs for restricted and extended He^+ conics. We also note that the pre-noon maximum in occurrence frequency is noticeably less for He^+ extended conic distributions.

Energetic ion conics identified by our algorithm are found most frequently above 70° INVL. O^+ conics occur poleward of He^+ conics, and the peak occurrence frequency of He^+ conics extends over a broader range of invariant latitude. There is also a very small poleward shift of extended conic distributions.

The results reported for the occurrence distribution of restricted ion conics in Table 1 and Figure 12 are mostly consistent with previous morphological studies. Because we have used a different technique to identify ion conics, we expect the present results to differ from those reported by Yau et al. [1984], Sagawa et al. [1987], and Kondo et al. [1990] which were also based on data from the DE -1 EICS instrument,but which were more coarsely binned in energy and angle. These authors used a one dimensional algorithm that identified angular peaks in three broad energy ranges. (10 eV - 1 keV, 1-4 keV, and 4-17 keV) for H^+ and O^+ ions. For O^+ Yau et al. [1984] found high occurrence frequencies (~20%) in the 10 eV - 1 keV energy channel, and frequencies comparable to those reported in Figure 12 in the higher energy channels. They found an altitude dependence for O^+ conics in the two high energy ranges similar to that of O^+ restricted conics shown in Figure 12. The Yau et al. O^+ invariant latitude distributions showed two relative maxima near 65° and 75°, compared to Figure 12 which shows a single maximum for O^+ conic occurrence above 75°. This difference arises from the different model dependent definitions of an ion conic distribution. In the present study we require that the figure of merit (defined above) of an ion conic distribution be at least 10 times that of any candidate conic distribution flowing in the opposite direction. The second, low latitude, maxima in conic occurrence frequency found by Yau et al. comes from the characterization of asymmetric counter streaming ion conic distributions noted by Sagawa et al. [1987] as ion conic distributions. The 6-12 MLT maxima and 15-20 MLT minima in O^+ restricted conic occurrence shown in Figure 12, are also clearly defined in the Kondo et al. [1990] results. Dayside maxima in ion conic occurrence has also been reported by Gorney et al. [1981], Thelin et al. [1990] and Miyake et al. [1993].

A side benefit of the identification of a conic path is the ability to take the flux values actually measured in each energy angle pixel along the path, $j(E,\theta)$, and use them to calculate characteristic fluxes and energies. We can then calculate average characteristic energies and fluxes and the variation of these average quantities over the altitude, invariant latitude and magnetic local time ranges available to us to provide further insight into the processes responsible for conic formation. There are many different quantities that we can calculate to characterize the flux and energy of ion conic events. Among them are the upward directed flux calculated using the measured fluxes in each pixel of the conic path, J_{PATH}, the flux weighted average energy associated with pixels in the conic path, E_{PATH}, and the net flux above the energy threshold used in our study, J_{NET}. These quantities are explicitly defined in equations 1-3.

$$J_{PATH} = \int_{PATH} \mathbf{j} \bullet \mathbf{n} \, dE \quad (1)$$

$$E_{PATH} = \int_{PATH} E \, \mathbf{j} \bullet \mathbf{n} \, dE \, / J_{PATH} \quad (2)$$

$$J_{NET} = \int_{LC}^{180} \int_{150\ eV}^{17\ keV} \mathbf{j} \bullet \mathbf{n} \, dE \, d\Omega \quad (3)$$

Where **j • n** is the vector product of the directed flux and a unit vector in the magnetic field direction. The flux values used in equations 1-3 have all been normalized to 1,000 km using a dipole magnetic field model. The path in Equations 1 and 2 includes only flux pixels in the conic path determined by the algorithm described above. In Equation 3 the integral is taken over all energies and angles excluding those angles in the loss cone and those energies below the threshold used in this study, 150 eV. The velocity space density in pixels along the conic path $f(E,\theta) \propto j(E,\theta)/E$. It is therefore possible to define a temperature (kT in keV) by fitting a Maxwellian to the measured velocity space densities along the conic path

$$f(E) = \chi\ e^{-E/kT} \tag{4}$$

In equation 4, χ and kT are the two constants determined from the fit of the velocity space densities $f(E_i, \theta_i)$ at the pixels located at energy and angle points E_i and θ_i along the conic path. Evaluation of the average values of the quantities defined in equations 1-4 from the thousands of conic events identified by our algorithm also requires making a correction for the false positive conic events discussed above. In this report we will present results for J_{PATH} and E_{PATH}. Presentation and discussion of the average values of J_{NET}, kT, and other quantities determined from the flux images and fluxes along the conic paths requires more space than is available here.

Figure 13 presents the average value of J_{PATH} normalized to 1,000 km in units of $(cm^2\text{-s-sr-keV})^{-1}$ as a function of altitude, invariant latitude and magnetic local time. The average values plotted in Figure 13 have been corrected for the contribution of false positive conics by using the value determined from the identified, false positive, downflowing conics using the relation:

$$J_{PATH}(\Psi) \equiv \{N_U(\Psi) \times J_{UPATH}(\Psi) - N_D(\Psi) \times J_{DPATH}(\Psi)\} \div \{N_U(\Psi) - N_D(\Psi)\} \tag{5}$$

Where Ψ refers to one of the intervals of altitude, INVL, or MLT in Figure 13, N is the total number of conic events found in the interval ψ, $J_{PATH}(\psi)$ is the average value in that interval, and the subscripts U and D refer to upward and downward directed conics identified by the algorithm. J_{PATH} values for O^+ restricted and extended conics and He^+ restricted conics are shown in Figure 13. J_{PATH} for He^+ extended conics and all conics at invariant latitudes greater than 80° or less than 65° have been omitted from Figure 13, because the statistical uncertainty in their values was comparable to their values. The uncertainty in the values for He^+ restricted conics and conic events at latitudes greater than 80° is the direct result of too few events. The large uncertainty in the values of J_{PATH} for latitudes less than 65° comes from the comparable number of upflowing and downflowing events found in this latitude range and the large relative uncertainties in the value determined by subtracting numbers of comparable magnitude.

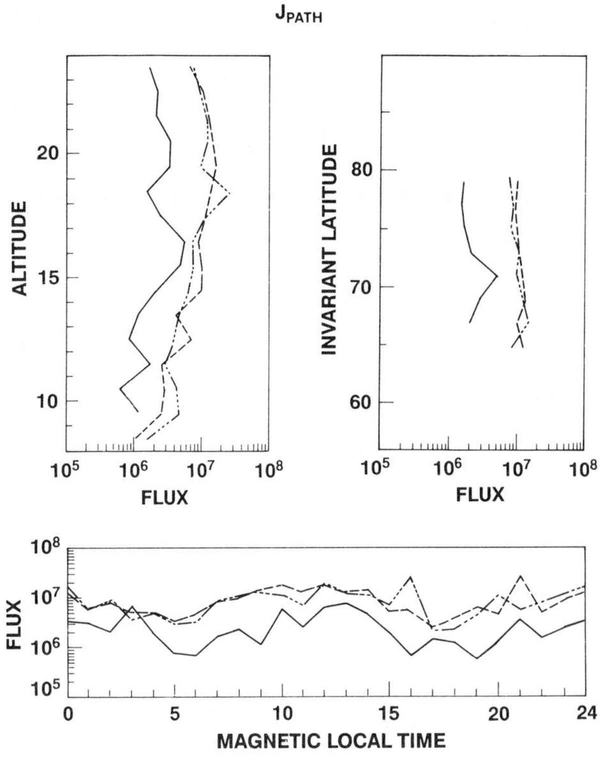

Fig. 13. Distribution in altitude, invariant latitude, and magnetic local time of the average value of J_{PATH} in units of $(cm^2\text{-s-sr-keV})^{-1}$. J_{PATH} is defined by Equations 1 and 5. The format is the same as Figure 12 except only three of the types of ion conic distribution identified by our algorithm are displayed. Key: ... O^+ restricted; _ .. _ O^+ extended; and ____ He+ restricted.

Figure 13 shows that the upward directed flux associated with restricted and extended O^+ events is comparable in all regions of the magnetosphere and that the He^+ conic fluxes are, on average ~20% of the O^+ fluxes. The altitude distribution shows a steady increase in J_{PATH} from 8,000 to ~17,000 km where it remains more or less constant at a value of 10^7 $(cm^2\text{-s-sr-keV})^{-1}$. The magnetic local time distribution shows the maximum J_{PATH} values at MLT's near the region of peak occurrence frequency shown in Figure 12. The invariant latitude distribution of J_{PATH} values is constant for O^+ conics.

Figure 14 presents the average value of E_{PATH} as a function of altitude, invariant latitude and magnetic local time. A correction similar to that shown in equation 5 to account for the contribution of false positive upflowing conics by using the value determined from the identified downflowing false positive conics has been applied to the data shown in Figure 14. E_{PATH} values for He^+ restricted conics, conic events at latitudes greater than 80°, less than 68 °, and altitudes less than 12,000 km are not shown because of the uncertainties in calculating the average value are large. Preliminary analysis shows that the source of the variation in the MLT distribution of E_{PATH} values of O^+ extended conics does not come from the statistical uncertainty of the measurement, a more detailed investigation to determine the source of the MLT variation has been undertaken. The working hypothesis for this investigation is that the variation is related to variations in geomagnetic activity at the times the data were acquired.

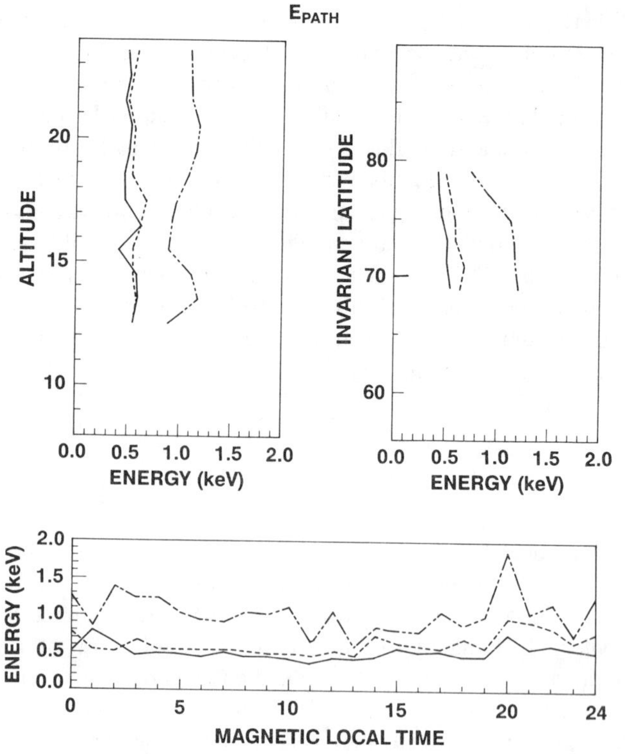

Fig. 14. Distribution in altitude, invariant latitude, and magnetic local time of the average value of E_{PATH} in units of keV. E_{PATH} is defined by Equation 2 and a second equation similar to Equation 5. The format is the same as Figure 12 except only three of the types of ion conic distribution identified by our algorithm are displayed. Key: ... O^+ restricted; _ .. _ O^+ extended; and ____ He+ restricted.

DISCUSSION

Above we have presented details of the two dimensional algorithm we have developed to identify and characterize energetic ion conic distributions. We have also compared restricted ion conic distributions identified by the algorithm with the results from earlier morphological studies. In this section we discuss the results we have obtained using our new algorithm.

The data presented in Table 1 and Figures 12-14 present the first detailed results from the application of an improved procedure to automatically identify, classify, and characterize extended energetic ion conic events. We have determined that at altitudes above 15,000 km conic distributions with an extended angular distribution occur more frequently than ion conic distributions with restricted angular distributions. We have determined that extended conic distributions are a non-negligible fraction of the conic distributions found at altitudes above 8,000 km, all invariant latitudes and magnetic local times sampled by the DE -1 spacecraft. We have also assembled some average plasma parameters calculated from the flux images and conic paths in order to learn more about the transverse ion energization processes acting at mid altitudes (8,000 to 25,000 km).

The distribution in space of the occurrence frequency, average flux, J_{PATH}, and energy, E_{PATH}, for restricted and extended O^+ conic distributions have some similarities and differences that place constraints on micro and/or macro physical mechanisms that generate O^+ conic distributions. (We defer discussion of the similarities and differences in He^+ conic parameters until we have binned them into altitude, magnetic local time, and invariant latitude bins broad enough to provide statistically significant samples over the entire volume of space sampled by DE -1.) As noted in the discussion of the data presented in Figure 12 above, the occurrence frequency of O^+ restricted and extended conics have almost identical distributions in invariant latitude, slightly different distributions in magnetic local time, and dissimilar distributions in altitude. The peak occurrence frequency at 77° INVL corresponds roughly to the peak of auroral activity in the 6:00 to 12:00 interval where the distribution in magnetic local time is maximum. The occurrence frequency distribution of extended conics has a much stronger altitude dependence than that for restricted conics. The broader distribution of restricted conics in magnetic local time and distinctly different distribution in altitude could arise from several sources including differences in the MLT distributions of the energy sources responsible for formation of the two types of conic. We also note that the increase in the occurrence frequency of extended ion conics with altitude is consistent with their generation by broad band, low frequency waves generated above 25,000 km and transported down magnetic field lines as discussed by Crew et al. [1990], and André et al. [1990].

The average J_{PATH} values for O^+ restricted and extended conics shown in Figure 13 are similar. Two other features of the average J_{PATH} values for O^+ restricted and extended conics are remarkable: 1) the independence of the average J_{PATH} value on invariant latitude; and 2) the steady fall off in the average J_{PATH} value at altitudes below 17,000 km. The equality of the average J_{PATH} values for O^+ restricted and extended conics and lack of variation in invariant latitude most probably reflects a common ionospheric source population. This is not an unanticipated conclusion. Peterson et al. [1993] used simultaneous observations of O^+ ion distributions obtained at two altitudes on auroral magnetic field lines to conclude that multiple heavy ion acceleration processes act on the same field line. The processes include an ionospheric pre-energization mechanism that provides sufficient energy to transport O^+ ions with energies in the range 1 - 10 eV to altitudes ~ 5,000 km; and transverse and parallel energization at altitudes above the ionosphere. The steady fall off of the average O^+ J_{PATH} values below 17,000 km could be a threshold effect as discussed below.

The average E_{PATH} values for O^+ restricted and extended conics have essentially the same distributions in altitude, invariant latitude, and magnetic local time. These distributions have less variations than those for occurrence frequency or J_{PATH}. The most remarkable feature of the E_{PATH} data displayed in Figure 14 is that extended O^+ conics have an average E_{PATH} value approximately twice that of restricted O^+ conic distributions. This difference in E_{PATH} values for O^+ restricted and extended conics and their weak variation in altitude, magnetic local time and invariant latitude could be an artifact of the algorithm or of the data. A preliminary investigation of variation in E_{PATH} and the average temperature determined by fits to Equation 4 about their average values showed, in some altitude, invariant latitude, and magnetic time ranges, variations comparable with the average values. Clearly a systematic investigation of the variation of parameters derived from conic paths and flux images must be performed before conclusions based on the data presented in Figure 14 are made. Such a study is planned, but has not yet been performed.

We now return to the the topic of the altitude dependance of J_{PATH} values. the most probable explanation for the slow, steady fall off of average O^+ J_{PATH} values below 17,000 km is a threshold effect, where the ion population acquires energy as it goes up the magnetic field line until, on average, at 17,000 km most ion distributions are above the energy threshold of our study. This suggests that there should be an detectable increase in a characteristic energy, such as E_{PATH}, as a function of altitude. As noted above there are large variations in E_{PATH} values at fixed altitudes which could mask a systematic increase in E_{PATH} with altitude. No clear altitude dependence of E_{PATH} is detectable in Figure 14.

The data presented above and in recent reports by Peterson et al., [1992], Miyake et al. [1993], and Peterson et al. [1993] provide some new insights into the distribution in altitude of transverse ion acceleration processes. The literature discussing the transverse ion energization mechanisms responsible for generating conic distributions is enormous, too large to review or even briefly summarize here. A good introduction, however, to this literature has been prepared by Klumpar, [1986].

Peterson et al. [1993] used simultaneous observations of O^+ ion distributions obtained at two altitudes on auroral magnetic field lines to conclude that multiple heavy ion acceleration processes commonly act on the same field line. Peterson et al. [1992] have examined the altitude dependence of restricted conic distributions presented in Figure 12 and reported the average pitch angle of the restricted conic events as a function of altitude. From these results and a report of a similar study made from data acquired below 10,000 km by the Akebono satellite [Miyake et al. 1993] Peterson et al. [1992] concluded that conic formation by localized, explosive, transverse energization is not the dominate mechanism responsible for producing the energetic conic distributions found above 8,000 km. The data presented in Figures 13 and 14 affirm the conclusion of Peterson et al. [1992] and further show that the upflowing flux associated with energetic conic events has less variation at a given altitude, invariant latitude or magnetic local time than a characteristic energy (E_{PATH}) of the distribution.

The picture that emerges from these recent observations is that transverse ion energization occurs in varying degrees at all altitudes above the ionosphere on auroral field lines.

CONCLUSIONS

We have developed and demonstrated an algorithm to identify and characterize energetic ion conic events and used the algorithm to classify and characterize a very large number of energetic ion conic events. We have noted slight differences in the conic events identified by our two dimensional algorithm and an earlier, one dimensional, algorithm used on the same data.

The data presented here are the first morphological study of the occurrence frequency distribution and average energy and flux associated with extended ion conic events. They show that, at altitudes above 15,000 km, conic distributions with an extended angular distribution occur more frequently than ion conic distributions with restricted angular distributions. We have determined that extended conic distributions are a substantial fraction of the conic distributions found at altitudes above 8,000 km, all invariant latitudes and magnetic local times sampled by the DE -1 spacecraft. These data and other recent reports affirm the conclusion that transverse ion energization occurs in varying degrees at all altitudes above the ionosphere on auroral field lines.

Acknowledgments. We gratefully acknowledge helpful discussions with Andrew Yau, Mats André, and Geoff Crew. We thank Daryl Carr for help with assembling the data base. Support was provided by NASA contract NAS5-33032 and the Lockheed Independent Research program.

REFERENCES

André, Mats, G.B. Crew, W.K. Peterson, A.M. Persoon, C.J. Pollock, and M.J. Engebretson, Ion heating by broadband low-frequency waves in the cusp/cleft, *J. Geophys. Res. 95, 20809,* 1990.

Chang, T., G.B. Crew, N. Hershkowitz, J.R. Jasperse, J.M. Retterer, and J.D. Winningham, Transverse acceleration of oxygen ions by electromagnetic ion cyclotron resonance with broad band left-hand polarization waves, *Geophys. Res. Lott. 13*, 636, 1986.

Crew, G.B., T. Chang, J.M. Retterer, W.K. Peterson, D.A. Gurnett, and R.L. Huff, Ion cyclotron resonance heated conics: Theory and observations, *J. Geophys. Res. 95*, 3959, 1990.

Davis, C., and N. Vaidta, Software for optical archive and retrieval (SOAR), users' guide, National Space Science Data Center, Greenbelt Maryland, *NSSDC/WDC-A-R&S, 89-16,* 1989.

Doherty, M.F., C.M. Bjorklund, W.K. Peterson and H.L. Collin, Automatic detection of mass-resolved ion conics, Lockheed Document Number *IEEE Transactions on Geoscience and Remote Sensing,* 31, 407 1993.

Gorney, D.J., A. Clarke, D. Croley, J.F. Fennell, J. Luhmann, and P.F. Mizera, The distribution of ion beams and conics below 8000 km, J. Geophys. Res. 86, 83, 1981.

Gorney, D.J. Y.T. Chiu, and D.R. Croley, Jr., Trapping of ion conics by downward parallel electric fields, *J. Geophys. Res. 90*, 4205, 1985.

Hoffman, R.A., and Schmerling, E.R., Dynamics Explorer program: An overview, *Space Sci. Instrum. 5,* 345, 1981.

Horwitz, J.L., Velocity filter mechanism for ion bowl distributions (bimodal conics), *J. Geophys. Res. 91*, 4513, 1986.

Klumpar, D.M., W.K. Peterson, and E.G. Shelley, Direct evidence for two-stage (bimodal) acceleration of ionospheric ions, *J. Geophys. Res. 89*, 10779, 1984

Klumpar, D.M., A digest and comprehensive bibliography on transverse auroral ion acceleration, in *Ion Acceleration in the Magnetosphere and Ionosphere,* Tom Chang, Ed., Geophysical Monograph #38, American Geophysical Union, Washington DC. p 389, 1986.

Kondo, T., B.A. Whalen, A.W. Yau, and W.K. Peterson, Statistical analysis of upflowing ion beam and conic distributions at DE -1 altitudes, *J. Geophys. Res.* 95, 12091, 1990.

Miyake, W., T. Mukai, and N. Kaya, On the evolution of ion conics along the field line from Exos -D observations, t*J. Geophys. Res.*, 98, 11127, 1993.

Peterson, W.K., E.G. Shelley, S.A. Boardsen, D.A. Gurnett, B.G. Ledley, M. Sugiura, T.E. Moore, and J.H. Waite, Transverse ion energization and low-frequency plasma waves in the mid-altitude auroral zone: A case study, J. Geophys. Res. 93, 11405, 1988.

Peterson, W.K., H.L. Collin, M.F. Doherty, and C.M. Bjorklund, O^+ and He^+ restricted and extended (bi-modal) ion conic distributions, Geophys. Res. Lett., 19, 1439, 1992.

Peterson, W.K., A.W. Yau, and B.A. Whalen, Simultaneous observations of H^+ and O^+ ions at two altitudes by the Akebono and Dynamics Explorer -1 satellites, *J. Geophys. Res.*, 98, 11177 1993.

Sagawa, E., A.W. Yau, B.A. Whalen, and W.K. Peterson, Pitch angle distributions of low-energy ions in the near-Earth magnetosphere, *J. Geophys. Res. 92*, 12241, 1987.

Sharp, R.D., R.G. Johnson, and E.G. Shelley, Observation of an ionospheric acceleration mechanism producing energetic (keV) ions primarily normal to the geomagnetic field direction, *J. Geophys. Res. 82*, 3324, 1977.

Shelley, E.G., D.A. Simpson, T.C. Sanders, E. Hertzberg, H. Balsiger, and A. Ghielmetti, The energetic ion composition spectrometer (EICS) for the Dynamics Explorer -A, *Space Sci. Instrum. 5,* 443, 1981

Temerin, M., Evidence for a large bulk ion conic heating region, *Geophys. Res. Lett. 13*, 1059, 1986.

Thelin, B., B. Aparicio, and R. Lundin, Observations of upflowing ionospheric ions in the mid-altitude cusp/cleft region with the Viking satellite, *J. Geophys. Res. 95*, 5931, 1990.

Yau, A.W., B.A. Whalen, W.K. Peterson, and E.G. Shelley, Distributions of upflowing ionospheric ions in the high-altitude polar cap and auroral ionosphere, *J. Geophys. Res.* 89, 5507, 1984.

W.K. Peterson, H.L. Collin, and C.M Bjorklund, Lockheed Palo Alto Research Laboratory, Palo Alto, California 94304.

M.F. Doherty, 913 Laguna Ave., Burlingame, California 94010

Nonlinear Wave-Particle Interactions in the Magnetosphere

Manju Prakash

Physics Department, State University of New York at Stony Brook

Stony Brook, New York.

The nonlinear interaction between the bounced motion of a proton and the kinetic Alfvén wave excited in the magnetosphere is studied. Using the Poincaré surface of section technique, the motion of the proton is traced numerically in the phase space. With suitable initial conditions, the particle trajectories are calculated under paraxial approximation, for several values of the wave amplitude. The trajectories show chaotic behavior at some finite amplitude of the wave. The threshold electrostatic field computed for the onset of chaos is less than the expected electric field associated with a kinetic Alfvén wave in the magnetosphere. This suggests that nonlinear wave-particle interactions can lead to chaos and subsequent instability of the bounced motion of the proton.

1. INTRODUCTION

The primary objective of the present work is to investigate wave–particle chaos near the Earth's equatorial region in the magnetosphere. The work was motivated by satellite observations of a large amplitude kinetic Alfvén wave (KAW) near the equatorial region and the encouraging progress made in the phenomenon of chaos in the field of plasmas in general[*Prakash*, 1989]. Chaotic aspects of particle dynamics can lead to a better understanding of various physical processes taking place in the magnetosphere [*Abdalla and Baker*, 1991]. Among other things, one can calculate transport coefficients in the magnetosphere[*Horton and Tajima*, 1991]. Valuable insights can be gained by comparing the results of calculations with observations of particle flux oscillations[*Chen and Palmadesso*, 1984; *Curran et al.*, 1986].

The present work investigates the stability of motion of a bouncing proton near the equatorial region of the Earth. Ring current proton with energy in the keV range can interact with a large amplitude KAW observed near the equatorial region[*Takahashi et al.*, 1985]. The trajectory of the proton becomes chaotic due to nonlinear wave particle interactions. The particle is scattered into its loss cone due to its random parallel velocity. It can then precipitate in the atmosphere. The emphasis of the work is to illuminate the role of the dynamical aspects of wave–particle chaos and not to perform the quantitatively rigorous calculation of the amount of precipitation.

We will study the dynamics of the proton and the KAW in the three dimensional phase space (p, q, t). Thus we have a system with 3/2 degrees of freedom. The Hamiltonian of the system (wave–particle) is time dependent. For small wave amplitudes, the dynamics can be studied using analytical methods. In this regime, the proton motion is predominantly regular. For large amplitudes of the wave, the invariants of motion break down completely due to nonlinear effects[*Lichtenberg and Lieberman*, 1983]. Nonlinearity arises due to the space dependence of the field quantities and leads to irregular particle motion. We will study particle dynamics by solving the nonlinear equations numerically. The stochastic aspects are studied in the Poincaré surface of section plot. *Liu* [1991] has studied wave–particle chaos due to a test electron and

Space Plasmas: Coupling Between Small
and Medium Scale Processes
Geophysical Monograph 86

standing kinetic Alfvén waves by relaxing the assumption of a parabolic potential.

2. HAMILTONIAN APPROACH

The KAW differs from the MHD Alfvén wave due to the presence of a sizeable parallel electric field [*Hasegawa and Chen*, 1976]. The parallel component is a consequence of the fact that the perpendicular wave vector is comparable to the inverse of ion–gyro radius. We consider a kinetic Alfvén wave of constant amplitude E_0 and frequency ω superimposed on the magnetic field of the Earth under the dipole approximation. Near the equatorial region, the magnetic field can be approximated by a parabolic potential well under the assumption of magnetic moment conservation. Under the paraxial approximation of the ambient magnetic field, a proton executes simple harmonic motion with bounce frequency ω_b. The amplitude of the oscillator is given by the distance z_m along the line of force measured from the magnetic equator to a mirror point. We study wave–particle interaction under the guiding center approximation[*Prakash*, 1989]. The equations of motion are:

$$\frac{dp_z}{dt} = -\mu\frac{\partial B}{\partial z} + eE(z,t)\,, \tag{1}$$

$$p_z = m_0\frac{dz}{dt}\,, \tag{2}$$

$$\mu = \frac{p_\perp^2}{2m_0 B} = \text{constant}\,. \tag{3}$$

In the above, the Earth's magnetic field is denoted by $B(z)$. The quantities e and m_0 represent the charge and the mass of the proton, respectively. The parallel and perpendicular components of the particle's momentum are denoted by p_z and $p_\perp$, respectively. Henceforth, the z coordinate is rescaled to a dimensionless quantity

$$z = \frac{z}{LR_E}\,, \tag{4}$$

where R_E is the radius of the Earth, and L is a dimensionless number. We choose the parallel electrostatic field of the wave as

$$E(z,t) = E_0\cos(kz - \omega t)\,, \tag{5}$$

where k is the parallel wave vector. Near the equatorial plane, the magnetic field $B(z)$ may be assumed to be parabolic

$$B(z) = B_0(1 + bz^2)\,, \tag{6}$$

$$b = \left.\frac{1}{2B_0}\frac{\partial^2 B}{\partial z^2}\right|_{z=0}. \tag{7}$$

The magnetic field $\vec{B}$ satisfies Maxwell's equation $\nabla\cdot\vec{B} = 0$ if we introduce a small component in the perpendicular direction. Substituting Eq. (6) in Eqs. (1)–(3), we obtain

$$\ddot{z} = -\omega_b^2 z + \frac{e}{m_0}E(z,t)\,. \tag{8}$$

The bounce frequency ω_b of the particle is given by

$$\omega_b^2 = \frac{2\mu b B_0}{m_0}\,. \tag{9}$$

3. PARTICLE DYNAMICS

The equation of motion of the proton Eq. (8) can be rewritten as

$$\ddot{z} = -\omega_b^2 z + \epsilon\,\omega_b^2 cos(kz - \omega t)\,. \tag{10}$$

where the perturbation parameter

$$\epsilon = \frac{eE_0}{m_0\omega_b^2}\frac{1}{LR_E}\,. \tag{11}$$

For $\epsilon << 1$, Eq. (10) can be integrated analytically over the unperturbed particle trajectory. This enables us to study linear but nonadiabatic processes such as wave–particle resonance [*Prakash*, 1989]. Nonadiabatic behaviour of the magnetic moment of a charge particle in a dipole magnetic field and in the presence of the external electric field has been studied by *Murakami et al.* [1990].

In the next section, we study the proton dynamics numerically with ϵ and q as parameters. The winding number q is defined by

$$q = \frac{\omega}{\omega_b}\,. \tag{12}$$

Our results will be displayed in the surface of section plot. For small values of the parameter ϵ, and for a rational value of q, the trajectory is overall periodic and closes upon itself. In the Poincaré surface of section plot, this is represented by a curve with a finite number of points. In the range of small ϵ, and irrational q, the particle trajectory is quasi–periodic and fills a curve densely. For larger values of ϵ, there is no integral of motion; the particle motion becomes irregular and fills up the phase space ergodically.

4. NUMERICAL ALGORITHM

To study particle dynamics for arbitrary values of ϵ, the equation of motion (Eq. (10)) can be integrated numerically using the Gear–Hindmarsh scheme. The scheme is accurate to fifth order in the time step. To check the stability of the numerical results with the total time of integration, calculations were performed for several values of T_{max} in the range $(100 - 400)$ T_ω, where T_ω is the wave period. The dynamical pattern remains invariant in the time range considered. We will present our results for $T_{max} = 400$ T_ω. For optimum transfer of energy, protons with bounce period T_b comparable to the wave period T_ω will be considered. This implies protons with energies in the keV range. The position z and the velocity v_z (instead of p_z) of the proton were computed at integral values of T_ω. The total time of integration employed was 500 times the characteristic time for bounce motion. This time is sufficiently long to study the time evolution of wave–particle dynamics.

5. CHAOS AND NONLINEAR DYNAMICS

The phase space structure of the perturbed system depends on the perturbation parameter ϵ, the winding number q and the initial conditions $z(0)$ and $v(0)$. The results presented in Figs. (1) through (12) focus on the approach to chaos for various values of ϵ.

In Figure 1, we show the surface of section plot for $\epsilon = 0.35$ and $q = 4/5$. The wave period $T_\omega = 100$ s and $T_b = 80$ s and the initial conditions are $z(0) = 0.015$, $v(0) = 0.0$ s^{-1}. The initial conditions are normalised with respect to the Earth's radius. The particle trajectory is bounded and integrable. The motion of the integrable orbit is constrained to the surface of an invariant torus and its intersection with the $z - v_z$ plane is a closed curve. This is known as the invariant KAM curve[*Arnold*, 1968] and is described by a finite number of points. For a given ϵ, the number of points is determined by the value of q. We find that particle motion is regular for $\epsilon < 0.55$. In this region, the salient features of particle dynamics do not change with changes in the initial conditions. This was verified by repeating the calculations for nearby initial conditions.

With further increase in ϵ (from 0.55), the trajectory is no longer confined to the invariant torus. Figure 2, demonstrates this for $\epsilon = 0.55$ when the invariant curve breaks down into a set of islands [*Arnold*, 1968]. The center of each island is an elliptic point and the islands correspond to stable regions of phase space with nearby trajectories staying close to them. Between the islands there are hyperbolic points [*Greene*, 1968, 1979]. This is shown in Figure 3, for $\epsilon = 0.6$. Nearby trajectories in this region diverge from each other. This is charac-

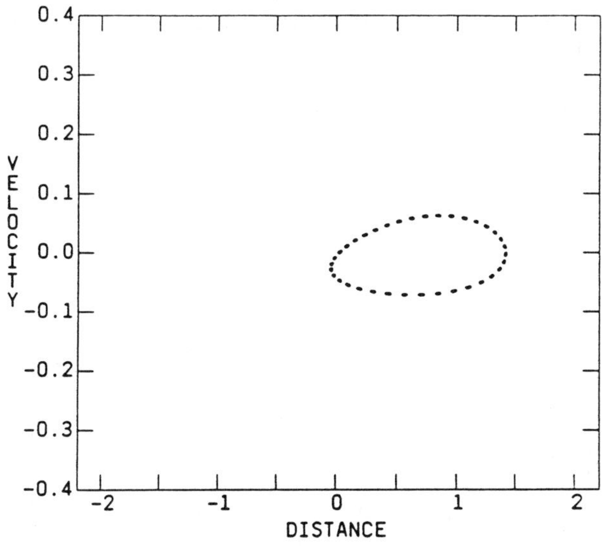

Fig. 1. Surface of section plot of a bouncing proton in the presence of a kinetic Alfvén wave with period $T_w = 100$ s and total time of integration $T_{max} = 400$ T_w. The wave vector k has the value $\pi/(3R_E)$. The bounce period is $T_b = 80$ s. The initial conditions are $z(0)$=0.015 and $v(0)$=0.0 s^{-1}. The dots represent a periodic orbit for a rational winding number, q=4/5. The perturbation parameter $\epsilon = 0.35$.

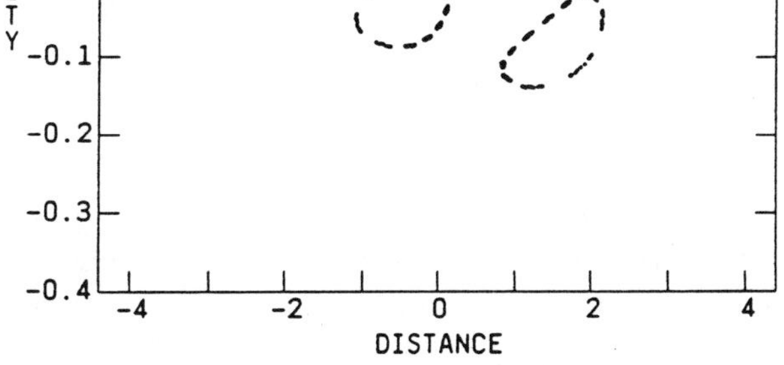

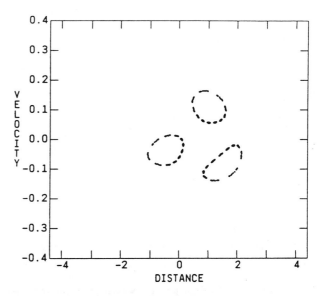

Fig. 2. Surface of section plot for a rational winding number q =4/5 and ϵ= 0.55. The initial conditions are as in Fig. 1. The trajectory breaks into a set of islands.

teristic of irregular dynamics. Thus with an increase in the energy of the system, the phase space is a mixture of regular and irregular regions. The size of the irregular region grows with the increase in perturbation parameter ϵ.

Figures 4, and 5, show the growth of the irregular region in between the islands for $\epsilon = 0.7$ and 1.0, respectively. This growth results in merging of the islands showing that particle motion is predominantly irregular. In this parameter range, we study the characteristics of the regular motion by repeating the calculations with nearby initial conditions. For example, with ϵ=1.0, $z(0) = 0.017$, and $v(0) = 0.0$ s^{-1}, Figure 6, (in comparison to Figure 5) illustrates the change in the pattern

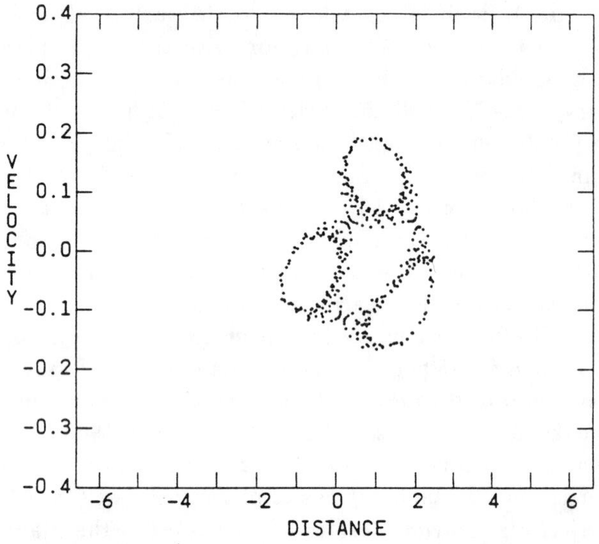

Fig. 3. Surface of section plot for a rational winding number q=4/5 and $\epsilon = 0.6$. The irregular region coexists with regions of regular motion.

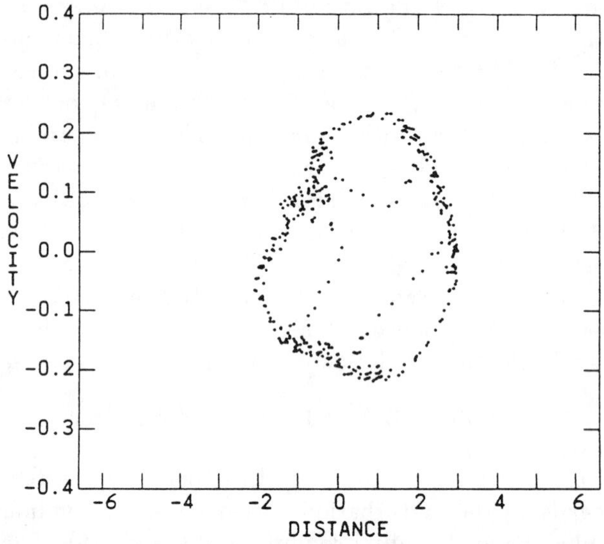

Fig. 4. Surface of section plot for a rational winding number q=4/5 and ϵ=0.7. The initial conditions are $z(0)$= 0.015 and $v(0)$= 0.0 s^{-1}. The irregular region grows due to the overlap of islands .

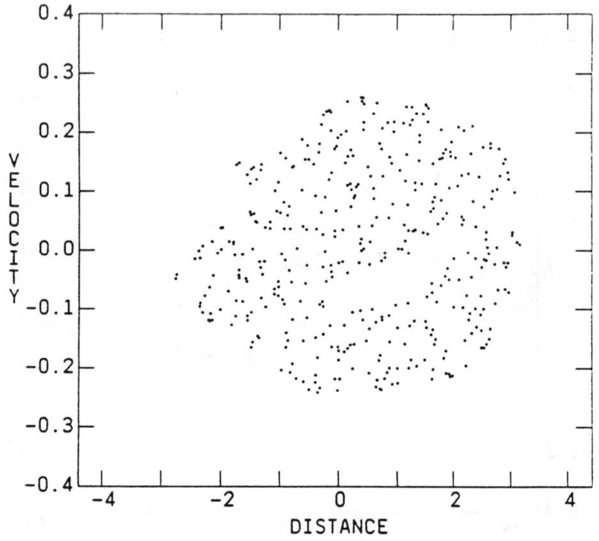

Fig. 5. Surface of section plot for a rational winding number, $q = 4/5$ and $\epsilon = 1.0$. The initial conditions are as in Fig. 1. The phase space is a mixture of regular and irregular regions.

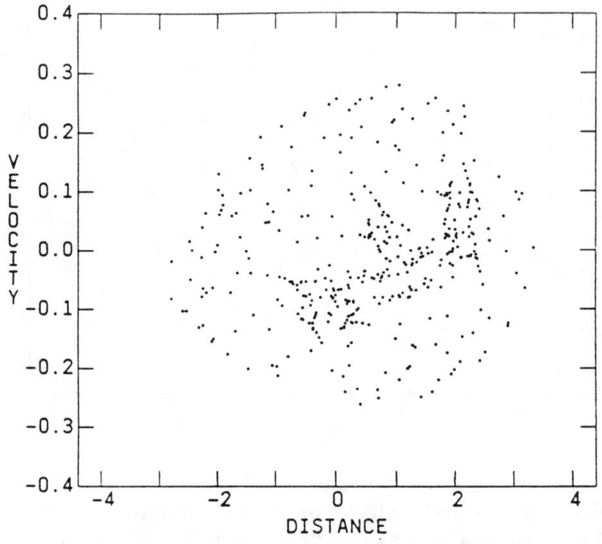

Fig. 6. Surface of section plot for a rational winding number $q = 4/5$ and $\epsilon = 1.0$. The initial conditions are $z(0) = 0.017$ and $v(0) = 0.0$ s^{-1}. Comparison with Fig. 5 illustrates the effects of small changes in the initial conditions.

of dynamics with changes in the initial conditions. In the linear region, the trajectories diverge linearly, and therefore they cover different regions of phase space with change of initial conditions. This explains why the dynamical pattern changes with change of initial conditions for $\epsilon < \epsilon_c$.

In Figure 7, results are shown for the onset of chaos which takes place for $\epsilon_c \simeq 1.2$. Nearby initial conditions reproduce the pattern of particle chaos in the surface of section plot. This implies that nearby trajectories diverge exponentially, i.e. the measure of regular motion disappears. Clearly, there is a progression from a KAM surface to islands to chaos as we increase the energy of the wave–particle system. For $\epsilon > \epsilon_c$, the salient features of particle dynamics do not change. However, the amplitude of motion in phase space increases.

Next we studied the evolution of particle dynamics for an irrational value of $q = 1/\sqrt{2}$. The initial conditions used were as in Figure 1. For $\epsilon << 1$, particle motion is quasi–periodic and fills a KAM curve densely. The curve is gradually distorted as we increase the value of ϵ, e.g. see the curve for $\epsilon = 0.6$ in Figure 8,. With further increase in ϵ, the curve breaks down into scattered dots which is shown in Figure 9, for $\epsilon = 0.8$. Particle dynamics has some measure of regular motion for $\epsilon < 1.2$. This was verified by repeating the calculations for nearby initial conditions. i.e. $z(0)$=0.017 and $v(0) = 0.0$ s^{-1}. For $\epsilon = 1.2$, the KAM curve breaks down completely into chaos (Figure 10,). We note that in this region, the pattern of chaos is not changed by varying initial conditions, or by further increase in the value of ϵ. Curves with rational and irrational winding numbers respond quite differently to perturbation; the former transform to islands, while the latter retain their character and distort gradually.

We also studied the variation of the threshold value ϵ_c with the parameter q (see Figures (11), and (12),). The details of the calculations are discussed in *Prakash*[1989]. Our conclusions are that there is no systematic behaviour of ϵ_c with the value of q for values of T_ω comparable to T_b. In the range studied, ϵ_c increases with T_b for a given T_ω

6. APPLICATION TO MAGNETOSPHERE

In this section, we will illustarate wave particle chaos with the suitable choice of the parameters of the wave and those of the bouncing proton near the equatorial region of the Earth. We will compute the quantity ϵ_c for such a system.

We consider a bouncing proton ($T_b \simeq 80$ s) near the Earth's equator. The proton can interact with the parallel electric field of the KAW. The wave with period $T_\omega \simeq$ 100 s has been observed near the Earth's equatorial region [*Takahashi et al.*, 1985; *Su et al.*, 1977; *Hughes et al.*, 1978] in connection with ULF pulsations. The KAW may be excited by the field line resonance[*Chen and Hasegawa*, 1974; *Southwood*, 1974] or by a bouncing elec-

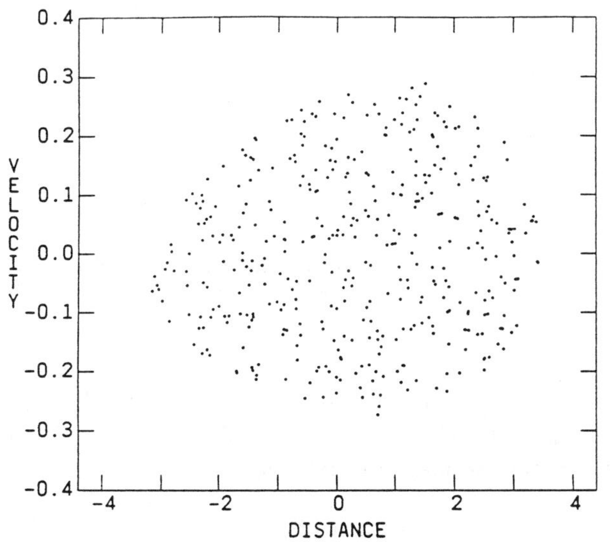

Fig. 7. Surface of section plot for a rational winding number, q =4/5 and $\epsilon_c = 1.2$. The initial conditions are as in Fig. 1. The system exhibits chaos .

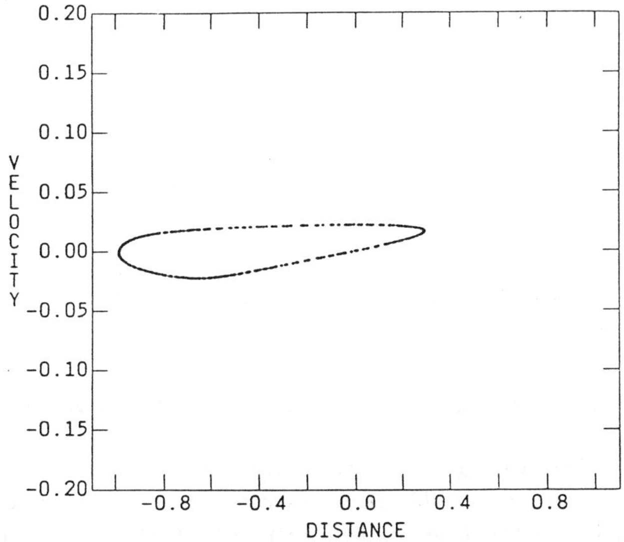

Fig. 8. Surface of section plot for $q = 1/\sqrt{2}$ and $\epsilon = 0.6$. The initial conditions are as in Fig. 1. The KAM curve is distorted.

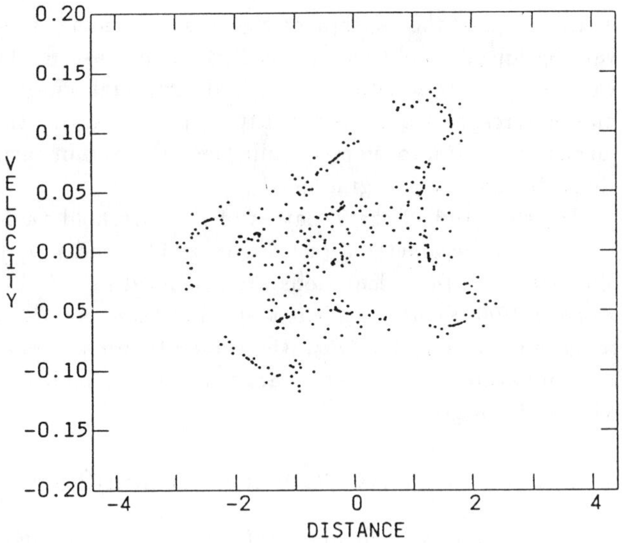

Fig. 9. Surface of section plot for $q = 1/\sqrt{2}$ and $\epsilon = 0.7$. The initial conditions are as in Fig. 1. The KAM curve breaks up into regular and irregular regimes.

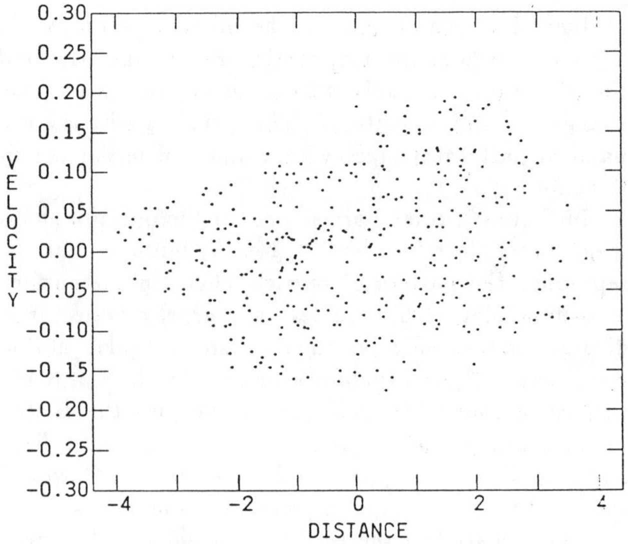

Fig. 10. Surface of section plot for $q = 1/\sqrt{2}$ and $\epsilon_c = 1.2$. The initial conditions are as in Fig. 1. The system exhibits chaos.

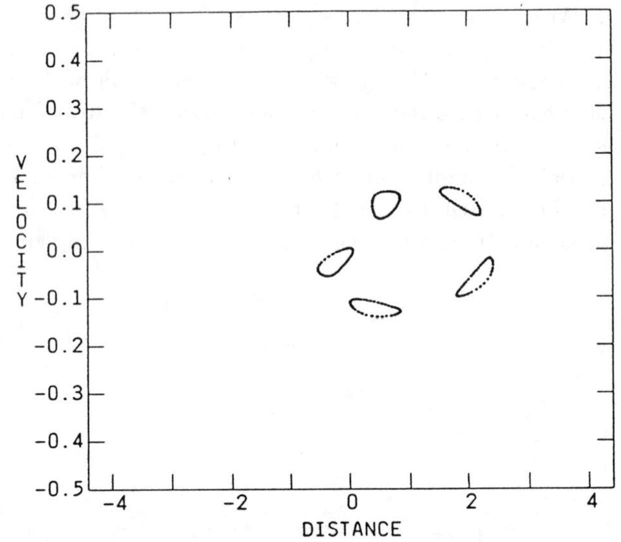

Fig. 11. Surface of section plot for $q= 9/10$ and $\epsilon = 0.4$. The initial conditions are as in Fig. 1. The KAM curve breaks into islands.

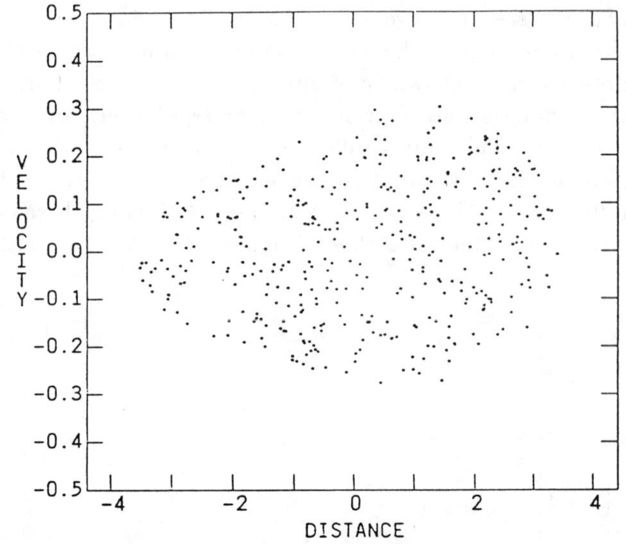

Fig. 12. Surface of section plot for $q= 9/10$ and $\epsilon_c = 1.4$. The initial conditions are as in Fig. 1. The islands overlap to yield chaos.

tron beam of high energy during a substorm[*Hasegawa*, 1979a].

We will show that the wave particle system with $q= 4/5$ can exhibit chaos for $\epsilon = \epsilon_c \simeq 1.2$. The parallel electric field of the kinetic Alfvén wave is given by[*Hasegawa*, 1979b]

$$E_z = k\phi(k_\perp \rho_i)^2 = \frac{k}{k_\perp} E_y (k_\perp \rho_i)^2 \, . \tag{13}$$

The quantity ρ_i denotes the gyro–radius of the ion. The perpendicular electric field E_y of the wave is related to the transverse magnetic field B_{y0} by

$$E_y = V_A B_{y0} \, , \tag{14}$$

where V_A is the velocity of the Alfvén wave and ϕ is the electrostatic potential. Equations (13) and (14) are valid when both $(k_\perp \rho_i)^2 << 1$ and $T_e << T_i$, where T_e and T_i are the electron and ion temperatures, respectively.

We consider ULF pulsation with transverse magnetic field $B_{y0} \cong 20$ nT and wave velocity $V_A \cong 1.5 \times 10^6$ m/s, respectively. These parameters yield a transverse electric field $E_y = 30$ mV/m. We take $k_\perp \rho_i = 0.5$ with $\rho_i = 0.2 R_E$. We choose $k = \pi/(3R_E)$ and $L = 6$. The parallel electric field E_z was estimated to be 3.1 mV/m. This yields a value of $\epsilon \cong 1.2$ (see Eq. (11)). This value is the same as the threshold value ϵ_c required for the onset of chaos (Fig. 7).

Therefore, protons with bounce period T_b equal to 80 s (with energy range keV) and interacting nonlinearly with the parallel component of the wave electric field (associated with ULF pulsations) will exhibit chaos. In the chaotic region, nearby particle trajectories diverge exponentially from each other. *Ho et al.* [1993] have calculated the Lyapunov exponent for such a system with various initial conditions. In the chaotic regime the bouncing proton acquires random parallel velocity. This results in the scattering of protons into the loss cone. The protons can therefore precipitate in the atmosphere.

7. CONCLUSIONS

Thus wave–particle chaos results in the loss of the mirroring protons and subsequent precipitation [*Ho et al.*, 1993, *private communication*] in the atmosphere.

Our conclusions are relevant to protons executing simple harmonic motion near the equatorial region. These conclusions may not extend to protons mirroring at higher latitudes where the approximations of a parabolic potential and the dipole field may not be valid. The model calculation presented here can be extended to study Arnold diffusion [*Chirikov*, 1979] and radial diffusion due to drift-bounce resonance when effects due to finite perpendicular wave vector are incorporated[*Chan et al.*, 1989].

Acknowledgements. I thank A. Hasegawa for suggesting this problem and for many discussions which stimulated this work. This work was supported by NSF contract EAR-0291.

REFERENCES

Abdalla, M. A., and D. N. Baker, Chaos and stochasticity in space plasmas, *Geophys. Res. Letts.*, *18*, 1573, 1991.

Arnold, A., *Ergodic Problems of Classical Mechanics*, W. A. Benjamin, New York, 1968.

Chan, A. A., L. Chen, and R. B. White, Nonlinear interaction of energetic ring current protons with magnetospheric hydromagnetic waves, *Geophys. Res. Letts.*, *16*, 1133, 1989.

Chen, J., and P. J. Palmadesso, Tearing instability in an anisotropic neutral sheet, *Phys. Fluids*, *27*, 1198, 1984.

Chen, L., and A. Hasegawa, A theory of long--period in magnetic pulsations, 1, Steady state excitation of field line resonance, *J. Geophys. Res.*, *79*, 1024, 1974.

Chirikov, B. V., A universal instability of many dimensional oscillator system, *Phys. Rep.*, *52*, 263, 1979.

Curran, D. B., C. K. Goertz, and T. A. Whelan, Ion distribution in two-dimensional reconnection geometry, *Eos, Trans. AGU*, *67*, 354, 1986.

Greene, J., M., Two dimensional measure preserving mappings, *J. Math. Phys.*, *9*, 760, 1968.

Greene, J., M., A method for determining a stochastic transition, *J. Math. Phys.*, *20*, 1183, 1979.

Hasegawa, A., Excitation of Alfvén waves by bouncing in electron beams: An origin of nightside magnetic pulsations, *Geophys. Res. Lett.*, *6*, 664, 1979a.

Hasegawa, A., Particle dynamics in low frequency electromagnetic waves in an inhomogeneous plasma, *Phys. Fluids 22*, 1988, 1979b.

Hasegawa, A., and L. Chen, Kinetic processes in plasma heating by resonant mode conversion of Alfvén wave, *Phys. Fluids 19*, 1924, 1976.

Ho, A. Y., S. P. Kuo, A. A. Chernikov and G. Schmidt, Chaotic ion motion in the magnetosphere, *Comments Plasma Phys. Controlled Fusion 15*, 129, 1993.

Hughes, W. J., R. L. McPherron, and J. N. Barfield, Geomagnetic pulsations observed simultaneously on three geostationary satellites, *J. Geophys. Res. 83*, 1109, 1978.

Lichtenberg, A. J., and M. A. Lieberman, *Regular and Stochastic Motion*, Springer--Verlag, New York, 1983.

Liu, W., Chaos driven by kinetic Alfvén waves, *Geophys. Res. Letts.*, *18*, 1611, 1991.

Murukami S., T. Sato and A. Hasegawa, Nonadiabatic behaviour of the magnetic moment of a charge particle in the dipole Field, *Phys. Fluids B*, *2*, 715, 1990.

Prakash, M., Particle precipitation due to nonlinear wave--particle interactions in the magnetosphere, *J. Geophys. Res. 94*, 2497, 1989.

Southwood, D. J., Some features of field line resonance in the magnetosphere, *Planet. Space Sci. 22*, 483, 1974.

Su, S.-Y., A. Konardi, and T. A. Fritz, On the propagation direction of ring current proton in ULF waves observed at ATS--6 at 6.6 R_E, *J. Geophys. Res. 82*, 1859, 1977.

Takahashi, K., P. R. Higbie, and D. N. Baker, Azimuthal in propagation and frequency characteristic of compressional Pc 5 waves observed at geostationary orbit, *J. Geophys. Res.*, *90*, 1473, 1985.

Manju Prakash, Physics Department, State University of New York at Stony Brook, Stony Brook, New York 11794-3800.

Transversely Accelerated Ions in the Topside Ionosphere

John M. Retterer[1]
Ionospheric Physics Div., Phillips Laboratory, Bedford, MA 01731

Tom Chang
Center for Space Research, MIT, Cambridge, MA 02139

John R. Jasperse[1]
Ionospheric Physics Div., Phillips Laboratory, Bedford, MA 01731

Data from the rocket campaigns *MARIE* and *TOPAZ* III, within regions of low-altitude transversely accelerated ions, are interpreted to explain the acceleration of the ions. Using the Monte Carlo kinetic technique to evaluate the ion heating produced by the simultaneously observed lower-hybrid waves, we find that their observed electric field amplitudes are sufficient to explain the observed ion energies in the *MARIE* event. Much of the uncertainty in evaluating the efficiency of a plasma wave-induced particle heating process which is dependent on a velocity resonance comes from the lack of information on the phase velocities of the waves. In the case of the *MARIE* observations, our modeling efforts show that features in the ion velocity distribution are consistent with the wave phase velocities inferred from interferometer measurements of wavelengths. The lower-hybrid waves with which low-altitude transversely accelerated ions are associated are frequently observed to be concentrated in narrow wave packets called 'spikelets'. We demonstrate through the scaling of the sizes of these wave packets that they are consistent with the theory of lower-hybrid envelope solitons that are a hallmark of the nonlinear wave-coupling processes that aid the transverse ion acceleration process.

Introduction

The rocket campaigns *MARIE* [*Kintner et al.*, 1986; *Yau et al.*, 1987; *LaBelle et al.*, 1986] and *TOPAZ* III [*Kintner et al.*, 1992; *Vago et al.*, 1992; *Arnoldy et al.*, 1992; and *Lynch et al.*, 1993] have given us detailed simultaneous wave and particle-flux observations within the regions of low-altitude transversely accelerated ions. Based on the excellent correlation between the ion energy flux and the amplitude of electric field fluctuations in the lower-hybrid frequency range demonstrated in the *MARIE* event [*Kintner et al.*, 1986], it is believed that these plasma waves are responsible for the transverse acceleration of the ions. The causal basis for this belief, however, has not been tested. The question can be expressed simply as: are the waves intense enough to explain the observed ion energies? This report offers a theoretical interpretation of the observations, based on the published data, using the model of lower-hybrid heating to describe the ion acceleration and answer this question. We will use the Monte Carlo technique [*Retterer et al.*, 1983; 1990] to solve the kinetic equation for the phase-space density of the ions interacting with lower hybrid waves and evaluate the ion heating produced by the observed waves. Our discussion here will focus on the *MARIE* observations; a complete analysis of the more recent *TOPAZ* III measurements will be presented in a later report.

Space Plasmas: Coupling Between Small
and Medium Scale Processes
Geophysical Monograph 86

Much of the uncertainty in evaluating the efficiency of wave-induced particle heating in situations where the resonance between wave and particle requires matching of velocities (the case for ion interaction with lower hybrid waves, see below) comes from the lack of information on the phase velocities of the waves. In the case of the *MARIE* event, wavelength measurements by an interferometer [*LaBelle and Kintner*, 1989] can possibly assist us in this problem. Recently, independent measurements of the average wavelength of auroral VLF emissions as a function of frequency were obtained on a sounding rocket using an orthogonal antenna technique [*Ergun et al.*, 1991]. The results were comparable to those of *MARIE*. We will show with our modeling efforts that features in the ion velocity distribution are consistent with the phase velocities inferred from the wavelength measurements.

TRANSVERSE ION HEATING BY LOWER HYBRID WAVES

A mechanism by which wave-particle interaction of ions with the intense turbulence near the lower-hybrid frequency leads to ion conic formation was advocated by Chang and Coppi [1981] : first, the turbulence is excited in the ambient plasma of the suprauroral region by the accelerated electrons precipitating to form the aurora. Wave-particle interaction of the ambient ion population with the turbulence near the lower-hybrid frequency then leads to ion acceleration nearly perpendicular to the field line, which is followed by the adiabatic folding of velocities as the ions mirror and travel up the field line, to create the conic velocity distributions observed by satellites.

Plasma simulations [*Retterer et al.*, 1986] have shown this mechanism to be effective not only for the transverse acceleration of ions, but also for the acceleration of the ambient electrons in the directions parallel to the geomagnetic field. The electron velocity distribution is predicted to have enhanced fluxes of energetic electrons both parallel and antiparallel to the geomagnetic field (a consequence of nonlinear mode coupling processes), a form which is indeed observed in space in conjunction with ion conics: counterstreaming electrons [*Lin et al.*, 1982]. In the *MARIE* event, also, there is a evidence of an enhanced suprathermal electron flux in the field-aligned direction out of the ionosphere [*Yau et al.*, 1987].

To develop our model for the wave-particle interaction process, we will first consider it alone, as it would act in a homogeneous medium. This model can then be incorporated into a mesoscale model of the more complicated geometry of the suprauroral region. In the presence of a turbulent electric field spectrum, the resonant (irreversible) effect on the ion velocity distribution function f is adequately described by a diffusion equation [*Davidson*, 1972]

$$\frac{\partial f}{\partial t} = \frac{\partial}{\partial \mathbf{v}} \cdot \left(\mathbf{D} \cdot \frac{\partial f}{\partial v} \right). \tag{1}$$

The velocity diffusion tensor $\mathbf{D}$ depends on the spectral density of the electric-field turbulence as a function of frequency and wave-vector. If we specialize to the study of ion interactions, and use the unmagnetized trajectory approximation, we have

$$\mathbf{D} = \left(\frac{q}{m}\right)^2 \int \frac{d\omega}{2\pi} \int \frac{d^3\mathbf{k}}{(2\pi)^3} \frac{\mathbf{kk}}{k^2} |E|^2(\mathbf{k},\omega)\pi\delta(\omega - \mathbf{k}\cdot\mathbf{v}) \tag{2}$$

where q and m are the ion charge and mass, ω is the frequency, $\mathbf{k}$ is the wave vector, and $|E|^2(\mathbf{k},\omega)$ is the spectral density of the magnitude of the electric field, which is assumed here to be electrostatic in nature. Observations of auroral hiss tell us something about the electric-field spectral density which allows us to simplify this picture. The largest amplitudes of the fields are found at frequencies which are only a few times larger than the lower-hybrid resonance frequency ω_{LHR}:

$$\omega_{LHR}^2 \equiv \frac{\omega_{pi}^2}{1 + \omega_{pe}^2/\Omega_e^2} \tag{3}$$

where ω_{pi} and ω_{pe} are the ion and electron plasma frequencies, and Ω_e is the electron cyclotron frequency. The whistler resonance-cone dispersion relation

$$\omega^2 = \frac{\omega_{pi}^2 + \omega_{pe}^2 \cos^2\theta}{1 + \omega_{pe}^2 \sin^2\theta/\Omega_e^2}, \tag{4}$$

where θ is the angle between the geomagnetic field and $\mathbf{k}$, tells us that the wave vectors of these waves are directed nearly perpendicular to the geomagnetic field. Thus, the largest contribution to the integral over the spectral density for $\mathbf{D}$ comes for the perpendicular component $D_\perp = \mathbf{k}_\perp \cdot \mathbf{D} \cdot \mathbf{k}_\perp/k^2$. This explains the essentially transverse heating of the ions observed in ion conic events, and means that we can write the local evolution equation for the ion velocity distribution keeping only the perpendicular contribution

$$\frac{\partial f}{\partial t} = \frac{1}{v_\perp}\frac{\partial}{\partial v_\perp}\left(v_\perp D_\perp \frac{\partial f}{\partial v_\perp}\right) \tag{5}$$

where

$$D_\perp = \left(\frac{q}{m}\right)^2 \int \frac{d\omega}{2\pi} \int \frac{d^3\mathbf{k}}{(2\pi)^3} \frac{k_\perp^2}{k^2} |E|^2(\mathbf{k},\omega)\pi\delta(\omega - \mathbf{k}\cdot\mathbf{v}) \tag{6}$$

Note the delta function in the integrand, imposing a resonance condition that states that the only waves which will cause ions of velocity **v** to diffuse in velocity are those waves for which the projection of the phase velocity in the direction of **v** equals the ion speed.

Some of the consequences of this resonance condition can be briefly outlined. Because the lower-hybrid waves initially excited by the linear instability of the energetic auroral electrons are of high phase velocity, they would resonantly interact only with energetic ions, which are few in number in the ionosphere. Other excitation mechanisms for lower hybrid waves, such as the instabilities due to the anomalous Doppler resonance [*Parail and Pogutse,* 1976], also face this high-phase-velocity problem. This situation can be resolved by nonlinear mode-mode coupling processes which can effectively aid the ion acceleration process by generating short wavelength modes with lower phase velocities. Evidence of these processes has been seen in computer simulations of the phenomena [*Retterer et al.,* 1986]. (Propagation of the resonance-cone waves through a transverse density gradient toward the lower-hybrid resonance can also reduce the phase velocities of the waves, as will linear mode conversion at the resonance, both aiding the access of ions to the waves.)

In a subsequent set of two-dimensional simulations [*Retterer and Chang,* 1988], a mixture of ions species was included, using both light ions representing the hydrogen of the ionosphere, and heavy ions representing oxygen. It was found in these simulations that the heating of the heavier species was negligible. This can be understood as a consequence of the velocity dependent resonance condition for interaction with the waves. The waves are excited at very high phase velocities by the electron beam, and then are subject to mode-coupling processes that reduce their phase velocities. These processes continue until the phase velocities of the waves are sufficiently degraded that the cool ambient plasma species can begin to interact with them. The species of particles with the largest velocities will be able to interact with the waves first. By absorbing the wave energy, this first ion species prevents the waves from undergoing further degradation in phase velocity and isolates the species with smaller velocities, namely species with heavier masses, from interaction with the waves. For the ion species that can interact with the waves, the effect of the wave-particle interaction is to produce high-velocity tails on their velocity distribution because lower velocity ions within the distribution are immune to the influence of the waves.

The assumption of unmagnetized ions may seem unrealistic in light of the occurrence of H^+ gyrofrequency structure in the spectrum of auroral hiss observed with the transversely accelerated ions [*Kintner et al.,* 1991]. Simulation studies [*Kamimura et al.,* 1978] of Bernstein modes, however, show that when many modes are excited, phase mixing among them can lead to the restoration of the unmagnetized Landau damping rate. Because the lower-hybrid frequency is generally much larger than the ion gyrofrequency, this phase mixing should be effective in the wave-particle interaction between ions and lower hybrid waves in the suprauroral region, allowing us to use the simplification of the unmagnetized trajectory formulation. Our simulations [*Retterer et al.,* 1986] of this interaction with and without magnetized ions show that the phenonena are indeed nearly the same in the two cases.

Although the plasma-simulation results serve as a qualitative guide to the phenomena observed in space, the situation in the auroral region is more complicated than we can practically study using self-consistent plasma simulations. Important phenomena occur on a wider range of spatial and temporal scales than self-consistent numerical models can accomodate simultaneously. For success in modeling the observations we must hope to extract from the simulations a description of the micro-physics of the wave-particle interaction, to incorporate into a model of the larger scale variations in the suprauroral region. For the heating of the ambient particle species, the model is the velocity diffusion equation, to which the simulations provide a model for the aspects of the spectral density of the electric fields which cannot be measured directly.

Because of the variety of circumstances to be found in the suprauroral region, we must also expect to employ wave observations in our specification of particle heating rates, to reflect the variability of the phenomena. Satellites rarely measure wavevector information for the electric field spectral density; the observed spectral density of any component of the electric field is a function only of frequency (which unfortunately includes the possibility of aliasing of spatial structure if the spacecraft is moving through an inhomogeneous medium). Thus, if the satellite antenna system were equally responsive to all wavelengths, the spacecraft receiver would measure

$$\int \frac{d^3\mathbf{k}}{(2\pi)^3} |E|^2(\mathbf{k},\omega) = |E|^2(\omega). \tag{7}$$

One can obtain a rough order of magnitude of the transverse velocity diffusion rate by using the observed value of $|E|^2(\omega)$ at some frequency near the lower hybrid frequency in the order-of-magnitude evaluation of Eq.(6)

$$D_\perp \sim \left(\frac{q}{m}\right)^2 |E|^2(\omega) \tag{8}$$

and this is frequently all that can be done to estimate the diffusion rate. One cannot, however, describe the velocity dependence of $D_\perp$ using a description this crude, and the high-energy tail structure of the particle fluxes seen both in space plasma and the simulations requires a velocity dependent diffusion rate. This is the point where our simulation results can suggest the form for a model of the unobservable wave vector specta of the resonance-cone waves[*Retterer et al.*, 1992].

Crew and Chang [1985] suggested the following form for the electric-field spectral density's dependence on $k_\perp$, to describe the general situation where $S_\perp$ peaks near a wave vector k_o

$$S_\perp(k_\perp) \propto \frac{(k_\perp/k_o)^\mu}{1+(k_\perp/k_o)^\nu}. \tag{9}$$

In this expression the free parameters are the peak wave vector k_o and the two spectral indices μ and ν. For the integral of $S_\perp$ to be defined, we need $\nu > \mu + 3$, and for $D_\perp$ to be nonsingular, $\mu > 0$. This form, it turns out, fits the body of the wave vector spectra measured in the one-dimensional simulations quite well, and we will use it to parametrize the wave vector spectra to calculate the velocity diffusion rate. Crew and Chang [1985] next showed that to a good approximation, the diffusion coeficient one obtains using this form has the following velocity dependence

$$D_\perp(v_\perp) \propto \frac{(v_\perp/v_o)^{\nu-\mu-3}}{1+\eta(\mu,\nu)(v_\perp/v_o)^{\nu-\mu}}. \tag{10}$$

Here η is a numerical constant of order unity, dependent on μ and ν, and v_o is a characteristic velocity near which the diffusion rate is a maximum; if the frequency spectrum peaks near a frequency ω_o, then $k_o \sim \omega_o/v_o$. We expect that v_o is of the order of the perpendicular phase velocity of the waves excited by the auroral electron beam, or smaller if mode-coupling processes have been especially effective in transferring wave energy to shorter wavelength, lower phase-velocity waves. As $v_\perp$ goes to zero, $D_\perp$ goes to zero as $v_\perp^{\nu-\mu-3}$; at velocities much larger than v_o, $D_\perp$ falls as $v_\perp^{-3}$.

The peaked nature of $D_\perp$ reflects the way in which wave-particle interaction leads to the formation of high-energy tails on the velocity distributions, rather than, say, simple bulk heating. Because $D_\perp$ falls to zero at small velocity, the bulk of the velocity distribution is unaffected by interaction with the waves. At higher velocity $D_\perp$ begins to become appreciable, and the diffusion process begins to affect particles in the distribution. Because there are initially more particles at lower velocity than at higher velocity, the net flux in velocity space is toward higher velocities, leading to the formation of high-energy tails. Crew and Chang [1985] demonstrated this by developing an asymptotic method for analytically solving the velocity diffusion equation with the peaked diffusion coefficient and presenting the solutions.

We must now consider how our local description of particle heating by wave-particle interaction with intense auroral hiss can be applied in the supraauroral region, where ion conics and counter-streaming electrons are observed in the vicinity of auroral arcs. In the absence of wave-particle interactions, the evolution along a geomagnetic field line of the ion velocity distribution, f, would be described by the Liouville equation [*Roederer,* 1970]. If we use the guiding center approximation and also neglect any DC electric fields, the equation is the following:

$$\frac{df}{dt} \equiv \frac{\partial f}{\partial t} + v_\parallel \frac{\partial f}{\partial s} + \frac{1}{2B}\frac{dB}{ds}\left(-v_\perp^2 \frac{\partial f}{\partial v_\parallel} + v_\parallel v_\perp \frac{\partial f}{\partial v_\perp}\right) = 0 \tag{11}$$

where s is a coordinate denoting position along a geomagnetic field line, B is the magnitude of the geomagnetic field, and q and m are the charge and mass of an ion. The motion of the ions is constrained to lie along a geomagnetic field line, with their magnetic moments and energies conserved.

In the presence of a transverse heating process, the effect on the ions of their interaction with the turbulence is adequately described by a diffusion equation,

$$\frac{df}{dt} = \frac{1}{v_\perp}\frac{\partial}{\partial v_\perp}\left(v_\perp D_\perp \frac{\partial f}{\partial v_\perp}\right), \tag{12}$$

where the left-hand side of the equation contains the terms of the Liouville equation and $D_\perp$ is the quasilinear velocity diffusion rate perpendicular to the geomagnetic field obtained earlier in this section. To predict the form of the distribution function, it is necessary to solve the kinetic equation in detail. Even in the simplest situation where a steady state solution is sought, the problem is a three dimensional one, with the independent variables position along the field line, perpendicular velocity, and parallel velocity. In addition, the specified fields in a problem will not necessarily be found in forms that make the problem amenable to analysis.

Simulation techniques offer a flexible and robust way of attacking such complicated problems. Because the number of ionospheric ions and electrons which are affected by this turbulence will be small—being those particles which can meet the velocity resonance conditions for interaction—their effect back on the fields may be small. A test-particle approach, in which the motion and heating of particles are studied in specified global models of the geomagnetic field, the steady electric fields, and the spectral density of the turbulent fluctuating electric fields, is then justified. We will solve the kinetic equation for the ion velocity distribution using a Monte Carlo model which was developed to investigate problems of this kind [*Retterer et al.*, 1983]. This model of mesoscale wave-particle interaction employs a Monte Carlo technique to follow the trajectories of many ions as they undergo the influences of wave-particle interaction and large-scale electric and magnetic fields in the supraauroral region. It has enjoyed considerable success [*Retterer et al.*, 1987] in modeling the ion velocity distribution of ion conics formed in other regions by interaction with turbulence in the ion cyclotron frequency range (See the review by Crew *et al.*[1990] for a discussion of this other form of ion conic.)

From an initial distribution in velocity and space, the calculation of the evolution of the distribution proceeds by following the trajectories of a large number of ions with time. Between the velocity perturbations caused by interaction with the waves, it is assumed that the ions travel in the static geomagnetic field with constant energy and magnetic moment. To introduce the effects of the wave-particle interaction, the velocities of the ions are perturbed randomly. The random velocity perturbations are chosen so that statistically they yield the appropriate quasilinear velocity diffusion [*Retterer et al.*, 1983].

LOW ALTITUDE ION CONIC

Now let us consider the transversely accelerated ions observed during the sounding rocket flight code-named *MARIE*. This rocket campaign at Fort Churchill, Canada has provided us with what is one of the most complete and detailed set yet available of observations within the regions of low-altitude transversely accelerated ions. The rocket passed through an auroral arc on its upleg and downleg flights; at each crossing a strong correlation was seen between the intensity of the auroral hiss electric fields at frequencies between 3 and 16 kHz and the flux of ions flowing up the field line at energies of 50 to 370 eV, with pitch angles near ninety degrees. The intense fluxes of both waves and particles were observed at altitudes between 500 and 800 kilometers. A mass spectrometer on board determined that the ions were primarily hydrogen, consistent with our expectation that the lighter ions are preferentially heated by the lower-hybrid turbulence.

To model this event, we first recognize that hydrogen is generally a minority species of the plasma at these altitudes, where the major constituent is expected to be oxygen. Thus the hydrogen ions observed here have probably been drawn down from higher altitudes. This affects the residence time of the ions in the vicinity of the intense turbulence, and thus the energies the ions achieve. Ions traveling down the field line enter the region of intense turbulence and begin to be transversely heated by interaction with the waves. At the same time, the ions's parallel motion slows down and eventually reverses as they mirror in the geometry of the geomagnetic field. Thus, the ions spend more time in the region of wave-particle interaction than they would if they were already traveling up the field line,increasing the effectiveness of the heating process.

For our simulation of the *MARIE* event, then, we begin with a hydrogen ion species with a temperature of 1 eV at the upper end of the simulated region, from 500 to 800 km altitude. To model the heating process, we specify a model for the velocity diffusion rate as a function of perpendicular velocity and position along the field line. For the velocity dependence we choose the rational form given above, Eq.(10), while for the spatial variation, we assume that $D_\perp$ is independent of position over the small range of altitude of the model. We can not make any claims about the uniqueness of this form; rather, one should conclude simply that it is a form that can produce conics with the observed shape. Altogether, the diffusion rate is written

$$D_\perp = \left(\frac{q}{m}\right)^2 S_{E0} \frac{(v_\perp/v_o)^\beta}{1+(v_\perp/v_o)^{\beta+3}} \qquad (13)$$

where S_{E0} is a reference value of the electric-field spectral density, v_o is the reference phase velocity, and β is the velocity power-law index ($\beta = \nu - \mu$ from the last section).

The peak spectral density of the electric field observed in the first pass through the arc, from about 05:29 to 5:31 UT, is around $(5\text{mV/m})^2$ per kHz, averaged over the range from 3 to 16 kHz. The fitting of the other parameters of the diffusion rate was done by trial and error, by comparing the calculated phase-space density integrated over the range of upflowing pitch angles, 90°—120°, to the data. Figure 1 shows the results of the simulation as two contour plots of the ion phase-space density as a function of $v_\perp$ and $v_\parallel$ at the altitudes 620 and 740 km. We see in both cases the characteristic near-ninety degree pitch angle of the conic, despite the fact that the heating has occurred over the course of several hundred kilometers. Figure 2 shows the phase-

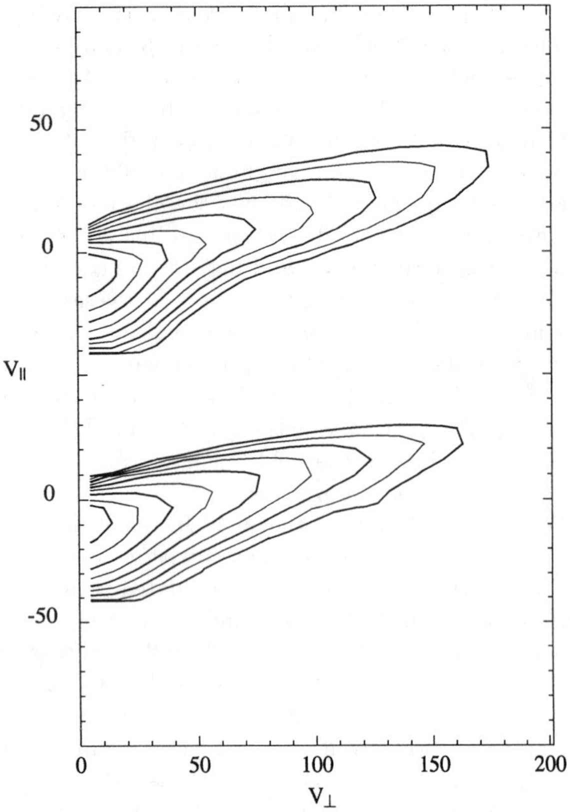

Fig. 1. Simulation of the *MARIE* ion conic event. Contour plots of the ion velocity distribution as a function of perpendicular and parallel velocities (in km/sec) at 620 km altitude (bottom) and 740 km (top). The contours are a half decade apart.

space density integrated over pitch angle at an altitude of 620 km. Yau *et al.*[1987] reported that the characteristic energy of the conic was about 22 eV, with a tail emerging from the body of the distribution at around 150 to 200 eV, extending out to a few hundred eV. We see here that in the simulation, the observed level of turbulence was able to produce the heating required to allow the ions to reach these energies. The dashed line in the figure is a Maxwellian distribution function with a temperature of 22 eV.

The shape of the high-velocity tail of the ion distribution can be fitted by adjusting the velocity dependences of the diffusion coefficient. A constant diffusion rate would produce an approximately Maxwellian tail on the velocity distribution. A flaring tail is produced by allowing $D_{\perp}$ to vary with velocity. Some trial and error showed us that the choice $\beta = 2$ produced the appropriate amount of flare for the *MARIE* event. We expect that the phase velocity of the peak spectral density would be the velocity corresponding to an energy above the point where the tails emerge from the body of the velocity distribution, around 200 eV. This energy corresponds to a velocity of about 200 km/s, which we selected as the value for the reference phase velocity, v_o. On the *MARIE* rocket, an electric field interferometer was flown to attempt to measure wavelengths of the observed plasma waves [*LaBelle and Kintner*, 1989]. When the rocket passed through the ion-acceleration region, the electric field interferometer detected a feature with measureable coherence which at a frequency of 9 kHz had a wavelength of about 20 m. This represents a phase velocity of 180 km/s, in agreement with the phase velocity of the peak spectral density inferred from the ion velocity-distribution measurement. The interferometric technique works only when the **k** spectrum is fairly coherent (dominated by a single wavelength); it is thus the peak of a broadband **k** spectrum that would be identified by the method.

Another order-of magnitude argument can give us the smallest phase velocity or shortest wavelength of the modes effective in the interaction with the ions. From

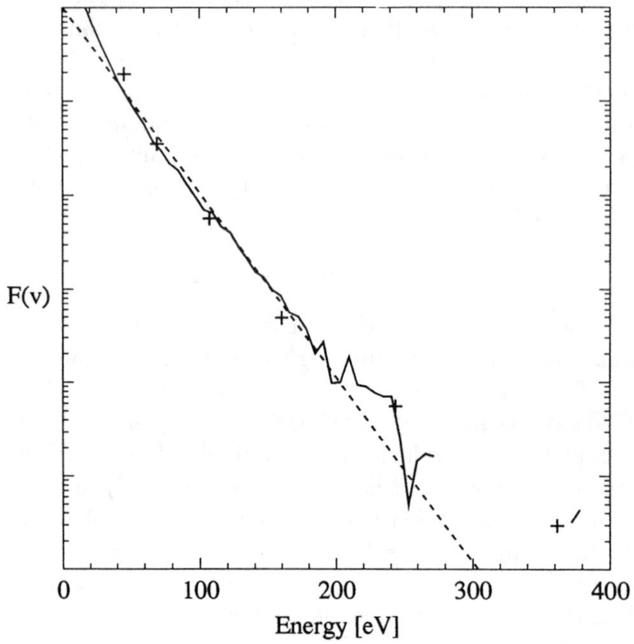

Fig. 2. The calculated phase-space density of the *MARIE* conic, integrated over a range of upflowing pitch angles. The velocity distribution at the altitude 620 km is plotted as a function of energy (in eV). The dashed line is a Maxwellian distribution function with a temperature of 22 eV. The pluses are data from the *MARIE* flight around 5:29 UT.

the ion flux measurements we can roughly estimate the density of the heated ions to be about 0.2cm^{-3}, if we assume a perpendicular temperature of 22 eV and a parallel temperature of 1 eV. If the ambient hydrogen density is 200cm^{-3} (consistent with the lower hybrid frequency and not unreasonable for the high latitude regions at 700-800 km altitude), then the heated ions represent a fraction of the ambient population which, if they came from the high velocity tails of the ambient distribution of temperature 1 eV, would correspond to all the ions above a velocity of 3.8 v_{ti} or 37 km/s, equivalent to 7 eV. With a characteristic frequency of 5 kHz for the turbulence, this minimum phase velocity implies that the shortest wavelength is about 7 m, and $k_\perp \rho_i$ is a little larger than unity, if we use 2 m for the 1 eV hydrogen gyroradius. This is the order of magnitude we expect for the short wavelength limit of the turbulence excitable in the plasma.

An application of the Monte Carlo model presented here was made [*Retterer et al.*, 1992] to a folded conic, which was observed above the region of transverse acceleration (the common situation when the ions are observed by satellite). By matching the observed ion mean energy and the shape of the energetic tail of the ion velocity distribution, we tried to place some constraints on the spectral density of the electric fields in the region of acceleration, which is remote from the point where the ions are observed. We studied the ion conic that was observed at high altitude (5700 km) by the S3-3 satellite on day 280, 1976, orbit 718. This conic has been described by Temerin *et al.*[1981] and by Gorney *et al.*[1982]. We found a good fit to the observed ion distribution when the velocity diffusion was described by the model of this section, using the following parameters:$S_{E0} = (20\text{mV/m})^2/\text{kHz}$, $v_o = 310$ km/s (corresponding to 500 eV), and $\beta = 3$. Note that this value of S_{E0} represents a fairly modest value of the total electric field in the turbulence, whose values have been observed to range up to 50 mV/m or more. The phase velocity of the peak of diffusion, v_o, corresponds to an energy of 500 eV, which is certainly of the right order of magnitude if the turbulence was generated by electron beams of a few keV energy and then subjected to degradation of phase velocity due to propagation and mode coupling effects.

Lower hybrid spikelets

Recent sounding rocket observations (*MARIE* and *TOPAZ* III) of the region of low-altitude transversely accelerated ions have also pointed out the correlation of lower-hybrid 'spikelets,' wave packets of intense electric fields oscillating near the lower hybrid frequency [Labelle *et al.*, 1986; Arnoldy *et al.*, 1992; Kintner *et al.*, 1992; Vago *et al.*, 1992], with the energetic ions. One possible mechanism for the generation of the spikelets is the nonlinear collapse of VLF waves excited by auroral electron precipitation into lower-hybrid envelope solitons[*Retterer et al.*, 1986]. One-dimensional and two-dimensional plasma simulations[*Retterer and Chang*, 1988] showed that whistler resonance-cone waves excited by an electron beam population decay into a broadband wavevector spectrum consistent with the collapse process. This same process of scattering of waves into shorter wavelengths also serves to facilitate the heating of the ionospheric plasma by reducing the phase velocities of the waves, leading to tail formation on both the electron and ion velocity distributions[*Retterer et al.*, 1986]. A theoretical model for the collapse process leads to an equation for the electric field evolution in the form of the nonlinear Schroedinger equation (NLSE) in one dimension[*Retterer et al.*, 1986], although the equation has a more complicated form in higher dimension, when electron $E \times B$ coupling is included[*Retterer and Chang*, 1988]. The kinetic effects of the exchange of energy between the waves and particles were described by simultaneously solving the velocity diffusion equations for the evolution of the particle velocity distributions.

This theoretical model was solved in one dimension by Retterer *et al.*[1986], and it was found that the model described well the phenomena observed in one-dimensional particle simulations of the plasma instability, due to the auroral electron beam, and ion acceleration. The spatial structure of the electric fields, however, was not examined then. We have recently rerun the model and plotted the amplitude of the electric field oscillating at the lower hybrid frequency in real space and time. Figure 3 illustrates the evolution of the electric field amplitude in a gray-scale image. We see at the beginning of the run (the bottom of the figure) that the amplitude of the electric field grows from low levels, more or less uniformly across the simulation region. About a quarter or a third of the way through the simulation, when the amplitude of the electric field has grown sufficiently, the electric field intensity begins to concentrate in narrow regions, which then begin to drift slowly (in the scale of the figure, a drift at the ion thermal velocity would correspond to a line inclined at 45 degrees), some to the left, but most to the right, in the direction of the beam which originally excited the long-wavelength waves in the plasma. These wave packets can also be observed to cross one another without

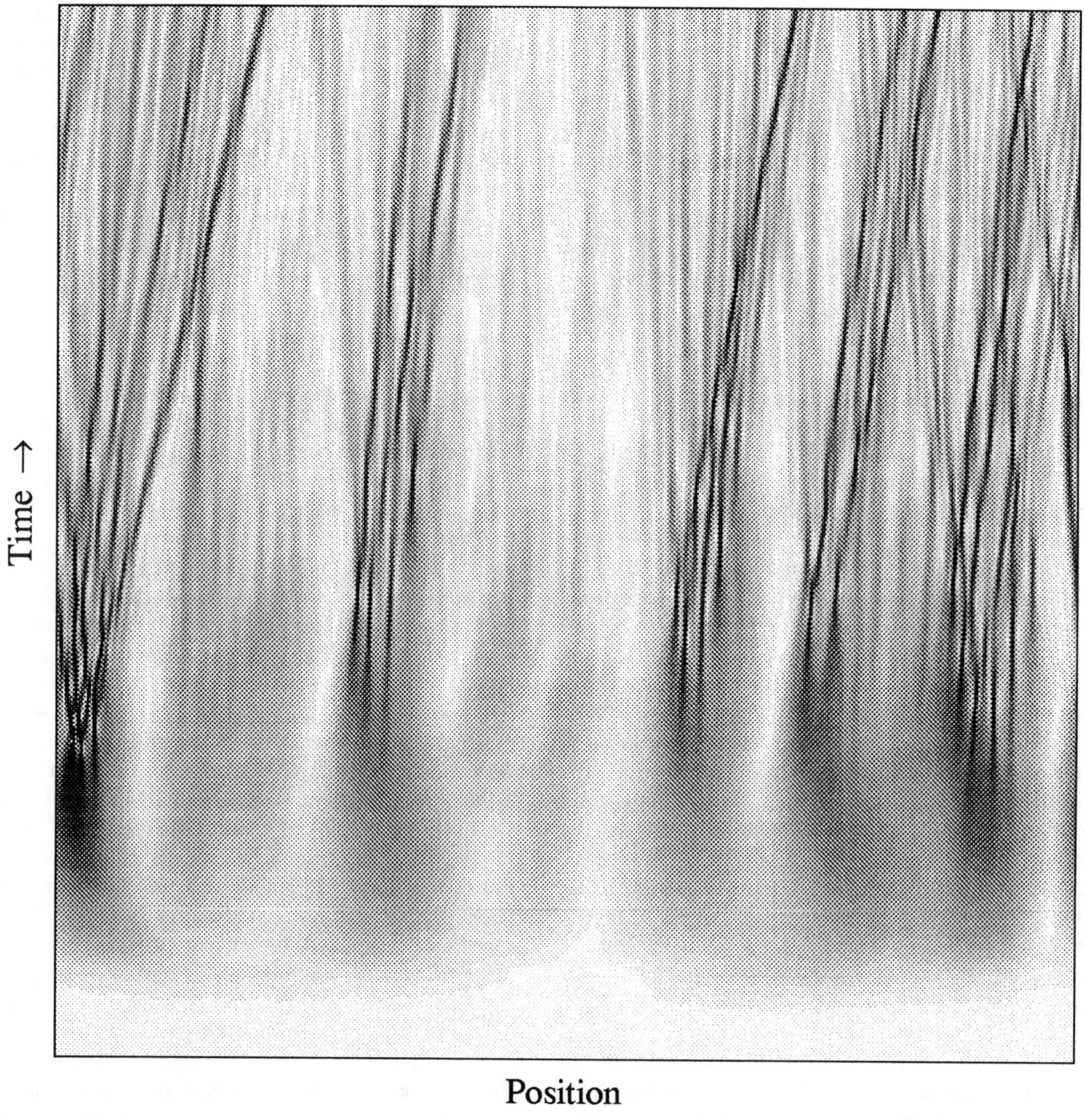

Fig 3. The evolution of the electric field amplitude in the NLSE/kinetic model of the suprauroral beam instability. Time runs from the bottom of the figure up; darker gray means more intense electric field.

alteration. These indeed are all characteristics of the envelope solitons which are solutions to the nonlinear Schroedinger equation[*Retterer et al.,* 1986].

Are the scales of these solitons consistent with the observed lower hybrid spikelets? Let us consider the scaling by analysing the one-dimensional NLSE:

$$\frac{i}{\omega_o}\frac{\partial E}{\partial t} + D\frac{\partial^2 E}{\partial^2 x} + C \mid E \mid^2 E = 0 \qquad (14)$$

where E is the amplitude of the electric field oscillating at the frequency ω_o in the lower-hybrid frequency range, and the coefficients C and D are given in Retterer *et al.*[1986] (they depend on plasma parameters such as the plasma temperature, the ion charge-to-mass ratio, and the ion plasma frequency, and the ratio of ω_o to the lower hybrid resonance frequency). A soliton solution of this equation is given by[*Goldman,* 1984]:

$$E(x,t) = E_o \, \text{sech}(x/L) \, e^{it/T} \quad (15)$$

where the time and length scales of the soliton wave packet are functions of the plasma parameters and the amplitude of the electric field:

$$T = \frac{2}{C \, E_o^2 \, \omega_o} \quad (16)$$

and

$$L = \sqrt{\frac{2D}{C \, E_o^2}} \quad (17)$$

In a typical *MARIE* spikelet[*LaBelle et al.*, 1986], the peak of the spectral density was around 10 kHz, the lower hybrid resonance frequency was 3 kHz, and the amplitude of the electric field was 200 mV/m. If the temporal duration of the event is interpreted to be due to the rocket motion through a spatial structure, the wave packet of the spikelet was 14 meters long. The rocket's velocity, however, was nearly along the geomagnetic field; if the temporal duration was limited strictly by the transverse dimension, the length was 2 meters. If we assume that the plasma temperature was 0.25 eV, the formula for the length scale above gives 19 meters, certainly of the correct magnitude for the phenomenon. The instrumentation on subsequent rockets was designed to measure the spikelets with even better resolution. In the spikelet events observed by *TOPAZ* III [*Vago et al.*, 1992], the frequencies were similar, but the electric-field amplitudes were smaller (75 mV/m) and the durations longer (50-70 meters). The formula for L does in this case give 50 meters. The modulation times T, Eq(16), are of the order of 1–6 seconds for the events observed on the two rockets. How do these one-dimensional calculations apply to what is undoubtedly a two or three dimensional phenomenon in the ionosphere? The soliton solutions in higher dimensions are unstable, but it is found[*Goldman*, 1984] that the natural breakup scale for solitons in higher dimension is given by the same formula as the one-dimensional soliton scale, Eq(17). Thus, we expect that L can still give us the characteristic scale of the solitons observed in the ionosphere, or at least a break point in the distribution of spatial scales of the solitons.

Discussion

To summarize the points on which the *MARIE* and *TOPAZ* III in-situ plasma and wave measurements agree well with the predictions of the model of ion interaction with intense electrostatic auroral hiss near the lower hybrid frequency, we have 1) a strong correlation between VLF wave intensity and energetic ion flux; 2) preferential acceleration of thermal hydrogen over thermal oxygen in the *MARIE* event but preferential acceleration of preheated oxygen in the *TOPAZ* III observations[*Arnoldy et al.*, 1992]; 3) enhancements of field-aligned electron fluxes; 4) the frequency spectrum of the VLF turbulence around the lower hybrid frequency shows structure spaced at the hydrogen gyro frequency, suggesting that ion damping of the waves is responsible for the acceleration; 5) the (average) amplitude of the VLF turbulence is several tens of mV/m, strong enough to account for the ion energies; 6) the form of the ion velocity distribution has the high energy tails that are characteristic of the acceleration observed in the simulation, and the measured phase velocities of the waves match the features of the ion distribution; and finally, 7) the observed lower hybrid spikelets correspond to the predicted results of the nonlinear wave-interaction processes that are invoked to generate the electric fields that accelerate the ions. Further, on both rocket flights, detailed correlations between the spikelets and bursts of accelerated ions were observed. Many of these points have been observed on other missions, but never perhaps as convincingly as observed on *MARIE* or *TOPAZ* III.

One of the longstanding controversies in the study of ion conics near auroral arcs has been the question of identifying the mechanism for the transverse ion heating, whether through wave particle interaction with the lower hybrid waves discussed here (see also Ashour-Abdalla *et al.*[1988]) or with electrostatic ion cyclotron waves [*Ashour-Abdalla and Okuda*, 1984] or through other mechanisms. Searches [*Kintner and Gorney*, 1984] for satellite observations of conics of pitch angle near ninety degrees (that were presumably in the process of being accelerated) were inconclusive but suffered because fast velocities and slow spin rates make satellites a poor platform from which to observe transverse ion acceleration near auroral arcs. The advantages of sounding rockets for these observations were demonstrated in the *MARIE* and *TOPAZ* III campaigns, in which the simultaneous observation of transverse ion acceleration and intense lower-hybrid waves provided strong evidence for the lower-hybrid heating mechanism.

To model the observations of ion conics and counter-streaming electrons near auroral arcs requires the consideration of the global evolution of both the particle populations and the wave spectra over ranges of distance and time that stretch our present models. The complete and self-consistent analysis of the fields and particles near auroral arcs remains still only a distant goal of space physics. For the calculation of ion conics, in particular, we are still hampered by the difficulty of measuring the relevant quantities for our theories, and thus considerable uncertainty remains in the empirical estimation of the rate of transverse ion acceleration because of the lack of information on the wave-vector spectrum of the VLF turbulence. Efforts to measure the wavelengths of the plasma waves [*LaBelle and Kintner,* 1989] are most welcome, although such results are still generally limited to the characteristic or dominant wavelength rather than the spectrum of wavelengths that would be needed to calculate wave-particle interaction rates. Such measurements certainly will help elucidate and discriminate among the variety of mechanisms that intervene between the linear excitation of plasma waves in the auroral arc environment and the heating and acceleration of ionospheric ions and electrons. Using the models we have developed, it may be possible to use the form of the ion conic velocity distribution to infer constraints on the spectral shape and magnitude of the VLF turbulence, a great benefit when the acceleration region is localized and remote from the platform observing the accelerated particles.

Acknowledgments. We thank P. Kintner, K. Lynch, R. Arnoldy, J. Vago, G.B. Crew, R. Ergun and R.L. Lysak for interesting discussions. This research is partially supported by the Air Force Office of Scientific Research and by Air Force Contracts numbered F19628-88-K-0008, F49620-86-C-0128, and F19628-86-K-0005. NASA provided support through grant NAGW-1532. The use of the supercomputer facilities at the National Center for Supercomputer Applications is gratefully acknowledged.

References

Arnoldy, R.L., K.A. Lynch, P.M. Kintner, J.L. Vago, S.W. Chesney, T. Moore, and C. Pollock, Bursts of transverse ion acceleration at rocket altitudes, *Geophys. Res. Lett.*, *19*, 413, 1992.

Ashour-Abdalla, M., and H. Okuda, Turbulent heating of heavy ions on auroral field lines, *J. Geophys. Res.*, *89*, 2235, 1984.

Ashour-Abdalla, M., D. Schriver, and H. Okuda, Transverse ion heating in multicomponent plasmas along auroral zone field lines, *J. Geophys. Res.*, *93*, 12,826, 1988.

Chang, T., and B. Coppi, Lower hybrid acceleration and ion evolution in the suprauroral region, *Geophys. Res. Lett.*, *8*, 1253, 1981.

Crew, G.B., and T. Chang, Asymptotic theory of ion conic distributions, *Phys. Fluids*, *28*, 2382, 1985.

Crew, G.B., Tom Chang, J.M. Retterer, W.K. Peterson, D.A. Gurnett, and R.L. Huff, Ion cyclotron resonance heated conics: theory and observation, *J. Geophys. Res.*, *95*, 3959–3985, 1990.

Davidson, R.C., *Methods in Nonlinear Plasma Theory*, Academic Press, New York, 1972.

Ergun, R.E., E. Klementis, C.W. Carlson, J.P. McFadden, and J.H. Clemmons, Wavelength measurement of auroral hiss, *J. Geophys. Res.*, *96*, 21,299, 1991.

Goldman, M.V., Strong turbulence of plasma waves, *Rev Mod. Phys.*, *56*, 709, 1984.

Gorney, D.J., S.R. Church, and P. Mizera, On ion harmonic structure in auroral zone waves: The effect of ion conic damping of auroral hiss, *J. Geophys. Res.*, *87*, 10,479, 1982.

Kamimura, T., T. Wagner, and J.M. Dawson, Simulation study of Bernstein modes, *Phys. Fluids*, *21*, 1151–1167, 1978.

Kintner, P.M., and D.J. Gorney, A search for the plasma processes associated with perpendicular ion heating, *J. Geophys. Res.*, *89*, 937–944, 1984.

Kintner, P.M., J. LaBelle, W. Scales, A.W. Yau, B.A. Whalen, and Craig Pollock, Observations of plasma waves within regions of perpendicular acceleration, *Geophys. Res. Lett.*, *13*, 1113, 1986.

Kintner, P.M., W. Scales, J. Vago, A. Yau, B. Whalen, R. Arnoldy, and T. Moore, Harmonic H+ gyrofrequency structures in auroral hiss observed by high altitude auroral sounding rockets, *J. Geophys. Res.*, *96*, 9627, 1991.

Kintner, P.M., J. Vago, S. Chesney, R.L. Arnoldy, K.A. Lynch, C.J. Pollock, and T. Moore, Localized lower hybrid acceleration of ionospheric plasma, *Phys. Rev. Lett.*, *68*, 2448, 1992.

LaBelle, J., P.M. Kintner, A.W. Yau, and B.A. Whalen, Large amplitude wave packets observed in the ionosphere in association with transverse ion acceleration, *J. Geophys. Res.*, *91*, 7113–7118, 1986.

LaBelle, J. and P.M. Kintner, The measurement of wavelength in space plasmas, *Reviews of Geophysics*, *27*, 495–518, 1989.

Lin, C.S., J.L. Burch, J.D. Winningham, and R.A. Hoffman, DE-1 observations of counterstreaming electrons at high altitudes, *Geophys. Res. Lett.*, *9*, 925, 1982.

Lynch, K.A., R.L. Arnoldy, P.M. Kintner, and J.L. Vago, Electron Distribution Function Behavior during Localized Transverse Ion Acceleration Events in the Topside Auroral Zone, *J. Geophys. Res.*, submitted, 1993.

Parail, V.V. and O.P. Pogutse, Instability of the runaway-electron beam in a tokamak, *Sov. J. Plasma Phys.*, *2*, 125, 1976.

Retterer, J.M., T. Chang, and J.R. Jasperse, Ion acceleration in the suprauroral region: a Monte Carlo model, *Geophys. Res. Lett.*, *10*, 583, 1983.

Retterer, J.M., T. Chang, and J.R. Jasperse, Ion acceleration by lower hybrid waves in the suprauroral region, *J. Geophys. Res.*, *91*, 1609, 1986.

Retterer, J.M., T. Chang, G.B. Crew, J.R. Jasperse, and J.D. Winningham, Monte Carlo modeling of ionospheric oxygen acceleration by cyclotron resonance with broadband electromagnetic turbulence, *Phys. Rev. Lett.*, *59*, 148–151, 1987. See also *Physics of Space Plasmas (1985-7)*, SPI Conference Proceedings and Reprint Series, vol. 6, Scientific Publishers, Inc.

Retterer, J.M. and T. Chang, Plasma simulation of intense VLF turbulence and particle acceleration in the supraauroral region, In *Physics of Space Plasmas (1988)*, SPI Conference Proceedings and Reprint Series, T. Chang, J.R. Jasperse, and G.B. Crew, editors, page 309, Scientific Publishers, Inc., Cambridge, MA, 1988.

Retterer, J.M., T. Chang, and J.R. Jasperse, Particle acceleration by intense auroral VLF turbulence, In *Physics of Space Plasmas (1989)*, SPI Conference Proceedings and Reprint Series, T. Chang, J.R. Jasperse, and G.B. Crew, editors, page 119, Scientific Publishers, Inc., Cambridge, MA, 1990.

Retterer, J.M., T. Chang, and J.R. Jasperse, The Ion Conic Observed by MARIE, submitted to *J. Geophys. Res.*, 1992.

Roederer, J.G., *Dynamics of Geomagnetically Trapped Radiation*, Springer-Verlag, Berlin, 1970.

Temerin, M.A., C.A. Cattell, R.L. Lysak, M.K. Hudson, R.B. Torbert, F.S. Mozer, R.D. Sharp, and P.M. Kintner, The small scale structure of electrostatic shocks, *J. Geophys. Res.*, *86*, 278, 1981.

Yau, A.W., B.A. Whalen, F. Creutzberg, and P.M. Kintner, Low altitude transverse ion acceleration: auroral morphology and in-situ plasma observations, In *Physics of Space Plasmas (1985-7)*, SPI Conference Proceedings and Reprint Series, T. Chang, J. Belcher, J.R. Jasperse, and G.B. Crew, editors, page 77, Scientific Publishers, Inc., Cambridge, MA, 1987.

Vago, J.L., P.M. Kintner, S.W. Chesney, Arnoldy, R.L., K.A. Lynch, T.E. Moore, and C.J. Pollock, Transverse ion acceleration by localized lower hybrid waves in the topside auroral ionosphere, *J. Geophys. Res.*, submitted, 1992.

J.M. Retterer and J.R. Jasperse, Ionospheric Physics Div., Phillips Laboratory, Bedford, MA 01731.

T. Chang, Center for Space Research, MIT, Cambridge, MA 02139.

Are Relativistic Effects Significant For The Analysis Of Whistler-Mode Waves In The Earth's Magnetosphere?

S.S. SAZHIN

Fluent Europe Ltd, Hutton's Buildings, 146 West Street, Sheffield S1 4ES, U.K.

A.E. SUMNER

Physics Department, University of Canterbury, Christchurch, New Zealand

N.M. TEMME

Centrum voor Wiskunde en Informatica, Kruislaan 413
1098 SJ, Amsterdam, The Netherlands

It is pointed out that relativistic effects lead to an increase in whistler-mode refractive index and to a decrease in the range of frequencies for which instability of these waves occurs. The latter effect is particularly important in a rarefied plasma in a strong magnetic field where the electron plasma frequency (Π) is of the order of or smaller than the electron gyrofrequency (Ω). The condition $\Pi \lesssim \Omega$ is satisfied in some regions of the Earth's magnetosphere (e.g. outside the plasmasphere near the magnetospheric equator) and in these regions the analysis of whistler-mode waves requires the weakly relativistic rather than the nonrelativistic approach even if the thermal velocities of magnetospheric particles are well below the velocity of light.

1. INTRODUCTION

Since the pioneering paper by Storey (1953) there have been different developments in the theory of whistler-mode waves. In particular, attempts have been made to take into account both non-zero electron temperature and anisotropy of the electron distribution function and their effects on whistler-mode propagation, growth or damping [e.g. *Sazhin*, 1993a]. In most cases the theory of whistler-mode propagation in a hot anisotropic plasma has been based on the non-relativistic approximation, i.e. on a self-consistent solution of the linearized Maxwell equations and the nonrelativistic Vlasov equation. This approximation was believed to be applicable with confidence to magnetospheric conditions, which satisfy

$$w \ll c, \tag{1}$$

where w is the thermal velocity of electrons, and c is the velocity of light. At the same time, most of those who developed this theory seem to have ignored the fact that the nonrelativistic approach to the analysis of waves in a hot plasma is not self-consistent, as it neglects relativistic effects in the Vlasov equation, but takes them into account in the Maxwell equations, which are relativistic by their nature.

An alternative approach to the theory of whistler-mode waves in a hot plasma has been based on the so called weakly relativistic approximation which uses condition (1) with terms of the order of w^2/c^2 taken into account, allowing substantial simplification of the general relativistic wave dispersion equation. However, even this relatively simple weakly relativistic whistler-mode dispersion equation is much more complicated than the nonrelativistic dispersion equation, and its applications have been very limited [e.g. *Robinson*, 1987].

In a series of papers reviewed by Sazhin (1993a) [see also the most recent publications: *Sazhin et al.*, 1993; *Sazhin*, 1993b] we attempted to develop an asymptotic theory of whistler-mode propagation, instability or damping in a weakly relativistic plasma. Restricting ourselves to a limiting, but practically important case of whistler-mode propagation strictly parallel to the magnetic field we obtained some approximate solutions to the weakly relativistic dispersion equation. These solutions are rather simple and we hope that they will be widely used in the plasma physics

Space Plasmas: Coupling Between Small
and Medium Scale Processes
Geophysical Monograph 86

community. The main restriction of these solutions is that we cannot always be sure of the range of their applicability, just as we cannot always be sure of the range of applicability of solutions to the nonrelativistic dispersion equation.

In order to specify the range of applicability of different approximate solutions to the weakly relativistic dispersion equation, as well as those of the nonrelativistic equation, we need to compare them with the results of a rigorous numerical analysis of the weakly relativistic whistler-mode dispersion equation. Such a comparison for parallel whistler-mode waves was made by Sazhin et al (1993a). In what follows we briefly discuss the main ideas of this paper with particular emphasis to magnetospheric applications of the obtained results.

2. BASIC EQUATIONS

The dispersion equation for parallel whistler-mode waves in a weakly relativistic (condition (1) is valid) anisotropic plasma can be written as [*Sazhin et al.*, 1992]:

$$N^2 = 1 - \frac{2X}{r}\left[\mathcal{F}_{1/2,2} - \frac{d\mathcal{F}_{3/2,2}}{dz}(A_e - 1)N^2\right], \quad (2)$$

where:

$$\mathcal{F}_{q,p} \equiv \mathcal{F}_{q,p}(z,a,b) = -i\int_0^\infty e^{izt - \frac{at^2}{1-it}}(1-it)^{-q}(1-ibt)^{-p}dt, \quad (3)$$

$z = 2(1-Y)/r$; $a = N^2/r$; $r = p_{0\parallel}^2/(m_e^2c^2)$; $b = A_e$; $X = \Pi_0^2/\omega^2$; $Y = \Omega_0/\omega$; $A_e = p_{0\perp}^2/p_{0\parallel}^2$; Π_0, Ω_0 and ω are the electron plasma frequency at rest, the electron gyrofrequency at rest and the wave frequency (which is complex in general) respectively, $N \equiv ck/\omega$ is the wave refractive index, m_e is the electron mass at rest, c is the velocity of light, k is the wave number.

When deriving (2) we assumed an electron distribution function of the form:

$$f(p_\perp, p_\parallel) = \left(\pi^{3/2} j! p_{0\perp}^{2j+2} p_{0\parallel}\right)^{-1} p_\perp^{2j} \exp\left(-\frac{p_\perp^2}{p_{0\perp}^2} - \frac{p_\parallel^2}{p_{0\parallel}^2}\right), \quad (4)$$

where $p_{0\perp(\parallel)}$ is the electron thermal momentum in the direction perpendicular (parallel) to the magnetic field, $p_\perp$ and $p_\parallel$ are the electron momenta in the corresponding directions.

In a nonrelativistic limit, $c \to \infty$, equation (2) can be simplified to:

$$N^2 = 1 + (A_e - 1)X + \frac{X}{N\sqrt{r}}\left[A_e + (1 - A_e)Y\right] Z(\xi), \quad (5)$$

where

$$Z(\xi) = i\sqrt{\pi}\exp(-\xi^2) - 2\int_0^\xi \exp(-\xi^2 + t^2)dt, \quad (6)$$

$\xi = z/2\sqrt{a}$.

Solutions to equations (2) and (5) are greatly simplified if we restrict our analysis to the case of weakly growing or weakly damped waves, i.e.

$$|\gamma| \equiv |\Im\omega| \ll \min(\Re\omega, \Omega_0 - \Re\omega). \quad (7)$$

In view of (7) we can present the complex equation (2) as a system of two equations:

$$N^2 = 1 - \frac{2X}{r}\left[\Re\mathcal{F}_{1/2,2} - \frac{d\Re\mathcal{F}_{3/2,2}}{dz}(A_e - 1)N^2\right], \quad (8)$$

$$\tilde{\gamma} = \frac{X\left[\Im\mathcal{F}_{1/2,2}(1 + (A_e - 1)N^2) - (A_e - 1)N^2\Im\mathcal{F}_{3/2,2}\right]}{r\left[1 + \frac{2X}{r^2}\sum_{i=0}^3 \tilde{a}_i \Re\mathcal{F}_{i-\frac{1}{2},2}\right]}, \quad (9)$$

where

$$\tilde{a}_0 = Y + N^2(1 + Y(A_e - 1)) + N^4(A_e - 1),$$

$$\tilde{a}_1 = (A_e - 1)N^2 r - Y - 2N^2(1 + Y(A_e - 1)) - 3N^4(A_e - 1),$$

$$\tilde{a}_2 = -(A_e - 1)N^2 r + N^2(1 + Y(A_e - 1)) + 3N^4(A_e - 1),$$

$$\tilde{a}_3 = -(A_e - 1)N^4,$$

$$\tilde{\gamma} \equiv \frac{\gamma}{\omega}.$$

To simplify the notation we hereafter assume that $\omega \equiv \Re\omega$.

In a similar way we can reduce the complex nonrelativistic dispersion equation (5) to the following system of equations:

$$N^2 = 1 + (A_e - 1)X + \frac{X}{N\sqrt{r}}\left[A_e + (1 - A_e)Y\right]\Re Z(\xi_0), \quad (10)$$

$$\tilde{\gamma} \equiv \frac{\gamma}{\omega} = \frac{-\sqrt{\pi}\left[A_e - (A_e - 1)Y\right]\exp(-\xi_0^2)}{\frac{2(1-Y)}{X\xi_0} + \kappa\xi_0 + \Re Z(\xi_0)\left[A_e + \kappa\xi_0^2\right]}, \quad (11)$$

where

$$\xi_0 \equiv \Re\xi, \quad \kappa = \frac{2\left[A_e + Y(1 - A_e)\right]}{Y - 1}.$$

Equations (8) and (10) describe wave propagation, while equations (9) and (11) describe wave growth ($\tilde{\gamma} > 0$) or damping ($\tilde{\gamma} < 0$).

In the cold plasma limit ($|\xi_0| \to \infty$) equation (10) reduces to:

$$N = N_{00} \equiv \sqrt{1 + \frac{X}{Y - 1}}, \quad (12)$$

while equation (11) gives in the same limit: $\tilde{\gamma} = 0$. This means that in a cold plasma whistler-mode waves propagate without damping or growth. In the limiting case $Y \gg 1$ and $X \gg 1$ equation (12) reduces to that derived and analized by Storey (1953).

In the limit $r \to 0$, but keeping terms of the order of r, we can write an approximate solution of (8) as (Sazhin, 1993a):

$$N = N_{00}\left[1 + \frac{\beta_e Y^3}{2(Y-1)^3} + \frac{\beta_e(1 + 4A_e)Y^2}{4(Y-1)^2 N_{00}^2} - \frac{\beta_e A_e Y^2}{2(Y-1)^2}\right], \quad (13)$$

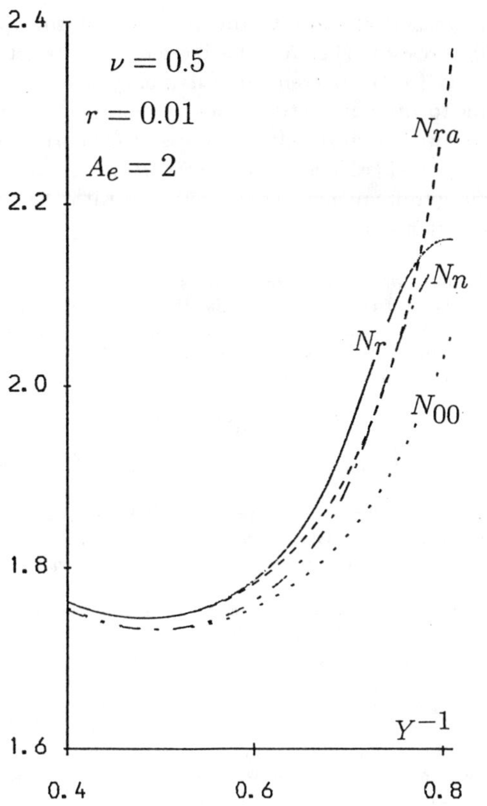

Fig 1. Plots of N_r versus $Y^{-1} \equiv \omega/\Omega_0$, solid, (see equation (8)), N_n versus $Y^{-1} \equiv \omega/\Omega_0$, — · — · — (see equation (10)), N_{ra} versus $Y^{-1} \equiv \omega/\Omega_0$, – – – –, (see equation (13)), N_{00} versus $Y^{-1} \equiv \omega/\Omega_0$, – – – (see equation (12)) for a plasma with the following parameters: $r = 0.01$, $\nu = 0.5$ and $A_e = 2$. Plots are shown only for those $Y^{-1} \equiv \omega/\Omega_0$ when $\tilde{\gamma}_{r,n,ns} \lesssim 0.2$ (condition (7) is satisfied).

where $\beta_e = 0.5\nu r$, $\nu = \Pi_0^2/\Omega_0^2$, N_{00} is defined by (12). In the limit $N_{00}^2 \to \infty$ this expression reduces to that which could be derived from the corresponding nonrelativistic dispersion equation (10). At $A_e = 1$ it reduces to the expression derived by it Imre (1962)

The nonrelativistic expression for $\tilde{\gamma}$ can be considerably simplified if we neglect the effects of nonzero electron temperature on wave propagation, it is then reduced to:

$$\tilde{\gamma} = \frac{\sqrt{\pi}\xi_{00}(Y-1)\left[A_e - (A_e - 1)Y\right]\exp(-\xi_{00}^2)}{\frac{2(1-Y)^2}{X} + Y}, \qquad (14)$$

where

$$\xi_{00} = \frac{1-Y}{N_{00}\sqrt{r}}.$$

The values of N which follow from equations (8), (10), (12), (13) and the values of $\tilde{\gamma}$ which follow from equations (9), (11) and (13) will be compared in the next section for different values of the plasma parameters (ν, r and A_e).

3. RESULTS

In this section we compare numerical values of N and $\tilde{\gamma}$, obtained from the approximations discussed in section 2 for different values of the parameters typical for conditions in the Earth's magnetosphere. Firstly we consider a relatively hot ($r = 0.01$, which corresponds to $T \approx 2.5$ keV) and rarefied ($\nu = 0.5$) plasma with anisotropy $A_e = 2$ (realistic conditions outside the plasmasphere). Plots of N versus $Y^{-1} \equiv \omega/\Omega_0$ and $\tilde{\gamma}$ versus $Y^{-1} \equiv \omega/\Omega_0$ for the different approximations for this plasma are shown in Figs 1 and 2. As can be seen from Fig 1, the values of N_{ra}, determined by (13), are closest to those of N_r, determined by (2) at $\omega/\Omega_0 \lesssim 0.78$. In any case approximation N_{ra} is better than the nonrelativistic approximation N_n, determined by (5) in this frequency range. The latter approximation is applicable for the qualitative analysis of N only. The curves N versus $Y^{-1} \equiv \omega/\Omega_0$ are hereafter presented only for those ω/Ω_0 at which $|\tilde{\gamma}| \lesssim 0.2$.

The curves $\tilde{\gamma}_{ns}$, determined by (14), and $\tilde{\gamma}_n$, determined by (11), shown in Fig 2, coincide within the accuracy of plotting, but the deviation between relativisrtic and nonrelativistic curves is quite noticeable. Relativistic theory predicts stronger instability but at a lower frequency. The frequency of marginal stability is approximately equal to

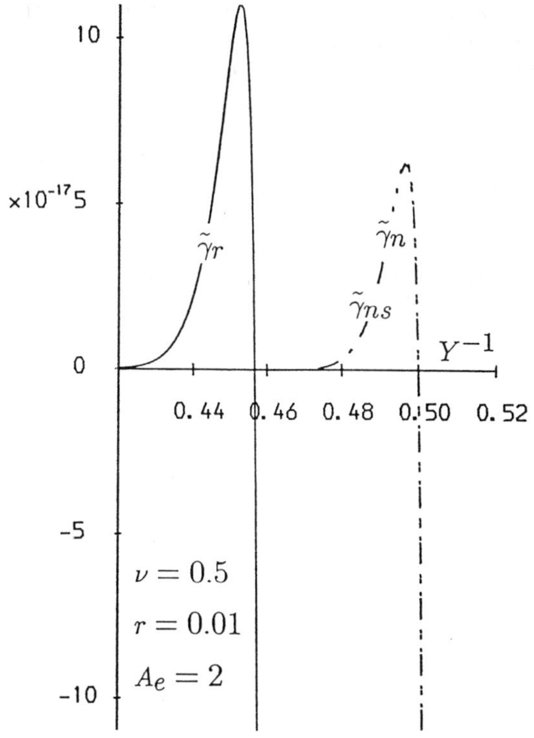

Fig 2. Plots of $\tilde{\gamma}_r$ versus $Y^{-1} \equiv \omega/\Omega_0$, solid, (see equation (9)), $\tilde{\gamma}_n$ versus $Y^{-1} \equiv \omega/\Omega_0$, — · — · -œ-, (see equation (11)), $\tilde{\gamma}_{ns}$ versus $Y^{-1} \equiv \omega/\Omega_0$, – – – –, (see equation (14)) for the same plasma parameters as in Fig 1.

$0.457\ \Omega_e$, which is consistent with the results of asymptotical analysis [*Sazhin et al.*, 1992).

Curves similar to those shown in Figs 1 and 2 were presented by Sazhin et al (1992) for other values of parameters ($\nu = 1$, $r = 0.01$, $A_e = 2$; $\nu = 5$, $r = 0.01$, $A_e = 2$; $\nu = 1$, $r = 0.004$, $A_e = 2$; $\nu = 1$, $r = 0.004$, $A_e = 1$; $\nu = 1$, $r = 0.004$, $A_e = 3$). We tried to choose values of the parameters r, ν and A_e, used in the computations, relevant to the conditions in the Earth's magnetosphere. However, these parameters vary there over such a wide range that we were bound to restrict ourselves to some illustrative examples only. The results of our analysis of these curves allowed us to draw the following conclusions.

1) Expression (13) for N is the best approximation for numerical values of N obtained in a weakly relativistic limit at relatively low frequencies (roughly $Y^{-1} \equiv \omega/\Omega_0 \lesssim 0.6$), but it breaks down rapidly as the wave frequency approaches the electron gyrofrequency (roughly $Y^{-1} \equiv \omega/\Omega_0 \gtrsim 0.6$). In a dense plasma, where $N_{00} \gg 1$, the contribution of the relativistic term in (13) (the term proportional to N_{00}^{-2} in square brackets) is negligible, while in a rarefied plasma, where N_{00} is close to unity, the contribution of this term can be of the same order of magnitude or larger than the contribution of other terms.

2) The nonrelativistic expression for N following from equation (10) can be used for qualitative analysis of whistler-mode propagation in a weakly relativistic plasma; this expression is the best approximation to the corresponding weakly relativistic expression when the wave frequency is close to the electron gyrofrequency, provided condition (7) is satisfied.

3) Relativistic effects lead to a decrease in the frequency range of instability, and the magnitude of this effect increases as ν decreases.

4) For moderate values of anisotropy ($T_\perp/T_\parallel \approx 2$) relativistic effects lead to an increase in the maximum value of the increment of instability (γ_{max}); γ_{max} increases rapidly with increasing electron temperature and anisotropy.

5) Expression (14) for the increment of whistler-mode instability (or the decrement of damping) is a good approximation to the nonrelativistic expression (11) when $\gamma > 0$, i.e. when we have instability; expression (14) predicts weaker damping of whistler-mode waves at frequencies close to the electron gyrofrequency, when compared with the prediction of expression (11).

Acknowledgements. The authors are grateful to the referees for their helpful comments (special thanks are due to one of the referees who drew our attention to the paper by Imre, 1962) One of the authors (S.S.) would like to thank NERC for financial support at the initial stage of this project.

REFERENCES

Imre, K., Oscillations in a relativistic plasma, *Phys. Fluids, 5*, 459-466, 1962.

Robinson, P. ., Thermal effects on parallel-propagating whistler-mode waves, *J. Plasma Phys., 37*, 149-162, 1987.

Sazhin, S. S., *Whistler-mode waves in a hot plasma.* Cambridge University Press, 1993a.

Sazhin, S. S., A relativistic theory of plasma cut-offs, *Astroph. Space Sci., 203*, 317-327, 1993.

Sazhin, S. S., A. E. Sumner, and N. M. Temme, Relativistic and non-relativistic analysis of whistler-mode waves in a hot anisotropic plasma, *J. Plasma Phys, 47*, 163-174, 1992.

Sazhin, S. S., A. E. Sumner, N. M. Temme and F. Gugic, An approximate solution of the parallel whistler-mode waves dispersion equation in a weakly relativistic plasma, *Plasma Phys. Control. Fusion, 35*, 117-126, 1993.

Storey, L. R. O., An investigation of whistling atmospherics, *Phil. Trans. Roy. Soc (London), A246*, 113-141, 1953.

S. S. Sazhin, Fluent Europe Ltd., Hutton's Buildings, 146 West Street, Sheffield, S1 4ES, U.K.

A. E. Sumner, Physics Department, University of Canterbury, Christchurch, New Zealand.

N. M. Temme, Centrum voor Wiskunde en Informatica, Kruislaan 413, 1098 SJ, Amsterdam, The Netherlands.

Quasi-Periodic Global Substorm Generated Flux Variations Observed at Geosynchronous Orbit

R D Belian, T E Cayton and G D Reeves

Los Alamos National Laboratory, Los Alamos, NM 87545

Giant, quasi-periodic, global flux variations with periods of 2 to 4 hours have been observed by the Los Alamos National Laboratory energetic particle detectors, CPA (Charged Particle Analyzer). These are not uncommon events, however, in most cases thus far analyzed, the number of cycles is limited to two or three. An episode on February 8, 1986, during a large magnetic storm associated with a large solar particle event is of particular interest because of the number and regularity of the variations. AE indices for February 8 show similar periodicities. Solar wind data from IMP-8, although very spotty, indicate that the solar wind speed became very high (>800 km/sec) towards the end of February 7, and remained high for the duration of the variations. The IMF data from IMP-8 indicate that the field was strong and southward throughout the event until 1800 UT when it became positive. Other events will be presented to show this is not an isolated case. We will discuss several possible mechanisms for the variations discussed here.

INTRODUCTION

Large geomagnetic pulsation structures (waves) have been observed for many years. Jacobs (1970), Nishida (1978), and McPherron (1972) have provided excellent reviews of such oscillations. Higbie et al., (1982) have reported simultaneously recorded oscillations (PC 5 waves) from a constellation of six spacecraft in an event of November 14 -15, 1979 during which oscillations were continuously recorded for 48 hours. They term these oscillations, seen in both charged particle and magnetometer measurements, as global because they were observed throughout the dayside. Takahashi, et al., (1987), have shown that the wavelength of the November 14 -15, 1979 oscillations had wavelengths up to 10000 km, with radial extents up to 1.7 Re.

We report here on quasi-periodic particle flux modulations, associated with substorm activity, with periods of hours, which are truly global in nature. These modulations were observed for a period of ~10 hours on February 8, 1986, and were recorded throughout the period at three geosynchronous orbit satellites which were nearly equally spaced in longitude.

Space Plasmas: Coupling Between Small
and Medium Scale Processes
Geophysical Monograph 86

INSTRUMENTATION

The data shown are from Charged Particle Analyzer (CPA) instruments on four geosynchronous orbit satellites, 1982-019, 1984-037, 1984-129, and 1977-007. The set of detectors comprising the CPA instrument has previously been described in Higbie et al., (1978), and the details will not be repeated here. Instead, we provide the following brief descriptions.

Data presented here are taken from two of the four detectors comprising the CPA. The LoE subsystem is a set of five, similar, solid state sensors arranged at angles of +/- 60, +/- 45 and 0 degrees to the normal to the satellite spin axis which is directed toward the Earth. Each element has a collimating aperture with a half angle of ~2.6 degrees, which provides a geometrical factor of 3.1E-03 cm sq sr. As the satellite spins, with a period of ~10 seconds, each telescope sweeps out a band of the unit sphere. During the course of one rotation, 200 samples of the unit sphere are recorded by the five LoE detectors for each energy band. There are six nested energy bands with approximate lower thresholds of 30, 45, 65, 95, 140, and 200 keV. Each band has an upper threshold of 300 keV.

The LoP subsystem is a one element, collimated, solid state sensor mounted perpendicular to the spacecraft spin axis with a field-of-view half angle of 3.7 degrees and a

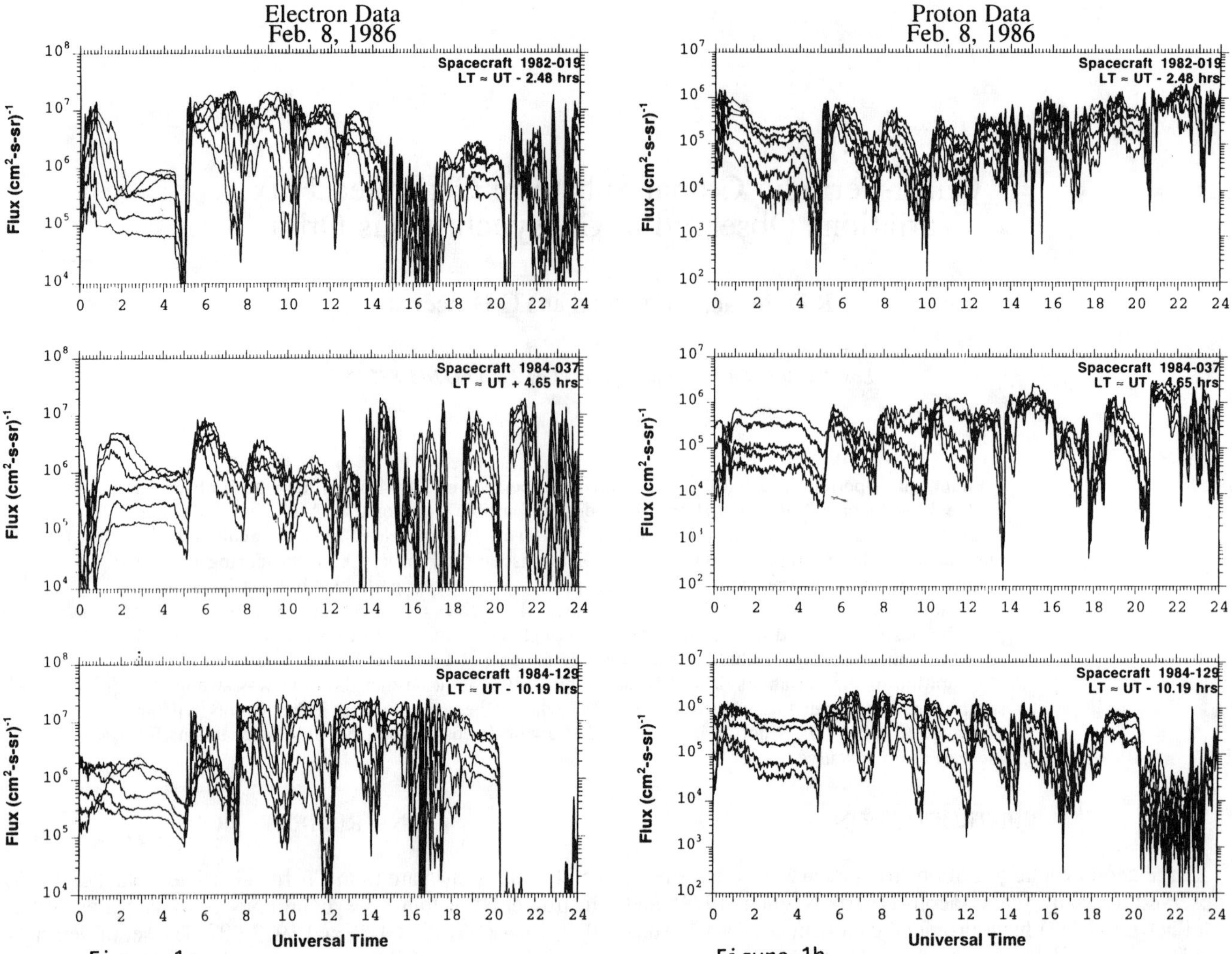

Figure 1a

Figure 1b

Figure 1. 24 hour plot of one minute averaged data. 1a - electron integral fluxes covering the energy range from 30 keV to 300 keV. 1b - proton integral fluxes covering the energy range from ~70 keV to ~600 keV. The data is for February 8, 1986, from three widely separated geostationary satellites. The scale at the bottom is Universal Time, that at the top of each panel indicates the local time of the location of each satellite.

geometrical factor of 4.0E-03 cm sq sr. The LoP has 10 nested energy bands with lower thresholds that vary from satellite to satellite, but which are about 80, 90, 110, 135, 175, 200, 240, 300, 365, 455 keV and an upper threshold of ~580 keV. The LoP records 40 samples of the particle distribution each rotation. Particles penetrating the sensor are vetoed by an anti-coincidence scintillation discriminator, and electrons are eliminated by a sweeping magnet which forms part of the front aperture.

OBSERVATIONS

We plot in Figures 1a and 1b 24 hours of data from three geosynchronous satellites for February 8, 1986. Figure 1a is the LoE electron plots, while Figure 1b is the LoP proton plots. Note the giant, pulsation-like character of the flux profiles between ~0500 and ~1500 UT at all three satellite locations. The set of numbers at the top of each panel are the local time of each satellite at the universal time shown below the third panel. Virtually all local times are represented at one or the other of the three satellites during the 10 hours of the event. These curves show the global aspect of the events and their occurrence at multiple energies and for both electrons and protons. Even though the form of the pulsation varies according to satellite location there is no mistaking their regularity at the vastly different locations. On the average these variations have a periodicity of about 2.3 hours.

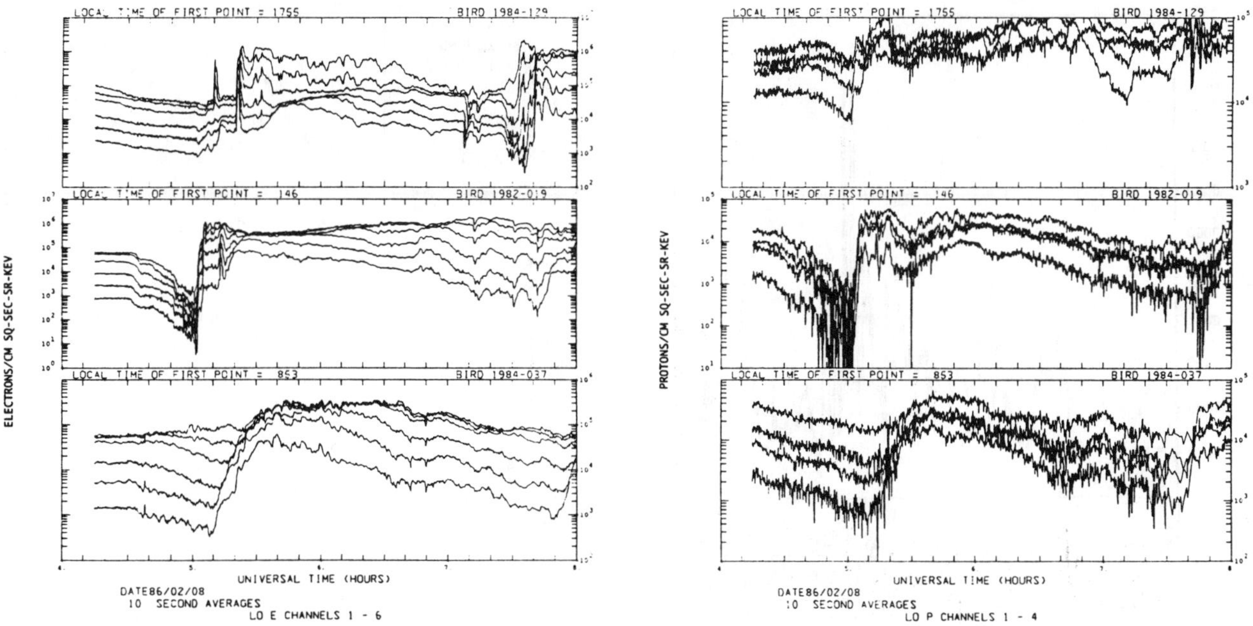

Figure 2. Same as figures 1a and 1b, except that data is only shown for a 4 hour period, encompassing one of the variations.

A sawtooth-like character to the envelope of flux profiles is easily discernible, especially in the proton data. The sawtooth character is brought about by alternating rapid flux enhancements followed by gradual flux declines. This is a truly remarkable event. We have looked at about 200 days of data in this 24 hour format, and the February 8, 1986 event is by far the most regular we have come across.

The apparent flux dropouts (seen especially around 0500 UT), causes one to suspect recurrent substorm activity to be the source of the variations. To investigate this further we have plotted Figures 2a and 2b. They represent 1:6 expansions of the data around the initial flux variation. Four hours of data are shown beginning at 0400. Figures 2, on first consideration, reinforce our thoughts of a substorm generation mechanism. First between 0400 and 0500 UT, we see the familiar one half to one hour flux decline usually associated with the substorm growth phase. Then at or beyond 0500 UT we observe the recovery to enhanced flux levels, often observed at geosynchronous, that we attribute to particle injection. And finally we observe the flux decrease normally seen after substorm injections during the substorm recovery phase. In addition, if one looks at the satellite nearest local midnight, 1982-019, one observes that the injection is without dispersion; this is another expected substorm characteristic. We also observe dispersed enhancements at the satellite near dawn, 1984-037. Again, this is just what one would expect to see in the substorm injection scenario.

There are other aspects of the situation on February 8 that strongly support the substorm injection scenario. The upper panel of Figure 3 shows the AE index plotted for the entire day. The lower panel is simply the upper panel of Figure 2b. One can discern a series of modulations in the AE index that are similar in form and duration to the flux variations. The modulations begin at 0500 UT, repeat approximately every 2 hours (like the flux variations) and quit after being interrupted by the strong shock that produced the 5000 gamma spike in AE index at ~1500 UT.

In addition to the AE index modulations and the form of the flux modulations, there are other substorm indicators. Figure 4 is a plot of IMP-8 interplanetary conditions for a few hours before February 8, 1986 through a few hours after February 8. The values were read by eye every 15 minutes from the normal moment plots of the IMP-8 data courtesy of S. Bame and M. Thomsen at the Los Alamos National Laboratory. The upper panel plots the solar wind bulk speed for the time period. Although the coverage is limited one sees that the bulk speed increased several hours before the event period. As far as we can tell from the available data the speed remained high for the duration and after. The lower panel of Figure 4 shows the latitude of the Interplanetary Magnetic Field (IMF) for the same period. The IMF data is read every 15 minutes from 15 second averaged plots of the IMF, kindly provided by R. P. Lepping, through the National Space Science Data Center at GSFC. In the 15 second data (not shown) the IMF became slightly southward at ~0500 UT, became gradually more and more southward, becoming strongly southward by about 0600 UT. There was a long IMF data gap before 0500 UT.

We have examined several other periods of solar particle

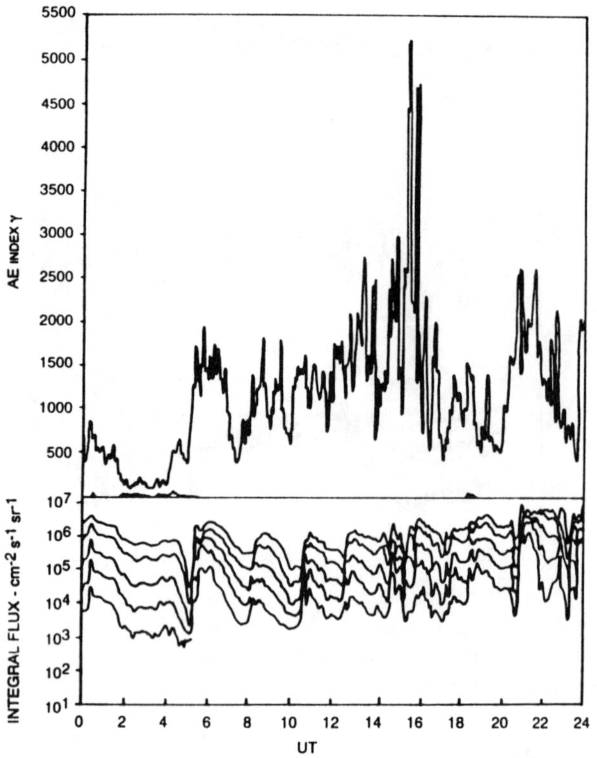

Figure 3. 24 hour plot of AE index for February 8, 1986, upper panel, and the same Lo P fluxes from 1982-019 as seen in Figure 1, lower panel.

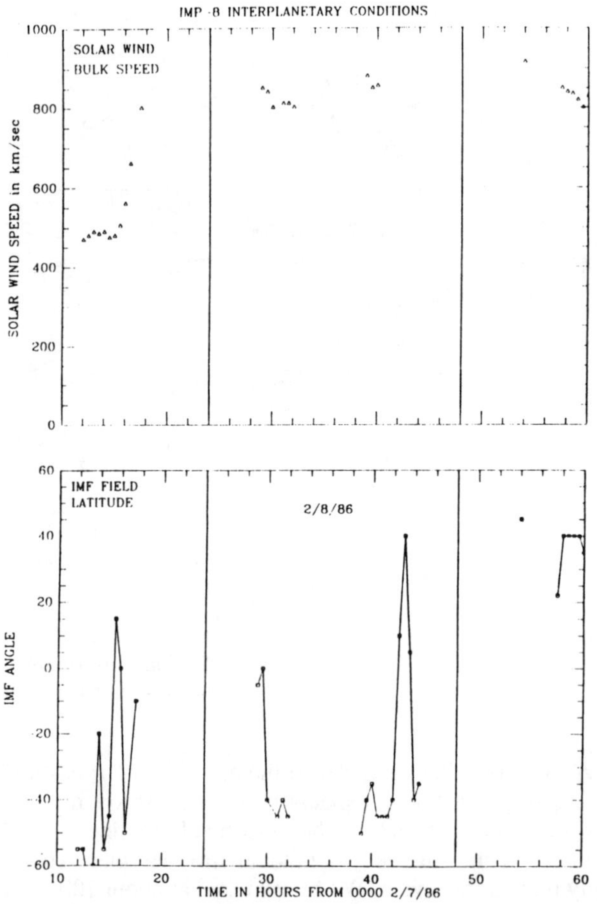

Figure 4. Interplanetary conditions. IMP-8 data read from plots of solar wind speed and IMF at 15 minute intervals, courtesy of M. Thomsen. The figure covers the period of time from 1000 UT on February 7, to 1200 UT on February 9, 1986 thereby bracketing the pulsation event. Upper panel: Solar wind bulk speed during times that that parameter is available. Lower panel: IMF field latitude during times that that parameter is available.

events during which we have good, continuous solar wind data, in order to identify similar events and indeed others have been found. Figure 5 is a composite of the type of data presented in Figures 1 through 4 for one of the events that occurred on January 31, 1982, while ISEE-3 was located at the sub-solar Lagrangian point. The panels are, from top to bottom, the flux profiles for satellite 1977-007 for electrons and protons, the AE index, bulk solar wind speed, and IMF latitude. First of all it is obvious, from the flux profiles, that this event was somewhat different in appearance than the first one discussed. In particular, the sawtooth nature of the variations is not so apparent for this event. However, the substorm like character (dropout, enhanced recovery, etc.) exists here also. The AE index tends to be modulated in similar fashion as the fluxes as was the case in the February 8, 1986 event. The solar wind speed and IMF latitude, much more continuous here, are high and strongly southward respectively, again as on February 8, 1986. These data reinforce our contention that the periodic variations occur when the IMF is strongly southward and the solar wind parameters are strong and constant for long periods of time. The fact that the bulk speed is considerably lower here, ~500 km, and not quite steady may account for the difference in flux profile between this and the previous event.

DISCUSSION

One can think of several causes of the phenomena shown in Figures 1 and 5. Hardening of the spectrum early on followed by softening as the event continues suggest the initial conclusions of Belian and Cayton, 1989, who have proposed global, non-adiabatic, periodic contractions of the magnetosphere.

We can also interpret these observations as samples of relaxation oscillations of a driven system, wherein the solar energy input remained essentially constant, the IMF remained southward and stresses built up at a constant rate. The stresses then relaxed by periodic substorm action.

These events are complex, and do not easily lend themselves to an understanding based on any existing substorm model. Given the constant solar wind and IMF

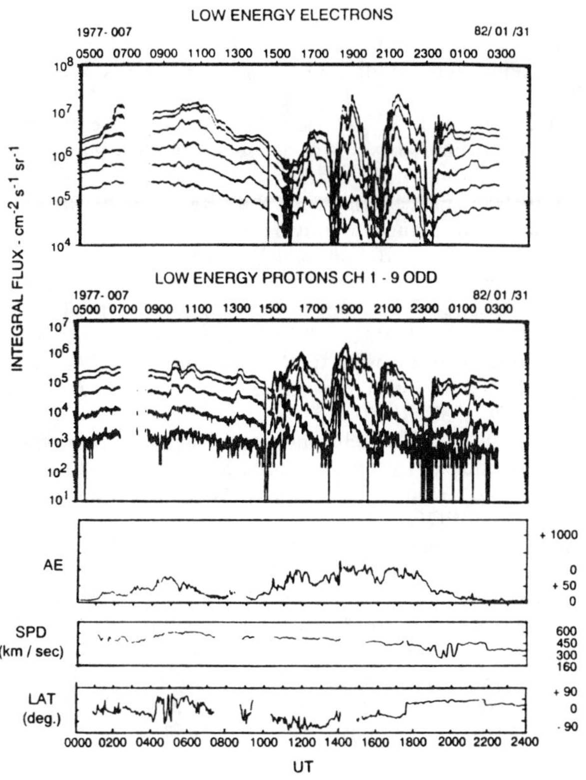

Figure 5. Data from an event that occurred on January 31, 1982. Composite of the types of data presented in Figures 1 through 4, except that flux profiles are presented only for one satellite. The AE index, bulk solar wind speed, and IMF latitude angle were taken from the WDC-C2 for Geomagnetism, Data Book, Special Issue (ISEE-3, AE, Dst, Kp), August 1986.

conditions, if a simple isolated substorm generated injection were to have occurred, one would expect to see a certain sequence of events, i.e. the classical substorm injection. This is not quite the case here. While some aspects of the classical injection are evident, and indeed, were discussed in the observations part of the paper, others are not. For example: The model predicts that with the southward turning of the IMF, the magnetosphere should quickly respond by stretching of the tail field lines, which then results in thinning of the plasma sheet and flux reductions at midnight zone geosynchronous orbit. This indeed is the case: witness the flux drop-out at the near midnight satellite, 1982-019 shown in figure 2. But notice also, that the fluxes diminish at all three locations, even at the satellite near 1000 LT. Indeed, in successive events, in every case, regardless of the satellite location, all three recorded flux reductions prior to every flux enhancement (see Figure 1). Flux drop outs near local- noon are not typical, expected substorm behavior.

The timing sequence at the various satellites seems to conform to that expected for drifting particle bunches. That is, for electrons the midnight satellite should record the onset, followed sequentially by satellites to the east. In this case, the enhancement beginning at ~0500 UT qualitatively follows the expected scenario: 1982-019 at ~0300 LT first recorded the enhancement, followed 7 minutes later by 1984-037 7 hours to the east and finally 11 minutes after that by 1984-129, 9 hours further east. But the appearance of the injected particle bunch at the three satellite locations does not display the expected dispersion. 1984-129 near 1915 LT recorded a sharp onset nearly 20 minutes after 1982-019 also recorded a sharp onset while near 0245 LT. Clearly sharp onsets at two widely separated locations with 20 minutes delay between them does not follow that expected from injection models. Also, the pulse shape at 1984-037, the second satellite to record the enhancement, is greatly dispersed and has a very long rise time, whereas the enhancement at 1984- 129 is sharp and non-dispersed. This is true even though 1984-129 is 11 hours east of 1984-037. Timing for the protons is another case. Satellites 1984-129 and 1982-019, both being in the evening sector, recorded the 'injection' at about the same time. This, of course, can happen in substorms when the injection front is wide so that both satellites are engulfed by the particle cloud at about the same time. In this case 1984-129 actually recorded the enhancement a moment before 1982-019. Other things could be discussed. However, it is not within the scope of this paper to discuss the details of individual flux enhancements observed. That may be the subject of a later paper.

One is drawn by the repeating flux drop out-flux enhancement sequence to consider the models of Baker et. al.,1990, 1991. In these models, which build upon the 'leaking faucet' model of Hones et al., 1979, the magnetosphere reacts to direct energy and particle input, first by energy storage and then the periodic release of the same. The result is a periodic tailward moving 'plasmoid' and the injection of energized particles earthward. In the cases shown here, possibly in response to a steady, high solar wind speed, and a steady, southward IMF the magnetosphere responded with a series of loadings and unloadings. However, these models suggest classic substorm behavior, whereas, as discussed above, some of the necessary substorm observables were not apparent.

Since the interplanetary conditions appear to remain constant prior to and during these events, and the events are delayed several hours after the magnetosphere had been hit by large changes in the solar wind conditions, it seems unlikely that the source of these large oscillations is directly related to the changes in those conditions. And they may in fact be the result of events taking place in the tail.

CONCLUSIONS

For a period of about 10 hours, the magnetosphere responded to the dual conditions of high energy input (at least for the event on February 8, 1986) and southward IMF by successive, periodic flux enhancements and decays which had a period of about 2.3 hours. The events were global, having been recorded at three widely separated geosynchronous locations. The sequence is highly reminiscent of relaxation oscillator operation. The events resemble successive magnetospheric substorm injections and occurred during times of IMF conditions conducive to substorm activity. Some aspects of the injection model are not followed. The 'growth phase' is apparent at all satellite locations. This is not typical substorm behavior. The 'injection' phases often continue for a large portion of the time (> 1 hour) between successive ' growth phases'. The pulse structure is often inconsistent with particle injections. This series of events resemble the expected 'loading- unloading' model of Baker, et al., but not all aspects thereof.

A list of possible mechanisms:

Giant global magnetospheric pulsations.

Large, regular variations in solar wind parameters.

Constant loading-unloading of the magnetosphere, i.e. "leaking faucet model".

Acknowledgments. This work was performed under the auspices of the United States Department of Energy.

REFERENCES

Baker, D. N., A. J. Klimas, R. J. McPherron, and J. Buchner, The evolution from weak to strong geomagnetic activity: an interpretation in terms of deterministic chaos, Geophys. Res. Letts., 17, 41, 1990.

Baker, D. N., A. J. Klimas, and D. A. Roberts, Examination of time-variable input effects in a nonlinear analog magnetosphere model, Geophys. Res. Letts., 18, 1631, 1991.

Belian, R. D., and T. E. Cayton, The effects of the major solar storm of February 1986 as seen by energetic particle detectors on three nearly equidistant geosynchronous satellites, Proceeding of the Research Institute of Atmospherics, Nagoya University, 36, 127, 1989.

Higbie, P. R., R. D. Belian, and D. N. Baker, High-resolution energetic particle measurements at 6.6 Re, 1, electron micropulsations, J. Geophys. Res., 83, 4851, 1978.

Higbie, P. R., D. N. Baker, R. D. Zwickl, R. D. Belian, J. R. Asbridge, J. F. Fennell, B. Wilken, and, C. W. Arthur, The global Pc 5 Event of November 14 - 15, 1979, J. Geophys. Res., 87, 2337, 1982.

Hones, E. W. Jr., Transient phenomena in the magnetotail and their relation to substorms, Space Sci. Rev., 23, 393, 1979.

Jacobs, J. A., Geomagnetic Micropulsations, Springer-Verlag, New York, 1970.

Nishida, A., Geomagnetic Diagnosis of the Magnetosphere, Springer-Verlag, New York, 1978.

McPherron, R. L., C. T. Russell, and P. J. Coleman Jr., Fluctuating magnetic fields in the magnetosphere, II, ULF waves, Space Sci. Rev., 13, 411, 1972.

Takahashi, K., J. F. Fennell, E. Amata, and P. R. Higbie, Field-aligned structure of the storm time Pc 5 wave of November 14-15, 1979, J. Geophys. Res., 92, 5857, 1987.

Observed Features in Current Disruption and Their Implications to Existing Theories

A. T. Y. Lui

The Johns Hopkins University Applied Physics Laboratory, Laurel, Maryland 20723-6099

Current disruption is a fast temporal and spatially localized phenomenon. Data from the Charge Composition Explorer are used to examine the evolution of the local plasma conditions leading up to and during current disruptions. Observations reveal that as disruption onset is approached, the plasma pressure becomes enhanced, with a tendency towards isotropy. The plasma beta becomes high (as much as ~70), and the observed plasma pressure gradient at the neutral sheet becomes large along the tail axis. The deduced local current densities prior to disruption onset range from ~27 to ~80 nA/m^2 and from ~85 to 105 mA/m when integrated over the sheet thickness. During current disruption, there are large changes in the local magnetic field strength, enhancements of magnetic noise over a broad frequency range, magnetic field-aligned counterstreaming electron beams, ion energization perpendicular to the magnetic field, and reduction in current similar to the amount built up during the growth phase. These observed features compare favorably with the expected signatures of the cross-field current instability.

INTRODUCTION

The investigation of the global flow of energy from the solar wind through the magnetosphere into the ionosphere is fundamental for understanding the coupling between the solar wind and the magnetosphere. It is no wonder that investigation of global energy flow is a primary objective of some major international space programs for this decade, such as the International Solar-Terrestrial Physics Program (ISTP), Solar-Terrestrial Energy Program (STEP), and the Geospace Environment Modeling (GEM) program. Energy from one region in the magnetosphere can be delivered efficiently to another by electrical currents. Therefore, the study of the substorm current wedge can contribute significantly towards our understanding of the spectacular energy flow occurring during substorms. The substorm current wedge can be viewed as a partial diversion or disruption of the cross-tail current which joins with the ionospheric currents through field-aligned currents in the region 1 sense, i.e., downwards to the ionosphere in the morning sector and upwards to the magnetosphere in the evening sector [Boström, 1964; Atkinson, 1967; Akasofu, 1972].

Data from the Charge Composition Explorer (CCE) of the Active Magnetospheric Particle Tracer Explorers (AMPTE) program reveal exciting evidence of large magnetic field fluctuations occurring in the near-Earth neutral sheet region at substorm onsets. The disturbance is accompanied by a reduction of the fringe field associated with the intense cross-tail current built up during the substorm growth phase. The north-south component of the magnetic field is often found to reverse in sign momentarily and repeatedly. These features have been interpreted as turbulent characteristics of current disruption detectable by a spacecraft embedded within the disruption site [Lui et al., 1988]. Large fluctuations of magnetic field within the neutral sheet region are not limited to the downstream distances within the CCE apogee (~8.8 R_E) as similar phenomenon has been previously reported at further down the magnetotail during substorms [Hruska and Hruskova, 1970; Lui and Meng, 1979; Coroniti et al., 1980].

In this paper, we present some characteristics in the current disruption region obtained by CCE measurements, with emphasis on the neutral sheet conditions leading up to and during current disruption. These results are then used to compare with predictions from existing theories for substorm initiation.

Space Plasmas: Coupling Between Small and Medium Scale Processes
Geophysical Monograph 86

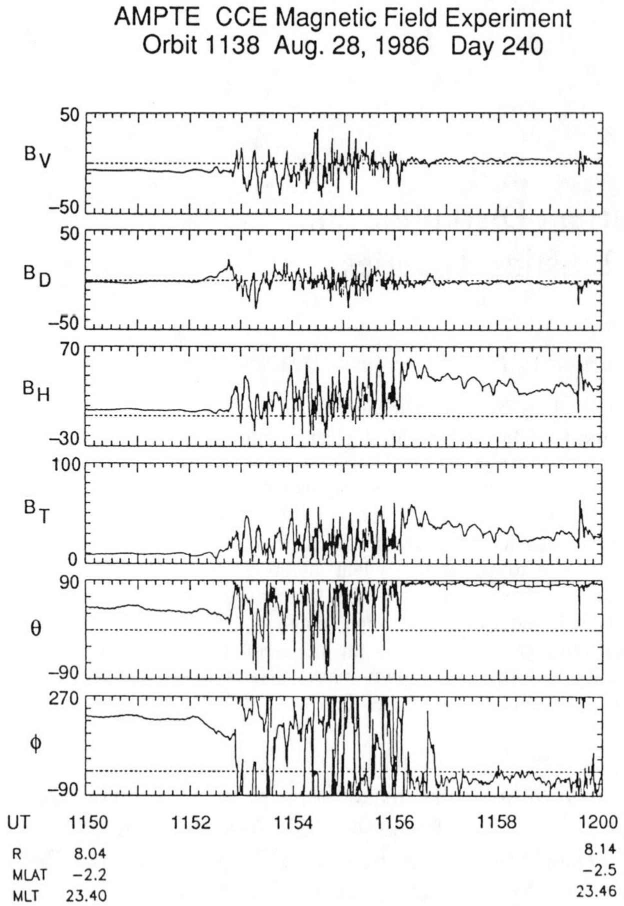

Fig. 1. CCE magnetic field measurements at high time resolution on August 28, 1986. The field strength was depressed to about one-seventh of the dipolar field value before the start of large field fluctuations at ~1153 UT.

CASE STUDY

The high level of magnetic field fluctuations which can be seen during current disruption is dramatically illustrated in Figure 1 by the event on August 28, 1986 [Takahashi et al., 1987]. The magnetic field measurements are given in the dipole VDH coordinate system. During this event, the spacecraft was located at a geocentric distance of ~8.1 R_E, slightly before the magnetic local midnight. The field magnitude prior to disruption onset was extremely small, ~8 nT, in comparison with the Earth's dipole field of ~58 nT at that distance. The low field strength indicates a buildup of an intense cross-tail current such that its fringe field canceled about six-sevenths of the dipole contribution. Large magnetic field fluctuations began at ~1152:40 UT, a time which corresponded to within 1 min of the onsets of magnetic activity and Pi 2 micropulsation on the ground in the CCE sector.

Another current disruption event which was detected closer to the neutral sheet than the previous event is shown in Figure 2 where the Medium Energy Particle Analyzer (MEPA) data for energetic ions are displayed with the anisotropy information. The proximity to the neutral sheet during this event is indicated by the latitude angle of the magnetic field being close to 90°. The "–" and "+" signs in the color spectrograms mark the sectors at which particles entered the detector from the dawn and dusk directions, respectively, at 90° with respect to the local magnetic field. Thus, for a northward field, the "–" and "+" sectors show the intensities of particles with gyrocenters at one gyroradius distance earthward and tailward, respectively, of the spacecraft location. Since all channels shared the same viewing aperture, the "+" and "–" sectors were the same for all channels at each sampling interval. The first indication of particle intensity enhancement for this event came from the "–" sector, implying that particles earthward of the spacecraft were energized first. This occurred before the increase of magnetic field strength at ~23:14:08 UT (which coincided closely with substorm westward electrojet intensification at Syowa at ~23:15 UT; see Lui et al. [1992] for details) and well before the H-component reversed in direction. Although only measurements of the lower-energy ions are shown here, the higher-energy ions behaved similarly. If the activity were propagating from tailward of the spacecraft, then ion energization would have been seen first in the "+" sector, then in sectors between "+" and "–" sectors, and finally in the "–" sector. That this temporal sequence was not seen demonstrates clearly the fact that the first activation occurred earthward of the spacecraft at ~8.8 R_E and was not due to earthward propagation of energization across the spacecraft from the tailward direction. After ~2315 UT, the site of particle intensification shifted tailward and earthward of the spacecraft repeatedly. This continual jump of intensification site relative to the spacecraft strongly suggests that the impulsive ion energization occurred all around the spacecraft and is not due to a single acceleration site moving back and forth across the spacecraft.

Can the north-south magnetic field reversal seen during current disruption be taken as the encounter of an X-type magnetic neutral line where particle energization takes place? The north-south component of the magnetic field reversed sign four times between 2314:30 and 2315:00 UT, which may be interpreted as a neutral line crossing over the spacecraft four times. However, a look at the simultaneous particle intensity indicates that there

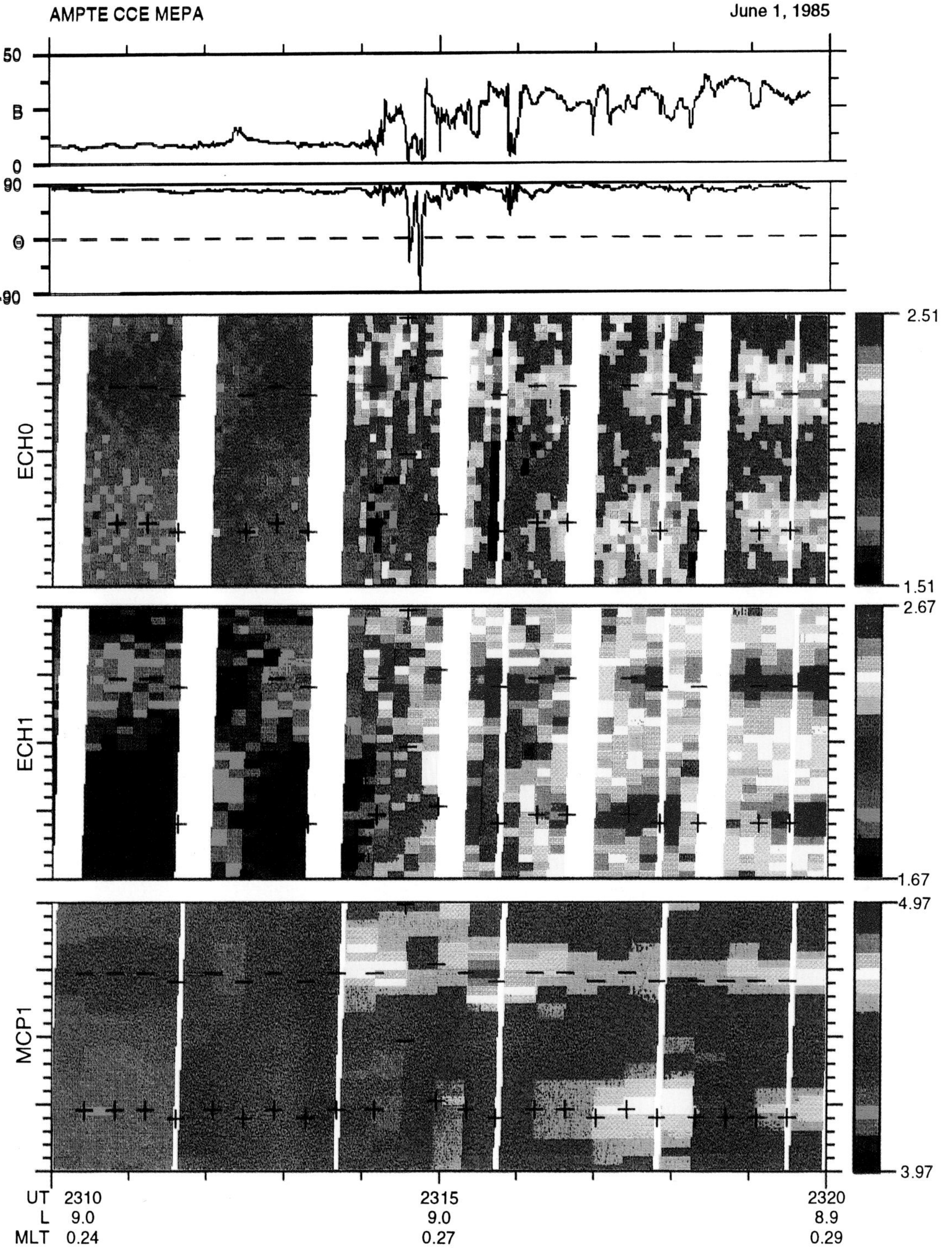

Fig. 2. Energetic ion and magnetic field measurements on CCE during the current disruption event on June 1, 1985. The three spectrograms correspond to ion measurements of 31-43 keV, 43-63 keV, and >10 keV from top to bottom. The first sign of particle energization occurred for particles with gyrocenters earthward of the spacecraft and it occurred before the increase of magnetic field strength.

is no dramatic enhancement during this interval of field reversals in comparison with the initial particle enhancement prior to this interval. As shown later in Figure 4, the electrons also did not show any dramatic energization in association with this field reversal. Therefore, it is difficult to justify that the north-south field reversal is related to an X-line with particle energization. Rather, the field reversal may be viewed as part of the manifestation of field turbulence bearing no special significance.

During this event, the CCE was in the neutral sheet region for an extended period, permitting a close examination of the temporal development of pressure components surrounding the disruption onset in the near-Earth neutral sheet region. The evolution of the particle pressures, pressure anisotropy, and plasma beta for this event are shown in Figure 3. The ion pressures are calculated from measurements of MEPA and the Charge-Energy-Mass (CHEM) spectrometer covering an energy range of 1 keV to ~4 MeV while the electron pressures are based on measurements from the Hot Plasma Composition Experiment (HPCE) covering an energy range of ~50 eV to ~25 keV. For typical thermal energies of ~10 keV for ions and ~2 keV for electrons, the energy ranges provided a coverage of ~90.5% and ~97.5% for ions and electrons, respectively, of the entire population with a stationary Maxwellian distribution (and even higher percentages for a Kappa distribution). The perpendicular ($P_\perp$) and parallel ($P_\parallel$) components of pressure for protons, oxygen, and electrons are given in panel (a) and (b), respectively. The total plasma pressure components shown in panel (c) are the sum from these three species. The plasma beta is displayed in panel (d). These parameters are calculated in intervals of ~3.2 min, which corresponds to an instrument cycle of CHEM. Times shown are measured with respect to the disruption onset time.

It can be seen from Figure 3 [panels (a) and (b)] that the perpendicular and parallel pressures from protons were highest among the species. Next were that of electrons. Oxygen ions contributed least to the pressure components among these three species. There was a general trend of increasing pressure in both perpendicular and parallel components toward the disruption onset. In general, the pressures of the three species as well as the total plasma pressure were nearly isotropic prior to the onset of activity. Near the end of the disturbance, we find that protons developed a pancake pitch angle distribution ($P_\perp > P_\parallel$) while the electrons were observed to show field-aligned bidirectional anisotropy ($P_\parallel > P_\perp$) as reported by Klumpar et al. [1988]. Furthermore, the sum of pressure components from all three species [panel (c) in Figure 3] were definitely enhanced relative to the pre-activity levels. It should be noted that although the level of magnetic field fluctuations was high during disruption, the field latitude angle was mostly near ~90° so that the pitch angle of particles sampled in each sector remained relatively unchanged. The plasma beta [shown in panel (d)] was extremely large throughout the interval, ranging from ~6 to ~70. Its maximum value was reached just prior to activity onset when the plasma energy density was high and the ambient magnetic field was low. Its minimum value occurred at the end of the activity when the magnetic field strength was highest.

The plot in Figure 4 shows a 2-min interval of electron data at the highest time resolution. The 2.9-4.9 keV electron channel sampled electrons near the thermal energy. The large increase in B_H, signifying current disruption, started at ~23:14:08 UT. Prior to the current disruption, there is an indication that electron fluxes showed a spin modulation with peak fluxes near a 90° pitch angle. Electrons thus had a pancake distribution prior to current disruption, which persisted well past the current disruption onset time until ~23:14:30 UT. This was followed by ~15 s of near-isotropic distribution when the magnetic field turbulence level was high and B_H reversed in sign for two brief intervals as noted before. As pointed out earlier, the north-south field reversals were not accompanied by significant electron energization. After this period of isotropy, spin modulation of electron pressure reappeared with enhanced values occurring near 0° and 180° pitch angles. This is due mostly to the development of field-aligned counterstreaming electron beams at the higher-energies as shown by 15-25 keV electrons. Similar field-aligned electron beams have been observed previously at geosynchronous altitude during substorms [McIlwain, 1975; Kremser et al., 1988].

The observed magnetic field fluctuations during current disruptions are generally fully developed such that the frequency band from which the fluctuations originate is not readily discernible. However, for weaker events, structures in the power spectra of the magnetic field fluctuations can be seen as shown in Figure 5 for the event on August 30, 1986. The power spectra of magnetic field fluctuations are given in all three components for three separate time intervals of pre-disruption, early disruption, and late disruption. Details on the procedure of extracting the power spectra and the times associated with these intervals were given in Lui et al. [1992]. The slight leveling off of power beyond 1 Hz frequency for the pre-disruption spectra is due to digitization sensitivity. The dashed line adjacent to the early disruption spectrum in the B_x panel is drawn to represent the slope of the late disruption spectrum. This reference line indi-

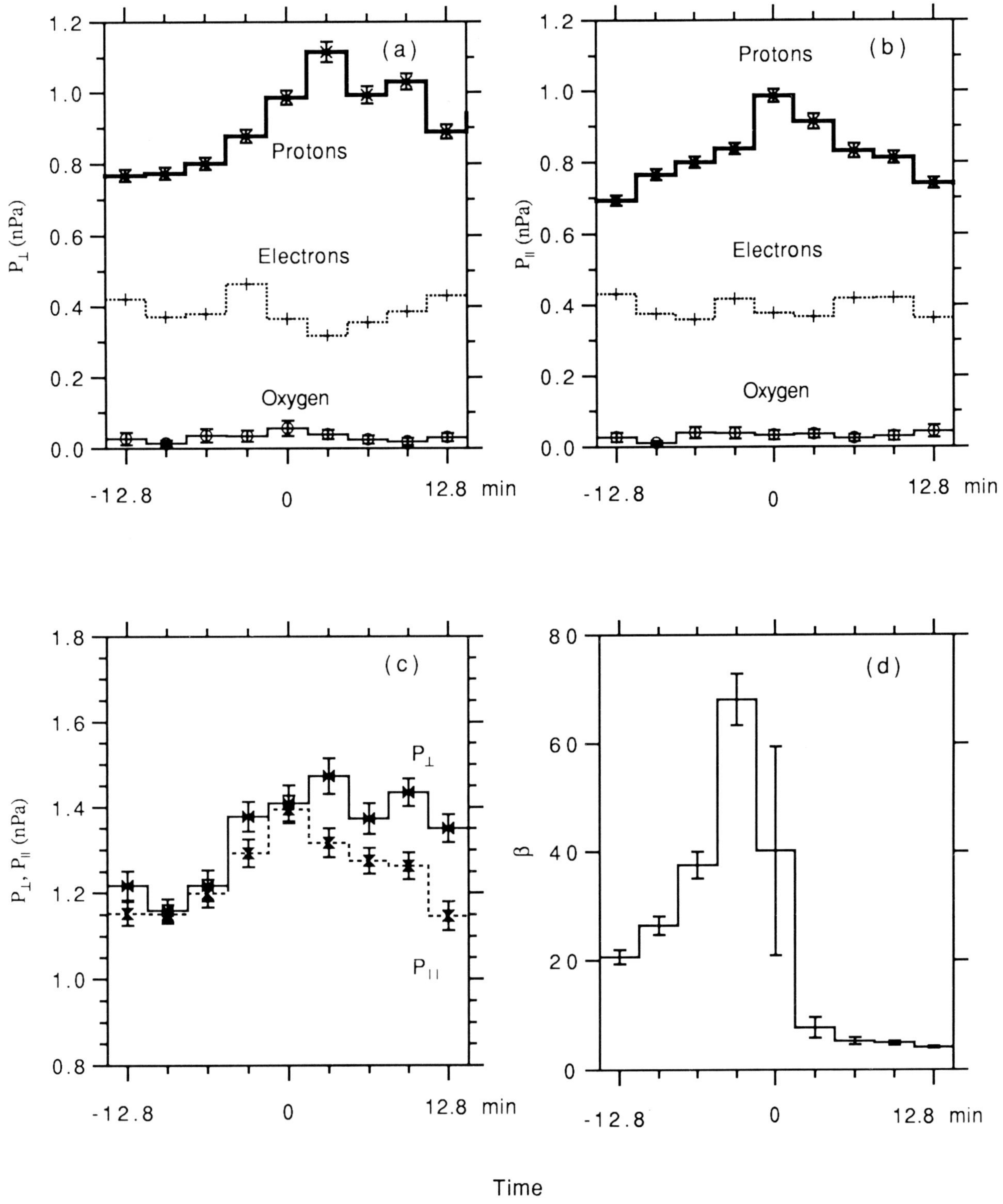

Fig. 3. Time development of pressure components during the current disruption event on June 1, 1985. The pressures from protons were highest, followed by electrons and then by oxygen ions. The total plasma pressure was nearly isotropic prior to the current disruption onset. Note also the extremely high value of the plasma beta just before current disruption.

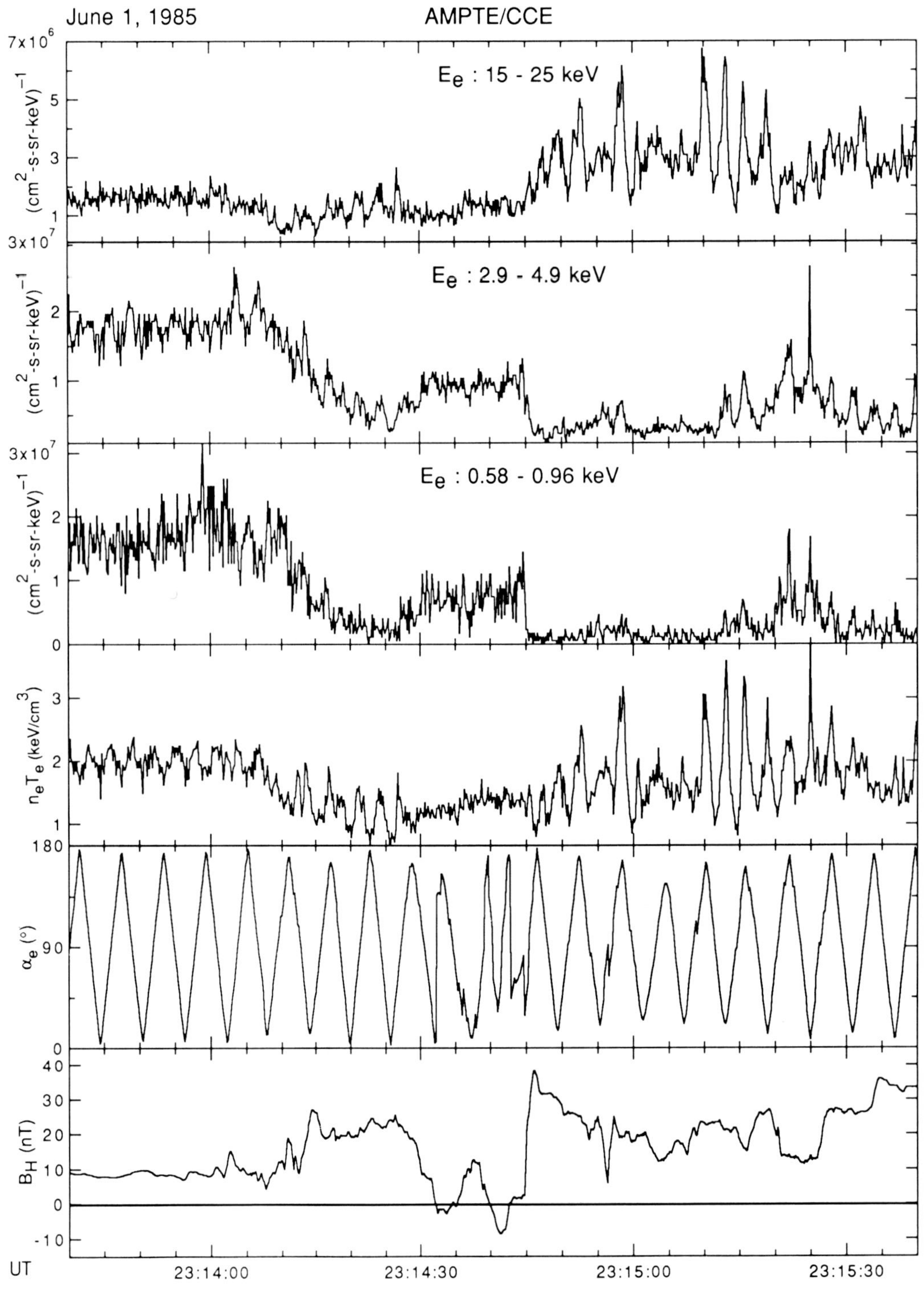

Fig. 4. High time resolution of electron data surrounding the onset time of current disruption. Note that trapped pancake pitch angle distribution of thermal electrons (represented by the 2.9-4.9 keV channel) persisted well after the onset. Nearly isotropic electron distribution occurred when the magnetic field fluctuations were high and gave rise to very brief negative B_z excursions. Afterwards, magnetic field-aligned counterstreaming beams were detected.

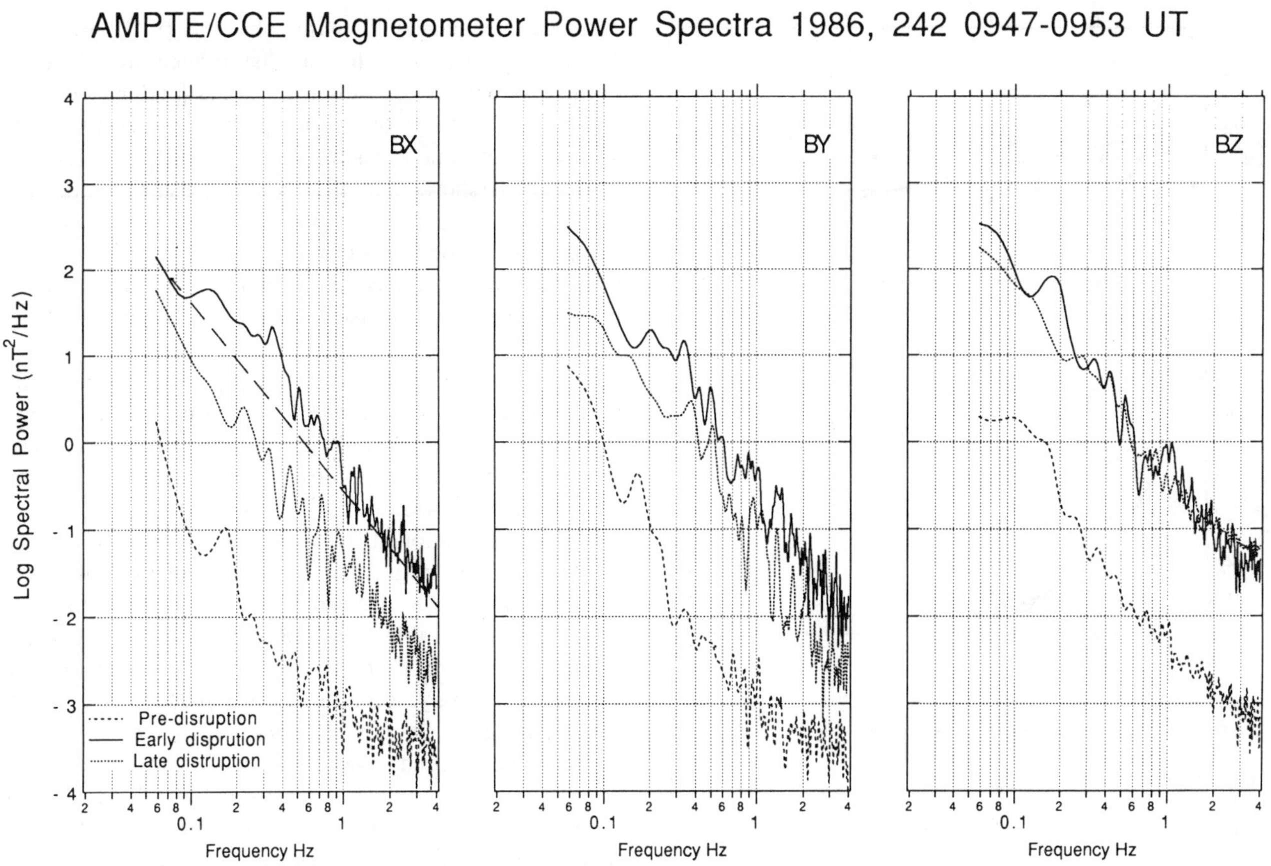

Fig. 5. Power Spectra of magnetic field fluctuations (after subtracting a sliding average; for details, see Lui et al., 1992) during three intervals of the current disruption event on August 30, 1986. Wave powers were generally two to three orders of magnitude above the pre-onset period.

cates the enhanced power to be in the frequency band of ~0.1-1 Hz seen during early disruption. This enhanced power was seen at frequencies near and above the proton gyrofrequency since the averaged proton gyrofrequency was ~0.59 Hz with a low value of ~0.09 Hz. The power spectral index for magnetic field fluctuations in this event varied from about −2.3 to −2.5.

STATISTICAL STUDY

Three general trends are discernible from a statistical study of 15 events of current disruptions from CCE observations. Figure 6(a) shows the anticorrelation between ΔB_T, the change in magnetic field magnitude between immediately before and after disruption, and the estimated distance from the neutral sheet $|dz|$ based on the formula from Lopez [1990]. The absolute magnitude of the magnetic field change ΔB_T is an indicator for the amount of current change through the Ampere's law. Although there is some degree of scatter, the plot suggests that the closer to the neutral sheet, the larger the field magnitude change. Another trend is shown in Figure 6(b) where the field change $\Delta B_T/B_T$ appeared to be anticorrelated with the amount of departure of the field strength from the quiet time level represented by the quantity, B_T/B_{dip}. We use the normalization factors to minimize the radial dependence of the quantities ΔB_T and B_T. The observed trend suggests that the stronger the current buildup, the larger the subsequent field change. The largest value of $\Delta B_T/B_T$ is ~6, indicating that the local field was seen to change by as much as a factor of ~7. Since the field change reflects the amount of current diverted into the ionosphere to form the substorm westward electrojet, there is a possible connection between the amount of current buildup prior to dipolarization and the strength of the substorm electrojet, i. e., the intensity of the substorm disturbance. This idea is tested in Figure 6(c) where the subsequent enhance-

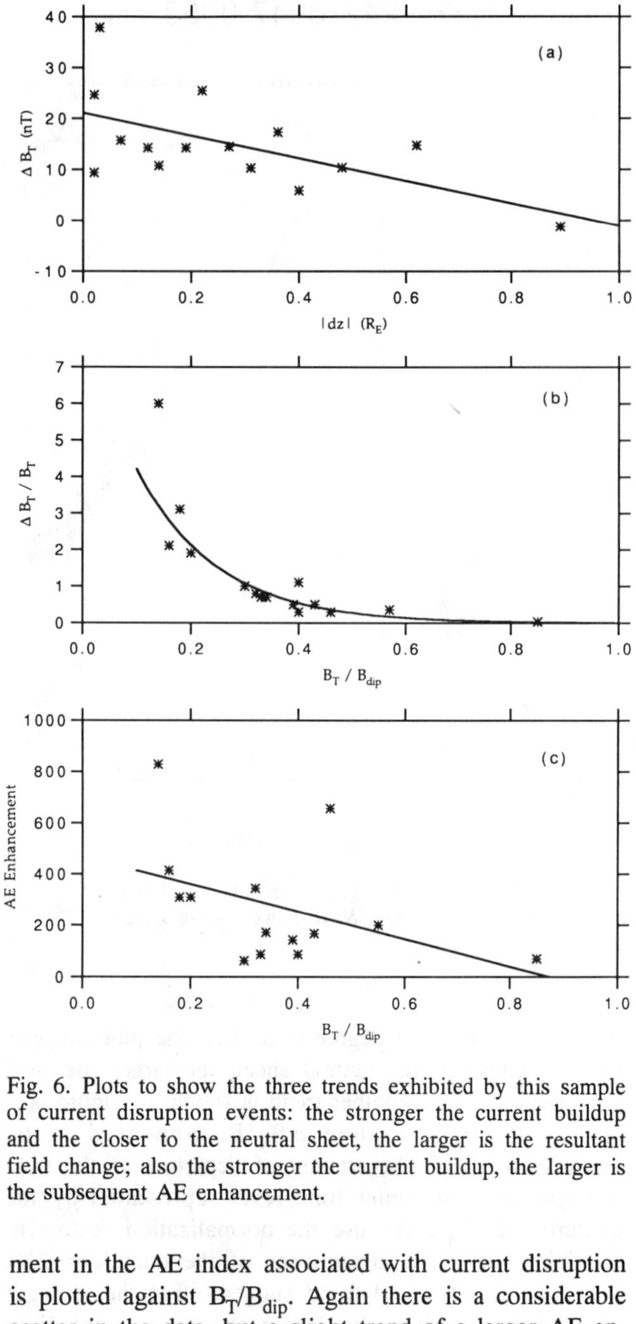

Fig. 6. Plots to show the three trends exhibited by this sample of current disruption events: the stronger the current buildup and the closer to the neutral sheet, the larger is the resultant field change; also the stronger the current buildup, the larger is the subsequent AE enhancement.

ment in the AE index associated with current disruption is plotted against B_T/B_{dip}. Again there is a considerable scatter in the data, but a slight trend of a larger AE enhancement associated with a smaller B_T/B_{dip} (stronger current buildup prior to disruption) is found. This association should be explored with larger samples in the future.

There are seven events altogether in which the pitch angle range covered was sufficiently complete to allow a reasonably accurate assessment of the pressure components at the time just prior to current disruption. Table 1 lists the pressure components from protons, oxygen ions, and electrons separately (with subscripts p, O, and e, respectively) just prior to the disturbance for all these events. As before, the subscripts $\perp$ and $\|$ refer to perpendicular and parallel to the magnetic field, respectively. The total plasma pressure anisotropy $P_{\perp,t}/P_{\|,t}$ is also tabulated. It shows that the proton pressure components are always the dominant ones for the total plasma pressures. The second most important contribution comes from the electrons, and the oxygen ions are the least important. These are persistent features exemplified by the June 1, 1985 event.

The plasma pressure anisotropy prior to current disruption is consistently above 1, indicating that in no case does the condition $P_{\|} > P_{\perp}$ hold for the near-Earth current sheet prior to the occurrence of current disruption in these events, as demonstrated in the June 1, 1985 event. It should be cautioned, however, that within the margin of statistical uncertainty, an anisotropy of less than ~5% is consistent with an isotropic pressure. In terms of individual species, Table 1 indicates that protons typically show a predominance of the perpendicular pressure. The electron anisotropy seems to favor a predominance of its parallel pressure component. From event to event as well as from one sampling interval to another within the same event, there was a significant variability of the electron anisotropy at a finer time-scale than the time-scale used here to obtain the proton anisotropy. These variations, however, do not affect the total plasma pressure anisotropy significantly in these events due to the dominance of proton pressure and the typically low electron anisotropy. For oxygen ions, there are two cases indicating strong anisotropy with the parallel pressure dominating and one case with the opposite anisotropy, while the other four cases are consistent with an isotropic pressure. The energy density of various species can be determined from the given pressure components. In particular, Table 1 gives the ratio of energy density of oxygen ions to protons W_O/W_p which ranges from ~0.5% to about 11%. This suggests that current disruption can occur in the neutral sheet environment with very low to moderate abundance of oxygen ions.

IMPLICATIONS

A puzzling dilemma noted by Kaufmann [1987] is that while the current is found to intensify in the near-Earth region prior to substorm onset, the particle fluxes measured by geosynchronous satellites are typically low. This raises the question of what constitutes the intensified current during the substorm growth phase. One should recall that the decrease in particle flux may be

Table 1. Pressure Components Prior to Current Disruption

Date	UT	$P_{\perp,p}$	$P_{\|,p}$	$P_{\perp,O}$	$P_{\|,O}$	$P_{\perp,e}$	$P_{\|,e}$	$P_{\perp,t}$	$P_{\|,t}$	$P_{\perp,t}/P_{\|,t}$	W_O/W_p
85/4/25	0112	0.89	0.71	0.076	0.074	0.37	0.38	1.34	1.16	1.16	0.091
85/6/1	2314	0.88	0.84	0.035	0.039	0.47	0.42	1.39	1.27	1.07	0.033
85/6/12	0950	0.87	0.81	0.039	0.034	0.24	0.28	1.15	1.02	1.03	0.044
86/8/28	1200	1.34	1.14	0.040	0.058	0.42	0.53	1.80	1.73	1.05	0.034
86/8/30	0949	1.39	1.04	0.150	0.130	0.26	0.29	1.80	1.46	1.23	0.113
86/10/19	2212	1.48	1.29	0.010	0.003	0.14	0.17	1.63	1.46	1.12	0.005
86/11/16	0447	0.88	0.84	0.012	0.035	0.48	0.44	1.43	1.32	1.08	0.069

due to thinning of the plasma sheet in the near-Earth tail prior to substorm onset causing the observing spacecraft to exit from the plasma sheet region rather than an actual decrease in plasma pressure. This indicates that the dilemma is best addressed by observations made near the neutral sheet without the complication brought about by plasma sheet thinning. Such a situation is realized for the June 1, 1985 event and the pressure evolution shown in Figure 3 provides an answer to the above dilemma. For that event, the plasma pressure was observed to increase as substorm onset approached. Therefore, the intensification of current in the near-Earth region prior to substorm onset is associated with plasma pressure enhancement. This trend of increasing pressure towards current disruption onset is not unique to this event but is also found in other events as well.

In a stationary plasma with anisotropic pressures $P_\perp$ and $P_\|$, the perpendicular current density $\mathbf{j}_\perp$ is given by

$$\mathbf{j}_\perp = \frac{\mathbf{B}}{B^2} \times \left[\nabla P_\perp + (P_\| - P_\perp) \frac{\mathbf{B} \cdot \nabla \mathbf{B}}{B^2} \right].$$

The first term is associated with the pressure gradient perpendicular to the field (which can be approximated as $P_\perp/(BL_g)$, where L_g is the scale-length for the pressure gradient) while the second term is associated with the product of the pressure anisotropy and the radius of field line curvature (which can be approximated by $(P_\| - P_\perp)/(BL_c)$, where L_c is the radius of field line curvature). If $P_\perp > P_\|$, then this second term opposes the first term. In this study, the plasma pressure anisotropy just prior to the activity typically showed a predominance of $P_\perp$ over $P_\|$. Therefore, a physical process which simply isotropizes the pitch angle distribution of these particles will in fact enhance the current density by eliminating the negative contribution from the curvature current term. This result contradicts the suggested scenario that the current intensification prior to substorm onset could be due to a positive feedback mechanism in which $P_\|$ is enhanced relative to $P_\perp$.

The June 1, 1985 event can be further used to estimate the development of current density and other relevant parameters just prior to current disruption. Using the ion-sounding technique on MEPA data (>10 keV ion channel which has a temporal resolution of ~24 s; note also the ion thermal energy was ~12 keV, close to the energy threshold of this channnel), we estimate the pressure gradient scale-length L_g perpendicular to the magnetic field to be ~1400 km just prior to the large magnetic field fluctuations. This pressure gradient scale-length is along the tail axis rather than in the north-south direction since the field latitude angle was close to 90°. This short gradient length-scale implies that the strong radial gradient must be rather localized along the tail axis. The perpendicular current density associated with this pressure gradient is $j_y \approx P_\perp/(BL_g) \approx 71$ nA/m^2. This is likely to be the peak current density of the current sheet since it is determined at the neutral sheet close to the field reversal plane. The number density was determined to be about 0.6 cm^{-3}, implying the net relative drift between the ions and the electrons v_d for this current density to be ~750 km/s. The proton thermal speed v_i was ~1300 km/s. Thus, the relative drift speed between protons and electrons was more than half of the proton thermal speed, which is a sizable fraction.

The magnetic field in the tail lobe needed to preserve the pressure balance in the north-south direction just prior to current disruption (before the dynamic activities exhibited by the particles and fields) is ~57 nT and the north-south integrated current density is ~91 mA/m. Together with the peak current density of 71 nA/m^2 as deduced above, we estimate the north-south current sheet thickness L_z to be ~1400 km, about twice the local gyroradius ρ_i of the thermal protons, if the current density is assumed to vary linearly with the distance from the field reversal plane.

Similar analysis can be made for other disruption events provided that the pitch angle coverage of particles was sufficiently good. We find the deduced current density to be ~27-80 nA/m^2, the north-south integrated current density to be ~85-105 mA/m, and the relative drift normalized to the ion thermal speed to be ~0.22 to 0.54. These current densities may be underestimated because (1) the temporal resolution of measurements is not high enough to capture the highest current density developed seconds before activity onset (Note that Ohtani et al. [1991] reported evidence for explosive enhancement of current just prior to disruption) and (2) the current density is obtained in most cases not at the center of the neutral sheet and is thus expected not to be the peak value in the current density profile. Nonetheless, the deduced current density values are quite consistent with other earlier estimates [Lui, 1978; Kaufmann, 1987; Tsyganenko, 1989; Mitchell et al., 1990].

These observations also indicate that current changes are substantial in the near-Earth region. No similarly large current density change is reported in the mid-tail or distant tail regions during the growth phase. The lack of observational evidence that the cross-tail current is reduced in the mid-tail or distant tail prior to substorm onset therefore suggests that the current intensification in the near-Earth region is due to an enhancement of the solar-wind-magnetosphere dynamo in the near-Earth region as deduced from stress balance argument by Siscoe and Cummings [1969] and Coroniti and Kennel [1972] and not due to a re-routing of the cross-tail current from further downstream in the tail.

One may question whether dipolarization events are simply a broadening of the current sheet or an actual reduction of current intensity. This can be answered from an examination of the plasma pressure profile given in Figure 3 for the current disruption event on June 1, 1985. Just prior to disruption, the tail lobe field B_o required to balance the total plasma pressure in the north-south direction is ~57 nT, while at the end of the disruption the required tail lobe field is ~43 nT. With these values and the Ampere's law, we obtain a reduction of the north-south integrated current density J_y $(=2B_o/\mu_o)$ from ~91 mA/m to ~68 mA/m., i. e. ~25 % reduction, in agreement with results from other earlier analyses [Lui, 1978; Jacquey et al., 1991]. Furthermore, the north-south plasma pressure (i.e., $P_{\|}$) well before disruption is about the same value as well after the disruption [Figure 3(c)], indicating that current (relatable to $P_{\|}$) in the near-Earth region was indeed enhanced during the substorm growth phase and was reduced to nearly the pre-growth phase value after disruption. From this, one can conclude that dipolarization events in the near-Earth region are not simply due to thickening of the current sheet without current reduction and that the amount of current reduction in the near-Earth region after current disruption is approximately the same as the amount of current buildup during the substorm growth phase.

COMPARISON WITH THEORIES

One theory for the substorm onset invokes the tearing instability [Schindler, 1974; Birn, 1980; Coroniti, 1985; Wang et al., 1990]. The occurrence of a magnetic field normal to the neutral sheet presents a difficulty in initiating this instability. The work of Lembége and Pellat [1982] has shown that the ion tearing instability is stabilized by the adiabatic response of the electrons. Utilizing the result of Chen and Palmadesso [1986] that some particle orbits in a neutral sheet geometry are chaotic, Büchner and Zelenyi [1987, 1989] explored the chaotization of electron orbits in a thin neutral sheet to eliminate the stabilizing effect of the electrons to tearing perturbations. They adopted a parameter κ to characterize the particle orbit in a neutral sheet environment. This parameter is basically the square root of the ratio of the minimum radius of field line curvature over the maximum gyroradius. For a slab geometry with a normal field B_z and a north-south scale length L_z, $\kappa \approx [B_z L_z/(B_o \rho_z)]^{1/2}$, where B_o is the magnetic field in the tail lobe and ρ_z is the gyroradius at the neutral sheet. The κ-parameter for the thermal ion κ_i and that for thermal electron κ_e are related by $\kappa_e = \kappa_i\,[m_iT_i/(m_eT_e)]^{1/4}$, where m_i and m_e are the mass of ions and electrons, respectively. Casting the result of Sergeev et al. [1983] in their κ-parameter specification, Büchner and Zelenyi [1989] noted that for $\kappa \approx 3.3$, nonadiabatic motion begins for particles with small pitch angles at the neutral sheet. Therefore, a gradual transition to chaotic regime occurs at $\kappa \approx 3$, as found in an earlier study by Imhof [1988] in his study of particle precipitation related to the trapping boundary. A value of $\kappa \approx 1$ is needed to suppress the stabilizing effect of the electrons for the ion tearing instability.

Although the effectiveness of chaotization of electron orbits in destabilizing the ion tearing instability has been challenged by Pellat et al. [1991], it is still useful to determine the characteristics of the electron orbits. The κ-parameter just prior to disruption for thermal ions and electrons for the seven events of Table 2 is computed assuming the same ratio of the north-south scale-length of the current sheet L_z to the gyroradius of the thermal ion as we have deduced for the June 1, 1985 event. Table 2 shows that the values for κ_i are found to be within ~0.51 to 0.93, i. e., the ratio of the minimum radius of field line curvature to maximum gyroradius is ~0.26 to 0.86.

Table 2. Current Sheet Parameters Prior to Current Disruption

Date	UT	j_y (nA/m^2)	J_y (mA/m)	v_d/v_i	κ_i	κ_e
85/4/25	0112	59	86	.38	0.51	4.3
85/6/1	2314	71	91	.54	0.56	4.4
85/6/12	0950	27	85	.22	0.76	6.8
86/8/28	1200	80	105	.41	0.93	8.6
86/8/30	0949	64	101	.39	0.81	9.5
86/10/19	2212	47	97	.24	0.77	9.1
86/11/16	0447	47	93	.39	0.58	5.0

This is consistent with the notion that the orbits of thermal ions are in the chaotic ($\kappa \sim O(1)$) or Speiser-orbit ($\kappa \sim O(0.1)$) regimes. On the other hand, for electrons, κ_e ranges from 4.3 to 9.5, i.e., the ratio of the minimum radius of field line curvature over the maximum gyroradius ranges from ~18 to 90. Under these circumstances, chaotization of electron orbits is unlikely to occur for these near-Earth current disruption events. This result is reinforced by the detailed examination of pitch angle distribution of electrons just prior to current disruption onset (Figure 4). The thermal electrons (in the 2.9-4.9 keV channel) exhibited trapped (pancake) distribution well after the onset of current disruption, indicating that the thermal electrons were not randomized prior to current disruption onset. One should note, however, that this result does not imply that chaotization of electron orbits (to possibly initiate ion tearing instability) will not occur further down the tail beyond the CCE apogee.

The range of κ_i obtained prior to onset at the disruption region carries an important implication. This relates to the question of whether or not the current disruption region maps to the discrete auroral arc region in the ionosphere. Substorm onset is marked by a breakup of a discrete auroral arc. At low altitudes, the discrete auroral arc region is almost exclusively confined to a region of isotropic ion precipitation [Lyons et al., 1988]. This is taken as an indication that the magnetic field lines containing discrete auroral arcs thread the tail current sheet where ion motion violates the guiding center approximation. The range of κ_i determined here at the disruption site prior to activity suggests that the ions there behave nonadiabatically and violate the guiding center approximation. This ion characteristic is consistent with mapping the current disruption region to the discrete auroral arc region at low altitudes. This implication is further reinforced by the occurrence in the disruption region of field-aligned counterstreaming electron beams which is regarded as another signature at the equatorial magnetosphere of the magnetic field line linking with a discrete auroral arc [Klumpar et al., 1988].

Ignoring the stabilizing influence of electrons, Baker et al. [1982] argued heuristically that the ion tearing can be triggered by an increase in oxygen concentration. This idea has found support in a recent analysis of one case study by Daglis et al. [1990] who claimed the predominance of parallel pressure over the perpendicular pressure for the oxygen species being sufficient to constitute the strong current required for the substorm growth phase. Their event was located near the plasma sheet boundary (during an extremely disturbed interval) and, therefore, mapped to the neutral sheet region further down the tail than the CCE location. As we have shown in Table 1, the contribution of oxygen ions to the total plasma pressure is consistently the least important among the three species. Furthermore, the ratio of energy density of oxygen ions to protons W_O/W_p ranges from ~0.5% to ~11%. Therefore, current disruption can occur in a neutral sheet environment with a very low concentration of oxygen ions. The global scenarios which stress the importance of O^+ ions, such as that discussed by Lopez et al. [1990] and Daglis et al. [1990], do not seem to be supported by the observations presented here.

Another mechanism for current disruption is the ballooning instability [Roux et al., 1991; Korth et al., 1991]. The condition for its onset is $L_p/R_c < \beta < 2L_p/R_c$, where L_p/R_c is the ratio of the plasma pressure radial gradient length-scale over the radius of field line curvature. This criterion arises from the consideration that the pressure gradient has to be large enough for the instability onset but at the same time small enough to have force balance in the radial direction. The disruption events analyzed here show L_p/R_c to be of the order of unity while β is typically well above 10. Therefore, the condition for ballooning instability onset does not appear to be satisfied. However, it is premature to rule out this instability since the above onset condition was derived on the basis of a fluid treatment which may be substantially modified when a kinetic treatment is carried out.

A cross-field current instability (CCI) has been proposed for current disruption [Lui et al., 1991]. The cross-tail current is visualized to intensify during the substorm growth phase to a point which exceeds the unstable threshold for this instability. A pre-requisite for this process is the drift associated with the current in the neutral sheet to be a substantial fraction of the ion thermal speed. Solution to the dispersion equation formulated for this environment shows that at least two modes can be excited. One mode is the ion Weibel instability (IWI) which causes current filamentation and electromagnetic turbulence. For the near-Earth plasma parameters, waves are excited over a broadband near and above the ion gyrofrequency. The other mode is analogous to the modified two stream instability (MTSI) which gives rise to ion acceleration perpendicular to the magnetic field and electron acceleration parallel to the field [McBride et al., 1972]. The excited waves for the near-Earth plasma parameters are also broadband at frequencies about one order of magnitude lower than the lower hybrid frequency. Numerical calculations demonstrate that for the near-Earth parameters, both modes can be excited with a high growth rate with the e-folding time-scale of ~5 to 50 s, which is quite comparable to the current disruption and substorm onset time-scales.

To compare the theoretical results with observations, we first examine the appropriateness of local approximation and the residence time of the wave in the relatively thin current sheet in comparison with the wave growth time. The former imposes $1 << k_z L_z$ and the latter requires

$$\int_{-L_z}^{L_z} dz/v_{gz} \approx 2L_z/v_{gz} >> \tau.$$

Here k_z is the wave number of the wave in the z-direction, L_z is the gradient scale length of the current sheet in the z-direction, v_{gz} is the group velocity of the wave in the z-direction and τ is the e-folding time of the wave growth. If we consider the June 1, 1985 event, the deduced ion drift speed with respect to the electrons is about half of the ion thermal speed and the estimated north-south gradient length-scale L_z is ~1400 km. For the excited wave at large propagation angles, we find that $\tau \approx 38$ s, $2L_z/v_{gz} \approx 70 - 350$ s, and $k_z L_z \approx 2.0 - 2.5$. Therefore, the first condition is quite well satisfied while the second one is marginally satisfied.

Table 2 shows that prior to current disruption, the ion drift at the spacecraft frame relative to its thermal speed ranges from 0.22 to 0.54. Keeping in mind that the drift speeds are probably underestimates of the actual values, we find these values to be consistent with what are needed for the CCI to operate. It is interesting to note from Table 2 that when the current sheet is thinned to a fraction of an earth radius in thickness, the high values of κ_e indicate that the current sheet is still below the unstable threshold for ion tearing instability even by assuming chaotization of electron orbits can indeed destabilize the mode, but the drift speed is already high enough to trigger the CCI. In other words, as the near-Earth plasma sheet thins during the substorm growth phase, the unstable threshold for CCI is reached first before that for the ion tearing instability assuming electron chaotization can destabilize it.

There are also encouraging agreements between the predictions of the CCI theory and observations. As pointed out in the presentation of spectra for the magnetic field turbulence seen during current disruption, there appears to be enhanced wave power near the ion gyrofrequency and above, consistent with the expected frequencies and the broadband nature of the excited waves. Furthermore, the observed excited waves are probably not associated with tearing instability. The anticipated waves from tearing instability are of hydromagnetic frequencies (well below the ion gyrofrequency) and long wavelength such that they are unlikely to be Doppler shifted to these higher frequencies.

SUMMARY AND CONCLUSIONS

Current disruption events observed by the CCE were examined and a summary of the observed features is shown in Figure 7. Several important conclusions can be drawn from the above findings concerning current disruptions in the near-Earth magnetotail within a geocentric distance of ~9 R_E.

(1) The current intensification during the substorm growth phase is associated with enhancement in the particle pressure at the neutral sheet. Therefore, the observed decreases in particle fluxes at the geosynchronous orbit during this substorm phase arise from thinning of the near-Earth plasma sheet at that time.

(2) The dipolarization events in the near-Earth region are associated with current reduction similar to the amount of current built up during the growth phase and not simply due to broadening of the current sheet.

(3) The lack of dominance of parallel plasma pressure over the perpendicular pressure in these current disruption events provides a strong evidence that an extremely thin current sheet requiring parallel pressure exceeding the perpendicular pressure does not develop at these distances prior to current disruption. Therefore, any current disruption theory requiring parallel pressure exceeding perpendicular pressure prior to disruption onset is not applicable to these events.

(4) The computation of the particle orbit parameter just prior to current disruption indicates that the thermal ion motion is nonadiabatic while the thermal electron motion is still in the adiabatic regime. The orbit characteristics of the thermal ions are consistent with the expected ion behavior in the tail region which connects magnetically to a discrete auroral arc. The adiabatic electron behavior indicates that the onset condition for the

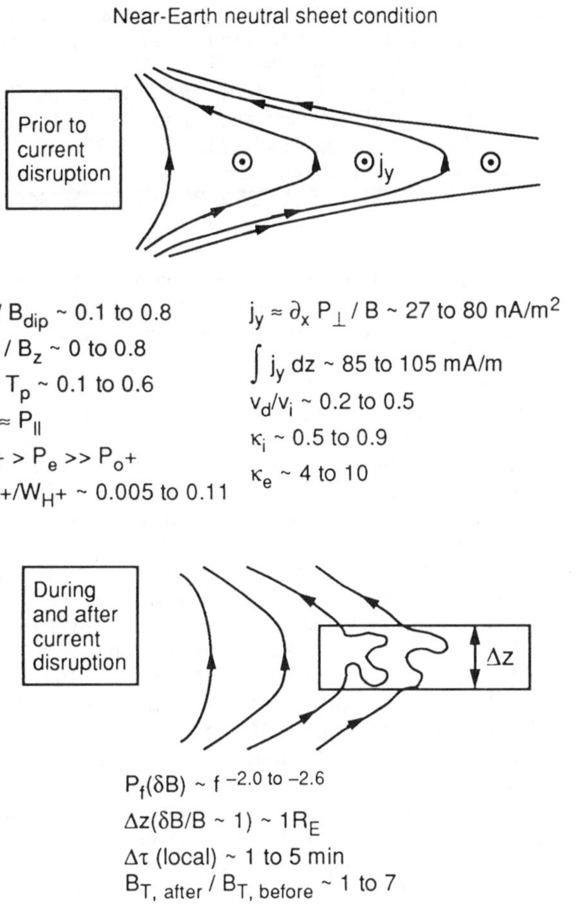

Fig. 7. A summary of several observed features in the near-Earth neutral sheet region prior to current disruption as well as during and after the disruption. For prior to current disruption, the symbols B, T, P, W, j, v, κ denote, respectively, the magnetic field, temperature, pressure, energy density, current density, velocity, and the kappa-parameter as adopted by Büchner and Zelenyi [1989]. For during and after current disruption, symbols P, Δz, $\Delta \tau$, B denote the power spectral density, the north-south thickness of large field fluctuation region associated with current disruption, the local duration of large magnetic field fluctuations, and the magnetic field, respectively.

ion tearing instability arising from chaotization of electron orbits is not reached prior to current disruptions identified within the CCE apogee. This feature is also reinforced by the fact that electron anisotropy (trapped pancake distribution) was observed to persist even after the current disruption onset.

(5) The observed parameters in the neutral sheet region indicate that the intense cross-tail current developed prior to current disruption is unstable to the cross-field current instability. The observed wave frequencies and the broadband nature in the magnetic field fluctuation spectra are consistent with the characteristics of the excited waves predicted by that theory. The current disruption time-scale is in agreement with the predicted growth rate. The characteristics of particle energization are compatible with the anticipated interaction between the excited wave modes and the particles. However, the theory needs to be extended to include nonlocal and nonlinear analyses in order to provide a more definitive comparison with observations.

Acknowledgments. This work was supported in part by NASA under Task I of contract N00024-85-C-8301 and in part by the Atmospheric Sciences Section of the National Science Foundation, grant ATM-9114316 to the Johns Hopkins University.

REFERENCES

Akasofu, S-I., Magnetospheric substorms: a model, in *Solar Terrestrial Physics, Part III*, ed. by Dyer, D. Reidel Publ. Co., Hingham, MD, U. S. A., p. 131, 1972.

Atkinson, G., An approximate flow equation for geomagnetic flux tubes and its application to polar substorms, *J. Geophys. Res.*, *72*, 5373, 1967.

Baker, D. N., E. W. Hones, Jr., D. T. Young, and J. Birn, The possible role of ionospheric oxygen in the initiation and development of plasma sheet instabilities, *Geophys. Res. Lett.*, *9*, 1337, 1982.

Boström, R., A model of the auroral electrojets, *J. Geophys. Res.*, *69*, 4983, 1964.

Birn, J., Computer studies of the dynamical evolution of the geomagnetic tail, *J. Geophys. Res.*, *85*, 1214, 1980

Büchner, J., and L. M. Zelenyi, Chaotization of the electron motion as the cause of an internal magnetotail instability and substorm onset, *J. Geophys. Res.*, *92*, 13,456, 1987.

Büchner, J., and L. M. Zelenyi, Regular and chaotic charged particle motion in magnetotaillike field reversals, 1. Basic theory of trapped motion, *J. Geophys. Res.*, *94*, 11,821, 1989.

Chen, J., and P. J. Palmadesso, Chaos and nonlinear dynamics of single-particle orbits in a quasi-neutral sheet with normal field, *J. Geophys. Res.*, *91*, 1499, 1986.

Coroniti, F. V., On the tearing mode in quasi-neutral sheets, *J. Geophys. Res.*, *85*, 6719, 1980.

Coroniti, F. V., and C. F. Kennel, Changes in magnetospheric configuration during the substorm growth phase, *J. Geophys. Res.*, *77*, 3361, 1972.

Coroniti, F. V., Explosive tail reconnection: the growth and expansion phases of magnetospheric substorms, *J. Geophys. Res.*, *90*, 7427, 1985.

Daglis, I. A., E. T. Sarris, and G. Kremser, Indications for ionospheric participation in the substorm process from AMPTE/CCE observations, *Geophys. Res. Lett.*, *17*, 57 1990.

Hruska, A., and J. Hruskova, Transverse structure of the Earth's magnetotail and fluctuations of the tail magnetic field, *J. Geophys. Res.*, *75*, 2449, 1970.

Imhof, W. L., Fine resolution measurements of the L-dependent energy threshold for isotropy at the trapping boundary, *J. Geophys. Res., 93*, 9743, 1988.

Jacquey, C., J. A. Sauvaud, J. Dandouras, Location and propagation of the magnetotail current disruption during substorm expansion: analysis and simulation of an ISEE multi-onset event, *Geophys. Res. Lett., 18*, 389, 1991.

Kaufmann, R. L., Substorm currents: growth phase and onset, *J. Geophys. Res., 92*, 7471, 1987.

Klumpar, D. M., J. M. Quinn, and E. G. Shelley, Counter-streaming electrons at the geomagnetic equator near 9 R_E, *Geophys. Res. Lett., 15*, 1295, 1988.

Korth, A., Z. Y. Pu, G. Kremser, and A. Roux, A statistical study of substorm onset conditions at geostationary orbit, *Magnetospheric Substorms*, ed. by J. R. Kan, T. A. Potemra, S. Kokubun, and T. Iijima, AGU, Washington, DC, p. 343, 1991.

Kremser, G., A. Korth, S. L. Ullaland, S. Perraut, A. Roux, A. Pedersen, R. Schmidt, and P. Tanskanen, Field-aligned beams of energetic electrons (16 keV $\leq$ E $\leq$ 80 keV) observed at geosynchronous orbit at substorm onsets, *J. Geophys. Res., 93*, 14,453, 1988.

Lembége, B., and R. Pellat, Stability of a thick two-dimensional quasineutral sheet, *Phys. Fluids, 25*, 1495, 1982.

Lopez, R. E., The position of the magnetotail neutral sheet in the near-Earth region, *Geophys. Res. Lett., 17*, 1617, 1990.

Lopez, R. E., D. G. Sibeck, R. W. McEntire, and S. M. Krimigis, The energetic ion substorm injection boundary, *J. Geophys. Res., 95*, 109, 1990.

Lui, A. T. Y., Estimates of current changes in the geomagnetotail associated with a substorm, *Geophys. Res. Lett., 5*, 853, 1978.

Lui, A. T. Y., and C.-I. Meng, Relevance of southward magnetic fields in the neutral sheet to anisotropic distribution of energetic electrons and substorm activity, *J. Geophys. Res., 84*, 5817, 1979.

Lui, A. T. Y., R. E. Lopez, S. M. Krimigis, R. W. McEntire, L. J. Zanetti, and T. A. Potemra, A case study of magnetotail current sheet disruption and diversion, *Geophys. Res. Lett., 15*, 721, 1988.

Lui, A. T. Y., C.-L. Chang, A. Mankofsky, H.-K. Wong, and D. Winske, A cross-field current instability for substorm expansions, *J. Geophys. Res., 96*, 11,389, 1991.

Lui, A. T. Y., R. E. Lopez, B. J. Anderson, K. Takahashi, L. J. Zanetti, R. W. McEntire, T. A. Potemra, D. M. Klumpar, E. M. Greene, and R. Strangeway, Current disruptions in the near-Earth neutral sheet region, *J. Geophys. Res., 97*, 1461, 1992.

Lyons, L. R., J. F. Fennell, and A. L. Vampola, A general association between discrete auroras and ion precipitation from the tail, *J. Geophys. Res., 93*, 12,932, 1988.

McBride, J. B., E. Ott, J. P. Boris, and J. H. Orens, Theory and simulation of turbulent heating by the modified two-stream instability, *Phys. Fluids, 15*, 2367, 1972.

McIlwain, C. E., Auroral electron beams near the magnetic equator, in *Physics of the Hot Plasma in the Magnetosphere,* ed. by B. Hultqvist and L. Stenflo, p. 91, Plenum, New York, 1975.

Mitchell, D. G., D. J. Williams, C. Y. Huang, L. A. Frank, and C. T. Russell, Current carriers in the near-Earth cross-tail current sheet during substorm growth phase, *Geophys. Res. Lett., 17*, 583, 1990.

Ohtani, S., K. Takahashi, L. J. Zanetti, T. A. Potemra, R. W. McEntire, and T. Iijima, Tail current disruption in the geosynchronous region, *Magnetospheric Substorms*, ed. by J. R. Kan, T. A. Potemra, S. Kokubun, and T. Iijima, AGU, Washington, DC, p. 131, 1991.

Pellat, R., F. V. Coroniti, and P. L. Pritchett, Does ion tearing exist?, *Geophys. Res. Lett., 17*, 143, 1991.

Roux, A., S. Perraut, A. Morane, P. Robert, A. Korth, G. Kremser, A. Pederson, R. Pellinen, and Z. Y. Pu, Role of the near Earth plasmasheet at substorms, *Magnetospheric Substorms*, ed. by J. R. Kan, T. A. Potemra, S. Kokubun, and T. Iijima, AGU, Washington, DC, p. 201, 1991.

Schindler, K., A theory of the substorm mechanism, *J. Geophys. Res., 79*, 2803, 1974.

Sergeev, V. A., E. M. Sazhina, N. A. Tsyganenko, J. A. Lundblad, and F. Soraas, Pitch-angle scattering of energetic protons in the magnetotail current sheet as the dominant source of their isotropic precipitation into the ionosphere, *Planet. Space Sci., 31*, 1147, 1983.

Siscoe, G. L., and W. D. Cummings, On the cause of geomagnetic bays, *Planet. Space Sci., 17*, 1795, 1969.

Takahashi, K., L. J. Zanetti, R. E. Lopez, R. W. McEntire, T. A. Potemra, and K. Yumoto, Disruption of the magnetotail current sheet observed by AMPTE/CCE, *Geophys. Res. Lett., 14*, 1019, 1987.

Tsyganenko, N. A., On the re-distribution of the magnetic field and plasma in the near nightside magnetosphere during a substorm growth phase, *Planet. Space Sci., 37*, 183, 1989.

Wang, X., A. Bhattacharjee, and A. T. Y. Lui, Collisionless tearing instability in magnetotail plasmas, *J. Geophys. Res., 95*, 15,047, 1990.

A. T. Y. Lui, The Johns Hopkins University Applied Physics Laboratory, Johns Hopkins Road, Laurel, Maryland 20723-6099.

Particle Acceleration Very Near an X-Line in a Collisionless Plasma

L. R. LYONS AND D. C. PRIDMORE-BROWN

Space and Environment Technology Center, The Aerospace Corporation,
El Segundo, California 90245

In a previous paper, we applied a simplified model for particle motion in the vicinity of a magnetic X-line that had been introduced by Dungey. We used the model to quantitatively show that an electric force along an X-line can be balanced by the gyroviscous force associated with the off-diagonal elements of the pressure tensor. Distribution functions near the X-line were shown to be skewed in azimuth about the magnetic field and to include particles accelerated to very high energies. In the present paper, we apply the previous model and use the distribution functions to evaluate the energization that results from particle interactions with the X-line. We find that, in general, this interaction gives a spectrum of energized particles that can be represented by a Maxwellian distribution. A power-law, high-energy tail does not develop. The thermal energy, K, of the Maxwellian can be expressed simply in terms of the field parameters and particle mass and charge. It is independent of the thermal energy, K_i, of the particle distribution incident upon the region of the X-line, provided that $K_i < K$. Significant energization is not found for $K_i > K$.

INTRODUCTION

Energy transfer in space plasmas is often associated with a magnetic X-line along which lies an electric field [e.g. Vasyliunas, 1975], and the physics of the forces and particle acceleration near such X-lines is an active research topic. In this paper, we consider the acceleration of particles very near an X-line using particle distributions that are consistent with the requirement for force balance near the X-line

Plasma dynamics near an X-line are often addressed within the framework of MHD under the assumption that some sort of resistivity maintains the force that opposes the electric force along the X-line. However, space plasmas are generally collisionless so that the applicability of the concept of resistivity is highly questionable.

It has been pointed out [Vasyliunas, 1975; Sonnerup, 1979, 1988] that force balance along an X-line in a collisionless plasma requires a balance between the electric force and the gradient of the off-diagonal elements of the electron pressure tensor (referred to as "gyroviscosity"); however, identification of a process that could lead to such gyroviscosity was not accomplished until recently. The breakthrough on this topic was by Dungey [1988], who evaluated collisionless particle motion near an X-line using a simplified model. He showed that the particle motion near an X-line will lead to a skewing of the velocity-space particle distributions and thus to significant off-diagonal pressure tensor elements. He suggested that the resulting gyroviscosity could be sufficient to balance the electric force along the X-line. Such gyroviscosity should be present in simulations of collisionless reconnection and should be identifiable by evaluation of the off-diagonal elements of the pressure tensor.

We [Lyons and Pridmore-Brown, 1990; hereafter referred to as paper 1] applied Dungey's model to evaluate particle distribution functions very near an X-line, and we used these distributions to quantitatively evaluate the force balance. We found that, within the limits imposed by Dungey's model, gyroviscosity can balance the electric force along an X-line in a collisionless plasma independent of other plasma and field parameters. The distribution functions we evaluated include the effects of particle energization from electric field acceleration in the weak magnetic field region near the X-line. Here we use these distributions to evaluate this energization.

Space Plasmas: Coupling Between Small
and Medium Scale Processes
Geophysical Monograph 86

DUNGEY'S MODEL

The magnetic field (**B**) model used to evaluate the particle trajectories and distribution functions is two-dimensional, having $\partial/\partial y = B_y = 0$. The X-line is taken to lie along the y axis and to be imbedded within a current sheet that is centered in the z = 0 plane. B_x reverses in sign across the current sheet. Symmetry is assumed about the X-line, so that $B_x(z) = -B_x(-z)$, and $B_z(x) = -B_z(-x)$. The normal component of **B** across the current sheet is expanded in a Taylor series about the X-line, so that $B_z = \beta x$. The electric field E_y is taken to be uniform and to be directed parallel to the current in the positive y-direction.

The basic motion of particles in the vicinity of a current sheet having a weak but uniform B_z was originally presented by Speiser [1965] and is described in paper 1. Particles that reach the midplane of the current sheet oscillate about the midplane and simultaneously gyrate about B_z for one-half a gyro-circle; the gyration begins with an x-component of velocity, v_{xo}, directed towards the X-line, and $v_{yo} = 0$. This gives a current in the positive y-direction, as is required to maintain the current of the current sheet. Only particles with $qv_y > 0$ remain in the current sheet. This is because particles with $qv_y < 0$ are accelerated away from the current sheet by the magnetic force due to B_x (which the electrons encounter as a result of their oscillatory motion about z = 0).

Following Dungey [1988] and paper 1, we take advantage of the decoupling that occurs between the x-y component and the z component of motion for particles as they oscillate about the z = 0 plane [Speiser, 1965, 1968]. This decoupling is assumed to hold as the x-y component of a particle trajectory crosses an X-line. Speiser did not specifically include this situation or the effects of $\partial B_z/\partial x \neq 0$ in his analysis, though particle behavior near an X-line has been considered by Stern [1979]. The evaluation of particle motion within the midplane of the current sheet can thus be simplified by calculating only the x-y component of the particle trajectory at z = 0. This motion is taken to begin with $v_y = 0$.

Under the above conditions, the equations of motion for particles of charge q and mass m are:

$$m(dv_x/dt) = q\beta x v_y \qquad (1)$$

$$m(dv_y/dt) = q(-\beta x v_x + E_y). \qquad (2)$$

Using the normalized temporal and spatial variables introduced by Dungey [1988],

$$x' = x/L, \qquad \text{where } L = |E_y m / (\beta^2 q)|^{1/3} \qquad (3)$$

$$t' = t/\tau, \qquad \text{where } \tau = L^2\beta / |E_y|, \qquad (4)$$

the equations of motion become

$$dv_x'/dt' = x'v_y'\text{sign}(q) \qquad (5)$$

$$dv_y'/dt' = (-x'v_x' \pm 1)\text{sign}(q), \qquad (6)$$

where ± refers to the sign of E_y, which we take to be positive in the present work. Note that E_y, m, β, and q all have a magnitude of unity in normalized units. Also, since the equations of motion are applicable to particles of any charge, we take sign(q) = +1. Thus, our results are applicable to electrons and ions provided the sign of v_y' is changed for electrons.

Representative particle trajectories in the vicinity of the X-line at x' = 0, obtained from solving equations (5) and (6) using a Runge-Kutta scheme, are shown in Figure 1. The trajectories shown are for particles that start their motion within the current sheet at values of x' ≥ 0. The distortion of the trajectories from semicircles results from the acceleration by the electric field and the variation of B_z with x'.

VELOCITY SPACE DISTRIBUTIONS

Trajectories such as shown in Figure 1 can be used to evaluate normalized distribution functions, f', in phase space for the z' = 0 plane if a distribution function, f_i', is specified for particles incident upon that plane. In paper 1, we illustrated the features of these distributions by taking a "top hat" form for f_i'. The value of f_i' was taken to be a constant for all initial velocities, v_{xo}', directed toward the X-line and having $0 \leq |v_{xo}'| \leq 1$, and f_i' was taken to be zero otherwise. Particles were assumed to be incident at all values of x'. With this incident distribution, the distribution at all x' in the z' = 0 plane can be described by an area in v_x',v_y' space within which f' is constant and outside of which f' = 0. It thus suffices to trace the boundary of this area in order to describe $f'(v_x',v_y')$ for an x' at z' = 0.

Boundaries in velocity space within which f' is constant, as obtained in paper 1 for selected values of x', are shown in Figure 2. The boundaries show that f' becomes increasingly skewed towards positive v_x' and v_y' as x' → 1. This is the skewing that gives rise to positive values for the off-diagonal elements of the pressure tensor and thus to the gyroviscosity. In paper 1, we used these distributions to evaluate moments of f' and the balance of forces near the X-line.

While all particles used to construct the distributions in Figure 2 entered the z' = 0 plane with |v'| ≤ 1, the distributions show tails extending to very high values of v'. These tails result from particles having trajectories that become nearly parallel to the X-line at very small values of x' (see trajectories in Figure 4 of paper 1), and the tails are associated with a net energization of particles that results from

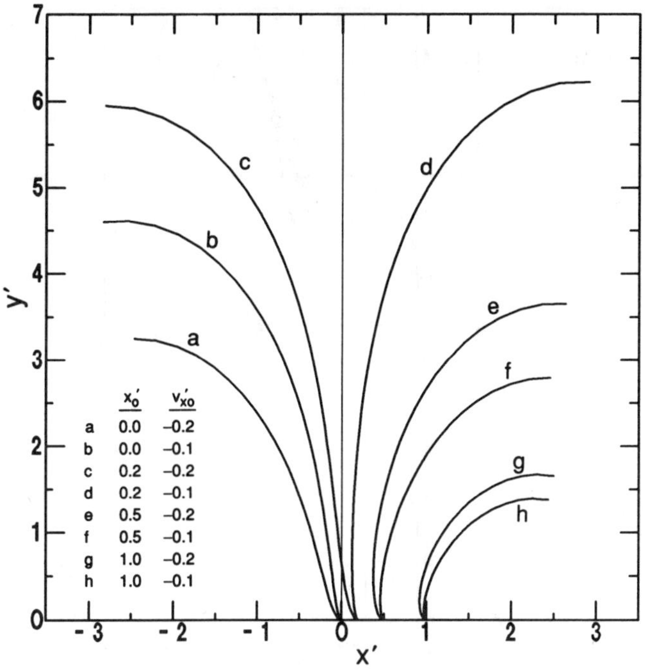

Fig. 1. Representative particle trajectories in the x'-y' plane near an X-line at x' = 0. The sign of y' is correct for positively charged particles. For negatively charged particles, the sign of y' must be changed.

their interaction with the X-line.

To evaluate the energization of particles from their interaction with the X-line, we assume a Maxwellian,

$$F_i'(v_{xo}') = \exp[-v_{xo}'^2/v_i'^2] , \qquad (7)$$

for the distribution of initial speeds of particles following their incidence upon the z = 0 plane. Under the assumption that initial velocities are directed towards the X-line, equation (7) corresponds to an incident distribution function in velocity space given by:

$$f_i'(v_{xo}',v_{yo}',v_{zo}') = \exp[-v_{xo}'^2/v_i'^2]\delta(v_{yo}')f_i'(v_{zo}'), \qquad (8)$$

$$\text{for } x'v_{xo}' \leq 0,$$

where $\int f_i'(v_{zo}')dv_{zo}' = 1$, and $f_i' = 0$ for $xv_{xo}' > 0$. Applying the assumption that the z and x-y components of particle motion are decoupled within the z = 0 plane, we take $\int f'(v_z')dv_z' = 1$ throughout the z' = 0 plane and ignore the v_z' dependence of f'.

Using equations (3) and (4), the normalized thermal velocity of incident particles can be expressed in terms of real parameters as

$$v_i' = [2(K_i q^{1/3}\beta^{2/3})/(E_y^{4/3}m^{1/3})]^{1/2}, \qquad (9)$$

where K_i is the thermal energy of the distribution expressed in eV. If we chose as nominal parameters for the distant geomagnetic tail K_i = 110 eV, β = 0.3 nT/R_e, and E_y = 0.25 mV/m, then equation (9) gives v_i' = 1.0 and 0.286 for electrons and protons, respectively.

With a Maxwellian distribution for $F_i'(v_{xo}')$, $f'(v_x',v_y')$ does not divide into regions where f = constant and where f = 0. However, the regions of maximum f remain similar in shape to the boundaries shown in Figure 2, so that the distributions in Figure 2 still qualitatively represent $f(v_x',v_y')$. To quantitatively evaluate $f'(v_x',v_y')$ at a given x', we use equations (5) and (6) to trace particle trajectories backwards in time from points in v_x',v_y' space until v_y' = 0. In this way, we find initial speeds, v_{xo}', corresponding to the points in v_x',v_y' space, and we obtain f' from equation (8).

Since our goal is to evaluate the energization from particle interactions with an X-line, we choose values of $v' = (v_x'^2 + v_y'^2)^{1/2}$ and determine f' over a range of values for f $= \tan^{-1}(v_y'/v_x')$ selected so as to allow the evaluation of $F'(v') = \int f'(v',\phi)d\phi/\pi$. This gives the distribution of speeds for particles within the current sheet after their interaction with the X-line, which can be compared directly with the incident distribution given by equation (7). As can be seen from Figure 2, the range of important ϕ values becomes increasingly narrow with increasing v', and, in fact, becomes several orders of magnitude less than 1° wide for the highest v' values of interest. This narrow range would be difficult to handle in a model that includes three-dimensions of particle motion; however, it does not present any problems here.

Comparisons between F' and F_i', plotted versus v'/v_i' for x' = 1, are displayed in Figure 3. In principal, the results are only dependent on v_i'. However, we show results for values of v_i' that correspond to specific real parameters. Results for the values of v_i' given above that correspond to electrons and ions for nominal tail parameters are shown in the middle panel (identified as E_y = 0.25 mV/m). Results are also shown for values of v_i' that have been adjusted, using equation (9), for E_y = 0.062 mV/m and for E_y = 1.0 mV/m with K_i and β unchanged. In each panel, the solid line gives F_i'. When plotted versus v'/v_i', F_i' is the same for all v_i', which facilitates comparison of the amount of energization between the different cases. Our evaluation of F'(v') for each E_y is given by filled circles for the value of v_i' corresponding to electrons and by filled triangles for the value of v_i' corresponding to protons.

Figure 3 shows significant energization for all cases having $v_i' \lesssim 1$, the energization increasing with decreasing v_i'. Such energization does not occur for $v_i'^2 \gg 1$, as can be seen for electrons with E_y = 0.062 mV/m, which corresponds to v_i' = 2.5. Keeping other parameters fixed, the energization is greater for protons than for electrons, since $v_i' \propto m^{-1/6}$,

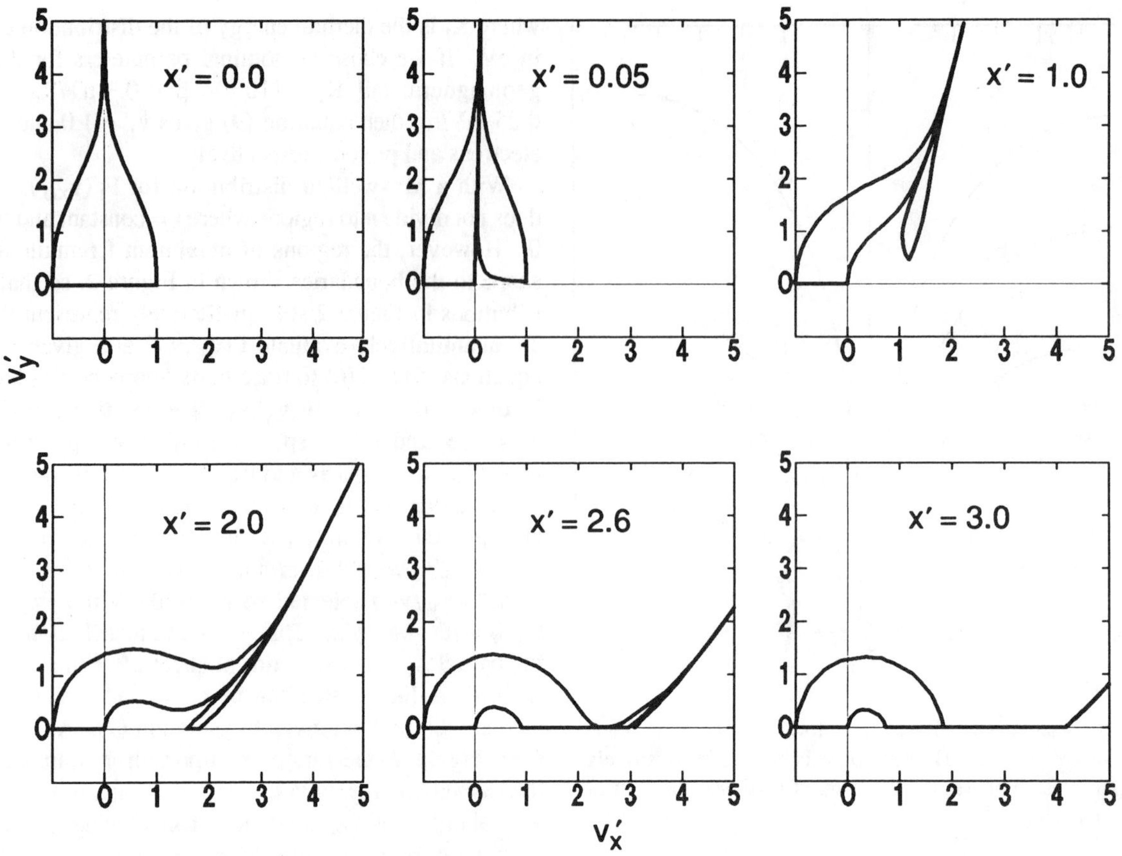

Fig. 2. Boundaries in normalized velocity space within which f' is constant at selected values of x' as obtained in paper 1 for the "top-hat" distribution of particles incident upon the current sheet.

and it increases with E_y since $v_i' \propto E_y^{-2/3}$.

The distributions in Figure 2 suggest that the energy distributions for the particles might show a high energy tail since some low-energy particles are accelerated to very high energies. However, Figure 3 shows that F'(v') falls off as a Maxwellian for $v' > v_i'$. In fact, we have found that F'(v') at $v' \gtrsim 1$ is very well represented by a Maxwellian having a normalized thermal velocity equal to two for all the cases examined that have $v_i' \lesssim 1$. Specifically, the Maxwellian

$$F'(v') = 0.5\, v'^{3/2} \exp[-v'/2]^2, \qquad (10)$$

which is shown in Figure 3 by dashed lines, can be seen to fit our results very well for $v' \gtrsim 1$. All our distributions have $F'(v') \rightarrow 0.5$ as $v' \rightarrow 0$; thus equation (10) does not fit F'(v') for $v' \ll 1$.

The results in Figure 3 are all for x' = 1. Results for the nominal tail parameters are shown in Figure 4 for x' = 0, 1, and 2, which show that F'(v') is essentially independent of x' near the X-line.

The results in Figures 3 and 4 show that the energy distribution of particles in a current sheet very near an X-line should have a thermal velocity of approximately two in normalized units, independent of the thermal energy of the particles incident upon the current sheet, provided the normalized thermal velocity of the incident particles $v_i' \lesssim 1$. This implies, from equations (3) and (4), that such particles will have a thermal energy K in eV given by

$$K = 2E_y^{4/3} m^{1/3} / q^{1/3} \beta^{2/3}, \qquad (11)$$

independent of K_i provided $K \gtrsim 4K_i$. No energization occurs if equation (11) gives $K \lesssim K_i$. For our nominal tail parameters (E_y = 0.25 mV/m), equation (11) give a thermal energy of 0.4 keV for electrons and 5.3 keV for protons. These are of the order of the thermal energies of particles observed within the tail plasma sheet; however, the effects of particle motion away from the X-line region would need to be evaluated for a direct comparison with plasma sheet observations to be meaningful.

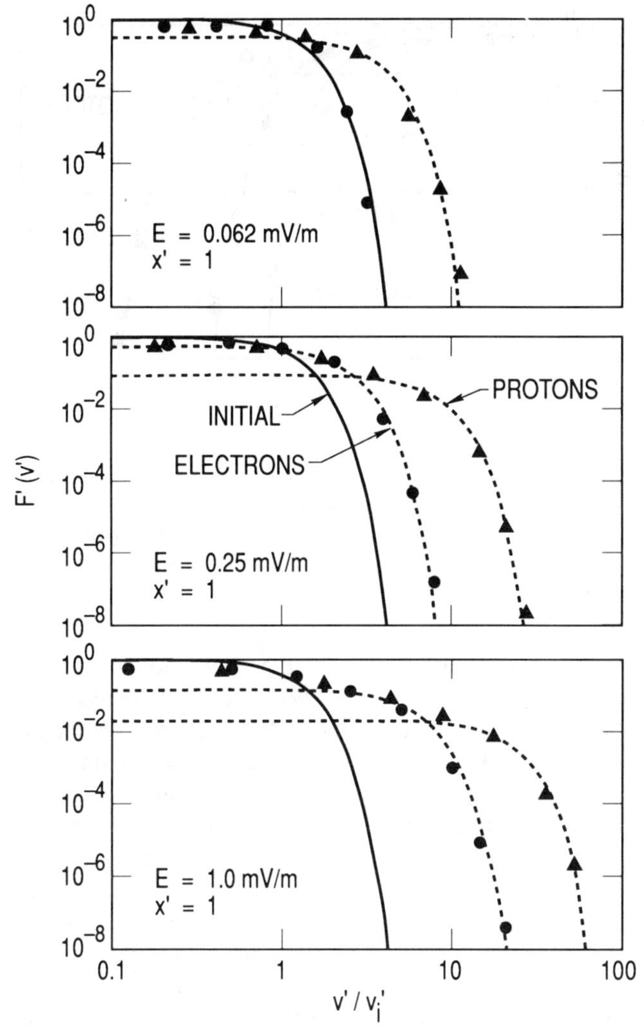

Fig. 3. Normalized distribution functions F'(v') versus v'/v$_i$'. In each panel, our evaluations of F' at x' = 1 are given by filled circles for electrons and by filled triangles for protons. The Maxwellian approximation for F'(v'), equation (10), is given by the dashed lines, and the distribution of incident particles, equation (7), is given by the solid lines. The values of v$_i$' used for electrons and ions, respectively, are 2.50 and 0.715 in the upper panel, 1.00 and 0.286 in the middle panel, and 0.400 and 0.113 in the bottom panel. These values correspond to the electric fields given in each panel and β = 0.3 nT/R$_e$.

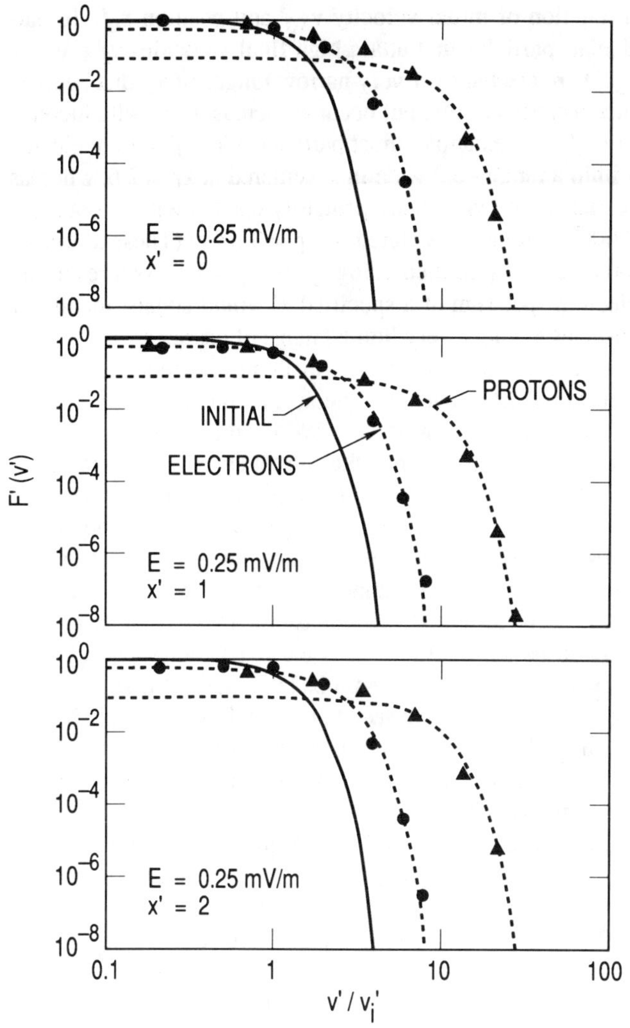

Fig. 4. Same as Figure 3, except that results are shown at x' = 0, 1, and 2 for the values of v$_i$' corresponding to the 0.25 mV/m electric field.

SPECTRA OF PARTICLES EJECTED FROM CURRENT SHEET

The energy spectra calculated in the previous section would only be directly observable near the mid-plane of the tail current sheet and within a normalized distance x' $\lesssim$ 2 from an X-line. This distance is quite small for electrons, and is given by 2L = 1700 km for our nominal tail parameters. For protons, 2L = 20,800 km. However, after traversing half a gyro-orbit about B_z, all particles in our model are presumed to be ejected from the current sheet. These particles will flow along field lines just within the boundary between open and closed field lines, where they are more likely to be observed than within the current sheet. It thus is worthwhile to consider the energy spectra of the ejected particles.

Within the context of our model, it is not sufficient to simply evaluate the distribution function of ejected particles as a function of energy at a specified x'. This is because of the strong variation of the location x_f' of particle ejection as a function of initial velocity v_{xo}' and position x_o'. In particular, particles that attain high final velocities v_f' (i.e., $v_f' \gtrsim 3$) are ejected over very narrow ranges of x_f' that become increasingly narrow, but occur at increasing x', with increasing v_f'. For example, most particles with $v_f' = 3$ are ejected within a range ~0.2 wide in x' centered at $x_f' = 2.6$, whereas particles with $v_f' = 6$ are primarily ejected within a range ~ 0.0002 wide in x' centered at $x_f' = 3.536$. (These x' ranges have been determined using f_i' for $v_i' = 1$.) As a result, an ejection spectrum at a specified x' would consist mostly of particles having a very limited range of energies.

We thus have decided to evaluate the total number of ejected particles that have interacted with the region near the X-line as a function of v_f'. We do this by assuming that particles are incident upon the current sheet within a range of x_o' centered about x' = 0, and evaluating $F_T'(v_f') = \int f'(v_f', x_f')dx_f'$ for various values of v_f'. This integration gives us an energy spectra that can be interpreted as the spectra averaged over the spatial extent of the particles' gyromotion near the X-line. Here we take the range of incident particles to be $|x_o'| \leq 2.25$. This particular range was chosen because 2.25 is the minimum $|x_f'|$ for incident particles having a speed of zero, the speed at which F'(v') maximizes, and we thus obtain a continuous spatial distribution for these particles. The precise size of this range, however, is not important to our results.

Energization of particles to high final velocities maximizes very close to x' = 0, whereas these energized particles are ejected beyond |x'| = 2. Thus, at higher speeds (i.e. $v_f' \gtrsim 2$), we expect the distribution of ejected particles $F_T'(v_f')$ to be similar to the distributions F'(v') within the current sheet at $x' \lesssim 2$ that are shown in Figures 3 and 4. Figure 5 shows this similarity for the values of v_i' that correspond to our nominal tail parameters. In Figure 5, $F_T'(v_f')$ is given by open circles for electrons and by open triangles for protons. The similarity between $F_T'(v_f')$ and F'(v') is also found for the other cases shown in Figure 3. (All particles are ejected with $v_f' \gtrsim 1$, so that $F_T'(v_f')$ goes to zero at low values of v_f'.) Thus, the energy spectra given by equation (10) is appropriate both within the current sheet and for the particles ejected from the X-line region. Again, however, an actual comparison with observations will require additional considerations of particle motion. In particular, the effects of particle mirroring and the resulting multiple interactions with the current sheet need to be evaluated.

COMPARISON WITH PREVIOUS RESULTS

Several studies have been performed that include the full three-dimensions of particle motion near an X-line. In this section, we compare our results to results from these studies. The comparisons provide a test of the assumptions applied in our model to limit the evaluation of particle motion to two-dimensions. When three dimensions of motion are included, it becomes difficult to evaluate the trajectories of a sufficient number of particles to determine the spectra of energized particles to high energies, where the value of F(v) is many orders of magnitudes below its peak near v = 0. To the best of our knowledge, such a study has not been carried out. Also, none of the three-dimensional studies have considered the possibility of significant azimuthal asymmetries about **B**, which lead to the gyroviscosity and are crucial to the physics near the X-line. Nevertheless, a number of interesting results have been obtained that are relevant to the present study.

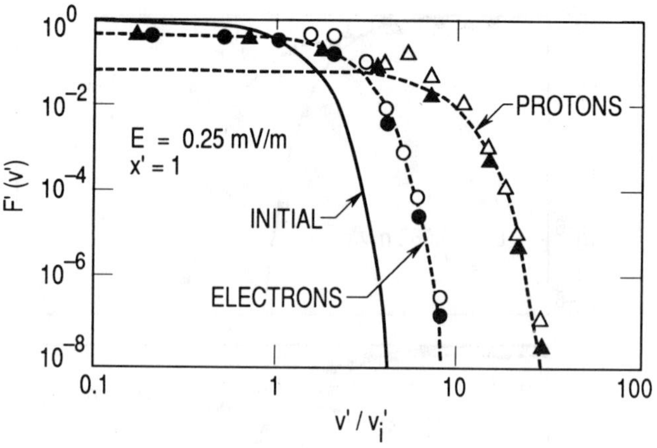

Fig. 5. Same as the middle panel of Figure 3, except the calculated distribution $F_T'(v_f')$ of particles ejected from the current sheet has been added. Values of the ejected distribution are given by open circles for electrons and by open triangles for protons.

Martin and Speiser [1988] evaluated particle distributions that have been averaged over the azimuthal angle around **B**. They found a ridge of enhanced values of f in velocity space that results from particle energization near the X-line in their model. This ridge is probably the same feature as the tail extending to high energies in Figure 2; however, the skewing that gives rise to the gyroviscosity cannot be determined from their azimuthally averaged distributions.

Burkhart et al. [1990] considered cold particle distributions incident upon the region near an X-line. They found that the velocity of the particle distribution flowing away from the X-line was proportional to $E_y^{2/3}q^{1/3}/m^{1/3}\beta^{1/3}$, which is the same scaling relation that we have obtained [equation (11)] for the thermal velocity. They found that the outflowing particles were magnetic field-aligned. This information cannot be obtained from our evaluation of particle motion. However, if we assume that particles ejected from the current sheet in our model are field aligned, then we obtain the same scaling as did Burkhart et al. for the outflow velocity. Thus, their results provide an excellent verification of the scaling that we have obtained.

Deeg et al. [1991] followed the three-dimensional trajectories of 14,000 cold (~1 eV thermal energy) protons near an X-line and were able to obtain a statistically significant energy spectrum of energized particles for one set of conditions. They assumed that particles are incident upon the X-line over a limited range of x near the X-line, and they counted the total number of particles ejected from the X-line region as a function of energy. This procedure is essentially the same as ours, so that their spectrum provides an excellent test for our two-dimensional results. Their spectrum is shown in Figure 6 as the total number of particles per unit energy N(K)dK [which is proportional to $v^2F_T(v)$] versus particle energy.

They used E_y = 0.3 mV/m and $\beta = 2.5 \times 10^{-15}$ T/m. From equation (11), our analysis predicts a Maxwellian distribution of energized particles having a thermal energy of 480 eV. The dashed line in Figure 6 shows a Maxwellian distribution having a thermal energy of 300 eV (which provided a better fit then did 480 eV), which can be seen to fit the simulated data very well. This comparison shows very good agreement between the general results from our two-dimensional model and the three-dimensional results for the particular set of parameters used by Deeg et al. [1991], the difference in thermal energies being not very significant. Also, using 500 particles, Deeg et al. investigated how the energization varies with E_y and β, and they found the same variation that we have found.

Note that even with 14,000 particles, Deeg et al.'s [1991] spectrum does not extend to high enough energies to distinguish a power-law high-energy tail from a Maxwellian. Our spectra do show this distinction, whereas their spectrum at energies above ~ 0.5 keV can be fitted as well by a power law as by a Maxwellian. However, the excellent agreement we find with their results gives strong support to the validity of our general two-dimensional result for the particle energization.

Goldstein et al. [1986] and Ambrosiano et al. [1988] evaluated the acceleration near X-lines and included the effects of magnetic turbulence obtained from MHD simulations. They found significant acceleration due to particle trapping in magnetic bubbles that developed in the simulations. Our results, however, are not directly comparable to their results because we did not allow for turbulence and because the basic physics of collisionless particle motion near an X-line (e.g., the gyroviscous force and the associated asymmetries in particle distribution functions about B) does not exist in MHD.

Finally, it is interesting to compare the results for particle energization near an X-line with those obtained by Speiser and Lyons [1984] for a weak, but constant, B_z across a current sheet. They found that initially cold particles were accelerated to a final velocity $v_f = 2E_y/B_z$. Our present results give a thermal velocity $v_{th} = 2E_y^{2/3}q^{1/3}/m^{1/3}\beta^{1/3} = 2E_y/B_z(x' = 1)$. Thus the present results give a thermal velocity that is the same as the final velocity in the B_z constant model, if B_z is evaluated one scale length L away from the X-line.

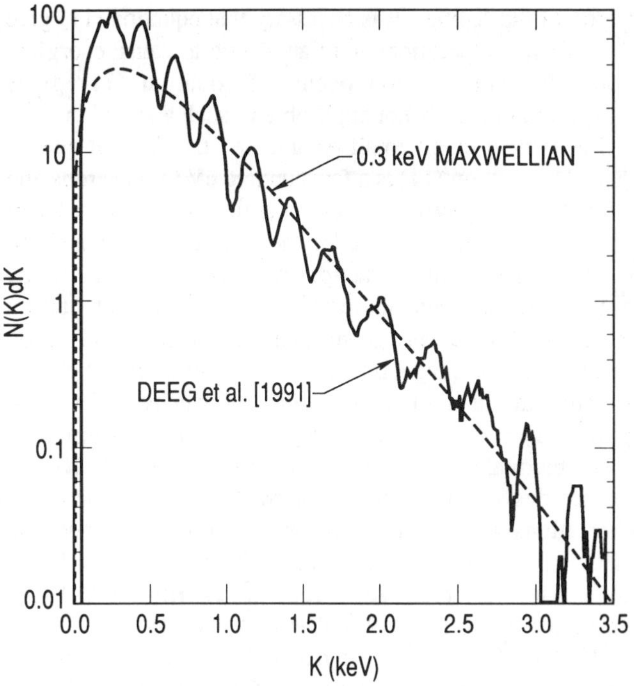

Fig. 6. Total number of particles per unit energy ejected from the region near an X-line as calculated by Deeg et al. [1991] from a three-dimensional simulation of particle motion. The dashed line gives a Maxwellian distribution for 0.3 keV thermal energy.

CONCLUSIONS

We have applied Dungey's [1988] two-dimensional model of fields and collisionless particle motion to evaluate the particle energization near a magnetic X-line. We find that this process results in an energy spectrum of particles that can be approximated as a Maxwellian. The same Maxwellian applies for both the particles within the current sheet in the vicinity of the X-line and for the particles ejected from that region of the current sheet. This Maxwellian has a thermal energy, K, given by equation (11), that depends upon field parameters but is independent of the thermal energy, K_i, of the particles incident upon the region of the X-line. It is necessary that equation (11) give $K \gtrsim K_i$ for interactions with an X-line to cause energization. No energization occurs if equation (11) gives $K \lesssim K_i$, and (11) is not applicable to such a situation.

For typical magnetotail parameters, the thermal energy given by equation (11) is a few hundred eV for electrons and a few keV for protons. These energies are of the order of magnitude of those observed in the distant tail plasma sheet, which suggests that interactions with the neutral line in the distant tail may contribute to the energization of plasma sheet particles. During geomagnetic disturbances, induction by time-varying magnetic fields may yield electric fields in the tail that are significantly enhanced over typical values. Taking $E_y = 4$ mV/m as perhaps an extreme example, we obtain thermal energies of 17 keV for electrons and 210 keV for protons. This is consistent with the suggestion that interactions with a neutral line may contribute to energetic particle phenomena that have been observed in the tail during substorms [e.g., Armstrong and Krimigis , 1968; Krimigis and Sarris, 1979; Baker at al., 1979].

We have compared our results to results from three-dimensional studies of particle motion near an X-line. These comparisons show our results to be consistent with the parameter scalings and the one detailed energy spectra obtained from the three-dimensional studies. Our results compliment the three-dimensional studies by giving an accurate energy spectrum to high energies that can be scaled to different magnetic and electric field parameters, and by providing details of the particle distribution functions such as the azimuthal asymmetries very near an X-line.

Acknowledgements. We are grateful to J. W. Dungey for stimulating discussions that motivated this study and to M. Schulz for valuable discussions throughout the course of the study. This work was supported by the NASA Space Physics Theory Program grant NAGW-2126.

REFERENCES

Ambrosiano, J., W. H. Matthaeus, M. L. Goldstein, and D. Plante, Test particle acceleration in turbulent reconnecting magnetic Fields, *J. Geophys. Res., 93*, 14,383, 1988.

Armstrong, T. P., and S. M. Krimigis, Observations of protons in the magnetosphere and magnetotail with Explorer 33, *J. Geophys. Res., 73*, 143, 1968.

Baker, D. N., R. D. Belian, P. R. Higbie, and E. W. Hones, Jr., High-energy magnetospheric protons and their dependence on geomagnetic and interplanetary conditions, *J. Geophys. Res., 84*, 7138, 1979.

Burkhart, G. R., J. F. Drake, J. Chen, Magnetic Reconnection in collisionless plasmas: Prescribed fields, *J. Geophys. Res., 95*, 18,833, 1990.

Deeg, H.-J., J. E. Borovsky, and N. Duric, Particle acceleration near X-type magnetic neutral lines, *Phys. Fluids B, 3*, 2660, 1991

Dungey, J. W., Noise-free neutral sheets, in *Proceedings of an International Workshop in Space Plasma,* held in Potsdam, GDR, ESA SP-285, Vol. II, p. 15, 1988.

Goldstein, M. L., W. H. Matthaeus, and J. J. Ambrosiano, Acceleration of charged particles in magnetic reconnection: solar flares, the magnetosphere, and solar wind, *Geophys. Res. Lett., 13*, 205, 1986.

Krimigis, S. M., and E. T. Sarris, Energetic particle bursts in the Earth's magnetotail, in *Dynamics of the Magnetosphere,* ed. S.-I. Akasofu, p. 599, D. Reidel, Dordrecht, Holland, 1979.

Lyons, L. R., and D. C. Pridmore-Brown, Force balance near an X-line in a collisionless plasma, *J. Geophys. Res., 95*, 20,903, 1990.

Martin, R. F., Jr., and T. W. Speiser, A predicted energetic ion signature of a neutral line in the geomagnetic tail, *J. Geophys. Res., 93*, 11521, 1988.

Sonnerup, B. U. Ö., Magnetic field reconnection, in *Solar System Plasma Physics III*, eds. L. J. Lanzerotti, C. F. Kennel, and E. N. Parker, North-Holland, Amsterdam, p. 45, 1979.

Sonnerup, B. U. Ö., On the theory of steady-state reconnection, *Computer Physics Communications, 49*, 143, 1988.

Speiser, T. W., Particle trajectories in model current sheets, 1, Analytical solutions, *J. Geophys. Res., 70*, 4219, 1965.

Speiser, T. W., and L. R. Lyons, Comparison of an analytical approximation for particle motion in a current sheet with precise numerical calculations, *J. Geophys. Res., 89*, 2177, 1984.

Stern, D. P., The role of O-type neutral lines in magnetic merging during substorms and solar flares, *J. Geophys. Res., 84*, 63, 1979.

Vasyliunas, V. M., Theoretical models of magnetic field line merging, 1, *Rev. Geophys. Space Phys., 13*, 303, 1975.

L. R. Lyons and D. C. Pridmore-Brown, The Aerospace Corporation, P. O. Box 92597, M2-260, Los Angeles, CA 90009.

Formation of the Macroscopic Tail Current Sheet in a Microscopic Distributed-Source Model

P. L. Pritchett[1] and F. V. Coroniti[2]

Department of Physics, University of California, Los Angeles

The macroscopic structure and stability of the tail current sheet in which the ions carry most of the current and momentum are investigated by means of a two-dimensional self-consistent microscopic model in which the current sheet is established from ion streams injected at the boundaries. For the case of a cold current sheet the characteristic half-width λ is given by $\lambda \approx (B_z/B_0)^{4/3} c/\omega_{p0}$, where B_z is the normal field component (assumed to be spatially uniform), B_0 is the asymptotic magnitude of the reversing field, and c/ω_{p0} is the collisionless ion skin depth based on the asymptotic density. While these idealized current sheets are found to be stable to the ion tearing mode, they are unstable to kink perturbations driven by the anisotropic pressure distribution produced by the chaotic nature of the particle orbits in a field-reversal region. The combined effects of chaotic scattering and finite thermal spread are likely to preclude the formation in the geomagnetic tail of stable current sheets with half-width less than about $0.25c/\omega_{p0}$ (~200 km).

1. INTRODUCTION

The central plasma sheet in the Earth's magnetic tail can be divided roughly into three characteristic regions. The first is the immediate vicinity of the distant neutral (or X) line. During quiet times this neutral line is located at about $100R_E$ down the tail from the Earth. The second consists of extended regions on either side of the X line where a one-dimensional (1D) current sheet description should form a valid first approximation. Here the field line tension is balanced by the dynamic pressure of the inflowing tail lobe plasma. The third is composed of the near-Earth region where the 1D approximation is broken due to the mirroring back into the current sheet of previously-accelerated ions which then thermalize to form the hot near-Earth plasma sheet. Here the field tension is balanced by the pressure gradient along the tail axis of the isotropized trapped hot plasma; the equilibrium is then at least two dimensional (2D) in nature.

In the present work we are concerned with the second of these regions, the 1D current sheet. It should be noted, however, that recent observations [*Mitchell et al.*, 1990; *Sergeev et al.*, 1990; *Lin et al.*, 1991] have indicated that current sheets with total thickness as small as ~400–800 km can occur in the near-Earth plasma sheet. It may thus be that the forced 1D current sheets we will discuss here are also of importance in the near-Earth region during substorm growth phase and leading to current disruption [*Burkhart et al.*, 1992*a*, 1992*b*].

The quantitative investigation of current sheet or field reversal regions dates back to the work of *Speiser* [1965, 1967, 1968], who obtained the properties of the nonadiabatic single-particle trajectories. *Eastwood* [1972, 1974] found the self-consistent solution for the Speiser trajectories and observed very narrow (2–4 km) current sheets. *Francfort and Pellat* [1976] demonstrated the existence of adiabatic solutions with no characteristic scale length provided that a magnetically trapped particle population exists within the current sheet. In more recent years it has been realized that stochastic effects play an important role in the particle trajectories [*Büchner and Zelenyi*, 1986; *Chen and Palmadesso*, 1986; *Martin*, 1986].

There remain a number of unsolved problems regarding the 1D current sheet structure which we shall address in this work. The first concerns the question of the length scale in z. If the marginal firehose stabil-

---

[1] Also at Institute of Geophysics and Planetary Physics, University of California, Los Angeles.

[2] Also at Department of Astronomy, University of California, Los Angeles.

Space Plasmas: Coupling Between Small
and Medium Scale Processes
Geophysical Monograph 86

ity condition is satisfied [*Rich et al.*, 1972; *Cole and Schindler*, 1972; *Eastwood*, 1972; *Hill*, 1975], then the general stress balance condition does not require any fundamental length scale. In addition, there has never been any self-consistent, dynamical demonstration of the formation of such a structure. Finally, the stability of such structures in the presence of weakly chaotic orbits remains unclear.

In section 2 we employ a cold model similar to that of *Eastwood* [1972] to determine the equilibrium structure by solving the coupled field and particle equations for the case where all particles follow the same trajectory. We then construct a 2D self-consistent simulation model in which the 1D current sheet is established from particle flows at the boundaries (section 3). This model allows us to address the question of stability of the current sheet equilibrium (section 4). The significance of the results is discussed in section 5.

2. STRUCTURE OF THE COLD 1D CURRENT SHEET

In this section we determine the structure of the self-consistent 1D current sheet following the model introduced originally by *Eastwood* [1972]. The magnetic field has the form

$$\mathbf{B} = (B_x(z), 0, B_z), \tag{1}$$

where B_x is an odd function of z and $B_z > 0$ is taken to be constant. The analysis is performed in a frame of reference in which the electric field E_y vanishes. This frame moves with a speed cE_y/B_z toward the Earth with respect to the neutral line rest frame. As shown by Eastwood, the electron contribution to the current J_y is smaller than that of the protons by a factor of the order of m_e/m_p. Thus the electron dynamics will be neglected; their role is simply to maintain charge neutrality. The incoming stream velocity $\mathbf{v}_0$ is assumed to be much greater than the protron thermal velocity. Thus protons flow in along the field lines and have identical trajectories. The initial velocity components satisfy

$$|v_{z0}|/v_{x0} = B_z/B_0, \tag{2}$$

where B_0 is the asymptotic value of $|B_x(z)|$. The condition for equilibrium obtained by *Eastwood* [1972] is just the marginal firehose stability condition, which takes the form

$$B_0^2/4\pi = \rho_0 v_{x0}^2, \tag{3}$$

where $\rho_0 = m_p n_0$ is the total mass density outside the current sheet. The ion plasma frequency based on the density n_0 is denoted by ω_{p0}.

The results for the self-consistent current sheet agree with those found by *Eastwood* [1972], but we shall cast them in a form which leads to useful scaling relations for the characteristic current sheet thickness and density enhancement as a function of the ratio B_z/B_0. Figure 1 shows the quantities $z(t)$, $v_z(t)$, and $v_y(t)$ for the case of $B_z/B_0 = 0.2$ for a particle that was started at $z = 2c/\omega_{p0}$. As the particle approaches the field-reversal region, the velocity components v_y and v_z undergo a small-amplitude oscillation at approximately the cyclotron frequency Ω_0 due to nonuniformities in the self-consistent $B_x(z)$. Upon approaching $z \approx 0$ for the first time, the particle becomes trapped and oscillates about $z = 0$, with the x energy converted into significant y and z energies. The y speed builds up to $v_y^{max} = (v_{x0}^2 + v_{z0}^2)^{1/2} = 1.02 v_{x0}$ and then decreases again as the particle is turned back in the opposite direction in x. The trapping time, as expected [*Speiser*, 1965], is given by $T \approx \pi/\Omega_z$. The frequency of the rapid oscillations while the particle is trapped is $\omega_t = 2.2\Omega_0$. Upon leaving the field-reversal region, the particle exhibits a large gyromotion with $v_\perp \approx v_{z0}$. This gyromotion is an indication of the scattering experienced

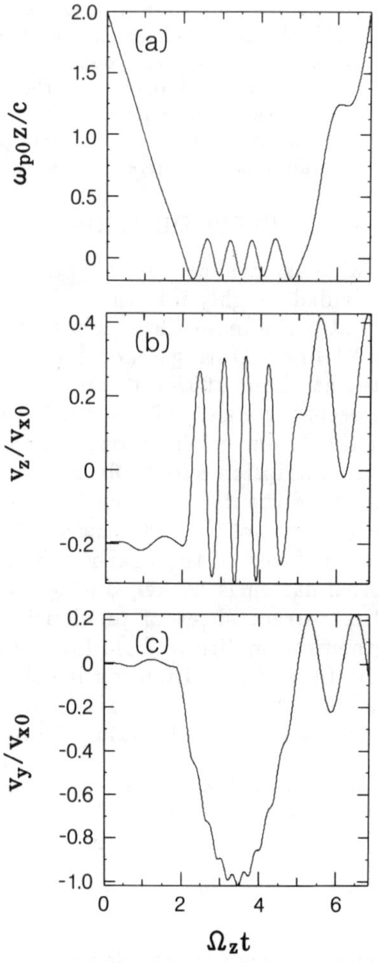

Fig. 1. Particle trajectories in the cold self-consistent current sheet for the case of $B_z/B_0 = 0.2$. (*a*) $z(t)$, (*b*) $v_z(t)$, and (*c*) $v_y(t)$.

by the particle during the traversal across the current sheet.

The trapping of the ions near the center of the current sheet, as illustrated in Figure 1, has a direct bearing on the structure of the density $n(z)$ and current density $J_y(z)$ profiles. Since each particle spends an increased time at the turning points of its z motion, these profiles exhibit a characteristic double-peaked structure (Figure 2). These peaks become increasingly pronounced as B_z/B_0 is decreased. The resulting self-consistent field profile $B_x(z)$ is shown in Figure 3. It is apparent that the current sheet thickness decreases considerably for the smaller values of B_z.

It is possible to obtain the analytic scaling for the half-thickness λ from a WKB solution to the z equation of motion [*Pritchett and Coroniti*, 1992]. Assuming that the profile $B_x(z)$ can be aprrroximated as linear, one finds the result that

$$\lambda/(c/\omega_{p0}) \approx (B_z/B_0)^{4/3}. \tag{4}$$

An alternative derivation of this result has been given by *Burkhart et al.* [1992*a*].

Once the scaling for λ has been obtained, it is straightforward to estimate the average density $\bar{n}$ in the current sheet by requiring that the particle flux injected into the sheet from one side ($n_0 v_{z0}/2$) be equal to the number of particles trapped in the half sheet ($\bar{n}\lambda$) divided by the particle lifetime in the sheet (π/Ω_z). In the harmonic approximation the central density at $z = 0$ should be roughly $n_c \approx (2/\pi)\bar{n}$, so that we find

$$n_c/n_0 \approx v_{x0}/\lambda\Omega_0 \approx (B_0/B_z)^{4/3}. \tag{5}$$

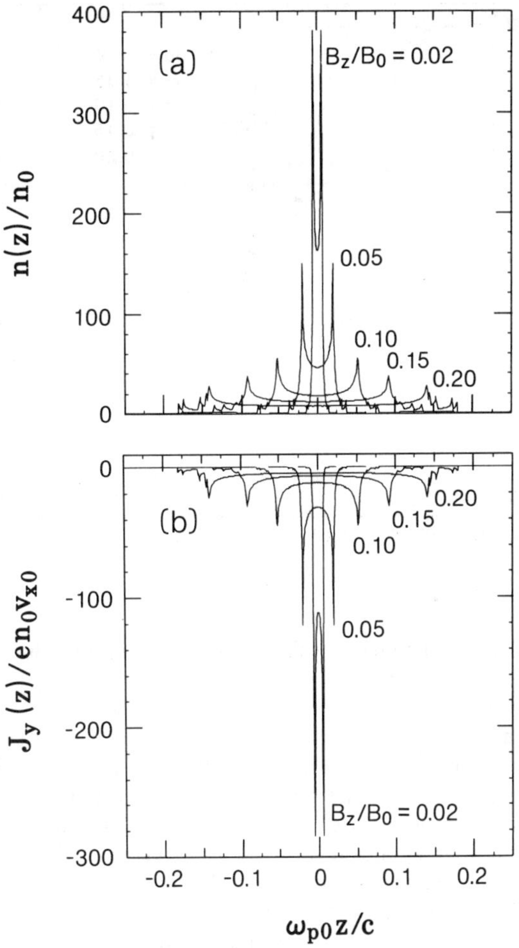

Fig. 2. The self-consistent profiles of (*a*) density $n(z)$ and (*b*) current density $J_y(z)$ for values of B_z/B_0 ranging from 0.02 to 0.20.

3. DISTRIBUTED-SOURCE MODEL

We have developed a 2D simulation model which allows us to investigate the formation of a current sheet from field-aligned ion streams (which may have a small thermal spread) injected at the boundaries. This is meant to model the inflow of tail lobe plasma into the central plasma sheet. The simulation is performed in a frame moving in the negative x direction (earthward) at a constant speed cE_y/B_z. The basic configuration is illustrated in Figure 4. The simulation is performed in the xz plane, with all quantities independent of y. The model solves Maxwell's equations in the Darwin approximation and assumes a single active species (ions) [*Dickman et al.*, 1969]. Thus the electrons are again assumed to form a neutralizing background. As indicated by the arrows at the boundaries, particles are injected into the system at three sides: along $x = 0$ and at $z = \pm L_z/2$ for $0 \leq x \leq x_3$. The x velocity of all particles is v_{x0}, while the z velocity is $\mp v_{z0} = \mp(B_z/B_0)v_{x0}$ depending upon whether the z coordinate is positive or negative. Initially, the magnetic field is kept fixed with $B_x(z) = -B_0 \tanh(z/\lambda)$ and $B_z =$ constant, where the appropriate value of λ is known from the results in the previous section. The particles are allowed to flow through the system until they reach any boundary, where they are then removed. The system is extended beyond the maximum injection point $x = x_3$ to $x = L_x$ to permit all particles to complete their trajectories across the current sheet. This is indicated schematically by the dashed line representing a gyroorbit in the region beyond $x = x_3$. The particles are advanced in time until a steady state is reached in which the outflow of particles is just balanced by the continued injection of new particles. This requires about one full gyroperiod in the B_z field. At this point one has established counter-streaming particle beams which, outside the field-reversal region, are nearly field aligned.

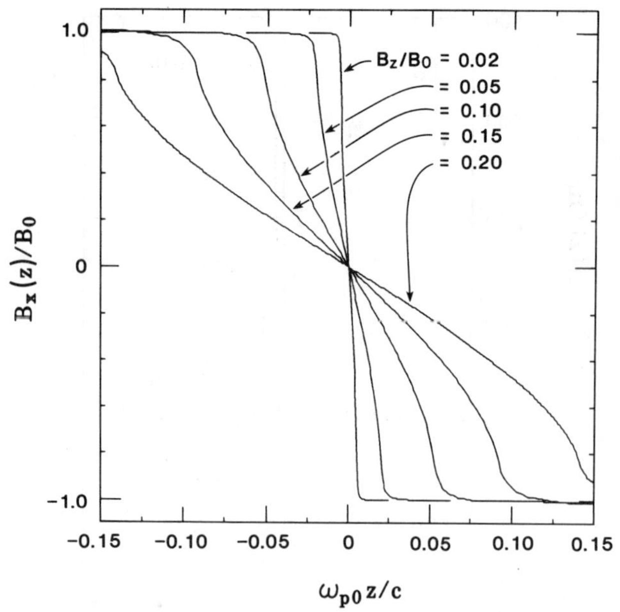

Fig. 3. The self-consistent field profiles $B_x(z)$ for values of B_z/B_0 ranging from 0.02 to 0.20.

The simulation then proceeds via the following two-step procedure [*Hamilton and Eastwood*, 1982]. In the first stage a self-consistent $B_x(z)$ is computed at each step while B_z is kept constant. This 1D field solution is obtained by accumulating the density and canonical momentum carried by the particles on the xz grid to form P_y, averaging over x values in the range $x_1 \le x \le x_3$, and then solving the field equation

$$(d^2/dz^2 - \omega_p^2(z)/c^2)A_y = -(4\pi e/mc)P_y \qquad (6)$$

for the perturbed vector potential A_{1y} by an iterative procedure [*Pritchett et al.*, 1991]. The boundary condition imposed is that $A_{1y}(\pm L_z/2) = 0$. The $B_x(z)$ profile is then obtained by a centered finite-difference approximation to $B_x = -dA_y/dz$. This stage allows an overall adjustment between the field profile and the particle distributions. The grid quantities are computed using an area-weighting scheme, and the particle advance is performed using the standard leapfrog algorithm [*Birdsall and Langdon*, 1985]. The particles are advanced in this 1D self-consistent field for typically another full gyroperiod in B_z.

In the final stage the particles are advanced in a 2D self-consistent field described by $A_y(x,z,t)$. This field is computed using the method described by *Pritchett et al.* [1991] applied to the active region $x_1 \le x \le x_2$, $-L_z/2 \le z \le L_z/2$ subject to the boundary conditions that the perturbed A_{1y} vanishes at the four edges of this region. During this stage B_z is no longer constant in either space or time. Here the particles are followed for typically some 3–5 gyroperiods in B_z.

The features of the steady state reached during traversal through the fixed magnetic field (stage 1) are illustrated in Figure 5 for the case of $B_z/B_0 = 0.2$. Figure 5*a* shows the v_x–x phase space for the range $0 \le x \le x_3 = 512\Delta$, corresponding to the range over which particles are injected. The large white area at the left represents the inaccessible region of phase space for particles injected with velocity $v_x = v_{x0}$. This region is never included in any of the spatially dependent calculations. Figures 5*b* and 5*c* show profiles of density and y current which have been integrated over all z as a function of x from $x = 0$ to $x = 512\Delta$. It is apparent that, apart from the inaccessible region, these profiles are essentially constant in x. Thus our numerical procedure does lead to the formation of a true 1D structure.

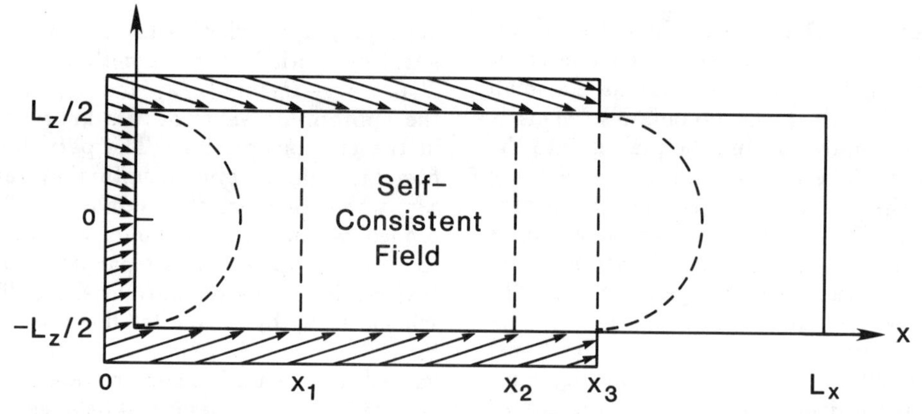

Fig. 4. Schematic diagram of the 2D distributed-source model. Particles are injected into the system at the boundaries on three sides as indicated by the arrows. The injection velocities are v_{x0} and $\mp v_{z0}$ depending upon whether z is positive or negative. The particles are allowed to flow through the system until they reach any boundary, where they are then removed. The 2D self-consistent fields are determined in the region $x_1 \le x \le x_2$ for all values of z.

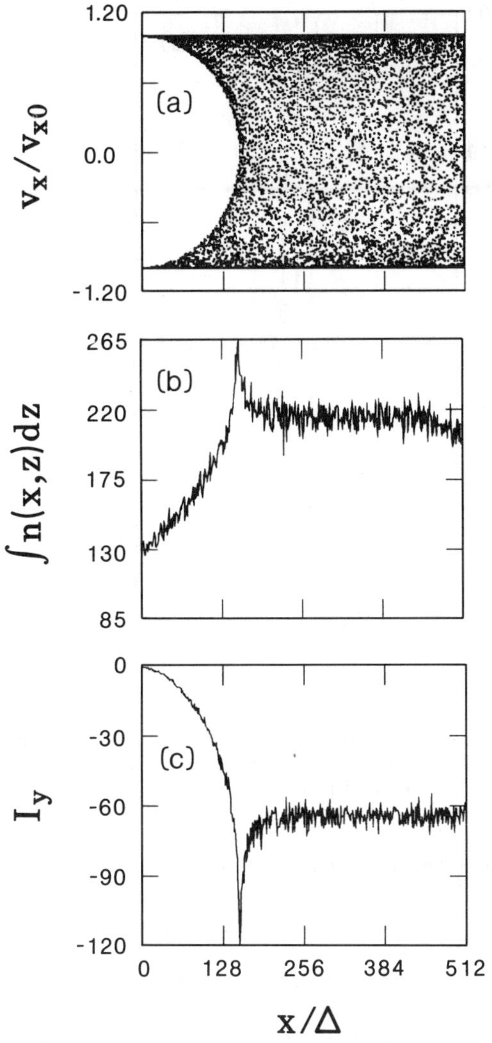

Fig. 5. Properties of the steady state reached for motion in a fixed magnetic field profile with $B_z/B_0 = 0.2$. (*a*) v_x–x phase space; profiles of (*b*) density and (*c*) current which have been integrated over all z as a function of x.

4. STABILITY OF THE 1D CURRENT SHEET

We shall discuss in detail the case where the initial value of $B_z/B_0 = 0.2$. The grid dimensions are varied from $N_x \times N_z = 896 \times 128$ to 1280×128; the grid spacing $\Delta = c/32\omega_{p0}$, so that the z boundaries are at $z = \pm L_z/2 = 2c/\omega_{p0}$. The value of x_1 is 200Δ, while those of x_2 and x_3 are either 456Δ and 512Δ (giving an active region of length 256Δ) or 712Δ and 768Δ (active region length $= 512\Delta$), respectively. The time step is $\Omega_0 \Delta t = 0.0125$ ($v_{x0}\Delta t = 0.4\Delta$). At each time step some 180–220 particles are injected into the system distributed according to the local flux $(n_0/2)v_0\Delta t$. This results in some 400,000–550,000 particles in the system when the steady state is reached.

Figure 6 shows density $n(z)$ and current $J_y(z)$ profiles averaged over the region $256\Delta < x < 384\Delta$ as determined at the end of each of the first two stages of the calculation: fixed hyperbolic tangent field (Figures 6*a*, 6*b*) and 1D self-consistent field (Figures 6*c*, 6*d*). The changes from the first to the second stage are relatively minor: the central density and current regions are reduced in width and increased in height somewhat, and there is some enhanced structure outside the field-reversal region associated with the enhanced scattering in the self-consistent fields. The basic double-humped structure, however, is clearly preserved. In the self-consistent results the peak density occurs at $z = \pm 0.13c/\omega_{p0}$, in close agreement with the earlier results of the single particle model (section 2). The values of $n(0)$ and $J_y(0)$ also agree with the earlier results, but the peak values ($n_{\max}/n_0 = 12$, $|J_y|^{\max}/en_0v_{x0} = 8$) are only about 40% as large as the corresponding cases in Figure 2. These decreased peak values are compensated by a larger total width for the current sheet; the enhanced density region extends out to about $0.22c/\omega_{p0}$ compared to the earlier value of about $0.17c/\omega_{p0}$. Thus the distributed 2D source leads to a somewhat broader current sheet.

Figure 7 shows v_y–z phase space plots averaged over x for each of the three stages in the distributed-source simulation. In the fixed-field calculation (Figure 7*a*) the particles emerge from the current sheet with only a small perturbation on the inflow field-aligned conditions; the final v_y component is very small ($\lesssim 0.03v_{x0}$). In the 1D self-consistent field (Figure 7*b*), however, the velocities acquire a substantial (as large as $0.28v_{x0}$) and irregular gyrocomponent; since the fields are independent of x, all particles follow the same trajectory in phase space. In the 2D self-consistent field (Figure 7*c*), there is a considerable smearing of this coherent structure in phase space.

In the 2D self-consistent simulation with an active region length of 256Δ, we observe an initial transient period associated with the establishment of the 2D fields which lasts for approximately one gyroperiod in B_z. After this period the perturbed field levels remain nearly constant in time to the end of the run at $\Omega_0 t = 200$. There is no indication of any tearing mode instability even though the gyroradius in the B_0 field due to the acquired gyromotion is comparable to or larger than the current sheet thickness. This is in agreement with the analytic result of *Burkhart et al.* [1992*b*] that the ion tearing mode in a forced current sheet equilibrium is stable in the absence of a diamagnetic current in the outer plasma sheet.

The chaotic nature of the particle orbits across the field-reversal region, which is manifested in the acquired perpendicular component of velocity, does have an important consequence, however. Figure 8*a* shows the anisotropy factor $A \equiv 4\pi(P_{\parallel} - P_{\perp})/B_0^2$ as a function of

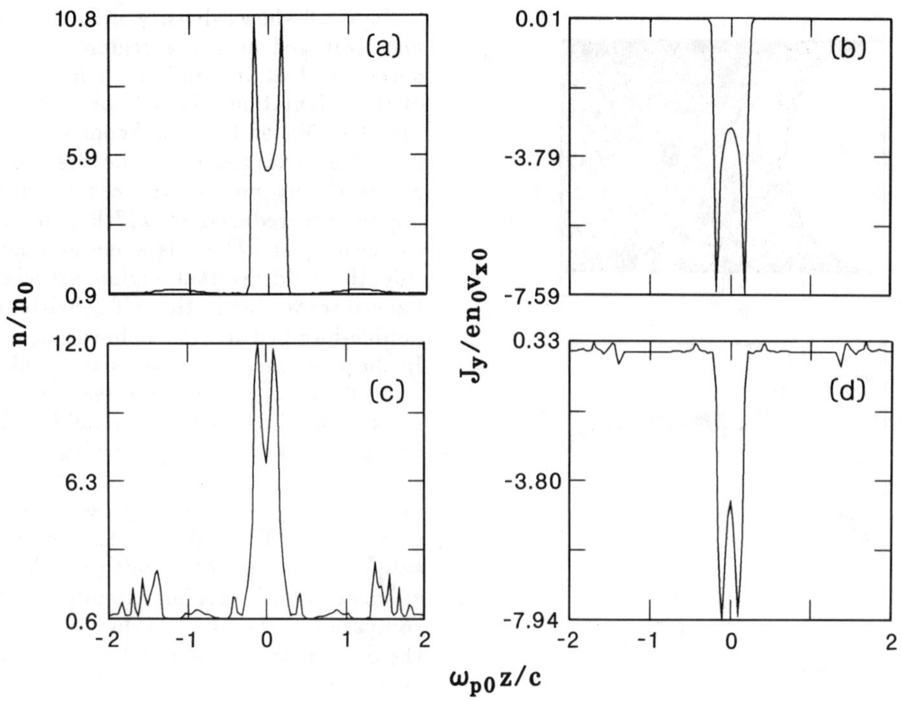

Fig. 6. Density $n(z)$ and current density $J_y(z)$ profiles averaged over the region $256\Delta < x < 384\Delta$ as determined at the end of each of the first two stages of the distributed-source simulation with $B_z/B_0 = 0.2$: (a,b) fixed hyperbolic-tangent field ($\Omega_z t = 10$), (c,d) 1D self-consistent field ($\Omega_z t = 10$).

z averaged over the region $256\Delta < x < 384\Delta$ at the end of the 1D self-consistent field stage. The field-aligned injection conditions used in the simulation correspond to $A = 1$. For $|z| \gtrsim 0.5c/\omega_{p0}$, however, A is about 0.5 or less. Near the edge of the current sheet A increases rapidly to greater than 3 as a result of the density spike; in the center of the current sheet A becomes strongly negative as the particle energy is converted primarily into y velocity. The marginal firehose stability condition is seen to be satisfied almost nowhere. In the presence of the 2D field perturbations, the anisotropy profile is substantially altered (Figure 8b). The marginal stability condition $A = 1$ is now well satisfied on average outside of the current sheet, the maximum at the edge of the current sheet is reduced, and the central minimum is much deeper and narrower.

When we increase the system size so that the active region is doubled to 512Δ, we observe a significant alteration in the structure of the current sheet. Figure 9a shows the time histories of the perturbed field intensity $|A_{1y,m}|^2$ integrated over all z for modes $m = 3$, 4, and 5. There is clear exponential growth for all 3 modes, with growth rates γ/Ω_0 in the range 0.08–0.12. As shown in Figure 10, the growth of these modes leads to a kinking of the current sheet. As the relative contribution of the various modes changes, the precise pattern of the deformation is altered. The effect of the instability on the structure of the current sheet is to smear out the double-humped nature of the current and density profiles, leaving an average structure which is peaked at $z = 0$ with a maximum density of about $6n_0$ and a broadened half-width of about $0.3c/\omega_{p0}$. The average anisotropy profile no longer has a peak with $A \gg 1$ at the edge of the current sheet, and the value of A for $|z| > 0.5c/\omega_{p0}$ is increased to ~0.9. It thus appears that the idealized cold, firehose marginally stable, 1D current sheet can never be realized due to the chaotic nature of the orbits in the field-reversal region.

An additional effect which would prevent attainment of the cold current sheet is the presence of a finite temperature. A typical mantle ion temperature of 10^6 K yields a thermal velocity $v_{Ti} = (T_i/m_p)^{1/2} = 91$ km/s, which is of the order of 25–50% of the inflow speed v_{x0}. We have thus repeated the distributed-source simulations for $B_z/B_0 = 0.2$ but now including a finite thermal velocity v_{Ti} for the inflow ion streams. Already for $v_{Ti}/v_{x0} = 0.1$, the effect is to smear out completely the coherent structure in phase space, even for the case of the fixed field. In addition, the thickness of the current sheet is broadened. During the 2D self-consistent field simulations with finite v_{Ti}, the growth of the $A_{1y,m}$ amplitudes is reduced considerably compared to the previous case with $v_{Ti} = 0$. Figure 9b shows the results for $v_{Ti}/v_{x0} = 0.1$. The maximum values of $|A_{1y,m}|^2$ are

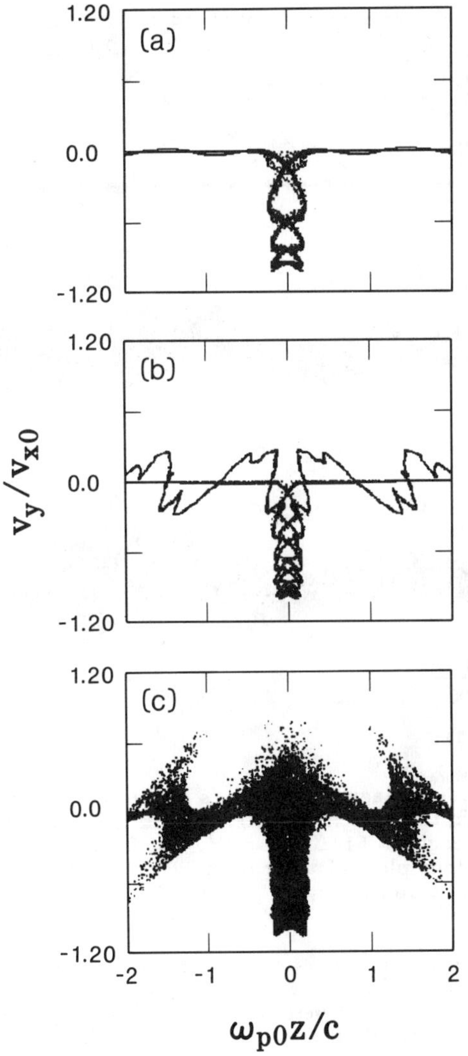

Fig. 7. v_y–z phase space plots as determined at the end of the three stages of the distributed-source simulation with $B_z/B_0 = 0.2$: (a) fixed hyperbolic-tangent field ($\Omega_z t = 10$), (b) 1D self-consistent field ($\Omega_z t = 10$), (c) 2D self-consistent field with active region length $= 256\Delta$ ($\Omega_z t = 40$).

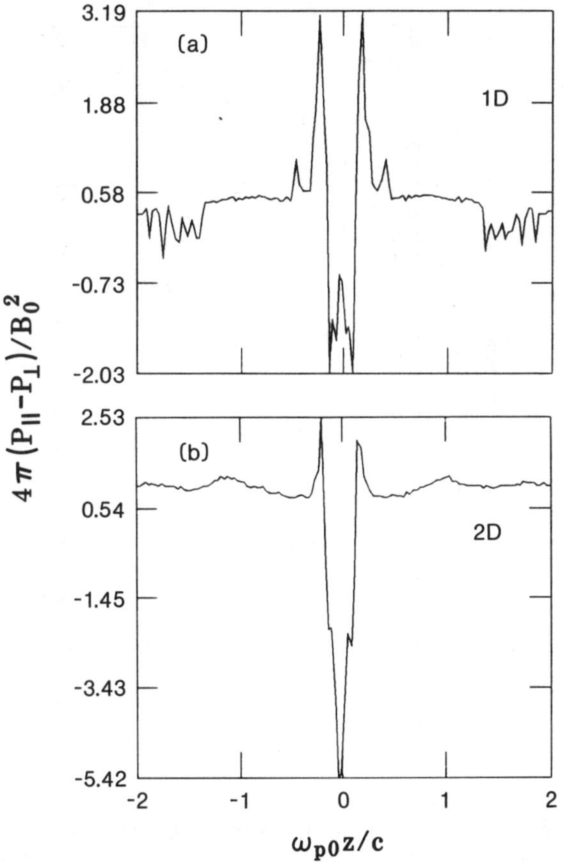

Fig. 8. Anisotropy factor $A \equiv 4\pi(P_{\|} - P_{\perp})/B_0^2$ as a function of z averaged over the region $256\Delta < x < 384\Delta$ for the distributed-source simulation with $B_z/B_0 = 0.2$ as determined (a) at the end of the 1D self-consistent field stage and (b) at the end of the 2D self-consistent field stage with active region length $= 256\Delta$.

smaller by at least an order of magnitude. For $v_{Ti}/v_{x0} = 0.2$ there is a further reduction by about a factor of 4. The anisotropy profiles for these thermally broadened current sheets possess no peak with $A \gg 1$ at the edge of the current sheet and satisfy $A \approx 1$ outside of the current sheet.

It would thus appear that, due to either or both of chaotic scattering and thermal effects, it is unlikely that a 1D current sheet can be formed with a half-width appreciably smaller than $0.3c/\omega_{p0}$.

5. SUMMARY

We have investigated the structure and stability of thin 1D current sheets in which the ions carry most of the current and momentum. The presence of a finite normal B_z component prevents the current sheet from collapsing to zero thickness in response to a uniform dawn-to-dusk electric field. Our analysis established that for the case of a cold current sheet this characteristic thickness is given by $\lambda \approx (B_z/B_0)^{4/3}c/\omega_{p0}$, where c/ω_{p0} is the collisionless ion skin depth based on the lobe density. During traversal of the current sheet, ions are trapped in the field reversal region for a period π/Ω_z. During this interval they perform fast oscillations between $z \approx \pm\lambda$. As a result of the increased time spent at the turning points of the oscillations, a characteristic double-humped density profile is established. The density enhancement at the center of the current sheet is given by $n_c/n_0 \approx c/\lambda\omega_{p0} \approx (B_0/B_z)^{4/3}$ and can thus be very substantial.

To address the questions of the formation and stability of these 1D current sheets, we have developed

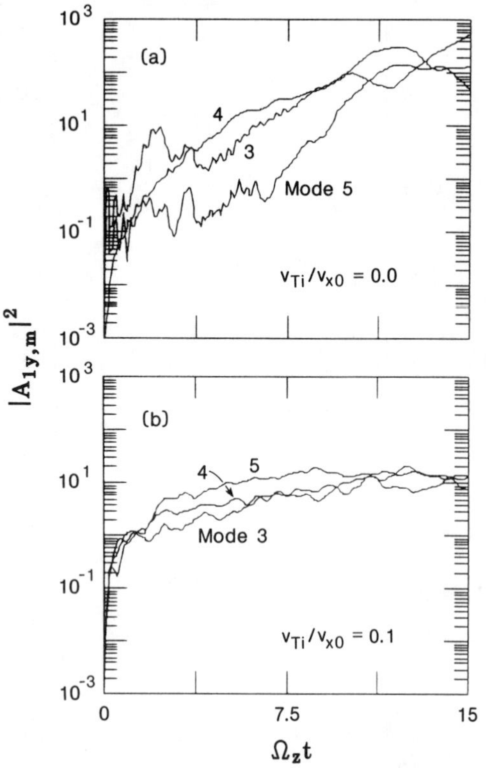

Fig. 9. Time histories during 2D self-consistent simulations with $B_z/B_0 = 0.2$ and active region length = 512Δ of the perturbed field intensity $|A_{1y,m}|^2$ integrated over all z for modes m = 3, 4, and 5 for the cases of (a) $v_{Ti}/v_{x0} = 0$ and (b) $v_{Ti}/v_{x0} = 0.1$.

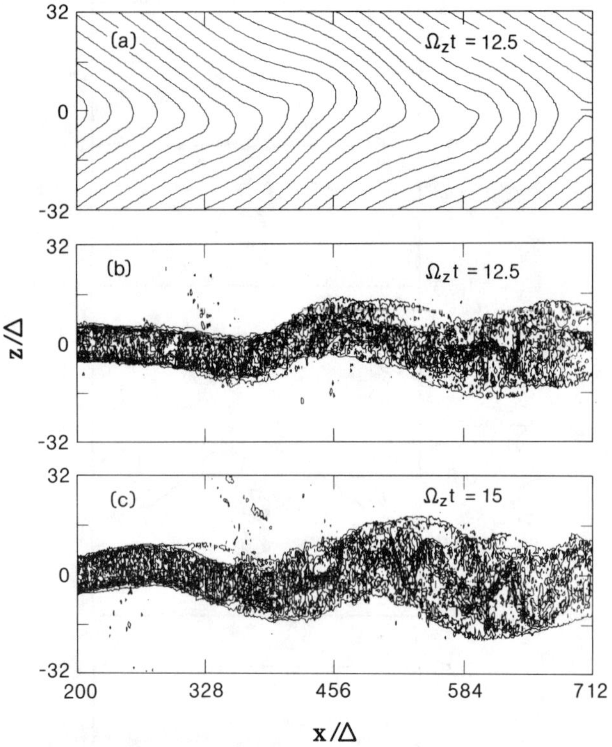

Fig. 10. (a) Magnetic field lines at time $\Omega_z t = 12.5$ as observed in the 2D self-consistent field simulation with active region length = 512Δ and $v_{Ti}/v_{x0} = 0$. (b,c) Corresponding contour plots of the current density $J_y(x, z)$ at times $\Omega_z t = 12.5$ and 15.

a 2D self-consistent dynamical simulation model. The current sheet equilibrium is formed in a frame moving toward the earth with a speed cE_y/B_z by injecting field-aligned streams on all field lines which leave the earthward and north-south boundaries of the simulation domain. In the central or active part of the simulation region the 2D self-consistent electromagnetic fields are solved continually in the Darwin approximation as the ions stream through the system. In the absence of the 2D fields, this distributed-source model does lead to the formation of a 1D double-humped current sheet. Due to the chaotic nature of the particle orbits, however, the firehose marginal stability condition is not satisfied. With the inclusion of the 2D fields, the 1D sheet proves to be stable to the ion tearing mode but unstable to kink-like perturbations which broaden the average current sheet thickness. The combined effects of the chaotic scattering and finite thermal velocity spread lead to a minimum half-thickness of the order of $0.25c/\omega_{p0}$ and produce a configuration satisfying the marginal firehose stability condition outside of the central current sheet.

Acknowledgments. This work was supported by National Science Foundation grant ATM 92-01662. The particle simulations were performed at the San Diego Supercomputer Center, which is supported by the National Science Foundation.

REFERENCES

Birdsall, C. K., and A. B. Langdon, *Plasma Physics via Computer Simulation*, McGraw-Hill, New York, 1985.

Büchner, J., and L. M. Zelenyi, Deterministic chaos in the dynamics of charged particles near a magnetic field reversal, *Phys Lett. A, 118*, 395, 1986.

Burkhart, G. R., J. F. Drake, P. B. Dusenbery, and T. W. Speiser, A particle model for magnetotail neutral sheet equilibria, *J. Geophys. Res., 97*, 13,799, 1992*a*.

Burkhart, G. R., J. F. Drake, P. B. Dusenbery, and T. W. Speiser, Ion tearing in a magnetotail configuration with an embedded thin current sheet, *J. Geophys. Res., 97*, 16,749, 1992*b*.

Chen, J., and P. J. Palmadesso, Chaos and nonlinear dynamics of single-particle orbits in a magnetotaillike magnetic field, *J. Geophys. Res., 91*, 1499, 1986.

Cole, G. H. A., and K. Schindler, On the equilibrium configuration of the geomagnetic tail, *Cosmic Electrodyn., 3*, 275, 1972.

Dickman, D. O., R. L. Morse, and C. W. Nielson, Numerical simulation of axisymmetric, collisionless, finite-β plasma, *Phys. Fluids, 12*, 1708, 1969.

Eastwood, J. W., Consistency of fields and particle motion in the "Speiser" model of the current sheet, *Planet. Space Sci., 20*, 1555, 1972.

Eastwood, J. W., The warm current sheet model and its implications on the temporal behavior of the geomagnetic tail, *Planet. Space Sci., 22*, 1641, 1974.

Francfort, P., and R. Pellat, Magnetic merging in collisionless plasmas, *Geophys. Res. Lett., 3*, 433, 1976.

Hamilton, J. E. M., and J. W. Eastwood, The effect of a normal magnetic field component on current sheet stability, *Planet. Space Sci., 30*, 293, 1982.

Hill, T. W., Magnetic merging in a collisionless plasma, *J. Geophys. Res., 80*, 4689, 1975.

Lin, N., R. L. McPherron, M. G. Kivelson, and R. J. Walker, Multipoint reconnection in the near-earth magnetotail: CDAW 6 observations of energetic particles and magnetic field, *J. Geophys. Res., 96*, 19,427, 1991.

Martin, R. F., Jr., Chaotic particle dynamics near a two-dimensional magnetic neutral point with application to the geomagnetic tail, *J. Geophys. Res., 91*, 11,985, 1986.

Mitchell, D. G., D. J. Williams, C. Y. Huang, L. A. Frank, and C. T. Russell, Current carriers in the near-earth cross-tail current sheet during substorm growth phase, *Geophys. Res. Lett., 17*, 583, 1990.

Pritchett, P. L., and F. V. Coroniti, Formation and stability of the self-consistent one-dimensional tail current sheet, *J. Geophys. Res., 97*, 16,773, 1992.

Pritchett, P. L., F. V. Coroniti, R. Pellat, and H. Karimabadi, Collisionless reconnection in two-dimensional magnetotail equilibria, *J. Geophys. Res., 96*, 11,523, 1991.

Rich, R. J., V. M. Vasyliunas, and R. A. Wolf, On the balance of stresses in the plasma sheet, *J. Geophys. Res., 77*, 4670, 1972.

Sergeev, V. A., P. Tanskanen, K. Mursula, A. Korth, and R. C. Elphic, Current sheet thickness in the near-earth plasma sheet during substorm growth phase, *J. Geophys. Res., 95*, 3819, 1990.

Speiser, T. W., Particle trajectories in model current sheets, 1, Analytical solutions, *J. Geophys. Res., 70*, 4219, 1965.

Speiser, T. W., Particle trajectories in model current sheets, 2, Applications to auroras using a geomagnetic tail model, *J. Geophys. Res., 72*, 3919, 1967.

Speiser, T. W., On the uncoupling of parallel and perpendicular motion in a neutral sheet, *J. Geophys. Res., 73*, 1113, 1968.

F. V. Coroniti and P. L. Pritchett, Department of Physics, University of California, Los Angeles, CA 90024-1547.

Vortex-Induced Reconnection and Turbulent Reconnection in Magnetospheric Boundary Regions

Z. Y. Pu and S. Y. Fu

Department of Geophysics, Peking University, Beijing

Z. X. Liu

Institute of Space Physics, Academia Sinica, Beijing

F. Li

Institute of Electronics, Academia Sinica, Beijing

Vortex induced magnetic reconnection (VIMR) at the dayside magnetopause is investigated and compared with bursty single X-line reconnection (BSXR) by performing 2-D MHD simulations. When the interplanetary magnetic field has a southward orientation, it is found that VIMR and BSXR occur when the flow-aligned Alfven Mach number is, respectively, larger than and less than ~0.5. Concentric flow vortices and magnetic islands, typically R_E in diameter, are formed by VIMR, while in the asymptotic quasi-steady state of BSXR, vortices are created around the X-line. If the magnetospheric boundary regions are open systems, and if the Reynolds number and the magnetic Reynolds number are greater than ~800, vortices and magnetic islands ranging from a few tens to a few hundreds of km are formed in the VIMR process. The field lines are irregularly broken and reconnected, leading to the formation of irregular meso and small-scale structures. The features of this type of turbulent reconnection are presented. In our study, no irregular meso and small-scale structures in the magnetic field and flow patterns are found to occur in association with BSXR.

1. INTRODUCTION

Magnetospheric boundary regions are the sites of significant energy, momentum and mass transfer from the solar wind to the magnetosphere. The transition region between the magnetosheath and the magnetosphere is a complicated layered structure, as well as a variety of meso and small-scale structures in different regions of the magnetopause. The processes that lead to the formation of such structures are believed to be fundamental to the energy, momentum and mass coupling between the solar wind and the magnetosphere. The flux transfer events (FTEs), of which the scale length is a few thousand kilometers, are the most frequently observed phenomena at the dayside magnetospheric boundary region (Russell and Elphic, 1978). There is considerable evidence that FTEs result from patchy and transient magnetic reconnection at the magnetopause when the interplanetary magnetic field (IMF) has a southward component (Lee, 1990). To explain the intermittent and spatial natures of the dayside reconnection associated with FTEs, several theoretical models have been proposed. These include the single connected flux tube model by Russell and Elphic (1978), the multiple X line magnetic reconnection model (MXMR) by Lee and Fu (1985), the bursty single X line reconnection model (BSXR) by Scholer (1988) and Southwood (1988), and the vortex induced magnetic reconnection model (VIMR) by Liu and

Space Plasmas: Coupling Between Small
and Medium Scale Processes
Geophysical Monograph 86

Pu (Liu and Hu, 1988; Hu et al., 1988; Pu et al., 1990a,b). Each of these models predicts specific patterns and structures in the magnetic and velocity fields. Currently, the available data are not sufficient to determine which model is the most appropriate. To get an idea how and where these models are different from each other, we recently compared the BSXR model and the VIMR model by doing a 2-D MHD simulation. In the third section of this paper we will present some of our results from this simulation.

ISEE 1 and 2 observations showed that small-scale magnetic structures with characteristic lengths of tens of kilometers occur within the magnetopause current layer when IMF has a southward orientation (Elphic, 1989). The fact that fluctuations are seen in all components of the magnetic field indicates that these small-scale structures manifest the presence of 3-D filamentary currents. It is not yet known what process is responsible for the occurrence of such current structures. In the fourth part of this paper we will propose a turbulent reconnection model, that provides a physical mechanism which may explain the creation of the filamentary current structures in the magnetopause current sheet. It will be seen that if the magnetospheric boundary regions are open systems and have a Reynolds number (R) and a magnetic Reynolds number (R_m) greater than ~800, vortices and magnetic islands with a scale size ranging from a few tens to a few hundreds of km appear in the VIMR process. The field lines are irregularly broken and reconnected, leading to the formation of irregular meso and small-scale structures. The larger R_m and R are, the more are the irregular meso and small-scale structures. The stronger the flow shear intensity is, the more quickly irregular meso and small-scale structures are produced. Meso and small-scale structures can co-exist with large-scale structure in the simulation domain for a very long time. No such Irregular meso and small-scale structures are found in connection with BSXR in our study.

2. SIMULATION METHOD

It is assumed that at the magnetopause there exist both a sheared velocity field and a reversed magnetic field together with inhomogeneities in the temperature and number density expressed by

$$\boldsymbol{V} = V_0 \tanh\frac{x}{l}\, \boldsymbol{e_z} \tag{1}$$

$$\boldsymbol{B} = ((B_1+B_2)+(B_1-B_2)\tanh\frac{x}{l})\, \boldsymbol{e_z} \tag{2}$$

$$N=((N_1+N_2)+(N_1-N_2)\tanh\frac{x}{l})/2 \tag{3}$$

$$T=((T_1+T_2)+(T_1-T_2)\tanh\frac{x}{l})/2 \tag{4}$$

where V_0, B_0, N_i (i=1,2), and T_i are all of dimensionless variables measured in units of, respectively, the characteristic Alfven speed, magnetic field, number density, and temperature, l represents the scale length of the shear layer, and subscript 1 (or 2) indicates the magnetosheath (or magnetosphere). Suppose all quantities satisfy $\partial/\partial y=0$, the incompressible MHD dimensionless equations can be written as (Pu et al., 1990a)

$$\frac{\partial \Omega}{\partial t} = -\boldsymbol{V}\cdot\nabla\Omega + \boldsymbol{B}\cdot\nabla J + \frac{1}{R}\nabla^2\Omega \tag{5}$$

$$\frac{\partial A}{\partial t} = -\boldsymbol{V}\cdot\nabla A + \frac{1}{R_m}\nabla^2 A \tag{6}$$

$$\frac{\partial N}{\partial t} = -\boldsymbol{V}\cdot\nabla N \tag{7}$$

$$N\frac{\partial E}{\partial t} = -\boldsymbol{V}\cdot\nabla E + \frac{J^2}{R_m} + \frac{1}{2R}\sum_{i,J}\left(\frac{\partial V_i}{\partial x_j}+\frac{\partial V_J}{\partial x_i}\right)^2 + \frac{1}{R}\nabla^2 E \tag{8}$$

where $\mathbf{V}=V_x\mathbf{e}_x+V_z\mathbf{e}_z$, $\mathbf{B}=B_x\mathbf{e}_x+B_z\mathbf{e}_z$, $\mathbf{V}=-\nabla x(\Phi\mathbf{e}_y)$, $\mathbf{B}=\nabla x(A\mathbf{e}_y)$, $\nabla^2\Phi=\Omega$, $\nabla^2 A=-J$, t is measured in units of the Alfven time $t_A = l/V_A$, V_A represents the Alfven speed, E indicates the internal energy density. In studying the VIMR, we set $R_m=R=R_0$ for simplicity; to simulate BSXR, we assume a local enhancement of anomalous resistivity to appear which takes the form

$$\eta = \eta_o[1+\alpha\exp(-\frac{(x-x_o)^2+(z-z_o)^2}{L^2})\exp(-\frac{t}{t_d})] \tag{9}$$

where η_0 is the normalized background dimensionless resistance, t_α indicates the decay time, and α the enhancement coefficient.

The simulation region is taken to be a square in the (x,z) plane defined by $-L/2 \le x \le L/2$, $-L/2 \le z \le L/2$. Perturbations in **B** and **V** are all assumed to vanish at the boundaries of $x=\pm L/2$ and $z=\pm L/2$. A mesh system of 64x64 is used. The differential derivatives in space are replaced by the central difference formulas, whereas for time evolution we use the 4th order Runge-Kutta scheme.

3. COMPARISON BETWEEN BSXR AND VIMR

BSXR may take place when there is a local enhancement of resistivity in the magnetopause current sheet, while VIMR may occur if a velocity shear and magnetic field reversal exist simultaneously in the magnetopause boundary regions.

For moderate values of R and R_m, there exists a critical magnitude (M_c) of the flow-aligned Alfven Mach number ($M_A=V_0/V_A$). When $M_A<M_c$, and no local resistivity

enhancement is present (α=0), reconnection appearing in the current sheet is called 'spontaneous' and also has a single X-line in the reconnection region. However, if α becomes non-zero somewhere in the center of the current sheet, BSXR takes place and develops much faster than the spontaneous single X-line reconnection. The X- line of BSXR is located where η is maximum. Figure 1 shows the asymptotic quasi-steady state of BSXR with M_A=0, N_1=N_2, and T_1=T_2. The reconnection rate of BSXR is much higher than that of the spontaneous single X-line reconnection. For moderate values of R_m=R, M_c is found to be ~0.5 in our simulation.

When M_A>M_c, VIMR occurs if there is no resistivity enhancement, of which the reconnection rate is also much higher than that of spontaneous single X-line reconnection. On the other hand, if local enhancement of resistivity exists, the BSXR type perturbation first appears around the resistivity maximum; and somewhat later, fluid vortices and their associated magnetic islands develop. They grow with time until VIMR dominates the system and reaches its asymptotic quasi-steady state (see Figures 2a and 2b).

The asymptotic quasi-steady state of BSXR is similar to that of single spontaneous X-line reconnection in which magnetic field lines do not form closed loops, but instead assume a bubble-like configuration. Meanwhile, vortices form around the X-line. Quite differently, in the asymptotic quasi-steady state of VIMR, vortices and magnetic islands are co-located; the streamlines, magnetic field lines, contour lines of vorticity and contour lines of T and N (if initially they are non-uniform) are all in alignment and take concentric vortical patterns, no matter what boundary conditions (periodic or non-periodic) are used in the flow-aligned direction. Figure 3 shows an example of patterns of the streamlines, magnetic field lines, and contour lines of the temperature and vorticity in a run of VIMR with N_1=N_2.

4. TURBULENT RECONNECTION AND MESO AND SMALL-SCALE STRUCTURES

Recently we developed a 2-D turbulent reconnection model which produces meso and small-scale structures in magnetopause boundary regions ranging from a few tens to a few hundreds of kilometers. The main assumptions of this model include (1) A reverse magnetic field and a velocity shear exist simultaneously at the magnetopause boundary. (2) Magnetopause boundary regions are open and dissipative systems. (3) The Reynolds and magnetic Reynolds number are both greater than some critical values. To simplify the problem, N and T are taken to be constant. Apparently our model is considerably different from that of Matthaeus and

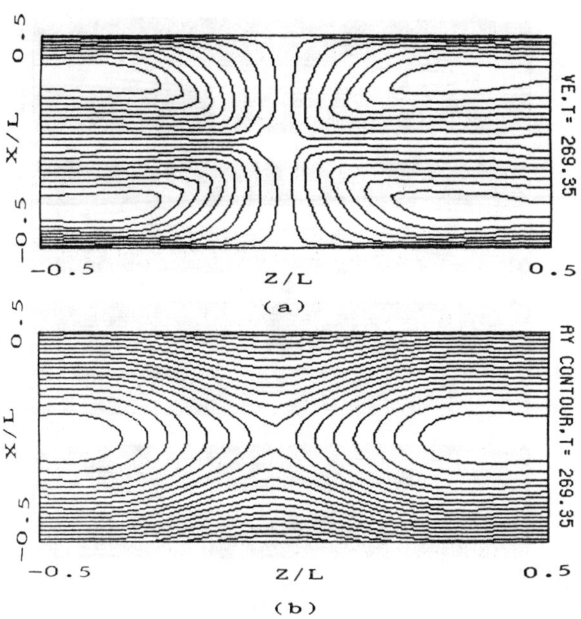

Fig. 1. The asymptotic quasi-steady state of BSXR with M_A=0: (a) the stream lines; (b) the magnetic field lines.

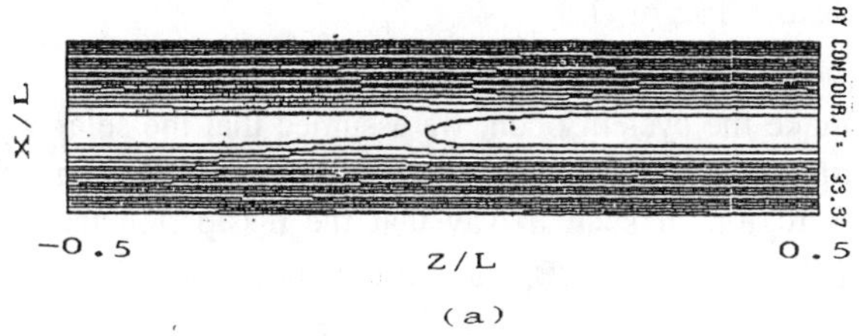

Fig. 2a. The magnetic field pattern for the case of α=3 and M_A=1.5 at t=33.

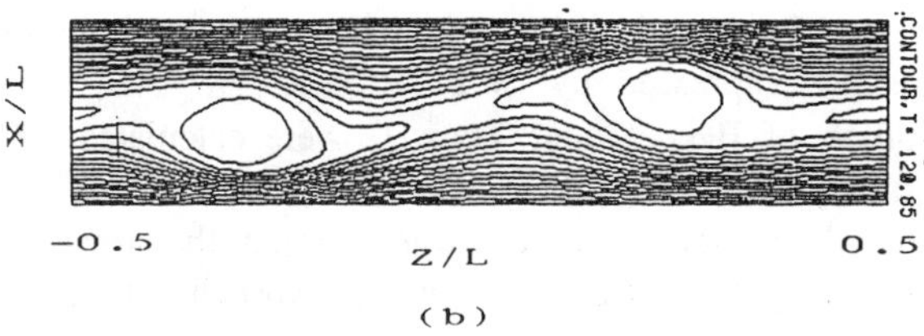

Fig. 2b. The magnetic field pattern for the case of α=3 and M_A=1.5 at t=121.

Lamkin (1985; 1986) which has neither energy input into the system nor the velocity shear; only high R_m and R and broadband initial fluctuations are assumed.

We performed a number of computer experiments with our turbulent reconnection model. The numerical methods used were basically those applied in studying VIMR and BSXR.

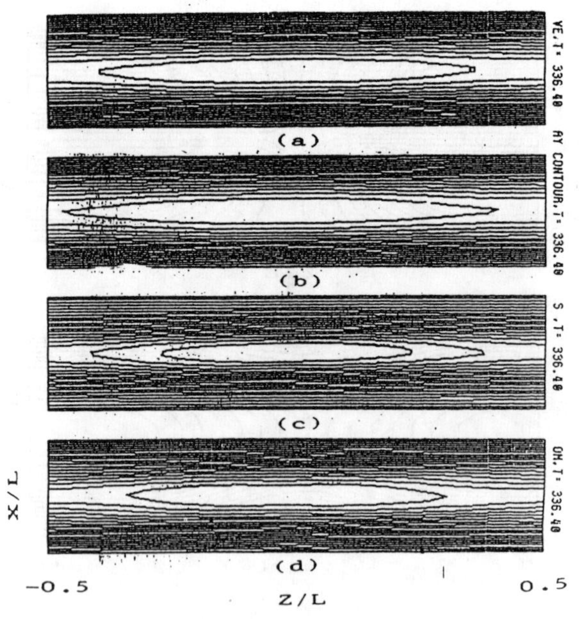

Fig. 3. The asymptotic quasi-steady state of VIMR with M_A=1.0: (a) stream lines, (b) magnetic field lines, (c) contour lines of T, and (d) contour lines of Ω.

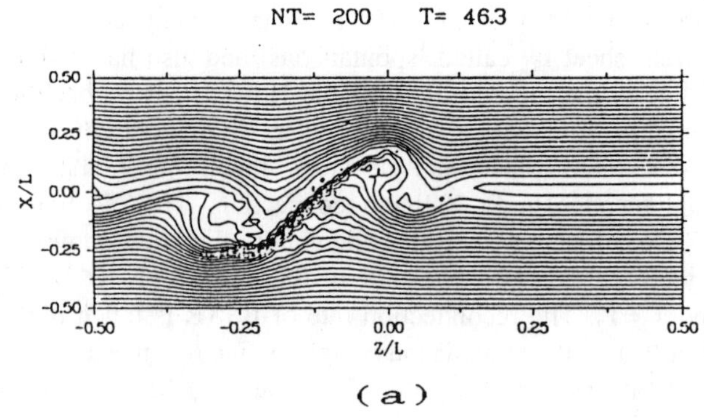

Fig. 4a. The magnetic field line pattern of turbulent reconnection at t= 46 for the case of R_m=R=1000 and M_A =2.0.

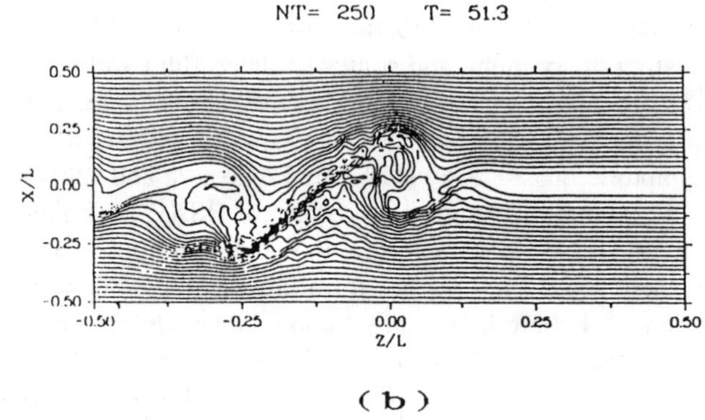

Fig. 4b. The magnetic field line pattern at t= 51.3 for the case of R_m=R=1000 and M_A =2.0.

To make the system open, we assumed that the solar wind continuously transports a certain amount of energy into the shear region in such a way that the dissipation terms of $\nabla^2\Omega_0/R$ and $\nabla^2 A_0/R_m$ are just compensated for by the energy input. In other words, if there were no disturbances, the velocity and magnetic field profiles expressed by (1) and (2) would be maintained. We then write $A(x,z;t)=A_0(x,z)+A'(x,z;t)$, $\mathbf{B}(x,z;t)=\mathbf{B}_0(x,z)+\mathbf{B}'(x,z;t)$, $\mathbf{V}(x,z;t)=\mathbf{V}_0(x,z)+\mathbf{V}'(x,z;t)$, and $\Omega(x,z;t)=\Omega_0(x,z)+\Omega'(x,z;t)$, and insert them into (5) and (6). The remaining dissipation terms are expressed by $\nabla^2\Omega'/R^{-1}$ and $\nabla^2 A'/R_{m-1}$. The evolution of $\mathbf{B}(x,z;t)$ and $\mathbf{V}(x,z;t)$ were calculated. Our major results are the following.

After the initial disturbance was imposed, the streamlines first reconnected to form an X-line and vortices, at the same time a wave-like perturbation in the magnetic field appeared. The field lines became twisted, and the vortex induced type of reconnection then took place. Meanwhile meso and small-scale magnetic structures start to occur in the vicinity of the X-line and near the separatrix line. These can be seen in Figures 4a and 4b in which the magnetic field profiles at two different times are plotted. With time, the field lines became stochastically broken and reconnected over an increasing area. A large number of irregular magnetic islands were produced; each of these islands corresponded to a current tube or filament. The scale sizes of such irregular meso and small-scale structures can be estimated to be a few tens to a few hundreds of kilometers. No dominant typical length could be found using Fourier transformation. Magnetic field variations obtained in this simulation can obviously be regarded as a type of turbulent reconnection.

High magnitudes of R_m and R and the existence of the velocity shear were both crucial to this kind of turbulent reconnection. For turbulent reconnection to appear, R_m and R must be greater than ~800. The larger R_m and R were, the more irregular meso and small-scale structures appeared. Figures 5a and 5b show the field lines for R_m=R=80 and 800, respectively. As we see irregular meso and small-scale structures only occur in the latter case. Figures 6a and 6b plot the magnetic field for the cases of V_o/l=0.38 and 0.75, respectively. These indicate that the stronger the flow shear intensity is, the more quickly the irregular meso and small-scale structures occur. Finally, it is worth noting that in our study, no visible irregular meso and small-scale structures were seen in the magnetic field and flow patterns in BSXR.

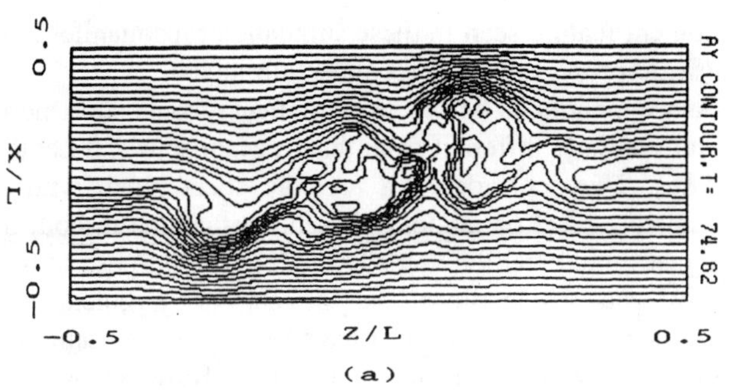

Fig. 5a. The magnetic field line pattern at t= 83 for the case of R_m =R=80 and M_A =2.0.

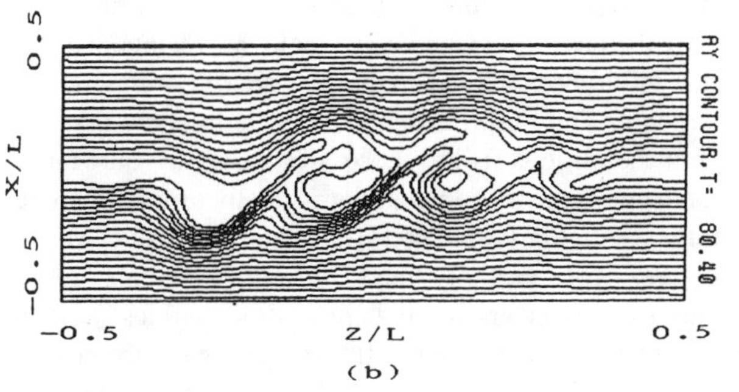

Fig. 5b. The magnetic field line pattern of turbulent reconnection at t= 60 for the case of R_m=R=800 and M_A =2.0.

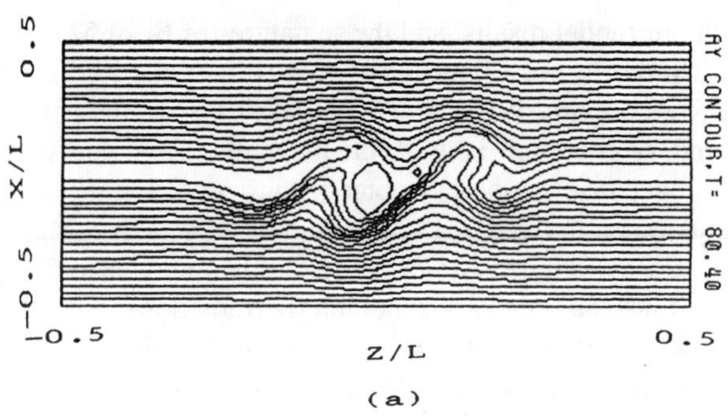

Fig. 6a. The magnetic field line pattern at t= 59 for the case of R_m=R=800 and V_0 /L =0.38.

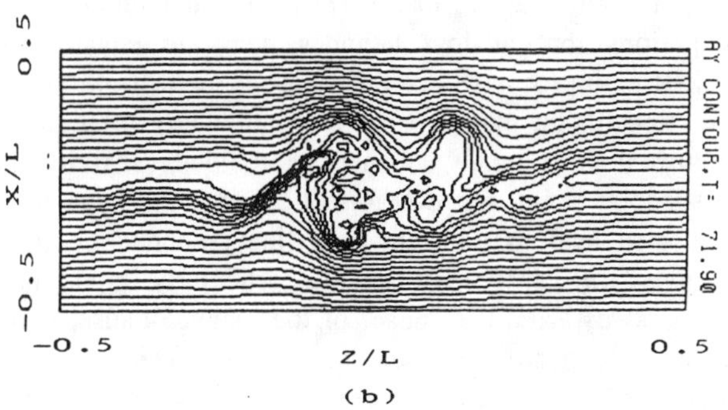

Fig. 6b. The magnetic field line pattern of turbulent reconnection at t= 45 for the case of R_m=R=800 and V/L=0.75.

5. DISCUSSION

It is known that MXMR is driven by an external cross-field flow, while BSXR is related to a local bursty enhancement of the resistivity; physically they are both associated with the development of the tearing mode instability. However, VIMR manifests an alternate mechanism. The Kelvin-Helmholtz instability excited in the center of the current sheet causes fluid vortices there. As these vortices grow, they cause the field lines to twist and merge to form magnetic islands (magnetic vortices). The instability associated with this process arises from coupling of the Kelvin-Helmholtz instability with the tearing mode, and was named the vortex induced tearing mode instability (Pu et al, 1990a,b). No matter what kind of boundary condition is imposed in the flow-aligned direction (periodic or non-periodic), the magnetic islands in the asymptotic quasi-steady state of VIMR must occur with vortices. In the three-dimensional situation with non-zero B_y , the vortex tube coincides with the magnetic flux tube, and results a MHD vortex tube (Liu et al., 1992). On the other hand, the field lines do not form closed loops in the BSXR case. While initially there were more than one local enhancement of the resistivity with the same value, only one X-line appears stochastically at one of the maxima. Plasmas on each side of the current sheet move towards the X-line and then turn back, just like what takes place in the conventional tearing mode case, hence producing a flow pattern topologically different from that of VIMR.

On the bases of VIMR, we proposed an FTE model (Liu and Hu, 1988; Hu et al.,1988; Pu et al., 1990a,b) in which the connected open flux tube assumed to be associated with FTEs is identified with the MHD vortex tube (Liu et al., 1992). Similar to the single connected flux tube model by Russell and Elphic (1978), the single connected flux tube in the VIMR model has a basically cylindrical symmetry, with several layers of different plasma and field regimes about the central core. The field lines and stream lines lying on the surface of the connected open flux tube and inside are helical, and are related to a field and stream-aligned current and vorticity. This geometry appears to be essential in predicting the observed signatures of FTEs, which depend on the direction that the connected open flux tube is moving, as well as on the position of the spacecraft relative to the passing flux tube. Recently we analyzed 61 FTEs data and found that the observed signatures of B_N are all consistent with our model results, and the signatures of B_L of 52 events are similar to our calculated shapes (Liu et al., 1992). We reported previously (Pu et al., 1990a) that when the periodic boundary condition is used, the magnetic island and vortex are concentric in the asymptotic quasi-steady state of our simulation provided $M_A > 0.4$. Now we see that M_c, the critical flow-aligned Alfven Mach number, may be as low as ~0.5 if a fixed boundary condition (perturbations in **B** and **V** vanish at $x=\pm L/2$, and $z=\pm L/2$) is applied. This means that for typical magnetopause conditions at times when FTEs were detected by ISEE 1 and 2 (Paschmann et al., 1982; Coroniti, 1985), so long as the magnetosheath flow passes over the magnetosphere at v>100-141 km/s, the VIMR will take place (Pu et al., 1990a,b). Thus, we concluded from our simulations that at low latitudes near the equator, the MXMR may proceed since there the magnetosheath plasma flows towards the magnetosphere. When micro-turbulence in the current sheet locally enhances anomalous resistivity near the subsolar point, the BSXR will become the most likely process to lead to transient field line reconnection there. However, the VIMR may become particularly important in regions away from the "nose" of the magnetopause, where the magnetosheath flow parallel to the interface has been accelerated to high speed. Regardless whether micro-turbulence produces locally enhanced anomalous resistivity, the VIMR appears to provide an effective mechanism for the formation of FTEs. Besides, it is of interest to recall that the asymptotic quasi-steady state of the VIMR may be regarded as the fixed attractor of an MHD system with a velocity shear and a reverse field (Pu et al., 1990b). Therefore, as the plasma flow convects to mid and high latitudes, the topology of the magnetic field and vortex structures originally created by either MXMR or BSXR near the equator may change to show patterns typically associated with the VIMR.

Turbulent reconnection has attracted much attention in the past ten years (Matthaeus and Montgomery, 1981; Matthaeus, 1982; Matthaeus and Lamkin, 1985; 1986). Matthaeus and Lamkin (1986) conducted a comprehensive study in a slab model with a periodic sheet pinch configuration. Their simulation showed that small-scale magnetic " bubbles" and vorticity structures developed in the vicinity of the X-line. A comparison of our Figures 4a-6b to their results clearly shows that the number of irregular meso and small-scale structures and the extent of irregularity are substantially higher in our simulation than in theirs. To check whether these irregular meso and small-scale structures are real, we made a series of calculations by reducing the time or spatial step to one half, one fourth, etc. The field line and stream line patterns thus obtained are qualitatively the same as the original ones. Therefore, the turbulent feature seen in these simulation runs manifests the intrinsic property of our reconnection model.

As we mentioned before, our turbulent reconnection model is substantially different from that of Matthaeus and Lamkin (1986). They considered a periodic sheet pinch with an initial low level of random phased fluctuations in both the magnetic field and the velocity modes. The purpose of their study was to explore the effects of turbulence on reconnection, and to confirm their belief that a certain level of turbulence is present in reconnection. However, we treat a transition region in which both a reversed magnetic field and a sheared flow exist with perturbations in **B** and **V** vanishing at the boundaries. We also allow an energy input into the system so that if there were no disturbances, the initial profiles of (1) and (2) would be maintained at all times. Initial broadband fluctuations are of no importance in our model. The purpose of this work is to show that under certain conditions (like those described in our model), reconnection may become an intrinsically turbulent process, yielding irregularly shaped fluid and magnetic vorticity structures (or vorticity and current filaments in 3-D situations). It is known that in hydrodynamics, a weakly dissipative turbulent flow is full of vortices with variety of scale sizes, which are often produced by a velocity shear. In our sheared MHD system, as VIMR takes place, meso and small-scale vortices are irregularly created close to the X-line, generating stochastic variations (motion) in the magnetic field in regions near the X-line and the separatrix. Turbulent reconnection is thus triggered with the field lines being randomly broken and reconnected. Similar to hydrodynamic turbulence, here very large values of R_m and R play an important role, because in order to retain small-scale structures, dissipation must be very weak. The system must be assumed to be open so that dissipation can be compensated for. We have also simulated a closed system with the same profiles as (1) and (2) and large values of R_m and R. No discernible irregular meso and small-scale structures are seen in the magnetic field and flow field with the present diagnostic resolution. So far we have not found similar irregular meso and small-scale structures in BSXR, either. It is worth noting that the interface between the solar wind and the magnetosphere away from the subsolar point manifests itself as both a current layer and a flow shear. The magnetopause currents are maintained by the presence of the jump from the IMF to the magnetospheric field across the boundary. It is reasonable to assume that the average kinetic energy of plasma in magnetopause transition regions originates in the solar wind flow. Weak turbulence theory has predicted very low levels of anomalous resistivity at the magnetopause (Coroniti, 1985). All these features are consistent in principle with the main assumptions of our model.

By a careful inspection of ISEE 1 and 2 data, Elphic (1989) pointed out that while the magnetopause current sheet thickness is typically many ion gyroradii, there is evidence of numerous small-scale structures within the boundary, suggesting the presence of filamentary or time-varying currents. On November 12, 1977 between 2258 and 2259 UT, ISEE 1 and 2 were crossed the magnetopause and were separated by only 15 km. The magnetosheath field at this time had a southward orientation. Two spacecraft observed significant differences in three components of the magnetic field which were due to local currents with scale sizes of the order of 10's of km. The difference field (ISEE 1 minus ISEE 2) was small and very quiet within the magnetosphere, but noisy within the magnetopause current layer. The fact that magnetic field fluctuations were present in all components indicates that these small-scale structures manifest the presence of 3-D filamentary currents. At that time the process responsible for the appearance of these current filaments was unknown. Based on our simulation study we suggest that in wide regions of the dayside magnetopause where l, the scale length of the current sheet, is many gyroradii of ions, turbulent reconnection may possibly provide a mechanism that can lead to the formation of current and vorticity structures with scale sizes ranging from a few tens to a few hundreds of km. The MHD approach is applicable in studying of the physics of the magnetopause when l is much larger than the average gyroradius of ions, r_i, which lies in the order of a few tens of km for typical magnetopause parameters. Therefore, it is unlikely that our simulation results can directly be used to explain the formation of filamentary currents with the scale size of 10 km (Elphic, 1989). To interpret the appearance of these filamentary currents in light of turbulent reconnection, a forward cascade process involving kinetic effects is required to cause the small structures to decay to smaller scales.

It should be noted that, in reality, anomalous resistivity may effectively be influenced by the occurrence of small-scale magnetic structure. If we assumed that η rises with increasing J(x,z;t) as J exceeds a critical current, the system will arrive at an asymptotic quasi-steady state in which irregular meso and small-scale structures can co-exist with meso and large-scale structures for a very long time. The size of small-scale magnetic islands in such a "final" state critically depends on what values of R_m and R are used. Furthermore, after attaining an asymptotic quasi-steady state, if we considerably reduce R_m an R to change the system , meso and small-scale magnetic islands may coalesce to form a larger-scale one; while a large (meso)-scale structure may break into smaller-scale ones provided R_m and R are substantially increased. In this regard, magnetic reconnection can also be considered as a breaking bifurcation process of a current sheet (Parker et al., 1990), and the relevant main bifurcation parameters are R_m and R. Magnetopause conditions change continuously with time and position. For instance, the thickness of the current layer lies between ~100 to ~1000 km (Elphic, 1989; Berchem and Russell, 1982). The tangential velocity of the magnetosheath flow outside the interface ranges from ~0 at the subsolar point to ~600 km at mid and high latitudes in high speed solar wind situations (Spreiter and Alksne, 1969). In addition, it has been widely accepted that micro-turbulence may cause anomalous resistivity, which is a number of orders higher than the classical one (Coroniti, 1985). Therefore, it is reasonable to expect that both R_m and R may vary with position and solar wind condition over quite a wide range. Hence we suggest that quasi-steady reconnection, patchy and transient reconnection, and turbulent or chaotic reconnection, may possibly appear at different time and in different areas of the magnetopause; a variety of magnetic field and vortex structures with different patterns and scale sizes may occur, depending on different reconnection processes involved. Under certain conditions, FTEs manifest a quasi-steady phenomena, while in other cases, they represent a transient state evolving to a turbulent one. Finally it should be emphasized that 2-D simulation along can not confirm such conjectures. To answer these outstanding questions, we must rely on the CLUSTER and other ISTP missions, along with 3-D simulations and advanced theoretical studies.

Acknowledgments. The authors wish to thank Dr.Z.F. Fu for his helpful assistance. This project is supported by CNSF.

REFERENCES

Berchem, J. and C. T. Russell, The thickness of the magnetopause: ISEE 1 and 2 observations, *J. Geophys. Res., 87*, 2108, 1982.

Coroniti, F.V., Space plasma turbulent dissipation: Reality or myth? *Space Sci. Rev., 42*, 399, 1985.

Elphic, R.C., Magnetic structure and dynamics, in *Geospace Environment Modeling, GEM report of the Workshop on Magnetopause and Boundary Layer Physics,* ed. by M. Ashour-Abdalla, p.13, 1989.

Hu, Y.D., Z.X. Liu, and Z.Y. Pu, Response of the magnetic field to the flow vortex field and reconnection in the magnetopause boundary region, *Sci. Sin. A., Engl. Ed., 10,* 1100, 1988.

Lee, L.C., The magnetopause : A tutorial review, in *Physics of Space Plasmas,* Proceedings of the 1990 Cambridge Workshop in Geoplasma Physics on Magnetic Fluctuations, Diffusion and Transport in Geoplasmas and the 1990 MIT Winter Symposium on the Physics of Space Plasmas, ed. by T. Chang, G.B. Crew, and J.R. Jasperse, p.33, Scientific

Publishers, Inc., 1990.

Lee, L.C., and Z. F. Fu, A theory of magnetic flux transfer events at the Earth's dayside magnetopause, *Geophys. Res. Lett., 12,* 105, 1985.

Liu, Z.X., and Y.D. Hu, Local magnetic reconnection caused by vortices in the flow field, *Geophys. Res. Lett., 15,* 752, 1988.

Liu, Z.X., Z.W. Zhu, F. Li, and Z.Y. Pu, Topology and signatures of a model for flux transfer events based on vortex-induced reconnection, *J. Geophys. Res., 97,* 19351, 1992.

Matthaeus, W.H., and D. C. Montgomery, Nonlinear evolution of the sheet pinch, *J. Plasma Phys., 14,* 283, 1981.

Matthaeus, W.H., Reconnection in two dimensions: Localization of vorticity and current near magnetic X-points, *Geophys.Res. Lett., 9,* 660, 1982.

Matthaeus, W.H., and S.L. Lamkin, Rapid magnetic reconnection caused by finite amplitude fluctuations, *Phys. Fluids, 28,* 303, 1985.

Matthaeus, W.H., and S.L. Lamkin, Turbulent magnetic reconnection, *Phys. Fluids, 29,* 2513, 1986.

Paschmann, G., G. Haerendel, I. Papamastorakis, N. Sckopke, S.J. Bame, J.T. Gosling, and C. T. Russell, Plasma and magnetic field characteristics of magnetic flux transfer events, *J. Geophys. Res., 87,* 2159, 1982.

Pu, Z.Y., M. Yei, and Z. X. Liu, Generation of vortex-induced tearing mode instability at the magnetopause, *J. Geophys. Res., 95,* 10559, 1990a.

Pu, Z.Y., P.T. Hou, and Z.X. Liu, Vortex-induced tearing mode instability as a source of flux transfer events, *J. Geophys. Res., 95,* 18861, 1990b.

Russell, C. T., and R. C. Elphic, Initial ISEE magnetometer results: Magnetopause observations, *Space Sci. Rev., 22,* 681, 1978.

Scholer, M.D., Magnetic flux transfer at the magnetopause based on single X-line bursty reconnection, *Geophys. Res. Lett., 15,* 291, 1988.

Southwood, D.J., C. J. Farrugia, and M.A. Saunders,What are flux transfer events? *Planet. Space Sci., 36,* 503, 1988.

Spreiter, J. R. and A. Y. Alksne, Plasma flow around the magnetosphere, *Rev. Geophysics, 7,* 11, 1969.

Magnetic Reconnection and Current-Sheet Formation at X-Type Neutral Points

R. S. Steinolfson

Department of Space Sciences, Southwest Research Institute, San Antonio, Texas

L. Ofman[1] and P. J. Morrison

Institute for Fusion Studies, University of Texas, Austin, Texas

Numerical solutions of the nonlinear, resistive magnetohydrodynamic (MHD) equations are used to study the evolution of a perturbed or stressed x-type neutral point. By performing individual simulations for both compressible and incompressible plasmas, we are able to demonstrate that the important physics for this problem involves just the interaction between the plasma flow velocity and the magnetic field and that the thermodynamics has a relatively passive effect. We have also done separate simulations for both solid, conducting wall boundary conditions at a fixed distance from the x-point and for open boundary conditions that adjust as required by the evolving solution within the boundaries. With solid, conducting wall boundary conditions, our solutions for azimuthally symmetric disturbances agree (for essentially linear perturbations) with those obtained in previous analytic linear studies. In this case the stressed x-point relaxes back to the unstressed state on a time scale somewhat shorter than the time scale for the linear resistive tearing mode. Perturbations that are not azimuthally symmetric can relax even faster than the symmetric modes. When the conditions at the boundary are free to adjust, the disturbances grow in amplitude on an Alfvén time scale with the eventual formation of a current sheet separating two y-points. This rapid growing behavior is, of course, in sharp contrast to the relatively slow decaying solutions obtained with closed boundaries. The growing solutions qualitatively agree with previous analytic x-point solutions that have been suggested as an explanation for the rapid energy conversion in flares and substorms.

1. INTRODUCTION

Localized regions within a plasma with an imbedded magnetic field containing points or lines at which the magnetic field strength vanishes have long been recognized as having important and unique topographical features. In a plasma for which the physical variables do not change in one dimension (the ignorable coordinate), one such region is the x-type neutral point (also referred to here as an x-point or neutral point), which becomes a neutral line along the ignorable coordinate. The x-point is of particular interest in solar physics since this may be the location of the rapid conversion of magnetic energy to other forms of energy in solar flares, as suggested by *Giovanelli* [1947]. *Dungey* [1953] quantified the suggestion by Giovanelli and showed that x-type neutral points were unstable and that a small perturbation initiated a rapid increase in the current at the neutral point. *Chapman and Kendall* [1963] used a more complete quantitative analysis than Dungey and were able to show that the instability grew on a time scale comparable to an Alfvén crossing time. In addition, for the Chapman and Kendall solution the x-point eventually changed to a current sheet with y-points at the ends. *Syrovatsky* [1966] later included this instability mechanism and the current sheet formation in a solar flare model. As the above references indicate, much of the early interest in perturbed x-type neutral points concentrated on the growing solutions and the potential application to explosive phenomena such as flares and auroral discharges [*Akasofu and Chapman*, 1961].

More recently, there has been some effort to examine solutions for perturbed x-type neutral points that do not grow

[1]Currently at NASA Goddard Space Flight Center, Greenbelt, MD

Space Plasmas: Coupling Between Small
and Medium Scale Processes
Geophysical Monograph 86

in time but rather decay back to the unperturbed configuration [*Craig and McClymont*, 1991; *Hassam*, 1991; *Craig and Watson*, 1992]. A primary interest in these studies has been to demonstrate that reconnection at the neutral point occurs at a rate faster than either the classical reconnection rate for the tearing instability [*Furth, Killeen, and Rosenbluth*, 1973] or the slightly faster rate predicted by *Sweet* [1958] and *Parker* [1963]. An important difference between these decaying solutions and the above growing solutions is the form of the outer boundary conditions at some fixed distance from the x-point. The perturbed magnetic flux must vanish at a solid, conducting wall surrounding the neutral point for the decaying solutions, while it must be allowed to adjust as required by conditions within the boundary for the growing solutions. In addition, for the two-dimensional cylindrical geometry often used in studies of the decaying solutions, the perturbed magnetic flux must depend only on radial distance from the neutral point in order to obtain the fast reconnection.

In the present study we numerically solve the two-dimensional, nonlinear, dissipative magnetohydrodynamic (MHD) equations for both a compressible and an incompressible plasma. We consider both the solid, conducting wall boundary conditions used in previous decaying solution studies as well as open or free boundary conditions for the growing solutions. We begin with solid, conducting wall boundary conditions, which should be applicable to fusion devices and laboratory plasmas, and simulate the relaxation of a perturbed or stressed x-point back to the initial potential state and show that our results are in quantitative agreement with those from earlier studies. When the boundaries are allowed to be open, which should have application to more general space plasmas and to flares and auroral discharges, the perturbation grows at a rate comparable to a relevant Alfvén time scale and the x-point evolves into a current sheet as predicted in the more approximate studies indicated above.

Our objectives are twofold. First of all, by applying both incompressible and compressible theory to the same problem, we are able to demonstrate, for this application, that the important physics lies not in the plasma thermodynamic properties but rather in the coupling between the inertial and magnetic (including resistivity effects) terms. Secondly, by using the same numerical codes with different outer boundary conditions, we show the dramatic influence the boundary conditions can have on the computed solutions; i.e., decaying solutions with solid, conducting walls and growing solutions when the plasma motion and magnetic field are not restricted at the outer boundary.

2. MHD EQUATIONS AND NUMERICAL SOLUTION

We make the usual assumption that MHD adequately describes the essential physics in this problem. By further assuming that the plasma resistivity is constant and isotropic and that gravitational and viscous effects are negligible, the compressible equations can be written in nondimensional form as follows:

$$\frac{\partial \rho}{\partial t} + \nabla \cdot (\rho \boldsymbol{u}) = 0 , \tag{1a}$$

$$\rho \frac{d\boldsymbol{u}}{dt} = -\frac{\beta}{2}\nabla p + (\nabla\times\boldsymbol{B})\times\boldsymbol{B} , \tag{1b}$$

$$\frac{\partial \boldsymbol{B}}{\partial t} = \nabla\times(\boldsymbol{u}\times\boldsymbol{B}) - \frac{1}{S}\nabla\times(\nabla\times\boldsymbol{B}) , \tag{1c}$$

$$\frac{d}{dt}\left(\frac{p}{\rho^{\gamma}}\right) = 0 , \tag{1d}$$

where d/dt indicates the convective derivative. The physical quantities are the density ρ, the velocity $\boldsymbol{u}$, the magnetic field $\boldsymbol{B}$, and the thermal pressure p. The specific heat ratio is represented by γ, which is taken here to be 5/3. The equations have been put in dimensionless form by normalizing time to a relevant Alfvén time $\tau_a = a(4\pi\rho_o)^{1/2}/B_o$, distance to the quantity a, which is a representative scale length for changes in the magnetic field, velocity to a/τ_a, thermodynamic quantities to their initial, uniform values (ρ_o and p_o), and magnetic field strength to the average value over the outer boundary B_o. The two parameters are the plasma beta ($\beta = 8\pi p_o/B_o^2$, the ratio of thermal pressure to magnetic pressure) and the Lundquist number ($S = \tau_r/\tau_a$, the ratio of the resistive diffusion time to the Alfvén time). The resistive diffusion time is defined as $\tau_r = 4\pi a^2/c^2\eta$, where c is the light speed and η is the plasma resistivity. We further make the assumption that the initial equilibrium magnetic field is externally maintained so it is not resistively dissipated.

The two-dimensional (2-D) incompressible MHD equations are obtained by writing the magnetic field in terms of a stream function using $\boldsymbol{B} = \nabla\psi \times \boldsymbol{e}_z$ and the velocity in terms of a flux function $\boldsymbol{u} = \nabla\phi \times \boldsymbol{e}_z$, where $\boldsymbol{e}_z$ is the unit vector in the ignorable direction z in a Cartesian coordinate system. By then taking the curl of the momentum equation

(1b) to eliminate the thermal pressure p and setting the constant density to unity, the equations can be written as:

$$\frac{\partial\psi}{\partial t} = -\frac{\partial\phi}{\partial y}\left(\frac{\partial\psi}{\partial x}+x\right) + \frac{\partial\phi}{\partial x}\left(\frac{\partial\psi}{\partial y}-y\right) - \frac{1}{S}J\,, \quad (2a)$$

$$\frac{\partial\omega}{\partial t} = -\frac{\partial\phi}{\partial y}\frac{\partial\omega}{\partial x} + \frac{\partial\phi}{\partial x}\frac{\partial\omega}{\partial y} + \left(\frac{\partial\psi}{\partial y}-y\right)\frac{\partial J}{\partial x} - \left(\frac{\partial\psi}{\partial x}+x\right)\frac{\partial J}{\partial y}\,, \quad (2b)$$

where $J = -\nabla\psi_\perp^{\,2}$ is the z-component of the current density, $\omega = -\nabla\phi_\perp^{\,2}$ is the z-component of the vorticity, and $\nabla_\perp^{\,2} = \partial^2/\partial x^2 + \partial^2/\partial y^2$ with $\partial/\partial z = 0$. Note that the dimensionless parameter β does not enter into the incompressible equations. The thermodynamics uncouples from the evolution of the plasma velocity and the magnetic field with this approximation.

The compressible equations are solved numerically using an explicit finite differencing scheme of the Lax-Wendroff type [*Rubin and Burstein*, 1967] with a smoothing term suggested by *Lapidus* [1967]. The equations are solved in the r,θ plane of a cylindrical coordinate system with a computational domain limited by $0 \le r \le r_m$, $0 \le \theta \le \pi/2$. All physical quantities are either symmetric or antisymmetric at the $\theta = 0$ and $\theta = \pi/2$ boundaries. The outer boundary conditions at $r = r_m$ will be discussed separately for each of the two studies in the following sections. The incompressible equations are solved numerically using an alternating-direction implicit method as discussed further by *Steinolfson and van Hoven* [1984]. These equations are solved in the x,y plane of a Cartesian geometry with a computational domain limited by $-x_m \le x \le x_m$, $-y_m \le y \le y_m$. As for the compressible case, the outer boundary conditions will be discussed in the following sections.

The equilibrium potential magnetic field for the compressible study is $\boldsymbol{B}_o = r(\sin 2\theta \boldsymbol{e}_r + \cos 2\theta \boldsymbol{e}_\theta)$. The initial equilibrium magnetic flux for the incompressible study can be written as $\psi_o = x^2 - y^2$, in which case the equilibrium potential magnetic field components are $B_{xo} = -2y$ and $B_{yo} = -2x$. The initial magnetic field for both cases contains an x-point, as shown for Cartesian geometry in Figure 1. The perturbation magnetic fields used in the separate studies will be discussed later.

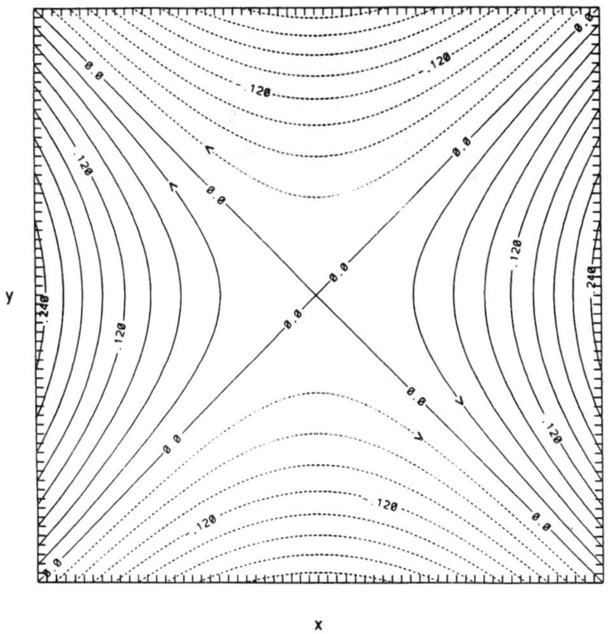

Fig. 1. The equilibrium magnetic field configuration containing an x-point. These are contour lines of the magnetic flux function ψ_o defined in the text.

3. DECAYING SOLUTIONS: RECONNECTION AT THE X-POINT

Hassam [1991] and *Craig and McClymont* [1991] investigated the behavior of perturbed x-points using linearized, compressible MHD with the low-beta approximation so thermal pressure gradients could be neglected. More importantly, in terms of the types of solutions they obtained, they assumed that the plasma was confined within solid, conducting walls. They solved the equations in the r,θ plane of a cylindrical coordinate system and showed that they can be reduced to a single, second-order partial differential equation for the perturbed magnetic flux function. Separation of variables, with an assumed time and angular dependence given by $\exp(\lambda t + im\theta)$, is used to reduce the equation to an ordinary differential equation in r. It is then argued that only the $m = 0$ (azimuthally symmetric) modes involve reconnection at the x-point, and therefore, the analysis is limited to the $m = 0$ modes. For these modes the approximate expressions for the oscillation and decay times of the fundamental radial mode ($n = 0$, the radial mode with longest wavelength) are derived as:

$$\tau_{osc} \approx 2\ \ln S\,, \qquad \tau_{dec} \approx 2(\ln S)^2/\pi^2\,. \quad (3)$$

Ofman [1992] has taken the above analysis one step further and has derived the general linear solution valid for any radial and azimuthal mode. His solution for the $m = n = 0$ mode is given by the solid curve in Figure 2, and the above approximate solution is given by the dashed line. As a check on the nonlinear incompressible code, the decay rate

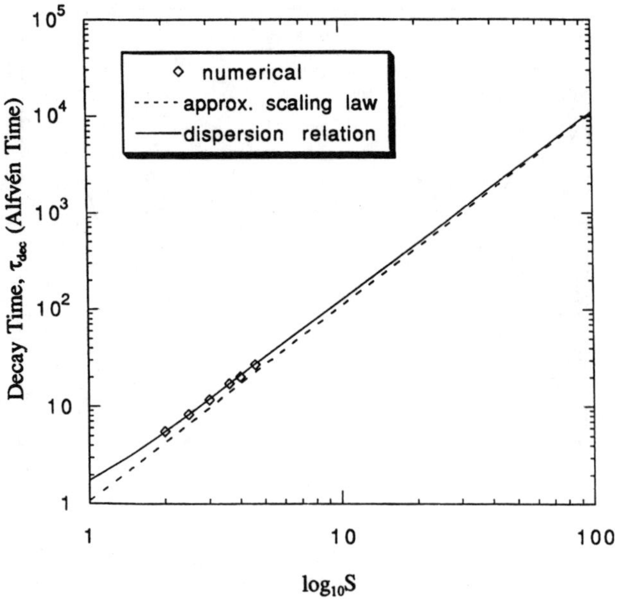

Fig. 2. The dependence of the decay time on S for the $m = n = 0$ mode. The dispersion relation is from *Ofman* [1992], the dashed curve is for τ_{dec} in equation (3), and the boxes indicate results from the numerical computation.

for small perturbations was computed for $10^2 \leq S \leq 4 \times 10^4$, and the results are compared with the analytic solutions in the figure. The perturbed magnetic flux function for this simulation is given by:

$$\psi_1 = A_i\,(x_m^{\,2}-x^2)\,(y_m^{\,2}-y^2)\,\exp(-5(x^2+y^2)) \qquad (4)$$

with $A_i = 10^{-2}$. The initial perturbation was selected to have approximate azimuthal symmetry ($m = 0$) near the origin, to vanish at the edges of the box ($x_m = y_m = 1$), and to decrease monotonically with distance from the origin ($n = 0$). Since the computational box is square, the disturbance will always have some azimuthal dependence at least near the edges of the box. Both the magnetic flux function ψ and the stream function ϕ are required to remain zero at the outer boundaries.

The relaxation of the perturbed x-point for the $S = 10^4$ case in Figure 2 is shown in Figures 3-5. The temporal decay and oscillation of the perturbed magnetic energy in the x-component of the field (curve A), the perturbed magnetic energy in the y-component of the field (curve B) and the total (magnetic plus kinetic, curve C) energies are given in Figure 3. These energies are computed by integrating over the entire region. The variation of the magnetic flux at the origin [$\psi(0,0,t)$] is shown in Figure 4. As discussed earlier, the magnetic field reconnects whenever $\psi(0,0,t)\neq 0$. This reconnection of the field when $\psi(0,0,t)\neq 0$ can be seen from plots of the magnetic field lines at three times during the oscillating decay in Figure 5. The results in Figure 5(a) are at the first minimum of $\psi(0,0,t)$ at $t \approx 7$, those in Figure 5(b) are at the second zero of $\psi(0,0,t)$ at $t \approx 12$, and those in

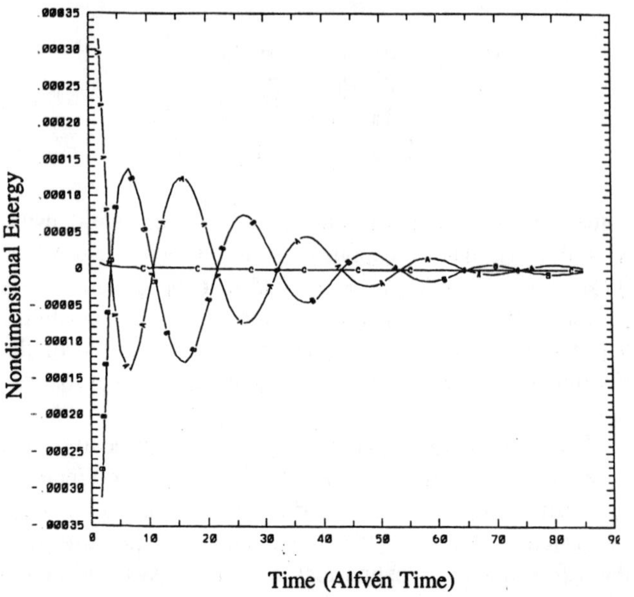

Fig. 3. The relaxation of the perturbed energies integrated over the numerical box for a disturbed x-point. The curves represent the x- and y-components of the magnetic energy (curves A and B) and the total energy (curve C).

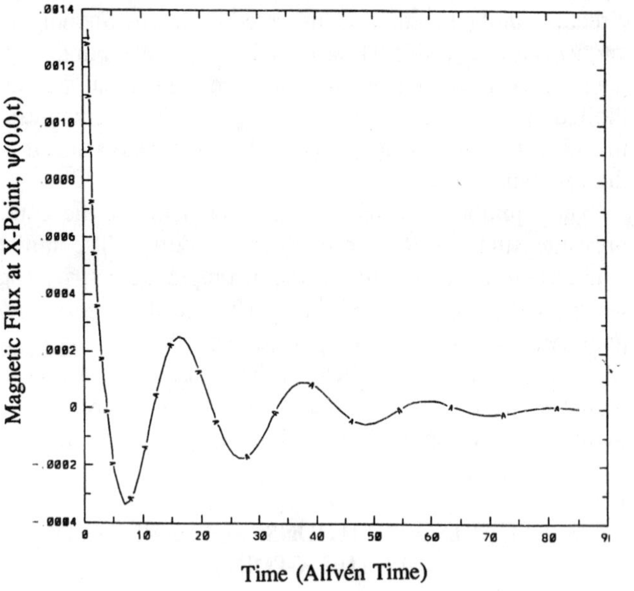

Fig. 4. The temporal variation of the magnetic flux at the x-point [$\psi(0,0,t)$].

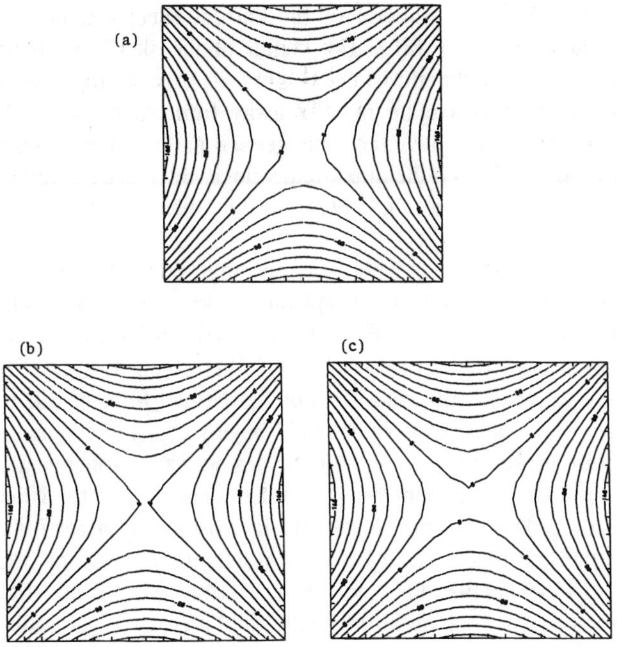

Fig. 5. Magnetic field lines at various times during the oscillating decay. The field lines in (a) are at the first minimum in $\psi(0,0,t)$ in Figure 4, those in (b) are at the second zero in $\psi(0,0,t)$, and those in (c) are at the time of the following maximum in $\psi(0,0,t)$.

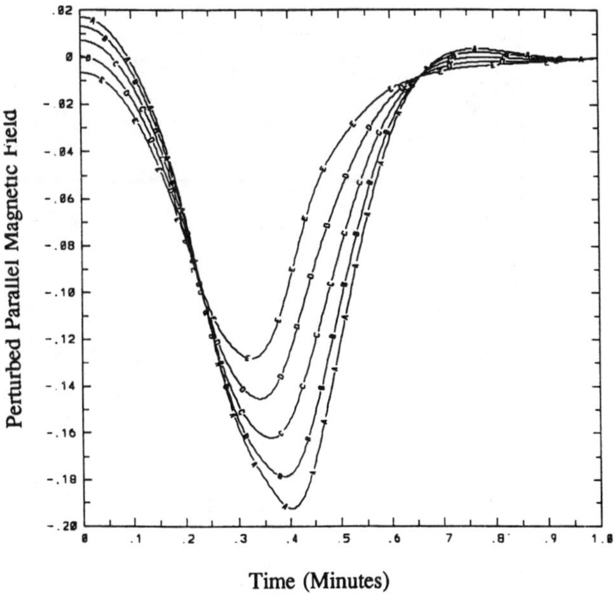

Fig. 6. The temporal decay of the perturbed component of the azimuthal magnetic field at $\theta = \pi/2$ and at several radial locations r_i near the x-point ($r_i = 0.02, 0.04, 0.06, 0.08, 0.10$).

Figure 5(c) are at the following maximum of $\psi(0,0,t)$ at $t \approx 16$. The reconnection oscillates between the field lines to the left and right of the x-point (at the center of the figure) and the field lines above and below the x-point. In Figure 5(b) the magnetic flux function vanishes at the origin, there is very little reconnection, and the magnetic configuration is nearly force-free. The oscillations are damped by the resistive reconnection, and the final state is the force-free x-point configuration given by ψ_o.

The decay and oscillation times computed with the compressible code also agree with the analytic results given above. The $m = 0$ perturbation (defined in terms of the magnetic flux function) used in this case is $B_\theta = A_c \sin \pi r$ for $0 \le r \le 1$, and the results in Figure 6 are for $A_c = 10^{-3}$ and $S = 10^4$. The outer boundary is located at $r = 1$ where the boundary conditions are such that the thermodynamic quantities remain fixed at their initial values, the flow is parallel to the boundary, and the component of the magnetic field parallel to the boundary remains fixed. The quantity shown in the figure is the perturbed component of the magnetic field at $\theta = \pi/2$, which is calculated as $\Delta B_\theta = [B_\theta(r_i,\pi/2,t) - B_\theta(r_i,\pi/2,0)]/B_\theta(r_i,\pi/2,0)$ where $r_i = 0.02, 0.04, 0.06, 0.08, 0.10$ (curves A-E, respectively). The minor phase differences at the various radii are due to nonlinear effects and can be made arbitrarily small by reducing the magnitude of the initiating perturbation (reduce A_c).

4.0 GROWING SOLUTIONS: CURRENT SHEET FORMATION

The boundary conditions at the outer boundary are now changed so that this boundary is open or free, as opposed to the closed boundary considered in the previous section. Another way to view the present boundary conditions is that they are now able to adjust to changes occuring within the region of interest rather than being fixed by some external means. The boundary values are computed by setting the perturbed value of each physical quantity at the boundary equal to the respective value at the grid point next to the boundary (zero-order extrapolation).

The results shown in Figure 7 after the solution has evolved for 5.3 Alfvén times are obtained from the incompressible code using the same initial perturbation as used in the above section. The transition to the sheet current seen in the plot of the contours of ψ in Figure 7(a) occurs approximately on an Alfvén time scale. The formation of the sheet current involves large flow vortices as shown in the contour plots of the velocity potential in Figure 7(b). The flow velocity is parallel to the contours with the dashed-line contours representing clockwise rotation and the solid-line contours indicating counter-clockwise rotation.

The formation of a current sheet in the compressible simlation is seen in the results presented in Figures 8-10.

(a)

(b)

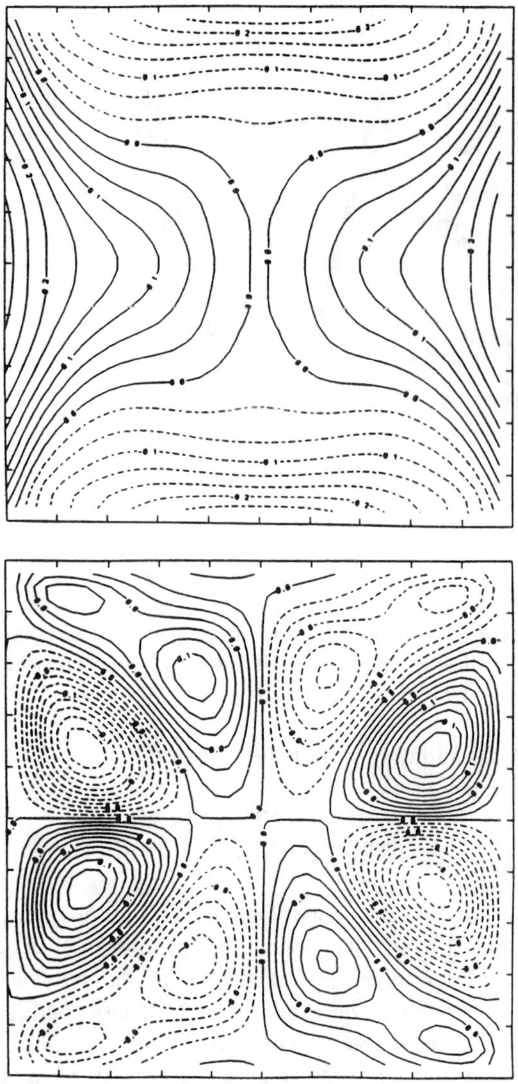

Fig. 7. The development of a sheet current in the incompressible MHD model. The magnetic field lines (contours of ψ) are shown in (a) and contours of the stream function (ϕ) are shown in (b). The flow velocity is parallel to the indicated ϕ contours with the dashed contours representing clockwise rotation and the solid contours representing counter-clockwise rotation.

The calculation extends out to $r = 2$, although the perturbation is the same as used in the previous section for the compressible solution (with $A_c = 0.05$) and only exists out to $r = 1$. The reason for extending the computation beyond the perturbation will be discussed later. The symmetries (or anti-symmetries) at $\theta = \pi/2$ (the vertical axis at the center of the panels in Figures 8 and 9) were used so the solution could be shown for $0 \le \theta \le \pi$. The magnetic field lines and velocity vectors are shown at the same three times in Figures 8 and 9. The formation of the current sheet can be seen clearly in Figure 8(b). The center of the sheet eventually becomes a magnetic island (Figure 8(c)) separating two x-points that propagate radially away from each other. The velocity vectors have the same general pattern at all times in Figure 9 although the maximum velocity increases by almost two orders of magnitude between the times used for the top and bottom panels.

The temporal behavior of the radial components of the perturbed magnetic field and the velocity at several radial locations (r_i = 0.2,0.4,0.6,0.8,1.0,1.2,1.4,1.6,1.8,2.0 for curves A-J) along a line at $\theta = 10°$ is given in Figure 10. These curves illustrate a couple of points. First of all, the large increase in the radial field begins near the original x-point location ($r = 0$) and propagates outward. Note also that the radial component peaks and then decreases as the island forms. Also, the radial velocity increases to a larger maximum before decreasing as the x-points travel outward since the reconnection then occurs in a stronger magnetic field region, since the initial magnetic field increases with radius.

As mentioned above, the computation extends one unit in radius beyond the maximum of the initial perturbation. If the calculation did only go out to $r = 1$, the results are almost identical to those shown here with the difference that the eruptive phase begins just a little earlier. The reason the eruption then begins earlier is that, after the transients from the initial perturbation decay away, the solution (primarily the velocity) must evolve into a particular pattern throughout the entire simulation region before the eruption begins. The initial transients can be seen at early times in the temporal plots in Figure 10. The time to set up this special initial state for the eruptive behavior naturally increases as the size of the numerical box increases. It now becomes easy to understand why the growing solutions do not develop when the radial velocity is forced to be zero at the outer boundary since the necessary initial state cannot be established.

It is important to realize that these growing solutions develop from physical conditions in the vicinity of the initial x-point. They are not a result of doing something special at the outer boundaries like pushing the field lines together or imposing an electric field. All that is required at the outer boundaries is that they be open so an appropriate state (after the transients due to the initial perturbation vanish) can be established. Once the eruption begins it continues to grow explosively with the x-points traveling outward into regions of stronger magnetic fields.

5.0 SUMMARY

The evolution of perturbed x-type neutral points has been studied using numerical solutions of both compressible and

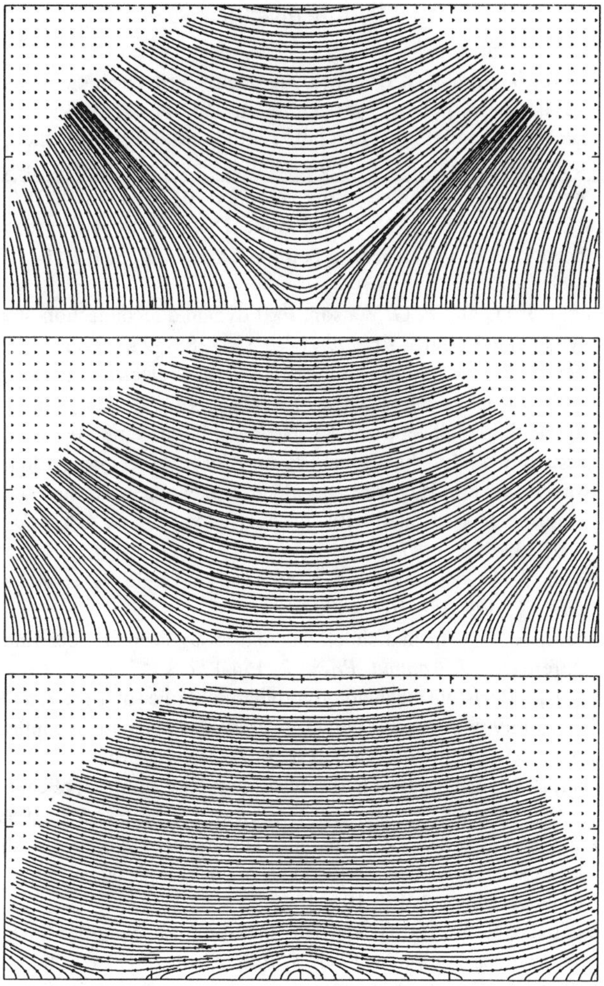

Fig. 8. Line segments drawn parallel to the magnetic field lines. The beginning and end of the lines are selected internally by the plot package, and consequently, the fact that the lines end suddenly does not imply the sudden appearance or disappearance of a magnetic field line. The solution in this and the following figure is symmetric about the lower axis in each panel and is shown at times of 5.4 τ_a, 8.8 τ_a, and 9.9 τ_a from top to bottom.

incompressible nonlinear MHD equations and for two distinctly different sets of outer boundary conditions. The qualitatively similar results obtained with the compressible and incompressible simulations demonstrates that the thermodynamics does not play a significant role in the evolution of x-points. When the boundary consists of a solid, conducting wall, the perturbed x-point relaxes back to the unstressed state on a time scale proportional to the resistivity. The decay time scale computed here for small perturbations agrees with earlier analytic linear studies and is somewhat faster than the time scale for reconnection in the resistive tearing instability. When the boundary is open, so conditions at the boundary can adjust to the evolution near the x-point, the perturbation grows in amplitude on an Alfvén time scale. The initial x-point evolves into a current sheet separating y-points at the ends of the sheet. The y-points propagate away from each other into regions of increasing magnetic field strength.

Acknowledgments. This work was supported by National Science Foundation Grant No. ATM-90-15705 and at the Institute for Fusion Studies by U.S. Department of Energy Contract No. DE-FG05-80ET-53088.

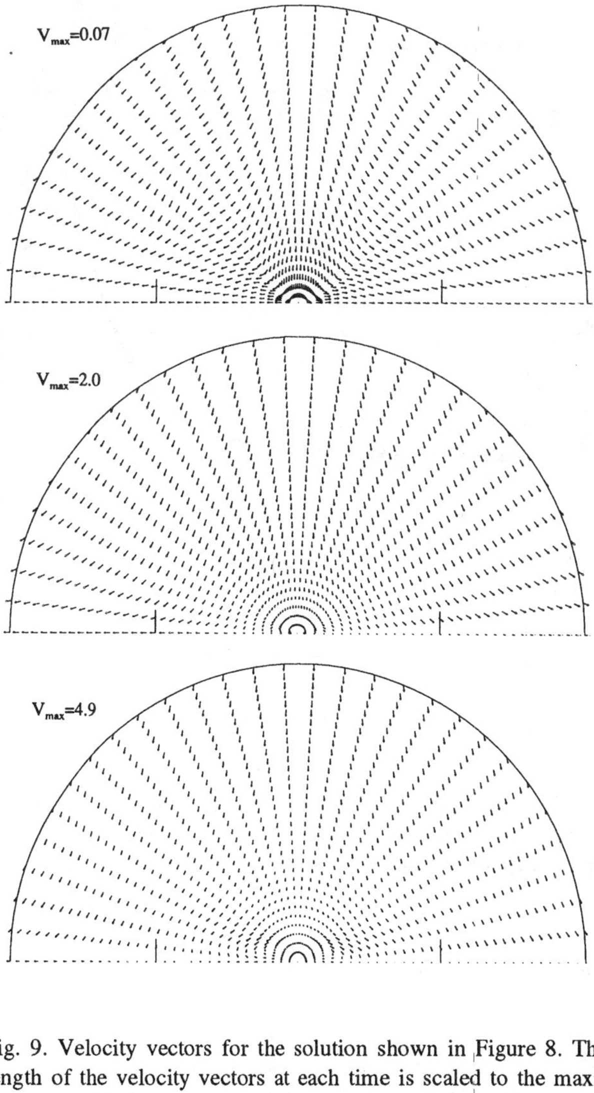

Fig. 9. Velocity vectors for the solution shown in Figure 8. The length of the velocity vectors at each time is scaled to the maximum speed throughout the grid, which is indicated by the given value of V_{max}.

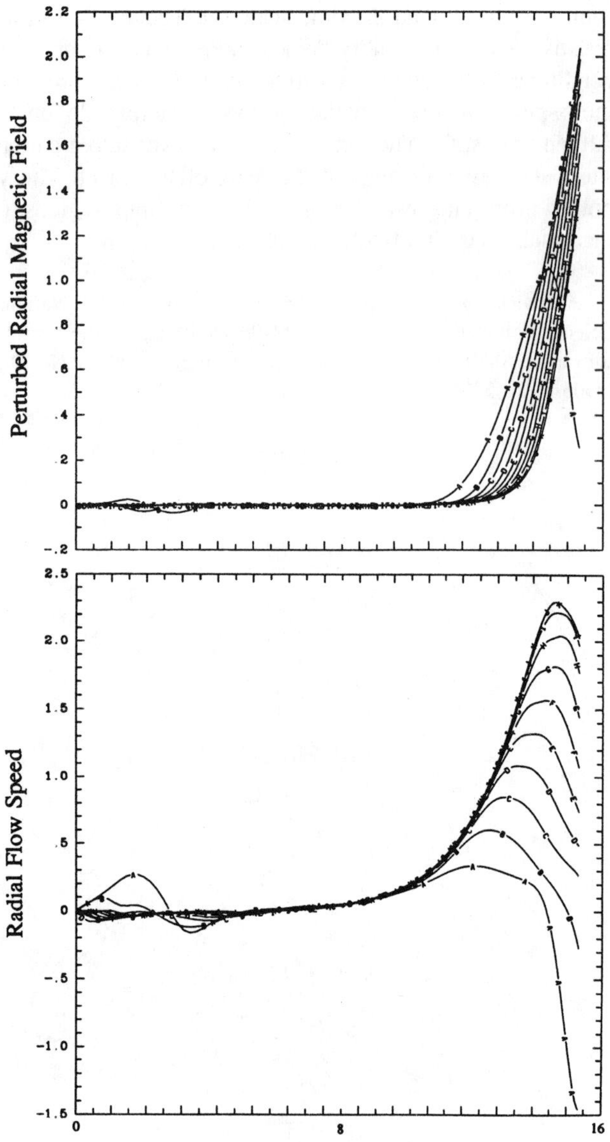

Fig. 10. The time variation of the radial components of the magnetic field and velocity at θ = 10° and at several radial locations spaced evenly between $r = 0.2$ and $r = 2.0$.

REFERENCES

Akasofu, S. -I., and S. Chapman, A neutral line discharge theory of the aurora polaris, *Phil. Trans. A, 253*, 44-49, 1961.

Chapman, S., and P. C. Kendall, Liquid instability and energy transformation near a magnetic neutral line: a soluble non-linear hydromagnetic problem, *Proc. Roy. Soc., A271*, 435-448, 1963.

Craig, I. J. D., and A. N. McClymont, Dynamic magnetic recon nection at an x-type neutral point, *Astrophys. J., 371*, L41-L44, 1991.

Craig, I. J. D., and P. G. Watson, Fast dynamic reconnection at x-type neutral points, *Astrophys. J., 393*, 385-395, 1992.

Dungey, J. W., Conditions for the occurrence of electrical discharges in astrophysical systems, *Phil. Mag., 44*, 725-738, 1953.

Furth, H. P., J. Killeen, and M. N. Rosenbluth, Finite-resistivity instabilities of a sheet pinch, *Phys. Fluids, 6*, 459-484, 1963.

Giovanelli, R. G., Magnetic and electric phenomena in the Sun's atmosphere associated with sunspots, *Mon. Not. R. Astro. Soc., 107*, 338-355, 1947.

Hassam, A. B., Reconnection of stressed magnetic fields, *Astrophys. J., 399*, 159-163, 1992.

Lapidus, A., A detached shock calculation by second-order finite differences, *J. Comput. Phys., 2*, 154-177, 1967.

Ofman, L., Resistive magnetohydrodynamic studies of tearing mode instability with equilibrium shear flow and magnetic reconnection, *IFSR 552*, Univ. of Texas, PhD. Thesis, 1992.

Parker, E. N., Kinematical hydrodynamic theory and its application to the low solar photosphere, *Astrophys. J., 138*, 552-575, 1963.

Rubin, E. L., and S. Z. Burstein, Difference methods for the inviscid and viscous equations of a compressible gas, *J. Comput. Phys., 3*, 178-196, 1967.

Steinolfson, R. S., and G. van Hoven, Nonlinear evolution of the resistive tearing mode, *Phys. Fluids, 27*, 1207-1214, 1984.

Sweet, P. A., The neutral point theory of solar flares, in *Electro magnetic Phenomenon in Cosmical Physics*, edited by B. Lehnert, pp. 123-140, Cambridge University Press, New York, 1958.

Syrovatskii, S. I., Dynamic dissipation of a magnetic field and particle dissipation, *Soviet Astron., 10*, 270-280, 1966.

L. Ofman and P.J. Morrison, Department of Physics and Institute for Fusion Studies, The University of Texas at Austin, austin, TX 78712.

R. S. Steinolfson, Department of Space Science, Southwest Research Institute, PO Drawer 28510, San Antonio, TX 78228-0510.

The Influence of Chaotic Particle Motion on Large Scale Magnetotail Stability

JÖRG BÜCHNER

Max-Planck-Institut für Extraterrestrische Physik, Außenstelle Berlin D-O-1199 Berlin, Germany F. R.

ANTONIUS OTTO

Geophysical Institute and Department of Physics, University of Alaska Fairbanks, AK, 99775-0800, U.S.A.

The stability of the collisionless plasma in the Earth's magnetotail depends on the microscopic processes taking place in its central current sheet. Recently we gained new insights into the nonlinear dynamics of the quasi–adiabatic, chaotic particle motion through current sheets. Here in this paper we report on our current investigation of the consequences of quasi–adiabatic electron behaviour in the plasma sheet for the large scale tail stability. We have developed a hybrid model were the large scale plasma and field dynamics is described by MHD simulation, while for the resulting parameters the non–adiabatic response of particles in the neutral sheet is considered kinetically. Utilizing results of the microscopic theory of chaotic particle motion the MHD equations are than modified locally, taking into account, therefore, the feedback of nonlinear microscopic particle motion to the large scale tail dynamics.

As a result we have found that in a typical quiet–time–tail equilibrium no tendency to instability quasi–adiabatic electron chaos exists inside $X_{GSM} = 70$ *Re* downstream the tail. Electrons become chaotic beyond this distance, but the corresponding growth time of a tearing mode instability would exceed the substorm time scale by far. Hence we conclude that a quiet–time–tail does not decay just due to chaotic particles. They are able, however, to support quasi–stationary reconnection in the distant tail. The situation completely changes after loading the initial equilibrium according to the tail stretching and neutral sheet thinning, observed 30 – 40 minutes before substorm onset in the mid–tail [*Baumjohann et al.*, 1992]. In this case a rapid instability starts by terminating the growth phase and forming a new neutral line in the mid–tail at a growth rate of the order of the time scale of minutes, the same as of substorm expansions.

1. Introduction

Magnetotail observations have provided evidence that during substorms a neutral line forms in the mid–tail at distances between 20 and 30 R_E from the Earth [*Hones*, 1979; *Nishida and Lyons*, 1988; *Baumjohann et al.*, 1989], which may lead to the formation of plasmoids [*Hones et al.*, 1984; *Scholer et al.*, 1984]. The most probable route to neutral line creation in the tail is a tearing mode instability [*Schindler*, 1974; *Galeev and Zelenyi*, 1976; *Coroniti*, 1985]. Two important facts have been established experimentally: The tail stability during quiet times and its transition to instability during substorms. It has been found that approximately 30–45 minutes before substorm onset the plasma sheet starts to thin and the tail magnetic fields to stretch at X_{GSM} distancies between 10 and 19 R_E from the Earth [*Baumjohann et al.*, 1992]. Major remaining questions are, therefore, the onset condition and location of the tail instability.

[1] *Geophysical Institute and Department of Physics, University of Alaska, Fairbanks, AK 99775-0800, U.S.A.*

Space Plasmas: Coupling Between Small and Medium Scale Processes
Geophysical Monograph 86

The tail destabilization observed during substorms is an essentially large scale, global phenomenon. Reconnection, as a possible candidate for it, on the other hand, destroys the plasma connectivity due to essentially localized effects. Consequently a reconnecting tail instability cannot be described in the framework of ideal MHD. One possible way to describe the stability of the tail is to apply resistive MHD [*Ugai and Tsuda*, 1977; *Sato*, 1979; *Sato and Hayashi*, 1989; *Birn*, 1980; *Lee et al.*, 1985; *Otto et al.*, 1990; *Scholer and Hautz*, 1991]. Due to the inefficiency of binary particle collisions these models assumed an anomalous resistivity, based on either current instabilities or on drift instabilities at the edge of the plasma sheet [*Huba*, 1985; *Hoshino*, 1991]. Such a dominant and primary role of microinstabilities in tail reconnection is, however, not obvious [*Büchner et al.*, 1988]. Also, the resulting threshold and growth of an anomalous–resistivity–based tearing mode instability were found to depend strongly on whether a localized or global finite resisivity was assumed, on the concrete dependence of the resistivity on the actual current strength et c. Generally it has been found that the assumed type of finite resistivity predetermines onset as well as the growth of the resisitive tearing mode instability (cf., e.g., *Otto et al.*, [1990]).

On the other hand, in dilute space plasmas a fundamental collisionless tearing mode instability can take place, based just on a direct Cherenkov interaction between plasma particles and the tearing mode field, without need in an intermediate scale microinstability caused anomalous resistivity [*Coppi et al.* 1966; *Schindler*, 1974; *Galeev and Zelenyi*, 1976]. Adiabatic electrons, however, stabilize the collisionless tearing mode [*Galeev and Zelenyi*, 1976; *Coroniti*, 1980; *Lembége and Pellat*, 1982]. Different lines have been proposed along which it might be able to remove the stabilizing influence of the adiabatic electrons, like non–adiabatic scattering et c. [*Coroniti*, 1980; *Goldstein and Schindler*, 1982; *Eseray and Molvig*, 1987; *Büchner and Zelenyi*, 1987]. One of these possible ways to instability is the chaotization of the electron motion in the current sheet. In this case $\kappa_e = \sqrt{R_{min}/\rho_{max}}$ (*Büchner and Zelenyi* [1986]) will be the relevant parameter for triggering a collisionless tearing mode instability. In the framework of the simple models used in analytical stability calculations this mode has currently been challenged by Pellat et al. [1991] and defended by Kuznetsova and Zelenyi [1991]. In fact, these calculations have been carried out for oversimplified models of the tail plasma system, like a parabolic magnetic field structure and other deviations from a tail equilibrium.

Here in this paper we develop a hybrid approach to the problem of large scale magnetotail stability including both the microscopic and the large scale aspects of tail instability. In our model the large scale magnetotail configuration is described by MHD equations. In contrast to previous MHD–simulations, however, we take the microscopic effects into account consistently, based on a kinetic derivation of the quasi–adiabatic electron response to a tearing mode perturbation. In fact the corresponding changes of macroscopic parameters are taken from the dispersion relation of the collisionless tearing mode, derived by Büchner and Zelenyi [1988] for the case of electron–κ smaller but comparable with unity. The non–ideal plasma response is limited to a region of effective interaction between field and particles. Inside this region we consider the plasma response using results from the kinetic theory, outside just the plain ideal MHD–equations. At the edge of the interaction region we match the two solutions.

At first we discuss in section 2 the microscopic model, used. Section 3 is devoted to a description of the MHD model (section 3.2) and the initial equlibrium configuration (section 3.1). So far we have considered two cases. The first corresponds closely to a magnetically quiet magnetosphere, which is found to stay stable although in its distant part electrons are already chaotic (section 4.1.). The second initial situation corresponds a typical pre–substorm loaded tail (section 4.2). The consequences of our model calculations for the magnetotail dynamics are discussed in section 5.

2. The Microscopic Model

Several suggestions exist for describing the consequences of the neutral sheet particle motion for magnetotail plasma. Following an earlier proposal of Speiser (1970) even an "Ohm's law" was derived [*Lyons and Speiser*; 1985], which does not apply, however, to tail stability considerations. The reason is that the effect discussed there does not correspond to a non–ideality or irreversibility of mode particle interactions which is neccesary for reconnection.

After the discovery of the transition to chaos in the tail current sheet [*Büchner and Zelenyi*, 1986; *Chen and Palmadesso*, 1986] the chaotic aspect of the particle motion came into consideration. Horton and Tajima (1991) calculated the correlation decay time for particles traversing the neutral sheet in order to determine the plasma response to neutral sheet perturbations. In contrast here in this paper we will follow the analysis of collisionles tearing mode dispersion relation for quasi–adiabatic electrons, carried out by Büchner and Zelenyi [1988]. Following them the threshold of the collisionless tearing mode instability is given by

$$\kappa_e = \frac{B_n}{B_o}\left(\frac{L}{\rho_{oe}}\right)^{0.5} \approx 1 \tag{1}$$

where ρ_{oe} is the electron gyroradius and Ω_{oe} the gyrofrequency in the tail–lobe magnetic field B_o:

$$\rho_{oe} = v_{te}/\Omega_{oe} \tag{2a}$$

$$\Omega_{oe} = e \cdot B_o/(m_e \cdot c) \tag{2b}$$

As it was shown by Büchner and Zelenyi, [1988] the dispersion relation (their equation (16)) of the collisionless tearing mode in case of quasi–adiabatic electrons compares to that of a resistive mode, with an effective conductivity σ_{eff}

$$\sigma_{eff} \approx \frac{n\,e^2}{m_e} \cdot \Omega_{ne}{}^{-1}, \tag{3}$$

concentrated to a limited (interaction) region. In equation (3) $\Omega_{ne} = \Omega_{ne} B_o/B_n$ is the electron gyrofrequency based on the local B_z component of the magnetic field B_n at the center of the neutral sheet. From the dispersion relation (16) in [*Büchner and Zelenyi*, 1988] one can estimate the following effective width of the interaction region:

$$\Delta_z \approx (B_n/B_o k L)^{0.5} \tag{4}$$

with k being the wavenumber of the instability and L the local width of the plasma sheet. This width becomes a crucial parameter after it does not exceed any more the thickness of the singular current layer in the framework of the resistive instability theory. Notice that in the derivation of the linear dispersion relation (16) in [*Büchner and Zelenyi*, 1988] the stabilizing influence of 10 $^o/_o$ of the electrons which move on ring–type orbits [*Büchner and Zelenyi*, 1989] have been neglected compared to the major part of about 90 $^o/_o$ electrons moving on cucumber orbits and having a destabilizing effect.

3. The Large Scale MHD Model

3.1. INITIAL TAIL EQUILIBRIUM

For the initial tail configurations we apply weakly two–dimensional magnetostatic equilibria of the kind, proposed by Birn et al. [1975]. The concrete shape of these equilibria is determined by a pressure function $\bar{p}(x)$ which prescribes the plasma pressure distribution at the neutral sheet ($Z = 0$) all along the tail. For these equilibria plasma pressure and magnetic field are given by

$$p(x,z) = \bar{p}(x)\,cosh^{-2}\left(\frac{z}{l}\right) + p_\infty \tag{5a}$$

$$b_x(x,z) = -\frac{1}{l} tanh\left(\frac{z}{l}\right) \tag{5b}$$

$$b_z(x,z) = 0.5\frac{d\,(ln\,\bar{p})}{d\,x}\left(\frac{z}{l} tanh\left(\frac{z}{l}\right) - 1\right) \tag{5c}$$

with the local width $l(x) = \bar{p}(x)^{-0.5}$ of the current layer. For the parameter κ one obtains the following expression

$$\kappa(x) = \kappa_o \frac{1}{2\sqrt{\bar{p}}}\left|\frac{d\,(ln\,\bar{p}(x))}{dx}\right| \tag{6}$$

It is appropriate to parametrize the pressure function $\bar{p}(x)$ as follows:

$$\bar{p}(x) = (1 - p_{dt})(1 + x/x_c)^{-q} + p_{dt}, \tag{7}$$

i.e. it is normalized to unity at $x = 0$, the earthward boundary of our tail model, located at, approximately, $X_{GSM} = 20\,R_E$. In the above presentation the pressure function decreases along the tail to a small but finite value at infinity. We have chosen a variation along the tail which by its upper and lower bounds was fitting best to values obtained from satellite observations [*Slavin et al.*, 1985].

3.2. MHD MODEL EQUATIONS

We solve the following set of MHD equations:

$$\frac{\partial \rho}{\partial t} = -\nabla \cdot \rho v \tag{8a}$$

$$\frac{\partial \rho v}{\partial t} = -\nabla \cdot \left(\rho v v + \frac{1}{2}(p + b^2)\underline{1} - bb\right) \tag{8b}$$

$$\frac{\partial b}{\partial t} = \nabla \times (v \times b - \eta \nabla \times b) \tag{8c}$$

$$\frac{\partial h}{\partial t} = -\nabla \cdot (hv) + \frac{\gamma - 1}{\gamma} h^{1-\gamma} \eta j^2 \tag{8d}$$

$$\text{with } h = (p/2)^{1/\gamma} \text{ and } j = \nabla \times b \tag{8e}$$

While in ideal MHD the normalisation is of minor importance, in our case it becomes important because we expect the macroscopic dynamical evolution being dependend on the specific properties of the microscopic processes taking place in the neutral layer. Hence we have chosen a normalisation of magnetic field, length and density by using

$$\hat{B} = B_o \times 10\,nT \tag{9a}$$

$$\hat{L} = L_o \times 1\,R_E \tag{9b}$$

$$\hat{n} = n_o \times 1\,cm^{-3}. \tag{9c}$$

Here B_o, L_o and n_o are dimensionless parameters which can be chosen to adjust the system to certain physical conditions. In this normalization the typical Alfvén velocity and – travel time are given by

$$\hat{v} = \frac{B_o}{\sqrt{n_o}} \times 218\,km\,s^{-1} \tag{10a}$$

$$\hat{\tau} = \frac{L_o\sqrt{n_o}}{B_o} \times 19.2\,s \tag{10b}$$

Specifying the ratio of electron to ion temperature by $\vartheta_e = T_e/T_i$ the parameter of chaotisation and the effective resistivity, discussed in section 2. become (in dimensionless units):

$$\kappa = \kappa_e = \kappa_o \times b_n \sqrt{l/b} \qquad (11a)$$

$$\eta_{eff} = \eta_{eff,o} \times b_n/\rho \qquad (11b)$$

with

$$\kappa_o = 34.6 \times L_o^{0.5} (n_o/\vartheta_e)^{0.25} \quad \eta_{eff,o} = 3.6 \times \frac{1}{L_o \sqrt{n_o}} \qquad (12)$$

where b, b_n, l, and ρ are the dimensionless lobe field, plasma sheet magnetic field, width of the current layer and mass density, respectively. Notice that we further will use just κ when we have in mind κ_e. It is evident that effective resistivity as well as the location where it becomes available depend on the chosen equilibrium model and the applied normalisation.

4. Numerical Calculations

We have calculated the evolution of our hybrid–model magnetotail starting with two different initial equilibria. One of them corresponds to a magnetically quiet magnetosphere (addressed as case 1), the other to a tail configuration typical for the late growth phase. In both cases we are consistently taking into account the microscopically determined non–adiabatic response of neutral sheet particles.

4.1. THE TAIL DURING QUIET TIMES

The quiet time configuration is described by

$$B_o = 2, \quad L_o = 2, \quad n_o = 0.25, \quad \vartheta = 0.25 \qquad (13a)$$

which yields

$$\kappa_o = 49, \quad \eta_o = 3.6 \times 10^{-2}, \quad \hat{\tau} = 14.6s \qquad (13b)$$

The corresponding pressure function is given by

$$p_{dt} = 0.1 \;\; x_c = 20 \;\; and \;\; q = 2 \qquad (13c)$$

The resulting quiet time pressure function, the b_z magnetic field component and the nonadiabaticity parameter κ are shown in Fig 1. as a function of $|X_{GSM}|$, given in R_E if one adds 20 to the numbers, indicated.

4.2. THE PRESTORM LOADED TAIL (LATE GROWTH PHASE)

We choose the following set of parameters in order to describe the stretched tail magnetic field and thin current sheet during the late growth phase:

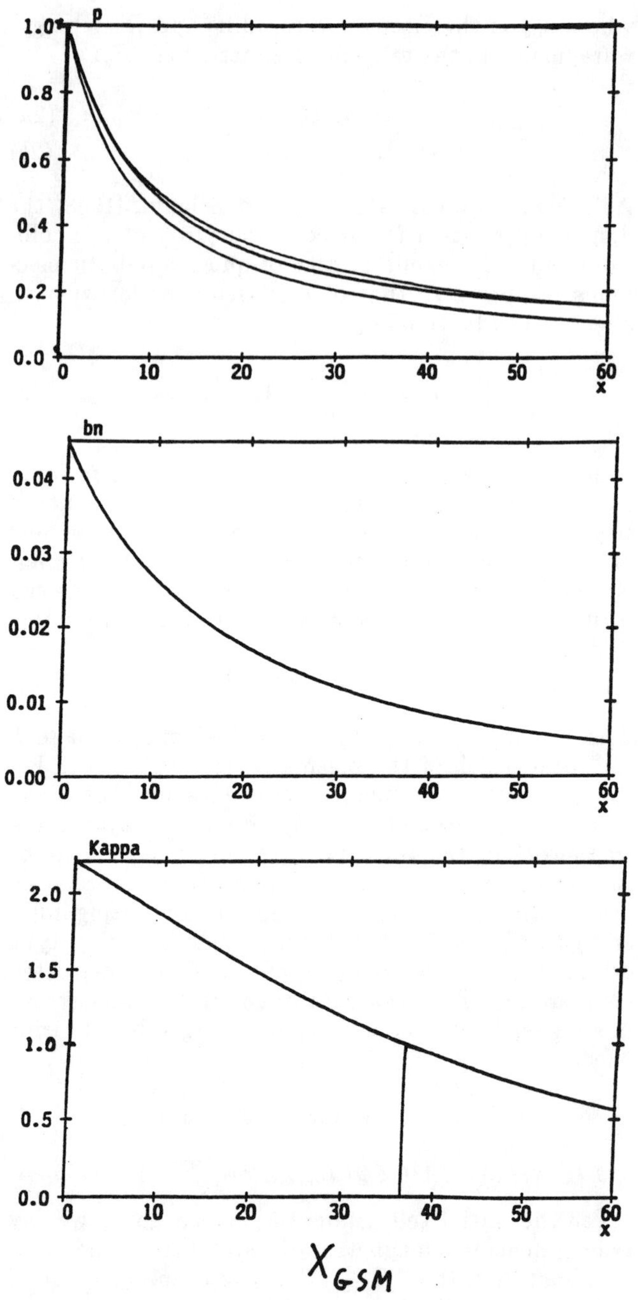

Fig. 1. The quiet time pressure function p, its upper and lower bound as well as the Slavin et al. [1985] fit in the middle, b_n magnetic field and κ parameter as a function of X_{GSM}.

$$B_o = 3, \quad L_o = 1.2, \quad n_o = 0.3, \quad \vartheta = 0.25 \qquad (14a)$$

$$\kappa_o = 40, \quad \eta_o = 5.5 \times 10^{-2}, \quad \hat{\tau} = 6.4s \qquad (14b)$$

and

$$p_{dt} = 0.1 \;\; x_c = 35 \;\; and \;\; q = 2 \qquad (14c)$$

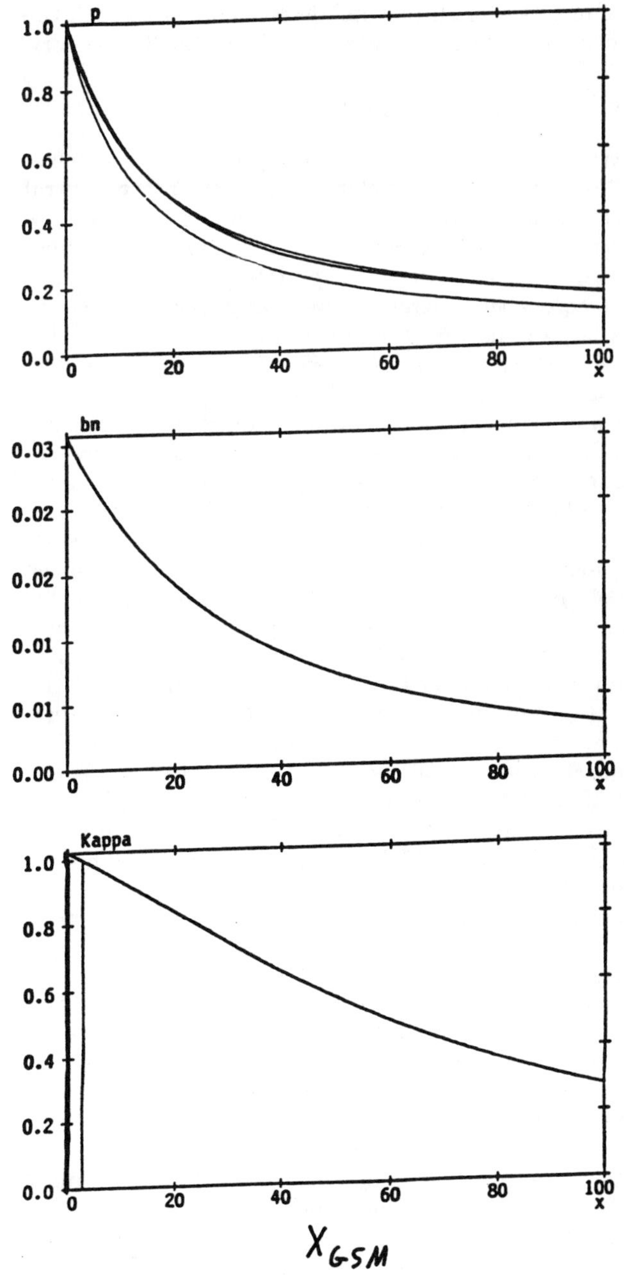

Fig. 2. Same as Fig. 1. for the loaded tail.

The results for $\bar{p}$, b_z, and κ are shown in Fig 2. The parameters (14a - 14c) have been adjusted in agreement with observational bounds on an average lobe field strenth which is indicated by the thin lines in the plots of the pressure function. It is important to note that these bounds do not give much freedom for the choice of the parameters in the pressure function.

Figs. 1. and 3. cover the range between $X_{GSM} = -20$ and -80 R_E, indicated as 0 and 60, i.e. one has to had 20 R_E to $|X_{GSM}$, Figs. 2. and 4. cover the range between $X_{GSM} = -20$ and -120 R_E, respectively. The thin vertical lines mark the location where κ becomes unity. From the comparison of the figures it is obvious that a loaded tail configuration leads to a much more earthward nonadiabatic electron motion. The effective resistivity at the corresponding locations is $\eta = 3.4 \times 10^{-4}$ for case 1 (magnetically quiet magnetospheric conditions) and $\eta = 1.3 \times 10^{-3}$ for the second case (loaded tail).

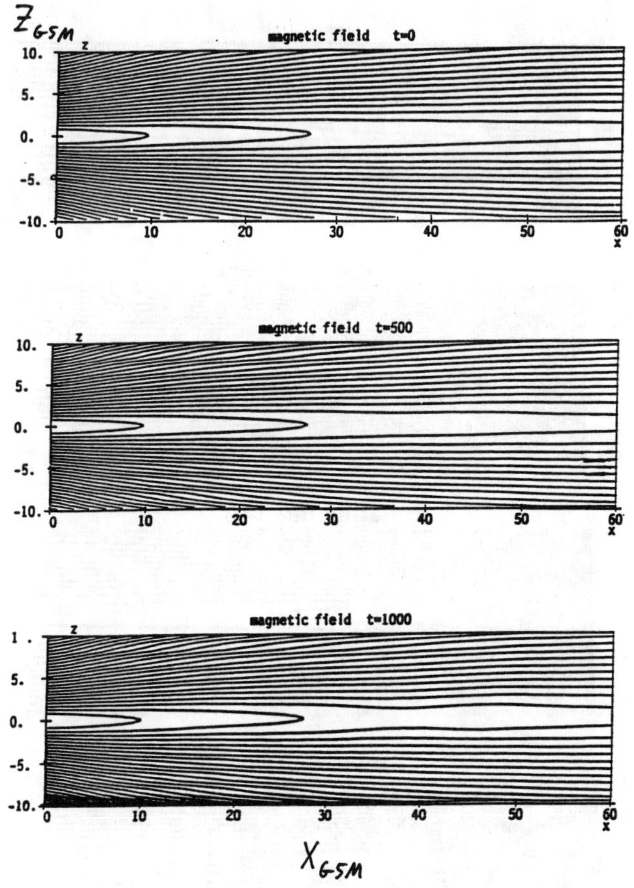

Fig. 3. Magnetic field lines, illustrating the dynamical tail evolution, for an initially quiet magnetosphere.

In order to compare our results with estimates from the resistive MHD instability theory we determine the local magnetic Reynolds number. The local width of the current layer and the local lobe magnetic field are for case 1: $l_l = 1.5$ and $b_l = 0.66$ and for the second case $l_l = 1.07$ and $b_l = 0.93$, respectively. Thus we arrive at a local magnetic Reynolds numbers of about $R_l = 6500$ for the quiet configuration and $R_l = 840$ in the late growth phase. Estimating for the initial phase of the dynamical tail evolution a growth time of a comparable resistive instability we obtain 2500 s for the quiet tail and 200 s for the loaded configuration.

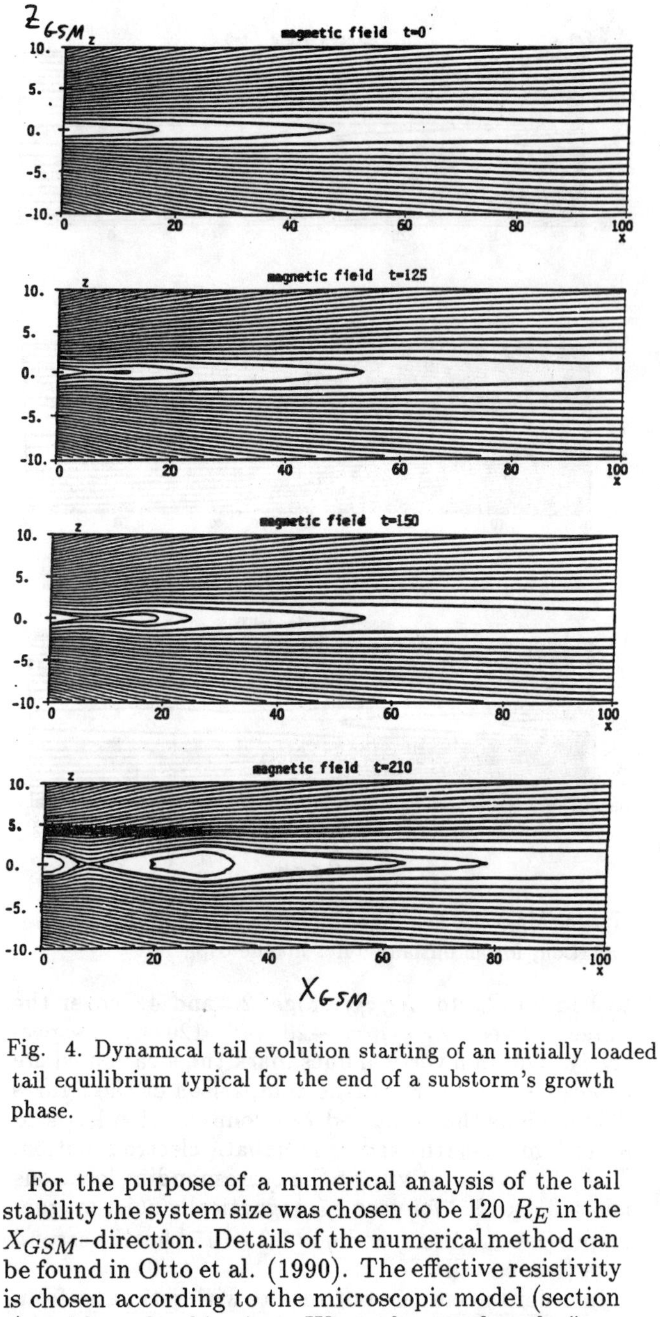

Fig. 4. Dynamical tail evolution starting of an initially loaded tail equilibrium typical for the end of a substorm's growth phase.

For the purpose of a numerical analysis of the tail stability the system size was chosen to be 120 R_E in the X_{GSM}–direction. Details of the numerical method can be found in Otto et al. (1990). The effective resistivity is chosen according to the microscopic model (section 2) and kept fixed in time. We made sure that the "numerical resistivity" resulting from the finite grid size, has been neglible small at any time step of the simulation. The results of these runs are shown in Figs. 3 and 4 for the quiet time and loaded configurations, respectively. Obviously the dynamical evolution is much faster for the loaded initial tail equilibrium. Reconnection onset for this run occurs at $116\hat{\tau}$ in contrast to $681\hat{\tau}$ in the case of the quiet tail. Considering the reconnection onset time as a rough measure for a physical growth rate we obtain a ratio of 13 for the growth rate in physical units which agrees quite well with the above mentioned estimate from linear resistive theory.

5. Summary And Discussion

We have investigated the large scale consequences of the quasi–adiabatic electron motion in the neutral sheet of the Earth's magnetotail for the tail stability. For this purpose we developed a hybrid model, where the knowledge of kinetic properties of the electron motion across the current sheet was applied to modify the large scale tail dynamics. In this paper we have presented the results of a two particular runs of this model, starting with two different initial tail equilibria, the first typical for a magnetically quiet conditions and another one for a prestorm–loaded tail. In our hybrid model the properties of charged particle motion near the tail neutral sheet are obtained consistently for any specific location along the tail. Considering the resulting modification of the macroscopic plasma response we obtain the feedback of the microscopic particle dynamics to the large scale tail dynamics. This allows us to analyse the stability of any initial tail equilibrium in dependence on a possible chaotization of the electron motion. As a result we have found that during magnetically quiet conditions (case 1) electrons become chaotic typically beyond $X_{GSM} = 70R_E$. The resulting growth rate of instability, however, tends to be very small. Hence chaotic electrons seem not to cause tearing mode instability of a quiet time tail. This result is in good accordance with the fact of a more or less stable existence of a quiet time Earth's magnetotail but allows some quasi–stationary reconnection of excess magnetic flux in the distant tail.

In a second run we calculated the large scale stability properties of a tail equilibrium with an initially more stretched magnetic field and thinner current sheet, as they were observed in the mid–tail prior substorm onset [*Baumjohann et al.*, 1992]. In this case we have found a quick tearing mode instability, growing on a time scale of hundreds of seconds, i.e. on the the typical substorm expansion time scale.

Our hybrid model has, therefore, revealed a consistent picture of both the quiet time quasi–stability and of a substorm related onset of a tail instability a typical growth phase current sheet thinning and magnetic field stretching. At the present moment our model does not extend to the further nonlinear evolution of the tail. It, nevertheless, provides an estimate of the instability growth time scale, which is of the order of the typical substorm expansion time. We want to emphasize that our model is not thought to verify the existing microscopic models but to investigate their consequences for the tail stability and the conditions of becominmg tearing mode unstable. Notice that one always finds locations in the tail, where thermal electrons are chaotic. Nevertheless only for a growth phase-like loaded tail the growth rate of the resulting tail instability becomes comparable with the time scale of substorm expansions. Indeed, in the quiet time tail equilibrium (case 1) the particle chaotization takes place in the distant tail (70 – 100 R_E) and the estimated growth time is about a few thousands of seconds. Under realistic conditions this is far too large for the development of a normal mode. In contrast to the late growth phase, loaded, tail equilibrium (case 2) electrons become chaotic relatively close to the Earth, perhaps even closer than 20 R_E (the edge of the scope of the applied equilibrium theory). We have found, however that even for a chatization slightly beyond 20 R_E the growth rate of the magnetotail instability becomes a few hundreds of seconds or less, i.e. well of the order of the substorm expansion time scale.

Our hybrid model provides an explanation why the magnetotail may stay stable for relatively long periods of time even in the presence of chaotic electrons in the distant tail, while in becomes unstable and a neutral line forms in the mid-tail after a growth phase related inward motion of the region of primary electron chaotization. The extension of this hybrid model to the nonlinear stage of the instability growth will be presented in a separate publication.

Acknowledgments. We gratefully thank Profs. K. H. Schindler and L. M. Zelenyi for discussions. This research was partially supported by the German Space Agency DARA, grant 50 QN 9102 3 and Deutsche Forschungsgemeinschaft DFG, grant BU 777/2-1.

REFERENCES

Baumjohann, W., Average plasma properties in the central plasma sheet, *J. Geophys. Res., 94,* 6597, 1989.

Baumjohann W., G. Paschmann, and T. Nagai, Thinning and expansion of the Substorm Plasma Sheet, *J. Geophys. Res., 97,* 1992, submitted.

Birn, J., Computer studies of the dynamic evolution of the geomagnetic tail, *J. Geophys. Res., 85,* 1214, 1980.

Birn, J., R. Sommer, and K. Schindler, Open and closed magnetospheric tail configurations and their stability, *Astrophys. Space Sci., 35,* 389, 1975.

Büchner, J., and L.M. Zelenyi, Deterministic chaos in the dynamics of charged particles near a magnetic field reversal, *Phys. Lett. A., 118,* 395, 1986.

Büchner J., and L. M. Zelenyi, Chaotization of the electron motion as the cause of an internal magnetotail instability and substorm onset, *J. Geophys. Res., 92,* 13,456, 1987.

Büchner J., and L. M. Zelenyi, Reconnection instability in collisionless plasma, *ESA SP-285, Vol. II,* 21, 1988.

Büchner, J., and L.M. Zelenyi, Regular and chaotic charged particle motion in a magnetotail-like field reversal, 1. Basic theory of trapped motion, *J. Geophys. Res., 94,* 11,821, 1989.

Büchner J., C. V. Meister, and B. Nikutowski, Reconnection and microturbulence, *ESA SP-285, Vol. II,* 29, 1988.

Chen J., and P. Palmadesso, Chaos and nonlinear dynamics of single-particle orbits in a magnetotaillike magnetic field, *J. Geophys. Res., 91,* 1986.

Coppi, B., G. Laval, and R. Pellat, A model for the influence of the Earth magnetic tail on geomagnetic phenomena, *Phys. Rev. Lett., 16,* 1207, 1966.

Coroniti F. V., On the tearing-mode in quasi-neutral sheets, *J. Geophys. Res., 85,* 6719, 1980.

Coroniti F. V., Explosive tail reconnection: The growth phases of magnetospheric substorms, *J. Geophys. Res., 90,* 7427, 1985.

Eseray E., and K. Molvig, A turbulent mechanism for substorm onset in the Earth's magnetotail, *Geophys. Res. Lett., 14,* 367, 1987.

Galeev A., A., and L. M. Zelenyi, Tearing instability in plasma configurations, *Sov. Phys. JETP, Engl. Transl., 42,* 1113, 1976.

Goldstein H., and K. Schindler, Large scale collision free instability of two-dimensional plasma sheets, *Phys. rev. Lett., 48,* 1468, 1982.

Hones, E. W. Jr., Transient phenomena in the magnetotail and their relation to substorms, *Space Sci. Rev., 23,* 393, 1979.

Hones, E. W. Jr., et al. Structure of the magnetotail at 220 R_E and its response to geomagnetic activity, *Geophys. Res. Lett., 11,* 5, 1984.

Horton, W., and T. Tajima, Collisionless conductivity and stochastic heating of the plasma sheet in the Geomagnetic tail, *J. Geophys. Res. 96* , 1991.

Hoshino H., Forced magnetic reconnection in a plasma sheet with localized resistivity profile excited by lower hybrid drift instability, *J. Geophys. Res., 96,* 11,555, 1991.

Huba J. D., Anomalous transport in current sheets, in *Proc. IAU Colloquium on unstable current systems and plasma instabilities in astrophysics,* 315, 1985.

Kuznetsova M., L. M. Zelenyi, The universality of the ion tearing mode, *Geophys. Res. Lett., 18,* 1991.

Lee L, C., Z. F. Fu, and S.-I. Akasofu, A simulation study of forced reconnection processes and magnetospheric storms and substorms, *J. Geophys. Res., 90,* 10,896, 1985.

Lembége and Pellat, Stability of a thick two-dimensional quasineutral sheet, *Phys. Fluids, 85,* 6719, 1982.

Lyons, L. R. and T. W. Speiser, Ohm's law for a current sheet *J. Geophys. Res., 90,* 8543, 1985.

Nishida A., L. Lyons, Assesment of the boundary layer model of the magnetospheric substorm, *J. Geophys. Res., 93,* 5579, 1988.

Otto, A., K. Schindler, and J. Birn, Quantitative study of the nonlinear formation of plasmoids in the earth's magnetotail, *J. Geophys. Res., 95,* 15,023, 1990.

Pellat R., F. V. Coroniti, P. Pritchett, Does the ion tearing exist?, *Geophys. Res. Lett., 18,* 1991.

Sato T., Strong plasma acceleration by slow shocks resulting from reconnection, *J. Geophys. Res., 79,* 7177, 1979.

Sato T., and T. Hayashi, Externally driven magnetic reconnection as a powerful magnetic energy converter, *Phys. Fluids, 22,* 1189, 1979.

Schindler, K., in: *Earth's Magnetospheric Processes, ed. by B.M. McCormac, D. Reidel,* 200, 1972

Schindler, K., A theory of the substorm mechanism, *J. Geophys. Res., 79,* 2803, 1974.

Scholer M., G. Gloeckler, B. Klecker, F.M. Ipavich, D. Hovestadt, and E. J. Smith, Fast moving plasma structures in the distant magnetotail, *J. Geophys. Res., 89,* 6717, 1984.

Scholer M., and R. Hautz, On acceleration of plasmoids in magnetohydrodynamic simulations of magnetotail reconnection, *J. Geophys. Res., 96,* 3581, 1991.

Scholer M., A. Otto, and G. J. Gadbois, Three–dimensional numerical simulations of magnetotail reconnection, in *Magnetospheric Substorms, Geophysical Monographs 64,* 171, 1991.

Slavin, J. A., E. J. Smith, D. G. Sibeck, D. N. Baker, R. D. Zwickel, and S.-I. Akasofu, *J. Geophys. Res., 90* , 10875, 1985.

Speiser, T., Conductivity without collisions or noise, *J. Geophys. Res., 75,* 1970.

Ugai, M, and T. Tsuda, Magnetic field line reconnection by localized enhancements of resistivity, 1: Evolution in a compressible MHD fluid, *J. Plasma Phys., 17,* 337, 1977.

Jörg Büchner, Max-Planck-Institut für Extraterrestrische Physik, Außenstelle Berlin, Rudower Chaussee 5, D-O-1199, Germany FR

Antonius Otto, Geophysical Institute and Department of Physics, University of Alaska, Fairbanks, AK, 99775-0800, U.S.A.

Global Consequences of Nonlinear Particle Dynamics in the Magnetotail

JAMES CHEN AND DANIEL L. HOLLAND [1]

Beam Physics Branch, Plasma Physics Division, Naval Research Laboratory, Washington, D. C.

Recent research efforts from the nonlinear dynamics point of view have yielded new insights into the relationships between the particle motion, global magnetic field geometry, and local plasma properties. Some novel theoretical predictions regarding local distribution functions and equilibrium structure of the current sheet have been made based on the dynamics of charged particles, yielding good agreement with observed data. A new framework of understanding kinetic physics of the magnetosphere is emerging. This paper gives a tutorial review of the nonlinear dynamics of charged particles in the magnetotail.

1. INTRODUCTION

The motion of charged particles is an essential ingredient for fully understanding plasma properties in a collisionless environment. In a system where the charged particle Larmor radii are much smaller than any spatial length scales, the particle motion can be well approximated by adiabatic theory. Adiabatic invariance may also be applicable if the time variation of the system is sufficiently slow relative to the gyroperiod. If the Larmor radii are much greater than any spatial scale lengths, one may be able to assume unmagnetized particle motion. However, if the Larmor radii and spatial scale lengths are comparable, the particle motion is nonadiabatic and is generally difficult to analyze. If, in addition, particle-particle or wave-particle interactions exist, collectively referred to as "collisions" in this paper, then the complexity of the particle motion is further increased.

The magnetosphere is made up of many distinct regions but exhibits globally coherent behavior governed by large-scale fields and collisionless charged particles. Figure 1 is a schematic drawing of the Earth's magnetosphere with various regions indicated. The coordinate system is shown. A region of particular interest is the nightside magnetotail which plays a central role in the storage and release of magnetic energy, thus controlling the dynamics of the entire magnetosphere. The magnetotail consists of the lobe and the central plasma sheet (CPS) which contains a thin current sheet at the midplane, $z = 0$, where the x-component of the magnetic field reverses sign. The internal structure of the magnetosphere is determined by the solar wind via large-scale momentum balance and energy and particle input. In this collisionless system, energy transport and coupling between different regions are critically determined by the charged-particle motion in the large-scale fields.

The current sheet represents magnetic energy, and its dynamics have been a long-standing topic of research. Numerous studies of the motion of particles in the magnetotail have been carried out [*Speiser*, 1965; *Alekseyev and Kropotkin*, 1970; *Sonnerup*, 1971; *West et al.*, 1978a, b; *Wagner et al.*, 1979; *Gray and Lee*, 1982; *Tsyganenko*, 1982; *Lyons and Speiser*, 1982; *Birmingham*, 1984; *Lyons*, 1984; *Chen and Palmadesso*, 1986; *Martin*, 1986; *Büchner and Zelenyi*, 1986, 1989; *Karimabadi et al.*, 1990; *Burkhart and Chen*, 1991]. The box on the right in Figure 1 shows the basic configuration of magnetic field models used in many theoretical studies. In the early works, individual orbits were studied [*Speiser*, 1965; *West et al.*, 1978a, b; *Wagner et al.*, 1979] using adiabatic invariance and deviations therefrom [*Sonnerup*, 1971; *Birmingham*, 1984]. More recently, *Büchner and Zelenyi* [1989] have extended the treatment of the action integral $I \equiv \oint \dot{z} dz$ [*Sonnerup*, 1971] to calculate the variation ΔI due to the nonadiabatic motion using an asymptotic matching approximation (the so-called slow separatrix crossing theory [*Cary et al.* [1986]).

A different approach is to consider families of orbits and phase space topology of the system rather than individual orbits. *Chen and Palmadesso* [1986] first showed that there exist only two global constants of the motion in involution in the magnetotail topology and that the particle motion is nonintegrable. Using the Poincaré surface of section technique, they found that the phase space is partitioned into disjoint regions occupied by dynamically distinct classes of orbits: unbounded stochastic (chaotic), unbounded transient and bounded integrable (regular) orbits. Based on the partitioning of the phase space, they proposed a process called "differential memory" (section 5) which can significantly influence the time-evolution of plasma

Space Plasmas: Coupling Between Small
and Medium Scale Processes
Geophysical Monograph 86

distributions. This process arises from the property that different phase space regions are occupied by orbits with widely separated orbital time scales so that the respective phase space regions can retain memory of existing population of particles for different lengths of time. The time scales referred to here are the accessibility time, the time for orbits far away from the midplane to reach (i.e., "access") various regions of the phase space. Differential memory has been used to model plasma distribution functions in the magnetotail [*Chen et al.*, 1990a, b; *Burkhart and Chen*, 1991; *Ashour-Abdalla et al.*, 1991], yielding results in agreement with observations [*Chen et al.*, 1990a]. The underlying partitioning of the phase space proved to be robust in the presence of collisional effects [*Holland and Chen*, 1991]. The same partitioning and differential memory also determine the current sheet equilibria [*Holland and Chen*, 1993]. Thus the nonlinear particle dynamics controls both local and global properties of the central plasma sheet (CPS).

Büchner and Zelenyi [1986] proposed a mapping to model the chaos in this system. The mapping is based on discrete jumps in the action, which were evaluated later using the separatrix crossing technique [*Büchner and Zelenyi* [1989]. In this approach, one divides a given trajectory into a sequence of segments and maps the action of one segment to that of the next using the variations in the action. Full solutions have not been generated for the modified Harris or parabolic field models using this mapping theory to date. A fully solved mapping has been applied to a model field in which different segments of a trajectory are calculated analytically and then matched [*Chen et al.*, 1990b]. The field is idealized but has the quasi-neutral sheet topology, and the basic dynamical properties are common to the more realistic models.

The magnetotail current sheet is a chaotic scattering system in which the outgoing conditions exhibit sensitive dependence on the incoming conditions [*Chen et al.*, 1990c; *Burkhart and Chen*, 1991]. The basic nonlinear dynamical properties of the system are now well established, and the reader is referred to recent reviews for more detail [*Chen*, 1992, 1993]. In another configuration of interest, *Martin* [1986] studied the nonlinear dynamics of charged particles in an X-type neutral line, using the maximal Lyapunov exponent to quantify the degree of chaos. Possible distribution function signatures [*Curran et al.*, 1987; *Curran and Goertz*, 1989; *Martin and Speiser*, 1988] and the kinetic structure of X-type reconnection regions [*Burkhart et al.*, 1990, 1991; *Moses et al.*, 1993] have been studied.

Much work has been done to explore the physical implications of nonlinear particle dynamics [e.g., *Curran et al.*, 1987; *Martin and Speiser*, 1988; *Curran and Goertz*, 1989; *Chen et al.*, 1990a, b; *Karimabadi et al.*, 1990; *Horton and Tajima*, 1990, 1991; *Burkhart and Chen*, 1991; *Ashour-Abdalla et al.*, 1990, 1991]. Some novel predictions based directly on the topological, i.e., geometrical, properties of the phase space have proved to be in good agreement with observed data [*Chen et al.*, 1990a; *Burkhart and Chen*, 1991]. In light of such successes, one can justifiably state that the study of particle motion and plasma properties in the magnetotail within the new framework of nonlinear dynamics has emerged from a *descriptive* stage and entered a *predictive* stage. This in turn has provided an impetus for further development of this subject. For example, attempts are being made to include time dependence in the understanding of particle dynamics [*Chapman and Watkins*, 1993; *Chapman*, 1994]. The present paper is intended to give an introductory overview of recent findings relevant to the Earth's magnetotail. It is based on the full-length review paper of *Chen* [1992] with additional information on the dynamics with $B_y \neq 0$ and chaotic conductivity. We will illustrate how the global field topology determines the locally measurable distribution functions, and how the current sheet is critically influenced by conditions in other regions such as the ionosphere. Because the charged particles are collisionless in the magnetosphere, processes occurring on micro- to macro-scales are coupled through the particle motion governed by the large-scale fields.

2. THE DYNAMICAL SYSTEM

A simple field model for the magnetotail is

$$\mathbf{B}(z) = B_0 f(z)\hat{\mathbf{x}} + B_y\hat{\mathbf{y}} + B_n\hat{\mathbf{z}}, \tag{1}$$

where $f(z) \rightarrow \pm 1$ as $z \rightarrow \pm\infty$ so that $B_0 = B_x(z \rightarrow \infty)$, B_n is the uniform field normal to the midplane ($z = 0$), B_y is a uniform field across the tail, and $B_0 f(z)$ is a field-reversed neutral sheet profile. (See Figure 1 for the coordinate system.) The function $f(z)$ is often chosen to be antisymmetric about $z = 0$ with $f(z = 0) = 0$. For the earth's magnetotail we typically consider $b_n \equiv B_n/B_0$ of the order of $0.05 \sim 0.1$ for the distant magnetotail, but b_n can be larger near the Earth. The B_y component is determined by the solar wind conditions, and $b_y \equiv B_y/B_z$ can have a range of values. In this article, we will use the gauge such that the vector potential is $\mathbf{A} = A_x(z)\hat{\mathbf{x}} + A_y(x, z)\hat{\mathbf{y}}$. Then,

$$A_x(z) = B_y z, \tag{2a}$$

$$A_y(x, z) = -B_0 F(z) + B_n x, \tag{2b}$$

where $F(z)$ is defined by $dF(z)/dz \equiv f(z)$, with the total asymptotic magnetic field $B = (1+{b_y}^2 b_n^2 + b_n^2)^{1/2} B_0$. The conclusions in this paper, however, are independent of gauge.

Two commonly used forms of $f(z)$ are the Harris model $f(z) = \tanh(z/\delta)$ and the parabolic model $f(z) = z/\delta$ where δ is the characteristic half-thickness of the neutral sheet. These two models give, respectively,

$$F(z) = \delta \ln[\cosh(z/\delta)], \quad F(z) = (1/2)z^2/\delta.$$

In the simplest case, $B_y = 0$ is used. In the vicinity of the midplane ($z/\delta \lesssim 1$), we have $\tanh(z/\delta) \sim z/\delta$, giving

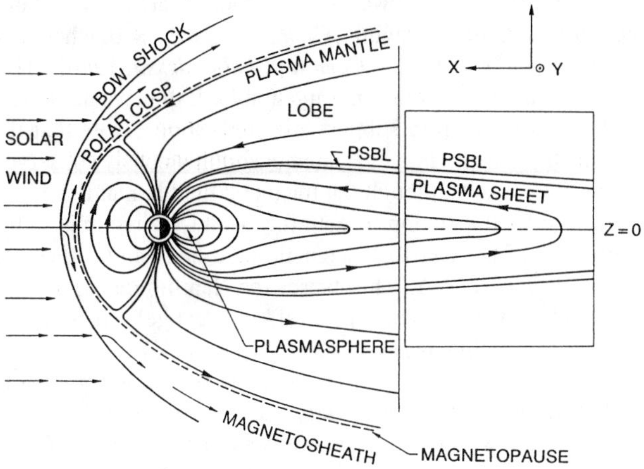

Fig. 1. Schematic drawing of the earth's magnetosphere. Magnetic field lines and various regions are depicted. The coordinate system is shown. (From *Chen* [1992].)

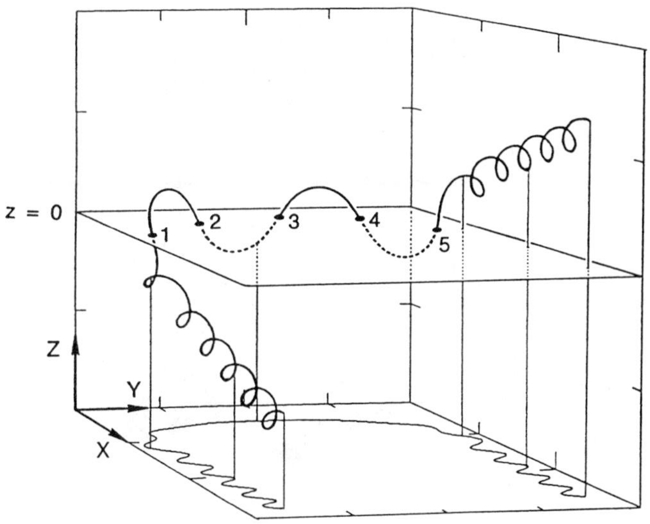

Fig. 2. A Speiser orbit. The crossing points at $Z = 0$ are indicated. $\hat{H} = 500$ and $b_n = 0.1$. The Z axis is expanded by 1.2. (Adapted from *Chen and Palmadesso* [1986].)

$\ln[\cosh(z/\delta)] \simeq (1/2)(z/\delta)^2$ so that the Harris and parabolic field models are similar for $z/\delta \lesssim 1$. If $z/\delta \gtrsim 1$, on the other hand, the two model fields diverge significantly. In particular, for the Harris model, if $|z|$ is greater than some $z_a \gg \delta$, we have $B_x \simeq B_0$, and the magnetic field is essentially uniform. We will refer to $|z| > z_a$ as the asymptotic regions. For the parabolic field model the magnitude of $f(z)$ continues to increase with increasing $|z|$, the current density is everywhere uniform, and there is no well defined asymptotic region. The essense of the asymptotic regions is the physical property that the current sheet is surrounded by regions with significantly less current, where the particle motion is qualitatively different from that inside the current sheet. Dynamically, this is manifested in the chaotic scattering nature of the system. From a modelling point of view, the dynamics in the asymptotic (or external) regions can be considered separately. Thus, the parabolic field should be "truncated" for application to any realistic system, for example, by setting $f(z) = 1$ for $|z| \gg \delta$. The system then becomes topologically equivalent to the modified Harris model. Thus, *Büchner and Zelenyi* [1990] heuristically incorporated separatrix crossing into a truncated parabolic field model. In both field models, the radius of curvature $R_c = B_n/(\partial B_x/\partial z)$ at $z = 0$ is $R_c = b_n\delta$. *West et al.* [1978b] found that the pitch-angle distributions calculated in a B_x field with "roll-off" provided by $\tanh(z/\delta)$ fit the observed data better than the models with power law dependence on z such as the parabolic field. Dynamically, these two configurations are qualitatively different. We will refer to these two models as the "modified Harris" and the "parabolic field" models, respectively. Both will be collectively referred to as quasi-neutral sheets as opposed to pure neutral sheet for which $B_n = 0$.

The single-particle motion can be described by the equation of motion

$$m\frac{d\mathbf{v}}{dt} = \frac{q}{c}\mathbf{v} \times \mathbf{B}. \tag{3}$$

In general, a nonzero E_y, albeit small during quiet times, exists. If we take $B_y = 0$ and $B_n =$ constant, a uniform E_y can be simply transformed away by an $\mathbf{E}_y \times \mathbf{B}_n$ translation. A nonuniform $\mathbf{E}$ field is more complicated and will not be considered.

There are two important time scales for the quasi-neutral sheet. One is $\Omega_0 \equiv eB_0/mc$ and the other is $\Omega_n \equiv eB_n/mc$. The corresponding Larmor radii are $\rho_0 \equiv v/\Omega_0$ and $\rho_n \equiv v/\Omega_n$ where v is the velocity.

It has been known that the particle motion in the magnetotail is nonadiabatic. *Speiser* [1965] first identified a class of orbits in the modified Harris configuration with $B_y = 0$, which are sometimes called Speiser orbits. Figure 2 shows such an orbit. This particle executes adiabatic motion far away from the midplane (noncrossing motion), enters the current sheet where it oscillates in the z direction, traverses the midplane while crossing the midplane at points 1, 2, 3, 4, 5, and is then ejected from the current sheet. The number of midplane crossings depends on the energy and is found to be roughly proportional to $\hat{H}^{1/4}$ [*Chen and Palmadesso*, 1986], where $\hat{H}^{1/4}$ is the normalized Hamiltonian (see equation (7) below). While traversing the midplane, the particle motion is a combination of "meandering" (midplane crossing) motion [*Sonnerup*, 1971] and gyration about the weak B_n field. For some time, the basic properties of the particle motion were understood in terms of the Speiser orbits and within the framework of adiabatic invariance and deviations therefrom [e.g., *Speiser*, 1965; *Sonnerup*, 1971; *Eastwood*, 1972; *Cowley*, 1978; *Lyons and Speiser*, 1982, 1985; *Speiser and Lyons*, 1984]. However, *West et al.* [1978a, b] found orbits whose

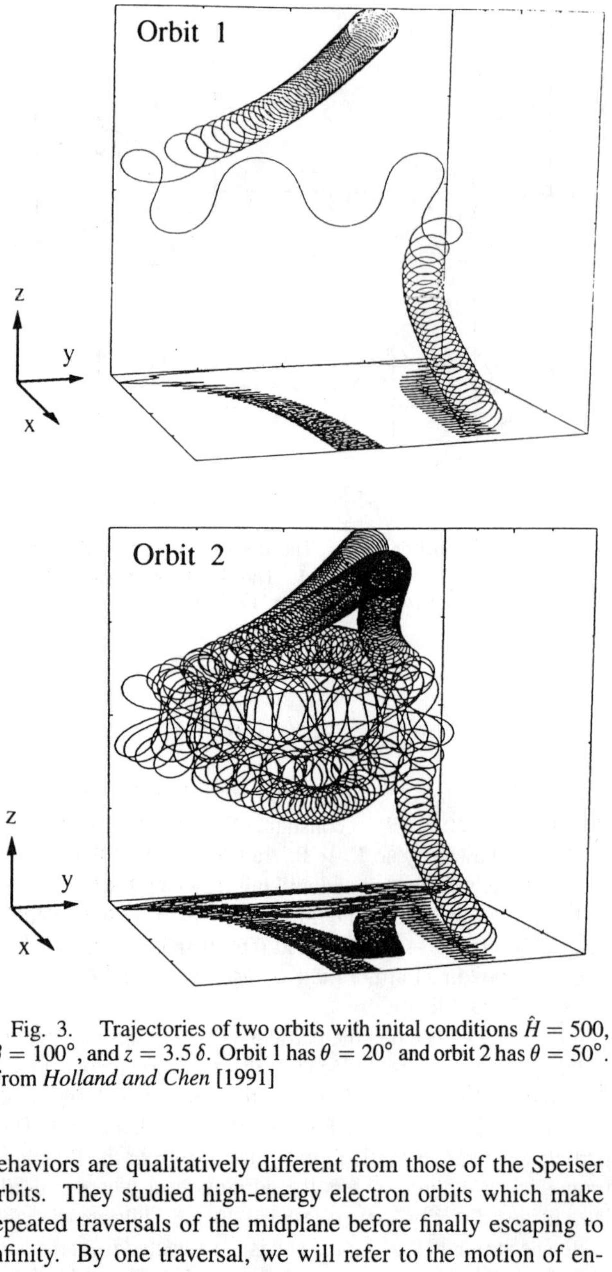

Fig. 3. Trajectories of two orbits with inital conditions $\hat{H} = 500$, $\beta = 100°$, and $z = 3.5\,\delta$. Orbit 1 has $\theta = 20°$ and orbit 2 has $\theta = 50°$. From *Holland and Chen* [1991]

behaviors are qualitatively different from those of the Speiser orbits. They studied high-energy electron orbits which make repeated traversals of the midplane before finally escaping to infinity. By one traversal, we will refer to the motion of entering and leaving the midplane once. One traversal generally has multiple midplane crossings. Figure 2 shows one traversal with five crossings. *Wagner et al.* [1979] also found similar ion orbits and noted that considerable variation in the motion of certain orbits occurs even with nearly identical initial conditions. This property is illustrated in Figure 3 which shows two orbits whose initial conditions differ only in the phase angle.

3. NONLINEAR DYNAMICS OF CHARGED PARTICLES

Equation (3) possesses three exact constants of the motion: $H = mv^2/2$, $P_y = mv_y + (q/c)A_y(x, z)$, and $C_x \equiv m(v_x - \Omega_n y + \Omega_n b_y z)$. However, it was found that these constants are not in involution, having $[C_x, P_y] = -m\Omega_n \neq 0$, where use was made of $[y, P_y] = 1$ [*Chen and Palmadesso*, 1986]. This suggests that the system is not integrable. Extensive numerical evidence was given to support this suggestion. Note that the canonical momenta P_x and P_y are not simultaneously conserved in this system even though the magnetic field is manifestly invariant under translation in both the x- and y-directions. This is because of the fact that magnetic field is given by a vector potential, $\mathbf{B} = \nabla \times \mathbf{A}$. This causes the Hamiltonian to depend on x and z (or on y and z in a different gauge). If an N dimensional system has N constants $\{C_1, \ldots, C_N\}$ in involution, i.e., $[C_i, C_j] = 0$ for all $i, j \leq N$, then the system is completely integrable, and there exists a canonical transformation to a coordinate system with N cyclic coordinates such that the first integrals C_i are the new canonical momenta. For $B_y \neq 0$, the above condition, $[C_x, P_y] = -m\Omega_n$, remains unchanged. Note that particle motion in the magnetic field $\mathbf{B}(z) = B_0 f(z)\hat{\mathbf{x}} + B_y(z)\hat{\mathbf{y}}$ is completely integrable because $C_x = P_x$, where P_x is the x canonical momentum, and $[P_x, P_y] = 0$.

It is convenient to rewrite equation (3) in a dimensionless component form [*Chen and Palmadesso*, 1986]:

$$\frac{d^2X}{d\tau^2} = \frac{dY}{d\tau} - b_y \frac{dZ}{d\tau}, \tag{4}$$

$$\frac{d^2Y}{d\tau^2} = \frac{d}{d\tau}\left\{b_n^{-2}\ln\left[\cosh\left(b_n Z\right)\right] - X\right\}, \tag{5}$$

$$\frac{d^2Z}{d\tau^2} = b_y \frac{dX}{d\tau} - b_n^{-1}\tanh\left(b_n Z\right)\frac{dY}{d\tau}, \tag{6}$$

where the normalization is $X \equiv \left(x - P_y/m\Omega_n\right)/b_n\delta$, $Y \equiv \left(y + C_x/m\Omega_n\right) b_n\delta$, $Z \equiv a/b_n\delta$, $\tau \equiv \Omega_n t$ and $\dot{X} \equiv dX/d\tau$. Here, we have assumed the modified Harris model of the magnetic field, $E_y = 0$, and $b_n \equiv B_n/B_0 \neq 0$. The normalized Hamiltonian is

$$\hat{H} \equiv H/\left(m b_n^2 \Omega_n^2 \delta^2\right). \tag{7}$$

With $B_y = 0$, equations (4)–(7) show that the modified Harris model requires two parameters to specify the dynamics: b_n and $\hat{H}$. In the parabolic field model, $b_n \neq 0$ can be normalized away and only $\hat{H}$ is needed. If $B_y \neq 0$, one additional parameter b_y need be specified in each case. The integrable limit $b_n \to 0$ cannot be taken directly in either field model. The limit $\hat{H} \to \infty$ corresponds to two distinct physical limits: $b_n \to 0$ or $H \to \infty$ where H is the unnormalized Hamiltonian. The former leads to the integrable neutral sheet while the latter is not integrable.

For $B_y = 0$, equations (4)–(6) are invariant under $Z \leftrightarrow -Z$ and under the simultaneous replacement $Y \to -Y$ and $\tau \to -\tau$. The latter symmetry operation implies that Y reversal (not invariant) and τ reversal are equivalent. With $B_y \neq 0$, the simultaneous replacement of $Z \leftrightarrow -Z$ and $b_y \leftrightarrow -b_y$ is an invariant symmetry operation as is the simultaneous replacement of

$Y \leftrightarrow -Y$, $\tau \leftrightarrow -\tau$, and $b_y \leftrightarrow -b_y$. The symmetry properties of the system are important for determining dynamical properties of the entire system.

The previous studies of particle dynamics in the magnetotail have been largely limited to the $B_y = 0$ case, and we will primarily discuss this case. However, we will briefly describe the generic phase space properties with $B_y \neq 0$ using the surface of section technique later in this section.

The fundamental dynamics of electrons is identical to that of ions. Aside from the trivial difference in the direction of motion, the main difference is the applicable regimes of $\hat{H}$. Using the normalization (7), a given value of H in physical units gives

$$\hat{H}_e = (m_e/m_i)\,\hat{H}_i, \qquad (8)$$

where $\hat{H}_e$ and $\hat{H}_i$ are H normalized for m_e and m_i, respectively.

In terms of the normalized quantities, we can write $\hat{H} = (1/2b_n^4)(\rho_0/\delta)^2$ with $\rho_n = \rho_0/b_n$. The normalized Larmor radii are $\hat{\rho}_0 = b_n(2\hat{H})^{1/2}$ and $\hat{\rho}_n = (2\hat{H})^{1/2}$. In this paper, we will use capital letters for dimensionless coordinates and lower case letters for unnormalized coordinates. Other normalized quantities will be denoted by carets.

Another frequently used parameter is $\kappa = (R_c/\rho_n)^{1/2} = (2\hat{H})^{-1/4}$, which has been used to parameterize pitch-angle scattering by the current sheet with $B_y = 0$ [*West et al.*, 1978b; *Sergeev and Tsyganenko*, 1982; *Sergeev et al.*, 1983]. *Büchner and Zelenyi* [1989] expanded the Hamiltonian for $\kappa \ll 1$ and proposed to use κ as a measure of chaos. This, however, proved to be inconsistent with the actual solutions (section 3).

The dynamical system represented by equations (4)–(7), or equation (3), was first described in detail using the Poincaré surface of section technique for $B_y = 0$ [*Chen and Palmadesso*, 1986]. For example, one can consider a plane at $z = 0$ and record the coordinates X and $\dot{X}$ as an orbit crosses the plane. The orbit in Figure 2 would leave five crossing points, $(X_1, \dot{X}_1), \ldots, (X_5, \dot{X}_5)$, in the midplane. One can then record these points and construct a constant-energy phase space plot which describes the long-time dynamical properties graphically. Figure 4a shows a surface of section plot constructed at $Z = 0$ for $\hat{H} = 500$ and $b_n = 0.1$ in the modified Harris model. This technique provides a graphic representation of the first return map. For example, in Figure 2, point 3 is the first return map of point 1, and point 5 is the mapping of point 3. Figures 4a–4b show all the generic phase space properties for this system. The most prominent feature is the sharp division of the phase space. The region denoted by A is an integrable (regular) region, and it is surrounded by region B referred to as a stochastic or chaotic region. The regions denoted by C are referred to as transient regions.

For $B_y = 0$, the equations of motion are symmetric in Z. As a result, the two directions of crossing at the midplane, $\dot{Z} > 0$ and $\dot{Z} < 0$, need not be distinguished. For $B_y \neq 0$, however, the Z-symmetry is lost, and the two directions of midplane crossing must be distinguished by using two phase space surfaces at $Z = 0$. For example, if $B_y \neq 0$, points 1, 3, 5 and points 2, 4 would be on different surfaces. Figures 4c–4d show the two return maps with $B_y = B_n = 0.1B_0$ and $\hat{H} = 500$. The regions A, B, and C are again occupied by integrable, stochastic, and transient orbits. Regions D are analogous to regions C and exist separately from regions C because of the lack of Z symmetry. These figures will be discussed in more detail later.

In the integrable region, a given trajectory is confined to an invariant surface, the so-called Kolmogorov-Arnol'd-Moser (KAM) surface, whose intersection with the surface $Z = 0$ is a closed curve. Several representative KAM curves are shown in region A of Fig. 4a. An integrable region consists of infinitely many nested closed curves. In the system given by equations (1) and (3), the z-extent of the integrable orbits is limited to $\sim d_j \equiv (2\rho_0\delta)^{1/2}$. In the absence of noise, integrable orbits remain on the KAM surfaces indefinitely without closing onto themselves (quasi-periodic). There exist integrable orbits which do close onto themselves. They are referred to as periodic or fixed-point orbits. *Büchner and Zelenyi* [1986, 1989] proposed that the action I can serve as the third invariant. Although I is approximately conserved in the current sheet [*Sonnerup*, 1971], it is easy to see that $[I, H] \neq 0$ so that I is not an invariant.

In region B of Figure 4a, nearby orbits which are initially arbitrarily close to each other can diverge exponentially during the time the orbits are in the vicinity of the current sheet [*Chen*, 1992]. The average exponential divergence rate is calculated in the same way as the standard Lyapunov exponent with the exception that the time used in the calculation is limited to the duration between the first and last crossings of the midplane.

In the modified Harris model, all nonintegrable orbits come from and escape to infinity ($z \to \infty$), and all orbits from infinity are nonintegrable, making their first midplane crossings in region $C1$. This is the entry region. All orbits escaping to infinity cross the midplane for the last time in region $C1'$ (not shown in Figure 2), where $C1'$ denotes the mirror image of $C1$ about $\dot{X} = 0$. The regions $C2$–$C5$ are the successive mapping of $C1$ by the equation of motion (3), or equivalently equations (4)–(7).

The regions C are shown as "blank" regions only to distinguish them from the surrounding stochastic region. Figure 4b shows the internal structures of the regions $C1$–$C5$. The orbits (i.e., points) shown are the transient orbits, making only one traversal of the midplane. For $\hat{H} = 500$, they have the appearance of the well-known Speiser orbits shown in Figure 2. All transient orbits enter the midplane through $C1$ and the majority of them pass through $C2$–$C5$, and escape to infinity. Some escape to infinity after crossing $C4$ just above $T4$, which is inside region $C1'$, the exit region. The transient orbits do not enter region B. The number N of the prominent transient regions is given by

$$N_t \sim (\hat{H}/2)^{1/4} + 1$$

for $\hat{H} \lesssim (1/2){b_n}^{-1/4}$ [*Chen and Palmadesso*, 1986]. (Note that N_t gives the number f as in *Cf.*) This is also the typical number of midplane crossings for an orbit of energy $\hat{H}$ as it makes one

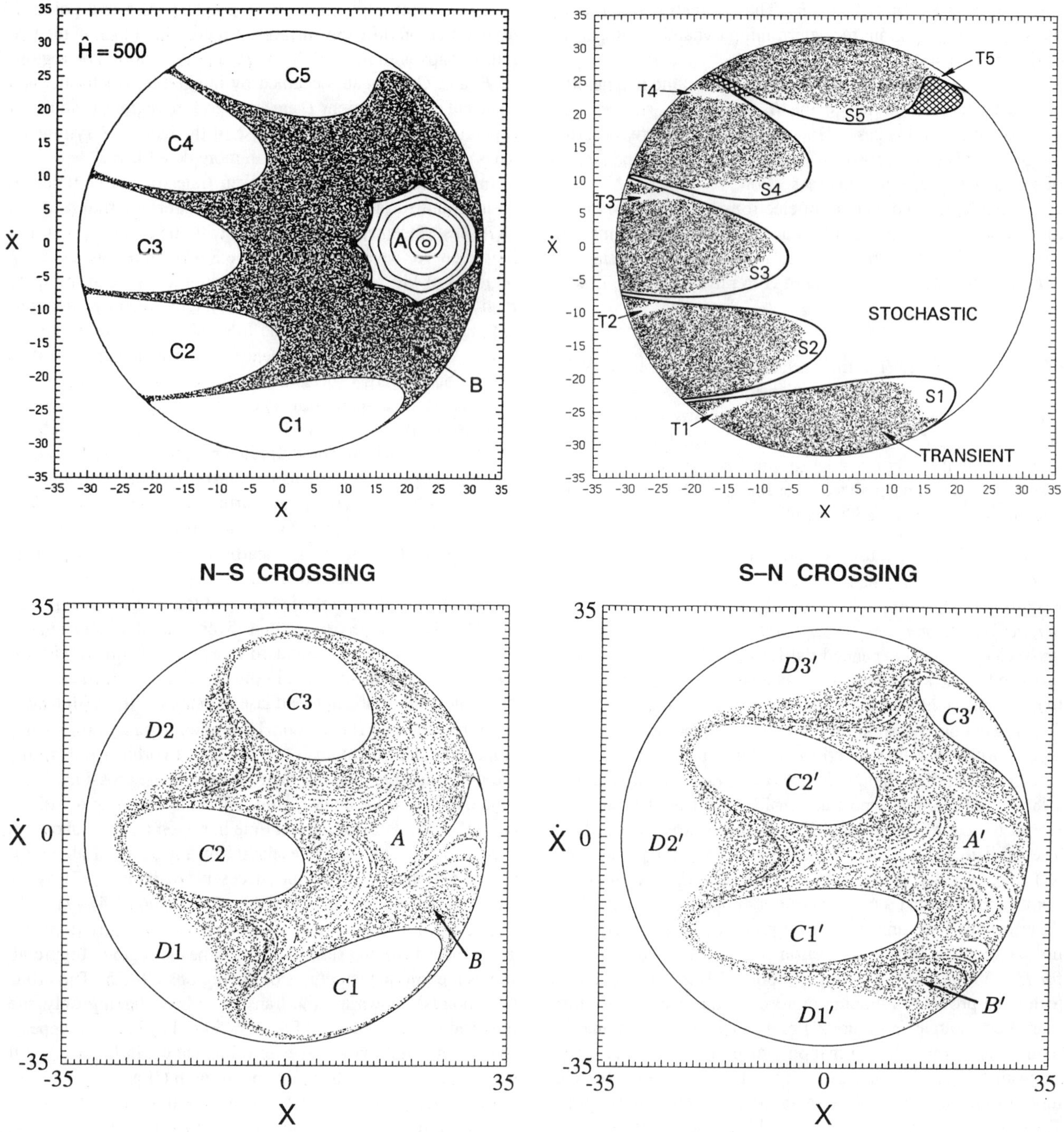

Fig. 4. A Poincaré surface of section map. $b_n = 0.1$. Adapted from *Chen and Palmadesso* [1986]. (a) $B_y = 0$. Region *A* is integrable and region *B* is stochastic. Regions *C1*–*C5* are transient regions. The boundaries $\partial(Cj)$ are shown. (b) Internal structure of the transient regions. *S*1 and *T*1 are the entry regions for stochastic orbits. The orbits shown are transient orbits. (c) $B_y = 0.1\ B_0$. Surface of section return map for north-to-south crossings. (d) $B_y = 0.1\ B_0$. South-to-north crossings.

midplane traversal. The scaling dependence is straightforward to obtain and is a consequence of the property $B_x(z) \simeq B_0 z/\delta$ for $z/\delta \lesssim 1$ and of the translational invariance in y. ($\rho_n \ll L_x$ is assumed.) Several important implications of this scaling dependence have been found recently, and we will discuss them in section 4. The orbits which enter the midplane through $S1$ or $T1$ subsequently cross the midplane through $S2$–$S5$ or $T2$–$T5$, respectively, and then enter region B. These orbits eventually escape through the cross-hatched regions, which are subregions of $C1'$, being the mirror images of $S1$ and $T1$. Generally, the entry region $C1$ consists of subregions corresponding to transient orbits of different midplane crossing numbers (e.g., 4 and 5 for $\hat{H} = 500$), and these subregions are separated by entry regions for stochastic orbits (e.g., $T1$).

It is instructive to understand the class of orbits comprising the boundaries of the transient regions. These boundaries, which will be denoted by $\partial(Cj)$, where $j = 1, 2, \ldots, f$, are the mapping forward in time of those orbits which have pitch angle $\beta = \pi/2$ at "infinity" [*Chen et al.*, 1990b]. The boundaries of the transient regions shown in Figure 4a are constructed by mapping such orbits at $z = 5\delta$ forward in time. The boundary $\partial(C1)$ is the first midplane crossing of these orbits, which are subsequently mapped to $\partial(C2)$–$\partial(Cf)$ by the equation of motion (3). For more detail regarding the structure and properties of $\partial(Cj)$, see *Chen* [1992, section 2.3].

In the $B_y \neq 0$ case the basic partitioning of phase space persists, albeit with some modifications. Figure 4c shows the return map at $Z = 0$ for north-to-south (N–S) crossings with $B_y = B_n = 0.1B_0$, and Figure 4d shows the return map for south-to-north (S–N) crossings with the same parameters, where north refers to $Z > 0$ and south to $Z < 0$. (Due to the symmetry of the equations of motion under the simultaneous replacement of b_y with $-b_y$ and Z with $-Z$, Figure 4c may also be considered to be the return map for south-to-north crossings and Figure 4d the return map for north-to-south crossings with $-B_y = B_n = 0.1B_0$.) Below we will give a brief discussion of the phase space structure for this case. The purpose here is not to discuss the detailed properties but to show that the concepts developed for the simpler $B_y = 0$ case can be similarly applied to obtain rigorous understanding of the more complicated system.

As in Figures 4a and 4b, regions A and A' are occupied by integrable (regular) orbits which are confined to KAM surfaces for all time in the absence of noise. All particles approaching the midplane from the north will first cross the midplane in region $C1$ (N–S crossing). The particles will then cross the midplane again in region $C1'$ (S–N crossing) followed successively by region $C2$, $C2'$, $C3$, and $C3'$. Some particles then escape to infinity (transient orbits), while others will reenter the phase space in region B (stochastic orbits). Similarly, those particles approaching the midplane from the south will first enter the surface of section in region $D1'$ (S–N crossing). The particles will then cross the midplane again in region $D1$ (N–S crossing), $D2$ and so forth. Subsequently, some particles will escape to infinity whereas others will reenter the phase space in region B and make multiple traversals of the midplane (stochastic orbits).

As in the $B_y = 0$ case, the entry regions ($C1$ and $D1'$) are the mapping of particles from infinity: $C1$ is the mapping of particles from $Z = +\infty$ while $D1'$ is the mapping from $Z = -\infty$. The escape regions are the mirror images of $C1$ and $D1'$ about $\dot{X} = 0$ in the respective surfaces (not shown but see *Chen* [1992] for the method of constructing phase space regions). Note that the mirror image of a region defined by a set of orbits is equivalent to the time reversed mapping of the same set of orbits. Unlike the $B_y = 0$ case, however, care must be used to ascertain the surface of exit if $B_y \neq 0$. That is, if orbits originating from the north escapes to the south, the escape region and entry region are in the same N–S surface. If the orbits escape to the north, then the entry region is in the N–S surface while the escape region is in the S–N surface.

Analogous to the $B_y = 0$ case, the entry regions are generally subdivided into subregions for transient and stochastic orbits. These subregions are the intersections between the entry region and the time reversed mapping of the regions C, C', D, and D'. Then, as mentioned above, the direction of the midplane crossing, i.e., N–S versus S–N, must be taken into account properly. Although this process is much more complicated than the $B_y = 0$ case, the subregions in the overall entry regions ($C1$ and $D1'$) can be mapped out geometrically and *exactly*.

The D regions have internal subdivisions just as in the standard transient regions (Figure 4b). The details of the phase space structure and their physical significance will be discussed in deatil elsewhere.

The phase space properties as represented graphically by the surface of section plots, which describe the dynamics of the system completely, are qualitativelysimilar for $B_y \neq 0$ and $B_y = 0$: the dynamical system consists of integrable, stochastic, and transient trajectories. In the remainder of the paper, we will primarily discuss the case with $B_y = 0$.

An important question is how collisions and noise can influence the dynamical properties which have been obtained in the absence of such effects. This is particularly relevant because realistic systems are generally "noisy", and stochastic orbits are sensitively affected by noise. *Holland and Chen* [1991] constructed an *ad hoc* scattering operator which parameterizes a collision by collision frequency ν, amplitudes of pitch-angle ($\Delta\beta$) and phase-angle ($\Delta\phi$) scattering, and amplitude of energy change (ΔE). It was found that the KAM surfaces are destroyed with the introduction of collisions while the transient region boundaries $\partial(Cj)$ are remarkably robust. Figure 5 shows a surface of section plot for $\hat{H} = 187.4$, and $b_n = 0.1$ with pure pitch-angle scattering of amplitude $\Delta\phi = \Delta\beta \leq 1°$ and frequency $\nu = \Omega_n/2$. The KAM curves in what would have been an integrable region without collisions are destroyed, showing that orbits can diffuse through the region. In contrast, the boundaries $\partial(C1)$–$\partial(C4)$ are essentially intact. This means that the separation between the stochastic and transient orbits remains sharp. Although the KAM surfaces are destroyed, the (former) integrable orbits are nearly trapped for substantially longer periods of time than the stochastic orbits. Thus, the partitioning of the phase space remains between all regions. For

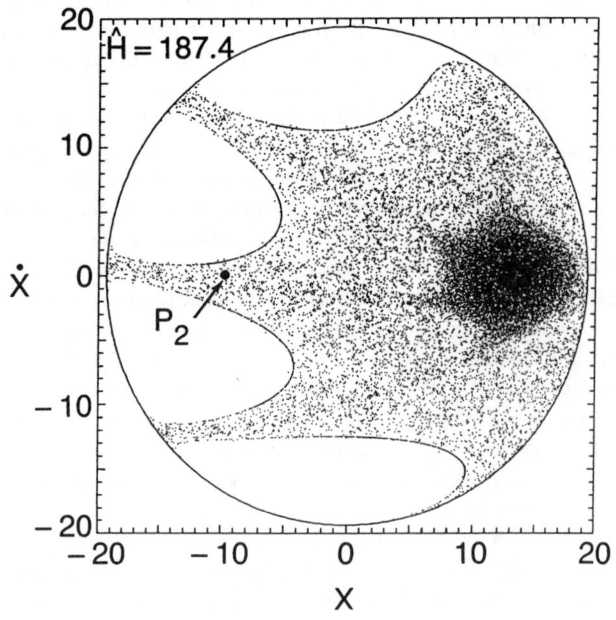

Fig. 5. Surface of section for $\hat{H} = 187.4$ including pure pitch-angle scattering. The scattering amplitude is 1° in pitch- and phase-angles with a frequency of $\nu = \Omega_n/2$. (From *Holland and Chen*, [1991].)

pure energy scattering, if the "collisions" are to have a significant effect on the phase space populations, particles must be transported onto energy surfaces with greatly differing phase space boundaries. If we assume that the "collisions" are due to wave particle interactions, we find that the observed frequencies and amplitudes will not produce a major change in the relative populations of transient and stochastic orbits.

The property that nonintegrable (stochastic and transient) orbits are unbounded distinguishes this system from the usual bounded dynamical systems. As Figures 2 and 3 show, the particle motion in the asymptotic regions is regular. Thus, it is possible to speak of well-defined incoming and outgoing conditions in the asymptotic regions. It was first shown [*Chen et al.*, 1990c] that the stochastic orbit entry regions (e.g., $S1$ of Figure 4b) are divided into subregions according to the outgoing states and that such subregions occur on a fractal set. Therefore the outgoing conditions of stochastic orbits depend sensitively on the incoming conditions while transient orbits exhibit no such sensitivity. As an example, the total number of midplane crossings between first and last crossings of the midplane can serve as an outgoing condition. The fractal dimension was found to be $d \simeq 1.8$ for $\hat{H} = 500$, indicating that the outgoing conditions are only poorly predictable from the initial conditions. Similarly the pitch angle is essentially constant for $|z| \gg \delta$ and can define asymptotic states. This type of system, which has regular asymptotic trajectories but whose outgoing conditions exhibit sensitive dependence on incoming conditions, are often called chaotic scattering systems. The physical importance of chaotic scattering is that the "external world" can influence and be influenced by the chaotic dynamics inside localized regions. For example, auroral region electric fields can modify the tailward flowing ion distributions which then determine the current sheet structure in the tail [*Holland and Chen*, 1993]. In this process, the incoming ions enter the current sheet through $C1$. The entry ($C1$) and exit ($C1'$) regions constitute the conduit between the current sheet and the "external" regions. As a result, the pitch-angle distribution of incoming particles and the internal structure of $C1$ determine the relative population of transient and stochastic orbits. Such structure of $C1$ and how the transient and stochastic orbits determine the current sheet structure will be discussed later (section 5). The term "chaotic scattering" has also been used [*Ashour-Abdalla et al.*, 1990, 1991; *Brittnacher and Whipple*, 1991] to refer to the variation ΔI in the action at the effective potential separatrix [*Büchner and Zelenyi*, 1989]. This usage is different from the standard terminology in nonlinear dynamics.

A quantitative measure of the degree of chaos is the Lyapunov exponent, which gives the average rate at which arbitrarily nearby orbits exponentiate away from a reference orbit. *Martin* [1986] computed the Lyapunov exponent for orbits in an X-type magnetic neutral line. For the quasi-neutral sheet configuration, *Chen* [1992] showed the exponential divergence rate λ^* for individual trajectories and defined a phase space average Λ of λ^*. The usual Lyapunov exponent is defined for infinite time, appropriate for bounded orbits. For chaotic scattering systems such as the quasi-neutral sheet configuration, the infinite time limit is not useful because all nonintegrable orbits are actually regular almost all the time. For individual orbits, a finite-time analogue of the Lyapunov exponent can be defined. Consider an orbit $\mathbf{x}$ and a nearby orbit $\mathbf{x} + \delta\mathbf{x}$. The difference vector $\delta\mathbf{x}$ lies in the tangent space at $\mathbf{x}$. Let its norm be $w(\mathbf{x}, t) \equiv |\delta\mathbf{x}|$. Then the exponential divergence rate for a given orbit is

$$\lambda^* = \frac{1}{L\Delta\tau}\sum_{i=1}^{L}\ln\left(\frac{w_i}{w_0}\right), \tag{9}$$

where w_0 is the norm of the initial tangent vector, w_i is the norm of the tangent vector after the i-th time step, and $\Delta\tau$ is the time step. The number L is the total number of time steps from the first crossing of the midplane (e.g., in $S1$ or $T1$ of Figure 4b) to the last point of crossing (e.g., in $S1'$ or $T1'$). The standard Lyapunov exponent corresponds to the limit $L \to \infty$. Recall that the dynamics (but not individual trajectories) in the modified Harris field can be parameterized by two quantities: $\hat{H}$ and b_n (plus b_y if $B_y \neq 0$). For given $\hat{H}$ and b_n, however, it is easy to show that the value of λ^* has a wide range, depending on the pitch and phase angles of the orbits. One can average over the pitch and phase angles to characterize the entire stochastic region for each energy surface. We define the average exponential divergence rate by

$$\Lambda(\hat{H}, b_n) \equiv \left(\frac{1}{M}\right)\sum_{m=1}^{M}\lambda_m^*, \tag{10}$$

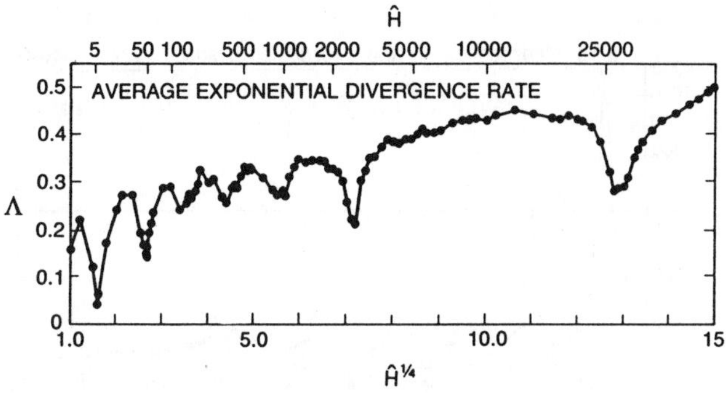

Fig. 6. Exponential divergence rate averaged over the phase space in the modified Harris model. The valleys occur at the resonance energies given by equation (15). (From *Chen* [1992].)

where λ_m^* is the divergence rate for the m-th orbit and M is the total number of orbits used in the calculation. The individual orbits are distributed uniformly in the stochastic orbit entry regions at $Z = 0$ (e.g., $S1$ and $T1$ of Figure 4b or Figure 8). The orbits are computed until they enter $S1'$ or $T1'$. Figure 6 shows the plot of Λ versus $\hat{H}$ for $b_n = 0.1$. We used $M =$ 40–100 for $\hat{H} \lesssim 10$ and $M \simeq$ 200–400 for higher values of $\hat{H}$. The rate λ_m^* for each orbit is computed from the first crossing to the last crossing of the midplane. We see that there are certain values of $\hat{H}$ where $\Lambda(\hat{H})$ is significantly reduced from the neighboring values. These $\hat{H}$ values are evenly spaced in $\hat{H}^{1/4}$, indicating the existence of resonance, and are related to the scaling behavior $N_t \simeq (\hat{H}/2)^{1/4} + 1$ previously mentioned. This resonance effect will be discussed in the next section. It can be shown that $\Lambda(\hat{H}, b_n)$ decreases linearly with b_n for fixed $\hat{H}$. This is to be expected because $b_n \rightarrow 0$ is a completely integrable limit.

It has been conjectured [*Büchner and Zelenyi*, 1987, 1989] that the degree of chaos increases as $\kappa \rightarrow 1$, attaining maximum at $\kappa = 1$ ($\hat{H} \rightarrow 0.25$) and that it decreases for $\kappa \rightarrow 0$ ($\hat{H} \rightarrow \infty$). The quantity κ was used as a small expansion parameter, but the conjecture was not based on quantitative relationship between κ and the rate of exponential divergence. Figure 6 shows that $\Lambda(\hat{H}, b_n)$ has no monotonic dependence on $\hat{H}$ (and for fixed b_n in the modified Harris model) and that the overall tendency of Λ is to increase with increasing $\hat{H}$. It has also been found that Λ is small near $\kappa \simeq 1$, reaching $\Lambda = 8 \times 10^{-2}$ as $\hat{H} \rightarrow 0.25$ (i.e., $\kappa \rightarrow 1$) [*Chen*, 1992, section 2.6]. These properties are contrary to the above conjecture. Another popular notion is that κ can distinguish between different classes of orbits. Actual solutions show that, at any given value of $\hat{H}$ or equivalently κ, there are integrable, stochastic and transient orbits. Recall that Figure 3 depicts two orbits whose initial conditions are identical except for the phase angle, i.e. they have exactly the same value of κ. When the particles leave the current sheet they have similar properties. Yet, one orbit is stochastic and the other is transient. In addition, there are integrable orbits with the same value of κ as shown in Figure 4a of *Chen and Palmadesso* [1986]. The exponential divergence rates λ^* for different individual orbits similarly do not have a single value for given $\hat{H}$ and b_n. For $\hat{H} \rightarrow \infty$, the integrable regions vanish, and all the orbits are of the transient type [*Chen et al.*, 1990b; *Burkhart and Chen*, 1991]. The transient orbits, however, do not exhibit sensitive dependence on initial conditions and are not integrable. Thus, the system does not become more integrable as $\kappa \rightarrow 0$. Despite its popularity, κ does not constitute a n objective measure of chaos. Note that there is an ambiguity in the limit $\kappa \rightarrow 0$ since this limit can be taken by either $\hat{H} \rightarrow \infty$ holding b_n fixed or $b_n \rightarrow 0$ holding $\hat{H}$ fixed. The latter corresponds to an integrable limit while the former does not. The mathematical difficulty arises from the fact that the normalization of equations leading to κ and equivalently to $\hat{H}$ diverges in the integrable limit $b_n \rightarrow 0$, as was pointed out earlier [*Chen and Palmadesso*, 1986]. The expansion based on $\kappa \ll 1$ is not meaningfully extendable to $\kappa \sim 1$. In contrast, it can be shown readily that Λ decreases linearly with $b_n \rightarrow 0$ for fixed $\hat{H}$. Similar ambiguities have also been noted in characterizing the particle motion in time-dependent models [*Chapman*, 1994].

The model given by equation (1) is independent of x. The neglect of x- and y-dependence within the current sheet is valid if $\rho_n \ll L_x, L_y$ where L_x, L_y are the gradient scale lengths in the x- and y-directions, respectively. *Karimabadi et al.* [1990] used an x-dependent model and found that the dynamics of charged particles can be significantly different from the x-independent models. In particular, they found a regime in $\hat{H}$ where the phase space area occupied by integrable orbits is much greater than that in the modified Harris field (the regime of enhanced integrability). They also found that the orbits travel substantially shorter distances in the x-direction before mirroring than in the x-independent model. In addition, with an imposed constant E_y, the particles gained unrealistically high energy. As a result, *Karimabadi et al.* [1990] questioned the applicability of existing x-dependent field models with uniform E_y. *Burkhart and Chen* [1992] further investigated the reason for the inapplicability. They found that the magnetic field as seen by the orbits in the regime of enhanced integrability does not resemble the presumed field geometry of the magnetotail. Specifically, they found that the field lines in this regime are "bulb-shaped" and that the x-excursion Δ of the orbits and the Larmor radius ρ_n are both comparable to the scale length L_x. This is different from the Earth's magnetotail where $\Delta, \rho_n \ll L_x \sim 50R_E$ [*Behannon*, 1968]. The apparent agreement between the predictions based on an x-independent model and observation in the magnetotail reported by *Chen et al.* [1990a] indicates that the particle dynamics in the distant magnetotail can be approximated by suitable x-independent models.

For $B_y \neq 0$, we must restrict ourselves to $E_y = 0$ or make $\mathbf{E} \perp \mathbf{B}$. Otherwise the field aligned electric field would accelerate particles to unrealistically high energies. It is found if the B_y field becomes sufficiently large ($B_y \gtrsim 2B_z$), the majority of the particles directly pass through the sheet and the resonance effects mentioned above and discussed below are negligible. This is a significant result in that it places a limit on the magnitude of the y-component of the magnetic field if the physical consequences of the resonance effects are observed.

4. PHASE SPACE RESONANCE

We have discussed how the dynamical properties of this system can be *graphically* described using Poincaré surface of section maps. We now consider how the phase space structures may be related to physical observables. A physically important effect is the phase space resonance referred to above. It was found [*Burkhart and Chen*, 1991] that, at certain resonance values of $\hat{H}$ satisfying

$$N \simeq \hat{H}_N^{1/4} - 0.6, \qquad (11)$$

where $N = 1, 2, 3, \ldots,$ the entry (Ci) and exit (Cf) regions become nearly symmetric. The value 0.6 is a numerical scale factor, and $\hat{H}_N$ denotes the N-th resonance energy. This equation refines the previously discussed expression $N_t = (\hat{H}/2)^{1/4} + 1$ with $N_t = N + 1$, and the resonance effect incorporates the additional ingredient of symmetry in Ci and Cf. The first few resonances occur at $\hat{H}_1 \simeq 6$, $\hat{H}_2 \simeq 50$, and $\hat{H}_3 \simeq 170$. For $b_n = 0.1$, the resonance terminates after $N = 6$. It is useful to note that there are $N_t = N+1$ prominent transient regions at resonance so that the majority of orbits traversing the current sheet return to the side from which they are incident ("backscattered") at odd-N resonances while the majority of incident orbits exit to the opposite side of the current sheet ("forward scattered") at even-N resonances. (Also see *Chen* [1992] for a detailed discussion.)

It is easy to infer from Figure 4b that the entry regions for stochastic orbits (e.g., $S1$ and $T1$ for $\hat{H} = 500$) decrease in relative phase space area as the degree of symmetry between Ci and Cf increases. By taking the mirror images of the entry regions, we can see that the area of the stochastic orbit exit regions (e.g., the cross-hatched regions in Figure 4b) decreases as Ci and Cf become more symmetric. To see the physical effect of this resonance, consider a population of orbits at a resonance energy, uniformly distributed in phase space in the asymptotic region. Because the relative area of the stochastic orbit entry regions decreases significantly at resonance, the proportion of stochastic orbits also decreases significantly in comparison with neighboring off-resonance energies. One physical consequence is that orbits in asymptotic regions have significantly decreased probability of entering stochastic regions near the midplane. A majority of orbits entering the current sheet are transient orbits at resonance energies. Conversely, stochastic orbits in the vicinity of the current sheet have significantly reduced probabilities of escaping to the asymptotic regions, leading to increased time during which the stochastic orbits are effectively "trapped" in the current sheet. *Burkhart and Chen* [1991] numerically calculated the time that stochastic orbits spend between the first

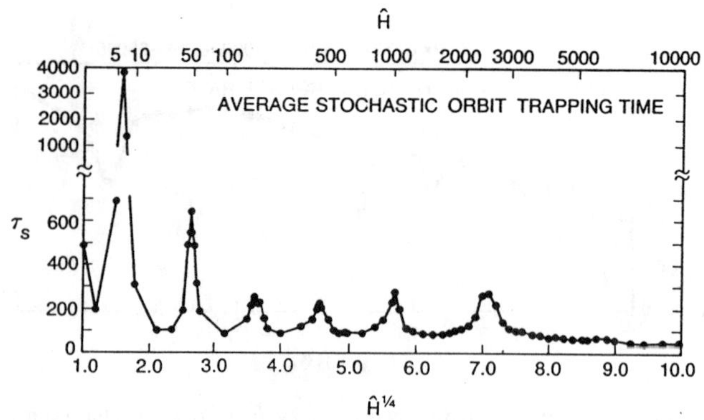

Fig. 7. Average stochastic orbit trapping time. (From *Burkhart and Chen* [1991].)

and the last crossings of the midplane. This time τ_s, referred to as the average stochastic orbit trapping time, is a measure of how long stochastic orbits spend in the neutral sheet and is a function of $\hat{H}$ and b_n. It was found that, at the values of $\hat{H}$ given by (11), $\tau_s(\hat{H})$ is "anomalously" long. Figure 7 shows $\tau_s(\hat{H})$ for $b_n = 0.1$. It is clear that τ_s has well-defined peaks evenly spaced in $\hat{H}^{1/4}$. By comparing Figures 6 and 7, it is easy to see that peaks in τ_s coincide with the valleys in Λ. Recall that the dynamics of the system depends on two independent parameters $\hat{H}$ and b_n. The quantity b_n is important because it is the measure of the nonintegrable perturbation.

A physical understanding of the resonance phenomenon can be obtained by carefully considering the phase space structures. Figure 8 shows the boundaries of the transient regions $\partial(Cj)$ which are the mapping of orbits with incoming pitch angle $\beta_1 = \pi/2$ at $z = 5\delta$ forward in time, as discussed earlier. These boundaries are shown by solid lines. Here, $j = 1, \ldots, f$ with $f = 5$ for $\hat{H} = 500$. Also shown are the time-reversed mapping of the same asymptotic orbits. They are shown as dashed lines. Because y- and time-reversals are equivalent, the dashed lines are simply the mirror images of the solid lines. The stochastic orbit entry regions $S1$ and $T1$ are identified. (Also see Figure 4b.) Formally, the stochastic orbit entry regions can be written as $C1 \cap \overline{Cf'} \cap C(f-1)' \cap \cdots$. Here $\overline{Cj}$ denotes the complement of the region Cj, and $\cap$ indicates the intersection of two regions. The exit regions are simply the mirror images of entry regions. Therefore, if $C1$ and Cf are completely symmetric, then $C1$ and Cf overlap completely and there is no entry or exit region for stochastic orbits. The value $\hat{H} = 500$ shown in Figure 8 is near the $N = 4$ resonance energy. As a result, $C1$ and $C5'$ have a large overlap. We have also shown several contours (— · —) of constant asymptotic pitch angles β_1 which are defined at $z = 5\delta$. These contours are the mapping of asymptotic orbits with $\beta_1 = 2°$, $30°$ and $60°$. We see that there is a critical value β_c such that all asymptotic orbits with pitch angle $\beta_1 < \beta_c$ are transient orbits which do not enter the stochastic region. For $\beta_1 > \beta_c$, some orbits enter stochastic regions and are scattered chaotically (section 3). For $\hat{H} = 500$, $\beta_c \simeq 30°$.

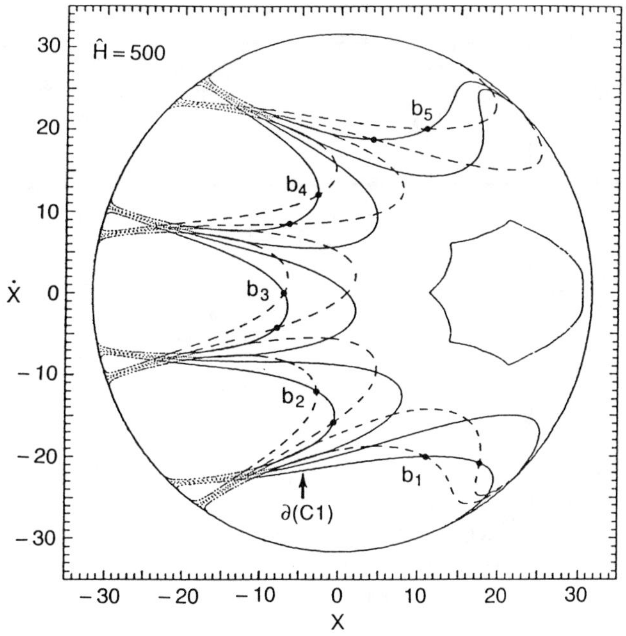

Fig. 8. Phase space boundaries for the modified Harris model. $\hat{H} = 500$ and $b_n = 0.1$. Solid lines are the boundaries $\partial(C1)$–$\partial(C5)$. The dashed lines are $\partial(C1')$–$\partial(C5')$. The regions $S1$, $T1$, $S5'$ and $T4'$ are shown. The mapping of orbits with the asymptotic pitch angle β_1. The contours (— · —) of $\beta_1 = 2°$, 30° and 60° are shown. The points a_1, b_1, and c_1 are mapped onto a_5, b_5, and c_5, respectively. The point d_1 is mapped to d_4 and then to ∞.

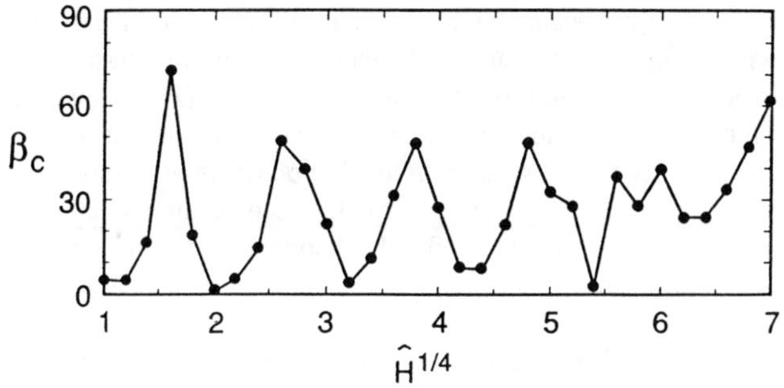

Fig. 9. β_c versus $\hat{H}^{1/4}$. Asymptotic incoming pitch angle $\beta_1 < \beta_c$ corresponds to regular scattering.

As discussed in section 3, the entry regions for stochastic orbits (e.g., $S1$ of Figure 4b or Figure 8) are divided by fractal boundaries corresponding to different outgoing conditions. In terms of pitch angle scattering, the outgoing asymptotic pitch angle β_2 depends sensitively on the incoming β_1 if an orbit enters the stochastic region. *Burkhart and Chen* [1992] found that asymptotically field-aligned orbits (i.e., $\beta_1 = 0$) undergo regular scattering at energies near resonance values, with the outgoing asymptotic pitch angles β_2 exhibiting no sensitive dependence on $\hat{H}$, while pitch-angle scattering is chaotic at off-resonance energies. This is because, at off-resonance energies, even the asymptotically field-aligned orbits enters stochastic regions so that $\beta_c \simeq 0$. The quantity β_c can be calculated numerically and is shown in Figure 9. We see that the peaks in β_c occur at nearly the same energies as the peaks in τ_s (Figure 7), which in turn are essentially the same as the valleys in the average exponential divergence rate Λ (Figure 6) and that $\beta_c \simeq 0$ approximately half way between neighboring resonance peaks. The numerical data points are simply connected by straight lines. Each point has been computed with 500 orbits, uniformly distributed over the phase space and initially located at $z = 5\delta$. We have compared a number of representative points of this figure with the corresponding surface of section plots and found that the results are consistent. This figure complements Figure 4 of *Burkhart and Chen* [1992].

Recall that the dynamics in the modified Harris model can be parameterized by two independent quantities $\hat{H}$ and b_n. The quantity b_n plays an important role in determining the dynamics because it describes the nonintegrable perturbation. We now discuss the dependence of the resonance phenomenon on b_n. Because of the normalization of $\hat{H}$ defined by (7), the resonance values $\hat{H}_N$ do not depend on b_n for $b_n \ll 1$ [equation (11)]. However, the termination point of the resonance effect and the number of resonance peaks are determined by b_n. The dependence of the resonance on b_n can be understood as follows [*Burkhart and Chen*, 1991]. There are two critical energies if the current sheet is confined to a finite thickness. The first occurs when ρ_0 is comparable to the radius of curvature $R_c = b_n\delta$ at $Z = 0$. Setting $\rho_0 = R_c$, we find

$$\hat{H} \simeq H_1 \equiv \frac{1}{2} b_n^{-2}. \tag{12}$$

This corresponds to $d_j = (2b_n)^{1/2}\delta < \delta$ so that $B_x \propto z$, and the analysis leading to $N \sim \hat{H}^{1/4}$ is applicable. As $\hat{H}$ increases, the orbits experience more deviations from $B_x \propto z$, and the resonance effect weakens, resulting in a decreased degree of symmetry between $C1$ and Cf. The resonance peaks in τ_s and τ_D are in fact strongest for $\hat{H} \lesssim H_1$, decreasing with increasing $\hat{H}$ (Figures 6 and 7). The second critical energy occurs when $\rho_0 \simeq \delta \gg R_c$, which is equivalent to $d_i \simeq \delta$. This energy is

$$\hat{H} \simeq H_2 \equiv \frac{1}{2} b_n^{-4}. \tag{13}$$

For $\hat{H} \geq H_2$, the typical midplane crossing orbits have z-excursions exceeding the current sheet thickness. The magnetic field as seen by such orbits deviates significantly from the linear dependence of $B_x \propto z/\delta$, and the resonance effect is no longer operative in general (but see the discussion in the next paragraph). Thus, for a given b_n, there are approximately $2^{-1/4} b_n^{-1}$ resonance energies. For $b_n = 0.1$, Figure 7 demonstrates these two properties clearly. Here, $H_1 = 50$ and $H_2 = 5000$, corresponding to $H_1{}^{1/4} \simeq 2.7$ and $H_2^{1/4} \simeq 8.4$.

In obtaining H_1 and H_2, we have used ρ_0 based on the total kinetic energy with no reference to the pitch angle of a trajectory. As a result, we expect that an ensemble of particles with a range of pitch-angle and energy distribution to show the expected cuttoff in the resonance indicated by (13). However, asymptotically field-aligned particles cross the midplane with small v_z, as can be seen from the 2° coontours in Figure 8, so that their z-excursions remain small. It can be shown that such trajectories continue to exhibit the resonance effect beyond H_2. This means that a field-aligned beam of particles can exhibit resonance beyond H_2.

The ratio R_1 of the area occupied by the entry region C1 to the total area of the energy surface is

$$R_1 \simeq b_n(2\hat{H})^{1/4}. \tag{14}$$

This is applicable to resonance as well as off-resonance $\hat{H}$ values. At resonance, it can be shown further that the ratio of the total area occupied by the transient regions to the total area of the energy surface is

$$R_T \simeq b_n 2^{1/4}\hat{H}^{1/2}, \tag{15}$$

which is simply $N_t R_1$. We have numerically verified that the scalings $R_1 \propto b_n \hat{H}^{1/4}$ is highly accurate (within a few percent) for $\hat{H} < H_2$. Physically, the explicit dependence on b_n is due to the finite thickness of the current sheet.

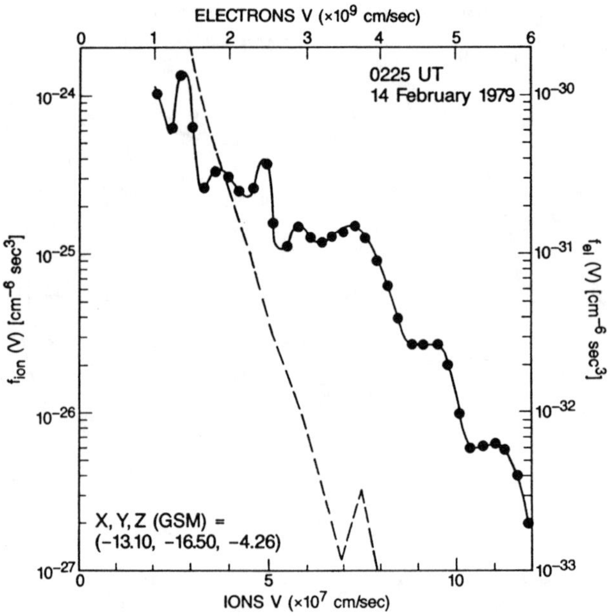

Fig. 10. A velocity distribution function in the central plasma sheet ($v_\parallel = 0$) measured by ISEE 1. The solid line is for ions (left and bottom axes) and the dashed line is for electrons (right and top axes.)

5. PHYSICAL CONSEQUENCES IN THE MAGNETOTAIL

One qualitatively new understanding from the dynamics point of view is the existence of disjoint phase space regions corresponding to distinct classes of orbits. The different regions are characterized by widely separated time scales. An implication is that the different regions are affected by external influences on different time scales in a time-dependent environment. The boundaries between the different regions do not allow uniform propagation of information or particle energy throughout the phase space. Suppose the system contains a population of charged particles initially in thermal equilibrium with external sources. If the conditions of the distant plasma distribution are changed, then the system cannot evolve or relax to conform to the new external conditions uniformly in time throughout the phase space. The dynamically different classes of particles retain the "memory" of the existing particle distribution for different lengths of time. This process was called "differential memory" [*Chen and Palmadesso*, 1986]. It was suggested that differential memory would generate non-Maxwellian distribution functions and associated free-energies. This process has been demonstrated in the trilinear magnetic field [*Chen et al.*, 1990b] and in the modified Harris field [*Chen et al.*, 1990a; *Burkhart and Chen*, 1991]. In these studies, the time-evolution of distribution functions following changes in source distributions was calculated using test particles in fixed magnetic fields. It was found that accessibility time scales of the different phase space regions play an important role in the evolution of distribution functions. In our discussion below, we will use the modified Harris model.

It was found that the reduced accessibility of the stochastic regions at resonance can be manifested as peaks and valleys in distribution functions. Applying the resonance condition (11) to the earth's magnetotail and constructing model distribution functions, it was found that observed ion distribution functions do exhibit peaks and valleys occurring at energies given by equation (11) [*Chen et al.*, 1990a]. Figure 10 shows an ion distribution function (the solid line) obtained by ISEE 1. Similar distribution functions were published earlier by *Huang et al.* [1989]. Following the model prediction, one can take the square root of the velocities (i.e., $H^{1/4}$) at which peaks and valleys occur, and the resulting values should be evenly spaced. In Figure 11, the circles on Line A are the $v^{1/2}$ of the peaks and valleys shown in Figure 10, and Line A is the theoretical prediction based on equation (11). The radius of the circles is roughly the size of error bars. Lines B and C correspond to two additional distribution functions reported by *Huang et al.* [1989]. We see that the agreement is good. The fit is obtained by demanding $N = -0.6$ at $v = 0$. In this figure, it was assumed that the incoming population of orbits from the northern side of the current sheet has a higher density than that of the southern source. Then, the valleys (peaks) correspond to even-N (odd-N) resonance energies with the first valley inferred to be at $N = 12$.

A second prediction is that electron distributions, corresponding to $\hat{H} \ll 1$ in the magnetotail [equation (8)], do not exhibit peaks and valleys obeying the $\hat{H}^{1/4}$ scaling law because the first resonance occurs at $\hat{H} = 6$. The dashed line in Figure 10 is the

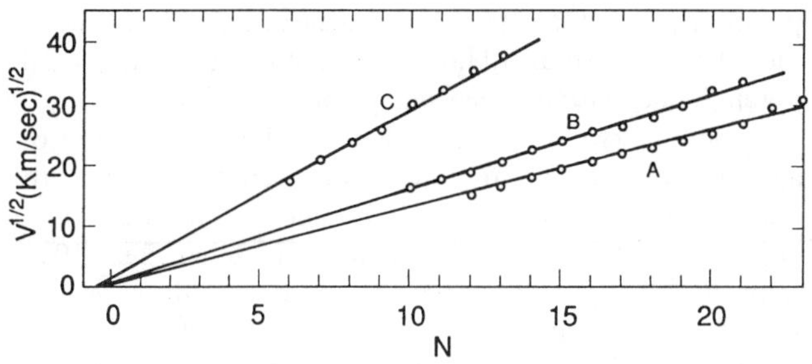

Fig. 11. Comparison between theoretical predictions and observations. Circles denote the locations of the valleys and peaks. The straight lines are the theoretical predictions given by equation (15). Line A corresponds to the distribution in Figure 9a. (From *Chen et al.* [1990a].)

electron distribution function taken at the same time as the ion distribution. We see that it is essentially structureless.

Based on the good agreement between the predicted and observed result, *Chen et al.* [1990a] suggested that the slope of the straight line could be used to infer the thickness of the current sheet using a single space craft. This is possible because the equilibrium current sheet thickness δ and the Larmor radius of thermal particles are related by

$$\delta = \alpha^{-2}(8/\pi)\Gamma^2(7/4)\rho_{\text{th}} \tag{16}$$

where α is a factor order unity to be determined, Γ is the gamma function and $\rho_{\text{th}} = v_{\text{th}}/\Omega_0$ with Ω_0 the Larmor frequency in the asymptotic magnetic field B_0. Using (16), we find that $b_n\Omega_n\delta = \left[(8/\pi)\Gamma^2(7/4)\right]\alpha^{-2}b_n{}^2v_{\text{th}}$, where $\Omega_0 = qB_0/mc = b_n^{-1}\Omega_n$ has been used. Thus, the slope and the thermal velocity of the distribution gives the quantity α/b_n. If α is computed separately, a task we are presently undertaking, then b_n and hence δ may be inferred from the observed distribution functions.

More generally, the phase space topology and the underlying dynamics are determined by global attributes of the system such as the field geometry. Local distribution function signatures which are manifestations of topological properties carry information on large-scale properties. Figure 11 and equation (16) are an example of how one can make use of observed distribution functions as diagnostic tools based on an understanding of the particle dynamics. In this regard, the scaling exponent 1/4 in equation (11) arises from the dependence $B_x(z) \propto z$ [*Chen*, 1992]. (Also see section 4.) Thus, the scaling exponent of observed distribution functions can be used to infer the average z-profile of $B_x(z)$ as seen by particles during midplane crossing. In numerical simulation studies, the scaling law can indicate the profile of the model field as seen by the simulation particles. The agreement between theory and data shown in Figure 11 indicates that the charged-particles see $B_x(z) \propto z$ near the midplane in the Earth's magnetotail.

The relationship between the phase space regions and distribution functions can be illustrated as follows. First, note that the majority of orbits at even-N resonance energies are scattered back toward the source by the current sheet while those at odd-N resonance energies are scattered to the other side of the current sheet. At off-resonance energies, incoming particles are scattered back toward the source and to the other side of the current sheet with nearly equal probability. This process was referred to as "coherent chaotic scattering" [*Chen et al.*, 1990a]. Consider now a distribution of incoming particles in the asymptotic region on one side of the current sheet, say, on the northern ($z > 0$) side. After traversing the current sheet, the particles near the even-N resonance energies are manifested as beams returning to the asymptotic source region while the particles near the odd-N resonances appear as depletions since the particles are coherently scattered to the other (southern) side of the current sheet. At off-resonance energies, backscattering and forward scattering occur with equal probability. A satellite would measure the superposition of particles coming directly from the source, the particles backscattered from the current sheet and the particles forward scattered from the other side of the current sheet, leading to distribution functions of the form given in Figure 10. (See *Chen et al.* [1990a] for a model distribution function based on this scenario.) The reason for the coherent scattering at resonance is the enhanced degree of symmetry between $C1$ and Cf.

The above scenario applies to $z \gg \delta$. In this regime, one possible interpretation of the result is that both peaks and valleys correspond to successive resonance energies. In the study of *Burkhart and Chen* [1991], the evolution of distribution functions was evaluated at the midplane with sources placed symmetrically about $z = 0$. In this case, distribution functions can develop either peaks or valleys at successive resonances depending on whether the source is depleted or enhanced, respectively. These are slightly different manifestations of the same resonance effect. There may be other manifestations. For example, *Ashour-Abdalla et al.* [1991] found beams in their test-particle simulations of the magnetotail, which are also a consequence of the resonance effect discussed above. The main difference is in the different magnetic field model used. We note here that if the magnetic field near the midplane varies as $(z/\delta)^m$ the separation between adjacent resonances should scale as $\hat{H}^\chi$, where $\chi = s/[2(s+1)]$. For $s = 1$ this reduces to the standard $\hat{H}^{1/4}$ scaling. The general scaling $H \propto \hat{H}^\chi$ has been verified for field aligned particles with $s = 3$ and $s = 5$ [*Chen*, 1993]. An implication is that the resonance in locally observed distribution functions in the tail can be used as a diagnositic tool for the field profile as seen by particles. Figure 9 suggests that pitch-angle resolved distributions would also show beams at resonance energies. It is interesting to note that the chaotic particle motion in the current sheet leads to more "ordered" (i.e., non-Maxwellian), rather than more random (i.e., Maxwellian) distributions.

The same dynamical properties also play a controlling role in the equilibrium cross-tail current $J_y(z)$ and particle density $n(z)$ [*Holland and Chen*, 1993]. A number of one-dimensional

(x-independent) self-consistent test-particle equilibrium models have been proposed [*Burkhat et al.*, 1992; *Pritchett and Coroniti*, 1992; *Holland and Chen*, 1993]. In these models, solutions of Vlasov equation satisfying $\mathbf{J} = (c/4\pi)\nabla \times \mathbf{B}$ and $(1/c)\mathbf{J}\times\mathbf{B} - \nabla\cdot\mathbf{P} = 0$ are sought. Some authors obtained equilibria in the regime $v_D \gtrsim v_t$ [*Eastwood*, 1972; *Burkhart et al.*, 1992; *Pritchett and Coroniti*, 1992] while others found solutions for $v_D \ll v_t$ [*Holland and Chen*, 1993]. Here $v_D = cE_y/B_n$ and v_t is the thermal velocity. In the Eastwood regime ($v_D \gg v_t$), both current and particle densities are peaked in the current sheet while in the $v_D \ll v_t$ regime, the particle density is nearly constant across the current sheet while $J_y(z)$ is peaked. The latter is consistent with quiet-time magnetotail observations [*McComas et al.*, 1986]. In the simulations of *Burkhart et al.* [1992], they could not find converged equilibria for v_D smaller than some critical value for given system parameters. In this work, the input distribution function is constrained to be a drifting Maxwellian parameterized by v_D. The samllest critical value was $v_D/v_t \simeq 0.35$ with another example giving 1.2. This was interepreted as a catastrophic loss of equilibrium as $v_D/v_t \to 0$. In contrast, *Holland and Chen* [1993] performed similar test-particle simulations in the $v_D/v_t \ll 1$ regime and obtained good equilibria down to $v_D/v_t < 0.1$. In obtaining equilibria in this regime, they allowed the input distribution functions to have high-energy tails. In this regard, a κ function (Maxwellian with a power-law tail), which is prevalent in magnetospheres [*Christon et al.*, 1989], was found to yield good equilibria at $v_D/v_t = 0.05$. *Holland and Chen* [1993] attributed the lack of equilibria to the constraint that the input distribution be drifting Maxwellians. Such distributions in the $v_D/v_t \ll 1$ regime results in strong diamgnetic currents which flow in the opposite direction to what is required by $\mathbf{J} = (c/4\pi)\nabla \times \mathbf{B}$. The high-energy tail or field-aligned distribution counters the diamagnetic effect. It should be noted that the catastrophe conclusion of *Burkhart et al.* [1992] is given in terms of a paramter κ_A, and a physical explanation in terms of force balance argument of *Francfort and Pellat* [1976]. Careful examination of the catastrophe result shows that (1) actual numerical convergence fails when v_D becomes small, and (2) neither κ nor forces of *Francfort and Pellat* [1976] is used in the numerical iteration process. Thus neither point is germane to the actual process of nonconvergence. One should show that diamagnetic current is weak near the point of suggested catastrophe for a demonstration for the physical applicability of the catastrophe to the magnetotail current sheet.

The above simulations show that the pressure tensor generally has off-diagonal elements to establish force balance in both x and z directions. As the input distribution function becomes more field-aligned, the current sheet thins. A thin current sheet is more unstable to the collisionless tearing mode [*Coppi et al.*, 1966] and may be more susceptible to other forms of disruption [e.g., *Lui et al.*, 1991]. The anisotropic pressure may also significantly increase the growth rate of the tearing mode [*Chen and Palmadesso*, 1984; *Burkhart and Chen*, 1989]. An interesting possibility is that the auroral region electric field may produce more field-aligned distribution of ions flowing tailward, resulting in thinner current sheets due to ion dynamics.

A non-Maxwellian distribution function such as the one shown in Figure 10 also represents free-energy that can drive plasma waves. *Huba et al.* [1992] studied linear stability properties of a model distribution function resembling Figure 10 and found that a low-amplitude broadband electrostatic noise can be generated in the quiet-time central plasma sheet of the earth's magnetotail. The linear dispersion properties proved to be quite consistent with observed wave spectra.

In a different vein, it has been suggested [*Martin*, 1986; *Horton and Tajima*, 1990, 1991; *Horton et al.*, 1991a,b] that the chaotic motion in the current sheet may constitute a form of collisionless "chaotic conductivity". *Martin* [1986] suggested that the Lyapunov exponent may provide the time scale of decorrelation of nearby orbits, taking the place of collision frequency. Using the fluctuation-dissipation theorem [*Horton et al.*, 1990b] and linearized Vlasov theory, [*Horton and Tajima*, 1990, 1991; *Horton et al.*, 1990a] argued that the zero frequency conductivity of the neutral sheet can be determined using the two-time velocity correlation function:

$$\sigma_{\alpha\beta} = \left(ne^2\right)/\left(Mv_{\text{th}}{}^2\right) \int dv f(v) \int_0^\infty d\tau\, C_{\alpha\beta}(\tau) \qquad (17)$$

$$C_{\alpha\beta}(\tau) = \lim_{T\to\infty} \frac{1}{T} \int_0^T dt\, v_\alpha(t) v_\beta(t+\tau). \qquad (18)$$

Calculation of $C_{\alpha\beta}(\tau)$ showed a power-law decay in time which is suggestive of a collision-like momentum scattering and dissipation in analogy with the Lorentz gas problem. For the magnetotail, σ_{yy} and C_{yy} are studied because the cross-tail current is of primary interest. The idea of chaotic conductivity is an appealing one in a collisionless environment. If such a quantity could be defined and calculated for the magnetotail in terms of quantities characterizing the entire system, the electric field E_y and current density J_y could be related by a *local* quantity σ such that $J_y = \sigma E_y$. In addition, by calculating how σ depends on the global system parameters, one could reduce global considerations to a local quantity.

Recently, *Holland and Chen* [1992] re-examined the concept of chaotic conductivity in the magnetotail. They found that since the magnetotail is a chaotic scattering system as opposed to a classical chaotic system, the underlying assumptions leading to the use of two-time correlation functions may be inapplicable to zero-frequency conductivity and that the two-time correlation functions do not lead to a unique value of "conductivity". Due to intrinsic ambiguities in determining the intial and final points in the time integrations they showed that the correlation function C_{yy} of a given orbit can decay differently depending on how C_{yy} is computed. As an illustration of this effect, they calculated the yy component of the correlation func-

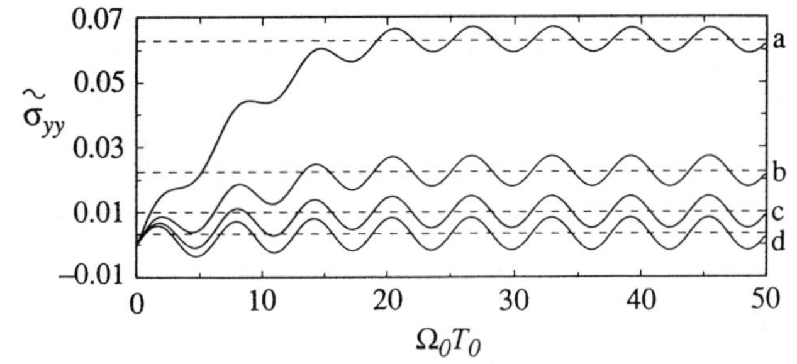

Fig. 12. $\tilde{\sigma}_{yy}$ versus $\Omega_0 T_0$: (the asymptotic mean value of $\tilde{\sigma}_{yy}$ is given in parentheses) (a) $\Omega_0 T = 274$, (0.0625); (b) $\Omega_0 T = 774$, (0.0224); (c) $\Omega_0 T = 1774$, (0.0099); and (d) $\Omega_0 T = 5774$, (0.00325).

tion for a typical transient orbit and the associated single particle conductivity,

$$\tilde{\sigma}_{yy}(T_0) = \int_0^{T_0} C_{yy}(\tau)\, d\tau$$

, for several values of the time T over which the correlation function is evaluated. They focused on transient orbits since these contribute the largest percentage of the current and hence would be expected to have the largest conductivity. The total conductivity is defined by $\sigma_{yy} = \int f(v)\, \tilde{\sigma}_{yy}\, d^3v$, with $T, T_0 \rightarrow \infty$. Figure 12 shows the calculated $\tilde{\sigma}_{yy}(T_0)$ for several values of T. For fixed T and large T_0, the conductivity oscillates with a constant amplitude about some mean value. As T go to infinity (equation(18)), the asymptotic mean value of $\tilde{\sigma}_{yy}$ approaches zero as $1/T$. Using an approximate analytical analysis, they were able to show that this is the expected behavior for a chaotic scattering system of this type. Thus, in practice σ_{yy} depends critically on the time over which C_{yy} is computed. In evaluating C_{yy} and $\tilde{\sigma}_{yy}$, the first crossing point was used as the starting point for the integrations since it is only in the vicinity of the midplane that the particles are expected to decorrelate. Additionally, they noted that different orbits with dissimilar correlation functions can have essentially the same energy gain. Recall that the orbits in Figure 3 have essentially the same initial and final conditions and therefore have essentially the same net energy gain. On the other hand, these two orbits obviously contribute very differently to the cross tail current since they have very different time histories. They gave an alternative expression to calculate the energy gain [*Cowley*, 1978]:

$$\Delta H = mc \left(\frac{E_y}{B_n}\right) \left(\frac{2H}{m}\right)^{1/2} \left(1 + b_n^2\right)^{-2} \left(|\cos\beta_1| + |\cos\beta_2|\right), \qquad (19)$$

where β_1 and β_2 are the asymptotic incoming and outgoing pitch angles, respectively. The chaotic motion of particles in the vicinity of the current sheet plays no role. Numerical calculations show that this expression gives a complete description of energy gain. They also pointed out that the current density $J_y(z)$ depends primarily on the source particle distribution and the magnetic field topology and that J_y and E_y are not related by simple relationships such as an Ohm's law. In this regard, we note that a nonzero E_y effectively modifies the source particle distribution functions and hence the current but it is in a nontrivial way. The relationship between β_1 and β_2 is, of course, complicated as evidenced by the existence of fractals and is determined by global magnetic topology. It is unlikely that the global magnetotail structure can be fully understood using local quantities such a "collisionless" conductivity.

6. DISCUSSIONS

The charged-particle motion in the magnetotail has been extensively studied for more than two decades because of its importance for understanding a wide variety of plasma phenomena in the geospace environment. Traditionally, the emphasis was to study the motion of individual particles. The adiabaticity and nonadiabaticity of particle orbits have received considerable attention. A qualitatively different approach is to treat the particle motion as a dynamical system, and geometrical (phase space) properties are exploited to obtain and explain physical observables. The important dynamical properties are the partitioning of the phase space and the chaotic scattering of incoming orbits by the current sheet. The underlying symmetry properties of the system are of principal importance. The significant phase space structures are robust in the presence of collisions, in contrast to properties of individual orbits. The basic dynamical properties of the magnetotail-like system have been firmly established from first principles. Moreover, the dynamical approach has yielded novel insights previously unavailable from the study of individual orbits. We emphasize that the geometrical approach yields qualitative but rigorously accurate for families of orbits. In contrast, detailed information regarding individual orbits or segments of orbtis is not generic to the dynamical system.

The ability of charged particles to carry information and energy over long distances ($\gg R_E$) in the essentially collisionless magnetotail means that the global particle dynamics can influence the distribution of internal energy and currents on a large scale. In view of its robustness, the partitioning of the phase space may play a significant role in determining such large scale properties. The magnetosphere may act as a self-organized critical system [*Bak et al.*, 1988]. Unlike the usual example of sandpiles, the magnetosphere is determined by globally coherent interaction of fields and collisionless particles. Thus, the nature of long-distance coupling within the magnetosphere is an essential question. If these conclusions prove to be correct, nonlinear dynamics may indeed provide a new framework for a deeper understanding of global magnetospheric processes. (See section 3.6 of *Chen* [1992].) Clarification of the relationships between particle dynamics and global properties is an important subject of future research. The importance of this line of inquiry is enhanced by the fact that the magnetotail is an observationally testable paradigm for a class of systems containing current sheets.

Acknowledgments. This work was supported by the National Aeronautics and Space Administration (W-16,991) and the Office of Naval Research.

REFERENCES

Alekseyev, I. I., and A. P. Kropotkin, Interaction of energetic particles with the neutral sheet in the tail of the magnetosphere, *Geomagn. Aeron., Engl. Transl., 10*, 615, 1970.

Ashour-Abdalla, M., J. Berchem, J. Büchner, and L. M. Zelenyi, Chaotic scattering and acceleration of ions in the Earth's magnetotail, *Geophys. Res. Lett., 17*, 2317, 1990.

Ashour-Abdalla, M., J. Berchem, J. Büchner, and L. M. Zelenyi, Large and small scale structures in the plasma sheet: A signature of chaotic motion and resonance effects, *Geophys. Res. Lett., 18*, 1603, 1991.

Bak, P., C. Tang, and K. Wiesenfeld, Self-organized criticality, *Phys. Rev. A, 38*, 364, 1988.

Behannon, K. W., Mapping of the Earth's bow shock and magnetic tail by Explorer 33, *J. Geophys. Res., 73*, 907, 1968.

Birmingham, T. J., Pitch angle diffusion in the Jovian magnetodisc, *J. Geophys. Res., 89*, 2699, 1984.

Brittnacher, M. J. and E. C. Whipple, Chaotic jumps in the generalized first adiabatic invariant in current sheets, *Geophys. Res. Lett., 18*, 1599, 1991.

Büchner, J., and L. M. Zelenyi, Deterministic chaos in the dynamics of charged particles near a magnetic field reversal, *Phys. Lett. A, 118*, 395, 1986.

Büchner, J., and L. M. Zelenyi, Chaotization of the electron motion as the cause of an internal magnetotail instability and substorm onset, *J. Geophys. Res., 92*, 13,456, 1987.

Büchner, J., and L. M. Zelenyi, Regular and chaotic charged particle motion in magnetotaillike field reversals, 1, Basic theory of trapped motion, *J. Goephys. Res., 94*, 11,821, 1989.

Büchner, J., and L. M. Zelenyi, The separatrix tentacle effect of ion acceleration to the plasma sheet boundary, *Geophys. Res. Lett., 17*, 127, 1990.

Burkhart, G. R. and J. Chen, Linear, collisionless, bi-Maxwellian neutral-sheet tearing instability, *Phys. Rev. Lett., 63*, 159, 1989.

Burkhart, G. R., and J. Chen, Differential memory in the Earth's magnetotail, *J. Geophys. Res., 96*, 14,033, 1991.

Burkhart, G. R., and J. Chen, Chaotic scattering of pitch angles in the current sheet of the magnetotail, *J. Geophys. Res., 97*, 6479, 1992a.

Burkhart, G. R., and J. Chen, Particle motion in *x*-dependent Harrislike Magnetotail models, in press, *J. Geophys. Res.*, 1992b.

Burkhart, G. R., J. F. Drake, P. B. Dusenbery, and T. W. Speiser, *J. Geophys. Res., 97*, 13,799, 1992.

Cary, J. R., E. F. Escande, and J. L. Tennyson, Adiabatic-invariant change due to separatrix crossing, *Phys. Rev. A, 34*, 4256, 1986.

Chapman, S. C., Properties of single particle dynamics in a parabolic magnetic reversal with general time dependence, *J. Geophys. Res., 99*, 5977, 1994.

Chapman, S. C. and N. W. Watkins, Parameterization of chaotic particle dynamics in a simple time-dependent field reversal, *J. Geophys. Res., 98*, 165, 1993.

Chen, J., Nonlinear particle dynamics in the magnetotail, *J. Geophys. Res., 97*, 15,011, 1992.

Chen, J., Kinetic physics of the magnetotail, *Phys. Fluids B 5*, 2663, 1992.

Chen, J. and P. J. Palmadesso, Tearing instability in an anisotropic neutral sheet, *Phys. Fluids, 27*, 1198, 1984.

Chen, J., and P. J. Palmadesso, Chaos and nonlinear dynamics of single-particle orbits in a magnetotaillike magnetic field, *J. Geophys. Res., 91*, 1499, 1986.

Chen, J., G. R. Burkhart, and C. Y. Huang, Observational signatures of nonlinear magnetotail particle dynamics, *Geophys. Res. Lett., 17*, 2237, 1990a.

Chen, J., H. G. Mitchell, and P. J. Palmadesso, Differential memory in the trilinear model magnetotail, *J. Geophys. Res., 95*, 15,141, 1990b.

Chen, J., J. L. Rexford, and Y. C. Lee, Fractal boundaries in magnetotail particle dynamics, *Geophys. Res. Lett., 17*, 1049, 1990c.

Christon, S. P., D. J. Williams, D. G. Mitchell, L. A. Frank, and C. Y. Huang, Spectral characteristics of plasma sheet ion and electron populations during undisturbed geomagnetic conditions, *J. Geophys. Res., 94*, 13,409, 1989.

Coppi, B., G. Laval, and R. Pellat, Dynamics of the geomagnetic tail, *Phys. Rev. Lett., 16*, 1207, 1966.

Cowley, S. W. H., A note on the motion of charged particles in one-dimensional magnetic current sheet, *Planet. Space Sci., 26*, 539, 1978.

Curran, D. B., and C. K. Goertz, Particle distributions in a two-dimensional reconnection field geometry, *J. Geophys. Res., 94*, 11,521, 1989.

Curran, D. B., C. K. Goertz, and T. A. Whelan, Ion distributions in a two- dimensional reconnection field geometry, *Geophys. Res. Lett., 14*, 99, 1987.

Eastwood, J. W., Consistency of fields and particle motion in 'Speiser' model of the current sheet, *Planet. Space Sci., 20*, 1555, 1972.

Francfort, Ph. and R. Pellat, Magnetic merging in collisionless plasmas, *Geophys. Res. Lett., 3*, 433, 1976.

Gray, P., and L. C. Lee, Particle pitch angle diffusion due to nonadiabatic effects in the plasma sheet, *J. Geophys. Res., 87*, 7445, 1982.

Holland, D. L., and J. Chen, Effects of collisions on the nonlinear particle dynamics in the magnetotail, *Geophys. Res. Lett., 18*, 1579, 1991.

Holland, D. L., and J. Chen, On chaotic conductivity in the magnetotail, *Geophys. Res. Lett., 19*, 1231, 1992.

Holland, D. L. and J. Chen, Self-consistent current sheet structures in the quiet-time magnetotail, *Geophys. Res. Lett., 20*, 1775, 1993.

Horton, W., and T. Tajima, Decay of correlations and the collisionless conductivity in the geomagnetic tail, *Geophys. Res. Lett., 17*, 123, 1990.

Horton, W., and T. Tajima, Collisionless conductivity and stochastic heating of the plasma sheet in the geomagnetic tail, *J. Geophys. Res., 96*, 15,811, 1991.

Horton, W., C. Liu, B. Burns, and T. Tajima, Collisionless plasma transport across loop magnetic fields, *Phys. Fluids B, 3*, 2192, 1991.

Horton, W., C. Liu, J. Hernandez, and T. Tajima, *Geophys. Res. Lett., 18*, 1575, 1991b.

Huang, C. Y., C. K. Goertz, L. A. Frank, and G. Rostoker, Observational determination of the adiabatic index in the quiet time plasma sheet, *Geophys. Res. Lett., 16*, 563, 1989.

Huba, J. D., J. Chen, and R. R. Anderson, Electrostatic turbulence in the Earth's central plasma sheet produced by multiple-ring ion distributions, *J. Geophys. Res., 97*, 1533, 1992.

Karimabadi, H., P. L. Pritchett, and F. V. Coroniti, Particle orbits in two- dimensional equilibrium models for the magnetotail, *J. Geophys. Res., 95*, 17,153, 1990.

Lembège, B. and R. Pellat, Stability of a thick two-dimensional quasineutral sheet, *Phys. Fluids, 25*, 1995, 1982.

Lui, A. T., C.-L. Chang, A. Mankofsky, H.-K. Wong, and D. Winske, A cross-tail current instability for substorm expansion, *J. Geophys. Res., 96*, 11,389, 1991.

Lyons, L. R., Electron energization in the geomagnetic tail current sheet, *J. Geophys. Res., 89*, 5479, 1984.

Lyons, L. R., and T. W. Speiser, Evidence of current sheet acceleration in the geomagnetic tail, *J. Geophys. Res., 87*, 2276, 1982.

Lyons, L. R., and T. W. Speiser, Ohm's law for a current sheet, *J. Geophys. Res., 90*, 8543, 1985.

Martin, R. F., Chaotic particle dynamics near a two-dimensional magnetic neutral point with application to the geomagnetic tail, *J. Geophys. Res., 91*, 11,985, 1986.

Martin, R. F., and T. W. Speiser, A predicted energetic ion signature of a neutral line in the geomagnetic tail, *J. Geophys. Res., 93*, 11,521, 1988.

McComas, D. J., C. T. Russell, R. C. Elphic, and S. J. Bame, The near-earth cross-tail current sheet: Detailed ISEE 1 and 2 case studies, *J. Geophys. Res., 91*, 4287, 1986.

Moses, R. W., J. M. Finn, and K. M. Ling, Plasma heating by collisionless magnetic reconnection: Analysis and computation, *J. Geophys. Res., 98*, 4013, 1993.

Neishtadt, A. I., Variation of adiabatic invariant accompanying transition through separatrix, *Sov. J. Plasma Phys.*, Engl. Transl., *12*, 568, 1986.

Pritchett, P. L. and F. V. Coroniti, Formulation and stability of the self-consistent one-dimensional tail current sheet, *J. Geophys. Res., 97*, 16,773, 1992.

Rogers, S. H., and E. C. Whipple, Generalized adiabatic theory applied to the magnetotail current sheet, *Astrophys. Space Sci., 144*, 231, 1988.

Sergeev, V. A., and N. A. Tsyganenko, Energetic particle losses and trapping boundaries as deduced from calculations with a realistic magnetic field model, *Planet. Space Sci., 30*, 999, 1982.

Sergeev, V. A., E. M. Sazhina, N. A. Tsyganenko, J. Å. Lundblad, and F. Sorras, Pitch-angle scattering of energetic protons in the magnetotail current sheet as the dominant source of their isotropic precipitation into the nightside ionosphere, *Planet. Space Sci., 31*, 1147, 1983.

Sonnerup, B. U. Ö., Adiabatic particle orbits in a magnetic null sheet, *J. Geophys. Res., 76*, 8211, 1971.

Speiser, T. W., Particle trajectories in model current sheets, 1, Analytical solutions, *J. Geophys. Res., 70*, 4219, 1965.

Speiser, T. W., Conductivity without collisions or noise, *Planet. Space Sci., 18*, 613, 1970.

Speiser, T. W., and L. R. Lyons, Comparison of an analytical approximation for particle motion in a current sheet with precise numerical calculations, *J. Geophys. Res., 89*, 147, 1984.

Swift, D., The effect of the neutral sheet on magnetospheric plasma, *J. Geophys. Res., 82*, 1288, 1977.

Timofeev, A. V., On the problem of constancy of the adiabatic invariant when the nature of the motion changes, *Sov. Phys. JETP, 48*, 656, 1978.

Tsyganenko, N. A., Pitch-angle scattering of energetic particles in the current sheet of the magnetospheric tail and stationary distribution functions, *Planet. Space Sci., 30*, 433, 1982.

Wagner, J. S., J. R. Kan, and S.-I. Akasofu, Particle dynamics in the plasma sheet, *J. Geophys. Res., 84*, 891, 1979.

West, H. I., R. M. Buck, and M. G. Kivelson, On the configuration of the magnetotail near midnight during quiet and weakly disturbed periods: State of the magnetosphere, *J. Geophys. Res., 83*, 3805, 1978a.

West, H. I., R. M. Buck, and M. G. Kivelson, On the configuration of the magnetotail near midnight during quiet and weakly disturbed periods: Magnetic field modeling, *J. Geophys. Res., 83*, 3819, 1978b.

Whipple, E., M. Rosenberg, and M. Brittnacher, Magnetotail acceleration using generalized drift theory: A kinetic merging scenario, *Geophys. Res. Lett., 17*, 1045, 1990.

Zwingmann, W., Self-consistent magnetotail theory: Equilibrium structures including arbitrary variation along the tail axis, *J. Geophys. Res., 88*, 9101, 1983.

Collisionless Resistivity and Velocity Power Spectrum for the Geomagnetic Tail

W. HORTON, J. HERNANDEZ, AND T. TAJIMA

Institute for Fusion Studies, University of Texas at Austin

The dissipative part of the finite, collisionless electrical conductivity for the current sheet plasma in the geomagnetic tail is obtained by launching ensembles of charged particles in tail-like magnetic field reversals, calculating the velocity power spectrum for each individual trajectory, and averaging the result over the ensemble of particles. The velocity power spectrum formalism can be derived from two alternative approaches: 1) the fluctuation-dissipation theorem and 2) linearized Vlasov theory. As a consistency check for the formalism, the dissipative part of the conductivity for an unmagnetized plasma is calculated. The formalism is then applied to study the effect of a dawn-dusk magnetic field component B_y on the conductivity. It is shown that B_y introduces a new term in the collisionless conductivity which scales as $(B_y/B_z)^2$. In contrast to the $B_y = 0$ case, the contribution of the electrons to the conductivity is very important. Even though the numerical values of the coefficients of the conductivity formula depend on the specific choice of magnetic field model $\mathbf{B}^{(0)}(\mathbf{x})$, the structure of the conductivity formula is model-independent and is due only to the tail-like, magnetic field reversal geometry of the different models. Several field models, including the self-consistent Voigt-Fuchs model, are discussed.

1. INTRODUCTION

Transport mechanisms in the current sheet plasma in the Earth's magnetotail play an important role in the dynamics of the large-scale magnetic disturbances known as magnetospheric substorms [*Akasofu*, 1980,1981]. A basic step in understanding the transfer of particles, momentum, and energy between the solar wind, the magnetosphere, and the ionosphere is the determination of the transport coefficients for the current sheet plasma. Here, in particular, we consider the origin of the irreversible dissipation in the perturbed current sheet plasma for various magnetic field models.

The role of dissipation in the current sheet plasma is made clear in the analogue models of the tippy bucket [*Akasofu*, 1981], the dripping faucet [*Baker et al.*, 1990], and the Faraday loop [*Klimas et al.*, 1991]. However, the origin of dissipation for the current sheet plasma has remained as a challenging problem due to the lack of collisions and of turbulent wave activity in the quasineutral layer. Our approach is based on the premise that transport in the current sheet is driven by particle dynamics rather than by collisional processes.

The particle dynamics driven transport view is implicitly and/or explicitly shared by several authors. For example, *Horton and Tajima* [1990, 1991a] have proposed that irreversible dissipation in the current sheet plasma is due to chaotic particle dynamics and *Martin* [1986] has proposed a chaotic conductivity for the X-point magnetic field. Furthermore, in particle simulations of the tearing mode instability it is not necessary to introduce some "anomalous resistivity," since dissipation should arise naturally from the particle dynamics [*Allen and Swift*, 1989]. This is in direct contrast with MHD simulations of the tearing mode for which it is necessary to introduce some form of artificial dissipation.

It has been shown by *Büchner and Zelenyĭ* [1986] and *Chen and Palmadesso* [1986] that in the current sheet region there are three kinds of orbits: 1) Speiser or transient orbits, 2) Cucumber, quasitrapped, or chaotic orbits, and 3) ring, trapped, or integrable orbits. Furthermore, *Chen and Palmadesso* [1986] have shown that the phase space can be partitioned into basically three regions, each region corresponding to one particular kind of orbit. The implications of phase space partitioning have been further explored by *Chen et al.* [1990], and

Space Plasmas: Coupling Between Small
and Medium Scale Processes
Geophysical Monograph 86

Burkhart and Chen [1991] and are reviewed by *Chen* [1992].

The phase space partitioning into three different regions and the orbit classification into Speiser, cucumber, and ring orbits are due to the magnetic field reversal geometry and thus we expect the structure of the conductivity formula to be weakly dependent on the details of the magnetic field model. The principal change in the conductivity is due to the change in the fractional populations of the three orbit types.

In the presence of electromagnetic perturbations the dawn-dusk current density is given by $j_y = j_y^{(0)} + \delta j_y$, where $j_y^{(0)}$ is the unperturbed current density and δj_y is the perturbed part. The unperturbed current density $j_y^{(0)}$ is entirely due to the particle dynamics and no electric field is necessary to maintain it. On the other hand, δj_y is causally related to the perturbing electric field E_y through the dawn-dusk component of the conductivity tensor $\sigma_{yy}(k_x, \omega)$.

A general procedure for determining the conductivity for collisionless plasmas is the method of characteristics, which consists of linearizing the Vlasov equation and then integrating over the unperturbed orbits [*Drummond*, 1958; *Rosenbluth and Rostoker*, 1958]. However, the problem in applying the method of characteristics to the current sheet plasma is that in general the unperturbed orbits are nonintegrable. *Horton and Tajima* [1991a,b] have followed two equivalent approaches for the calculation of the dissipative part of the collisionless mobility: 1) from the linearized Vlasov equation with the integration over the complex chaotic orbits carried out numerically, and 2) from the fluctuation-dissipation theorem assuming the system is sufficiently close to the local Maxwell-Boltzmann distribution.

The fluctuation-dissipation theorem formalism has been used by *Hernandez et al.* [1993] to assess the effect of a constant dawn-dusk magnetic field component B_y on the collisionless mobility. These authors have found an expression for the low frequency conductivity which consists of the modified *Lyons-Speiser* [1985] conductivity plus a new term with a strong $(B_y/B_z)^2$ scaling.

In Section 2 the magnetic field models used in the calculation of the collisionless conductivity are discussed. In Section 3 the fluctuation-dissipation theorem or velocity power spectrum formalism is presented and illustrated for the case of an unmagnetized plasma. In Section 4 the effect of B_y on the mobility is discussed. Throughout this work we use the geo-solar magnetospheric coordinate system with the x-axis pointing towards the sun, the y-axis going from dawn to dusk, and the z-axis in the conjugate direction such as to form a right-handed system.

2. Magnetic Field Models

Because of the ever-changing interplanetary magnetic field and solar wind conditions, it is reasonable to assume that the Earth's magnetosphere never reaches a true self-consistent equilibrium. Therefore, it is important to check the consistency of the results for different field models. For $X < -10R_E$ and $|Y| \leq 10$ to $15R_E$ translational invariance in the dawn-dusk direction for the magnetic flux function $\psi(X, Z)$ can be assumed. In the Y-independent approximation, the magnetic field is given by

$$\mathbf{B}(X, Z) = \widehat{e}_y \times \nabla\psi + B_y \widehat{e}_y \ , \tag{1}$$

where a constant magnetic field component B_y is assumed. In the case where the cross-tail field component can be neglected, $B_y \approx 0$, the magnetic field lines are the lines of constant $\psi(X, Z)$.

The single particle dynamics in the magnetic field model (1) is determined by the Hamiltonian

$$H = \frac{P_z^2}{2m} + \frac{1}{2m}\left(P_x - qB_y Z\right)^2 + \frac{1}{2m}\left(P_y + q\psi(X, Z)\right)^2 \ . \tag{2}$$

Because of the time independence of the fields and because of the translational invariance in the dawn-dusk direction, the constants of motion are the energy $H = mv^2/2$ and the y-component of the canonical momentum P_y. If the flux function is independent of X, that is if $B_z = 0$, then P_x is conserved and the motion is integrable. Adiabatic particle motion in the $B_z = 0$ approximation is discussed by *Sonnerup* [1971].

A useful magnetic field model is given by

$$\psi = \left[L_z\, B_{x0}\left(1 - \cos\left(\frac{Z}{L_z}\right)\right) - L_x\, B_{z0}\right] e^{X/L_x} \ , \tag{3}$$

where L_x and L_z are constant. The magnetic flux (3) has been used by *Horton et al.* [1993] to study pressure anisotropies, $P_\perp/P_\parallel \neq 1$, in the geomagnetic tail.

An interesting limit of the flux (3) is the *Voigt-Fuchs* [1979] model

$$\psi = -L_z\, B_{x0} \cos\left(\frac{Z}{L_z}\right) e^{X/L_x} \ , \tag{4}$$

which is obtained from (3) by choosing L_x such that

$$\frac{L_x}{L_z} = \frac{B_{x0}}{B_{z0}} \ . \tag{5}$$

The Voigt-Fuchs model is obtained solving the Grad-Shafranov equation for the equilibrium problem $\nabla^2\psi = \mu_0 j_y = -\mu_0 p'(\psi)$ with an isotropic pressure profile

$$p(\psi) = \frac{k^2\,\psi^2}{2\mu_0}\ ,$$

where $k^2 = 1/L_z^2 - 1/L_x^2$. Model (4) has a broad current profile

$$j_y = -k^2\psi\ ,$$

which must be terminated at $|Z| \leq \pi L_z/2$ to avoid negative j_y. The magnetic field components for the Voigt-Fuchs model are given by

$$B_x \;=\; B_{x0}\sin\left(\frac{Z}{L_z}\right)e^{X/L_x}\ ,$$

and

$$B_z \;=\; B_{z0}\cos\left(\frac{Z}{L_z}\right)e^{X/L_x}\ .$$

A widely-used model in the study of nonlinear dynamics effects in the central plasma sheet is the modified Harris sheet model

$$\psi = B_{x0}\,L_z\,\ell n\left[\cosh\left(\frac{Z}{L_z}\right)\right] - B_z\,X\ , \qquad (6)$$

with the corresponding magnetic field

$$\mathbf{B} = B_{x0}\tanh\left(\frac{Z}{L_z}\right)\widehat{\mathbf{e}}_x + B_z\,\widehat{\mathbf{e}}_z\ . \qquad (7)$$

It is important to note that (6) is a self-consistent Vlasov equilibrium model only in the limit of vanishing B_z [*Harris*, 1962; *Lembége and Pellat*, 1982]. Model (6) can be considered as the $L_x \gg L_z$ limit of (3).

In the $|Z| \ll L_z$ and $|x| \ll L_x$ limit, both model (3) and the modified Harris model (6) reduce to the local parabolic field model

$$\mathbf{B} = B_{x0}\,\frac{Z}{L_z}\,\widehat{\mathbf{e}}_x + B_z\,\widehat{\mathbf{e}}_z\ . \qquad (8)$$

This model applies in the interior of the current sheet where the current density $\mu_0 j_y = dB_z/dx = B_{x0}/L_z$ is constant.

Another important model is the *Kan* [1973] model. This model is a two-dimensional, self-consistent, Harris-like equilibrium model which contains as special case the one-dimensional Harris sheet model

$$\mathbf{B} = B_{x0}\tanh\left(\frac{Z}{L_z}\right)\widehat{\mathbf{e}}_z\ . \qquad (9)$$

The parameters for the Kan model are chosen by considering the observed variation of the magnetic field tailwards in the x-direction. The flaring in the magnetic field lines described above for the Voigt-Fuchs model is also present in the Kan model and the thickness of the current sheet plasma is of the same order as the thickness of the central plasma sheet. However, because the magnetic field has to be evaluated at every time step, Kan's model is numerically expensive and will be considered elsewhere.

From the particle dynamics point of view, the different magnetic field models differ only in the fractional populations of orbits in the three classes. Thus, the general structure of the conductivity formula should be the same for the different field models.

In all the models described above the relevant parameters are

$$b_z \equiv \frac{B_{z0}}{B_{x0}}\ ,\ b_y \equiv \frac{B_y}{B_{z0}}\ , \qquad (10)$$

the finite Larmor radius parameter

$$\epsilon \equiv \frac{\rho}{L_z}\ , \qquad (11)$$

where $\rho = v_{th}/\omega_{cx0}$, $v_{th} = (2H/m)^{1/2}$, and $\omega_{cx0} = qB_{x0}/m$ and the ratio of characteristic scale lengths along the z and x-directions

$$\delta = \frac{L_z}{L_x}\ . \qquad (12)$$

Another important parameter is the chaos parameter of Büchner and Zelenyĭ

$$\kappa \equiv \frac{\omega_{cz}}{\omega_{bz}} = \frac{b_z}{\epsilon^{1/2}}\ , \qquad (13)$$

where $\omega_{cz} = qB_{z0}/m$ and ω_{bz} is the north-south oscillation frequency. The chaos parameter κ is related to the $\widehat{H}$ parameter of Chen and Palmadesso by $\widehat{H} = 1/2\kappa^4$. In the local limit of Eq. (8), the scaled Hamiltonian depends on the single parameter κ. In the general case, the problem depends on all four parameters ϵ, δ, b_y, and b_z.

3. Collisionless Mobility

The problem of deriving a collisionless irreversible resistivity for the current sheet plasma is very difficult because of the spatial inhomogeneity of the system and because of the nonlinear particle dynamics in the current sheet. Since there is a net cross-tail current density j_y, which leads to the irreversible Joule heating of the current sheet plasma when perturbed by an electric field E_y, we will focus on the dawn-dusk component of the conductivity tensor $\sigma_{yy}(k_x,\omega)$ and consider tearing-like perturbations of the form

$$\delta\,A_y(x,z,t) = \delta\,A_y(z)e^{i(k_x x-\omega t)} + \text{c.c.}\ . \qquad (14)$$

At this point we remark that the conductivity relates

perturbations in the currents to perturbations in the fields, that is

$$J_y = J_y^{(0)} + \delta J_y = J_y^{(0)} + \sigma_{yy}(k_x,\omega)E_y \ , \quad (15)$$

and that the existence of a current does not imply that the current is produced by an electric field, as illustrated by the dawn-dusk current $J_y^{(0)}$ which can be determined purely in terms of the Speiser orbits [*Holland and Chen*, 1992].

Before introducing the collisionless conductivity, we define the two-time single particle velocity correlation function $C(\tau, k_x; \mathbf{X}_0)$ by

$$C(\tau,k_x;\mathbf{X}_0) \equiv \lim_{T\to\infty}\int_{-T/2}^{T/2}\frac{dt}{T}v_{y,k_x}(t;\mathbf{X}_0)v_{y,-k_x}(t-\tau;\mathbf{X}_0) \ , \quad (16)$$

where

$$v_{y,k_x}(t;\mathbf{X}_0) \equiv v_y(t;\mathbf{X}_0)e^{-i\,k_x x(t;\mathbf{X}_0)} \ , \quad (17)$$

and where $\mathbf{v}(t;\mathbf{X}_0)$ and $\mathbf{x}(t;\mathbf{X}_0)$ are the velocity and position at time t of a particle with initial conditions $\mathbf{X}_0 = (\mathbf{x}_0, \mathbf{v}_0)$ and whose motion is obtained by integration of the equations of motion,

$$\frac{d\mathbf{x}}{dt} = \mathbf{v} \ , \quad \text{and} \quad \frac{d\mathbf{v}}{dt} = q\,\mathbf{v}\times\mathbf{B}(\mathbf{x}) \ , \quad (18)$$

in the unperturbed field $\mathbf{B}(\mathbf{x})$. For integrable particle motion the correlation function oscillates without decay. On the other hand, for chaotic orbits the correlation function decays. The later behavior is similar to the behavior of the correlation function in collisional systems. In these systems, the single particle velocity correlations decay due to collisions with other particles and the collisional time is the characteristic decorrelation time.

In general, the conductivity is proportional to some effective "collisional" time τ_c. In the highly collisionless environment of the geomagnetic tail the effective collisional time $\tau_c(k_x,\omega)$ is the ensemble-averaged decay time of the velocity correlations [*Horton and Tajima*, 1990]

$$\tau_c(k_x,\omega) \equiv \frac{1}{v_{th}^2}\left\langle\int_0^\infty d\tau\, C(\tau,k_x;\mathbf{X}_0)e^{i\omega\tau}\right\rangle \ , \quad (19)$$

where $\langle\ldots\rangle$ means average over the ensemble of particles $v_{th} \equiv (T/m)^{1/2}$, and the collisionless conductivity is given by

$$\sigma_{yy}(k_x,\omega) = \frac{nq^2}{m}\,\tau_c(k_x,\omega) \ . \quad (20)$$

Wave propagation in the current sheet plasma is governed by the dielectric tensor $\epsilon_{\alpha\beta}$, which is related to the conductivity tensor $\sigma_{\alpha\beta}$ through the equation

$$\epsilon_{\alpha\beta}(k_x,\omega) = \delta_{\alpha\beta} - \frac{\sigma_{\alpha\beta}(k_x,\omega)}{i\omega\epsilon_0} \ , \quad (21)$$

where $\delta_{\alpha\beta}$ is the Kronecker delta.

The dissipative or real part of the conductivity tensor corresponds to the imaginary part of the dielectric tensor, and from (20)–(19) we have that

$$\mathrm{Re}\,\sigma_{yy}(k_x,\omega) = \frac{n_0\,q^2}{2m\,v_{th}^2}\langle I(k_x,\omega;\mathbf{X}_0)\rangle \ , \quad (22)$$

where $I(k_x,\omega;\mathbf{X}_0)$ is the Fourier transform of the single-particle velocity correlation function,

$$I(k_x,\omega;\mathbf{X}_0) \equiv \int_{-\infty}^{\infty} d\tau\, C(\tau,k_x;\mathbf{X}_0)e^{i\omega\tau} \ , \quad (23)$$

which is also known as velocity power spectrum or spectral correlation function.

The origin of the irreversible dissipation or Joule Heating for the current sheet plasma can be understood in the following way. The velocity power spectrum for integrable particle motion vanishes everywhere except for some spikes (δ-functions) at the orbital frequencies. The spiky nature of the power spectrum for integrable or reversible motion is clearly exemplified by the motion of a particle in the presence of a constant uniform magnetic field B_z. In this case the power spectrum has a spike at the cyclotron frequency ω_{cz}, as shown in Figure 1, and energy absorption by the particles from the waves is possible only when the resonance condition $\omega = n\omega_{cz} + k_x v_x$ is met. On the other hand, chaotic particle motion is characterized by a continuous "grassy" power spectrum, which is the same kind of spectrum as that obtained for a collisional gas. To convey the message from this nonlinear theory more broadly, we may state that the chaotic particle dynamics is a "delta function buster." That is to say the complex time series generated by the orbits, in for example $v_y(t)$, yields a continuous frequency spectrum rather than the sharp delta function spectrum characteristic of integrable motion. The continuous "grassy" structure of the single particle velocity power spectrum for chaotic orbits is shown in Figure 2a, which corresponds to the chaotic orbit, in the modified Harris model, shown in Figure 2b.

Averaging the single particle power spectrum $I(k_x,\omega;\mathbf{X}_0)$ over all the particles, the ensemble averaged power spectrum $< I(k_x,\omega;\mathbf{X}_0) >$ exhibits a

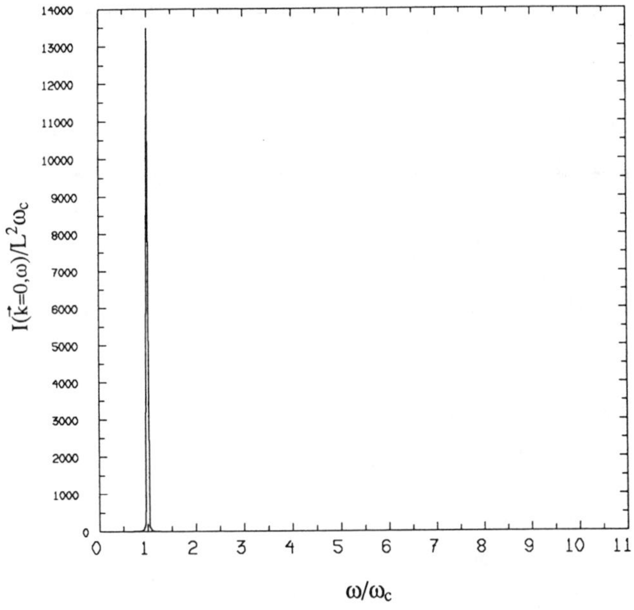

Fig. 1. Single Particle v_y-power spectrum for a charged particle in the presence of a constant, uniform B_z. The wave number is $\mathbf{k} = 0$, the unit of length is L, and the unit of time is ω_c^{-1}. The delta function at the cyclotron frequency demonstrates the discrete nature of the velocity power spectrum for integrable motion.

continuous, broad band structure. In systems where chaotic and integrable orbits coexist, the broad band structure of the ensemble-averaged power spectrum is due to both phase mixing and the broad band structure of the single particle power spectrum for chaotic orbits. The broadening of the ensemble-averaged power spectrum due to phase mixing is illustrated below, where we consider the case of a uniform unmagnetized plasma.

In previous work by *Horton and Tajima* [1991a], and *Doxas et al.* [1990], the conductivity formula was derived indirectly by integrating the equations of motion in the perturbed fields and using the results to calculate the dissipative part of the conductivity via

$$Re\, \sigma_{yy}(\omega, k_x) = \frac{\langle \delta J_y\, E_y \rangle}{\langle E_y^2 \rangle} \;. \tag{24}$$

Here we calculate directly the dissipative part of the conductivity using the information obtained by computing the velocity power spectrum. The practical difference is that in the first method using Eq. (24) the orbits are in the total $\mathbf{B}_0 + \delta \mathbf{B}$, δE_y field, while in the second method, based on the fluctuation-dissipation relation (FDR), they are computed in $\mathbf{B}(\mathbf{x}_0)$. We are currently using the FDR method to calculate complete expressions not only for the conductivity tensor $\sigma_{\alpha\beta}$ but also for the dielectric tensor $\epsilon_{\alpha\beta}$.

3.1 Unmagnetized Plasma

In order to illustrate the velocity power spectrum formalism, consider a uniform unmagnetized plasma. Assume that initially no electric and magnetic fields are present and that the unperturbed distribution function is a Maxwellian $f_M(v)$.

In the absence of unperturbed fields the particle motion consists only of straight lines

$$\begin{aligned} \mathbf{x}(t) &= \mathbf{x}_0 + \mathbf{v}t \;, \\ \mathbf{v}(t) &= \mathbf{v} = \text{const} \;. \end{aligned} \tag{25}$$

Assume now a perturbation of the form (14). From (16) and (25) we have that the single-particle velocity correlation function is given by

$$C(\tau, k_x; \mathbf{X}_0) = v_y^2\, e^{i\, k_x\, v_x \tau} \;, \tag{26}$$

Substituting (26) into (23) we find the v_y-power spectrum

$$I(k_x; \omega; \mathbf{X}_0) = 2\pi\, v_y^2 \delta(\omega - k_x\, v_x) \;. \tag{27}$$

Finally, averaging (27) over the Maxwellian distribution and substituting the result in (22), we get the dissipative part of σ_{yy}:

$$\text{Re}\; \sigma_{yy}(k_x, \omega) = \sqrt{\frac{\pi}{2}}\, \frac{nq^2}{m|k_x|\, v_{th}}\, e^{-\omega^2/(2k_x^2\, v_{th}^2)} \;. \tag{28}$$

Expressions for the other components of the conductivity tensor are calculated in a similar way. Figure 3 is a plot of the dissipative part of the conductivity. The smooth curve in Figure 3 is calculated from the theoretical expression (28), and the grassy curve is obtained by numerically calculating the velocity power spectrum (23) for each particle and averaging the result over the $N = 1000$ particles. The "spiky" appearance of the numerical curve is due to the delta function of Eq. (27), no to the chaotic orbits. Increasing the number of particles in the simulation, the spikes in the numerical curve are smeared out.

3.2 Green-Kubo Formulae

The justification of the spectral correlations approach for calculating the collisionless mobility is given by the Green-Kubo formulae [*Kubo*, 1985; *Sitenko*, 1982; *Klimontovich*, 1991; *Horton and Tajima*, 1991b; *Tajima et al.*, 1992; *Tajima and Cable*, 1992], which are obtained from linear response theory for many-body systems close to thermodynamic equilibrium, and are general relations between microscopic fluctuations and the transport coefficients. In particular, the mobility tensor

Fig. 2. (a): Velocity power spectrum for a charged particle executing chaotic motion in the modified Harris model (6). (b): Orbit of the particle used in (a). The complexity of the motion is due to the continuum mixing of frequencies. The unit of length is $(\rho L)^{1/2} \approx 1800\, km$, and the unit of time is $\omega_{bz}^{-1} \approx 1.8\, s$. The parameters of the simulation are $\kappa = 0.18$, $b_y = 0$, and $b_z = 0.05$. The initial conditions are $x = 0.6$, $P_x = 0.1$, $y = 0$, $P_y = 0.18$, $z = 0$, and $P_z = 0.99$.

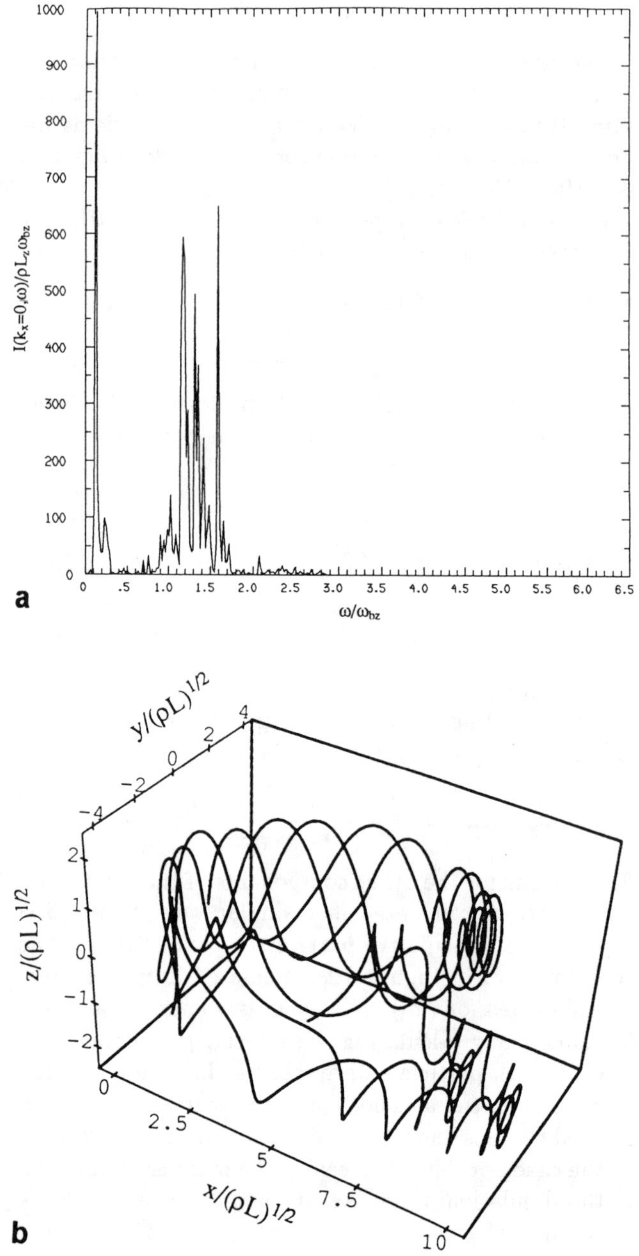

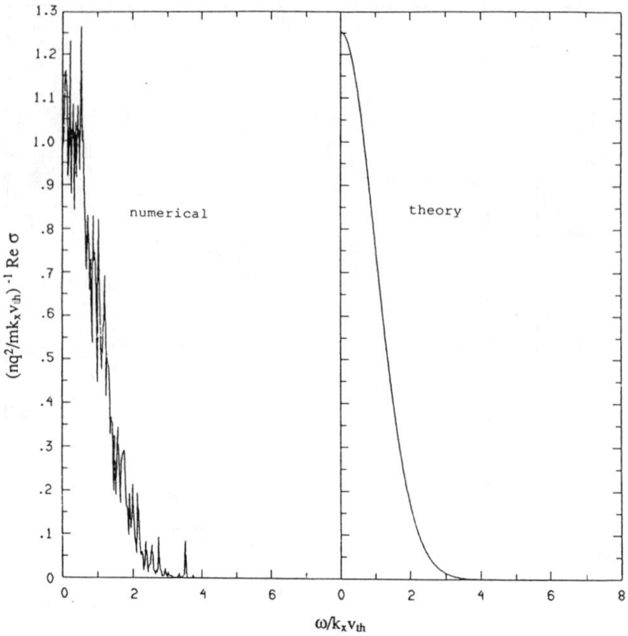

Fig. 3. Dissipative part of the y-component of the conductivity tensor, $\operatorname{Re} \sigma_{yy}(k_x, \omega)$, for an ensemble of particles in a field free region. The left side curve is $\operatorname{Re}\sigma_{yy}$ obtained numerically by computing the v_y-power spectrum and averaging the result over all the particles. The right side curve is the theoretical conductivity (28). The unit of length is $1/k_x$, the unit of time is $1/k_x v_{th}$. At $t = 0$ the particles were uniformly distributed from $x = 0$ to $x = 2\pi$ and their velocities were chosen according to a Maxwellian distribution.

for the plasma species "s" is given by

$$\mu_{\alpha\beta}^s(\mathbf{x}-\mathbf{x}', t-t') = \beta_s \langle v_\alpha(\mathbf{x},t) v_\beta(\mathbf{x}',t') \rangle \, , \quad (29)$$

where $\beta_s \equiv 1/k_B T_s$, and the velocity field

$$\mathbf{v}(\mathbf{x},t) = \sum_{i=1}^{N_s} \mathbf{v}_i(t) \delta(\mathbf{x} - \mathbf{x}_i(t)) = \sum_{\mathbf{k}} \mathbf{v}_{\mathbf{k}}(t) e^{i\mathbf{k}\cdot\mathbf{x}} \, , \quad (30)$$

with

$$\mathbf{v}_{\mathbf{k}}(t) = \sum_{i=1}^{N_s} \mathbf{v}_i(t) e^{-i\mathbf{k}\cdot\mathbf{x}_i(t)} \, , \quad (31)$$

where N_s is the number of particles of the species "s", and $\mathbf{x}_i(t)$ ($\mathbf{v}_i(t)$) is the position (velocity) of the ith particle at the instant t.

The electrical conductivity $\sigma_{\alpha\beta}$ is related to the mobility $\mu_{\alpha\beta}^s$ by $\sigma_{\alpha\beta} = \sum_s n_s q_s^2 \mu_{\alpha\beta}$ and the dissipative part of the diagonal elements of the electrical conductivity is given by

$$\operatorname{Re} \sigma_{\alpha\alpha}^s(\mathbf{k},\omega) = n_s q_s^2 \operatorname{Re} \mu_{\alpha\alpha}^s(\mathbf{k},\omega) \, , \quad (32)$$

where

$$\mathrm{Re}\,\mu^{s}_{\alpha\alpha}(\mathbf{k},\omega) = \beta_s \int_{-\infty}^{\infty} d\tau e^{i\omega\tau} \left\langle v_{\alpha,-\mathbf{k}}(t-\tau) v_{\alpha,\mathbf{k}}(t) \right\rangle . \tag{33}$$

Equations like (33), which relate spectral correlations to the dissipative part of the transport coefficients, are called fluctuation-dissipation theorems.

Holland and Chen [1992] and *Chen* [1992] have argued that the dc part of the collisionless conductivity, $\sigma^{dc} \equiv \lim_{\substack{\mathbf{k}\to 0\\ \omega\to 0}} \sigma(\mathbf{k},\omega)$, cannot be calculated on the basis of the fluctuation-dissipation theorem, which assumes, according to those authors, that the fluctuation frequencies are faster than the characteristic time scales. We question the validity of this criticism on the basis of the example for the unmagnetized plasma, in the last subsection, and on the basis of the example for cyclotron motion discussed by *Horton and Tajima* [1991b].

3.3 Dissipative Conductivity and the Dielectric Tensor

From Eq. (21), the dissipative (real) part of the conductivity corresponds to the imaginary part of the dielectric tensor. Once the dissipative part of the conductivity is determined, the full dielectric tensor can be derived using the Kramers-Kronig relations [*Ichimaru*, p. 64, 1992]

$$\mathrm{Re}\ \epsilon(\mathbf{k},\omega) - \mathbf{I} = \frac{1}{\pi} \int_{-\infty}^{\infty} d\omega' \frac{\mathcal{P}}{\omega' - \omega}\ \mathrm{Im}\ \epsilon(\mathbf{k},\omega') , \tag{34}$$

and

$$\mathrm{Im}\ \epsilon(\mathbf{k},\omega) = \frac{1}{\pi} \int_{-\infty}^{\infty} d\omega' \frac{\mathcal{P}}{\omega' - \omega} \left[\mathbf{I} - \mathrm{Re}\ \epsilon(\mathbf{k},\omega')\right] , \tag{35}$$

where $\mathbf{I}$ is the identity tensor.

The use of the Kramers-Kronig relations to calculate Re $\epsilon(\mathbf{k},\omega)$ and Im $\epsilon(\mathbf{k},\omega)$ from the velocity power spectrum is a well-known procedure, which can be carried out analytically for a uniformly magnetized plasma [*Tajima*, p. 46, 1989].

3.4 Numerical Calculation of the Spectral Correlations and the Conductivity

From the definition of the single particle velocity correlation function (16) and from (23) we have that

$$I(k_x,\omega;\mathbf{X}_0) = \frac{\omega_{\min}}{2\pi} \left|v_{y,k_x}(\omega;\mathbf{X}_0)\right|^2 \geq 0 , \tag{36}$$

where $v_{y,k_x}(t;\mathbf{X}_0)$ is defined in (17) and

$$\omega_{\min} = \frac{2\pi}{T} , \tag{37}$$

with T the total integration time in the numerical simulations. Typical values used in the simulations are $B_{x0} = 20\,nT$, $B_z = 1\,nT$, $B_y = 0 - 15\,nT$, $\rho = v_{th}/\omega_{cx0} = 500\,km$, $L_z = 1\,R_E$, total integration time $T = 1200\omega_{bz}^{-1} \approx 40$ min, and total number of particles $N = 500 - 1000$.

In the simulations we launch N particles uniformly distributed in the current sheet and satisfying a local Maxwellian distribution. The details of the initial distribution in configuration space are not really important as long as the particles tend to fill the magnetic flux tube after their launching. The velocity power spectrum is calculated for each particle according to (36), and the ensemble-averaged power spectrum is calculated according to the following rule

$$\langle I(k_x,\omega;\mathbf{X}_0)\rangle = \frac{1}{N} \sum_{j=1}^{N} I(k_x,\omega;\mathbf{X}_{0j}) ,$$

where the subscript j denotes the particle label and the sum runs over the N particles.

The procedure that we follow to determine the mobility is to generate $\langle I(k_x,\omega;\mathbf{X})\rangle$ for a range of values of k_x and ω keeping b_z and ϵ fixed, and then vary b_z and ϵ to determine how the mobility scales for fixed k_x and ω.

4. Effect of B_y on the Collisionless Mobility

From the observations [*Fairfield*, 1979; *Akasofu et al.*, 1978; *Cattel and Mozer*, 1982], the existence of a non-vanishing dawn-dusk magnetic field component B_y is well established. The B_y field is the dawn-dusk component of the IMF which soaks into the magnetosphere and can be comparable to the B_z component of the magnetic field.

The simplest model to study the effect of B_y on the mobility is to assume a constant B_y. The constant-B_y model, its effect on the mobility, and the probable stabilization of the collisionless tearing modes are discussed in *Hernandez et al.* [1993]. Here we give a brief account of the results. Surface of section plots for field reversal models including a constant B_y component are given by *Karimabadi et al.* [1990].

The constant-B_y field introduces a tilt on the magnetic field lines away from the meridian and increases the effective field line length in the current sheet, which is the region where the particles are coherently accelerated by the E_y field. The motion of the charged particles in the current sheet can be roughly decomposed into the cyclotron motion about B_z and the streaming motion of the particles along the tilted field lines. Each of these motions contributes separately to the conductivity and, for the low frequency part of the conductivity, we can write

$$\sigma(\omega; b_z, b_y, \epsilon) = a_1(\omega)\,\sigma_\perp(b_z, \epsilon) + a_2(\omega)\,\sigma_\parallel(b_y, b_z, \epsilon)\ , \tag{38}$$

where $\sigma_\perp$ is the contribution to the conductivity arising from the cyclotron motion about B_z, $\sigma_\parallel$ is the contribution to the conductivity due to the tendency of the particles to move along the tilted field lines, and a_1 and a_2, which are determined numerically, are weighting factors with a weak dependency on the frequency.

The local perpendicular conductivity is the *Lyons-Speiser*, [1985] conductivity

$$\sigma_{\rm LS} = \frac{nq^2}{m}\,\frac{2}{\pi|\omega_{cz}|}\ . \tag{39}$$

However, the Lyons-Speiser conductivity acquires its value only in a region of thickness $\Delta = (\rho L_z)^{1/2}$ about the equatorial plane, not in the whole central plasma sheet, whose half-thickness is L_z. Therefore, we need to multiply the Lyons-Speiser conductivity by a factor $\Delta/L_z = \epsilon^{1/2}$, and then

$$\sigma_\perp = \epsilon^{1/2}\,\sigma_{\rm LS} = \frac{nq^2}{m}\,\frac{2\epsilon^{1/2}}{\pi|\omega_{cz}|}\ . \tag{40}$$

The scaling of $\sigma_\perp$ with $\epsilon^{1/2}$ and $1/b_z$ is confirmed in *Horton and Tajima* [1991a] by using the full perturbed orbits in Eq. (24) and a more accurate numerical coefficient is given.

The parallel conductivity must be proportional to the extra time spent by the particles in the acceleration region because of the motion along the tilted field lines. The result is [*Hernandez et al.*, 1993]

$$\sigma_\parallel = \frac{nq^2}{m}\,\frac{\kappa^2 b_y^2 \epsilon^{1/2}}{|\omega_{cz}|}\ , \tag{41}$$

where again the $\Delta/L_z = \epsilon^{1/2}$ factor has been introduced for the same reasons as in the expression for $\sigma_\perp$ Eq. (40). From (38), (40), and (41) we get the low frequency conductivity formula

$$\sigma = \frac{nq^2}{m}\,\frac{\epsilon^{1/2}}{|\omega_{cz}|}\left[a_1 + a_2\,\kappa^2 b_y^2\right] \tag{42}$$

where, for the modified Harris sheet model (6), $a_1 \approx 0.05$ and $a_2 \approx 0.6$ The b_y^2 scaling of the conductivity has been thoroughly confirmed by the numerical simulations [*Hernandez et al.*, 1993].

From (42) we have that when an electric field is applied, the perturbed current density δj_y is enhanced with respect to its value for $B_y = 0$. For $b_y \ll 1$ the dominant contribution to the conductivity is given by $\sigma_\perp$; however, as b_y is increased the contribution from $\sigma_\parallel$ becomes more important, until, for some critical value of b_y, b_{yc}, $\sigma_\parallel$ overcomes $\sigma_\perp$. The critical value of b_y is found equating the Speiser and parallel contributions in (42) and is given by

$$b_{yc} = \left(\frac{a_1}{a_2}\right)^{1/2}\frac{1}{\kappa}\ . \tag{43}$$

Using (43) we can compare the critical b_y for the electrons, b_{yc}, with the critical b_y for the ions

$$b_{yce} = \left(\frac{\kappa_i}{\kappa_e}\right) b_{yci} = \left(\frac{\rho_e}{\rho_i}\right)^{1/2} b_{yci}\ . \tag{44}$$

Thus, due to the smallness of the electron Larmor radius, $\sigma_{\parallel e}$ dominates over $\sigma_{\perp e}$ for small values of b_y.

Another interesting quantity is the value of b_y for which the electron and ion conductivities coincide. Denoting this critical value by b_{yie}, and using (42) we have that

$$b_{yie} = \left(\frac{a_1}{a_2}\right)^{1/2}\left(\frac{1}{\kappa_i\,\kappa_e}\right)^{1/2}\ . \tag{45}$$

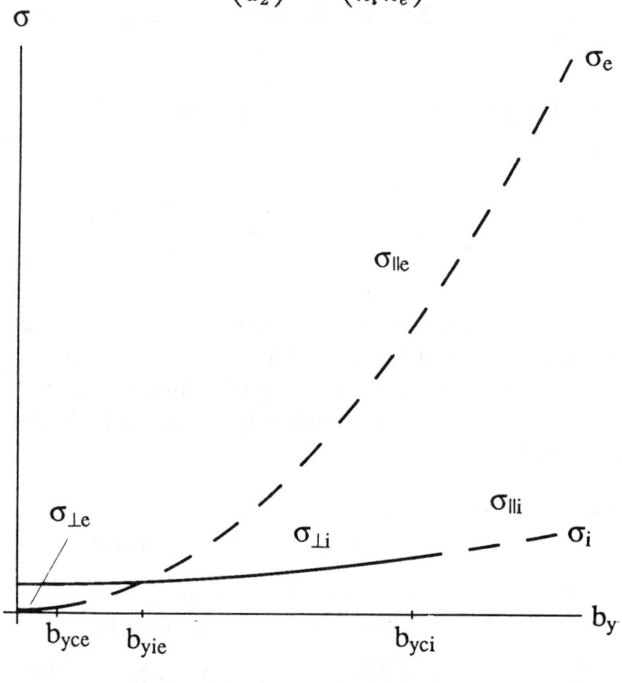

Fig. 4. Qualitative behavior of the conductivity (42) for both ions and electrons as a function of $b_y \equiv B_y/B_{z0}$. The solid (dashed) portions of the curves are the $\sigma_\perp$ ($\sigma_\parallel$)-dominated parts of the conductivities. The critical values b_{yce}, b_{yie}, and b_{yci} are obtained by setting $\sigma_{\perp e} = \sigma_{\parallel e}$, $\sigma_e = \sigma_i$, and $\sigma_{\perp i} = \sigma_{\parallel i}$.

From (44) and (45) we have that

$$b_{yci}:b_{yie}:b_{yce} = \left(\frac{\kappa_e}{\kappa_i}\right)^{1/2} > 1 \,. \qquad (46)$$

Choosing, for example, $\kappa_i = 0.2$, and $\kappa_e = 2.0$, we find that $b_{yce} = 0.14$, $b_{yie} = 0.44$, and $b_{yci} = 1.4$.

Figure 4 is a qualitative plot of σ_i, and σ_e as a function of b_y. The solid portions of the curves represent the conductivities σ_e, and σ_i, when the perpendicular conductivities $\sigma_{\perp e}$ and $\sigma_{\perp i}$ constitute the main contributions. The dashed curves represent the $\sigma_{\|}$-dominated phase.

Equation (42) was first obtained by considering the scaling of $\langle I(k_x,\omega)\rangle$ with respect to κ, b_y, and b_z. This is an example on how the velocity power spectrum can be used to obtain the dissipative part of the transport coefficients.

Acknowledgments. This work was supported by the National Science Foundation Grant ATM-88-11128, and by the NASA Grant NAGW-2590.

REFERENCES

Akasofu, S.-I., A.T.Y. Lui, C.I. Meng, and M. Haurwitz, Need for a three-dimensional analysis of magnetic fields in the magnetotail during substorms, *Geophys. Res. Lett.*, *5*, 283, 1978.

Akasofu, S.-I., The solar wind-magnetosphere energy coupling and magnetospheric disturbances, *Planet. Space Sci.*, *28*, 495, 1980.

Akasofu, S.-I., Energy coupling between the solar wind and the magnetosphere, *Space Sci. Rev.*, *28*, 121, 1981.

Allen, C.W., and D.W. Swift, A particle simulation of the tearing mode instability at the day side magnetopause, *J. Geophys. Res.*, *94*, 6925, 1989.

Baker, D.N., A.J. Klimas, R.L. McPherron, and J. Büchner, The evolution from weak to strong geomagnetic activity: An interpretation in terms of deterministic chaos, *Geophys. Res. Lett.*, *17*, 41, 1990.

Büchner, J. and L.M. Zelenyĭ, Deterministic chaos in the dynamics of charged particles near a magnetic field reversal, *Phys. Lett. A.*, *92*, 13,395, 1986.

Burkhart, G.R., and J. Chen, Differential memory in the earth's magnetotail, *J. Geophys. Res.*, *96*, 14,031, 1991.

Cattel, C.A., F.S. Mozer, E.W. Hones, Jr., R.R. Anderson, and R.D. Sharp, ISEE observations of the plasma sheet boundary, plasma sheet and neutral sheet, 1, electric field, magnetic field, plasma and ion composition, *J. Geophys. Res.*, *91*, 5663, 1986.

Chen, J., Nonlinear dynamics of charged particles in the magnetotail, to be published in *J. Geophys. Res.*, October, 1992.

Chen, J. and P.J. Palmadesso, Chaos and nonlinear dynamics of single-particle orbits in a magnetotail-like magnetic field, *J. Geophys. Res.*, *91*, 1499, 1986.

Chen, J., H.G. Mitchell, and P.J. Palmadesso, Differential memory in the trilinear model magnetotail, *J. Geophys. Res.*, *95*, 15,141, 1990.

Chen, J., G.R. Burkhart, and C.Y. Huang, Observational signatures of nonlinear magnetotail particle dynamics, *Geophys. Res. Lett.*, *17*, 2237, 1990.

Doxas, I., W. Horton, K. Sandusky, T. Tajima, and R. Steinfolson, Numerical study of the current sheet and plasma sheet boundary layer in a magnetotail model, *J. Geophys. Res.*, *95*, 12033, 1990.

Drummond, J.E., *Phys. Rev.*, *110*, 293, 1958.

Fairfield, D.H., The average configuration of the geomagnetic tail, *J. Geophys. Res.*, *84*, 1950, 1979.

Harris, E.G., On a plasma sheath separating regions of oppositely directed magnetic field, *Nuovo Cim.*, *23*, 115, 1962.

Hernandez, J., W. Horton, and T. Tajima, Low frequency mobility response functions for the central plasma sheet with application to tearing modes, *J. Geophys. Res.*, *98*, 5893, 1993.

Holland, D.L., and J. Chen, On chaotic conductivity in the magnetotail, *Geophys. Res. Lett.*, *19*, 1231, 1992.

Horton, W. and T. Tajima, Decay of correlations and the collisionless conductivity in the geomagnetic tail, *Geophys. Res. Lett.*, *17*, 123, 1990.

Horton, W. and T. Tajima, Collisionless conductivity and stochastic heating of the plasma sheet in the geomagnetic tail, *J. Geophys. Res.*, *96*, 15811, 1991a.

Horton, W., and T. Tajima, Transport from chaotic orbits in the geomagnetic tail, *Geophys. Res. Lett.*, *18*, 1583, 1991.

Horton, W., L. Cheung, J.-Y. Kim, and T. Tajima, Self-consistent pressure tensors from the Tsyganenko magnetic field models, to be published *J. Geophys Res.*, *98*, 17,327, 1993.

Ichimaru, S., Statistical plasma physics, Addison-Wesley Publishing Co., 1992.

Karimabadi, H., P.L. Pritchett, and F.V. Coroniti, Particle orbits in two-dimensional equilibrium models for the magnetotail, *J. Geophys. Res.*, *95*, 17,153, 1990.

Klimas, A.J., D.N. Baker, D.A. Roberts, D.H. Fairfield, and J. Büchner, A nonlinear analogue model of substorms, In *Magnetospheric Substorms*, Kan, J.R., T.A. Potemra, S. Kokubun, and T. Iijima, eds., pages 449-459, American Geophysical Union, 1991.

Klimontovich, Yu.L., Turbulent motion and the structure of chaos, Kluwer Academic Publishers 1991.

Kubo, R., M. Toda, and N. Hashitsume, Statistical Physics II, Springer-Verlag, New York, pages 167-174, 1985.

Lémbege, B., and R. Pellat, Stability of a thick two-dimensional quasineutral sheet, *Phys. Fluids*, *25*, 1995, 1982.

Lyons, L.R. and T.W. Speiser, Ohm's law for a current sheet, *J. Geophys. Res.*, *90*, 8543, 1985.

Martin, R.F., Chaotic particle dynamics near a two-dimensional magnetic neutral point with applications to the geomagnetic tail, *J. Geophys. Res.*, *91*, 11985, 1986.

Rosenbluth, M.N., and R. Rostoker, *Proc. 2nd Intern. Conf., Geneva*, *31*, 144, 1958.

Sitenko, A.G., Fluctuations and nonlinear wave interaction in plasmas, Pergamon Press, 1982.

Sonnerup, B.U., Adiabatic particle orbits in a magnetic null sheet, *J. Geophys. Res.*, *76*, 8211, 1971.

Tajima, T., Computational plasma physics: with applications to Fusion and Astrophysics, Addison-Wesley Publishing Co., 1989.

Tajima, T., S. Cable, K. Shibata, and R.M. Kulsrud, On the origin of cosmological magnetic fields, *The Astrophys. J.*, *390*, (1992).

Tajima, T., S. Cable, and R.M. Kulsrud, On zero-frequency magnetic fluctuations in plasmas, *Phys. Fluids B*, *4*, 2338, 1992.

W. Horton, J. Hernandez, and T. Tajima, Department of Physics and Institute for Fusion Studies, University of Texas, Austin, TX 78712.

Modeling Particle Distributions With Noninteracting Particles

R. F. Martin, Jr.[1]

Physics Department, Illinois State University, Normal, Illinois

Following the motion of individual particles has been a mainstay of magnetospheric modeling since the beginning of the field. We review recent applications of the single particle motion to the prediction of velocity distribution functions and discuss the observational evidence for them. New results are presented for the ion signature of an magnetotail X-type neutral line in the presence of a cross-tail magnetic field, B_y.

1. INTRODUCTION

The method of single particle dynamics has been used in Magnetospheric physics since its beginning. In the early decades of this century, Störmer followed the motion of particles in a magnetic dipole field in an attempt to understand the origin of the aurora (see [*Störmer*, 1955] for a review of his work). Störmer found that analytic solutions for particle dynamics in complex magnetic fields are difficult to come by (in fact, impossible, as we now know), and he was the first to use numerical integration to follow orbits. He was also able to obtain qualitative information on orbits, such as allowed and "forbidden" regions in phase space. These methods were used most fruitfully in analysis of cosmic rays.

Alfvén [1950] pioneered the guiding center approximation for motion in magnetic fields whose variations are small compared to a gyroradius. Such an approximation includes motion conserving the three adiabatic invariants (magnetic moment, bounce invariant, and drift invariant) in a dipole field, opening up a wide array of applications in the inner magnetosphere [see e.g. *Roederer*, 1970 and *Lyons and Williams* 1984].

When the field varies significantly over gyroradius scales, the invariants can be violated. In the magnetosphere, this occurs, for example, in the tail current sheet and the magnetopause, and these regions have been investigated in other approximations and numerically. Motion in these nonadiabatic regions can be very complex, and even chaotic under certain conditions [see e.g. *Lyons and Williams* 1984 and *Chen* 1992 for reviews].

In this paper we will consider the application of single particle dynamics to the magnetosphere, in particular its relatively recent use to model velocity distribution functions. Although the approximation of ensembles of noninteracting single particles has its drawbacks (lack of self-consistency being the major one), this method has nonetheless enjoyed a high degree of success. This *a posteriori* justification indicates that, at least in several regions of the magnetosphere, the collisionless nature of the plasma combines with an apparent lack of importance of collective effects, allowing the noninteracting particle approach to succeed.

We begin with a brief review of representative results, over the last decade or so, of single particle applications outside the inner magnetosphere. We concentrate on regions of dynamic interest: the magnetopause, the polar region, the current disruption region, and the tail current sheet. This will be followed by a presentation of new results on the effect of a cross-tail magnetic field on ion distributions near an X-line in the current sheet.

2. RECENT APPLICATIONS

Application of the noninteracting particles approach tends to take two different forms. First, one can study simple model systems in order to elucidate the basic physics underlying the interaction of particles with specific electromagnetic structures. Such studies may not bear a strong resemblance to reality, but the understanding resulting from them may be invaluable in interpreting

[1]Currently on leave at CETP/CNRS, Saint-Maur-des-Fossés, France

Space Plasmas: Coupling Between Small
and Medium Scale Processes
Geophysical Monograph 86

observations or more complex simulation results. The other approach is to begin with a field model that is designed to emulate the actual system (e.g. semi-empirical models), in order to more accurately explain observed distributions. The drawback in the latter case is that realistic models are often complex, with several adjustable parameters, making the basic physics less accessible. For both approaches, one is not limited to static field models, as one can choose explicitly time dependent fields or sequences of static field models. Nor is one limited to non-selfconsistent calculations since, with an iterative approach, one can obtain self-consistent steady state solutions.

2.1. *The Magnetopause*

Pioneering work by *Speiser et al* [1981] introduced the distribution function mapping method to the field. This method uses Liouville's Theorem, which states that the distribution function $f(\mathbf{v})$ is constant along particle trajectories. This allows one to map f from a source region, where f is assumed known (typically taken from observations), to the region of interest. In this way the effect of the intervening electric and magnetic fields on f can be understood, and results compared with observations (see Figure 1). This paper and subsequent work [*Speiser & Williams*, 1982] showed that quasi-trapped plasma was possible on open field lines indicating a magnetospheric source for some magnetosheath plasma. Agreement with ISEE-1 data was good for electric fields below about 0.5 mV/m. Figure 1 shows an example.

Curran et al [1987] and *Curran and Goertz* [1989] used the same mapping approach with a field model that included an X-type neutral line. Besides verifying the magnetospheric source, they obtained better agreement to ISEE-1 observations at higher electric fields, up to 1.8 mV/m. They were also able to show qualitative agreement with observations of the energy dependence of magnetosheath ion excursions (Figure 2).

2.2. *The Polar Region*

Horwitz [1986a] modeled ion distributions above the auroral acceleration region using a guiding center single particle simulation. He found good agreement between observed "bowl shaped" or conical distributions of transversely accelerated ions using a velocity filter method (Figure 3). The "filter" basically restricts the source distribution, and can be adjusted to provide a good fit to the data. Note in Figure 3 that very similar modeled distributions can be obtained with very different physics. There are other mechanisms proposed for producing transversely accelerated ion distributions involving wave-particle interactions that can produce similarly good agreement with observations(see e.g. Kintner [1991]). This points out a fundamental nonuniqueness problem in all such modeling: showing one mechanism that fits the data does not prove it is the correct mechanism, or the only one. One must consider other factors as well.

Using similar techniques, *Horwitz* [1986b] and *Horwitz and Lockwood* [1985] have studied the cleft "ion fountain" and the mass spectrometer effect of the polar magnetic field, with good agreement with observations. *Chappel et al* [1987] and *Horwitz et al* [1989] extended this work into the plasma sheet indicating that the ionosphere could be a major source of plasma sheet ions.

2.3. *The Current Disruption Region*

Time-dependent studies of the near-earth plasma sheet region have been done using a sequence of equilibrium magnetic field models, with the induction electric field computed from Maxwell's equations [*Mauk*, 1986; *Delcourt et al*, 1990; *Delcourt and Sauvaud*, 1991]. It was noted that magnetic moment conservation was violated due to rapid temporal changes in magnetic field, and that the boundary between adiabatic and nonadiabatic motion moved toward smaller L shells during substorm expansion phase (Figure 4). Identifying this adiabatic/nonadiabatic boundary as a substorm injection boundary, this effect was used to model the convection surge with qualitative agreement with observations. Ion banding due to the mass spectrometer effect was also modeled. *Ashour-Abdalla et al* [1992] find a similar dynamical boundary in a fixed field model, which they term the "wall", where ion chaos becomes important. Noting strong chaotic acceleration which enhances the cross-tail current, they speculate that this could lead to a current disruption.

2.4. *The Magnetotail Current Sheet*

The magnetotail is the most intensively studied region, using noninteracting particles. Its low density and highly collisionless nature make it an ideal candidate for such studies. Pioneering work by *Lyons and Speiser* [1982; henceforth LS] using the Liuoville mapping technique provided theoretical and numerical explanations for particle distributions in the plasma sheet boundary layer (PSBL). LS showed that a simple current sheet acceleration mechanism could effectively model parallel beams observed by ISEE spacecraft in the PSBL. Their comparison with data from *DeCoster and Frank* [1979] is shown in Figure 5. A Maxwellian distribution with a power law tail,

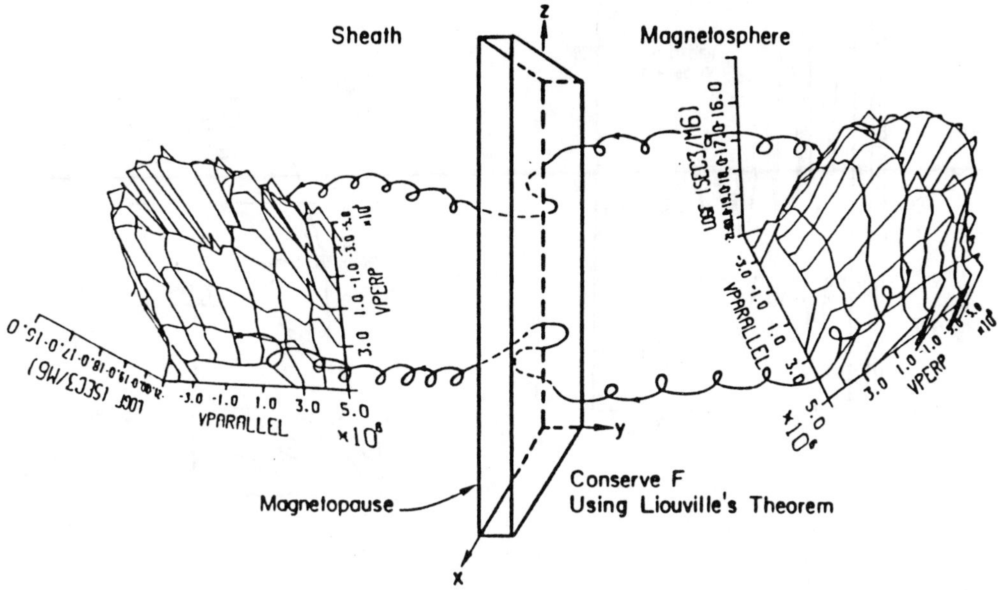

Fig. 1. Schematic representation of the Liouville mapping technique for modeling distribution functions (from *Speiser et al* [1981]). Here particles were traced backwards in time from the magnetosheath throught a one-dimensional model of the magnetopause, until they reached the magnetosphere. Then trajectories were tagged with an assumed source distribution obtained from observations, which is mapped back to the initial observation point.

modeling observed plasma mantle distributions was assumed for the source plasma. The parallel beam, then, is a signature of a constant B_Z linear current sheet magnetic field used in their simulation. These authors also presented a physical description and an analytic approximation [*Speiser and Lyons*, 1984] for this signature. *Moghaddam-Taheeri et al.* [1989] have suggested an alternate mechanism for production of the field-aligned beam: the interaction of ions with kinetic Alfvén waves.

One aspect of the PSBL beams evident in many observations is their "kidney bean" shape, which LS modeled with an enhanced tailward flow at higher energies. *Martin and Speiser* [1988] noted that an X-line magnetic field can produce similarly flattened beams without the enhanced flow. Better agreement has been obtained by *Onsager* et al [1991] using a combination of adiabatic deformation and velocity filter techniques.

Martin and Speiser [1988] (henceforth MS) and *Speiser and Martin* [1992] identified "ridges" in the $v_{\parallel}$ - $v_{\perp}$ distribution function as a signature of an *X* -type neutral line (Figure 6). The ridges arise from the topology of orbits: the *X*-line effectively splits the "observed" distribution into two parts with different source regions: earthward and tailward of the neutral line. In fact, the ridge is an observable consequence of the stable and unstable manifolds near the unstable fixed point at the X-line [*Martin*, 1994]. The fact that the source distribution was assumed to have a tailward bulk velocity amplifies the ridge: the distribution below the ridge is reduced since it's initially tailward (but earthward flowing) source is down in the tail of the source distribution, while above the ridge, tailward flowing particles from an earthward source are turned by B_z as in a constant B_z current sheet, thus coming from a higher probability region of the source distribution. Observational evidence for the ridge has been presented from ISEE [*Speiser et al.*, 1991] and AMPTE [*Speiser and Martin*, 1994] data.

Curran and Goertz [1989] have also investigated the *X* -line geometry. Employing a "double Harris" type *X*-line and Liuoville mapping, they observed a minimum in the omnidirectional flux, $j(E,z)$, at $z = 0$ as a possible signature. These authors suggested that neutral lines may be rare, since this signature is not observed.

Martin et al [1991] have investigated a similar model to MS including an *O* -type neutral line, in order to model a "magnetic island." Again a strong, nonequilibrium signature was observed: large jumps in the distribution function similar to the MS ridges. The *O* -line ridges, however, occur approximately at constant $v_{\perp}$ values, as opposed to *X* -line ridges which are closer to constant pitch angle. Moreover, multiple ridges are often seen for *O*-lines. Again the source of the jumps lies in the particle orbits:

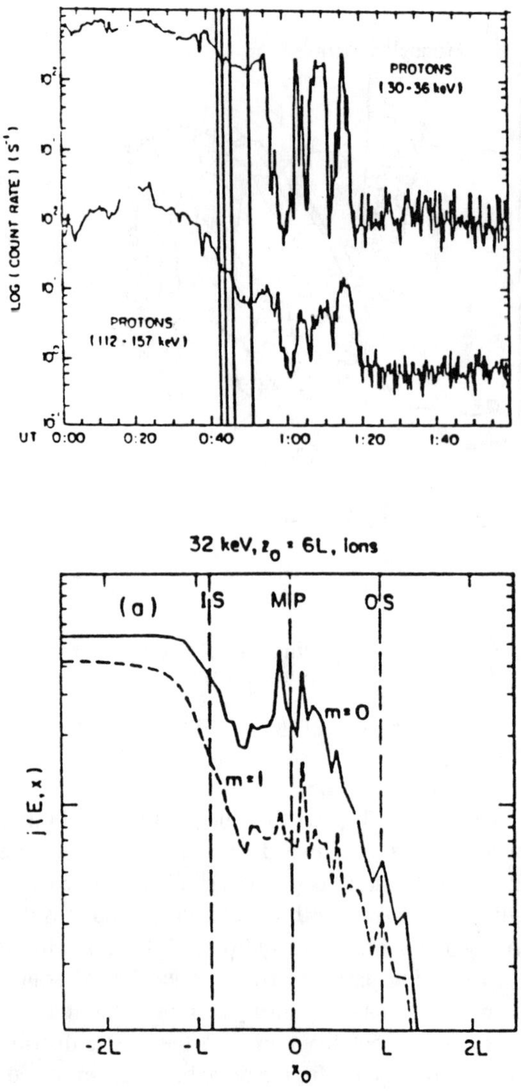

Fig. 2. Energy dependence of the omnidirectional ion flux through a two-dimensional model of the magnetopause (lower panel), to be compared with data from ISEE-1 (upper panel). Vertical lines in the upper panel denote magnetopause crossings. from *Curran and Goertz* [1989].

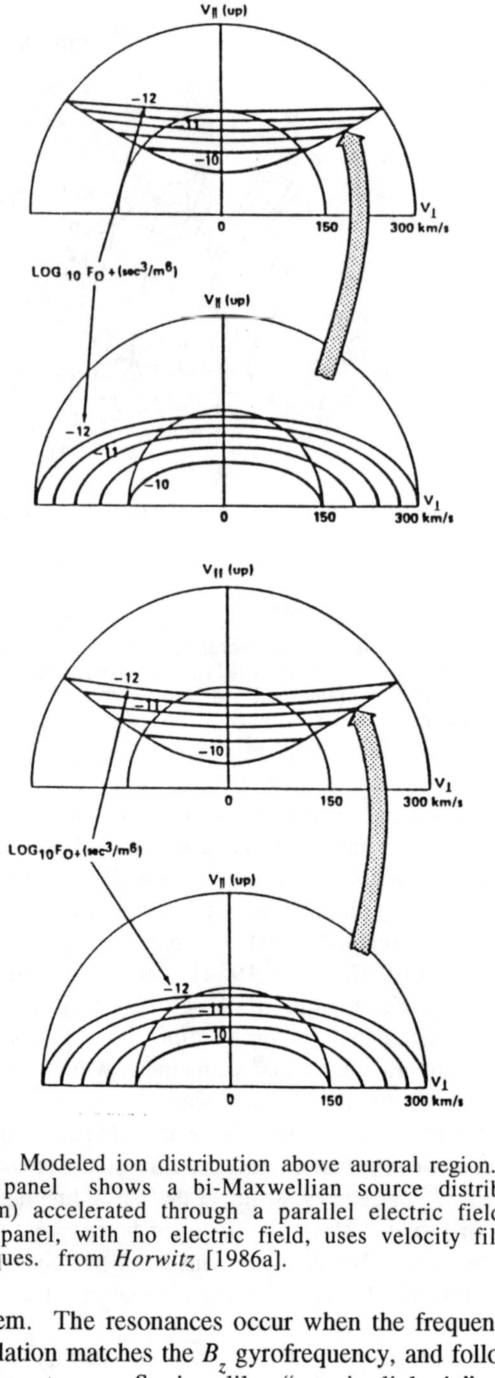

Fig. 3. Modeled ion distribution above auroral region. The upper panel shows a bi-Maxwellian source distribution (bottom) accelerated through a parallel electric field; the lower panel, with no electric field, uses velocity filtering techniques. from *Horwitz* [1986a].

each time the particle reverses its x velocity, a ridge ensues. With a tailward flowing source distribution, reversals yielding initially earthward ions cause a reduction in f, while those yielding initially tailward ions result in an increase in f.

Dips in the energy distribution $f(E)$ at resonant energies at $z = 0$ are predicted by *Burkhart and Chen* [1991]. Using a modified (constant B_z) Harris sheet, these authors start particles outside the current sheet with a Maxwellian velocity distribution and observe the subsequent velocity-space density at $z = 0$, also employing Liouville's Theorem. The resonances occur when the frequency of z oscillation matches the B_z gyrofrequency, and follow an $E^{-1/4}$ spectrum. Speiser-like "quasi-adiabatic" orbits dominate at resonant energies, while chaotic, quasi-trapped orbits occur at nonresonant energies. Resonant particles take a much longer time to reach the central plane, yielding gaps in the distribution. Good observational evidence for these dips was found in ISEE observations [*Chen et al.*, 1990]. Figure 7 shows an Example. Similar results were seen in a full tail simulation by *Ashour-Abdalla et al*

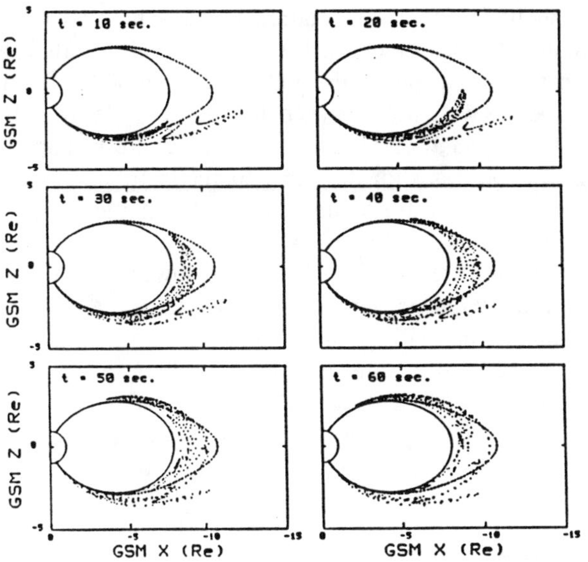

Fig. 4. Polar wind protons moving in an dipolarizing magnetic field. The dashed curve represents the initial adiabatic threshold while the solid curve is the final threshold. From *Delcourt and Sauvaud* [1991].

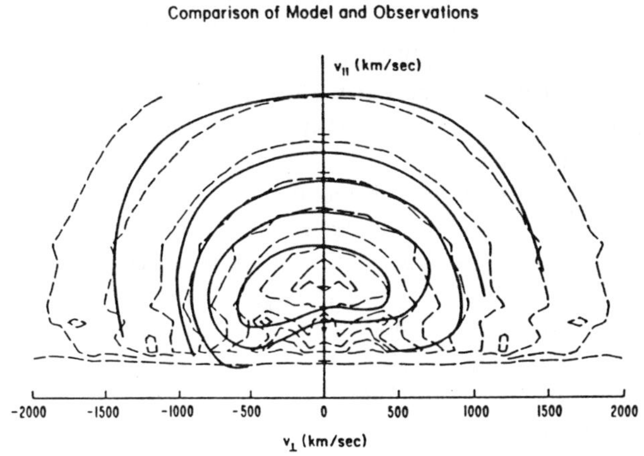

Fig. 5. Comparison of model results (solid contours) with ISEE data (dashed contours) in the plasma sheet boundary layer, showing the classic PSBL beam. From *Lyons and Speiser* [1982].

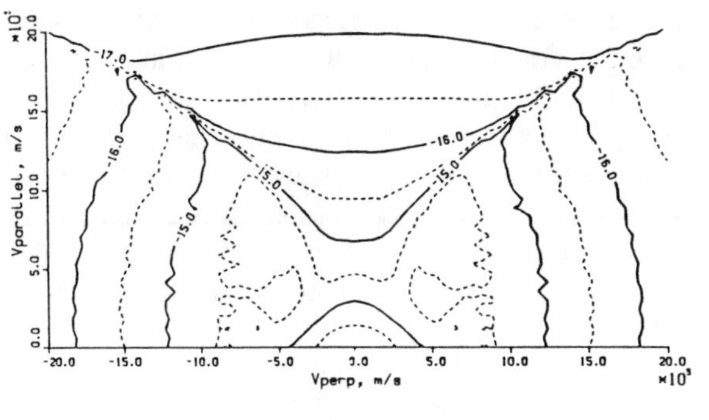

Fig. 6. Contours of the phase-averaged ion velocity distribution, showing the "ridge" caused by the source separation effect of the X-line. From *Martin and Speiser* [1988].

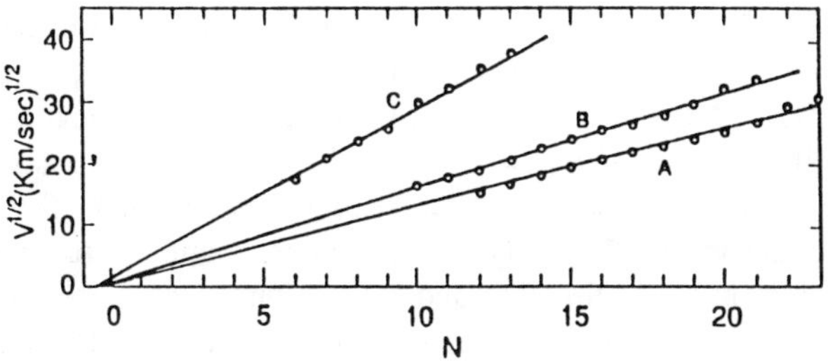

Fig. 7. Comparison of model results and ISEE-1 data showing agreement for the resonant energy dips in the energy distribution. The circles denote locations of valleys and peaks in the observed energy distribution, the lines are model calculations. The vertical axis is proportional to energy and the horizontal axis is the resonance number. (from Chen et al, [1990]).

[1991], who saw "beamlets" in the PSBL, and in the Tsyganenko model by *Kaufman* [1992] who dubbed the effect "correlated chaotic scattering". Theoretical justification has been given by *Burkhart and Chen* [1991] and *Buchner*, [1991].

Other authors have investigated more complex field models. *Doxas et al.* [1990] use a time-dependent tearing mode field which begins as a modified Harris sheet and develops an X-line and an O-line. No specific signature is observed, but a $T_{\perp}/T_{\parallel}$ asymmetry is observed at the CS center in accordance with observations. The MS ridge signature is also observable with this model [*Doxas and Speiser*, 1991; see also *Martin and Speiser*, 1992]. *Ashour-Abdalla et al.* [1990], using a source in the distant tail, were able to effectively model the formation of the plasma sheet boundary layer (PSBL), including the Lyons and Speiser beams. Subsequent work [*Ashour-Abdalla et al.*, 1991] with a mantle source and a distant neutral line also show the PSBL "beamlets" referred to above, which can reproduce some observed features of broadband noise in the region [*Shriver and Ashour-Abdalla*, 1991]. *Scholer and Jamitzky* [1987] have followed test particles in more complex, time-varying fields from an MHD simulation of the tail, including both X-lines and O-lines. They observed a high energy cutoff in the ion distribution function and a dawn-dusk asymmetry.

As the model complexity increases, it becomes more difficult to determine physical mechanisms for various signatures, even though correspondence with data can be achieved . For this reason, we will continue to study a simple model in order to elucidate the underlying physics, gradually adding complexity as our understanding grows. In this paper we continue the investigation of the *X* line model of *Martin and Speiser* [1988] and *Speiser and Martin* [1992, henceforth "SM"], investigating the effect that a magnetic field in the cross-tail *y*-direction has on the distribution.

2. THE EFFECT OF B_y

2.1. *Simulation Model*

We use the same *x* and *y* components of **B** as in SM: a magnetic field model consisting of hyperbolic field lines in a *T* by *L* region surrounding the neutral line, with constant normal field current sheets on its earthward and anti-earthward sides. *T*, here, is the Current sheet half-thickness while *L* is given by $L = T/b_n$ with $b_n = B_{z0}/B_{x0} = B(x = L, z = 0)/B(x = 0, z = T)$. The two-dimensional magnetic field is specified by $\mathbf{B_{xz}} = B_z\hat{x} + B_x\hat{z}$, where $\hat{x}$ and $\hat{z}$ are unit vectors in the *x* direction and *z* direction, respectively, in magnetospheric coordinates with origin at the neutral line (positive *x* toward Earth). To this is added a constant B_y component, tilting the field lines in the y-direction. In this study we used a uniform cross-tail electric field E_y, which has a disadvantage of producing parallel electric fields. However, preliminary work with a model in which $\mathbf{E} \cdot \mathbf{B} = 0$ give similar results.

In order to determine scaling behavior we write the fields and dynamical quantities in normalized variables defined by

$$B' = \frac{B}{B_{zo}}, \quad x' = \frac{x}{T}, \quad t' = \omega_{zo}t, \quad v' = \frac{v}{\omega_{zo}T}$$

where the maximum gyrofrequency about B_z is given by $\omega_{zo} = |q|B_{zo}/m$. Dropping the primes, the *x* and *z* component fields are

$$B_x = \begin{cases} b_n^{-1}z, & |z| < 1 \\ b_n^{-1}sign(z), & |z| \geq 1 \end{cases} \quad B_z = \begin{cases} b_n x, & |x| < b_n^{-1} \\ b_n sign(x), & |x| \geq b_n^{-1} \end{cases}$$

where $b_n = B_{zo}/B_{xo}$. We also define the parameter $b_y = B_y/B_{xo}$ as the relative strength of the B_y field.

In our simulation, orbits were started at an observation point at $z = L$, i.e. at the edge of the current sheet in the PSBL. Orbits were numerically integrated backwards in time until they leave the current sheet: exit conditions are 1) in *z*, at least two gyroradii above or below $|z| = T$; or 2) in *y* the equivalent of 40 R_E in the cross-tail direction. A time limit on the integration is also enforced to catch singular cases of trapped orbits. On exit, orbits can be tagged with a source distribution. A total of 15,625 initial velocities were used for each distribution: a 25 by 25 by 25 velocity grid in Cartesian velocity space. Results presented in section 3 are for the same source distribution used by MS, a tailward flowing Kappa distribution modeling plasma mantle ions (see MS for details).

Results are given in the following sections in unnormalized variables for definiteness and ease of comparison. Plots are shown for protons with the following tail parameters: $B_{z0} = 1$ nT, $B_{x0} = 20$ nT, $E_y = 2.5\times10^{-4}$ V/m, and $2T = 1000$ km, and tailward flow velocity, $U_{\perp x} = -350$ km/s (the same as the nominal parameters of MS and LS. For these values, $L = 10{,}000$ km.

2.2. *Results: PSBL Beam*

Figure 8 shows a sequence of distributions in the v_x- v_y plane (for $v_z = 0$) for an observation point at $x = 4L$, i.e. well earthward of the *X*-Line. The Lyons and Speiser beam is evident in panel (a) with $b_y = 0$, and remains well defined as b_y is increased to $b_y = 2$. Near $b_y = 4$ the beam begins to break up, but is still visible. As b_y is increased above 4, the beam disappears and the distribution becomes more isotropic, except for a "drop-out" at larger v_y values, seen clearly in panel (e).

The isotropization of the distribution is due to the increasing B_y field; the magnetic moment becomes more nearly conserved at $b_y = 20$, while it is not conserved for smaller values. The drop-out is a signature of the X-line: in the reduced *f* area, ions come from a source tailward of the *X*-line, while the isotropic portion above the drop-off is from earthward of the *X*-line (Figure 9). Thus, with a large B_y field, the *X*-line can be remotely sensed from a larger distance.

2.3. *Results: X-line Ridge and Drop-off*

The effect of B_y on the MS neutral line ridges is shown in Figure 10, taken at an $x = L$ observation point along the *X*-line separatrix. The ridge signature is evident up to $b_y = 4$, then fades away into a drop-off much like that observed at $x = 4L$, above. The source separation effect of the X-line is again the cause.

2.4. *Results: O-line Drop-Off*

Figure 11 shows the O-line drop-off signature presented in *Martin et al* [1991] for the same b_y values as in the

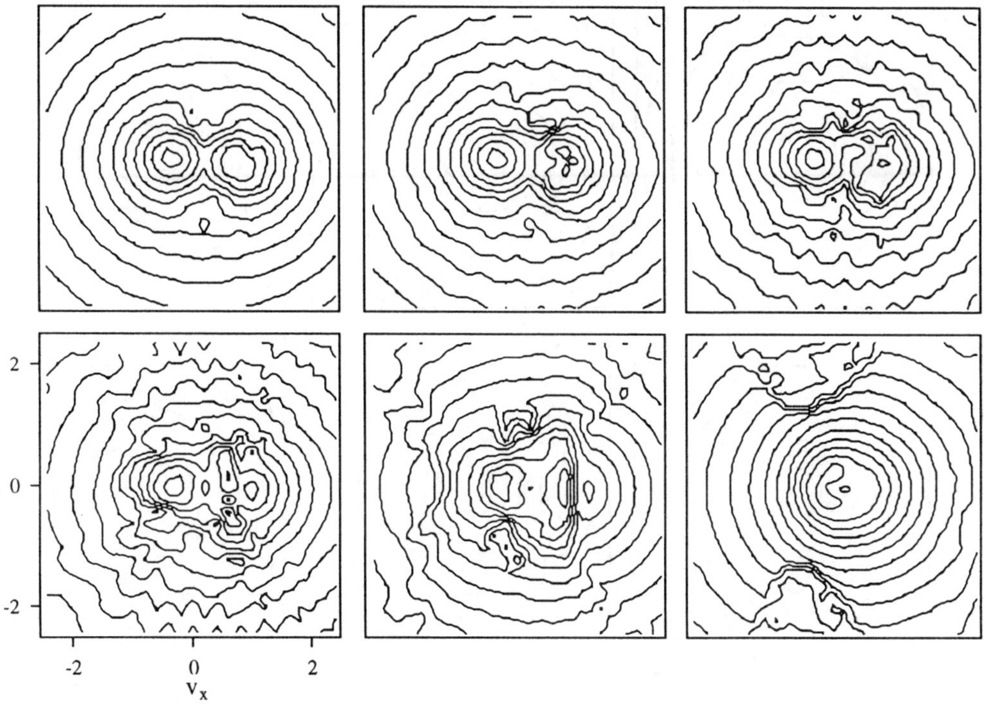

Fig. 8. Modeled distribution at $x = 4L$, well earthward of an X-type neutral line (the X-line is at $x = z = 0$), showing the evolution of the Lyons and Speiser PSBL beam as the field ratio b_y changes: a) $b_y = 0$, b) $b_y = 1$, c) $b_y = 2$, d) $b_y = 4$, e) $b_y = 8$, f) $b_y = 20$. Velocity axes are in units of 10^3 km/s.

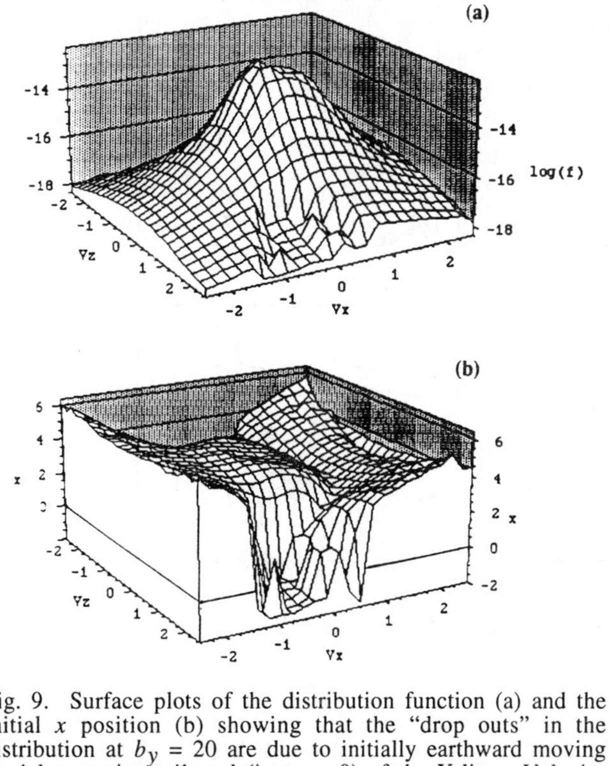

Fig. 9. Surface plots of the distribution function (a) and the initial x position (b) showing that the "drop outs" in the distribution at $b_y = 20$ are due to initially earthward moving particles starting tailward (i.e. $x < 0$) of the X-line. Velocity units are x 10^3 km/s and distance units are x10^4 km.

previous Figures. As b_y is increased the drop-off persists almost unchanged at $b_y = 4$, and even remaining visible at $b_y = 20$ at larger v_y values. Also evident at $b_y = 20$ is a steep decrease in f with "noisy" contours to the left. This appears to be a artifact of trapped orbits in the simulation which never leave the current sheet. In our simple model we stop such orbits at a fixed integration time, tagging them with the source distribution when stopped, even though they have not left the current sheet. Hence, to better understand this region, a more realistic source distribution for trapped orbits will be needed.

3. DISCUSSION

We have reviewed some recent results obtained from noninteracting particle simulations in the magnetosphere. The noninteracting particle approach has been found to be an effective method to study kinetic effects on particle distributions in several regions, yielding results which compare favorably to observations. The method is computationally efficient compared to Vlasov and self-consistent simulations allowing much larger scale regions to be studied, including full magnetosphere models. The full particle dynamics is used, including nonadiabatic effects and chaotic motion, which cannot be studied with Magnetohydrodynamics methods. Moreover, analytic

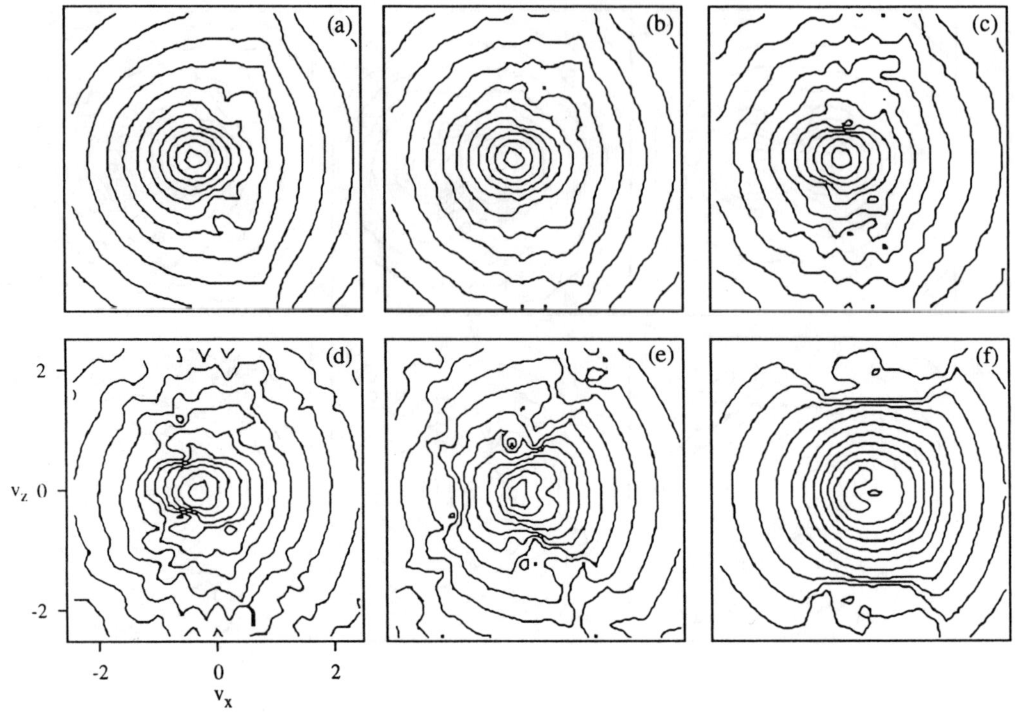

Fig. 10. Modeled distribution at $x = L$, earthward of an X-type neutral line, along the separatrix. The evolution of the neutral line ridge is shown, as the field ratio b_y changes: a) $b_y = 0$, b) $b_y = 1$, c) $b_y = 2$, d) $b_y = 4$, e) $b_y = 8$, f) $b_y = 20$. Units as in Fig. 8.

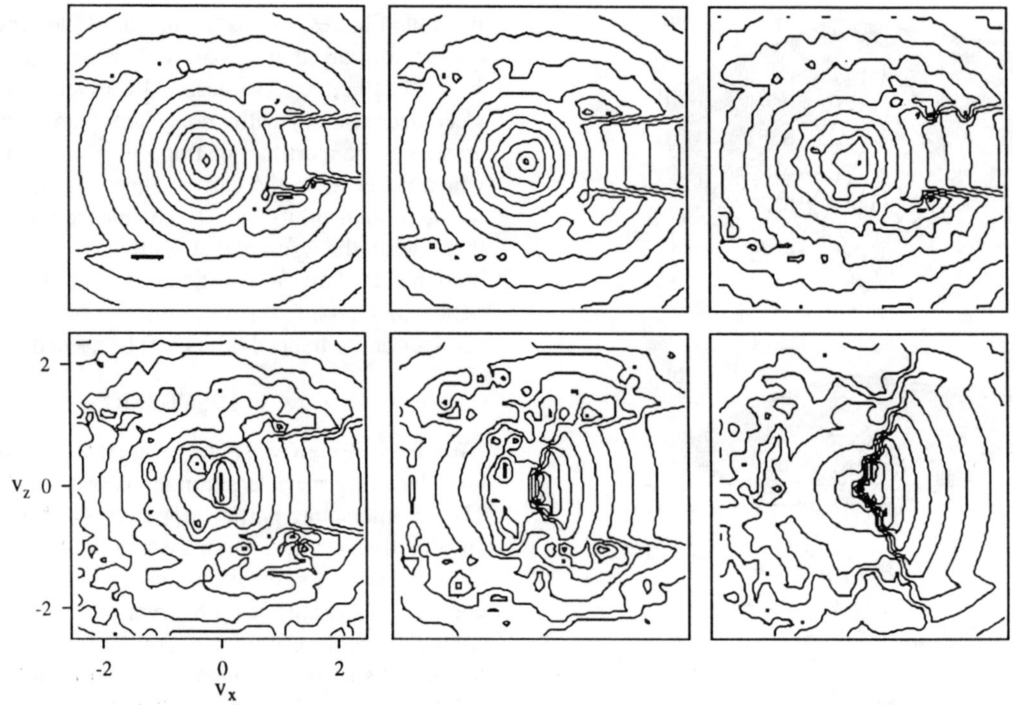

Fig. 11. Modeled distribution at $x = -L$, tailward of an O-type neutral line, showing the evolution of the O-line drop-offs as the field ratio b_y changes: a) $b_y = 0$, b) $b_y = 1$, c) $b_y = 2$, d) $b_y = 4$, e) $b_y = 8$, f) $b_y = 20$. Units as in Fig. 8.

results have been obtained in many cases, as well (see e.g. Chen, 1992). The major drawback is that collective effects cannot be included so that regions with strong wave-particle interactions must be studied with other methods.

We have also presented new results on the effect of a cross-tail B_y magnetic field on PSBL beams and X-line and O-line distribution ridges and drop-offs. In all cases, we find the respective signatures remarkably unaffected for $B_y \leq 4B_z$, showing that the dynamical source-separation property of neutral lines to be rather robust. For large b_y an X-type neutral line can be sensed by a drop-off in f at a larger distance than for small b_y, while the effect is masked in the O-line case due to trapped orbits.

Acknowledgments. The author wishes to thank Karen Klamczynski for assistance with the numerical simulations, and appreciates helpful discussions with T. Speiser. This research has been supported by the National Science Foundation (ATM-9002447) and by Illinois State University.

REFERENCES

Alfvén, H., *Cosmical Electrodynamics*, 1st edition, Oxford University Press, London and New York, 1950.

Ashour-Abdalla, M., J. Berchem, J. Büchner, and L. M. Zelenyi, Chaotic scattering and acceleration of ions in the Earth's magnetotail, *Geophys. Res. Lett., 17*, 2317, 1990.

Ashour-Abdalla, M., J. Berchem, J. Büchner, and L. M. Zelenyi, Large and small scale structures in the plasma sheet: A signature of chaotic motion and resonance effects, *Geophys. Res. Lett., 18*, 1603, 1991.

Ashour-Abdalla, M., L. M. Zelenyi,, J.-M. Bosqued, V. Peroomian, Z. Wang, D. Schriver, and R. L. Richard, The formation of the wall region: consequences in the near earth magnetotail, *Geophys. Res. Lett., 19* 1739, 1992.

Büchner, J., Correlation-modulated chaotic scattering in the earth's magnetosphere, *Geophys. Res. Lett., 18*, 1595, 1991.

Burkhart, G. R. and J. Chen, Differential memory in the Earth's magnetotail, *J. Geophys. Res., 96*, 14,033, 1991.

Chappel, C. R., T. E. Moore, and J. H. Waite, Jr., The ionosphere as a fully adequate source of plasma for the earth's magnetosphere, *J. Geophys. Res., 92*, 5896, 1987.

Chen, J., Nonlinear dynamics of charged particles in the magnetotail, *J. Geophys. Res., 97,* 11539, 1992.

Chen, J., G. R. Burkhart, and C. Y. Huang, Observational signatures of nonlinear magnetotail particle dynamics, *Geophys. Res. Let., 17*, 2237, 1990.

Curran, D. B., C. K. Goertz, and T. A. Whelan, Ion distributions in a two-dimensional reconnection field geometry, *Geophys. Res. Lett., 14*, 99, 1987.

Curran, D. B., and C. K. Goertz, Particle distributions in a two-dimensional reconnection field geometry, *J. Geophys. Res., 94*, 272, 1989.

DeCoster, R. J. and L. A. Frank, Observations pertaining to the dynamics of the plasma sheet, *J. Geophys. Res., 84*, 5099, 1979.

Delcourt, D. C., J. A. Sauvaud, and A. Pedersen, Dynamics of single-particle orbits during substorm expansion phase, *J. Geophys. Res.*, 95, 20853, 1990.

Delcourt, D. C., and J. A. Sauvaud, Generation of energetic proton shells during substorms, *J. Geophys. Res.*, 96, 1585, 1991.

Doxas, I., and T. W. Speiser, Evolution of the particle distribution function in a neutral line model (abstract), *Eos Trans. AGU, 72*, p. 251, 1991.

Doxas, I., W. Horton, K. Sandusky, T. Tajima, and R. Steinolfson, Numerical study of the current sheet and plasma sheet boundary layer in a magnetotail model, *J. Geophys. Res., 95*, 12,033, 1990.

Horwitz, J. L. and M. Lockwood, The cleft ion fountain: a two dimensional kinetic model, *J. Geophys. Res., 90*, 9749, 1985.

Horwitz, J. L., Velocity filter mechanism for ion bowl distributions (bimodal conics), *J. Geophys. Res., 91*, 4513, 1986a.

Horwitz, J. L., The tail lobe ion spectrometer, *J. Geophys. Res., 91*, 5689, 1986b.

Horwitz, J. L., K. Swinney, G. Wilson, T. Reyes, D. C. Delcourt, and T. E. Moore, "Kinetic" modeling of the transport of ionospheric ions into the magnetosphere, in *Physics of Space Plasmas (1988), SPI Conference Proceedings and Reprint Series*, Number 8, (Scientific Publishers, Inc., Cambridge, MA, 1989).

Kaufman, R. L., D. J. Larson, P. Beidl, and C. Lu, Mapping and energization in the magnetotail 1. Magnetospheric boundaries, *J. Geophys. Res., 98*, 9307, 1993.

Kintner, P. M., Plasma waves and transverse ion acceleration: a tutorial, in *Physics of Space Plasmas (1991), SPI Conference Proceedings and Reprint Series*, Number 11, edited by T. Chang, G. B. Crew, and J. R. Jasperse, Scientific Publishers, Inc., Cambridge, MA, 1992.

Lyons, L. R., and T. W. Speiser, Evidence for current sheet acceleration in the geomagnetic tail, *J. Geophys. Res., 87*, 2276, 1982.

Lyons, L. R. and D. J. Williams, *Quantitative Aspects of Magnetospheric Physics*, D. Reidel, Hingham, Mass., 1984.

Martin, R. F., Jr., and T. W. Speiser, A predicted energetic ion signature of a neutral line in the geomagnetic tail, *J. Geophys. Res., 93*, 11,521, 1988.

Martin, R. F., Jr., D. F. Johnson, and T. W. Speiser, The energetic ion signature of a plasmoid in the geomagnetic tail, *Adv. in Space Sci.*, vol. *11*, pp. 203-206, 1991.

Martin, R.F., Jr. and T.W. Speiser, "Nonlinear Dynamics of the Neutral Line Hamiltonian", in *Physics of Space Plasmas (1991), SPI Conference Proceedings and Reprint Series*, Number 11, edited by T. Chang, G. B. Crew, and J. R. Jasperse, pp. 311-334, Scientific Publishers, Inc., Cambridge, MA, 1992,

Martin, R.F., Jr., "Observable consequences of the stable-unstable manifold in a magnetic neutral line model", *Physics of Space Plasmas (1993), SPI Conference Proceedings and Reprint Series*, Number 13, edited by T. Chang, G. B. Crew, and J. R. Jasperse, Scientific Publishers, Inc., Cambridge, MA, in press, 1994.

Moghaddam-Taheeri, E., C. K. Goertz, and R. A. Smith, Ion beam generation at the plasma sheet boundary layer by kinetic Alfvén waves, *J. Geophys. Res., 94*, 10,047, 1989.

Mauk, B. H., Quantitative modeling of "convection surge" mechanism of ion acceleration, *J. Geophys. Res., 91*, 13423, 1986.

Onsager, T. G., M. F. Thomsen, R. C. Elphic, and J. T. Gosling, Model of electron and ion distributions in the plasma sheet boundary layer, *J. Geophys. Res., 96*, 20999, 1991.

Roederer, J. G., *Dynamics of Geomagnetically Trapped Radiation*, Springer-Verlag, Heidelberg, 1970.

Scholer, M., and F. Jamitzky, Particle orbits during the development of plasmoids, *J. Geophys. Res., 92*, 12,186, 1987.

Speiser, T. W., D. J. Williams and H. A. Garcia, Magnetopause modeling: flux transfer events and magnetosheath energetic ions, *J. Geophys. Res., 87,* 2177, 1982.

Speiser, T. W. and D. J. Williams, Magnetospherically trapped ions as a source of magnetosheath energetic ions, *J. Geophys. Res., 86,* 723-732, 1981.

Speiser, T. W., and L. R. Lyons, Comparison of an analytical approximation for particle motion in a current sheet with precise numerical calculations, *J. Geophys. Res.*, *89*, 147, 1984.

Speiser, T. W., P. B. Dusenbery, R. F. Martin, Jr., and D. J. Williams, Particle orbits in magnetospheric current sheets: Accelerated flows, neutral line signature, and transitions to chaos, in *Modeling Magnetospheric Plasma Processes*, *Geophys. Monogr. Ser.,* vol. *62*, edited by G. R. Wilson, pp. 71-79, AGU, Washington, D. C., 1991.

Speiser, T.W., and R.F. Martin, Jr., "Energetic ions as remote probes of neutral lines in the geomagnetic tail", *J. Geophys. Res.*, 97, 10775, 1992.

Speiser, T.W., and R.F. Martin, Jr., "Neutral line energetic ion signatures in the geomagnetic tail: comparison with AMPTE observations", in*Micro and Meso Scale Phenomena in Space Plasma Physics,* edited by Maha Ashour-Abdalla and T. Chang, AGU, (this volume), 1994.

Störmer, C., *The Polar Aurora*, Clarendon Press, Oxford, 1955.

R. F. Martin, Jr., Physics Department, Illinois State University, Normal, IL 61790-4560.

Neutral Line Energetic Ion Signatures in the Geomagnetic Tail: Comparisons with AMPTE Observations

T. W. Speiser

Astrophysical, Planetary and Atmospheric Sciences Department, University of Colorado, and Space Environment Laboratory, NOAA, Boulder, Colorado

R. F. Martin, Jr.

Physics Department, Illinois State University, Normal, Illinois

Neutral lines in the geomagnetic tail current sheet are thought to form near the earth near the onset time of substorms. Energetic ions which pass through the current sheet structure near a neutral line are more easily accelerated by the cross-tail electric field and thus impart a distinctive signature on the ion distribution function. The signature in the plasma sheet boundary layer is mainly a ridge in the distribution function plotted in $v_\perp/v_\parallel$ space. In the center of the current sheet, we project the distributions into three mutually perpendicular planes, v_x/v_y, v_x/v_z, v_y/v_z, in order to facilitate comparison with AMPTE/IRM observations [*Nakamura et al.*, 1991]. The model results for this time near the onset of a substorm, tell us that the source plasma was pre-accelerated earthward in the distant tail, that a secondary source was due to mirroring primaries, that a z-component of the electric field with magnitude of about 5 to 25 *mV/m* must have existed at the time of the observations, and that AMPTE/IRM was near but earthward of a near-earth neutral line.

1. INTRODUCTION

The observational evidence for magnetic neutral lines in the geotail is incomplete and indirect. Neutral lines have not been observed directly, but only inferred from other measurements, for example a change in direction of B_z, enhanced tailward plasma flow, tailward jetting of energetic ions and electrons, and observation of a plasmoid (itself indirect). The CDAW 6 substorm is a classic example where such indirect evidence has been used to infer the existence of a near-earth neutral line [*Fritz et al.*, 1984; *Paschmann et al.*, 1985]. Indirect energetic particle and plasmoid observations in the CDAW 7 [*Hones et al.*, 1986; *Kettmann et al.*, 1990] events are also interpreted as implying a near-earth neutral line.

To aid in the identification of such magnetic structures several neutral line signatures have been suggested from model results. Ridges, with dropoffs below, in the velocity space distribution function occur due to the neutral line separating source regions [*Martin and Speiser*, 1988; *Speiser and Martin*, 1992 (hereafter referred to as *MS* and *SM*, respectively)]. These authors have presented preliminary observational evidence for the ridges from the CDAW 6 energetic ion data [*Speiser et al.*, 1991]. Similar features have been seen by *Lyons and Pridmore-Brown* [1990]. *Curran and Goertz* [1989] predicted a dropoff in the omnidirectional flux at the center of the current sheet as a neutral line signature, suggesting that neutral lines might not exist, since this signature is not observed. *Doxas et al.* [1990] noted a rapid decrease in ion density before the formation of a neutral line in their model, and *Burkhart et al.* [1990, 1991] found sharp increases in both temperature and density just downstream from the boundary between inflow and outflow near the neutral line.

In this paper we provide further evidence for the observation of a neutral line using the ridge signature of Martin and Speiser. Through detailed comparisons with

Space Plasmas: Coupling Between Small
and Medium Scale Processes
Geophysical Monograph 86

AMPTE/IRM ion data in the central current sheet, we interpret a strong distribution function feature seen by *Nakamura et al.* [1991] (hereafter referred to as NPBS) as a neutral line ridge. Additionally, the model puts constraints on the nature of the source plasma.

2. OBSERVATIONS

NPBS have recently published observations of AMPTE/IRM three-dimensional ion distribution functions near and inside the tail current sheet (CS). The data were classified according to substorm phase. Their figure 4 is an example of a fast flow in the current sheet closely following substorm onset. This figure is reproduced here as Figure 1. This event is for 1 Mar 85, 02:42:40 - 02:45:19 *UT*, with substorm onset at 02:42, when AE jumped from 100 to 400 *nT*. At 02:45:19 *UT*, the fastest flow measured was 522 *km/s*, in earthward and dawnward directions, and the magnetic field was 1.6 *nT* northward. Figure 1a shows the magnetic field and plasma parameters. Three cuts through the three-dimensional distribution function are shown in Figure 1(b) in the (v_x,v_z), (v_y,v_z), and (v_x,v_y) planes in GSM coordinates. Projections of the magnetic field, $\boldsymbol{B}$, the bulk velocity, v, and the electric field, $\boldsymbol{E} = -\mathrm{v}\times\boldsymbol{B}$, on each plane are shown. The range of the observation is up to 40 *keV*, with the velocity scale given at the right of the plots. The phase space density scale is indicated at the bottom in s^3cm^{-6}. The distribution function shows a partial shell whose center of gravity is shifted earthward and dawnward. In the following section, we show model results of distribution function mappings for various initial distributions, and we compare the model results with the NPBS observations in section 4.

3. THE MODEL AND PREVIOUS RESULTS

We follow individual ions in a model of the tail current sheet, as discussed in *MS* and *SM*. The current sheet model is shown in Figure 2. In this model, a 2D neutral line is attached to a constant B_z current sheet with B_z at the center northward on the earthward ($x > 0$) side of the neutral line, and southward on the tailward ($x < 0$) side. The current sheet thickness is $2T$ and the neutral line width in the x-direction is $2L$. T and L are related by $T/L = B_{z0}/B_{x0}$, where B_{z0} is the magnitude of the normal component and B_{x0} is the magnitude of the B_x component for $|x| \geq L$, $|z| > T$.

For our simulations, we specify T, B_{x0}, B_{z0}, and E_y, the assumed cross-tail electric field. Note that for $E_y > 0$ (our general assumption), the model mimics reconnection, i.e., $\boldsymbol{E}\cdot\boldsymbol{j} > 0$ throughout the CS. Also note that E_y cannot be transformed away, as B_z is not constant, i.e., there is no unique DeHoffman-Teller frame, in contrast to some earlier studies. For each simulation run, the Lorentz force equations are numerically integrated backwards in time, with orbits started at an observation point and followed until they leave the CS. On exit, orbits can be tagged with a source distribution, and the source distribution then mapped back to the observation point, using the Liouville Theorem. A total of 29,400 initial velocities were used for each satellite position: a 35 by 70 velocity grid in the $v_\perp$-$v_\parallel$ plane, for each of 12 phase angles from 0° to 330°. See *MS* and *SM* for details and examples of mapped distributions. In *MS*, for example, it was shown that the model produces a beam in the plasma sheet boundary layer (PSBL) when the observation point is far from the neutral line. This beam essentially reproduces the model results of *Lyons and Speiser* [1982] and those model results compared favorably with *DeCoster and Frank's* [1979] observations. *MS* then found that the beam degenerated and was replaced by a "ridge-like" structure as the observation point approached the neutral line along the PSBL. *SM* showed mappings along the PSBL, in a north-south cross-section at $x = L$, and at the CS center ($z = 0$). The ridge structure was found throughout the current sheet, and it was found to be accentuated either earthward or tailward of the neutral line for a flowing (asymmetric) source distribution.

Mapped distribution functions were also shown for the PSBL in cartesian (v_x,v_y), (v_y,v_z), (v_x,v_z) coordinates by *SM*. In these coordinates, the phase dependency is not averaged out as it is in ($v_\perp$,$v_\parallel$) space, but nevertheless, the ridge structure was still apparent, especially in (v_x,v_y) (*SM* figures 6a and 6d). These simulations use a grid of 25x25x25 on v_x,v_y,v_z, i.e., 15,625 particle orbits for each run. As all dynamic variables are stored at the end of each run, it becomes possible to tag the orbits with a different source distribution without rerunning the computationally intensive particle orbit routine.

In the following section, we present mapped distribution functions in cartesian velocity space for an observation point near the CS center, and compare the model results with the AMPTE/IRM observations of section 2.

4. MODEL RESULTS IN THE CENTRAL CURRENT SHEET AND COMPARISONS TO OBSERVATIONS

We show in Figure 3(a) a modeled distribution function (in the v_x,v_y plane) using parameters based on the initial study of *Lyons and Speiser* [1982]. Here, $B_{x0} = 20\ nT$, $B_{z0} = 1\ nT$, $E_y = 1/4\ mV/m$, $T = 500\ km$ (CS half thickness), $L = 10{,}000\ km$, and the parameters of the assumed source distribution are: a tailward flowing kappa

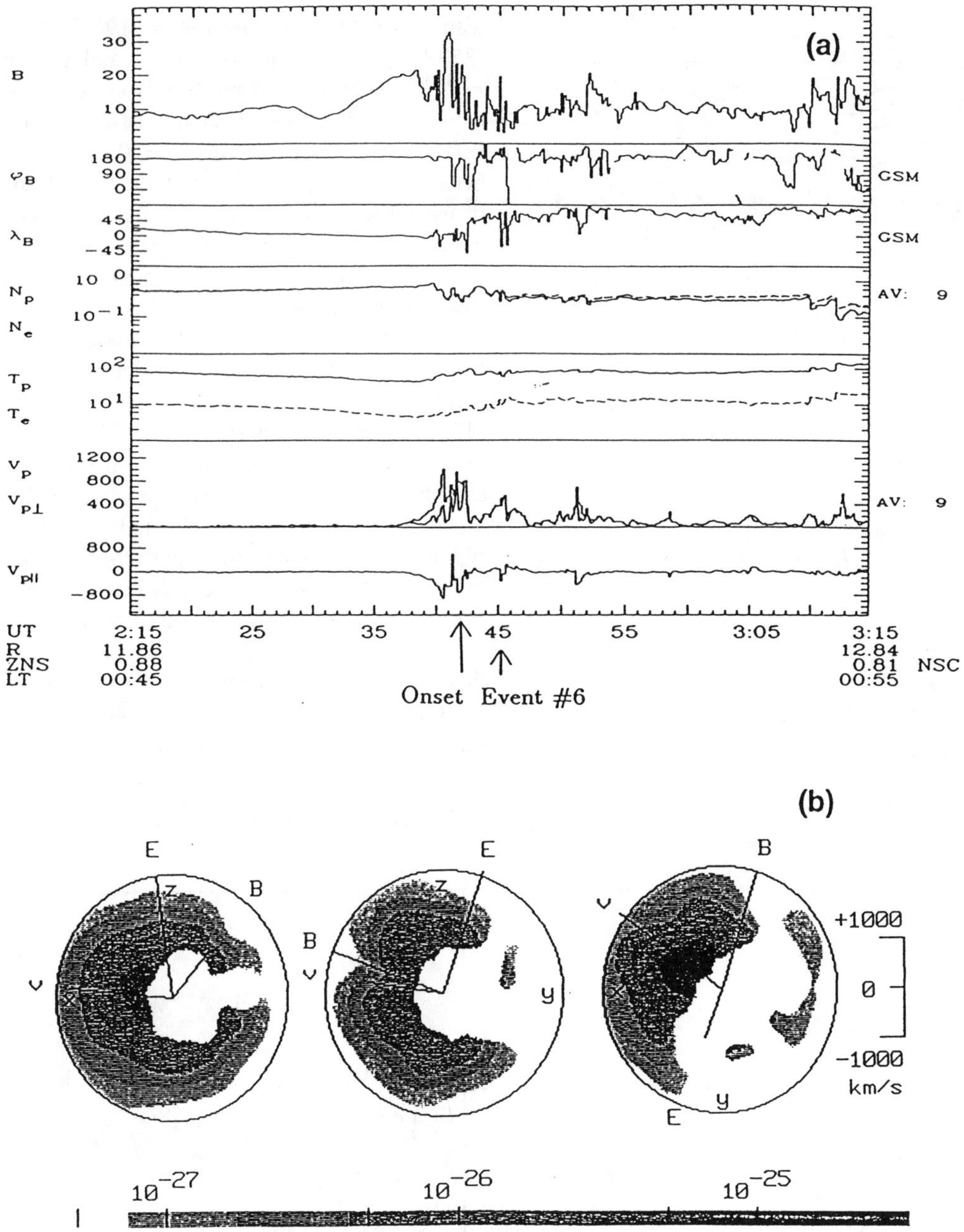

Fig. 1. (a) Plasma parameters within the magnetotail on 1 March 85, 02:15-03:15 *UT*. From top to bottom, the figure shows: the magnetic field strength (B), azimuth angle of the magnetic field (φ_B), elevation angle of the magnetic field (λ_B), proton density (N_p) and electron density (N_e), proton temperature (T_p) and electron temperature (T_e), proton bulk speed (v_p) and its perpendicular component to the magnetic field ($v_{p\perp}$), and parallel component of the ion velocity ($v_{p\parallel}$). The electron density N_e and the electron temperature T_e are plotted as broken lines. The units are nT in B, degree in φ_B and λ_B, cm^{-3} in N_p and N_e, 10^6 K in T_p and T_e, km/s in v_p, $v_{p\perp}$, and $v_{p\parallel}$. Universal time *UT*, distance from the earth R, distance from the theoretical neutral sheet $ZNS = Z_{NS}$ and local time *LT* are given along the bottom. Substorm onset time and the time of the event are shown by arrows. (b) Three cuts through the three-dimensional ion distribution at 02:45:19 *UT* on 1 March 85. From left to right, the cuts refer to the (v_x,v_z), (v_y,v_z), and (v_x,v_y) planes, in GSM coordinates. Projections of the magnetic field (**B**), the bulk velocity (v), and the electric field ($\boldsymbol{E} = -\mathrm{v} \times \boldsymbol{B}$) on each plane are shown. The range of the observation is up to 40 *keV* and the scale of the velocity is given at the right side of the plots. The phase space density scale in s^3cm^{-6} is given at the bottom. The cuts were obtained from an interpolation of the "high-resolution" 3D distribution [from *Nakamura, et al.*, 1991, their Figure 4].

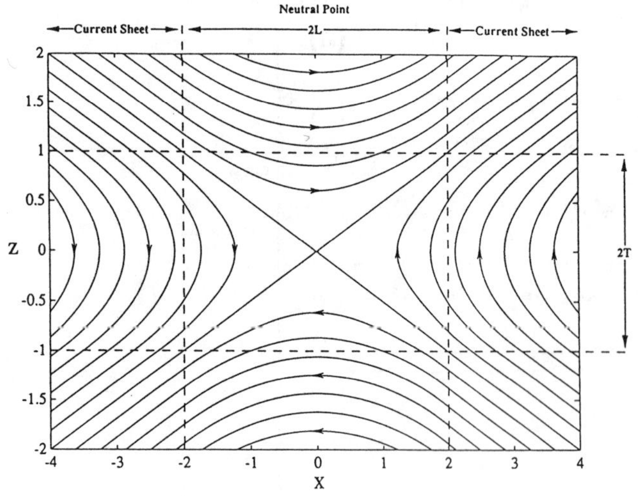

Fig. 2. Magnetic field model, current sheet with neutral line. North/south thickness = $2\ T$, $L = T/b_n$, $b_n = B_{z0}/B_{x0}$. For this diagram, $b_n = 0.5$.

distribution [e.g., *Krimigis et al.*, 1983], with flow of U_x = -350 *km/s*, temperature of 600 *eV*, and density of 1.04×10^4 m^{-3}. The observation position is at $z_0 = 50\ km = T/10$, $x_0 = +L$, i.e., 10,000 *km* earthward of the neutral line, and $y_0 = 0$. We did not start at $z_0 = 0$, the exact CS center, because there is a singularity in the particle dynamics at $z = 0$; particles are stuck at $z = 0$ and enter the CS from the edges of the tail. These are limiting cases (*SM* type 6) and were discussed in *SM*. The contours of the log of the distribution function are labeled in the figure and the units are $s^3 m^{-6}$, i.e., we multiply the observed distribution function (Figure 1) by 10^{12} in order to compare.

In Figure 3(b), we overlay as heavy dashed contours the -13, -14, and -15 contours from Figure 1(b) (the v_x,v_y observations), onto the model from Figure 3(a). We see that the centroid of the model is offset and does not well represent the data. Instead of a tailward flow of the source plasma, an earthward flow moves the centroid to regions of positive v_x (compare for example *SM* figures 6a and 6d). We experimented with various flows and found that we could fairly well match the observed contours with a flow that was strongly earthward and duskward (U_x = +1200 *km/s*, U_y = +800 *km/s*). Figures 4(a) and (b) show these model results and comparisons with the AMPTE

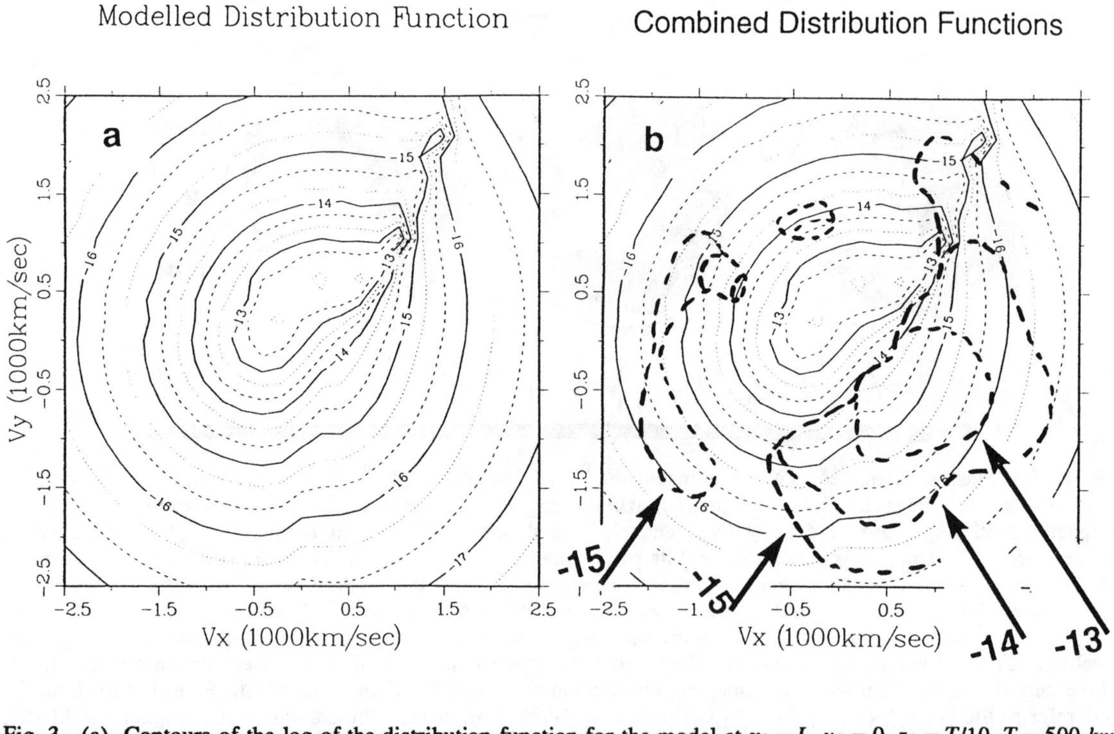

Fig. 3. (a) Contours of the log of the distribution function for the model at $x_0 = L$, $y_0 = 0$, $z_0 = T/10$, $T = 500\ km$, $L = 10{,}000\ km$, $E_y = 1/4\ mV/m$, initial distribution a tailward flowing kappa with $E_{th} = 600\ eV$, $n = 1.04 \times 10^4\ m^{-3}$, $U_x = -350\ km/s$, $U_y = U_z = 0$. (b) Only the -13, -14, -15 contours from the AMPTE observations (Figure 1(b) overlaid as dashed contours onto the model of (a). Observations from Figure 1(b) rotated by 180°.

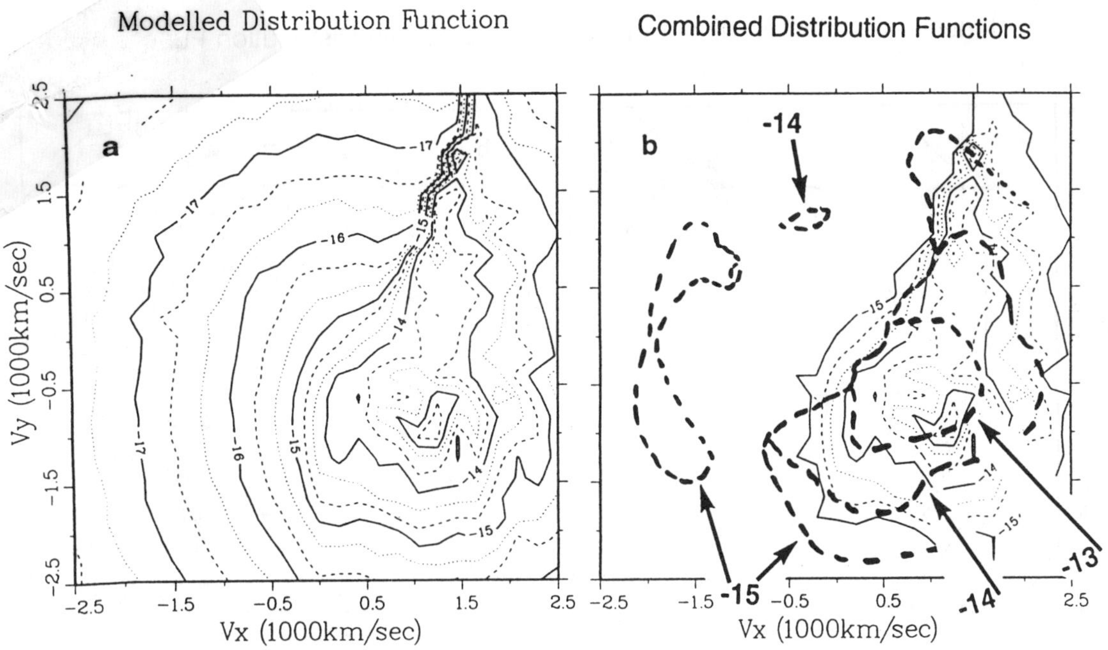

Fig. 4. (a) Same parameters as in Figure 3, (v_x,v_y) projection, but with earthward, duskward flow of $U_x = +1200$ *km/s*, $U_y = +800$ *km/s*. (b) Contours with distribution function less than 10^{-15} s^3m^{-6} are deleted, and AMPTE observations overlaid as dashed contours.

observations. Here (Figure 4(b)), model contours corresponding to values of the logarithm of the distribution function, f, less than -15, are deleted in order to match the sensitivity of the AMPTE results and to facilitate comparison. The match of the major contours is seen to be quite good for the cut through the distribution function in the v_x,v_y plane. In Figure 4(a), we identify the upper part of the distribution ($1000 \lesssim v_x \lesssim 1600$ *km/s*, $0 \lesssim v_y \lesssim 2500$ *km/s*), where the contours are squeezed together as the ridge signature in the central CS (see also Figures 7 and 8). The observed distribution function (Figure 4(b)) shows a similar ridge-like feature in the same region of velocity space.

We also ran cases where $x_0 = -L$, i.e., the observation point was tailward of the neutral line. In those cases, the ridge flips from the $v_x > 0$ sector to the $v_x < 0$ sector similar to Figure 2 of *SM*. The comparison with the AMPTE observations is thus not good, and so we conclude that the AMPTE observation point is earthward of the neutral line. There may be other mechanisms which are able to reproduce the centroid of the distribution. However, we do believe the model along the ridge approximates the observations fairly well, and we know of no other mechanism which can produce this signature.

In Figures 5(a) and (b) and 6(a) and (b), we show model results and data comparisons for the other two planes (v_x,v_z) and (v_y,v_z). In Figures 5(a) and (b), we see that the model roughly matches the centroid of the observations, but does not do well on the wings. Similarly, on Figures 6(a) and (b) the centroid is approximated but the wings, and the patch for $v_y > 0$, are not represented by the model at all.

In order to try to understand the source of these discrepancies, we show in Figure 7 for the (v_x,v_y) plane a mapping of some of the stored dynamic information: a) the number of current sheet crossings, b) the particle's initial position in x, and c) the particle's initial position in y. Figure 8 (a,b,c) overlays the observation contours onto the panels from Figure 7.

In Figures 7(a) and (b), the importance of the neutral line becomes apparent: those particles which have initial positions near $x = 0$ (the neutral line) form the velocity space ridge (for $v_y > 0$) and have a large number of oscillations across the CS. In Figures 8(a) and (b), we see that the particles making up the observations along the ridge ($1000 \lesssim v_x \lesssim 1600$ *km/s*, $0 \lesssim v_y \lesssim 2500$ *km/s*) were incident on the current sheet very close to the neutral line, and they made a large number of oscillations about the CS. From Figure 7(c), we see that those particles found along the ridge have also come the farthest in y, thus being most energized by the cross-tail E_y. We also see that most of the particles that make up the centroid of the observed distribution have come from distances 5 to 20 thousand *km* tailward of the neutral line (Figure 8(b)). (Recall that the

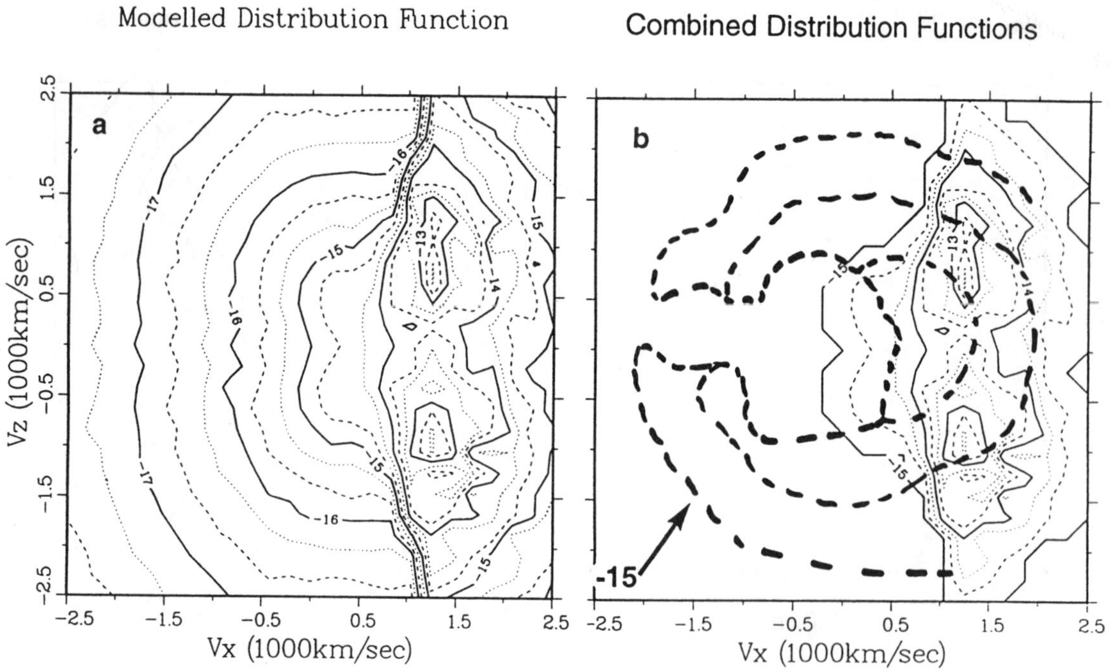

Fig. 5. Same parameters as in Figure 4, (v_x,v_z) projection, and AMPTE data overlay (-15, -14, -13 contours). Observations from Figure 1(b) rotated about the *z*-axis by 180°.

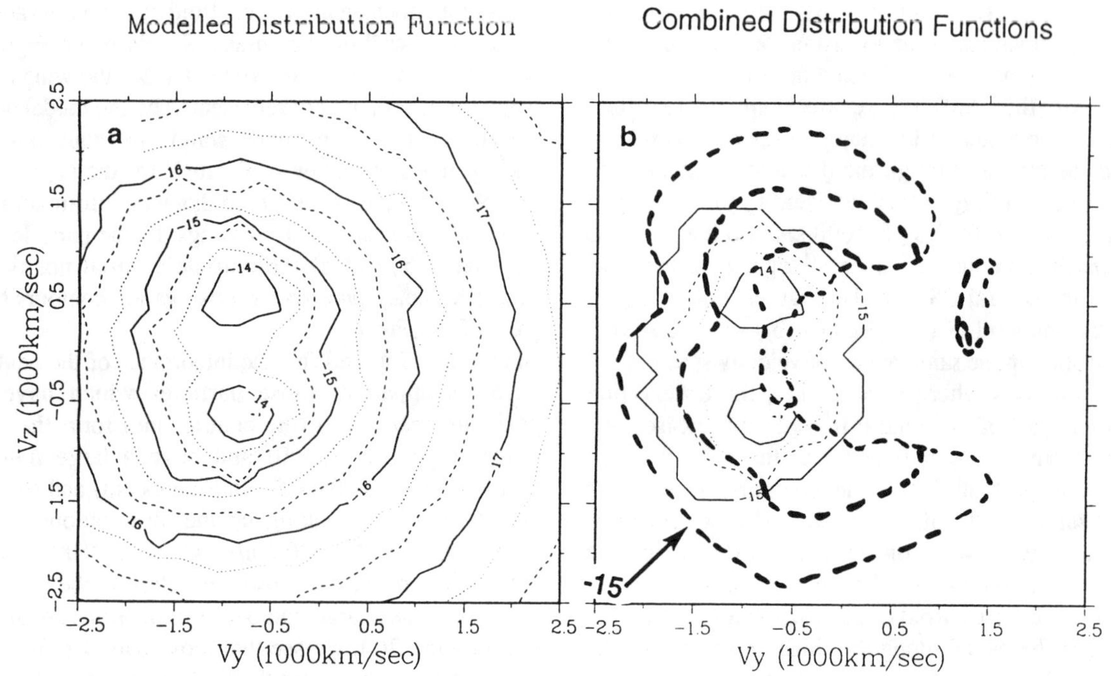

Fig. 6. Same parameters as in Figure 4, (v_y,v_z) projection, and AMPTE data overlay (-15, -14, -13 contours).

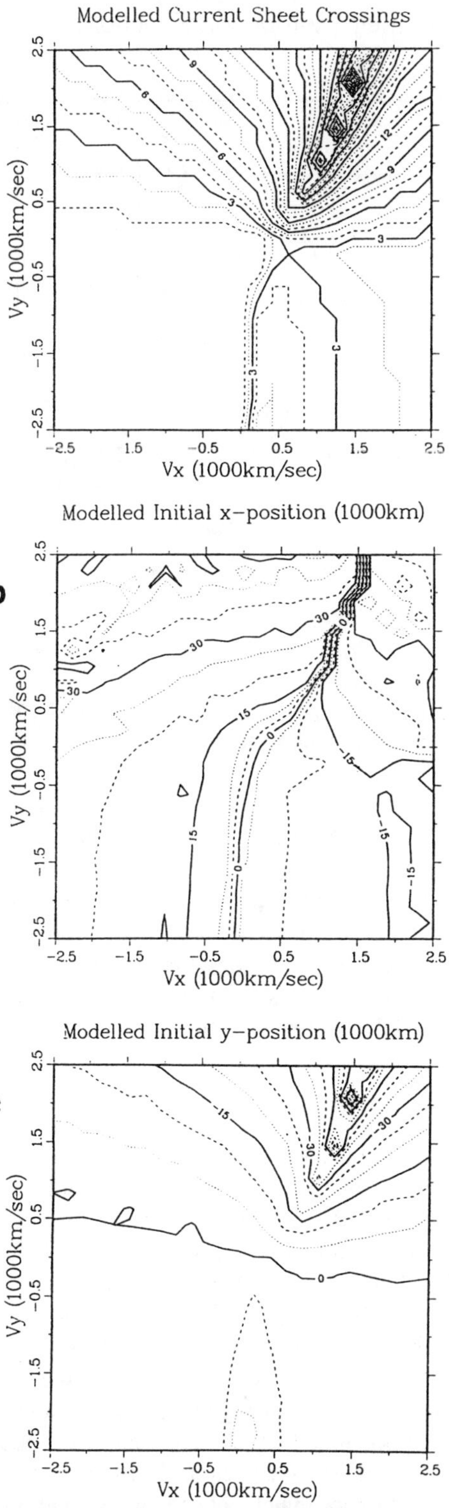

Fig. 7. (v_x, v_y) projections for the same parameters as Figure 4: (a) number of current sheet crossings; (b) initial *x*-position (source) of the particles; (c) initial *y*-position of the particles. Contour spacing for (b) and (c) is 5000 *km* increasing levels denoted by "solid", "dashed", then "dotted" lines.

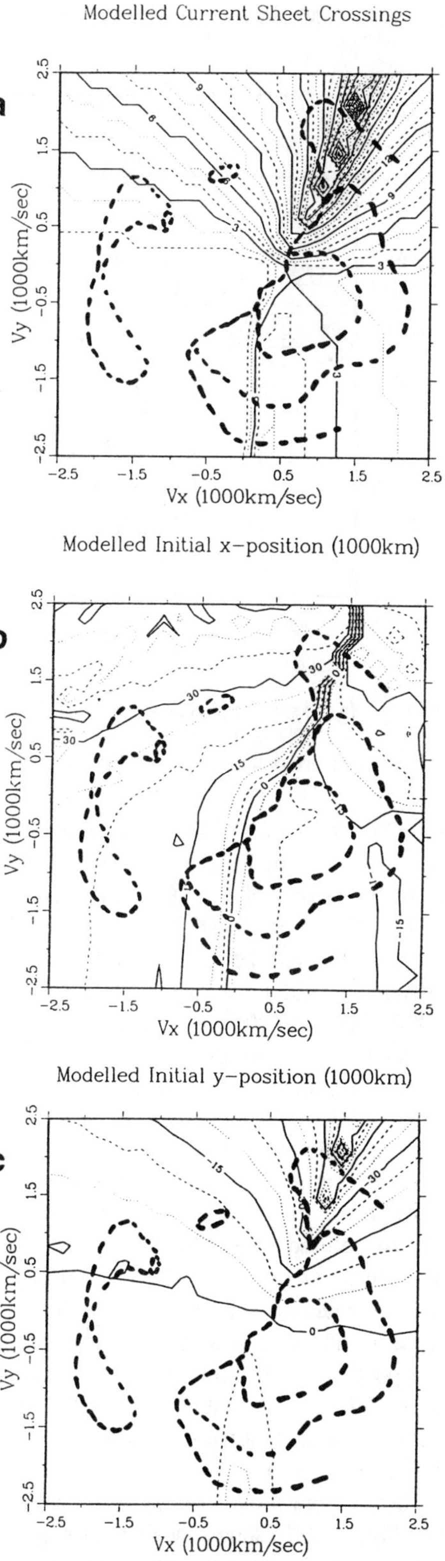

Fig. 8. Same as Figure 7, but with -13, -14, -15 contours of AMPTE observations overlaid.

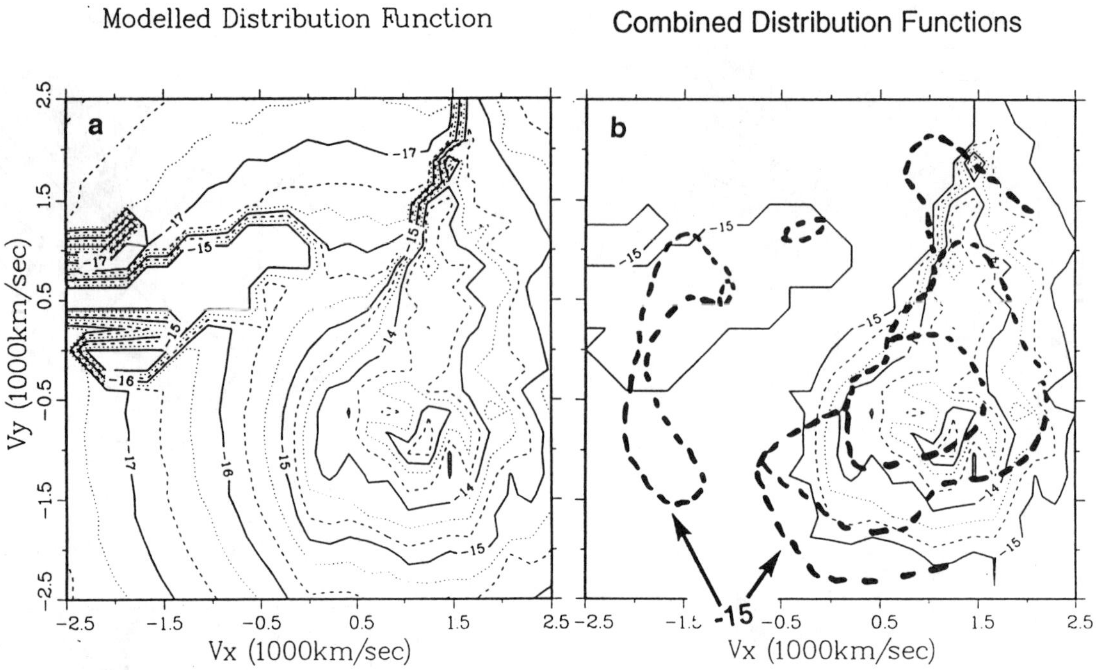

Fig. 9. (a) Same as Figure 4 (in (v_x,v_y) projection), but with addition of secondary source of magnitude 10^{-15} s^3m^{-6} for $20{,}000 \leq x \leq 30{,}000$ and $-10{,}000 \leq y \leq 1$ *km*. (b) Model contours with distribution function less than $10^{-15}s^3m^{-6}$ are deleted, and AMTPE observations overlaid as dashed contours.

observation point is 10,000 *km* earthward of the neutral line.) We also see in Figure 8(b) that those particles observed in the region of $v_x < 0$ would have come from distances $20{,}000 \lesssim x \lesssim 30{,}000$ *km* in the model. From Figure 8(c), we see that the model would predict that those same particles would have mainly come from initial *y* values $5000 \gtrsim y \gtrsim -30{,}000$ *km*. Therefore, we assume a secondary source with strength $|f| \simeq 10^{-15}\ s^3m^{-6}$, located in the *x-y* plane at a location $20{,}000 \lesssim x \lesssim 30{,}000$ *km* and $-30{,}000 \leq y < 5000$ *km*. We then make a new mapping with this additional secondary source, and the results of the model and the overlaid observations are shown in Figures 9(a) and (b). (In Figure 9(b), again the model contours less than -15 are deleted to facilitate comparison with the AMPTE observations.) The secondary source approximately matches the observations for $v_x < 0$, as it should since we added particles in these regions specifically for that purpose. But what does the same secondary source do to the model and observation comparisons in the other two planes? These comparisons are illustrated in Figure 10 (a,b) and Figure 11 (a,b).

Figure 10(a) is the model (v_x,v_z) plot with secondary source and Figure 10(b) shows the data overlay. The model wings for $v_x < 0$ are now extended, giving a very good qualitative and good quantitative match. Similarly, Figure 11 (a,b) extends the model into velocity space regions approximating the observations. Even the small observed "patch" at $v_y \approx 1.2$, $v_z \approx 0.4$ is now approximated by the model. Clearly, more permutations of the model using other sources could yield even better matching of the observations. The intent, however, is to show that we can fairly well match these 3D AMPTE/IRM observations using a model with few assumptions.

5. DISCUSSION

We have used the *MS* neutral line model to try to model a central CS, three-dimensional distribution function, measured by AMPTE/IRM [*Nakamura et al.*, 1991], at a time of fast flows near substorm onset. In contrast to earlier studies [*MS*; *SM*; *Speiser et al.*, 1991; *Lyons and Speiser*, 1982], it was found that an initially tailward flowing plasma produced poor agreement with the observed distribution function after interaction with the CS. Instead, a strongly earthward flowing source plasma was required to produce a good match for the centroid of the observations. AMPTE/IRM must have been just earthward of a near-earth neutral line (a poor match between model and observations resulted if AMPTE had been tailward of the neutral line). If this primary source were supplemented by an earthward, dawnward secondary source, then even many asymmetries in the observations could be explained. A sketch of these

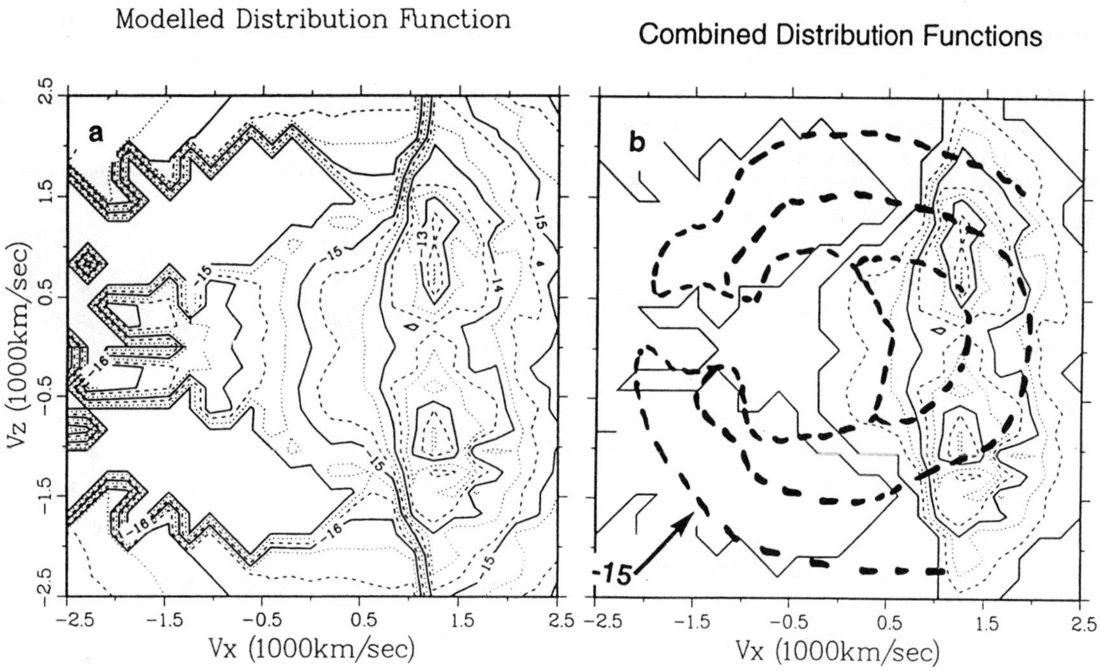

Fig. 10. Same as Figure 9, but in (v_x, v_z) projection.

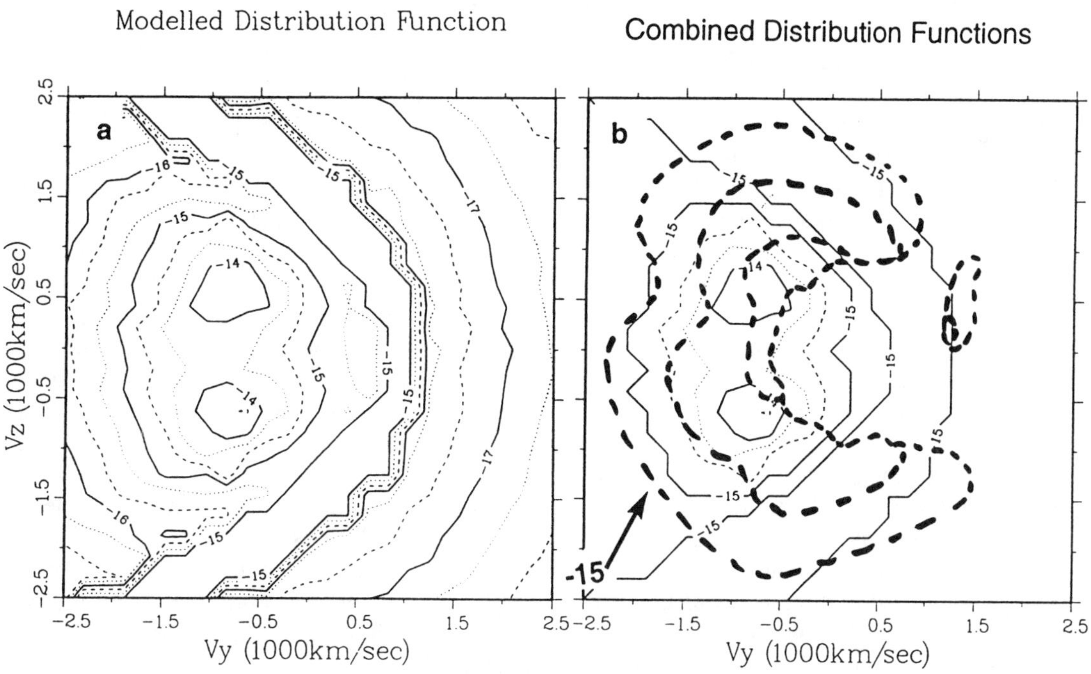

Fig. 11. Same as Figure 9, but in (v_y, v_z) projection.

presumed sources and the AMPTE/IRM location is shown in Figure 12.

In Figure 12, we indicate the location and flow of the primary source (U_x = 1200 *km/s*, U_y = 800 *km/s*) necessary for matching the main centroid features of the observations, and the location and flow of the secondary source which allows us to model asymmetries in the observed distribution function. This secondary source was chosen primarily to provide a match with the observations in the (v_x,v_y) plane (Figure 9), but it also provides a reasonable match in the other two planes. The secondary source was arbitrarily given a magnitude of 10^{-15} s^3m^{-6} (corresponding to the lowest measured distribution function), and was confined to $20 \le x \le 30$, $-10 \le y \le 1$ thousand *km*, i.e., 3 to 5 R_e earthward and 0 to 2 R_e dawnward of AMPTE/IRM (x = -12, y = -3 R_e). While the direction of the particles in the primary source is mainly earthward, from velocity plots (not shown), the particles in the secondary source are mainly moving in the tailward and duskward direction (secondary source: $-2000 \le v_x \le -1000$ *km/s*, $0 \le v_y \le 1300$ *km/s*). The tailward initial motion is consistent with the idea that some of the primary particles mirror close to the earth and return. We could imagine that the lobe field lines have some positive B_y which would explain the necessity for $U_y \approx 800$ *km/s* (primary source) if the primary source were essentially field aligned. However, the measured lobe field of 6-30 *nT* was almost entirely in the earth-sun direction, indicating little B_y lobe (note φ_B = 180° before onset in Figure 1(a)). In addition, if B_y lobe were positive, then a mirrored secondary source would have a v_y mostly negative, in disagreement with the model results. On the other hand, the required U_y of 800 *km/s* could be an electric drift due to a *z*-component of the electric field. Taking B_{xlobe} = 6 to 30 *nT*, and U_y = 800 *km/s*, gives E_z = 5 to 25 *mV/m*, northward above the current sheet and southward below it. This electric field would then also act on mirroring secondary particles, resulting in initial positive values of v_y. Such an electric field is not divergenceless and implies a positive charge density in the current sheet. In fact, *Li and Speiser* [1991] inferred a similar E_z, positively correlating with B_x for another time interval, namely 22 March 1979 - the CDAW 6 interval. (The data used was from Williams' medium energy particles experiment on ISEE 1. Integrations over the 3D distribution function were used to infer perpendicular and parallel bulk flows.) Using Li and Speiser's linear regression (their equation 2; E_z $(mV/m) = 0.475\ B_x\ (nT) - 2.01$; cc = .91) gives $-16 < E_z < -5$ *mV/m* for $-30 < B_x < -6$ *nT*, very close to the range we have inferred above. Referring back to the AMPTE measurements of Figure 1(a), we see that there is little evidence of any measurable component of the perpendicular flow ($v_{p\perp}$) before the substorm onset time (02:42). However, at 02:42, $v_{p\perp} \approx$ 400 *km/s* (and $B \approx 20$ *nT*); at 02:44, $v_{p\perp} \approx$ 200 *km/s* (and $B \approx 16$ *nT*); and at 02:51, $v_{p\perp} \approx$ 300 *km/s* (and $B \approx 12$ *nT*). So the measurements are not inconsistent with strong perpendicular flows near the edge of the plasma sheet after substorm onset, although we do not know that the direction of $v_{p\perp}$ is consistent with the value of U_y required in this analysis. An E_z (z) has not been included in the simulations and could well alter our distribution function mappings. Its effect could be minimal on the average *x-y* motion of the CS ions, as it should go to zero in the CS center. It would definitely increase the oscillation amplitude of the ions and thus the CS thickness. We plan to investigate the effects of E_z on the particle dynamics in a future study.

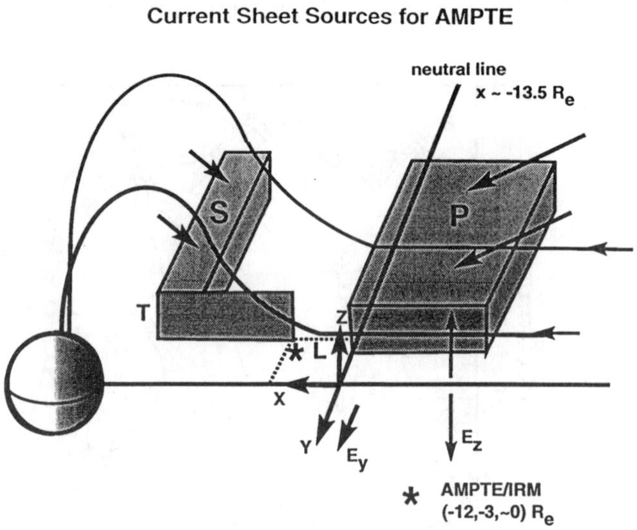

Fig. 12. Sketch of possible primary and secondary source regions in the geomagnetic tail. The primary source (p) is suggested to be due to pre-acceleration in the distant current sheet, while the secondary source (s) would be due to partial mirroring of the primaries. For the primary source, bulk flow of (+1200, +800, 0) *km/s* incident on $-18.2 < x < -13.5$ R_e, $-7.7 < y < -2.2$ R_e; for the secondary source, incident particles are in the velocity range $-2000 < v_x < -1000$ *km/s*, $0 < v_y < +1300$ *km/s*, and at distances $-8.9 < x < -7.3$ R_e and $-4.6 < y < -2.8$ R_e. Current sheet parameters: E_y = 1/4 *mV/m*, T = 500 *km*, L = 10,000 *km*, B_{x0} = 20 *nT*, B_{z0} = 1 *nT*. If the y-component drift for p and s is due to an $E_z(z)$, then $|E_z| \sim$ 5 to 25 *mV/m* (see text). AMPTE/IRM position (at *) of (-12, -3, ~0) R_e, see Figure 1. For the model, the neutral line is 10,000 *km* tailward of AMPTE, i.e., at $x \approx -13.5$ R_e.

In summary, we believe the *Nakamura et al.* [1991] 3D observations in the central CS near the time of substorm onset imply that the source of the particles was mainly a pre-accelerated earthward flow either from the distant current sheet or a distant neutral line. This primary beam fell upon the CS a few R_e tailward and dawnward of AMPTE/IRM and was further accelerated in the CS as the population moved toward the AMPTE/IRM position. A near-earth neutral line was about 1.5 R_e tailward of AMPTE. A secondary source was provided from mirroring particles near the earth, returning to the tail CS a few Re earthward and dawnward of AMPTE/IRM. An electric field component in the z-direction is implied, agreeing with the conclusions of *Li and Speiser* [1991]. While there may be other mechanisms capable of reproducing the centroid of the observations, no other mechanism has been suggested which can also reproduce the ridge-like features of the observed distribution.

Acknowledgments. We wish to thank M. Nakamura and N. Sckopke for their help in interpreting the AMPTE/IRM data. We also thank P. Bornman and R. Zwickl for useful discussions. This research has been supported by the National Science Foundation (ATM-9002447, ATM-9112954), the National Aeronautics and Space Administration (NAGW-1176, NAG5-1506, NAGW-2968), and Illinois State University.

REFERENCES

Burkhart, G. R., J. F. Drake, and J. Chen, Magnetic reconnection in collisionless plasmas and prescribed fields, *J. Geophys. Res., 95*, 18,833, 1990.

Burkhart, G. R., J. F. Drake, and J. Chen, Structure of the dissipation region during magnetic reconnection in collisionless plasma, *J. Geophys. Res., 96*, 11,539, 1991.

Curran, D. B. and C. K. Goertz, Particle distributions in a two-dimensional reconnection field geometry, *J. Geophys. Res., 94*, 272, 1989.

DeCoster, R. J. and L. A. Frank, Observations pertaining to the dynamics of the plasma sheet, J. Geophys. Res., 84, 5099, 1979.

Doxas, I., W. Horton, K. Sandusky, T. Tajima, and R. Steinolfson, Numerical study of the current sheet and plasma sheet boundary layer in a magnetotail model, *J. Geophys. Res., 95*, 12,033, 1990.

Fritz, T. A., D. Dn. Baker, R. L. McPherron, and W. Lennartsson, Implications of the 1100 UT March 22, 1979 CDAW 6 substorm event for the role of magnetic reconnection in the geomagnetic tail, in *Magnetic Reconnection in Space and Laboratory Plasmas, Geophys. Mono. Ser., 30*, 203, 1984.

Hones, E. Q., Jr., T. A. Fritz, J. Birn, J. Cooney, and S. J. Bame, Detailed observations of the plasma sheet during a substorm on April 24, 1979, J. Geophys. Res., 91, 6845, 1986.

Kettmann, G., T. A. Fritz, and E. W. Hones, Jr., CDAW 7 revisited: Further evidence for the creation of a near-earth substorm neutral line, *J. Geophys. Res., 95*, 12,045, 1990.

Krimigis, S. M., J. F. Carbary, E. P. Keath, T. P. Armstrong, L. J. Lanzerotti, and G. Gloeckler, General characteristics of hot plasma and energetic particles in the saturnian magnetosphere: results from the voyager spacecraft, *J. Geophys. Res., 88*, 8871, 1983.

Li, X. and T. W. Speiser, An Estimation of the Electric Field in the Magnetotail Current Sheet, *Geophys. Res. Lett., 18*, 1967, 1991.

Lyons, L. R. and D. C. Pridmore-Brown, Force balance near an x line in a collisionless plasma, *J. Geophys. Res., 95*, 20,903, 1990.

Lyons, L. R. and T. W. Speiser, Evidence for current sheet acceleration in the geomagnetic tail, *J. Geophys. Res., 87*, 2276, 1982.

Martin, R. F., Jr. and T. W. Speiser, A predicted energetic ion signature of a neutral line in the geomagnetic tail, *J. Geophys. Res., 93*, 11,521, 1988.

Nakamura, M., G. Paschmann, W. Baumjohann, and N. Sckopke, Ion distributions and flows near the neutral sheet, *J. Geophys. Res., 96*, 5631, 1991.

Paschmann, G., N. Sckopke, and E. W. Hones, Jr., Plasma measurements during the 1054 UT substorm of March 22, 1979 (CDAW 6), *J. Geophys. Res., 90*, 1217, 1985.

Speiser, T. W. and R. F. Martin, Jr., Energetic ions as remote probes of x-type neutral lines in the geomagnetic tail, *J. Geophys. Res., 97*, 10,775, 1992.

Speiser, T. W., P. B. Dusenbery, R. F. Martin, Jr., and D. J. Williams, Particle orbits in magnetospheric current sheets: Accelerated flows, neutral line signature, and the transitions to chaos, in *Modeling Magnetospheric Plasma Processes, Geophys. Mono. Ser., 62*, 71, 1991.

R. F. Martin, Jr., Physics Department, Illinois State University, Normal, IL 61761-6901.

T. W. Speiser, Astrophysical, Planetary and Atmospheric Sciences Department, University of Colorado, Campus Box 391, Boulder, CO 80309-0391.

Fine-Scale Structures in Auroral Arcs: An Unexplained Phenomenon

JOSEPH E. BOROVSKY

Space Plasma Physics Group, Los Alamos National Laboratory, Los Alamos, New Mexico

An investigation into the fine-scale structures within auroral arcs reported by *Maggs and Davis* [1968] finds that the structures are an unexplained phenomenon. The fine-scale structures are sheets of airglow that have thicknesses of about 100 m in the auroral ionosphere. The fine-scale structures are considerably brighter than the background intensities within auroral arcs. When mapped along the magnetic-field lines to the plasma sheet, they have sizes of a few km. Utilizing existing theories to obtain estimates for auroral-arc-structure thicknesses, it is found that none of these theories can account for the 100-m structures. Several possible origins of the fine-scale structure are considered and a call for new observations is made.

1. INTRODUCTION

Auroral arcs are east-west-oriented, magnetic-field-aligned sheets of airglow in the upper atmosphere. Energetic electrons precipitating into the atmosphere are observed above the arcs [*McIlwain*, 1960] and the height profile of the airglow is consistent with energy deposition by the energetic electrons [*Belon et al.*, 1966]. Typically, the bottom of the airglow sheet is about 100 km in altitude and the sheet can be 100's of km tall [*Belon et al.*]. The auroral arcs may be 1000's of km long in the east-west direction.

In the north-south direction, there are several spatial scales associated with arcs [*Davis*, 1977, 1978, 1979; *Borovsky et al.*, 1991; *Borovsky and Suszcynsky*, 1992]. These are outlined in Table 1. On the largest scale there is the auroral oval, which is the region in which the auroral arcs reside and which is more-or-less the zone where the Region-I currents flow into the dusk ionosphere [*Kamide and Akasofu*, 1976; *Klumpar*, 1979; *Feldstein and Galperin*, 1985; *Friis-Christensen and Lassen*, 1991]. The auroral zone is typically 2° - 6° wide in latitude [*Feldstein and Starkov*, 1970; *Hones et al.*, 1991; *Elphinstone and Hearn*, 1993], which corresponds to 200 - 700 km in the ionosphere. The next-largest spatial scale is the quiescent channel of diffuse airglow associated with a single arc, which is 10's of km wide. Next is the narrow, bright, dynamic channel of emission that resides within the broad channel. This is a kilometer or so wide. Finally the smallest-known spatial scale is the fine-scale structure [*Davis*, 1967; *Maggs and Davis*, 1968; *Borovsky et al.*, 1991; *Borovsky and Suszcynsky*, 1992], which are 100-m or so wide. Typical widths W_{ion} in the ionosphere of these four spatial scales are recorded in Table 1.

There is much evidence that auroral-arc magnetic-field lines connect to the near-earth plasma sheet [*McIlwain*, 1975; *Meng et al.*, 1979; *Klumpar et al.*, 1988; *Shepherd and Shepherd*, 1985; *Feldstein and Galperin*, 1985 *Elphinstone et al.*, 1991]. During geomagnetically quiet times, these field lines may be dipolar; when they are, the a radial width W_{mag} in the equatorial magnetosphere can be calculated from the north-south width W_{ion} in the ionosphere with the use of the expression

$$\frac{W_{mag}}{W_{ion}} = L^{3/2}\left(4 - \frac{3}{L}\right)^{1/2} , \qquad (1)$$

obtained from expressions (2A.5) and (2A.7) of *Borovsky* [1992a], where L is the McIlwain parameter for the magnetic-field line. Taking $L = 8$, expression (1) yields $W_{mag} = 43\ W_{ion}$. In Table 1, this factor is used to estimate the radial widths in the equatorial plasma sheet of the various auroral spatial scales.

The focus of this report is the fine-scale structures that were first clearly reported by *Maggs and Davis* [1968]. The report is organized as follows. In Section 2 the observations of auroral-arc fine-scale structure are reviewed. In Section 3, various theories of auroral arcs are used to obtain predictions for the minimum thickness in the ionosphere that an auroral structure can have. From a comparison of the theoretical

Space Plasmas: Coupling Between Small
and Medium Scale Processes
Geophysical Monograph 86

TABLE 1. Typical north-south widths W_{ion} of the characteristic auroral spatial scales in the ionosphere and the radial widths W_{mag} of those regions mapped to the equatorial plasma sheet at $L = 8$.

STRUCTURE	W_{ion}	W_{mag}
Auroral Zone	100 km	4300 km
Broad Quiescent Channel	15 km	650 km
Narrow Dynamic Channel	1.5 km	65 km
Fine-Scale Structure	100 m	4.3 km

thicknesses with the observed thicknesses, it is concluded in Section 4 that no existing auroral-arc theory explains the fine-scale structures. In Section 5 several possible causes of the fine-scale structure are discussed. Future observations are is called for in Section 6 and the report is summarized in Section 7.

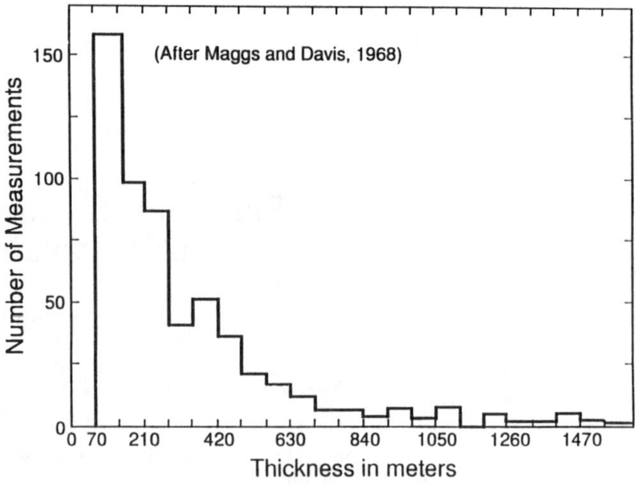

Fig. 1. The distribution of thicknesses of auroral-arc structures obtained by *Maggs and Davis* [1968].

2. Observations of Fine-Scale Structures

Although it is a straightforward task, measurements of the optical thicknesses of auroral arcs have rarely been made. Thickness measurements are performed by imaging arcs with a ground-based camera when the arcs are in the magnetic zenith, *i.e.*, when they are edge on. Some observers have reported thicknesses of larger than a kilometer [*Stormer*, 1955; *Kim and Volkman*, 1963]: these measurements may pertain to the larger-spatial-scale features of an arc (see Table 1), or the arcs may not have been in the magnetic zenith when they were imaged. *Elvey* [1957] obtained a value of 250 m for the thickness of an auroral arc and *Akasofu* [1961] obtained values of 144 m and 336 m for the thickness of one arc. These measurements probably pertained to auroral-arc fine-scale structures. *Davis* [1967] reported that auroral arcs are composed of parallel structures with widths of 100 to 300 m. This report was followed up with an extensive survey of arc-structure thicknesses [*Maggs and Davis*, 1968] that has yet to be surpassed.

The statistical findings of *Maggs and Davis* [1968] are reproduced in Figure 1. The arc thicknesses were measured from images obtained with an image-orthicon television system with one line of video resolution corresponding to 70 m at the inferred height of the arcs (100 km). As can be seen in Figure 1, the most-probable thickness was that of the system resolution, which indicates that arc structures could have been thinner than 70 m. Because the structures observed are so thin, the measurements of *Maggs and Davis* have profound implications for theories of auroral arcs. However, they have been largely ignored. The main criticism of their results rests on the fact that image-orthicon systems can have very nonlinear responses, and so they can produce a high-contrast image of a structure that is merely a few-percent enhancement over the background intensities [*Davis*, 1978]. Accordingly, some members of the research community thought that the *Maggs and Davis* small-scale arcs were only low-amplitude ripples in the background intensity of a broad auroral arc, hence they were not significant.

To answer the question of whether auroral-arc structures thinner than 70 m exist and to address the criticism that the fine-scale structures observed by *Maggs and Davis* [1968] might not be robust, a different type of imaging system was constructed and fielded [*Borovsky et al.*, 1991; *Borovsky and Suszcynsky*, 1992]. That imaging system had a spatial resolution of 2.1 m at auroral altitudes (as opposed to 70 m for *Maggs and Davis*) and did not suffer from a nonlinear response. With four nights of data in April, 1990 and zero nights of data in October, 1990, insufficient observations were obtained to repeat the statistical curve of *Maggs and Davis* (Figure 1). However, two results were obtained: (1) the observations confirmed the existence of fine-scale structures with thicknesses that were typically 100 m and (2) the observations demonstrated that the fine-scale structures are robust, with intensities that are about an order of magnitude higher than the background intensities with the arcs. The conclusion of *Borovsky and Suszcynsky* is that the fine-scale structures reported by *Maggs and Davis* are a real and important phenomenon.

A summary of the known and unknown properties of auroral-arc fine-scale structures appears in Table 2. Very little is firmly known about the structures — only their thicknesses have been quantified, and these thickness values are uncertain owing to the unconfirmed assumption that the fine-scale structures are located at a height of 100 km. When looking into the magnetic zenith, the fine-scale structures are seen to be elongated: it was determined [*Borovsky et al.*, 1991; *Borovsky and Suszcynsky*, 1992] that the structures were at least 500 m long (east-west), which was the east-west field of view (at 100-km heights) of the telescope

TABLE 2. Known and unknown properties of auroral-arc fine-scale structure

Known Properties	
	reside within auroral arcs
	thicknesses of about 100 m*
	east-west aligned
	sheet shaped
	drift in north-south direction**
	multiple and parallel
Unknown Properties	
	height and vertical extent
	total emitted power
	east-west extents
	emission-line ratios***
	velocity of structures relative to arc
	lifetimes
	time evolution
	relation to larger-scale arc structures

* The thickness values depend on the assumption that the fine-scale structures reside at altitudes of 100 km.

** Drifts have been observed relative to the ground, it is not known whether the structures drift with respect to the ambient plasmas.

*** A determination of the line ratios will indicate whether the structures arise from high-energy electron impact or from a low-energy electical discharge.

that was used. The fine-scale structures are probably sheet shaped, since they are only seen in the magnetic zenith when they are presumably edge on. Measured drift velocities of the structures are in the range 100 - 500 m/sec relative to the ground cameras (assuming a 100-km height). The fine-scale structures can also be multiple [*Davis*, 1967; *Borovsky et al.*], with separations of a hundred meters or so between the individual structures.

Also in Table 2 is a list of some unknown properties of fine-scale structures that could be determined from ground-based observations. (These will be discussed further in Section 6.) Note that the intensity ratios of the airglow emission lines have not been measured for the fine-scale structures, therefore it has not been determined for certain that the fine-scale structures are caused by energetic-electron precipitation, as is the emission from the major portion of an arc [*e.g. Belon et al.*, 1966].

In examing the possible origns of fine-scale structures (see Sections 3, 4, and 5), several questions arise that could be answered with further ground-based observations of the structures. Needed observations are outlined in Section 6.

3. Theoretical Thicknesses of Auroral-Arc Structures

The one well-characterized property of auroral-arc fine-scale structures is their thicknesses. By comparing this one observed property with predictions from theoretical models, it will be concluded that auroral-arc fine-scale structures are an unexplained phenomenon.

In this section auroral-arc models are briefly overviewed and minimum thicknesses of auroral-arc structures as predicted by each model are collected in Table 3. The details of these thickness calculations can be found in *Borovsky* [1992a]. The models are divided into accelerator mechanisms (Section 3A) and generator mechanisms (Section 3B). A calculation of the minimum-possible thickness for an airglow region in the upper atmosphere produced by the precipitation of energetic electrons is outlined in Section 3C.

A. Accelerator Mechanisms

Accelerator-mechanism theories (listed in Table 3) are concerned with explaining how the electrons that produce auroral arcs become energized. Most of the accelerator mechanisms assume that there is a potential drop $\Delta\phi$ between the magnetospheric plasma and the ionosphere on an auroral-arc magnetic-field line that accounts for the energy gain of precipitating electrons, and the various models assume that the potential drop takes various forms (*e.g.* anomalous resistivity, strong double layers). One accelerator model (kinetic Alfven waves) assumes the electrons gain their energy by passing through an electric field at the front of an Alfven wave (which can be both inductive and electrostatic), two of the models (electromagnetic ion-cyclotron waves and lower-hybrid Landau resonance) assume that auroral electrons gain their energy by means of Landau resonances with plasma waves, and one mechanism (AKR-induced precipitation) assumes that the electrons already have sufficient energy and are pitch-angle scattered by waves. All acceleration mechanisms are assumed to operate at altitudes of 1 - 2 R_E, where electrons are observed to be energized [*Gorney et al.*, 1981; *Bennett et al.*, 1983].

Eleven accelerator models are outlined in the following 11 paragraphs, and the minimum thickness that each theory predicts (from *Borovsky* [1992a]) is listed in Table 3.

PARTICLE ANISOTROPIES: In this model the accelerating potential along the magnetic field is assumed to be driven by differing pitch-angle distributions between the hot electrons and hot ions in the magnetosphere, giving the two species different mirror points, which results in a charge separation [*Alfven and Falthammer*, 1963; *Persson*, 1963; *Chiu and Schultz*, 1978]. In this model the hot magnetospheric ions drive the potential, hence the size of the potential drop $\Delta\phi$ is limited to approximately $k_B T_i/e$.

THERMOELECTRIC CONTACT POTENTIAL: In this model the contact between the hot magnetospheric plasma and the cool ionospheric plasma forms a potential drop at the interface [*Hultqvist*, 1971]. Wave-particle scat-

TABLE 3. Theoretical Predictions for the Minimum Widths of Auroral Structures in the Ionosphere

Accelerator Mechanisms		
MECHANISM	W_{ion}	TYPE
Static magnetosphere-ionosphere coupling	340 km	potential drop
Particle anisotropies	1.0 km	potential drop
Thermoelectric contact potentials	1.5 km	potential drop
Strong double layers	2.2 km	potential drop
Electrostatic shocks	1.4 km	potential drop
Wave-broadened electrostatic structures	???	potential drop
Anomalous resistivity	$\gg$1.5 km	potential drop
Weak double layers	$\gg$0.87 km	potential drop
Kinetic Alfven waves	1.7 km	induction E field
Electromagnetic ion-cyclotron waves	1.7 km	Landau resonance
Lower-hybrid-wave Landau resonances	$\gg$0.35 km	Landau resonance
AKR-induced precipitation	6.9 km	pitch-angle scattering
Generator Mechanisms		
MECHANISM	W_{ion}	LOCATION
Shear in low-latitude boundary layer	130 km	low-latitude boundary layer
Shear in plasma sheet	51 km	plasma sheet
Poynting-flux absorption	4.2 km	plasma sheet
Magnetic-field-line reconnection	3.5 km	plasma-sheet-boundary layer
Ionospheric-conductivity feedback	17 km	ionosphere
Plasma flow into high-conductivity zone	8.8 km	plasma sheet
Pressure gradients	2.6 km	plasma sheet
Electrostatic fluid turbulence	120 km	plasma sheet
Earthward ion streams	13 km	plasma-sheet boundary layer

tering is envisioned to reduce the mobility of particles at the interface. Some problems with the model are that contact potentials tend to be of the wrong sign (accelerating electrons upward) and the potential drops tend to be greatly reduced in the presence of cool electrons.

STRONG DOUBLE LAYERS: Here the major part of the field-aligned potential drop between the ionosphere and the magnetosphere is assumed to occur in a strong plasma double layer [*Carlqvist and Bostrom*, 1970; *Block*, 1972; *Borovsky*, 1992b]. The drawback to this model is that it predicts strong localized parallel electric fields, which have yet to be observed.

ELECTROSTATIC SHOCKS: In this model the potential drop between the magnetosphere and the ionosphere is assumed to occur within a structure called an electrostatic shock [*Swift*, 1975, 1976; *Hudson and Potter*, 1981]. Conceptually, electrostatic shocks are the same as strong double layers, both being potential-drop solutions to the Poisson-Vlasov equations, but electrostatic shocks are envisioned to have ion-gyroradius thicknesses [*Swift*, 1979] whereas magnetized double layers have Debye-length thicknesses [*Borovsky*, 1983]. However, the reality of electrostatic shocks is doubtful: *Swift* [1979] points out that electrostatic-shock solutions to Poisson's equation that satisfy the needed boundary conditions have not been found (as was also the case for the *Hudson and Potter* calculations). Therefore, it can be argued that electrostatic shocks and strong double layers are actually the same thing, and that the creation of the electrostatic-shock category was premature.

WAVE-BROADENED ELECTROSTATIC STRUCTURE: In this model the difference between the $\vec{E} \times \vec{B}$-drift velocities of electrons and ions owed to gradients in $E_\perp$ and finite ion gyroradii drive lower-hybrid waves via the modified-two-stream instability [*Smith*, 1986a, 1986b, 1987]. The driving of the instability by the electron-ion drift produces a viscosity between the two species, which results in a force-cross-B drift of the plasma, carrying along the charge density and field-aligned currents. This results in a broadening of the arc. The model suffers from the drawback that it requires high plasma densities ($\sim$ 100 cm^{-3}) in the auroral acceleration region, and such high densities have not been observed. Also, the model has not been advanced sufficiently to allow an auroral-structure thickness calculation to be made.

ANOMALOUS RESISTIVITY: Here, the field-aligned potential drop between the magnetosphere and ionosphere is envisioned to be a resistive drop in the auroral-arc current owed to the scattering of current carriers off electrostatic waves that are driven by the electron-ion relative drift [*Palmadesso et al.*, 1974; *Ionson et al.*, 1976; *Galeev*, 1983]. The most-often-discussed wavemode is the electrostatic ion-cyclotron mode: for the thickness estimate put into Table 3, the arc structure is taken to be much thicker than one perpendicular wavelength of an electrostatic ion-cyclotron wave.

WEAK DOUBLE LAYERS: Similar to the anomalous-resistivity model, the potential drop between the ionosphere and the magnetosphere is assumed here to be the sum of many small drops, each drop occurring across an ion-acoustic (weak) double layer [*Lotko and Kennel*, 1983; *Hudson et al.*, 1987; *Lotko*, 1986]. The question of how, in a three-dimensional geometry, a large-scale large-amplitude potential drop is constructed from the sum of individual, isolated charge separations has yet to be fully addressed. The width estimate inserted into Table 3 reflects this critical absence of a full picture: the estimate comes from the fact that the full potential structure must be much wider than one individual weak double layer.

KINETIC ALFVEN WAVES: The kinetic Alfven wave model has auroral electrons being accelerated by an electric field at the front of an Alfven wave [*Hasegawa*, 1976; *Goertz and Boswell*, 1979; *Lysak and Carlson*, 1981], rather than by falling through an electrostatic potential drop. The Alfven wave needed for the acceleration has a transverse width of an electron-plasma collisionless skin depth $2\pi\, c/\omega_{pe}$, which sets the minimum width of an auroral structure produced by kinetic Alfven waves.

ELECTROMAGNETIC ION-CYCLOTRON WAVES. This model assumes that auroral electrons are accelerated by Landau resonance with electromagnetic ion-cyclotron waves [*Temerin et al.*, 1986; *Temerin and Cravens*, 1990]. The electromagnetic ion-cyclotron wave of interest is similar to a kinetic Alfven wave in that it has a perpendicular wavelength of about 2π electron-plasma collisionless skin depths, however the ion-cyclotron wave has a frequency approaching the ion-cyclotron frequency, whereas the Alfven wave has a much-lower frequency. In this model, electrons are picked up in large amounts in the wave troughs and are released in surges, producing downward electron fluxes that are modulated at the wave period.

LOWER-HYBRID-WAVE LANDAU RESONANCE: In this model electrons are stochastically accelerated by means of Landau-resonance interactions with lower-hybrid waves [*Bingham et al.*, 1984; *Bryant et al.*, 1991]. The width of an arc structure corresponds to the width of the region of wave turbulence, which may in fact be set by the generator mechanism. For Table 3 the estimated thickness is written as being much greater than the perpendicular wavelength of a single lower-hybrid wave.

AKR-INDUCED PRECIPITATION: This model differs significantly from the other models that are grouped into accelerator models in that electrons are not energized. Instead, the auroral-arc airglow is envisioned to arise when energetic magnetospheric electrons are pitch-angle scattered into the atmospheric loss cone by interactions with electromagnetic waves known as auroral kilometric radiation (AKR) [*Calvert*, 1982, 1987]. In the model, the AKR intensity builds up on auroral-arc magnetic-field lines because of partial reflections of the AKR off density gradients. One drawback to the model is that it requires AKR reflection coefficients that may be unrealistically high [see, *e.g.*, *Pritchett*, 1986].

B. Generator Mechanisms

In the different generator models outlined below (and listed in Table 3), very different mechanisms are invoked to supply power, current, and charge to auroral-arc magnetic-field lines. The various mechanisms are thought to act in various regions of the magnetosphere and ionosphere. Some of these regions do not map to auroral arcs if the mapping discussed in Section 1 is accepted. The generator mechanisms that are inconsistent with this mapping will be so noted.

SHEAR IN THE LOW-LATITUDE BOUNDARY LAYER: In this model the inertia of a plasma flow powers an auroral current system, with the driving charge building up in the observed cross-field velocity shear in the low-latitude boundary layer [*Eastman et al.*, 1976; *Lundin and Evans*, 1985; *Siscoe et al.*, 1991]. A criticism of this model is that the Region-1 currents do not map to the low-latitude boundary layer except perhaps near the cusp; rather they map to the plasma sheet. So, any generator in the low-latitude boundary layer cannot directly power auroral-arc currents.

SHEAR IN THE PLASMA SHEET: Again, the inertia of a moving plasma acts as a generator, with charge and currents emanating from a shear zone [*Rostoker and Bostrom*, 1976; *Hasegawa and Sato*, 1979; *Birn and Hesse*, 1991]. The shear flows in the plasma sheet could be associated with earthward plasma-sheet convection deflecting around the dipole, with the breakdown of co-rotation, or with substorm current-wedge flows accompanying magnetotail dipolarizations.

POYNTING-FLUX ABSORPTION: In this model power is transferred from the solar wind to auroral field lines because the solar wind drives compressive Kelvin-Helmholtz waves on the magnetosphere, the Kelvin-Helmholtz waves coupling to kinetic Alfven waves on resonance layers in the plasma sheet [*Goertz*, 1983, 1984, 1990; *Harrold et al.*, 1990; *Samson et al.*, 1992]. The resonance layers are characterized by having Alfven velocities equal to the velocity of the Kelvin-Helmholtz waves. Electrons can be accelerated by the electric fields of the Alfven waves [*e.g. Hasegawa*, 1976; *Goertz*, 1983] or electrostatic structures that accelerate electrons can be produced by pumping ions off the field lines to leave them negatively charged [*e.g. Hasegawa*, 1987; *Moghaddam-Taaheri et al.*, 1991].

MAGNETIC-FIELD-LINE RECONNECTION: This model assumes that auroral arcs are powered by the energy that is released at an X-type neutral line during magnetic-field reconnection [*Akasofu et al.*, 1967; *Sato and Iijima*, 1979; *Atkinson et al.*, 1989]. The auroral arcs are envisioned as residing on magnetic-field lines that connect to the X-lines (separatrix field lines). The charge and current for the arc may be driven by Hall electric fields set up by the plasma flow near the X-line [*Atkinson*, 1992] or by space-charge along the neutral line [*Hoshino*, 1987; *Deeg et al.*, 1991]. A drawback to the model is that it cannot explain multiple quiet arcs, since during quiet times there is only one long-lasting neutral line (the distant neutral line). Another drawback is that the distant neutral line is located at the interface between the lobe and the plasma-sheet boundary layer: this location does not map to the auroral zone.

IONOSPHERIC-CONDUCTIVITY FEEDBACK: By means of this generator mechanism, north-south ionospheric electric fields (which may be powered by magnetospheric convection) drive auroral arcs by producing polarization charge on the edges of east-west-aligned high-conductivity channels in the ionosphere [*Ogawa and Sato*, 1971; *Sato*, 1978; *Watanabe and Sato*, 1988]. The positive polarization charge in the ionosphere attracts magnetospheric electrons down to increase the ionization level in the ionsophere, which maintains the high-conductivity channel.

PLASMA FLOW INTO HIGH-CONDUCTIVITY ZONE: This generator works when the cross-tail (east-west) electric field in the magnetosphere (which is powered by an earthward plasma flow) produces a north-south charge separation across a high-conductivity channel in the ionosphere under the arc [*Rothwell et al.*, 1991]. The north-south charge separation is produced by the Hall current and the high conductivity is produced by auroral-arc-electron precipitation (see also *Rothwell et al.*, [1988, 1989]). In this model, the earthward flow of plasma in the plasma sheet is slowed down and deflected toward the dawn as it crosses auroral-arc magnetic-field lines.

PRESSURE GRADIENTS: When a plasma pressure gradient in the plasma sheet is perpendicular to a gradient in the magnetic induction, then the diamagnetic-drift current can build up and maintain regions of space charge that can drive field-aligned auroral currents [*Vasyliunas*, 1970; *Stasiewicz*, 1985; *Galperin and Volosevich*, 1989].

ELECTROSTATIC FLUID TURBULENCE: In this model coherent flow structures arise in a plasma that contains turbulent $\vec{E} \times \vec{B}$-drift motions, with velocity-shear zones building up charge densities, which can drive field-aligned auroral currents [*Swift*, 1977; 1981; *Lotko and Schultz*, 1988]. One drawback to the model is that structures tend to become aligned in the north-south direction, rather than in the east-west direction as are auroral arcs. When the model is used to drive auroral currents from turbulence in the low-latitude boundary layer, the model is inconsistent with the mapping of auroral arcs into the magnetosphere discussed in Section 1: when the model is used to drive the currents from turbulence in the plasma sheet, it is consistent with that mapping.

EARTHWARD ION STREAMS: Here, the energy to generate an auroral arc comes from the kinetic energy of ions streaming earthward along the magnetic field from the magnetotail [*Kan*, 1975; *Lui et al.*, 1977; *Lyons and Evans*, 1984]. These earthward-streaming ions reside in the plasma-sheet boundary layer [*e.g. Takahashi and Hones*, 1988; *Zelenyi et al.*, 1990], which is a drawback to the model, since auroral arcs do not magnetically connect to the plasma-sheet boundary layer. Another drawback to the model is that it does not explain multiple quiet arcs, since only one ion beam is envisioned.

C. Minimum-Possible Structure Thickness

If an airglow structure is caused by energetic electrons hitting the upper atmosphere, then the of airglow structure has a minimum thickness because of Coulomb scattering of the precipitating electrons. Even if an infinitely thin sheet beam of electrons were producing a structure, the sheet beam will spread as it enters the atmosphere owing to to the scattering, and hence the resulting airglow region will be broadened. *Borovsky et al.* [1991] numerically calculated this broadening and found that infinitely thin sheet beam resulted in airglow regions with Gaussian intensity profiles: 5-keV electrons resulted in a 10-m-wide airglow structure and 25-keV electrons resulted in an airglow structure that was 22-m wide.

4. Comparison between Observation and Theories

The distribution of arc-structure thicknesses measured by *Maggs and Davis* [1968] is plotted again in Figure 2. Also plotted on that figure are the 20 theoretical minimum thicknesses from Table 3, where the plotted distribution is the number of theories, rather than the number of structures. Finally, the result quoted in Section 3C that the minimum-possible arc-structure thickness (if the airglow is produced by energetic-electron precipitation) is 10 - 22 m is plotted on Figure 2.

Examining Figure 2, it is clear that none of the auroral-arc thicknesses predicted by the theories can match the thicknesses of auroral-arc fine-scale structures as measured by *Maggs and Davis* [1968]. Therefore it is concluded that none of these theories explains the cause of the fine-scale structure observed by *Maggs and Davis* and by *Borovsky et al* [1991] and *Borovsky and Suszcynsky* [1992]. *Maggs and Davis* appear to have been observing a new phenomenon that has not been explained in 25 years.

As can be seen in Figure 2, the thicknesses of the fine-scale structures are greater than the minimum-possible thicknesses for energetic-electron-produced airglow. This means that it is possible that the fine-scale pattern of airglow is caused by a pattern of electron precipitation, although this

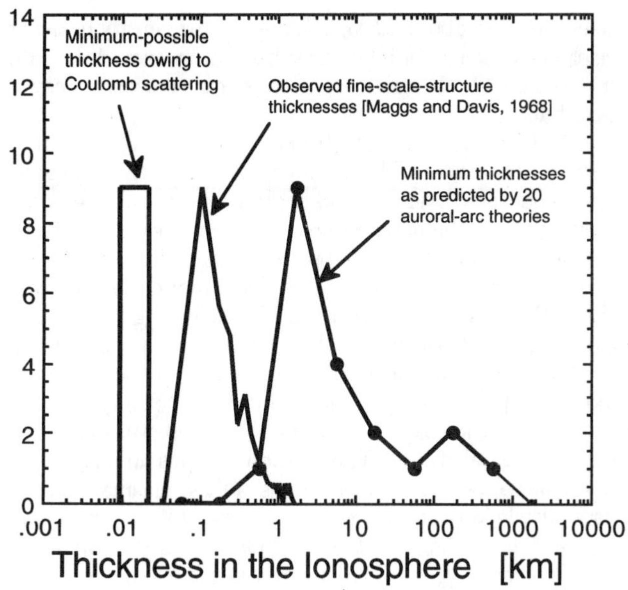

Fig. 2. The minimum thicknesses of auroral-arc structures as predicted by 20 auroral-arc theories compared with the thicknesses observed by *Maggs and Davis* [1968].

TABLE 4. Auroral-arc fine structure is mapped along dipole magnetic-field lines from the ionosphere to the acceleration region at 1.5 R_E altitude and on into the equatorial plasma sheet at $L = 8$. For those three regions of space, characteristic plasma spatial scales are calculated. For the auroral ionosphere $n = 10^5$ cm^{-3}, $T_e = 0.2$ eV, $T_i = 0.2$ eV, and $B = 0.6$ gauss; for the acceleration region $n = 0.3$ cm^{-3}, $T_e = 500$ eV, $T_i = 2.5$ keV, and $B = 3.3 \times 10^{-2}$ gauss; and for the equatorial magnetosphere $n = 0.3$ cm^{-3}, $T_e = 500$ eV, $T_i = 2.5$ keV, and $B = 6 \times 10^{-4}$ gauss. For the ion and electron gyroradii, energies of 5 keV were used since the potential drop of interest is $e\Delta\phi \sim 5$ keV.

Quantity Mapped	Auroral Ionosphere	Acceleration Region	Equatorial Magnetosphere
Fine Scales	100 m	430 m	4.3 km
$20\ \lambda_{De}$	21 cm	6 km	6 km
$2\ r_{gi_{energetic}}$	240 m	4.4 km	240 km
$\pi\ c/\omega_{pe}$	53 m	30 km	30 km
$2\ r_{ge_{energetic}}$	8 m	140 m	8 km
c/ω_{pi}	720 m	410 km	410 km
$4\ v_{ti}/\omega_{ci}$	300 cm	6.2 km	310 km

has not been observationally confirmed. This possibility and other possible origins for the fine-scale structure are discussed in Section 5, and measurements that can help to determine the nature of the fine-scale structures are discussed in Section 6.

5. Possible Origins of Fine-Scale Structure

The nature of auroral-arc fine-scale structure is a mystery. Some information about the possible causes of fine-scale structure and their locations can be obtained by comparing the thickness of a structure with characteristic spatial scales in a plasma. To do this, a typical (100-m) fine-scale structure is mapped along the dipole magnetic-field lines from the auroral ionosphere to the acceleration region at 1.5 R_E altitude and to the equatorial plasma sheet at $L = 8$. The mapped scale size appears in Table 4 and in Figure 3. Also included in Table 4 and Figure 3 are characteristic plasma scale lengths for these regions. The scale length $20\lambda_{De}$ (20 times the Debye length of the plasma) is the minimum width that an electrostatic-potential structure can have if the structure has a potential drop larger than the electron temperature of the plasma in which it resides. It is also approximately the smallest perpendicular wavelength for Langmuir waves, ion-acoustic waves, and lower-hybrid waves, and it is approximately equal to the transverse width of an ion-acoustic double layer. The scale length $\pi c/\omega_{pe}$ (π times the electron-plasma collisionless skin depth of the plasma) is the minimum thickness that an Alfven-wave structure can have and still be guided along the magnetic field. The scale length $2r_{gi_{thermal}}$ (the gyrodiameter of a thermal ion) is the minimum transverse thickness that a density gradient or a pressure gradient or a shear layer in the plasma can have, and it is the minimum thickness that a structure formed by mirroring ions can have. The scale length $4v_{Ti}/\omega_{ci}$ is approximately the perpendicular wavelength of an electrostatic ion-cyclotron wave in a warm-ion plasma (see, *e.g.* Figure 1b of *Kindel and Kennell* [1971]). It is also the minimum thickness of a Hartmann-flow viscous layer [*e.g. Jackson*, 1975], obtained by optimizing the collsision frequencies to maximize the perpendicular (Pederson) conductivity. The scale length c/ω_{pi} (the ion-inertial length) is the scale thickness of a slow-mode shock [*Coroniti*, 1971]. The final size considered is $2r_{ge_{energetic}}$ (the gyrodiameter of an energetic electron), which is not a characteristic scale size of the plasma, but rather is an indicator of how narrow a sheet of energetic electrons can be. As can be seen in Figure 3, in the acceleration region and in the equatorial magnetosphere the fine-scale-structure thickness is less than all of the characteristic plasma scale lengths. This has two related implications. First, it will be difficult to envision a mechanism in the acceleration region or plasma sheet that can supply a significant amount of kinetic energy to electrons to produce these narrow airglow structures. Second, an ionospheric or near-ionospheric origin for the fine-scale structures seems likely.

In the following three sections, three possibile locations for the source of fine-scale structures are considered and implications are drawn for each location. They are (1) the fine-scale structures originate from a phenomenon in the acceleration region at 1 - 2 R_E altitude, (2) the fine-scale structures originate from a phenomenon acting between the

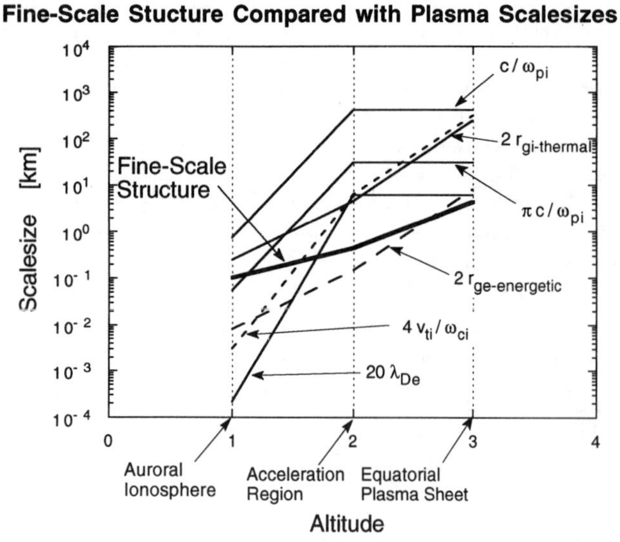

Fig. 3. The thickness of a fine-scale structure (100 m) in the ionosphere is mapped along dipole magnetic-field lines to the acceleration region (1 - 2 R_E altitude and to the equatorial plasma sheet at $L = 8$. The characteristic scale sizes for the plasmas at those three locations are plotted for comparison (from Table 4).

acceleration region and the upper atmosphere, and (3) the fine-scale structures originate from a process in the the upper atmosphere. The fine-scale structures originating in the plasma sheet is ruled out since a generator at the equator must deliver a current along the magnetic-field lines and the field-aligned-current channels are set up by Alfven-wave processes that result in current channels that can be no thinner than $\pi c/\omega_{pe}$, which yields a current channel much wider than the mapped fine-scale structure.

A. The Origin Is in the Acceleration Region

If the fine-scale east-west-aligned structures within an east-west-aligned auroral arc originate in the acceleration region, then two possibilities exist.

(1) *The Electron Accelerator Is Finely Structured.* The first possibility is that the electrons are accelerated downward in the acceleration region by a process that involves an electric field that is localized to about 400 m transverse to the magnetic field. From this 400-m-wide region a slab of energetic electrons emanates and this slab maps along the dipole field lines to produce a 100-m-wide airglow structure in the atmosphere (see Table 4). This possibility is problematic, since it requires an acceleration mechanism that has a spatial extent that is smaller than all of the characteristic spatial scales of the ambient plasma. Steady-state electrostatic structures that accelerate the electrons can certainly be ruled out, since the fine-scale structure maps to widths of about 1 λ_{De} at these altitudes and electrostatic structures in a plasma are excluded by Poisson's equation from having sizes less than about 20 λ_{De}. If the plasma density in the accelerator zone can build up to substantially exceed the value $n = 0.3$ cm^{-3}, then narrow electrostatic structures may be possibile.

(2) *The Flow of Electrons into the Accelerator Is Finely Structured.* The second possibility is that the accelerator operates over a broad spatial region, but the flux of electrons into the accelerator is somehow modulated in the north-south direction. One possible cause for a modulated electron flow is the presence of electrostatic plasma waves above the accelerator, with thc flow bcing higher though a trough in the wave than it is through a crest. This would produce a pattern of airglow emission in the atmosphere that moves in the north-south direction with a velocity that corresponds to the mapping down of the phase velocity of the wave at high altitudes. This possiblity seems unlikely, owing to the large transverse wavelengths of most plasma modes — for example ion-acoustic waves and lower-hybrid waves have $\lambda_\perp \gtrsim 20\lambda_{De}$ and electrostatic ion-cyclotron waves have $\lambda_\perp \gtrsim 2\pi r_{gi_{\rm thermal}}$. The electron-Bernstein mode [*Bernstein*, 1958; *Alexandrov et al.*, 1984], however, can have a very short perpendicular wavelength in the acceleration-region plasma: for $\omega_{pe} \ll \omega_{ce}$ the mode can have $\lambda_\perp < \lambda_{De}$, which can be compable to the width of a fine-scale structure. For $\omega_{pe}^2 \ll \omega_{ce}^2$ and $\lambda \gg r_{ge_{\rm thermal}} = (k_B T_e/m_e)^{1/2}/\omega_{ce}$, the dispersion relation for the $N = 1$ electron-Bernstein mode is (from expression 8.12.9 of *Krall and Trivelpiece* [1971]) $\omega^2 = \omega_{ce}^2 + \omega_{pe}^2\left(1 - k^2 r_{ge_{\rm thermal}}^2\right)$, which becomes the upper-hybrid mode in the long-wavelength limit. For $\lambda \sim 400$ m in the acceleration region, this mode has a phase and group velocity of $\omega/k \approx \omega_{ce}/k \sim 3.6 \times 10^9$ cm/sec and $\partial\omega/\partial k \approx \omega_{pe}^2 r_{ge_{\rm thermal}} 2k/\omega_{ce} \sim 7.0 \times 10^5$ cm/sec. The phase velocity of the mode is very large, which makes it unlikely that the moving fine-scale structures are mappings of the motions of the wave troughs. However, the group velocity of the mode is not so large (and it is very sensitive to the value of ω_{pe}), so it is possible that the fine-scale structures are caused by an interference of the electron flow by electron scattering in a slowly moving high-frequency electron-Bernstein wave packet. However, such wave scattering would seem to be better at making dark regions in an arc by reducing the electron flux rather than making the brighter fine-scale structures.

B. The Origin Is between the Acceleration Region and the Upper Atmosphere

If the origin of the fine-scale structure is between the acceleration region and the upper atmosphere, then two possibilities exist: the high-energy electron beam becomes finely structured before it reaches the upper atmosphere or a mechanism operates well below the acceleration region that energizes other electrons. It is not likely that ions are energized above the upper atmosphere to produce the fine-scale structures, since ions will not follow magnetic field lines down into the upper atmosphere owing to charge-exchange collisions that neutralize them [*e.g. Davidson*, 1965].

(1) *The Electron Sheet Beam Becomes Structured.* The first possible origin of the fine-scale structures is the structuring of the beam of electrons as they propagate downward along the magnetic-field lines from the acceleration region to the upper atmosphere. A candidate mechanism that will structure the broad slab-shaped sheet of keV electrons into thinner slabs that map to the fine-scale structures is not known. The beam-structuring mechanisms that are discussed in the literature will not transform a thick slab into a thinner slab: the charged-sheet (diocotron) instabilities form windings (curls) on a sheet beam [*Hallinan and Davis*, 1970; *Wagner et al.*, 1983], the current-sheet instabilities form windings (spirals) on a sheet beam [*Hallinan*, 1976], and the tearing-mode instabilities in a sheet current form windings on the sheet and bunch the sheet current into filaments [*Seyler*, 1988; *Otto and Birk*, 1992].

(2) *A Mid-Altitude Electron-Acceleration Mechanism Operates.* The second possible origin of the fine-scale structures is the precipitation of electrons that are energized somewhere above the ionosphere and below the main acceleration region at 1 - 2 R_E altitude. Whether they are isotropic or field aligned, the electrons accelerated at lower altitudes need to gain energies of on the order of 1 keV to make fine-scale structures that have extents along the magnetic field of less than about 50 km [*e.g. Berger et al.*, 1970; *Omholt*, 1971; *Banks et al.*, 1974]. The electrons that produce auroral arcs are believed to gain their kinetic energies in the acceleration region 1 - 2 R_E in altitude [*e.g. Mizera and Fennell*, 1977; *Paschmann et al.*, 1983; *McFadden et al.*, 1990], with little evidence that keV electrons originate at substantially lower altitudes. The electron accelerator mechanism may fall into one of three forms: a steady-state electrostatic-potential structure, an Alfven wave with a parallel electric field, or a spatial region containing electrostatic plasma waves. Strong d.c. electric fields are seen in the upper levels of the ionosphere [*e.g. Carlson and Kelley*, 1977; *Smiddy et al.*, 1977; *Maynard et al.*, 1982; *Chmyrev et al.*, 1985]. *Maynard et al.* have reported 200 mV/m perpendicular electric fields with spatial scales of 500 m at 940 km altitude in the auroral zone and *Chmyrev et al.* report an auroral-zone pass where a 200 mV/m electric-field structure with a spatial extent of 4 km was seen at an altitude of 811 km. These observed electric fields are insufficient to accelerate electrons to keV energies: an electrostatic field of 200 mV/m with an extent of 400 m only has a potential drop of 80 V, which is enough to accelerate electrons to energies sufficient to produce airglow, but the airglow produced by such low-energy electrons will appear at very high altitudes and will be very extended along the magnetic field. Thus, these weak fields are not the origin of the fine-scale structure. An additional argument against them is that strong-electric-field structures with widths cpomparable to 100 m have not been seen. The possibility that Alfven waves trapped at low altitudes within the ionosphere could accelerate electrons to produce multiple structures has been put forth [*Haerendel*, 1983; *Trakhtengerts and Feldstein*, 1991]. An analysis of Alfven-wave propagation that appears in the Appendix of *Borovsky* [1992a] found that Alfven waves narrow enough to produce the fine-scale structures would not propagate, but instead would resistively damp owing to the finite parallel conductivity of the upper ionosphere. Electrostatic plasma-wave acceleration of electrons above the ionosphere is a promising possibility, particularly since very-small-scale regions of intense plasma-wave activity have been observed there. Bursts of Langmuir waves with parallel electric fields up to 500 mV/m have been seen at rocket altitudes within auroral arcs [*McFadden et al.*, 1986; *Ergun et al.* 1991a, 1991b], with burst periods of about 0.1 sec. With a rocket velocity on the order of 1 km/sec across the magnetic field, this time corresponds to a width of 100 m. The appearance of the Langmuir-wave bursts sometimes coincided with the appearance of keV electrons moving down the magnetic-field lines (—it e.g. Figure 2 of *Ergun et al.* [1991a]). Bursts of Lower-Hybrid waves with amplitudes up to 200 mV/m have been seen at rocket altitudes within auroral arcs [*La-Belle et al.*, 1986; *Kintner et al.*, 1992], the burst durations corresponding to as little as 50 m of rocket flight. These Langmuir-wave and lower-hybrid-wave packets above the ionosphere are the only structures known to the author that are narrow enough to map to the fine-scale structures.

C. The Origin Is in the Upper Atmosphere

Looking for an origin in the upper atmosphere (ionosphere) is appealing because the thicknesses of the fine-scale structures are larger than the characteristic plasma scale lengths in the ionosphere (see Table 4 and Figure 3). Two possibilities are explored.

(1) *An Electrical Discharge at Low-Altitude Heats Electrons.* The possibility that the fine-scale structures are patches of airglow produced by few-eV electrons accelerated by ionospheric electric fields was explored by *Borovsky* [1992a]. It was found that airglow was unlikely to be produced in this manner in the lower ionosphere but that it could be produced in the middle or upper ionosphere. At altitudes of 100 km the very large cross section (which turn on at ~1 eV) for electron excitation of N_2 vibrational states makes it unlikely that an electron can be accelerated from ionospheric energies to the energies required to produce airglow (~4 eV), unless dc parallel electric-field strengths can exceed 70 mV/m at 100 km. At higher altitudes, the electric-field-strength restriction is less stringent: at 180 km a 0.1 mV/m parallel electric field could accelerate electrons to a few eV before they lost their energies to the N_2 vibrational states, hence a weak parallel electric field could result in airglow. A mechanism that will set up narrow regions with parallel electric fields in the ionosphere is not known.

(2) *Ionospheric Ions Are Heated.* A possible cause of fine-scale structures is the heating of ions in the ionosphere. Ions that precipitate from higher altitudes cannot make narrow airglow structures owing to charge-exchange collisions that they undergo in the upper atmosphere, which neutral-

ize them so they no longer follow the magnetic field lines, hence they spread [*Davidson*, 1965]. For ions that are energized at low altitudes, this spreading effect reduces. If ions are heated to 10's of eV in the upper atmosphere, then charge-exchange collisions will transform the ions into neutrals, some of which that in excited states, and the charge-exchange collisions will transform air atoms and molecules into ions, some of which are in excited states [*e.g. Wehrenberg and Clark*, 1973; *Torr et al.*, 1974; *Hoffman et al.*, 1981]. If the neutral density is high enough and the charge exchange cross sections are large enough, then this will result in characteristic line spectra emitted from the region of ion heating. If proton heating is the cause, a narrow structure emitting the hydrogen lines Hα at 6563Å and Hβ at 4861Å should be seen. If O^+ heating is the cause and charge exchanging with the atmospheric neutrals occurs, a narrow structure emitting the short-lifetime lines of molecular ions (such as the 3914-Å line of N_2^+ and the 4278-Å line of O_2^+) should be seen with an isotropic halo of emission in the 0.7-sec-lifetime atomic-oxygen 5577-Å line surrounding the narrow structure, the molecular-ion lines being emitted from air molecules that become excited molecular ions after the charge-exchange collisions and the 5577-Å emission coming from fast-moving excited oxygen atoms produced by neutralizing the fast-moving oxygen ions. It is not known whether or not it is possible to heat ions deep in the ionosphere.

D. Assessment of the Possible Origins

Of the several origins for the fine-scale structures considered in Sections A - C, only three are promising candidates: (1) wave acceleration of electrons to keV energies just above the ionosphere, (2) electrical discharges at altitudes of 150 km or so that heat electrons to energies of a few eV, and (3) the heating of ions at altitudes of 100 km or so. Other mechanisms that would operate at altitudes higher than the upper ionosphere are unlikely owing to the very large characteristic spatial scales of the plasmas at those higher altitudes (see Table 4 and Figure 3). As is discussed in Section 6, the three favored candidates each provide unique sets of signatures that can be seen with ground-based detectors.

6. Critically Needed Measurements

Table 2 contains a list of unknown properties of auroral-arc fine-scale structures that could be discerned from ground-based observations. A subset of these properties are critically needed, if the origin of the fine-scale structures is to be determined. In this section, these critical properties are discussed.

LINE RATIOS: It is very important to determine the relative brightness of various emission lines from the airglow of the fine-scale structures. Determining if the line ratios are consistent with energetic-electron precipitation will eliminate electrical-discharge origins and ion-heating origins. If the fine-scale airglow appears in hydrogen line emission, then ion heating is almost certainly their cause. If the 3914-Å emission line of N_2^+ is absent but fast-emission bands of N_2 and O_2 are present, then an electrical-discharge origin (very-low-energy electrons) is likely.

HEIGHTS: Determining the heights of the fine-scale structures by triangulation will allow a more-accurate determination of the thicknesses and velocities of the fine-scale structures. A height determination will also provide information as to whether the structures are at altitudes consistent with an electron-precipitation origin: if not, then electrical-discharge or ion-heating mechanisms would be suspected.

MOTIONS RELATIVE TO ARC MOTIONS: If the fine-scale structures have relative motions with respect to the moving larger-scale auroral arcs in which they reside, then wave mechanisms would be suspected for their origin. These wave mechanisms might be modulating the flow of electrons into the accelerator at 1 - 2 R_E altitudes, they might be producing energetic electrons between the accelerator altitudes and the ionosphere, they might be structuring the energetic-electron sheet beam as it propagates down toward the ionosphere, or they might be heating ions in the upper atmosphere. For each of these four possibilities, the waves would be powered by the auroral arc.

BRIGHTNESSES: Determinations of the absolute brightnesses of the fine-scale structures would provide an estimate of the power dissipated by the mechanism that makes the structures and determinations of the brightnesses of the fine-scale structures relative to the brightnesses of the larger-scale arcs would determine how important energetically the fine-scale structures are to auroral arcs and to solar-wing/magnetospheric coupling.

7. Summary

Fine-scale structures within auroral arcs are real and they are unexplained. At ionospheric altitudes they have thicknesses of about 100 m (Figure 1). The known and unknown properties of fine-scale structures are summarized in Table 2.

Existing auroral-arc theories (listed in Table 3) do not explain the fine-scale structures because none of these theories will allow a structure as narrow as the observed fine-scale structures (as summarized in Figure 2).

Several possible origins for the fine-scale structures were considered in Section 5: only three of those seemed likely. They are (1) the acceleration of electrons to keV energies within localized wave packets in the upper ionosphere (Section 5.B.2), (2) the heating of ionospheric electrons to energies of a few eV by electrical discharges acting at altitudes of 150 km or so (Section 5.C.1), and (3) the heating of ions by an unknown mechanism at altitudes of 100 km or so.

Ground-based observations that could help to determine the origin of the fine-scale structures were suggested in Section 6..

Acknowledgments. The author wishes to thank Jim Maggs, Peter Gary, and Dave Suszcynsky for useful conversations. This work was supported by the U.S. Department of Energy.

The author is grateful to Christine Hoffman (now Christine Borovsky) for choosing to vacation in Kauai at the time of the AGU Chapman Conference.

REFERENCES

Akasofu, S.-I., Thickness of an Active Auroral Curtain, *J. Atmos. Terr. Phys.*, *21*, 287, 1961.

Akasofu, S.-I., S. Chapman, and P. C. Kendall, The Significance of the Multiple Structure of the Auroral Arc, in *Aurora and Airglow*, B. M. McCormac (ed.), Reinhold, New York, 1967.

Alexandrov, A. F., L. S. Bogdankevich, and A. A. Rukhadze, *Principles of Plasma Electrodynamics*, sect. 5.8.3, Springer-Verlag, New York, 1984.

Alfven, H., and C.-G. Falthammar, *Cosmical Electrodynamics*, sect. 3.2.1, Oxford, Clarendon, 1963.

Atkinson, G., Mechanism by Which Merging at X Lines Causes Discrete Auroral Arcs, *J. Geophys. Res.*, *97*, 1337, 1992.

Atkinson, G., F. Creutzberg, and R. L. Gattinger, Interpretation of Complicated Discrete Arc Structure and Behavior in Terms of Multiple X Lines, *J. Geophys. Res.*, *94*, 5292, 1989.

Bennett, E. L., M. Temerin, and F. S. Mozer, The Distribution of Auroral Electrostatic Shocks below 8000-km Altitude, *J. Geophys. Res.*, *88*, 7107, 1983.

Berger, M. J., S. M. Seltzer, and K. Maeda, Energy Deposition by Auroral Electrons in the Atmosphere, *J. Atmos. Terr. Phys.*, *32*, 1015, 1970.

Belon, A. E., G. J. Romick, and M. H. Rees, The Energy Spectrum of Primary Auroral Electrons Determined from Auroral Luminosity Profiles, *Planet. Space Sci.*, *14*, 597, 1966.

Bernstein, I. B., Waves in a Plasma in a Magnetic Field, *Phys. Rev.*, *109*, 10, 1958.

Bingham, R., D. A. Bryant, and D. S. Hall, A Wave Model for the Aurora, *Geophys. Res. Lett.*, *11*, 327, 1984.

Birn, J., and M. Hesse, The Substorm Current Wedge and Field-Aligned Currents in MHD Simulations of Magnetotail Reconnection, *J. Geophys. Res.*, *96*, 1611, 1991.

Block, L. P., Potential Double Layers in the Ionosphere, *Cosmic Electrodynamics*, *3*, 349, 1972.

Borovsky, J. E., The Scaling of Oblique Plasma Double Layers, *Phys. Fluids*, *26*, 3273, 1983.

Borovsky, J. E., Auroral-Arc Thicknesses As Predicted by Various Theories, submitted to *J. Geophys. Res.*, 1992a.

Borovsky, J. E., The Strong-Double-Layer Model of Auroral Arcs: An Assessment, submitted to *Auroral Plasma Dynamics*, edited by R. Lysak, American Geophysical Union, Washington, 1992b.

Borovsky, J. E., and D. M. Suszcynsky, Optical Measurements of the Fine-Scale Structure of Auroral Arcs, submitted to *Auroral Plasma Dynamics*, edited by R. Lysak, American Geophysical Union, Washington, 1992.

Borovsky, J. E., D. M. Suszcynsky, M. I. Buchwald, and H. V. DeHaven, Measuring the Thicknesses of Auroral Curtains, *Arctic*, *44*, 231, 1991. Bryant, D. A., A. C. Cook, Z.-S. Wang, U. de Angelis, and C. H. Perry, Turbulent Acceleration of Auroral Electrons, *J. Geophys. Res.*, *96*, 13829, 1991.

Calvert, W., A Feedback Model for the Source of Auroral Kilometric Radiation, *J. Geophys. Res.*, *87*, 8199, 1982.

Calvert, W., Auroral Precipitation Caused by Auroral Kilometric Radiation, *J. Geophys. Res.*, *92*, 8792, 1987.

Carlqvist, P. and R. Bostrom, Space-Charge Regions above the Aurora, *J. Geophys. Res.*, *75*, 7140, 1970.

Carlson, C. W., and M. C. Kelley, Observation and Interpretation of Particle and Electric Field Measurements inside and Adjacent to an Active Auroral Arc, *J. Geophys. Res.*, *82*, 2349, 1977.

Chiu, Y. T., and M. Schulz, Self-Consistent Particle and Parallel Electrostatic Field Distributions in the Magnetospheric-Ionospheric Auroral Region, *J. Geophys. Res.*, *83*, 629, 1978.

Chmyrev, V. M., V. N. Oraevsky, S. V., Bilichenko, N. V. Isaev, G. A. Stanev, D. K. Teodosiev, and S. I. Shkolnikova, the Fine Structure of Intensive Small-Scale Electric and Magnetic Fields in the High-Latitude Ionosphere as Observed by INTERCOSMOS-BULGARIA 1300 Satellite, *Planet. Space Sci.*, *33*, 1383, 1985.

Coroniti, F. V., Laminar Wave-Train Structure of Collisionless Magnetic Slow Shocks, *Nucl. Fusion*, *11*, 261, 1971.

Davidson, G. T., Expected Spatial Distribution of Low-Energy Protons Precipitated in the Auroral Zones, *J. Geophys. Res.*, *70*, 1061, 1965.

Davis, T. N., Cinematographic Observations of Fast Auroral Variations, in *Aurora and Airglow*, B. M. McCormac (ed.), Reinhold, New York, 1967, pg. 133.

Davis, T. N., Observed Characteristics of Auroral Forms, *Space Sci. Rev.*, *22*, 77, 1978.

Davis, T. N., Observed Microstructure of Auroral Forms, in *Auroral Processes*, C. T. Russell (ed.), Japan Scientific Societies Press, Tokyo, 1979, pg. 171.

Deeg, H.-J., J. E. Borovsky, and N. Duric, Particle Acceleration near X-Type Magnetic Neutral Lines, *Phys. Fluids B*, *3*, 2660, 1991.

Eastman, T. E., E. W. Hones, S. J. Bame, and J. R. Asbridge, The Magnetospheric Boundary Layer: Site of Plasma, Momentum and Energy Transfer from the Magnetosheath into the Magnetosphere, *Geophys. Res. Lett.*, *3*, 685, 1976.

Elphinstone, R. D., and D. J. Hearn, The auroral distribution and its relation to magnetospheric processes, *Adv. Space Res.*, *13*(4), 17, 1993.

Elphinstone, R. D., D. Hearn, J. S. Murphree, and L. L. Cogger, Mapping Using the Tsyganenko Long Magnetospheric Model and Its Relationship to Viking Auroral Images" *J. Geophys. Res.*, *96*, 1467, 1991.

Elvey, C. T., Problems of Auroral Morphology, *Proc. Nat. Acad. Sci.*, *43*, 63, 1957.

Ergun, R. E., C. W. Carlson, J. P. McFadden, J. H. Clemmons, and M. H. Boehm, Langmuir Wave Growth and Electron Bunching: Results from a Wave-Particle Correlator, *J. Geophys. Res.*, *96*, 225, 1991.

Ergun, R. E., C. W. Carlson, J. P. McFadden, D. M. TonThat, J. H. Clemmons, and M. H. Boehm, Observation of Electron Bunching during Landau Growth and Damping, *J. Geophys. Res.*, *96*, 11371, 1991.

Feldstein, Y. I., and Y. I. Galperin, The Auroral Luminosity Structure in the High-Latitude Upper Atmosphere: Its Dynamics and Relationship to the Large-Scale Structure of the Earth's Magnetosphere, *Rev. Geophys.*, *23*, 217, 1985.

Feldstein, Y. I., and G. V. Starkov, The auroral oval and the boundary of closed field lines of geomagnetic field, *Planet. Space Sci.*, *18*, 501, 1970.

Friis-Christensen, E., and K. Lassen, Large-Scale Distribution of Discrete Auroras and Field-Aligned Currents, in *Auroral Physics*, edited by C.-I. Meng, M. J. Rycroft, and L. A. Frank, pg. 369, Cambridge University Press, Cambridge, 1991.

Galeev, A. A., Anomalous Resistivity on Auroral Field Lines and Its Role in Auroral Particle Acceleration, in *High-Latitude Space Plasma Physics*, edited by B. Hultqvist and T. Hagfors, pg. 437, Plenum, New York, 1983, pg. 437.

Galperin, Y. I., and A. V. Volosevich, The ARCAD-3 Project and the Theory of Auroral Structures, *Can. J. Phys.*, *67*, 719, 1989.

Goertz, C. K., Nonstationary Coupling between the Magnetosphere and Ionosphere, in *Sixth ESA Symposium on European Rocket and Balloon Programmes*, pg. 221, European Space Agency SP-183, 1983.

Goertz, C. K., Kinetic Alfven Waves on Auroral Field Lines, *Planet. Space Sci., 32*, 1387, 1984.

Goertz, C. K., Heating and Cooling of the Earth's Plasma Sheet, *Adv. Space Res. Suppl., 10*, 29, 1990.

Goertz, C. K., and R. W. Boswell, Magnetosphere-Ionosphere Coupling, *J. Geophys. Res., 84*, 7239, 1979.

Gorney, D. J., A. Clarke, D. Croley, J. Fennell, J. Luhmann, and P. Mizera, The Distribution of Ion Beams and Conics below 8000 km, *J. Geophys. Res., 86*, 83, 1981.

Haerendel, G., An Alfven Wave Model of Auroral Arcs, in *High-Latitude Space Plasma Physics*, edited by B. Hultqvist and T. Hagfors, pg. 515, Plenum, New York, 1983.

Hallinan, T. J., Auroral Spirals 2. Theory, *J. Geophys. Res., 81*, 3959, 1976.

Hallinan, T. J., and T. N. Davis, Small-Scale Auroral Arc Distortions, *Planet. Space Sci., 18*, 1735, 1970.

Harrold, B. G., C. K. Goertz, R. A. Smith, and P. J. Hansen, Resonant Alfven Wave Heating of the Plasma Sheet Boundary Layer, *J. Geophys. Res., 95*, 15039, 1990.

Hasegawa, A., Particle Acceleration by MHD Surface Wave and Formation of Aurora, *J. Geophys. Res., 81*, 5083, 1976.

Hasegawa, A., Beam Production at Plasma Boundaries by Kinetic Alfven Waves, *J. Geophys. Res., 92*, 11221, 1987.

Hasegawa, A., and T. Sato, Generation of Field Aligned Current During Substorm, in *Dynamics of the Magnetosphere*,

Hoffman, J. M., G. J. Lockwood, and G. H. Miller, Emission of 3914-Å N_2^+ Radiation from Charge-Transfer Excitation, *Phys. Rev. A, 23*, 2983, 1981. edited by S.-I. Akasofu, pg. 529, Reidel, Dordrecht, 1979.

Hones, E. W., A. B. Galvin, and P. R. Higbie, Poleward motions of auroral structures, in *Auroral Physics*, edited by C.-I. Meng, M. J. Rycroft, and L. A. Frank, pg. 299, Cambridge University Press, Cambridge, 1991.

Hoshino, M., The Electrostatic Effect for the Collisionless Tearing Mode, *J. Geophys. Res., 92*, 7368, 1987.

Hudson, M. K., and D. W. Potter, Electrostatic Shocks in the Auroral Magnetosphere, in *Physics of Auroral Arc Formation*, edited by S.-I. Akasofu and J. R. Kan, pg. 260, American Geophysical Union, Washington, D.C., 1981.

Hudson, M. K., T. L. Crystal, W. Lotko, and C. Barnes, Weak Double Layers in the Auroral Ionosphere, *Laser Part. Beams, 5*, 295, 1987.

Hultqvist, B., On the Production of a Magnetic-Field-Aligned Electric Field by the Interaction Between the Hot Magnetospheric Plasma and the Cold Ionosphere, *Planet. Space Sci., 19*, 749, 1971.

Ionson, J. A., R. S. B. Ong, and E. G. Fontheim, Anomalous Resistivity of the Auroral Plasma in the Top-Side Ionosphere, *Geophys. Res. Lett., 3*, 549, 1976.

Jackson, J. D., *Classical Electrodynamics*, second edition, sect. 10.4, Wiley, New York, 1975.

Kamide, Y., and S.-I. Akasofu, The Location of the Field-Aligned Currents with Respect to Discrete Auroral Arcs, *J. Geophys. Res., 81*, 3399, 1976.

Kan, J. R., Energization of Auroral Electrons by Electrostatic Shock Waves, *J. Geophys. Res., 80*, 2089, 1975.

Kaufmann, R. L., D. J. Larson, and C. Lu., Mapping and Distortions of Auroral Structures in the Quiet Magnetosphere, *J. Geophys. Res., 95*, 7973, 1990.

Kim, J. S., and R. A. Volkman, Thickness of Zenithal Auroral Arc over Fort Churchill, Canada, *J. Geophys. Res., 68*, 3187, 1963.

Kindel, J. M., and C. F. Kennel, Topside Current Instabilities, *J. Geophys. Res.*, 76, 3055, 1971.

Kintner, P. M., J. Vago, R. Arnoldy, and K. Lynch, Localized Lower Hybrid Acceleration of O^+ Ions in the Source Region, *EOS, Trans. Amer. Geophys. Soc., 73*(14), 262, 1992.

Klumpar, D. M., Relationships between Auroral Particle Distributions and Magnetic Field Perturbations Associated with Field-Aligned Currents, *J. Geophys. Res., 84*, 6524, 1979.

Klumpar, D. M., J. M. Quinn, and E. G. Shelley, Counter-Streaming Electrons at the Geomagnetic Equator near 9 R_E, *Geophys. Res. Lett., 15*, 1295, 1989.

Krall, N. A., and A. W. Trivelpiece, *Principles of Plasma Physics*, McGraw-Hill, New York, 1973.

LaBelle, J. A., P. M. Kintner, A. W. Yau, and B. A. Whalen, Large Amplitude Wave Packets Observed in the Ionosphere in Association with Transverse Ion Acceleration, *J. Geophys. Res., 91*, 7113, 1986.

Lotko, W., Diffusive Acceleration of Auroral Primaries, *J. Geophys. Res., 91*, 191, 1986.

Lotko, W., and C. F. Kennel, Spiky Ion Acoustic Waves in Collisionless Auroral Plasma, *J. Geophys. Res., 88*, 381, 1983.

Lotko, W., and C. G. Schultz, Internal Shear Layers in Auroral Dynamics, in *Modeling Magnetospheric Plasma*, edited by T. E. Moore, J. H. Waite, pg. 121, American Geophysical Union, Washington, D.C., 1988.

Lui, A. T. Y., E. W. Hones, F. Yasuhara, S.-I. Akasofu, and S. J. Bame, Magnetotail Plasma Flow during Plasma Sheet Expansions: VELA 5 and 6 and IMP 6 Observations, *J. Geophys. Res., 82*, 1235, 1977.

Lundin, R., and D. S. Evans, Boundary Layer Plasmas as a Source for High-Latitude, Early Afternoon, Auroral Arcs, *Planet. Space Sci., 32*, 1389, 1985.

Lyons, L. R., and D. S. Evans, An Association between Discrete Aurora and Energetic Particle Boundaries, *J. Geophys. Res., 89*, 2395, 1984.

Lysak, R. L., and C. W. Carlson, The Effect of Microscopic Turbulence on Magnetosphere-Ionosphere Coupling, *Geophys. Res. Lett., 8*, 269, 1981.

Maggs, J. E., and T. N. Davis, Measurements of the Thicknesses of Auroral Structures, *Planet. Space Sci., 16*, 205, 1968.

Mauk, B. H., and C.-I. Meng, The Aurora and Middle Magnetospheric Processes, in *Auroral Physics*, C.-I. Meng, M. J. Rycroft, and L. A. Frank (eds.), Cambridge University Press, Cambridge, 1991, pg. 223.

Maynard, N. C., J. P. Heppner, and A. Egeland, Intense, Variable Electric Fields at Ionsopheric Altitudes in the High Latitude Regions As Observed by DE-2, *Geophys. Res. Lett., 9*, 981, 1982.

McFadden, J. P., C. W. Carlson, and M. H. Boehm, Structure of an Energetic Narrow Discrete Arc, *J. Geophys. Res., 95*, 6533, 1990.

McIlwain, C. E., Direct Measurement of Particles Producing Visible Auroras, *J. Geophys. Res., 65*, 2727, 1960.

McIlwain, C. E., Auroral Electron Beams near the Magnetic Equator, in *Physics of Hot Plasma in the Magnetosphere*, B. Hultqvist and L. Stenfo (eds.), Plenum, New York, 1975, pg. 91.

Meng, C.-I., B. Mauk, and C. E. McIlwain, Electron Precipitation of Evening Diffuse Aurora and Its Conjugate Electron Fluxes near the Magnetospheric Equator, *J. Geophys. Res., 84*, 2545, 1979.

Mizera, P. F., and J. F. Fennell, Signatures of Electric Fields from High and Low Altitude Particle Distributions, *Geophys. Res. Lett., 4*, 311, 1977.

Moghaddam-Taaheri, E., C. K. Goertz, and R. A. Smith, Plasma Convection and Ion Beam Generation in the Plasma Sheet Boundary Layer, *J. Geophys. Res., 96*, 1569, 1991.

Ogawa, T., and T. Sato, New Mechanism of Auroral Arcs, *Planet. Space Sci., 19*, 1393, 1971.

Omholt, A., *The Optical Aurora*, Fig. 2.2, Springer-Verlag, New York, 1971.

Otto, A., and G. Birk, The Dynamical Evolution of Small-Scale Auroral Arc Phenomena Due to a Resistive Instability, *J. Geophys. Res., 97*, 8391, 1992.

Palmadesso, P. J., T. P. Coffey, S. L. Ossakow, and K. Papadopoulos, Topside Ionosphere Ion Heating Due to Electrostatic Ion Cyclotron Turbulence, *Geophys. Res. Lett., 1*, 105, 1974.

Paschmann, G., J. Papamastorakis, N. Sckopke, G. Haerendel, and E. G. Shelley, Altitude and Structure of an Auroral Arc Acceleration Region, *J. Geophys. Res., 88*, 7121, 1983.

Persson, H., Electric Field along a Magnetic Line of Force in a Low-Density Plasma, *Phys. Fluids, 6*, 1756, 1963.

Pritchett, P. L., Cyclotron Maser Radiation from a Source Structure Localized Perpendicular to the Ambient Magnetic Field, *J. Geophys. Res., 91*, 13569, 1986.

Rostoker, G., and R. Bostrom, A Mechanism for Driving the Gross Birkeland Current Configuration in the Auroral Oval, *J. Geophys. Res., 81*, 235, 1976.

Rothwell, P. L., L. P. Block, M. B. Silevitch, and C.-G. Falthammar, A New Model for Substorm Onsets: The Pre-Breakup and Triggering Regimes, *Geophys. Res. Lett., 15*, 1279, 1988.

Rothwell, P. L., L. P. Block, M. B. Silevitch, and C.-G. Falthammar, A New Model for Auroral Breakup During Substorms, *IEEE Trans. Plasma Sci., 17*, 150, 1989.

Rothwell, P. L., M. B. Silevitch, L. P. Block, and C.-G. Falthammar, Prebreakup Arcs: A Comparison Between Theory and Experiment, *J. Geophys. Res., 96*, 13967, 1991.

Samson, J. C., D. D. Wallis, T. J. Hughes, F. Creutzberg, J. M. Ruohoniemi, and R. A. Greenwanld, Substorm Intensifications and Field Line Resonances in the Nightside Magnetosphere, *J. Geophys. Res., 97*, 8495, 1992.

Sato, T., A Theory of Quiet Auroral Arcs, *J. Geophys. Res., 83*, 1042, 1978.

Sato, T., and T. Iijima, Primary Sources of Large-Scale Birkeland Currents, *Space Sci. Rev., 24*, 347, 1979.

Seyler, C. E., Nonlinear 3-D Evolution of Bounded Kinetic Alfven Waves Due to Shear Flow and Collionless Tearing Instability, *Geophys. Res. Lett., 15*, 756, 1988.

Shepherd, M. M., and G. G. Shepherd, Projection of Auroral Intensity Contours into the Magnetosphere, *Planet. Space Sci., 33*, 183, 1985.

Siscoe, G. L., W. Lotko, and B. U. O. Sonnerup, A High-Latitude, Low-Latitude Boundary Layer Model of the Convection Current System, *J. Geophys. Res., 96*, 3487, 1991.

Smiddy, M., M. C. Kelley, W. Burke, F. Rich, R. Sagalyn, B. Shuman, R. Hays, and S. Lai, Intense Poleward-Directed Electric Fields near the Ionospheric Projection of the Plasmapause, *Geophys. Res. Lett., 4*, 543, 1977.

Smith, R. A., Simulation of Double Layers in a Model Auroral Circuit with Nonlinear Impedance, *Geophys. Res. Lett., 13*, 809, 1986a.

Smith, R. A., Effects of Anomalous Transport on the Potentials of Discrete Auroral Arcs, *Geophys. Res. Lett., 13*, 889, 1986b.

Smith, R. A., Anomalous Transport in Discrete Arcs and Simulation of Double Layers in a Model Auroral Circuit, *Laser Part. Beams, 5*, 381, 1987.

Stasiewicz, K., Generation of Magnetic-Field Aligned Currents, Parallel Electric Fields, and Auroral Zone Structures by Plasma Pressure Inhomogeneities in the Magnetosphere, *Planet. Space Sci., 33*, 1037, 1985.

Stormer, C., *The Polar Aurora*, chap. 8, Oxford, London, 1955.

Swift, D. W., On the Formation of Auroral Arcs and Acceleration of Auroral Electrons, *J. Geophys. Res., 80*, 2096, 1975.

Swift, D. W., An Equipotential Model for Auroral Arcs 2. Numerical Simulations, *J. Geophys. Res., 81*, 3935, 1976.

Swift, D. W., Turbulent Generation of Electrostatic Fields in the Magnetosphere, *J. Geophys. Res., 82*, 5143, 1977.

Swift, D. W., An Equipotential Model for Auroral Arcs: The Theory of Two-Dimensional Laminar Electrostatic Shocks, *J. Geophys. Res., 84*, 6427, 1979.

Swift, D. W., The Generation of Electric Potentials Responsible for the Acceleration of Auroral Electrons, in *Physics of Auroral Arc Formation*, edited by S.-I. Akasofu and J. R. Kan, pg. 288, American Geophysical Union, Washington, D.C., 1981.

Takahashi, K., and E. W. Hones, ISEE 1 and 2 Observations of Ion Distributions at the Plasma Sheet–Tail Lobe Boundary, *J. Geophys. Res., 93*, 8558, 1988.

Temerin, M. A., and D. Cravens, Production of Electron Conics by Stochastic Acceleration Parallel to the Magnetic Field, *J. Geophys. Res., 95*, 4285, 1990.

Temerin, M. A., J. McFadden, M. Boehm, C. W. Carlson, and W. Lotko, Production of Flickering Aurora and Field-Aligned Electron Flux by Electromagnetic Ion Cyclotron Waves, *J. Geophys. Res., 91*, 5769, 1986.

Torr, M. R., J. C. G. Walker, and D. G. Torr, Escape of Fast Oxygen from the Atmosphere during Geomagnetic Storms, *J. Geophys. Res., 79*, 5267, 1974.

Trakhtengerts, V. Y., and A. Y. Feldstein, Turbulent Alfven Boundary Layer in the Polar Ionsophere 1. Excitation Conditions and Energetics, *J. Geophys. Res., 96*, 19363, 1991.

Vasyliunas, V. M., Mathematical Models of Magnetospheric Convection and Its Coupling to the Ionosphere, in *Particles and Fields in the Magnetosphere*, B. M. McCormac (ed.), pg. 60, Reidel, Dordrecht, 1970.

Wagner, J. S., R. D. Sydora, T. Tajima, T. Hallinan, L. C. Lee, and S.-I. Akasofu, Small-Scale Auroral Arc Deformities, *J. Geophys. Res., 88*, 8013, 1983.

Watanabe, K., and T. Sato, Self-Excitation of Auroral Arcs in a Three-Dimensionally Coupled Magnetosphere-Ionosphere System, *Geophys. Res. Lett., 15*, 717, 1988.

Wehrenberg, P. J., and K. C. Clark, Photon-Particle Coincidence Measurement of Charge-Transfer Excitation of the N_2^+ First Negative 3914-ÅBand by Protons, *Phys. Rev. A, 8*, 173, 1973.

Zelenyi, L. M., R. A. Kovrazhin, and J. M. Bosqued, Velocity-Dispersed Ion Beams in the Nightside Auroral Zone: AUREOL 3 Observations, *J. Geophys. Res., 95*, 12119, 1990.

J. E. Borovsky, Space Plasma Physics Group, Los Alamos National Laboratory, Los Alamos, NM 87545.

Coupling Between Mesoscale and Microscale Processes in the Cusp and Auroral Plasmas

J. L. Burch and C. S. Lin

Southwest Research Institute, San Antonio, Texas

J. D. Menietti

Department of Physics and Astronomy, University of Iowa, Iowa City, Iowa

R. M. Winglee

Geophysics Program, University of Washington, Seattle, Washington

Mesoscale phenomena in the Earth's magnetosphere are in many cases easily detectable with single spacecraft, while the macroscale, or global, phenomena, such as substorms, must be studied with multiple spacecraft and ground stations. The microscale processes, which are often responsible for the mesoscale phenomena we observe and sometimes result from them, are difficult to study experimentally or theoretically because of the fine spatial and temporal scales upon which they operate. The study of coupling between micro- and mesoscale processes thus requires a combination of experiment, theory, and numerical simulation. Four examples of mesoscale phenomena: electron conics, auroral kilometric radiation, cusp-region ion conics, and suprathermal electron bursts, are discussed in terms of the ongoing search for the microscale processes associated with them. In some cases (e. g., electron conics) the processes themselves have not been satisfactorily identified, while in others (e. g., AKR) a process (the cyclotron maser instability) is acknowledged to be important, but the source of free energy is not agreed upon. Final resolution of these questions will require measurements with higher resolution, simultaneous multi-point measurements, and/or more complete measurements.

1. INTRODUCTION

Most observations of the magnetosphere, whether from the ground or from spacecraft, are interpretable in terms of space plasma phenomena that can be considered as part of the mesoscale. Many examples come to mind, including longitudinally-extended sheets of field-aligned current, inverted-V structures in auroral electron precipitation, ion velocity dispersions in the polar cusps, westward-traveling surges and drifting omega bands along the auroral oval, substorm injection fronts at geosynchronous orbit, resonantly oscillating magnetic field lines, and flux transfer events. Historically, as multi-point measurements became available on the ground and in space, collections of mesoscale phenomena were recognized as part of global or macroscale phenomena. Good examples of macroscale phenomena that have been identified in the Earth's magnetosphere include the geomagnetic storm, the auroral oval, and the magnetospheric substorm.

Space Plasmas: Coupling Between Small and Medium Scale Processes
Geophysical Monograph 86

Coordinated observations of mesoscale phenomena throughout the magnetosphere can trace the flow of mass, momentum, and energy from the solar wind through the magnetosphere and into the ionosphere and upper atmosphere. Moreover, MHD models of the magnetosphere and its constituent parts can link the observations together into phenomenological models with predictive capabilities. Nonetheless, crucial questions about the sources of the mesoscale phenomena remain unanswered. The controlling mechanisms that underlie the mesoscale phenomena typically occur at the microscale in spatial and temporal domains that have generally not been explored with direct measurement or, except locally, with single-particle computer simulations because of practical resource limitations on spacecraft and computers, respectively.

The sketch in Figure 1 shows a number of macroscale, mesoscale, and microscale phenomena that occur in the magnetosphere in a space-time coordinate system. Although the space-time spectrum occupied by the various magnetospheric phenomena is continuous, consideration of the mesoscale as the domain that can be explored by single spacecraft or single or localized chains of ground stations tends to distinguish it from the other two scales. As mentioned before, global observing networks are needed to characterize the macroscale phenomena, while exploration of the microscale requires very high data rates and closely-spaced multiple spacecraft, which have only been approached with sounding rockets to date.

Coupling between mesoscale and microscale phenomena can be regarded as occurring at least in the following two ways: (1) microscale processes occurring in regions such as the reconnection diffusion region or thin neutral sheets lead to mesoscale phenomena such as flux transfer events, plasmoids, and earthward-streaming ion populations; and (2) mesoscale processes such as field-aligned current systems, downward-flowing cusp ions, or inverted-V events couple energy into the upper ionosphere where microscale processes generate other mesoscale phenomena such as auroral kilometric radiation [AKR], auroral hiss, and large-scale regions of upward-flowing conical electron and ion distributions. An example of a phenomenon that lies at the interface between the mesoscale and microscale is the individual auroral arc (see Figure 1). Full understanding of auroral arc physics will require much higher resolution measurements of the microscale than are currently available along with single-particle models of the individual plasma processes that occur in the arc.

Figure 2 shows several examples of the mesoscale-microscale-mesoscale coupling sequence that is referred to above. In each case the mesoscale phenomena in the first and last columns can be measured directly by spacecraft, while the

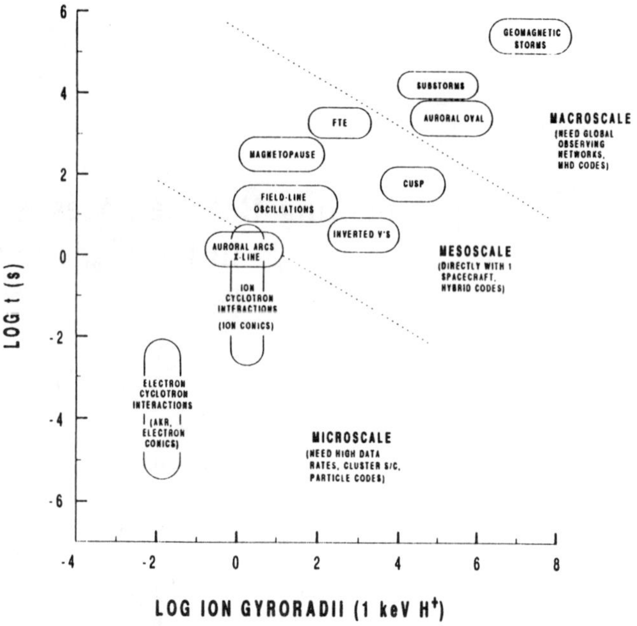

Fig. 1. Space-time representation of some magnetospheric macroscale, mesoscale, and microscale phenomena.

microscale phenomena in the middle column need to be studied by theory and simulation. Consideration of such coupling chains leads one to the visualization of a magnetosphere that is driven by the solar wind and by its own internal disruptions, which ultimately lead to large-scale energy dissipation and output, all of which is mediated by microscale processes of many different types. The ability of the magnetosphere to transmit electrical energy via field-aligned currents and the associated Alfvén waves quickly over large distances and its ability to store large amounts of energy are what makes it an intrinsically fascinating environmental system. It is natural to compare this environment to the meteorological environment of Earth, which is also so large and complex that only short-term predictions are yet possible. It was pointed out by Siscoe [1991] that there is no truly global meteorological disturbance that is comparable to the magnetospheric substorm.

Rossby planetary waves represent the only global meteorological mode, although they are a propagating rather than a truly global disturbance. However, at the macroscale, phenomena such as hurricanes and weather fronts are analogous to the macroscale magnetospheric phenomena in that imaging and multi-point measurements are needed to characterize them. For example, weather fronts were not even known until the deployment of large arrays of weather balloons during World War I. Mesoscale weather phenomena such as thunderstorms and tornados share characteristics

MESOSCALE ⇒	MICROSCALE ⇒	MESOSCALE
Plasma Sheet Thinning	Ion Tearing	Plasmoids
Inverted V	Cyclotron Maser Interactions	AKR
Cusp Ion Flows	Ion Heating	Large-Scale Ion Conic Regions
Birkeland Currents	Wave-Particle Interactions	Field-Aligned Electron Bursts
Neutral Sheet Current	Meandering Ion Orbits	Earthward Streaming Ion Beams
[*Measured on S/C*]	[*Theory and Simulation*]	[*Measured on S/C*]

Fig. 2. Examples of coupling of mesoscale and microscale processes in the magnetosphere.

of their magnetospheric counterparts in that they are easily observable from a single observing location and involve the dissipation of significant amounts of energy. Finally, the meteorological microscale (microbursts, lightning, etc.) is just as elusive as the magnetosphere's, with localized intense energy conversion that is difficult to observe and characterize. In meteorology, then, as in magnetospheric physics, the microscale represents the last frontier of measurement technology; however, in the magnetosphere the microscale processes possess the added potential of controlling larger-scale phenomena through the electromagnetic coupling that occurs efficiently throughout the plasma medium and through the release of stored plasma and magnetic-field energy.

In the following paragraphs a number of examples of coupling between mesoscale and microscale phenomena will be discussed. Although many such examples can be found throughout the magnetosphere (for example, microscale processes act in various ways to produce mesoscale phenomena such as the polar wind, the diffuse aurora and the bow shock), this discussion will be limited to phenomena that are commonly observed in the cusp and auroral plasmas.

The first example of mesoscale-microscale coupling that will be addressed is the generation of conical electron distributions in the auroral regions. Since their identification by Menietti and Burch [1985], electron conics have been the subject of numerous theoretical and experimental investigations. After the initial suggestion by Menietti and Burch [1985] that a perpendicular heating process similar to that proposed for ion conics was responsible for electron conics, numerous other ideas have been tested. Temerin and Cravens [1990] have shown that parallel heating of the electron population over an extended altitude range can produce a conical electron distribution that matches many observations. In a different approach Hultqvist et al. [1991] have addressed the question of the simultaneous appearance of upward-moving electron beams and ion conics and showed that a fluctuating double layer could produce both phenomena. André and Eliasson [1992] showed further that a similar process can lead to conical electron distributions.

The second example to be discussed is auroral kilometric radiation (AKR)—a readily observable mesoscale phenomena that has its source within another mesoscale phenomenon, the auroral acceleration region. The conversion of free energy from the energetic electron distribution is thought to occur through a resonant cyclotron interaction [the cyclotron maser] at the microscale. Present experimental limitations have frustrated the search for the source of free energy although several candidate velocity-space gradients have been identified. In this case the critical limitation is the time resolution of the available measurements of the energetic electron distributions within the auroral acceleration region. Complete 2-D electron distributions need to be made in less than one second [Melrose, 1986] in order to identify the free energy source for AKR. With present measurements, the strongest positive perpendicular gradients in velocity space ironically have to be discounted in favor of plateaus because if the observed gradients did constitute the free energy source they would have been smoothed out or eliminated by velocity-space diffusion during the time between measurements of the electron distribution if strong wave growth were to occur.

A third example of mesoscale-microscale coupling to be discussed occurs in the polar cusps, where downward-flowing magnetosheath ions and upward-flowing conical ion distributions are observed together. Ion conics are commonly detected throughout the auroral regions, but the cusp region allows them to be studied in relative isolation from other more turbulent phenomena that are typically present in the nightside auroral oval. Ion conics are most likely produced by the perpendicular heating of the topside ionospheric ion population by resonant cyclotron interactions. However, it has been shown that the conical distributions can result from an interaction region that is localized in altitude as well as by one that is greatly extended in altitude [Temerin, 1986]. Moreover, while the necessary wave modes and intensities have been observed along with the ion conics, it is often unknown whether similar waves were occurring within the actual heating region.

A final phenomenon that will be discussed is the field-aligned electron burst [Arnoldy, 1974; Johnstone and Winningham, 1982]. Both upward and downward electron beams are commonly observed throughout the auroral oval, with the upward beams generally seen at higher altitudes within and above the auroral acceleration region. It was shown by Burch et al. [1983] that the upward electron beams are the primary charge carriers for the morning-side Region-1 Birkeland currents and that the observed distributions are consistent with parallel potential drops of a few tens of eV located at altitudes near one Earth radius. Similarly, Menietti and Burch [1987] showed that such electron beams are the primary charge carriers of the Birkeland currents near and within theta auroras in the polar cap. Although the beam-like nature of the electron bursts suggests acceleration by parallel potential drops, as suggested by Burch et al. [1983], the fact that the beams are often field-aligned over a wide range of energies has led to several suggestions that a wave-particle interaction is responsible for these impulsive phenomena [Temerin et al., 1986; McFadden et al., 1987].

The field-aligned electron bursts are invariably associated with Birkeland currents and often with inverted-V events, both of which can be considered to be mesoscale phenomena. Thus, the suprathermal electron bursts are also phenomena that appear to result from the coupling between mesoscale and microscale processes along the auroral oval. Here again our knowledge is limited by the time resolution and completeness of both the particle and wave measurements in the interaction region.

In the following sections the current status of research on each of the four topics discussed above (electron conics, auroral kilometric radiation, polar cusp ion conics, and field-aligned electron bursts) is reviewed, and some new results from DE-1 and from computer simulations are presented.

2. CONICAL ELECTRON DISTRIBUTIONS

Electron conics were observed in the DE-1 data by Menietti and Burch [1985], who reported them to be associated with trapped particles and parallel electric fields. The authors suggested perpendicular heating via wave-particle interactions as a source mechanism, in analogy with ion conic formation. Lundin et al. [1987] and Hultqvist et al. [1988] later reported observations of electron conics in the Viking data and suggested that a parallel potential difference that varies in magnitude over a fraction of an electron bounce period might explain electron conics that were observed to be associated with ion conics. For plasma parameters characteristic of those observed by DE-1 in the mid-altitude nightside auroral region, Wong et al. [1988] have shown that upper hybrid waves generated by a loss-cone distribution in a background cold plasma can preferentially heat the electron population in a magnetically oblique direction by cyclotron resonance. They pointed out that any particle distribution that produces $df/dv_\perp > 0$ might act as the free energy source. Since then, numerical simulations of the production of electron conics by upper hybrid waves using a loss cone [Roth et al., 1989] and a loss cone with a ring distribution [Lin et al., 1990] have been performed. Roth et al. [1989] suggested that the heating is due to wave trapping instead of quasilinear diffusion, as held by Wong et al. [1988].

Wong et al. [1988] also pointed out that because of the dependence of the diffusion coefficient on energy and pitch angle, upper hybrid waves will heat the electrons in both the parallel and perpendicular directions. By contrast, for the case of parallel acceleration only, the ratio of upgoing to downgoing electrons at any specific perpendicular energy should always be unity except in the case of lower energy backscattered electrons. For all of the observed electron conical distributions presented by Menietti and Burch [1985], however, more particles were detected moving up the field line than down, with the ratio as large as 1.45.

Roth et al. [1989], and most recently Temerin and Cravens [1990] (extending the ideas of Lundin et al., 1987), have demonstrated that an electron conical distribution can result from parallel heating of the electrons via electrostatic or acoustic mode waves. Upon mirroring, this heated electron distribution resembles the conical distributions presented by Menietti and Burch [1985]. Roth et al. [1989] suggested that downgoing electron beams excite parallel wave modes (electron acoustic or Langmuir waves), which can provide parallel acceleration of electrons. These electrons are partly mirroring at lower altitudes, and the reflected electrons have an enhancement at the edge of the loss cone.

Temerin and Cravens [1990] showed by numerical simulation that electron conics can be generated by stochastic acceleration parallel to the magnetic field and identified Alfvén ion cyclotron waves generated within inverted-V regions as the most likely wave mode. An example of a conical electron distribution produced in the Temerin and Cravens [1990] simulation study of stochastic parallel heating by Alfvén electron cyclotron waves is shown in Figure 3. The difference between conical distributions produced by stochastic parallel heating and stochastic perpendicular heating over extended regions of altitude is rather subtle, as shown in Figure 4. The first adiabatic invariant is conserved with parallel heating but not with perpendicular heating, and this difference becomes evident when mirrored electrons return to the heating region producing the flared conic branches in the distribution function for perpendicular heating as shown in Figure 4.

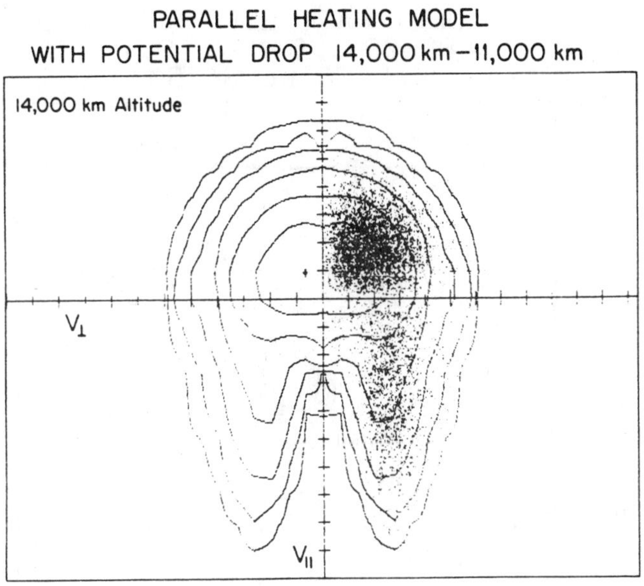

Fig. 3. Results of a simulation of stochastic parallel acceleration of electrons to form a conical distribution [Temerin and Cravens, 1990].

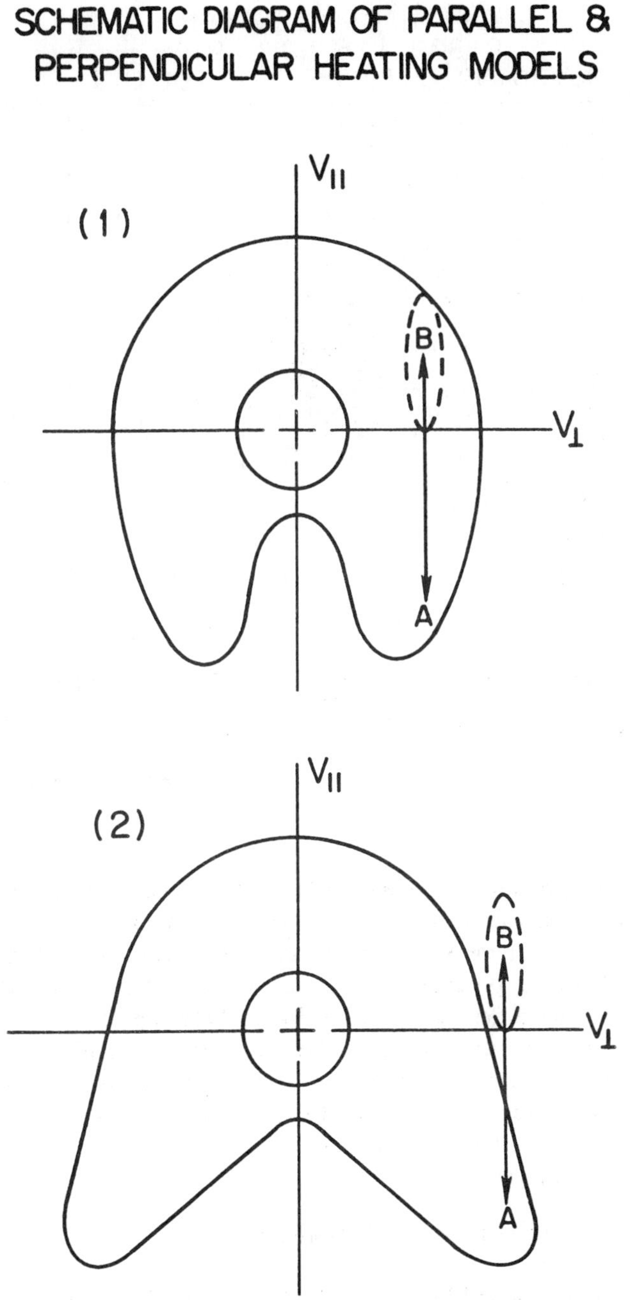

Fig. 4. Comparison of electron conics produced by parallel (top) and perpendicular (bottom) heating processes [Temerin and Cravens, 1990].

Therefore, as shown in Figure 4 Panel (1), for parallel heating the electrons in region A come from region B. On the other hand, for perpendicular heating [Figure 4 Panel (2)] there are insufficient electrons in region A to populate region B.

Recently Menietti et al. [1992a] have reported observations by both particle and plasma wave instruments on board DE-1 and the Swedish Viking satellite that demonstrate the presence of intense upper hybrid waves at mid-altitudes of the polar magnetosphere in association with electron conical distributions. This work suggests the possibility that upper hybrid waves heat electrons perpendicular to the magnetic field and are responsible for at least some electron conical distributions. Of 12 DE-1 orbits on which electron conics were observed in the Menietti et al. [1992] study, waves near the upper hybrid frequency were seen (within about 30 seconds of the time of one frequency sweep of the plasma wave instrument, PWI) on 10 of them. As an example, Figure 5 shows a contour plot of the electron distribution function obtained for a pass on day 81309 (November 4, 1981) for the time interval indicated. The contour plot clearly displays the electron conical signature of enhanced electron flux just outside the upgoing loss cone. Figure 6 displays a plot of the power spectral density obtained from the PWI data for a time interval that includes the electron conic shown in Figure 5. A distinct peak occurs just above the electron gyrofrequency, which was determined from the Magsat model to be about 78 kHz.

The above studies have thus far demonstrated by theory and simulation that electron conics can be generated by upper hybrid waves (perpendicular heating) and/or by electron acoustic waves, Langmuir waves, or Alfvén-ion cyclotron waves (parallel heating). Recently André and

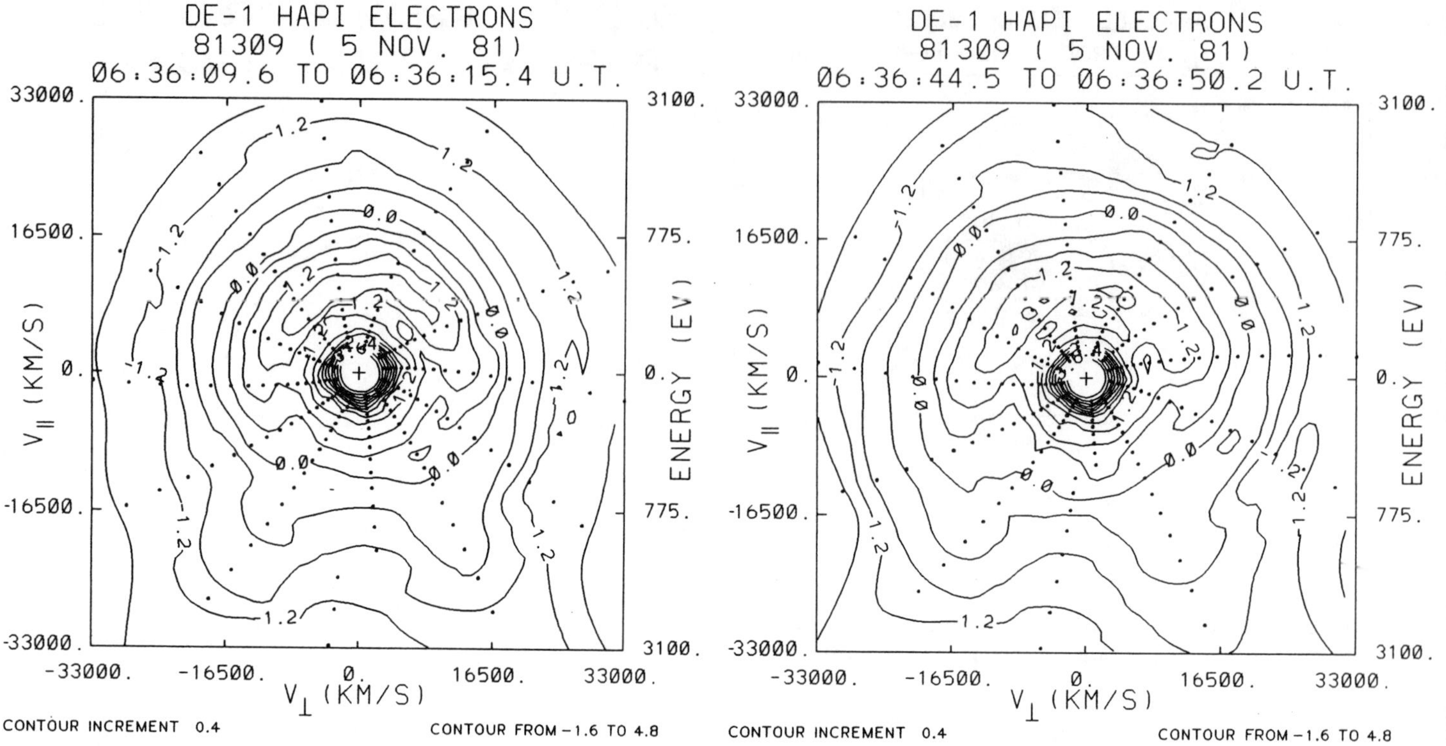

Fig. 5. Conical electron distribution measured by DE 1 [Menietti et al., 1992].

Eliasson [1992] have shown that low-frequency (~ 1 Hz) wave turbulence with parallel electric field components as observed on the Viking spacecraft by Block and Fälthammer [1990] can produce electron conical distributions, but that again the differences produced by parallel and perpendicular heating are very difficult to distinguish experimentally. It is obvious, then, that much more experimental and theoretical research remains to be done before the process or processes responsible for electron conical distributions can be identified with certainty.

3. AURORAL KILOMETRIC RADIATION

According to Wu and Lee [1979] the most likely source of free energy for the generation of AKR is the loss-cone distribution within a region of depleted plasma density where the ratio of electron plasma frequency to gyrofrequency is less than 0.3. This distribution provides the requirement of the cyclotron maser instability that $df/dv_{\perp} > 0$, and the theory has had success in explaining the observations to date. Pottelette et al. [1992] have recently shown how upper hybrid waves resulting from electron loss-cone distributions can interact with lower hybrid solutions generated by electron beams to produce AKR.

Omidi and Gurnett [1982] have demonstrated that f(v) distributions obtained by the S3-3 satellite within a source region of AKR seem to provide the large growth rate necessary to explain the observations with a cosmic noise background wave source and typical ray path lengths of 100 km. This analysis points out a fundamental problem in trying to resolve the source of free energy for AKR. That is, the AKR growth rate is calculated to be $\leq 10^4 s^{-1}$, and no satellite has been capable of obtaining a complete distribution in phase space on this time scale. Typically the best satellites require several seconds to obtain a complete distribution in phase space. During this time, the waves within an AKR source region would have grown and already altered the existing plasma distribution. This point was also made by Melrose [1986].

Observations of electron distributions in the AKR source region have recently been reported by investigators of the Viking satellite. Louarn et al. [1990] have reported simultaneous measurements of electromagnetic fields and particle distributions measured during a crossing of an AKR source region. They have determined that trapped electrons may play a more important role in the generation of AKR than loss-cone distributions. This conclusion was based on an instability analysis of $df/dv_{\perp} > 0$ in the "forbidden" region

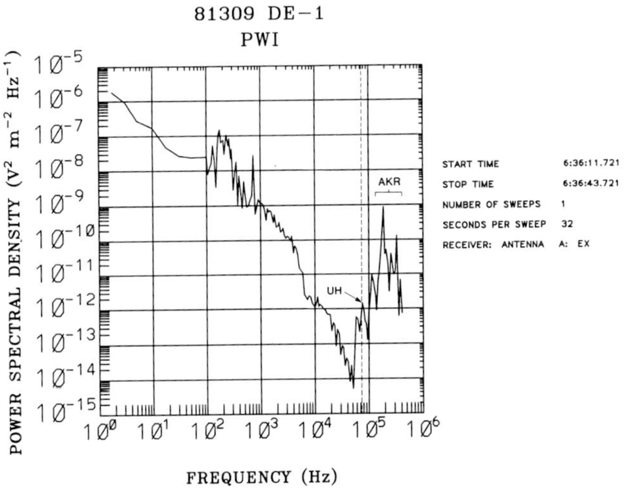

Fig. 6. Power spectral density of the electric field for a time period including that of the electron conic observation in Figure 5 [Menietti et al., 1992].

of velocity space [Knight, 1973], which lies outside the loss cone and between the ionospheric and magnetospheric electron populations. Although a plateau was actually observed along the $v_\perp$ axis, it was supposed by Louarn et al. [1990] that the velocity space gradient necessary for wave growth was smoothed out by the cyclotron maser instability in a time less than the measurement cycle of the Viking plasma instrument, thus underscoring the measurement time-scale problem pointed out above. Ungstrup et al. [1990] also used Viking data within an AKR source region to search for the source of free energy, but they favor the loss-cone distribution because the loss-cone gradients were steeper just outside the source region than within it, again supposing that the gradient was smoothed out by the cyclotron maser instability. This result suggests that the steep loss-cone distributions analyzed by Omidi and Gurnett [1982] may have been obtained just outside the AKR region rather than within it.

Menietti et al. [1993] recently reported DE-1 particle and wave data for a pass of the nightside auroral region near an AKR source center. They compared their results to those reported by Viking and also to recent numerical simulation studies of Winglee and Pritchett [1986], which show a great similarity to the observations. Figure 7, which is taken from Menietti et al. [1993] shows an electron energy-vs-time spectrogram for the nightside auroral region on day 285 (October 21) of 1981. The bottom panel displays PWI data, which show intense AKR extending down to frequencies near the local gyrofrequency (indicated by the black dotted line), in the time interval 00:28 $< t <$ 00:30. In fact, the AKR extends to frequencies below the gyrofrequency, which is to be expected for high-energy electrons satisfying the gyroresonant condition.

As noted in Figure 7, adjacent to and generally poleward of the AKR region was a region of electron distributions with $T_\perp/T_\parallel >> 1$. An example of such a distribution is shown in Figure 8. The electron distribution in Figure 8 is an example of a beam and temperature anisotropy of the kind Newman et al. [1988], Wong and Goldstein [1990] and Winglee et al. [1992] have shown is unstable to bursty radio emission via a modified kinetic Weibel instability (or temperature anisotropic beam instability). The elongated distribution-function plateaus along $\pm v_\perp$ in Figure 8 are similar to the trapped distributions presented by Louarn et al. [1990]. Of most interest to the question of the free-energy source for AKR is the possibility that a distribution with $df/dv_\perp > 0$ existed before the onset of the instability. Louarn et al. [1990] suggested fluctuating double layers as a process for moving electrons into the forbidden region. Menietti et al. [1993] have presented DE-1 data that could indicate that a mechanism suggested by Winglee and Pritchett [1986] is responsible for producing a distribution with $df/dv_\perp > 0$ outside the forbidden region and that the cyclotron maser instability itself leads to population of the forbidden region as well as to the AKR wave growth. An example of this type of distribution is shown in the contour plot in Figure 9. The small black dots in Figure 9 indicate the data points used in the contour plot; the spiral structure is produced by the energy scan of the electron detector (from high to low energy) and the spin of the spacecraft, which is counterclockwise in Figure 9 beginning near the top of the figure.

In Figure 9 the distribution function in the $-v_\perp$ half plane is clearly of a different character from that in the $+v_\perp$ half plane, which was sampled a few seconds later. The former distribution resembles the distribution proposed by Winglee and Pritchett [1986] to result from the spreading of a downward-directed beam in velocity space as it moves to lower altitudes, resulting in positive velocity-space gradients along both the parallel and perpendicular velocity axes. The simulation studies of Winglee and Pritchett [1986] demonstrated how the competition between the bump-in-tail instability and the cyclotron maser instability can then produce diffusion along both the parallel and perpendicular directions, with the resulting distribution depending on the initial parameters such as the electron density and the wave propagation angle. Note the similarity between the distribution in the left-hand portion of Figure 9 and that in Figure 10a (from Winglee and Pritchett, 1986), both of which have $df/dv_\perp > 0$. In the right-hand portion of Figure 9, as in Figure 10b, the positive velocity space gradient along the

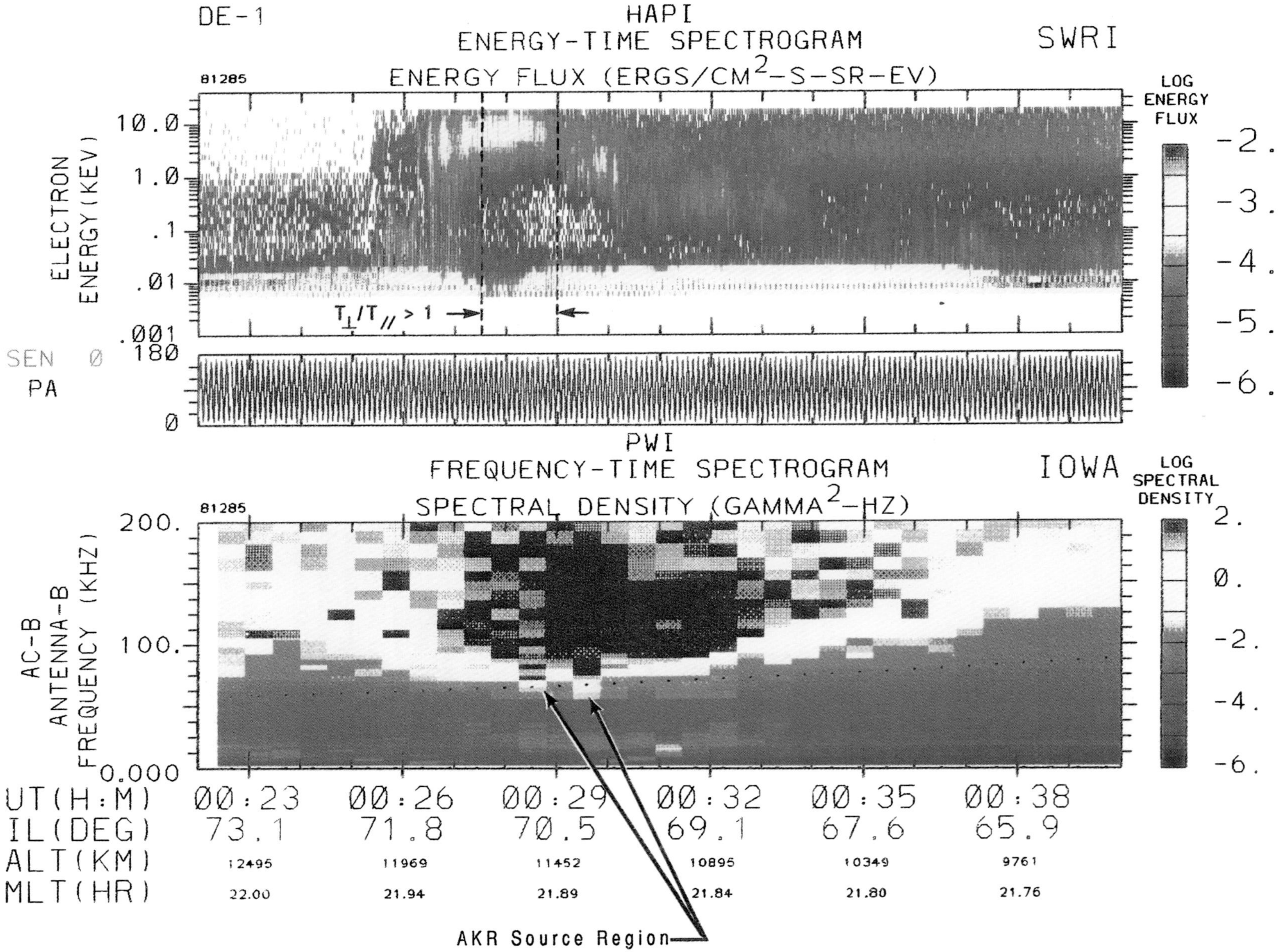

Fig. 7. Energy-time spectrogram of electrons (top panel); electron pitch angles (middle panel); and frequency-time spectrogram of the magnetic field (bottom panel) as measured by DE 1. The black dots in the wave spectrogram denote the electron (top) and ion (bottom) gyrofrequencies. AKR emissions extend down from the top of the wave spectrogram to just below the electron gyrofrequency between 00:28 and 00:30 UT.

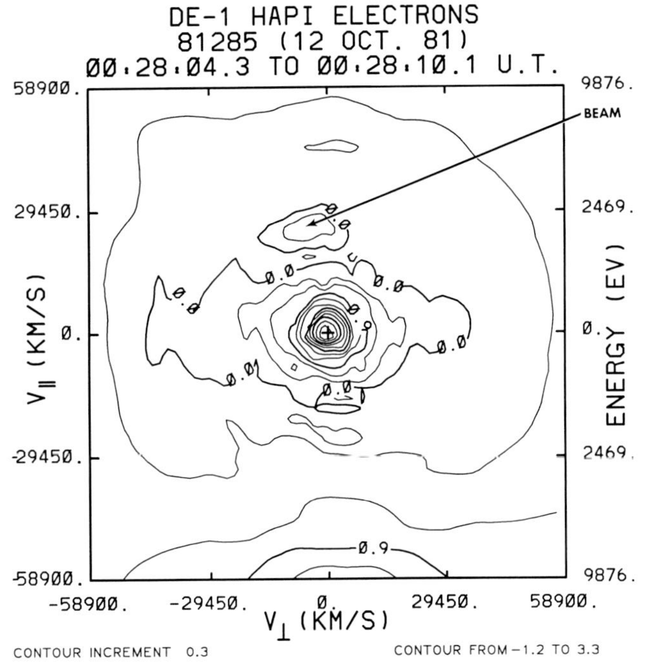

Fig. 8. Contour plot of the electron distribution function for a time period just after 00:28 UT during the DE 1 pass shown in Figure 7 [Menietti et al., 1993].

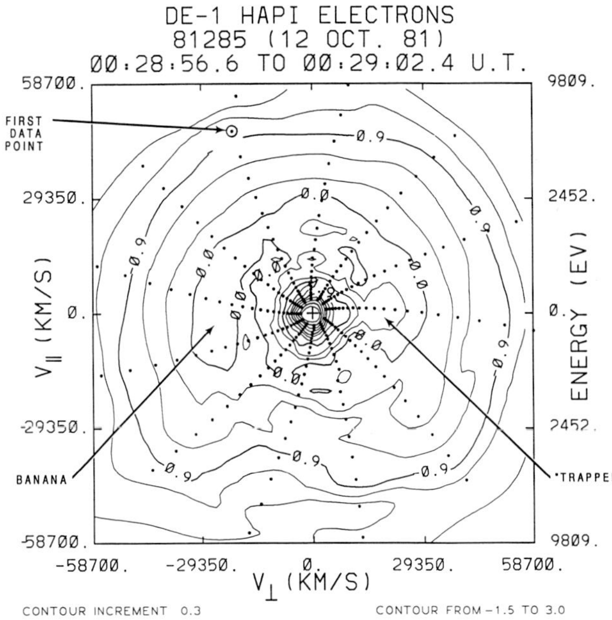

Fig. 9. Contour plot of the electron distribution function for a time period around 00:29 UT during the DE 1 pass shown in Figure 7 [Menietti et al.,1993].

perpendicular axis is absent, and the "forbidden" region is populated with electrons. The distribution in Figure 9 may indicate a temporal evolution of the distribution function as AKR is generated and the forbidden region is populated, but much more experimental evidence with data acquired at higher rates will be necessary to confirm this possibility.

4. CUSP ION FLOWS AND ION CONICS

The polar cusp is the only region of the magnetosphere through which solar-wind plasma has been observed to enter the magnetosphere directly. In addition to the direct penetration of solar-wind plasma, there is also propagation of waves that are excited at the magnetopause by processes such as the Kelvin-Helmholtz instability. Both phenomena carry energy down the polar cusp into the upper ionosphere where it can produce strong heating of ionospheric plasma. As a result of this heating, there is an enhanced outflow of ionospheric plasma known as the cleft ion fountain [Lockwood et al., 1985].

It has been suggested by Hudson and Roth [1986] that downflowing ions in the cusp can generate waves around the lower hybrid frequency and multiples of the ion gyrofrequencies, and that such waves can cause ion conics. More recently Viking and DE-1 data have been used to study cyclotron resonance heating and double cyclotron absorption in the cusp [André et al., 1988; Peterson et al., 1989; André et al., 1990; Ball and André, 1991]. Along these same lines, Winglee et al. [1993a,b] have considered the injection of energetic magnetosheath plasma through the polar cusp and the ion heating and upward ion conics that may result. Their approach was to use Dynamics Explorer plasma measurements to characterize the two mesoscale phenomena that are easily observed in the cusp (the cusp ion flows and the large-scale ion conic population) and to explore the microscale processes responsible for the ion heating with particle simulations.

Observations from Dynamics Explorer 1 have shown that the injected magnetosheath ions have very distinct characteristics, forming a "V" shape on energy/pitch-angle spectrograms [Burch et al., 1982]. This unique spectral feature arises from the fact that the transit time from the injection region to the point of observation in the mid-altitude cusp is a function of ion pitch angle. Both impulsive injection and injection through a restricted region in a convection electric field can lead to the "V" signatures. Superimposed on the "V" energy/pitch-angle signatures is an energy-time or energy-latitude signature (a velocity dispersion), which has the same cause as the "V" signatures. Reiff et al. [1977] interpreted the velocity dispersions observed in low-altitude

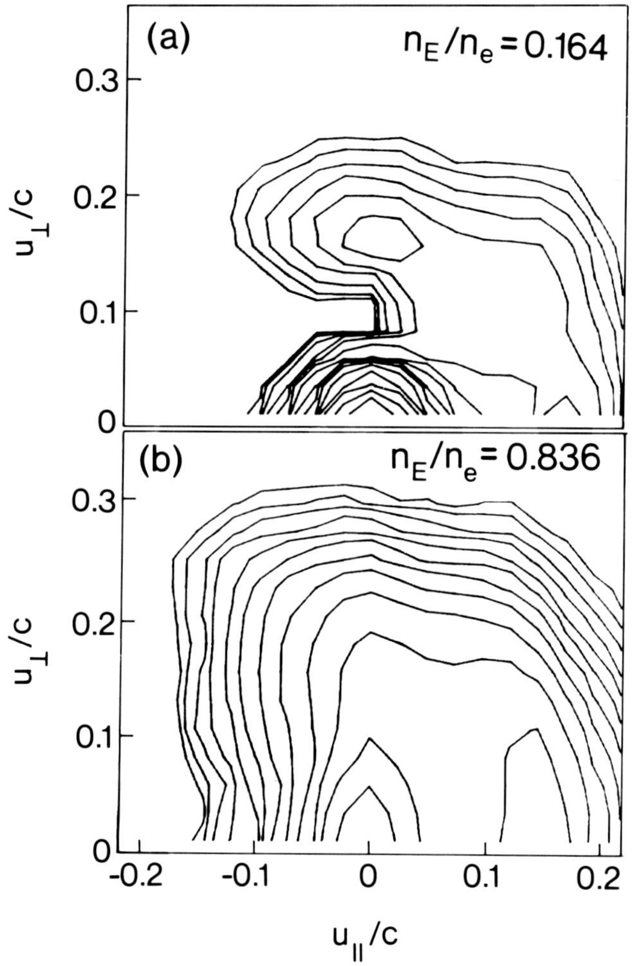

Fig. 10. Contour plots of the electron distribution from two-dimensional simulations presented by Winglee and Pritchett [1986].

satellite data as resulting from injection through a narrow region followed by poleward convection, while Carlson and Torbert [1980] interpreted the data from a much more slowly moving sounding rocket in terms of impulsive injection with the velocity dispersion being a temporal effect. Both effects probably occur, but measurements from multiple spacecraft will be needed to separate the spatial effects from the temporal ones.

Winglee et al. [1993a] have recently shown from DE 1 data that cusp ion injection is subject to short period modulation (of about 10 - 18 seconds with an injection duration of only a few seconds). This modulation is shown to be associated with modulation of the characteristics of the outflowing ionospheric plasma owing to the interaction of magnetosheath and ionospheric plasma. Figure 11 shows an example of a DE-1 cusp crossing. The example in Figure 11 shows the V-shaped energy/pitch-angle signatures of the cusp ion injections at energies above and below ~1 keV along with concurrent "tuning fork" signatures of upgoing ion conic distributions at lower energies. There is clearly an association between these two mesoscale phenomena, which suggests that they are coupled by microscale processes occurring at temporal scales not observable with DE 1. Even with the available data, however, Winglee et al. [1993a] have noted a correlation between the cusp ion injections and the ion conics along with concurrent periodicities in upgoing and downgoing cusp electrons. The possibility that the magnetosheath ions can provide a net downward momentum transfer to the ambient ions as well as enhancing their heating rate was investigated by Winglee et al. [1993b] with electrostatic particle simulations.

The results of the simulations performed by Winglee et al. [1993b] are consistent with the DE-1 data in that the magnetosheath ion injections initially suppress the upward ion conics, as downward momentum is imparted to the ambient ions, but the conics strengthen as the ion injection diminishes as a result of a heating process, which is a slow speed ion-ion drift instability between the outflowing heavy ions and the H^+ ions. Figure 12 (from Winglee et al., 1993a) shows the simulation results at low altitudes (left-hand column) and high altitudes (right-hand column). In the right-hand column of Figure 12, the cusp ion injection begins after $\omega_{pe}t$ = 800 (panel g), grows linearly to a maximum intensity at 2000, and then diminishes linearly to zero at 3600. The top panel in Figure 12 shows the ambient cold H^+ distribution; the second panel (1600) shows the additional distribution of injected magnetosheath H^+ in the positive-$v_\parallel$ half plane. The resulting ion conic appears strongly in the bottom two panels in the left-hand column as the injection has subsided.

5. SUPRATHERMAL ELECTRON BURSTS

In addition to the acceleration of auroral primary electrons, which produces inverted-V events with the characteristics of acceleration by parallel potential differences, there is a second class of auroral acceleration mechanisms involving nonadiabatic, or diffusive, processes that lead to burst-type electron distributions. The electron bursts are typically field-aligned over a broad range of energies and are often associated with conical ion distributions [Klumpar, 1979]. Electron distributions with designations such as field-aligned bursts, counterstreaming electrons, suprathermal bursts, and edge precipitation belong to this category. The burst-type distributions have been identified as primary charge carriers for major Birkeland current elements such as the morning-side region-1 currents [Burch et al., 1983], the currents associated with polar-cap theta arcs [Menietti and Burch,

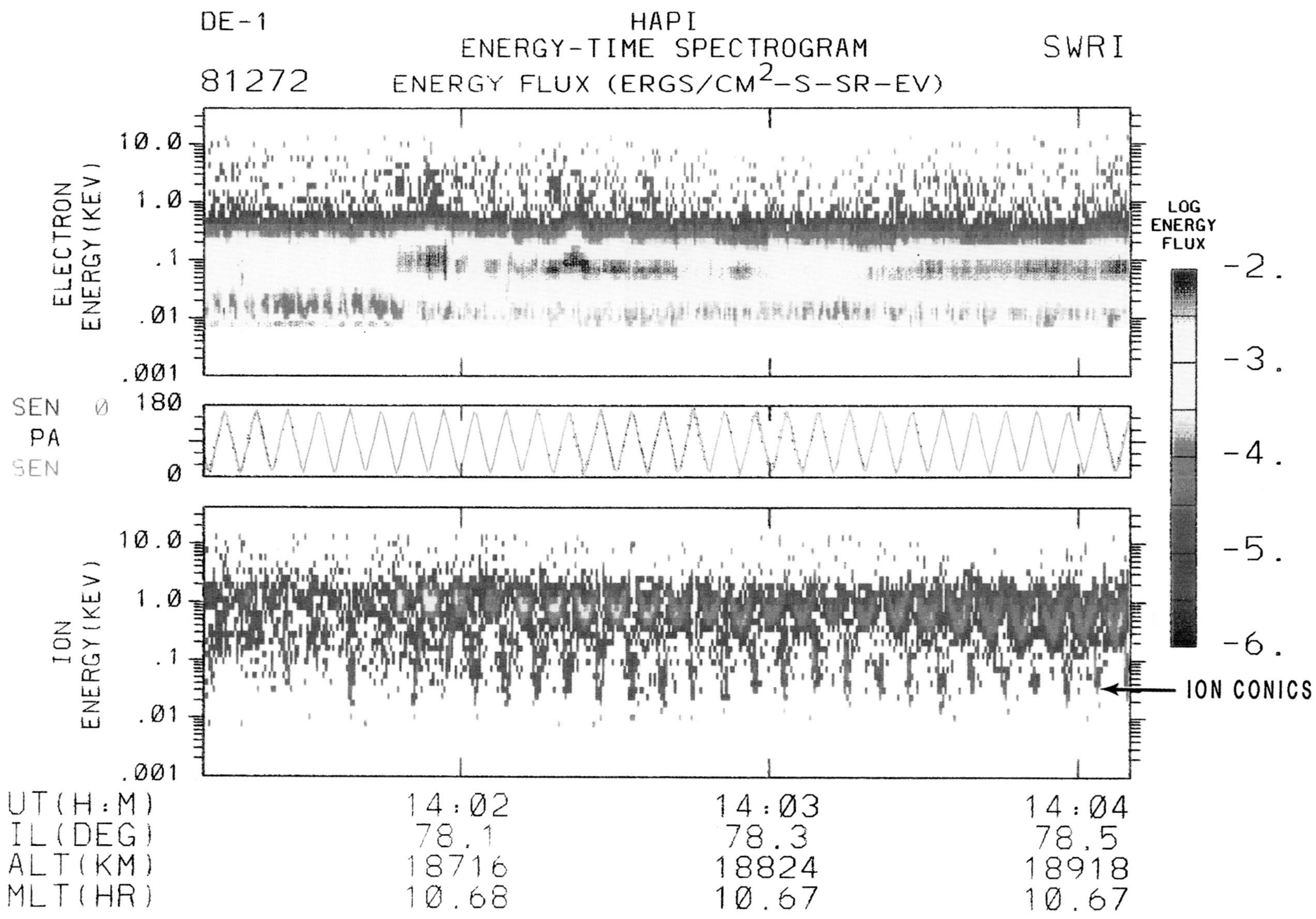

Fig. 11. Electron (top panel) and ion (bottom panel) energy-time spectrograms for a DE 1 pass through the mid-altitude cusp. The middle panel contains the particle pitch angles. [Winglee et al., 1993a].

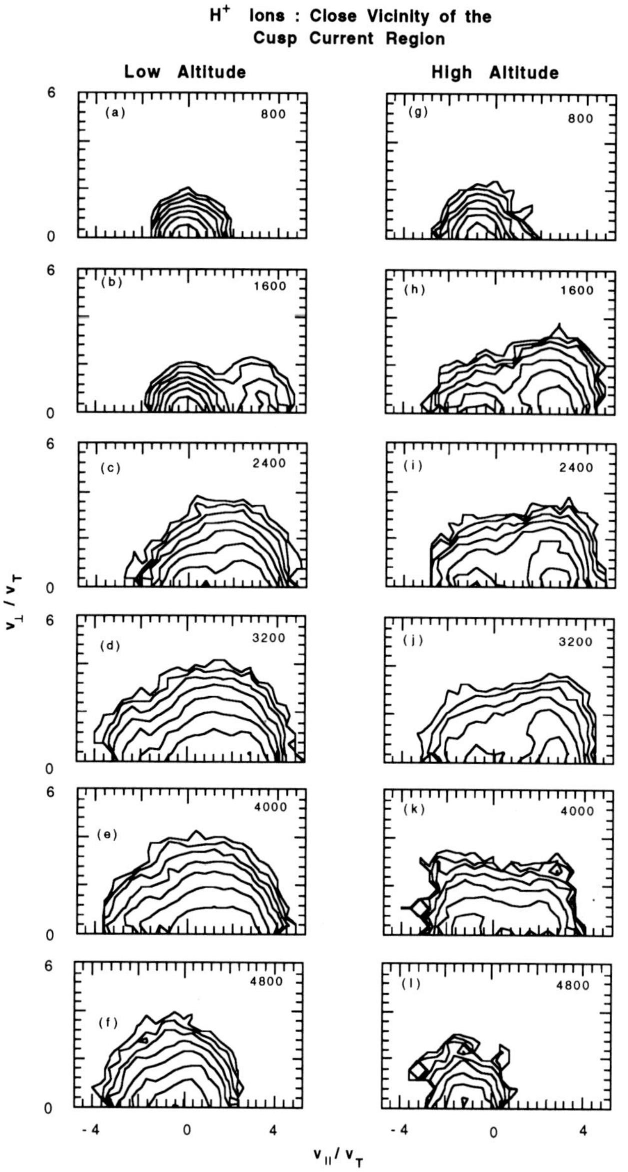

Fig. 12. Results of an electrostatic particle simulation of the interaction between cusp ions and ambient ions. Normalized times ($\omega_{pe}t$) are indicated in each panel. [from Winglee et al., 1993b].

1987], and the evening-side region-2 currents [Klumpar and Heikkila., 1982]. The burst distributions have also been associated with the flickering aurora because of the periodicities they have exhibited with periods in the range of a few seconds [McFadden et al., 1987].

After their initial discovery by Hoffman and Evans [1968] and further analysis by Arnoldy [1974], Johnstone and Winningham [1982] investigated field-aligned electron bursts with the ISIS 2 data and referred to the distributions as suprathermal bursts. Although it is not known for sure whether the electron bursts are primarily spatial or temporal effects, Johnstone and Winningham noted similar time durations for bursts observed by satellites and sounding rockets, which would favor the temporal effect. The suprathermal bursts observed at low altitudes with ISIS were primarily directed downward, although some upward bursts were also reported. Measurements at higher altitudes within and above the auroral acceleration regions have now shown an essentially equal probability of observing upward and downward bursts, which are associated with downward and upward Birkeland currents, respectively [Marshall et al., 1991]. Often, however, the upward and downward bursts are observed in very close proximity to one another or together as counterstreaming electrons [Sharp et al., 1980] with their relative intensities determining whether the net Birkeland current is upward or downward [Marshall et al., 1991].

Johnstone and Winningham [1982] noted in the suprathermal bursts a strong field alignment over a broad range of energies and a tendency for the bursts to occur in the central plasma sheet (CPS) region or in the BPS, but in the regions surrounding inverted-V structures rather than within them. Nonetheless, they considered the suprathermal burst process to be closely related to the inverted-V mechanism because the upper energy limit of the bursts seemed to rise and then fall again with latitude or, when an inverted V was present, to fall off with increasing distance from the edge of a nearby inverted V. In an earlier sounding rocket experiment, however, Arnoldy [1974] reported field-aligned electron burst distributions that seemed to fill in the energy regime below the peak in the primary auroral electron energy spectrum. Similarly, Burch et al. [1979] and McFadden et al. [1987] observed field-aligned electron bursts within inverted-V structures with the field-aligned electrons having a wide energy range extending up to, but not above the primary electron beam energy. Figure 13, from McFadden et al. [1987], shows the simultaneously measured field-aligned electron bursts and auroral primary distributions above a flickering auroral arc.

The close association that has been observed between field-aligned electron burst distributions and both Birkeland current systems and inverted-V events suggests that the bursts are another example of phenomena that result from the coupling of mesoscale phenomena into microscale processes in the upper ionosphere. The mechanism responsible for the electron bursts has not yet been determined, although it has been noted that the bursts are often accompanied by upward-moving ion conics and that the field- aligned motion of the suprathermal electrons and the cyclotron motion of the conic ions have similar time scales. This

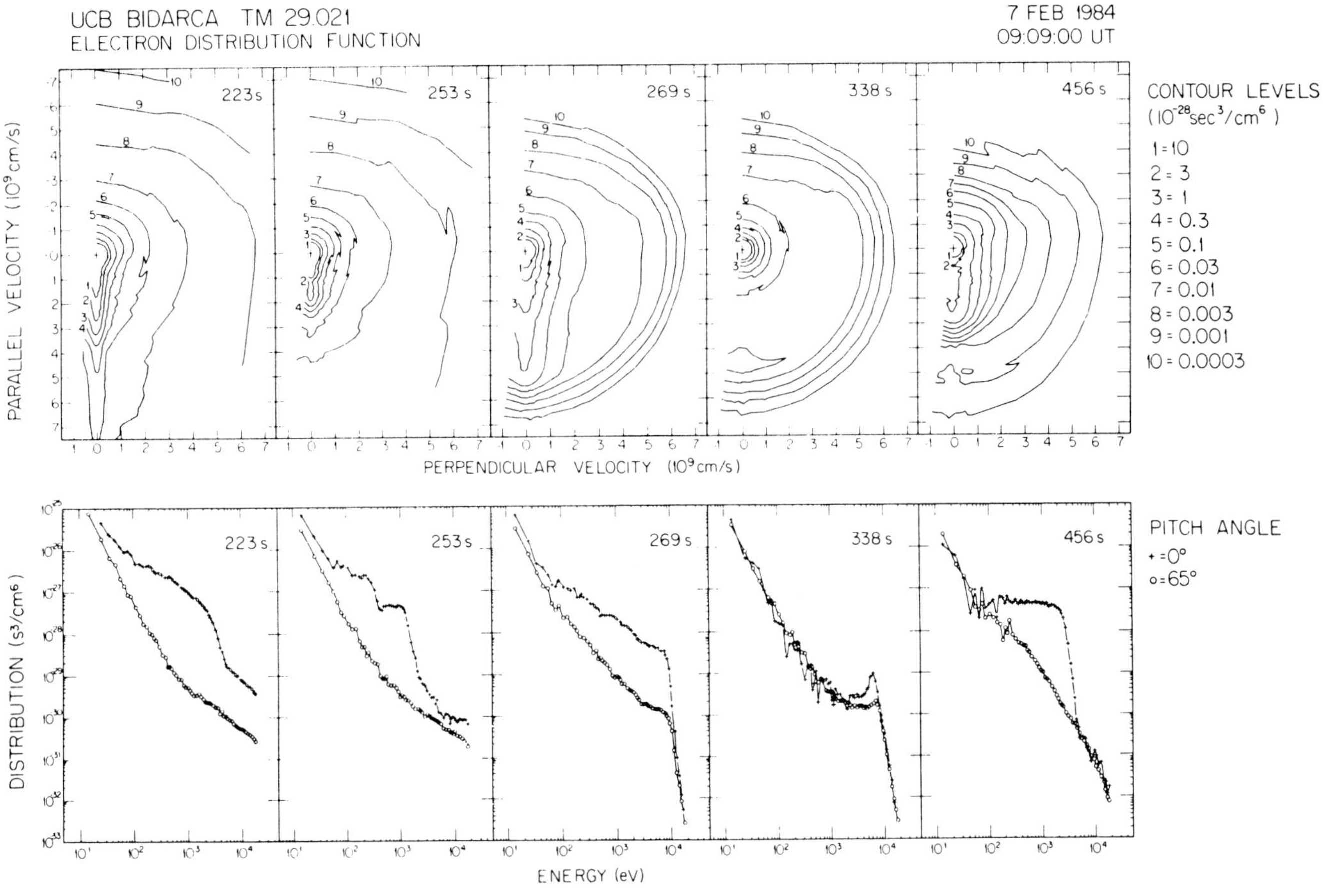

Fig. 13. Contour plots (top panels) and line plots (bottom panels) of auroral primary electrons and suprathermal electron bursts at lower energies measured by McFadden et al. [1987] on a sounding rocket in a flickering aurora.

notion has led to several models of the electron bursts, all of which involve waves in the VLF frequency range that are generated within an auroral acceleration region [Temerin et al., 1986; Lotko, 1986; McFadden et al., 1987]. Another class of models for the field-aligned electron bursts involves flickering or moving double layers [Sharp et al., 1980; Hultqvist, 1991].

6. SUMMARY AND CONCLUSIONS

In the preceding paragraphs we have considered several examples of coupling between mesoscale and microscale phenomena in the cusp and auroral region plasmas. The four phenomena discussed above—electron conics, auroral kilometric radiation, cusp region ion conics, and field-aligned electron bursts—are typical in the sense that full understanding will depend on measurements with higher resolution in either space or time than is presently achievable. In each case either the specific plasma process or the free energy source is indeterminate with existing data sets, so progress is being made by using numerical simulations to investigate the microscale processes and their coupling with the mesoscale phenomena, for which extensive data sets exist. However, only when the resolution of the measurements allows direct probing of the microscale processes will the unique identification of the specific coupling modes be possible.

It is significant, or at least interesting, that fluctuating double layers have been proposed in connection with the four phenomena we have discussed. André and Eliasson [1991] suggested fluctuations of parallel electric fields in low frequency waves observed by Viking as a cause of electron conics; Louarn et al. [1990] proposed similar fluctuations as a mechanism for populating Knight's forbidden region of velocity space [Knight, 1973] with trapped electrons; Sharp et al. [1990] proposed such fluctuations as the source of counterstreaming field-aligned electrons; and Hultqvist et al [1991] proposed them as an explanation for the simultaneous appearance of upward-moving ion and electron distributions. On the other hand, other instabilities and wave-particle interactions have also been shown to be consistent with the observations. The theoretical and numerical simulation results that are being obtained should prove valuable in guiding the development of the higher-resolution plasma and wave measurements that are needed to identify uniquely the microscale processes that are associated with the well-known mesoscale phenomena that characterize the cusp and auroral regions. The present capabilities and future needs with regard to investigation of mesoscale-microscale coupling processes can be summarized as follows:

Present Capabilities:

- Mesoscale phenomena can be directly detected with contemporary plasma and field instrumentation in missions such as Geotail, Polar, and Cluster.
- Theory and modeling at the microscale can be used to guide the most effective measurements of the mesoscale (mission-oriented theory).
- Velocity space (except for cold electrons) and the frequency domain are adequately characterized, but temporal resolution and resolution in configuration space are inadequate for investigation of microscale processes.

Future Needs:

- Measurements of wave vectors are needed to identify wave-particle interactions uniquely.
- Temporal resolution of plasma distribution function measurements needs to be less than the expected
- Temporal resolution of plasma distribution function measurements needs to be less than the expected growth rates of instabilities (<1 s or better).
- Multi-point measurements are needed within mesoscale phenomena so that spatial and temporal effects can be separated and the microscale processes identified.

These needs are being addressed by some of the new missions such as Freja (launched in October 1992), Cluster (scheduled launch in 1996) and the Fast Auroral Snapshot (FAST) explorer (scheduled launch in 1994). Rapid advances in computer technology will at the same time advance markedly the spatial and temporal domains over which particle simulations of the microscale processes can be applied.

REFERENCES

André, M., H. Koskinen, L. Matson, and R. Erlandson, Local transverse ion energization in and near the polar cusp, *Geophys. Res. Lett.*, *15*, 107-110, 1988.

André, M., G. B. Crew, W. K. Peterson, A. M. Persoon, C. J. Pollock, and M. Engebretson, Ion heating by broadband low-frequency waves in the cusp/cleft, *J. Geophys. Res.*, *95*, 20,809-20,823, 1990.

André, M., and L. Eliasson, Electron acceleration by low frequency electric field fluctuations: electron conics, *Geophys. Res. Lett.*, *19*, 1073-1076, 1992.

Arnoldy, R., Auroral particle precipitation and Birkeland currents, *Rev. Geophys. Space Phys.*, *12*, 217-231, 1974.

Ball, L., and M. André, Heating of O^+ ions in the cusp/cleft: Double-cyclotron absorption versus cyclotron resonance, *J. Geophys. Res.*, *96*, 1429-1437, 1991.

Block, L., and C.-G. Fälthammer, The role of magnetic-field-aligned electric fields in auroral acceleration, *J. Geophys. Res.*, *95*, 5877, 1990.

Burch, J. L., S. A. Fields, and R. A. Heelis, Polar cap electron acceleration regions, *J. Geophys. Res.*, *84*, 5863-5874, 1979.

Burch, J. L., P. H. Reiff, R. A. Heelis, J. D. Winningham, W. B. Hanson, C. Gurgiolo, J. D. Menietti, R. A. Hoffman, and J. N. Barfield, Plasma injection and transport in the mid-altitude polar cusp, *Geophys. Res. Lett.*, *9*, 921-924, 1982.

Burch, J. L., P. H. Reiff, and M. Sugiura, Upward electron beams measured by DE-1: a primary source of dayside region-1 Birkeland currents, *Geophys. Res. Lett.*, *10*, 753-756, 1983.

Carlson, C. W., and R. B. Torbert, Solar wind ion injections in the morning auroral oval, *J. Geophys. Res.*, *85*, 2903, 1980.

Hoffman, R. A., and D. S. Evans, Field-aligned electron bursts at high latitudes observed by OGO 4, *J. Geophys. Res.*, *73*, 6201-6214, 1968.

Hudson, M. K., and I. Roth, Ion heating in the cusp, in *Ion Acceleration in the Magnetosphere and Ionosphere*, ed. by T. Chang, 271-278, AGU Geophysical Monograph 38, Washington, D.C., 1986.

Hultqvist, B., R. Lundin, K. Stasiewicz, L. Block, P.-A. Lindqvist, G. Gustafsson, H. Koskinen, A. Bahnsen, T. A. Potemra, and L. J. Zanetti, Simultaneous observation of upward moving field aligned energetic electrons and ions on auroral zone field lines, *J. Geophys. Res.*, *93*, 9765, 1988.

Hultqvist, B., H. Vo, R. Lundin, B. Aparacio, P.-A. Lindqvist, F. Gustafsson, and B. Holback, On the upward acceleration of electrons and ions by low-frequency electric field fluctuations observed by Viking, *J. Geophys. Res.*, *96*, 11,609-11,615, 1991.

Johnstone, A. D., and J. D. Winningham, Satellite observations of suprathermal bursts, *J. Geophys. Res.*, *87*, 2321-2329, 1982.

Klumpar, D. M., Transversely accelerated ions: an ionospheric source of hot magnetospheric ions, *J. Geophys. Res.*, *84*, 4229-4237, 1979.

Klumpar, D. M., and W. J. Heikkila, Electrons in the ionospheric source cone: evidence for runaway electrons as carriers of downward Birkeland currents, *Geophys. Res. Lett.*, *9*, 873-876, 1982.

Knight, S., Parallel electric fields, *Planet. Space Sci.*, *21*, 741, 1973.

Lin, C. S., J. D. Menietti, and H. K. Wong, Perpendicular heating of electrons by upper hybrid waves generated by a ring distribution,, *J. Geophys. Res.*, *95*, 12,295-12,299, 1990.

Lotko, W., Diffusive acceleration of auroral primaries, *J. Geophys. Res.*, *91*, 191, 1986.

Louarn, P., A. Roux, H. de Féraudy, D. Le Quéau, M. André and L. Matson, Trapped electrons as a free energy source for the auroral kilometric radiation, *J. Geophys. Res.*, *95*, 5983-5995, 1990.

Lockwood, M., M. O. Chandler, J. L. Horwitz, J. H. Waite, Jr., T. E. Moore, and C. R. Chappell, The cleft ion fountain, *J. Geophys. Res.*, *90*, 9736-9748, 1985.

Lundin, R., L. Eliasson, B. Hultqvist, and K. Stasiewicz, Plasma energization on auroral field lines as observed by the Viking satellite, *Geophys. Res. Lett.*, *14*, 443, 1987.

Marshall, J. A., J. L. Burch, J. R. Kan, P.H. Reiff, and J. A. Slavin, Sources of field-aligned currents in the auroral plasma, *Geophys. Res. Lett.*, *18*, 45-48, 1991.

McFadden, J. P., C. W. Carlson, M. H. Boehm, and T. J. Hallinan, Field-aligned electron flux oscillations that produce flickering aurora, *J. Geophys. Res.*, *92*, 11,133-11,148, 1987.

Melrose, D. B., A phase-bunching mechanism for fine structures in auroral kilometric radiation and Jovian decametric radiation, *J. Geophys. Res.*, *91*, 7970-7980, 1986.

Menietti, J. D., and J. L. Burch, Electron conic signatures observed in the nightside auroral zone and over the polar cap, *J. Geophys. Res.*, *90*, 5345, 1985.

Menietti, J. D., and J. L. Burch, DE 1 observations of theta aurora plasma source regions and Birkeland current charge carriers, *J. Geophys. Res.*, *92*, 7503-7518, 1987.

Menietti, J. D., C. S. Lin, H. K. Wong, A. Bahnsen, and D. A. Gurnett, Association of electron conical distributions with upper hybrid waves , *J. Geophys. Res.*, *97*, 1353-1361, 1992.

Menietti, J. D., J. L. Burch, R. M. Winglee, and D. A. Gurnett, DE-1 particle and wave observations in AKR source regions, *J. Geophys. Res.*, 98, 5865-5880, 1993.

Newman, D., R. M. Winglee, and M. V. Goldman, Theory and simulation of electromagnetic beam modes and whistlers, *Phys. Fluids*, *31*, 1515, 1988.

Omidi, N., and D. A. Gurnett, Growth rate calculations of auroral kilometric radiation using the relativistic resonance condition, *J. Geophys. Res.*, *87*, 2377-2383, 1982.

Peterson, W. K., M. André, G. B. Crew, A. M. Persoon, M. J. Engebretson, C. J. Pollock, and M. Temerin, Heating of thermal ions near the equatorward boundary of the mid-altitude polar cleft, in *Electromagnetic Coupling in the Polar Clefts and Caps*, ed. by A. Egeland and P. E. Sandholt, 103, Kluwer Academic Press, Boston, 1989.

Pottelette, R., R. A. Treumann, and N. Dubouloz, Generation of auroral kilometric radiation in upper hybrid wave - lower hybrid soliton interaction, *J. Geophys. Res.*, *97*, 12,029-12,044, 1992.

Reiff, P. H., T. W. Hill, and J. L. Burch, Solar wind plasma injection at the dayside magnetospheric cusp, *J. Geophys. Res.*, *82*, 479, 1977.

Roth, I., M. K. Hudson, and M. Temerin, Generation Models of Electron Conics, *J. Geophys. Res.*, *94*, 10,095-10,102, 1989.

Sharp, R. D., E. G. Shelley, R. G. Johnson, and A. G. Ghielmetti, Counterstreaming electron beams at altitudes of 1 R_E over the auroral zone, *J. Geophys. Res.*, *85*, 92-100, 1980.

Siscoe, G. L., What determines the size of the auroral oval?, in *Auroral Physics*, ed. by C.-I. Meng, M. J. Rycroft, and L. A. Frank, 159-176, Cambridge University Press, 1991.

Temerin, M., Evidence for a large bulk ion conic heating region, *Geophys. Res. Lett.*, *13*, 1059-1062, 1986

Temerin. M., and D. Cravens, Production of electron conics by stochastic acceleration parallel to the magnetic field, *J. Geophys. Res.*, *95*, 4285-4290, 1990.

Temerin, M., J. McFadden, M. Boehm, C. W. Carlson, and W. Lotko, Production of flickering aurora and field-aligned electron flux by electromagnetic ion cyclotron waves, *J. Geophys. Res.*, *91*, 5769-5792, 1986.

Ungstrup, E., A. Bahnsen, H. K. Wong, M. André and L. Matson, Energy source and generation mechanism for auroral kilometric radiation, *J. Geophys. Res.*, *95*, 5973-5981, 1990.

Winglee, R. M., and P. L. Pritchett, The generation of low-frequency electrostatic waves in association with auroral kilometric radiation, *J. Geophys. Res.*, *91*, 13,531-13,541, 1986.

Winglee, R. M., J. D. Menietti, and H. K. Wong, Numerical simulations of bursty radio emissions from planetary magnetospheres, *J. Geophys. Res.*, *97*, 17,131-17,140, 1992.

Winglee, R. M., J. D. Menietti, W. K. Peterson, J. L. Burch, J. H. Waite, Jr., and B. Giles, Magnetosheath-ionospheric plasma interactions in the cusp/cleft: 1. observations of modulated injections and upwelling ion fluxes, *J. Geophys. Res.*, *98*, 19,315-19,330, 1993a.

Winglee, R. M., J. D. Menietti, and C. S. Lin, Magnetosheath-ionospheric plasma interactions in the cusp/cleft: 2. mesoscale particle simulations, *J. Geophys. Res.*, *98*, 19,331-19,348, 1993b.

Wong, H. K. and M. L. Goldstein, A mechanism for bursty radio emission in planetary magnetospheres, *J. Geophys. Res.*, *17*, 2229, 1990.

Wong, H. K., J. D. Menietti, C. S. Lin, and J. L. Burch, Generation of electron conical distributions by upper hybrid waves in the Earth's polar region, *J. Geophys. Res.*, *93*, 10025, 1988.

Wu, C. S.. and L. C. Lee, A theory of the terrestrial kilometric radiation, *Astrophys. J.*, *230*, 621, 1979.

clear in Auroral [illegible], edited by [illegible] Kelley, CMB, and Cogger, [illegible] A. [illegible] and P. [illegible], [illegible] Academic Press, [illegible]

Rosenberg, [illegible] Tokugawa, Indian [illegible] Tone [illegible] [illegible] and [illegible] waves [illegible] [illegible] interaction [illegible] Res., [illegible]

Roth, R. [illegible] W. [illegible] and [illegible] [illegible] [illegible] [illegible] Res., [illegible] 22, [illegible]

Roth, [illegible] K. [illegible] and [illegible] Electric Observations, [illegible]. Proc. [illegible] IDAS [illegible]

Shelley, E. G., [illegible] Sharp, [illegible] and C. Chappell, [illegible] [illegible] of [illegible] observations of [illegible] [illegible] Res., [illegible]

[illegible]

[illegible] Res. [illegible]

[illegible] J. Geophys. Res., [illegible] 1990.

[illegible] [illegible] [illegible] [illegible] [illegible] and [illegible] Geophys. Res., [illegible], 1990.

Wahlund, [illegible] and [illegible] [illegible] waves [illegible] with [illegible] Geophys. Res. [illegible] 1988.

Winglee, R. M., J. D. Menietti, and H. K. Wong, Numerical simulations of [illegible] [illegible] from [illegible] J. Geophys. Res., 97, 17,131–17,147, 1992.

Winglee, R. M., [illegible] D. Menietti, [illegible] [illegible] Berthelier, [illegible] [illegible] Geophys. Res., [illegible] 19,315–19,330, 1992a.

Winglee, R. M., J. D. Menietti, and H. K. Wong, Mechanism for [illegible] [illegible] J. Geophys. Res., 98, 19,331–19,344, 1993.

Wong, H. K., and [illegible] [illegible] [illegible] Geophys. Res., [illegible]

Wong, H. K., D. [illegible] Menietti, C. S. Lin, and J. L. Burch, Generation of electron conical distributions by upper hybrid waves in the Earth's polar region, J. Geophys. Res., 93, 10,025–10,036, 1988.

Wu, C. S., and [illegible] [illegible] of the [illegible] [illegible] mechanism, Astrophys. J., 230, [illegible]

Kaufmann, R. L., Thermally accelerated ions [illegible] the [illegible] J. Geophys. Res., [illegible] 1982.

Kintner, P. M., [illegible] W. J. Heikkila, [illegible] [illegible] [illegible] [illegible] [illegible] Geophys. Res. Lett., [illegible] 1982.

Kuijpers, J., [illegible] Space Sci. Rev., [illegible]

[illegible] J. [illegible] and [illegible] [illegible] [illegible] Geophys. Res., [illegible] 1988.

Lotko, W., [illegible] Res., [illegible]

[illegible] A., [illegible] [illegible] and [illegible] [illegible] Res., [illegible] 1990.

Lockwood, M., [illegible] Chandler, [illegible] T. [illegible] [illegible] [illegible] Geophys. Res., [illegible]

[illegible] [illegible] [illegible] [illegible] [illegible] [illegible] observed by the Viking [illegible] Geophys. Res., [illegible]

Mahmood, [illegible] J. [illegible] Burch, [illegible] Kelly, and [illegible] Source of [illegible] currents in the auroral [illegible] Geophys. Res. Lett., [illegible]

McFadden, J. [illegible] [illegible] [illegible] [illegible] Geophys. Res., 92, [illegible]

Melrose, D. B., [illegible] [illegible] [illegible] [illegible] Geophys. Res., [illegible] 1986.

Menietti, J. D., and J. L. Burch, [illegible] [illegible] J. Geophys. Res., [illegible] 1987.

Menietti, J. D., and [illegible] J. Geophys. Res., [illegible] 1987.

Menietti, J. D., [illegible] Lin, H. K. Wong, [illegible] and D. A. Gurnett, Association of electron conical distributions with upper hybrid waves, J. Geophys. Res., 97, 1353–1361, 1992.

Menietti, J. D., [illegible] Burch, R. M. Winglee, and D. A. Gurnett, DE-1 particle and wave observations in [illegible] [illegible] regions, J. Geophys. Res., [illegible] 1993.

Newman, [illegible] Winglee, and M. V. Goldman, [illegible] of [illegible] [illegible] [illegible]

Omidi, N., and D. A. Gurnett, Growth rate calculations of auroral kilometric radiation using the relativistic resonance condition, J. Geophys. Res., 87, 2377, 1982.

Peterson, W. K., [illegible] G. [illegible] A. M. Persoon, H. L. Collin, C. J. Pollock and M. Temerin, Heating of thermal ions near the equatorward boundary of the [illegible] [illegible] [illegible]

Auroral Plasma Dynamics in the Presence of a Finite-Width Current Filament and V-Shaped Potential Drop

Supriya B. Ganguli
Science Applications International Corporation, McLean, Virginia
H. G. Mitchell
NASA Goddard Space Flight Center, Greenbelt, Maryland
P. J. Palmadesso
Naval Research Laboratory, Washingon, D.C.

We have simulated plasma dynamics driven by a finite width current filament in ionosphere-magnetosphere coupling. The model uses the reduced set of sixteen moment system of equations and simultaneouly solves coupled continuity and momentum equations and equations describing the transport along the magnetic field lines of parallel and perpendicular thermal energy and heat flows for each species. The lower end of our simulation is at an altitude of 800 km, in the collisional topside ionosphere, while the upper end is at 10 R_E, well into the magnetosphere. The plasma consists of hot electrons and protons of magnetospheric origin and low-energy electrons, protons and oxygen ions of ionospheric origin. The dynamical interaction of the current filament with ionospheric and magnetospheric plasma generates a potential structure in the horizontal direction and a kV potential drop along the field line. In the presence of this potential structure and parallel electric field the ionospheric plasma density is depleted and velocity is reduced, while density enhancement and increased velocity is observed in magnetospheric plasma. The ionospheric and magnetospheric electron temperatures increase below 2 R_E due to magnetic mirror force on converging geomagnetic field lines.

INTRODUCTION

The auroral region is the most active region in the ionosphere-magnetosphere (I-M) system in terms of particle energizations, generation of parallel electric fields and field-aligned potential drops. Currents driven in the magnetosphere by the solar wind-magnetophere dynamo close in the ionosphere by means of magnetic-field-aligned currents in the high latitudes. Such currents give rise to kilovolt-sized potential structures along the magnetic field, leading to particle energization and plasma waves in this region. In order to understand the creation and maintenance of these field-aligned potentials, we have for a number of years been involved in the creation of one-dimensional dynamic simulations of this region [*Ganguli et al.*, 1988, 1991, *Mitchel et al.*, 1990, 1992].

DE 1/DE 2 observations reveal that parallel electrostatic fields of 1-10 kV potential drop at 1-2 R_E altitude are an important source for auroral particle acceleration [*Reiff et al.*, 1988; *Marshall et al.*, 1991; and *Lu et al.*, 1991]. The process by which kilovolt-sized potentials parallel to the magnetic field are maintained in collisionless plasma is one of the important topics of ionosphere-magnetosphere (I-M) coupling. *Knight* [1973] and later *Lyons* [1981] pointed out that these potentials are a natural consequence of upward field-aligned currents where high-energy electrons from the magnetosphere flow down the converging magnetic field

Space Plasmas: Coupling Between Small
and Medium Scale Processes
Geophysical Monograph 86

to the ionosphere against the resistance of the magnetic mirror. The upward-flowing low-energy electrons from the ionosphere do not experience the resistance of the magnetic mirror force. Numerical simulations of *Mitchell et al.,* [1990, 1992] indeed illustrate this diode-like response of the field-line plasma to the parallel currents coupling the ionosphere and magnetosphere. It is shown that while upward currents produce kilovolt-sized potential drops along the field, return currents produce only few volts.

The formation of auroral kV potential drops along field lines was also investigated by *Lysak and Hudson* [1987] using a combination of a simplified large scale model and inputs from two small-scale processes, namely double layers and anomalous resistivity. The study concluded that double layers contribute significantly toward the formation of potential drop than anomalous ressistivity. There are kinetic studies of potential structures using adiabatic assumptions for particle distributions [*Chiu and Schultz,* 1987; *Chiu and Cornwall,* 1980; *Newman, et al.,* 1986]. Kinetic simulations of horizontal potential structures have also been performed using two-dimensional particle-in-cell (PIC) techniques [*Singh, et al.,* 1983; 1987; *Wagner et al.,* 1980; 1981; and *Winglee et al.,* 1988] and references therein. Large scale plasma dynamics associated with the potential structure were not studied by these models.

In this paper, we investigate auroral plasma dynamics in the presence of a finite-width current filament using our generalized fluid model. The model uses separate magnetospheric and ionospheric electrons and ion species and a robust approximation of the heat flow. Generalized fluid formalisms based on Grad-type closure approximations fail in a collisionless plasma when the heat flow exceeds a threshold value. In the absence of a separate magnetospheric electron species, the large temperature difference between the magnetospheric and ionospheric electrons drives an unphysical and very large heat flow which make the simulation unstable. These improvements to the model have allowed us to directly simulate the large field-aligned currents (10 $\mu A/m^2$) and potentials one expects in the presence of downward electron currents.

The current filament self-consistently generates field aligned kV potential drops and its horizontal distribution. The dynamical interplay of plasma from ionospheric and magnetospheric origin is studied. The generation of electric fields, formation of V-shaped potential drops, evacuation of the flux tube plasma in response to currents, formation of density cavities/enhancements in I-M coupling are discussed.

THE MODEL

We have simulated two-dimensional plasma dynamics by bunching a number of one-dimensional flux tubes. A potential structure is produced self-consistently in the presence of a finite width current filament. The current filament is gaussian in shape with upward current density of 10 $\mu A/m^2$, the highest current density at the center of the filament, tapering off to .01 $\mu A/m^2$ at both sides.

The model [*Ganguli et al.,* 1992; 1993] uses the reduced set of sixteen moment system of equations and simultaneouly solves coupled continuity and momentum equations and equations describing the transport along the magnetic field lines of parallel and perpendicular thermal energy and heat flows for each species. The lower end of our simulation is at an altitude of 800 km, in the collisional topside ionosphere, while the upper end is at 10 R_E, well into the magnetosphere. The plasma consists of hot electrons and protons of magnetospheric origin and low-energy electrons, protons and oxygen ions of ionospheric origin [*Mitchell et al.,* 1992]. The collision terms used in our model are *Burgers'* [1965] Coulomb collision terms with corrections for finite species velocity differences.

All plasma species s are assumed to be gyrotropic, to have no significant variation perpendicular to $\mathbf{B}$ and no significant chemical interactions, and to be described by four quantities; number density n_s, velocity v_s, temperature $T_{s\|}$ parallel to $\mathbf{B}$, and temperature $T_{s\perp}$ perpendicular to $\mathbf{B}$. From these moments we construct the flux tube quantities $N_s = An_s$, $M_s = An_sv_s$, $P_{s\|} = An_skT_{s\|}/m_s$, and $P_{s\perp} = An_skT_{s\perp}/m_s$, and their associated equations as follows:

$$\frac{\partial N_s}{\partial t} = -\frac{\partial M_s}{\partial r} + \frac{\delta N_s}{\delta t} \tag{1a}$$

$$\frac{\partial M_s}{\partial t} = \frac{1}{e_s}(-D_s + \frac{1}{4\pi}E_{\|}\omega_{ps}^2) \tag{1b}$$

$$\frac{\partial P_{s\|}}{\partial t} = -\frac{\partial P_{s\|}v_s}{\partial r} - \frac{\partial Q_{s\|}}{\partial r} - 2P_{s\|}\frac{\partial v_s}{\partial r} + \frac{\delta P_{s\|}}{\delta t} + \frac{2Q_{s\perp}}{A}\frac{\partial A}{\partial r} \tag{1c}$$

$$\frac{\partial P_{s\perp}}{\partial t} = -\frac{\partial P_{s\perp}v_s}{\partial r} - \frac{\partial Q_{s\perp}}{\partial r} + \frac{\delta P_{s\perp}}{\delta t} - \frac{P_{s\perp}v_s + Q_{s\perp}}{A}\frac{\partial A}{\partial r} \tag{1d}$$

where,

$$D_s = e_s \left(\frac{\partial M_s v_s}{\partial r} + \frac{\partial P_{s\parallel}}{\partial r} - \frac{P_{s\perp}}{A} \frac{\partial A}{\partial r} + \frac{g N_s}{r^2} - \frac{\delta M_s}{\delta t} \right) \tag{1e}$$

represents terms of the momentum equation that arrive when we solve for the parallel electric field. We will discuss the physical meaning of D_s later in the context of parallel electric field equation (4). Here A is the area of the flux tube, m_s and e_s are the mass and charge of species s, $\omega_{ps} = (4\pi N_s e_s^2/m_s)^{1/2}$ is the plasma frequency of species s, $E_\parallel$ is the electric field parallel to $\mathbf{B}$, $\delta/\delta t$ represents the change of a quantity due to collisions, g is the constant of gravitational acceleration, and $Q_{s\parallel}$ and $Q_{s\perp}$ are the parallel heat fluxes of parallel and perpendicular energy for species s. Specifically, $Q_{s\parallel}$ and $Q_{s\perp}$ are N_s times the species' averages of the quantities $(v_\parallel - v_s)^3$ and $(v_\parallel - v_s)v_\perp^2$, respectively.

The above set of equations have two types of undetermined quantities: the parallel electric field $E_\parallel$ and the heat flows $Q_{s\parallel}$ and $Q_{s\perp}$. The manner in which the parallel electric field is determined defines the electrodynamic character of the model. The specification of the heat flows determines the character of the fluid approximation, since the heat flow equations close the fluid equations for each species.

The equation for the electric field is obtained by multiplying (1a) and (1b) by e_s and summing the results over all species s to obtain equations for the total charge $Q = \sum_s N_s e_s$ and total current $J = \sum_s M_s e_s$:

$$\frac{\partial Q}{\partial t} = -\frac{\partial J}{\partial r} + \frac{\delta Q}{\delta t} \tag{2}$$

$$\frac{\partial J}{\partial t} = -\sum_s D_s + \frac{1}{4\pi} E_\parallel \sum_s \omega_{ps}^2 . \tag{3}$$

Fluctuations in Q and J propagate parallel to $\mathbf{B}$ by means of Alfvén waves oblique to $\mathbf{B}$, which cannot be simulated in a one-dimensional model. For this reason, and since the Alfvén speed is of the order of 1 R_E/s in the central region of the simulation, we neglect finite Alfvén transit time effects and assume quasi-neutrality, even though this may be a questionable assumption at the upper end of the simulation. For this reason, our simulation is electrostatic.

We apply quasi-neutrality ($Q = 0$) to (2) and note that charge conservation implies that $\delta Q/\delta t = 0$. Therefore (2) becomes the current conservation equation $\delta J/\delta r = 0$; that is, J is always constant along the flux tube, even though it may vary in time. Equation (3) may be written in the form

$$E_\parallel = \frac{4\pi \left(\frac{\partial J}{\partial t} + \sum_s D_s \right)}{\sum_s \omega_{ps}^2} \tag{4}$$

where $\partial J/\partial t$ is independent of r. The first term in the numerator of (4) defines the time-dependent polarization electric field due to the time variation of the total current, while the second term defines the ambipolar electric field due to the imbalance in the sum of the momentum equation drivers D_s. This equation for the parallel electric field may be substituted into (1b) to yield a momentum equation in which the momentum transfer between species due to $E_\parallel$ is explicit, that is,

$$\frac{\partial M_s}{\partial t} = \frac{F_s}{e_s} \frac{\partial J}{\partial t} + \frac{1}{e_s} \sum_{s'} (F_s D_{s'} - F_{s'} D_s) \tag{5}$$

where $F_s = \omega_{ps}^2 / \sum_{s'} \omega_{ps'}^2$ is a number between 0 and 1. Since $\partial J/\partial t$ is an externally determined quantity in our simulation, (5) is a well-defined equation for M_s. Further, if quasi-neutrality and current conservation are both initially satisfied, (5) guarantees that they remain satisfied during the entire simulation.

For heat flows we use the equations

$$Q_{s\parallel} = -3 \frac{l_s N_s v_{s\parallel} k}{m_s} \frac{\partial T_{s\parallel}}{\partial r} \tag{6}$$

$$Q_{s\perp} = -\frac{l_s N_s v_{s\parallel} k}{m_s} \left(\frac{\partial T_{s\perp}}{\partial r} + (1 - \frac{T_{s\perp}}{T_{s\parallel}}) \frac{T_{s\perp}}{A} \frac{\partial A}{\partial r} \right) \tag{7}$$

where, $l_s = \min(5\ v_{s\parallel}/4\nu_{ss}, \lambda)$. To be precise, λ should be the local wavelength for perturbations in $T_{s\parallel}$ or $T_{s\perp}$. The value of λ that we use is the simulation grid size (which is always less than the local perturbation wavelength), which provides sufficient thermal diffusion to remove the spurious dynamics inherent in the moment approximation without the necessity of performing a nonlocal calculation of the perturbation wavelength. Since the simulation grid size multiplied by the gradient of a quantity is equal to the change of that quantity from one simulation cell to the next, our approximation effectively defines the saturated heat flow as being proportional to ΔT between two adjacent simulation cells rather than $\partial T/\partial r$.

SIMULATION RESULTS

Simulations are initialized with an equilibrium in which $J = 0$. This equilibrium has two primary components: the low-energy ionosphere and the high-energy magnetosphere. The low-energy ionosphere consists of the three species e^-, H^+, and O^+ in a plasma outflow configuration at a temperature of 2000 K. The dominant ion species O^+ has a density n_o of 6×10^4 cm^{-3} at the lower end of the simulation and is gravitationally bound. The e^- density n_e in this region must match n_o, and the resulting small-scale height produces an ambipolar electric field large enough to accelerate the minority species H^+ to supersonic outflow velocities by the time it becomes the dominant ion species at about 5000 km altitude. Above this altitude the H^+ velocity v_H and linear density N_H are roughly constant, while n_o is negligible. The e^- density and velocity match those of H^+ in this region so as to maintain quasi-neutrality and zero current. The plasma is maintained by an H^+ flux of 10^8 $cm^{-2}s^{-1}$ into the simulation at the lower end (representing the production rate of H^+ in the upper ionosphere), and an outward flux from the upper end (representing an effectively open field line). The high-energy magnetosphere is composed of the two species e^* and H^*, both at a parallel temperature of 1 keV and a perpendicular temperature of 3 keV [*Mitchell et al.*, 1992]. The equilibrium ionospheric electron density is shown in figure 1a and magnetospheric electron density in figure 1b.

These set of simulations were ran for 90s. At t = 0 second, the current was turned on in all fluxtubes with upward current density of 10 $\mu A/m^2$, the highest current at the center of the filament tapering off to .01 $\mu A/m^2$ at both sides. The velocities of ionospheric and magnetospheric electrons increase at the same rate (downward in this case), until the pressure imbalance in the magnetospheric electron momentum equation acts to reduce their velocity back to negligible levels. At 1 s, the e^* velocity has already been reduced relative to the e^- velocity, and at 2 s, the e^* distribution has "rebounded", i.e., the initial downward impulse imparted to the e^* distribution by the current onset has been reflected by the converging magnetic field, producing a transient upward flux of e^* below 3 R_E. The flux tube carrying the highest current achieved the highest electron velocity.

The current is carried almost entirely by the ionospheric electrons after the first few seconds. This upward current implies a source of cold electrons in the magnetosphere. Since this is unrealistic, our simulation prohibits an inward flux of ionospheric electrons at the upper boundary of the model. The upward current carried by downward-flowing cold electrons can therefore only persist at a certain altitude until all of the cold electrons above that altitude have been depleted. In our simulation the cold electrons at the upper end of the model (10 R_E) are depleted at about 17 s after current onset, after which the depleted region travels toward the ionosphere.

The e^- density cavity is formed at the upper boundary. The parallel electric field increases sharply at the upper boundary in order to force the e^* distribution to carry the current there. Below the H^+/O^+ cross over the precipitating e^- distribution encounters a dense and relatively stationary e^- population. The counterstreaming of these two populations appears as a sharp increase in the parallel e^- temperature in this region.

The potential structure is seen to form at the upper end of the simulation region at this time (17 s). The potential drop is strongest on the flux tube with highest current density. At 38 s after the current onset the structure is more defined and takes the form of a V-shaped potential drop (figure 2). The ionospheric electron density depletion at this time is shown in figure 3a. The magnetospheric electron density is shown in figure 3b. The corresponding velocities are shown in figures 4a and 4b.

At 60 s after current onset, the entire region above 2 R_E altitude has been depleted of cold electrons and will remain depleted as long as the current is maintained. This is similar to the results of a steady state study by *Newman et al.* [1985] in which the upward motion of cooler electrons is effectively curtailed by the introduction of a sizeable potential drop along an auroral field line. Our model clearly shows dynamically a path by which such a potential may evolve in the presence of a field-aligned current.

The potential structure moves downward to topside ionosphere with time. The parallel electric field at 72 s is shown in figure 5a and the corresponding potential drop in figure 5b. The evacuation of ionospheric plasma in this region in response to upward current is shown in figure 6a. The density cavity is deeper in the region of high current density. The corresponding velocity is shown in figure 6b. The ionospheric electron temperatures shown in figures 6c and 6d, is cool in most of the region except at the lower ionospheric boundary. The e^- temperatures have become very large below 2 R_E due to the precipitating e^- population.

In the depleted region, only magnetospheric electrons remain to carry the field-aligned current. As a result, the electric field must be large enough to maintain a sustained downward hot electron flux against the mirror action of the converging magnetic field. Magnetospheric electron density is enhanced. The density and velocity profiles for magnetospheric electrons are shown at 72 s in figures 7a and 7b. Since at this time there is very few ionospheric electrons in this region, the magnetospheric electrons are forced to maintain quasi-neutrality with all ion species in the depletion region. The e^* parallel and perpendicular temperatures are shown in figures 7c and 7d. The e^* temperature is hotter at the ionospheric boundary due to compression of e^* fluid elements as they move downward along converging magnetic field lines.

Before ionospheric electron density depletion begins, the field-aligned potential drop within the model is 1.5 V. This value increases dramatically when the magnetospheric electrons are forced to sustain the current, and the field-aligned potential drop reaches a value of 1 kV at 60 s after current onset.

The large parallel electric field above 2 R_E accelerates both ionospheric and magnetospheric H^+ ions. Upward acceleration of H^+ below 1 R_E due to the large, low-altitude ambipolar electric field was also observed. The oxygen ions are the dominant ion species 1 R_E altitude, and the ion density increases rapidly below that point. The time constant of electon dynamics is few seconds, where as ion dynamics takes much longer about 30 minutes.

DISCUSSIONS

We have simulated two-dimensional plasma dynamics driven by a finite width current filament in ionosphere-magnetosphere coupling. The dynamical interaction of the current filament with ionospheric and magnetospheric plasma generates a potential structure in the horizontal direction and a kV potential drop along the field line. The simulation displays the dynamical path by which the potential structure evolves in the presence of the field-aligned current filament. In the presence of this potential structure the ionospheric electron density is depleted and velocity is reduced, while density enhancement and increased velocity is observed in magnetospheric plasma. The ionospheric and magnetospheric electron temperatures increase below 2 R_E due to magnetic mirror force on converging geomagnetic field lines.

The highest current density in the current filament is 10 $\mu A/m^2$, a value which is realistic but difficult to simulate with conventional approaches. Our model uses separate ionospheric and magnetospheric plasma population and a robust approximation of the heat flow. These improvements to the model have allowed us to directly simulate the large field-aligned currents (10 $\mu A/m^2$) and potentials one expects in the presence of downward electron currents. In the absence of a separate magnetospheric electron species, the large temperature difference between the magnetospheric and ionospheric electrons drives an unphysical and very large heat flow which make the simulation unstable. Most previous plasma outflow models use a low current density of 1 or 2 $\mu A/m^2$ [see *Mitchell et al.*, 1992 for references].

The potential structures are three-dimensional in nature. Therefore, we need to include cross-field plasma dynamics in a realistic way. This will be the subject of a future report. The cross-field dynamics is important and may even be crucial. For example, kV potential drops generated along one field-line can be moderated by cross-field transport thereby altering both the magnitude and stability of the potential. However, as a first step it is important to understand the plasma dynamics along the geomagnetic field-lines. Effects of wave-particle interactions can also be incorporated in our model via anomalous collision terms [*Ganguli and Palmadesso*, 1988]. But this must be done only after large-scale dynamics is well understood.

Acknowledgements. The research at SAIC is supported by NSF grant ATM-9020655, ATM-9307834, and NASA contract NASW-4601. The work at NRL was supported by the Office of Naval Research.

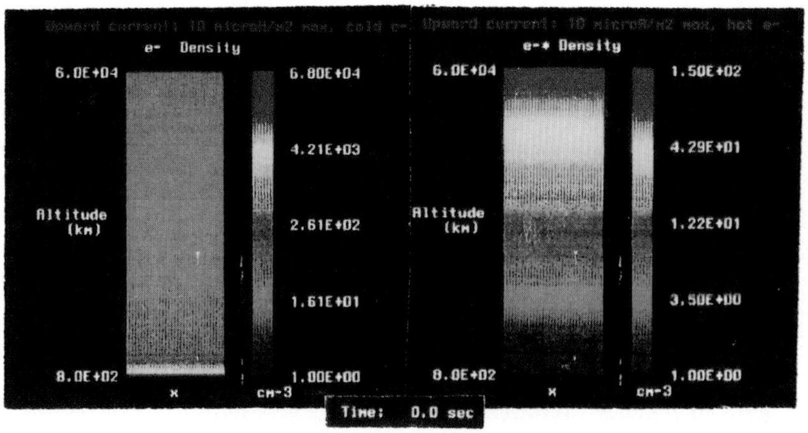

Fig. 1a and 1b. Ionospheric and magnetospheric electron density distribution at t = 0 s.

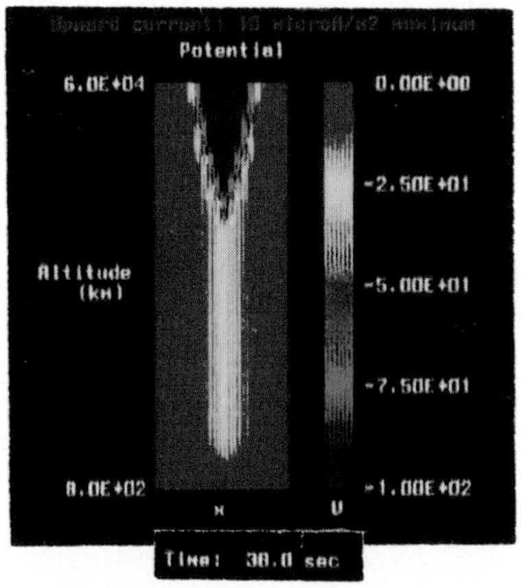

Fig. 2. Formation of the potential structure at t = 38 s.

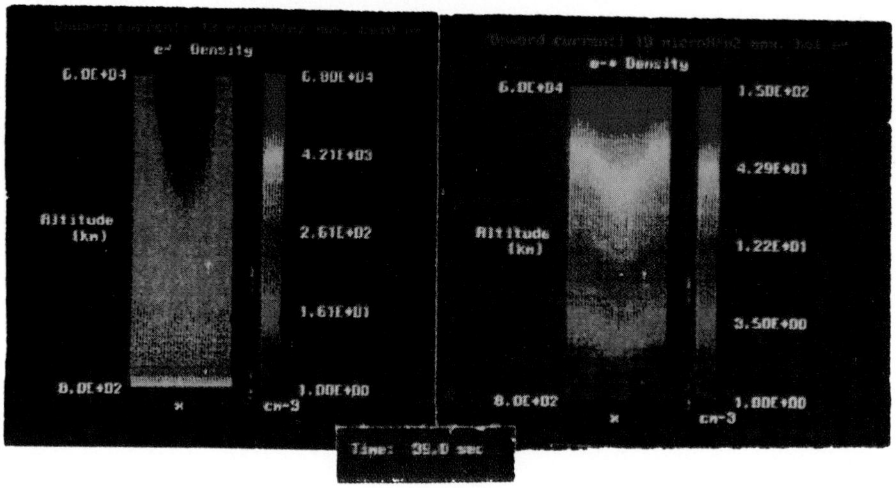

Fig. 3a and 3b. Ionospheric and magnetospheric electron density distribution at t = 39 s.

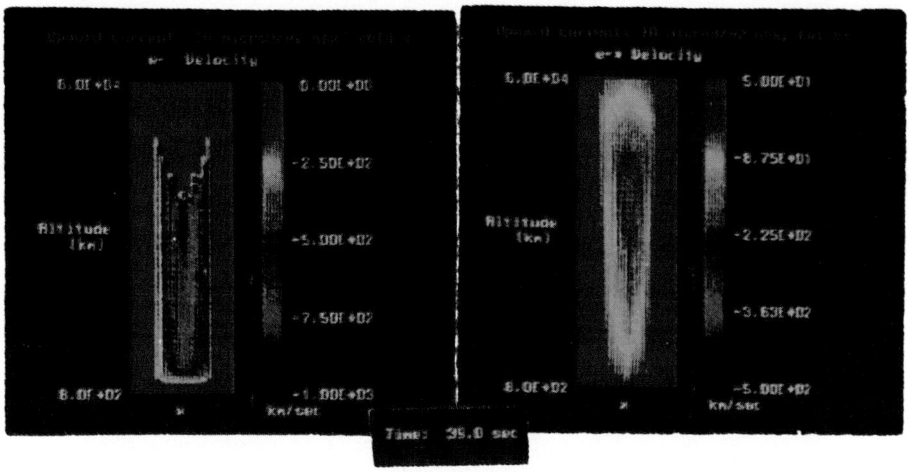

Fig. 4a and 4b. Ionospheric and magnetospheric electron velocity distribution at t = 39 s.

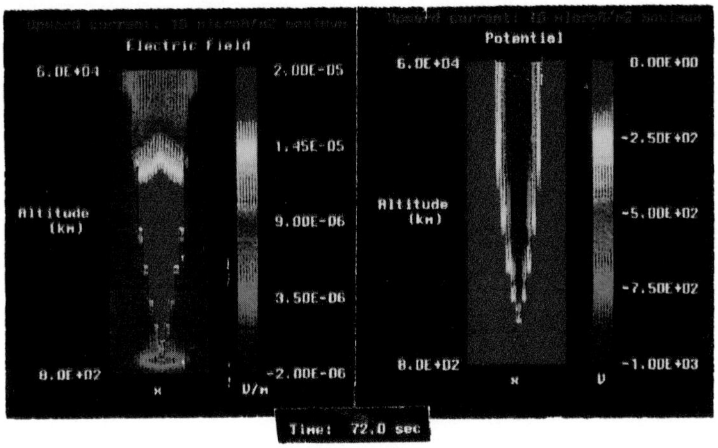

Fig. 5a and 5b. The electric field and the potential structure at t = 72 s.

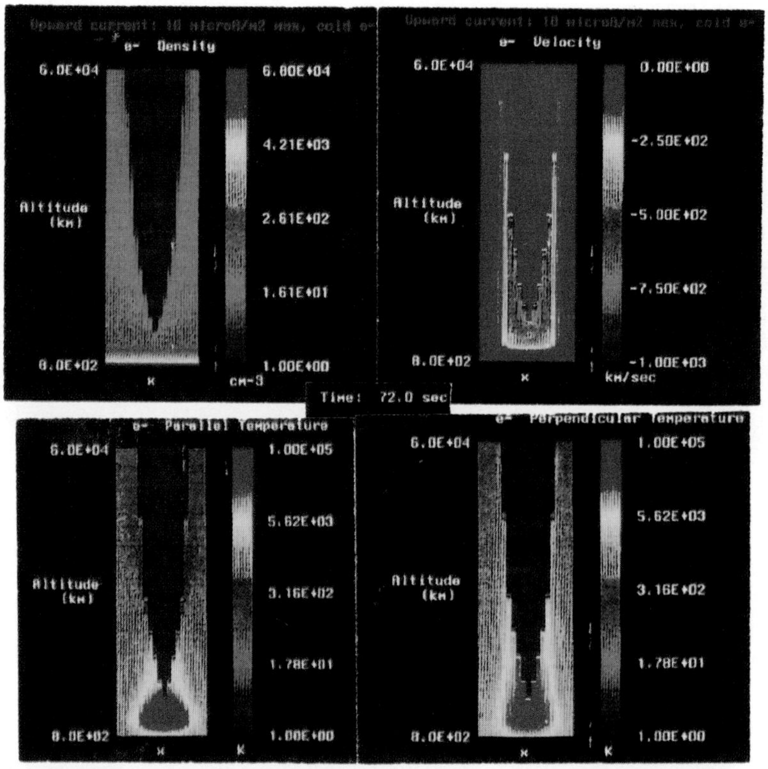

Fig. 6 the ionospheric electron (a) density (b) velocity (c) parallel temperature and (d) perpendicular temperature distribution at t = 72 s.

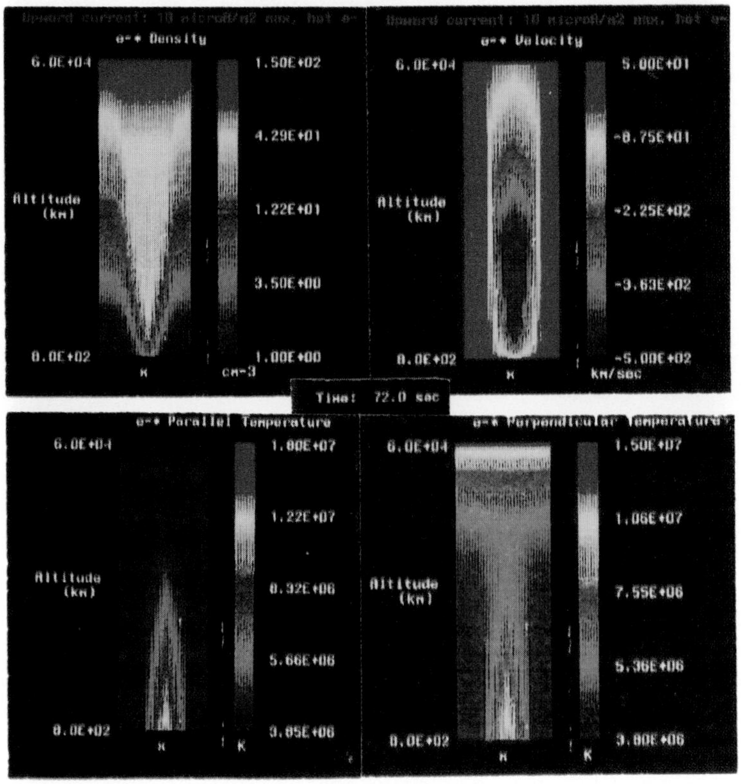

Fig. 7 the magnetospheric electron (a) density (b) velocity (c) parallel temperature and (d) perpendicular temperature distribution at t = 72 s.

REFERENCES

Burgers, J. M., *Flow equations for composite gases*, Academic Press, New York, 1969.

Chiu, Y. T., and M. Schultz, *J. Geophys. Res.*, 83, 629, 1987.

Chiu, Y. T., and J. M. Cornwall, *J. Geophys. Res.*, 85, 543, 1980.

Ganguli, S. B., and P. J. Palmadesso, *J. Geophys. Res.*, 92, 8673, 1987.

Ganguli, S. B., P. J. Palmadesso, and H. G. Mitchell, *Geophys. Res. Lett.*, 15, 1291, 1988.

Ganguli, S. B., P. J. Palmadesso, and H. G. Mitchell, in *SPI conference proceedings*, T. Chang and G. B. Crew editors, 197, 1991.

Ganguli, S. B., H. G. Mitchell and P. J. Palmadesso, Auroral plasma dynamics in the presence of a finite-width current filament and V-shaped potential drops, *AGU Chapman Conference on Micro and Meso Scale Phenomena in Space Plasmas*, Hawaii, 1992.

Ganguli, S. B., H. G. Mitchell and P. J. Palmadesso, *Geophys. Res. Lett.*, 20, 975, 1993.

Knight, S., *Planet Space Sci.*, 21, 741, 1973.

Lu, G., et al., *J. Geophys. Res.*, 96, 3523, 1991.

Lyons, L. R., in *Physics of Auroral Arc Formation*, Geophys. Mono. Ser., vol. 35, 252, AGU, Wash. D.C, 1981.

Lysak, R. L., and M. K. Hudson, *Laser and Particle Beams*, 5, 351, 1987.

Marshall, J. A., et al., *Geophys. Res. Lett.*, 18, 45, 1991.

Mitchell, H. G., S. B. Ganguli, and P. J. Palmadesso, *Geophys. Res. Lett.*, 1, 1873, 1990.

Mitchell, H. G., S. B. Ganguli, and P. J. Palmadesso, *J. Geophys. Res.*, 97, 12045, 1992.

Newman, A. L., Y. T. Chiu, and J. M. Cornwall, *J. Geophys. Res.*, 91, 3167, 1986.

Reiff, P. H., et al., *J. Geophys. Res.*, 93, 7441, 1988.

Singh, N., H. Thiemann, and R. W. Schunk, *Geophys. Res. Lett.*, 10, 745, 1983.

Singh, N., H. Thiemann, and R. W. Schunk, *Planet Space Sci.*, 3, 353, 1987.

Wagner et al., *Phys. Rev. Lett.*, 45, 803, 1980.

Wagner et al., in *Physics of Auroral Arc Formation*, S. -I. Akasofu edited, p. 304, 1981.

Winglee, R.W., et al., *J. Geophys. Res.*, 93, 14567, 1988.

[1] Science Applications International Corporation, 1710 Goodridge Drive, McLean, VA 22102

[2] NASA Goddard Space Flight Center, Greenbelt, MD 20771

[3] Naval Research Laboratory, Washington, D.C. 20375

Micro and Meso Scale Measurements by the Freja Satellite

Rickard Lundin
Swedish Institute of Space Physics, Kiruna, Sweden
Gerhard Haerendel
Max Planck Institut für extraterrestrisches Physik, Garching, Germany

Freja, a joint Swedish and German scientific satellite launched on October 6 1992, is designed to give high temporal/spatial resolution measurements of the auroral plasma characteristics. A high telemetry rate (520 kbits/s) and ≈15 Mbyte distributed on board memories (give ≈2 Mbits/s during one minute) enables Freja to resolve meso and micro scale phenomena in the 100 m range for particles and 1-10 m range for electric and magnetic fields. The on-board UV imager will resolve auroral structures of kilometer size with a time resolution of one image per 6 s. Novel plasma instruments give Freja the capability to increase the spatial/temporal resolution orders of magnitudes above that achieved on satellites before.

The scientific objectives of Freja is to study the interaction between the hot magnetospheric plasma with the topside atmosphere/ionosphere. This interaction leads to a strong energization of magnetospheric and ionospheric plasma and an associated erosion, and loss, of matter from the terrestrial exosphere. Freja orbits with an altitude of ≈ 600 -1750 km, thus covering the lower part of the auroral acceleration region. This altitude range hosts processes that heat and energize the ionospheric plasma above the auroral zone, leading to the escape of ionospheric plasma and the formation of large density cavities.

1. INTRODUCTION

The Freja mission represents a continuation of a line of magnetospheric research that commenced with Viking, the first Swedish satellite launched in 1986. Like on Viking, participation in the instrumentation for Freja is spread over a number of countries, but the satellite is mainly funded by Sweden (≈75%) and Germany (≈25%). The extreme low cost approach on Freja follows a highly successful tradition, such as Viking and AMPTE (Germany). A piggyback launch (on the Chinese Long March II) and a very streamlined project organization is utilized to keep the total costs of the project to within 100 million Swedish kronor (corresponding to ≈15 million US dollars). Although this cost may seem unreasonably low, the project is not suffering from any significant compromises on system safety or payload quality. For instance, all critical housekeeping systems such as transponders etc are doubled. Even more important from the science point of view is that Freja is equipped with 8 PI-instruments for particles, fields and auroral imaging that are capable of providing fine-structure plasma measurements with an hitherto unprecedented temporal/spatial resolution from satellites.

The scientific objective of Freja is to explore fine structure plasma properties within the low-altitude portion of the auroral acceleration region and to study the physical processes whereby ionospheric plasma is being heated/accelerated and subsequently ejected out into the magnetosphere. The altitude range traversed by Freja (≈600 km - 1750 km), constitute the topside ionosphere and the low-altitude part of the auroral energization region. This altitude range is in general considered well explored by numerous polar orbiting satellites. However, a high telemetry rate and state-of-the-art design of the instrumentation, enables more than an order of magnitude increased temporal/spatial resolution compared to its predecessors, which makes Freja even exploratory in some aspects. Results from the Freja mission are expected to improve our understanding of a number of processes and

Space Plasmas: Coupling Between Small
and Medium Scale Processes
Geophysical Monograph 86

TABLE 1: FREJA SCIENTIFIC PAYLOAD AND SCIENCE COORDINATION

Experiment	Measurement Technique	Principal Investigator
F1 Electric Fields	3 pairs of Wire booms, 20 tip-to-tip 2 comp. of E up to 6000 samples/s	*Göran Marklund*, Alfvén Lab., Royal Inst. of Technology, Stockholm, Sweden
F2 Magnetic Fields	Triaxial flux-gate on 2 m boom 3 comp. of B, 128 samples/s	*Lawrence Zanetti*, Johns Hopkins Univ., Applied Physics Lab., Laurel, MD, USA
F3H Particles, Hot Plasma:	2D Magnetic electron spectrometer 2D distr. 0.1-115 keV, 100 samples/s 2D Ion composition spectrometer => 3D distr. 0.001 - 10 keV in 3 s.	*Lars Eliasson* Swedish Institute of Space Physics, Kiruna, Sweden
F3C Particles, Cold Plasma:	2D ion/electrons on 2 m boom 3D distr. of cold plasma (<300 eV) >100 samples/s	*Brian Whalen*, National Research Council, Ottawa, Canada
F4 Waves	Wire booms + 3 axis search coil E, B, Δn Waves, 1 Hz - 2 MHz	*Bengt Holback* Swedish Institute of Space Physics, Uppsala-Div., Uppsala,Sweden
F5 Auroral Imager	2 UV CCD cameras Auroral images every 6 s.	*John S Murphree*, University of Calgary, Calgary, Canada
F6 Electron Beam	Three electron guns 3 components of E, 100 samples/s	*Götz Paschmann*, Max Planck Institut für extraterr. Physik, Garching, Germany
F7 Correlator	2D electron spectrometer 0.01 - 20 keV, correlation F4	*Manfred Boehm*, Max Planck Institut für extraterr. Physik, Garching, Germany
European Ground. Based Coordinator	European network of ground-based observations.	*Hermann Lühr*, Institut für Geophysik and Meteorologi, Braunschweig, Germany
North America Ground Based Coordinator	North American network of ground-based observations	*Gordon Rostoker*, University of Alberta, Edmonton, Alberta, Canada

phenomena in space plasma physics, such as:

* processes responsible for the transverse heating/energization and loss of ions and electrons in the topside ionosphere,
* plasma density cavities, their importance and consequences for the interaction between hot and cold plasma,
* low-altitude ion and electron acceleration,
* processes leading to the formation of fine-structure in auroral forms,
* Wave-particle interaction and the great variety of wave phenomena on auroral field lines.

The Freja payload comprises a full complement of high-resolution plasma diagnostic instruments and a fast auroral imager. Table 1 lists the individual instruments, their key elements, performance parameters, and Principal Investigators. Measurements performed by Freja are for instance:

- The electrostatic vector field (by both active and passive means),
- the magnetic vector field,
- hot and cold plasma density and distribution,
- hot and cold ion composition,
- electric, magnetic and plasma density waves,
- wave-particle correlations,
- ultraviolet auroral images.

Freja was launched as a piggy-back satellite on a Chinese Long-March II rocket in October 6, 1992. After launch into a 63° inclination, Freja was lifted to a higher perigee (≈600 km) and apogee (1750 km) by means of two separate solid fuel boost motors. The relatively low inclination means that good auroral oval coverage is obtained mainly over the American continent. Thus the Canadian Prince Albert ground station, is an important ground segment for the direct transmission of data. However, Esrange in Sweden will be the main operations centre where all uplink communications are performed.

2. SCIENTIFIC BACKGROUND

The sun represents not only the main source of energy but also a strong source of matter interacting with the planets in our solar system. The solar wind, the loss of matter from the sun, corresponds to about 5 Earth masses per billion year. This wind of solar plasma interacts with e.g. planets and comets in a complicated way that is highly dependent on their intrinsic magnetic properties. An important common consequence of the interaction is, however, the erosion of matter from planets and comets as a result of the energy and momentum transfer from the solar wind. The Earth is more effectively protected from this interaction by a strong intrinsic magnetic field which provides a "magnetic umbrella" shielding its topside atmosphere. Despite this the Earth is loosing matter at a rate of 2-4 kg/s (Chappell et al., 1987) as a result of the heating and acceleration of plasma along the auroral oval topside ionosphere. The loss rate may appear high, but it is nevertheless insignificant on cosmogonic time scales. It will for instance take some 100 billion years to evacuate the present terrestrial atmosphere at that rate. Other planets, such as Venus and Mars may have been less fortunate in retaining a hydrosphere and a "habitable" atmosphere due to their lack of a strong intrinsic magnetic field. There, the solar wind interacts directly with the entire frontside ionosphere and the momentum transfer and relative atmospheric loss rate become correspondingly larger.

The Sun-Earth relationship is a complex involving direct electromagnetic forcing from radiation as well as indirect electromagnetic forcing from the solar wind plasma interacting with the terrestrial magnetic field. Because the solar radiation input is 4 to 5 orders of magnitude higher than the solar wind energy input, one would expect solar radiation to dominate the dynamics of the topside ionosphere. However, contemporary space research have demonstrated that the solar wind forcing may become more important than the solar radiation for the redistribution and loss of ionospheric plasma.

There is a strong coupling between the ionization of the upper atmosphere by solar UV and EUV and the electromagnetic forcing induced by solar wind plasma interacting with the Earth´s magnetic field. An increased conductivity in the upper atmosphere due to ionization means that electromagnetic energy more easily dissipates there and a complex chain of transport dissipation and loss can be set up as a consequence of the magnetosphere-ionosphere coupling. In fact, dissipation of solar wind electromagnetic energy to the topside atmosphere and ionosphere of the Earth requires a finite electrical conductivity, the solar wind electromagnetic energy being dumped to the ionosphere as waves/electric fields (providing Joule heating) or via charged particle precipitation. The former (Joule heating) can be described by a current circuit analogy where the field-aligned/Birkeland currents connects the ionospheric load to the solar wind dynamo. Although Birkeland currents and their connection to the solar wind dynamo were extensively studied during the 70:ies and 80:ies (e.g. Ijima and Potemra, 1976) our knowledge of the properties of the source (dynamo) region remains relatively poor. The implications of the Birkeland currents are better known, though. The most important one being their relation to the auroral energization process.

Auroral energization processes

Aurora, in particular its most spectacular and intense form, the discrete aurora, manifests itself over the polar regions as a result of plasma energization processes primarily taking place in the altitude range 1 000 to 10 000 km. This has been a well accepted fact for about 15 years now - gradually evolving through an era of rising maturity of space explorations. The first in-situ measurements, performed in the sixties by sounding rockets, of monoenergetic beams of electrons over auroral arcs (e.g. McIlwain, 1960; Albert 1967; and Evans, 1968) and later experiments with actively injected Barium jets (Haerendel et al., 1976) became qualitative evidence for the hypothesis by Alfvén (1958) of electric potentials being established along the magnetic field lines.

During the seventies various models for the acceleration region above the auroral oval were moulded from numerous measurements by sounding rockets and low-altitude orbiting satellites. For instance, field-aligned pitch angle distributions and monoenergetic beams of electrons with rising and subsequently falling energies versus latitude, denoted inverted V:s, were interpreted as regions with parallel electric fields (e.g. Frank and Ackerson, 1971, Gurnett and Frank, 1973). The prime obstacle for establishing parallel electrostatic fields as proposed by Alfvén, the apparent infinite conductivity along magnetic field lines, was removed by introducing e.g. “anomalous” resistivity due to wave-particle interaction (Kindel and Kennell, 1971, Papadopolous and Coffey, 1974).

The first modelling attempt by Knight (1973) and

Lennartsson (1976) introduced a scheme whereby a nonresistive acceleration may occur simply because of the magnetic mirroring geometry. A quite successful model of the electron distribution resulting from such a nonresistive acceleration, including the backscattering of electrons, was introduced by Evans (1974). Many other models also followed based on the concept by Evans (see e.g. Fälthammar, 1983 for a review). Further indirect means of establishing a current-voltage relationship in the nonresistive acceleration region were made from sounding rocket measurements (e.g. Lundin and Sandahl, 1978; Lyons et al., 1979). On the basis of the current-voltage relationship one could also infer the altitude of the auroral acceleration region as being below ≈2 R_E (Lundin and Sandahl, 1978).

Conclusive evidence for the mid-altitude energization of ions and electrons came from the S3-3 satellite, traversing for the first time with appropriate instrumentation the core of the acceleration region. The S3-3 particle and electric field data (e.g. Shelley et al., 1976; Mizera and Fennell, 1977) frequently displayed many of the characteristic features expected from an electrostatic acceleration process with e.g. electrons accelerated in one direction (downward) and ions accelerated in the opposite direction (upward). A comparison between measured and simulated particle distribution functions within a field-aligned electrostatic potential (e.g. Chiu and Schultz, 1978) also showed good quantitative agreement with theory. The discovery of perpendicular electric "shocks" (Mozer et al., 1977) and weak double layers (Temerin et al., 1982) introduced further complexity to the perhaps too primitive conjecture of electrostatic field-aligned potentials along magnetic field lines. Instead of constituting a smooth potential well structure the electric field measurements implied considerably more temporal/spatial variations acting within the acceleration region. DE-1 and in particular Viking confirmed this considerably more complex picture of the auroral energization region.

The launch of DE-1 in 1981 into a somewhat higher orbit (apogee ≈23 000 km) than S3-3 (apogee ≈8 000 km) added further discoveries from the auroral energization region. For instance, so called "electron conic" signatures (Menietti and Burch, 1985) are features of the electron distribution function that cannot be attributed to a simple electrostatic acceleration. On the other hand, the high-low aspect introduced by the co-orbiting DE-1 and DE-2 indicated that at least the time averaged properties of the acceleration region were in good agreement with mid-altitude acceleration constituting a quasi-electrostatic potential well structure (Reiff et al., 1988).

Similarly, the launch of Viking in 1986 and the very detailed measurements of particles and fields carried out during the about 1 year lifetime of the spacecraft marks another landmark for the understanding of the mid-altitude auroral energization region. The apogee of Viking (≈13 500 km) was ideal because that appears to mark an upper bound of at least the dayside mid-altitude energization. Most of the traversals of the AKR generation region are below ≈10 000 km, indicating that the main field-aligned potential is below that altitude (e.g. Bahnsen et al., 1989). This upper altitude limit is consistent with Viking ion data of field aligned beams (Thelin et al., 1990). However, Viking data showed that the transverse energization of ions may continue up to at least to the Viking apogee.

Fig 1 illustrates the meso-scale structure of the parallel acceleration region driven by a magnetospheric dynamo which converts (magnetospheric) plasma kinetic energy into electromagnetic energy that drives the auroral current circuit. Energy dissipation within the current circuit takes place in primarily two regions, the topside ionosphere as plasma heating/acceleration and the mid-altitude acceleration region as field aligned accelerated beams of electrons and ions. The figure illustrate an important consequence of the auroral acceleration processes, the energization and evacuation of the topside ionospheric plasma that leads to the formation of plasma cavities. In what follows, the two types of energization processes are discussed.

Field aligned electron and ion acceleration.

A number of theories for the field aligned acceleration processes have been suggested, a few of them listed here:

1. Double layers (e.g. Block, 1972) are produced as a local plasma instability when the plasma contains an insufficient number of charge carriers. They constitute cavities with polarization on each side, frequently with a net voltage across. They have been studied in laboratories for more than 30 years and have now also been observed in space (Temerin et al., 1982; Boström et al., 1987).
2. Magnetic mirroring (e.g. Alfvén and Fälthammar, 1963) utilizes the magnetic mirroring force as a current limiter. For instance, when the current carried by electrons in the loss-cone is insufficient for closing the current loop driven by the magnetospheric dynamo, part of the dynamo voltage will progress to lower altitudes. This will both open up the loss-cone and accelerate the source electrons.
3. Collisionless thermoelectric fields (Hultqvist, 1971) were proposed to exist as a result of the temperature difference between the hot magnetospheric and cold ionospheric plasma.

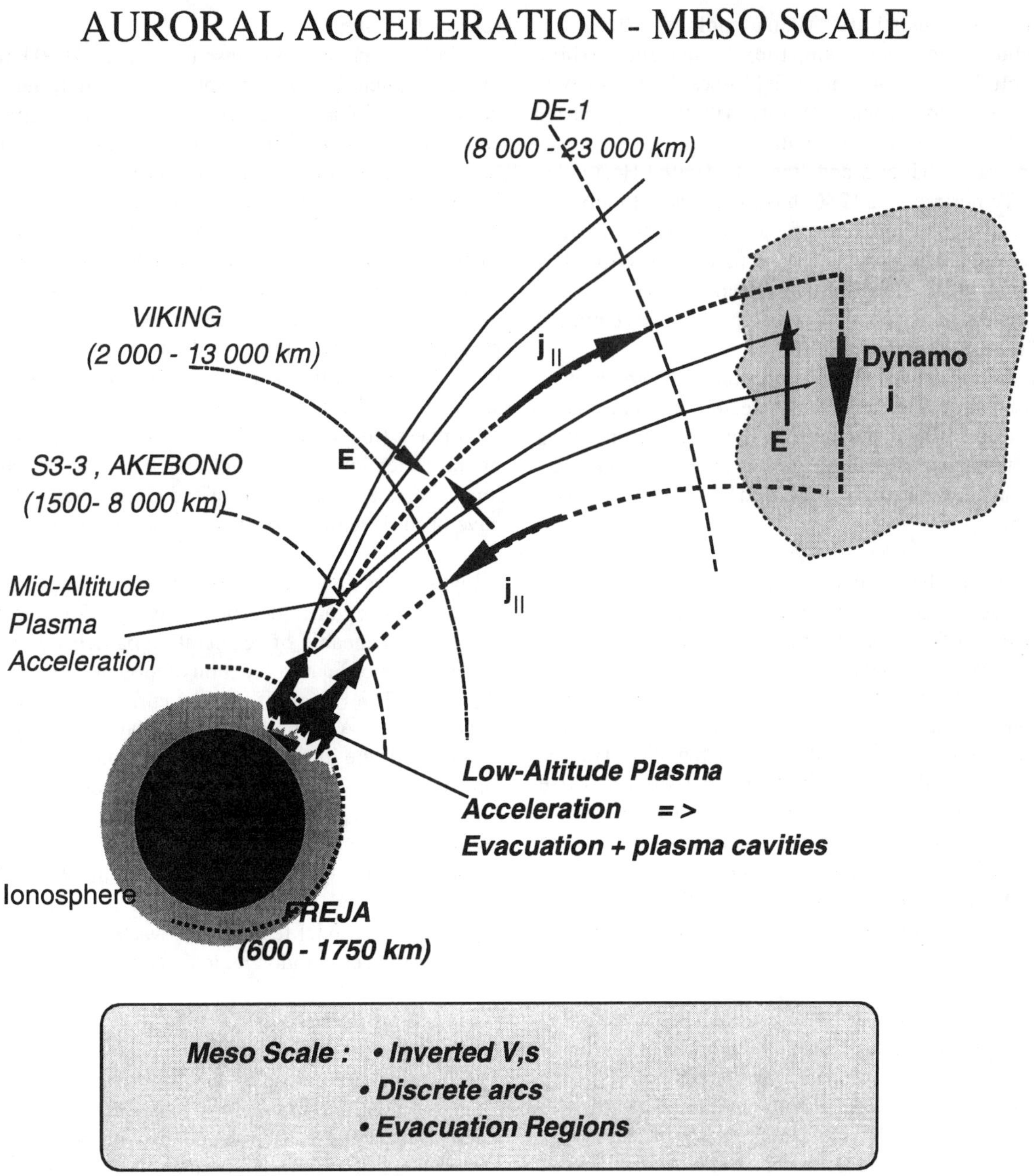

Figure 1 A diagrammatic illustration of the auroral acceleration region and its associated current circuit driven by a magnetospheric dynamo. The acceleration/energization of ionospheric plasma leads to evacuation and the formation of plasma density cavities.

4. "Anomalous resistivity" (Papadopolous, 1977) was introduced to overcome the apparent infinite conductivity along magnetic field lines. The resistivity should occur as particle collisions with waves produced as a result of a current driven instability.
5. Alfvén waves (Goertz and Boswell, 1979; Haerendel, 1980, Temerin et al., 1986) have been used to model a temporal/spatial variation of the field-aligned particle acceleration. The process can be used to describe time variations of parallel potentials (Haerendel, 1983) or can be considered a generically independent acceleration process (e.g. Temerin et al., 1986).
6. Lower hybrid waves (Bingham et al., 1984) have been proposed as an alternative to the electrostatic acceleration of electrons. Waves generated by downward propagating magnetospheric ions are here assumed to accelerate electrons downward in a phased or quasi randomly fashion.

Notice that all these "rivaling" field aligned acceleration theories have in common that it is a parallel electric field ($E_{||}$) that accelerates the electrons. They also require (in accordance with observations) the existence of intense Birkeland currents in the acceleration region. The disagreement between the theories is mainly related to the cause of $E_{||}$ and the temporal/spatial characteristics. The last two theories (5 and 6) differ from the others in that these processes can act resonantly. It is not clear from the wave acceleration theories how they can accelerate both ions and electrons field aligned in opposite directions as is frequently observed. Wave energization is likely to belong to a set of observations with field aligned electrons and transversely accelerated ions.

Fig 2 gives a cartoon showing the three most popular concepts for the downward acceleration of electrons, by quasi-static electric fields, together with a fourth way of energizing electrons by resonant wave interaction (Fig 2d). Fig 2a illustrates the electric equipotential structure with double layers, Fig 2b the anomalous resistivity caused by current instability and Fig 2c the electric potential derived from the requirement of current continuity in a magnetic mirror field geometry.

More distributed voltages parallel to B may be ascribed to the existence of anomalous resistivity introduced by plasma wave turbulence due to current instability. On the microscopic level these waves may turn out to have the structure of double layers, i.e. small potential jumps concentrated in solitary waves (Temerin, 1982, Boström et al., 1987). Thousands of such potential jumps of order kT/e may constitute the large scale voltages of several kV suggested to exist above auroral arcs and extending over several 1000 km.

Parallel potential drops, inserted at a few 1000 km altitude into a magnetospheric-ionospheric current system, are the consequences of a self-generated resistance, an extra mode of energy conversion, additional to the ohmic resistance of the lower ionosphere where the currents close. Currents interchanged between the outer magnetosphere and the ionosphere-atmosphere transport momentum and energy. They propagate presumably with the Alfvén speed, mostly parallel to the magnetic field. When such waves hit the ionosphere, they are partially reflected. Geomagnetic pulsations of a few minutes duration (Pi2) are witnesses thereof (e.g Baumjohann and Glassmeier, 1984). If the associated Birkeland currents become dissipative above the ionosphere (for one or several reasons as sketched in Fig 2), partial reflection and transmission occur already above the ionospheric end resistance of the circuit. Several concepts have been developed for this process. They are sketched in Fig. 3.

Fig 3 shows how incident Alfvén waves when modified in speed at lower altitudes develops parallel electric field components capable of accelerating particles. Fig 3c gives the envelope of $E_{||}$ as a function of altitude (after Temerin et al., 1986). The region of parallel current may also act as a wave emitter. A standing wave situation between the $E_{||}$ region and the magnetosphere, on one hand, and the ionosphere on the other hand, has been proposed by Haerendel (1980, 1989). Fig 3d shows how the acceleration region propagates into the current circuit from which it extracts stored magnetic energy. Besides the standing Alfvén wave pattern, by which the flow of magnetic energy is organized, the model features the evacuation of ionospheric plasma in the acceleration region (bell shaped area).

Transverse ion acceleration

Besides being affected by the same processes as those for electrons mentioned in the previous section, ions are more readily affected by transverse energization processes because of their larger Larmor radii. Furthermore, the large variety of ion species may cause other effects e.g. ion-ion two-stream instabilities.

Transverse energization may occur as a result of a resonant or nonresonant process. Resonant acceleration of particles occur when the force-(electric) field varies with the characteristic frequencies of the plasma, e.g. the cyclotron frequency. Particles may then gain energy even for a zero time-average electric field (E). Nonresonant acceleration of particles takes place in a conservative force field when the particles gain kinetic energy at the expense of their potential energy. In a field-aligned potential, $U_{||}$, the charged particles

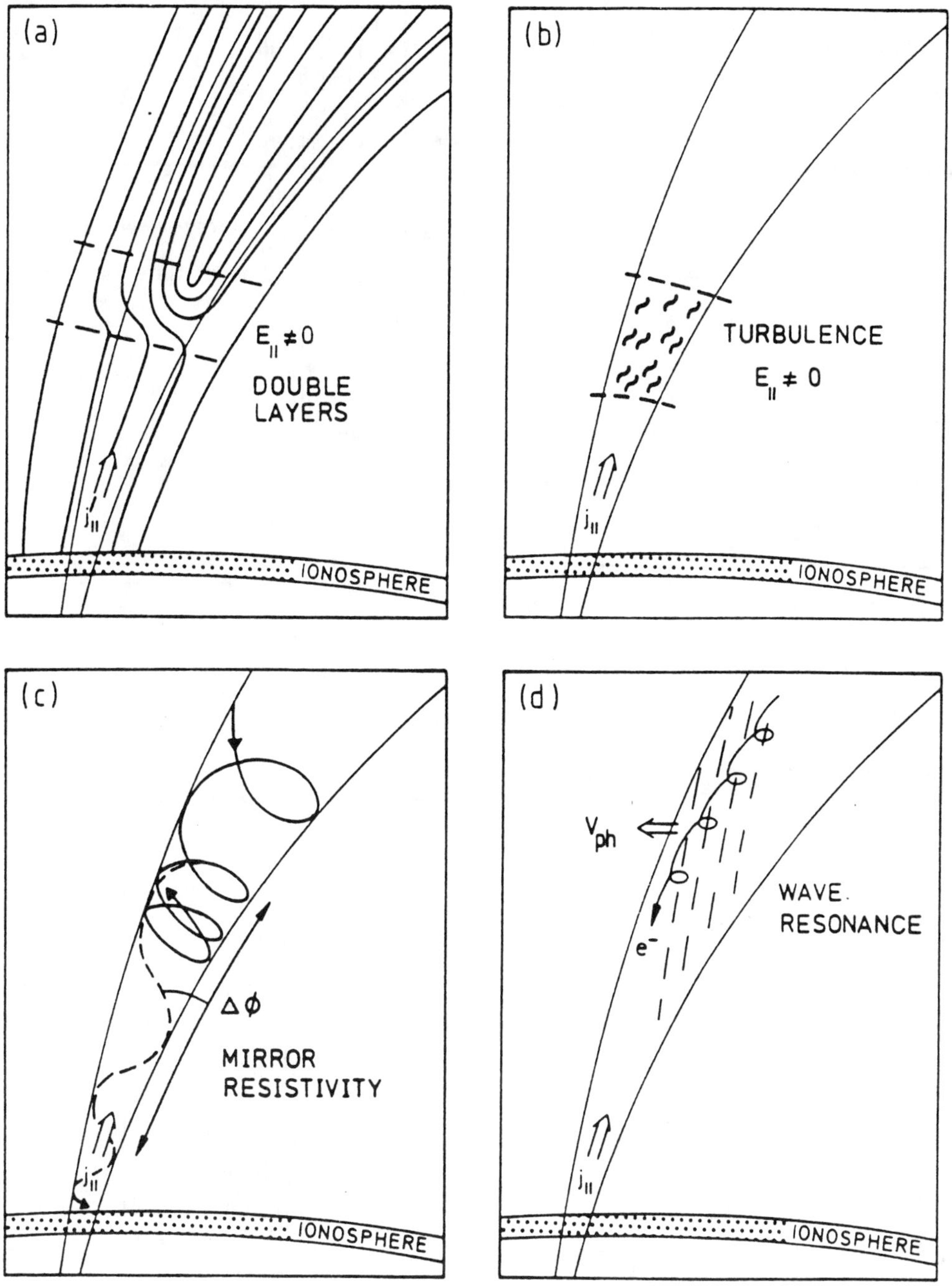

Figure 2 Four concepts of the auroral acceleration process: (a) - (c) show acceleration by parallel electric fields, (d) acceleration by wave-particle interaction. The electric equipotential structure with double layers is illustrated in (a): (b) shows the anomalous resistivity caused by current instability; (c) gives the electric fields derived from the requirement of current continuity in a mirror field geometry.

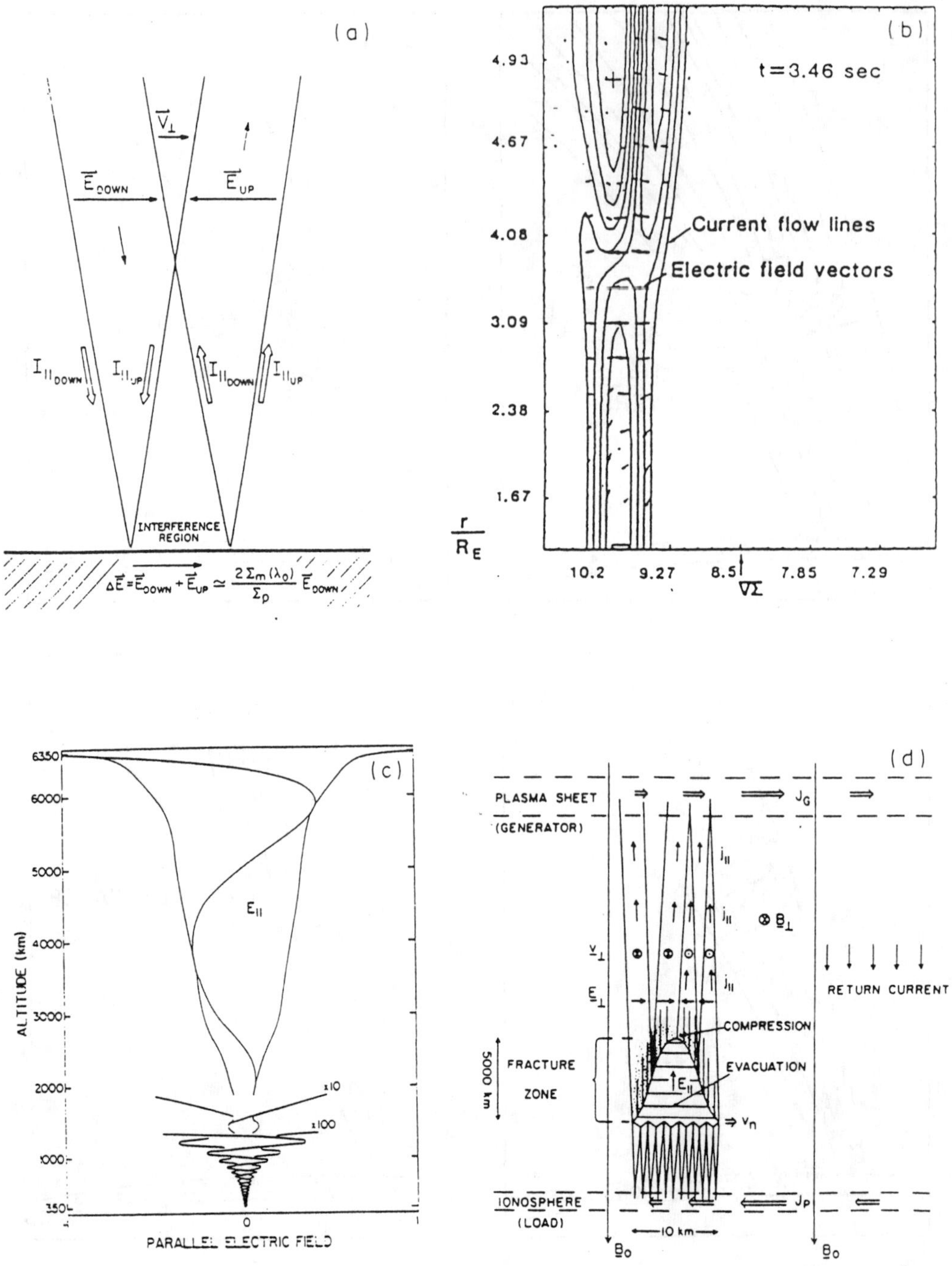

Figure 3 (a) Reflection of Alfvén waves at the ionosphere in a dissipation-free propagation medium (Mallinckrodt and Carlson, 1978). (b) Simulation of currents and electric fields in a region of instability (Lysak and Dum, 1983). (c) Height variation of Ell in an incident kinetic Alfvén ion-cyclotron wave (temerin et al., 1986). (d) standing Alfvén wave pattern between the acceleration region and the ionosphere and generator regions, respectively (Haerendel, 1989).

(especially the ones with pitch angles close to 0°) can easily gain a kinetic energy up to $qU_{||}$. However, a steady state transverse potential $U_{\perp}$ limits the acceleration of particles transverse to the magnetic field to velocities less than $2E_{\perp}/B$ ($V_{average} = E_{\perp}/B$ = convection velocity). To increase the kinetic energy above the convection speed some form of elastic scattering is required, e.g. in transverse temporal/spatial gradients of the electric potential ($\Delta E_{\perp}$). The nonresonant energization may act in two ways: either as a stochastic energization or as a non-gyrotropic walk in transverse potential gradients. In both cases the energization acts as an ion pickup process that "pumps up" the magnetic moment of the ions. The general principles for a time-dependent increment of the magnetic moment was introduced by Cole (1976). Later, Lennartsson (1980) and Borovsky (1984) proposed a transverse energization by non-gyrotropic jumps in electrostatic shocks /oblique double layers. Arguments for a stochastic transverse energization were proposed by Temerin (1986) to explain the ion bowl distribution. In fact it was demonstrated by Lundin and Hultqvist (1989) that the transverse electric fluctuations observed by Viking were alone sufficient to explain a massive outflow from the topside ionosphere.

Resonant heating has been proposed due to e.g. electrostatic ion cyclotron waves (e.g. Ungstrup et al., 1979; Klumpar, 1979), lower hybrid waves (Chang and Coppi, 1981), and Alfvén waves (Chang et al., 1986). The ion bowl distribution is proposed to be generated by an electromagnetic ion-cyclotron resonance (Retterer et al., 1987; Chang et al., 1986). As pointed out by Ball and André (1991) the rarely observed enhancement at the ion cyclotron frequency of the wave spectrum is not a necessary condition for resonant heating, wave absorption within a broad-band spectrum may be sufficient. There is a rich literature on this topic and it would be too far-reaching to fully survey the various physical processes involved. There are, however, strong arguments for resonant heating as playing a very important role in the energization of ionospheric ions.

The outflow of energized ionospheric ions appears in a form that varies from transversely accelerated ions (conics) to pure parallel accelerated ions (beams). Pure parallel acceleration appears to be very rare, though. In fact, the ion beams are almost always characterized by a higher perpendicular than parallel temperature, implying that transverse energization is basically always operating. This is still an idealized picture of an acceleration process that is likely to be highly dynamical, at least at times such as those illustrated in Fig. 4 which shows Viking electron and ion data taken from a pass over the dayside oval at altitudes of about 12 000 km. Although one may here also distinguish "traditional" field-aligned and transverse ion energization regions, the electrons data shows such immense structure and variability that it becomes virtually impossible to single out individual processes.

While the altitude of strong parallel acceleration of ions is usually considered to be above a few thousand kilometers, i.e. in the region of downward accelerated electrons, transverse acceleration of ions evidently starts already at altitudes just above ≈500 km (Whalen et al., 1978), with transverse energization in the hundreds of eV-range found above 1000 km (Klumpar et al., 1979). The good correlation between the occurrence frequency of ion conics and upgoing electron beams (Thelin et al., 1990) also implies that the process responsible for the transverse energization of ions are related with the same process energizing electrons. In fact, narrow upgoing electron beams coincident with transversely accelerated ions were observed to altitudes below 2000 km from Viking (Lundin and Eliasson, 1991). The production of very dynamic wide energy downward beams of electrons observed at rocket altitudes (e.g. Temerin et al., 1986) is possibly related with the same physical process.

3. SCIENTIFIC CONTRIBUTIONS FROM FREJA

The apogee of Freja, at ≈1750 km, implies operations at the lower edge, or below, the mid-altitude auroral energization region.Thus, the Freja mission differs from e.g Viking and other mid-altitude mission that emphasizes in situ measurements within the field-aligned acceleration region. The capability of Freja lies elsewhere. It is uniquely suited to study the impact of the auroral acceleration process on the structure of the topside ionosphere. Large holes in plasma density, created by cavitation processes in the lower edge of the auroral energization region (e.g. Haerendel, 1989), are expected to be seen regularly along Freja´s trajectory at auroral latitudes (Fig 5). It appears likely that at least some of these cavitations are due to the heating and upward loss of electrons and ions simultaneously, possibly due to the strong low-frequency electric turbulence observed within these cavities (Hultqvist et al., 1988, Lundin et al., 1990).

Plasma density cavities have strong effects on the auroral processes. They establish and maintain the conditions for current instability, they constitute a "memory" in (or imprint on) the magnetosphere-ionosphere interface of the recent occurrence of accelerations, they modify the transmission of energy carried by e.g. Alfvén waves towards the ionosphere, and they are the source region of auroral kilometric radiation. The high-density walls surrounding the cavities may efficiently trap plasma waves and lead to the

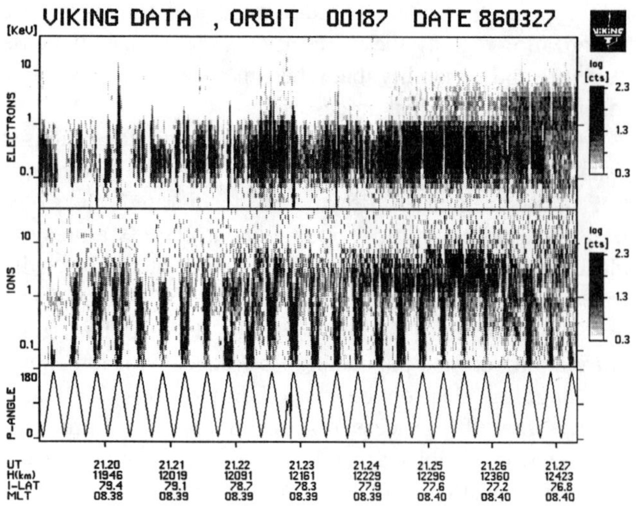

Figure 4 An example of a Viking energy-time spectrogram of electrons (upper panel) and ions (mid panel) versus pitch angle (lower panel) illustrating how extremely variable the auroral acceleration process may be.

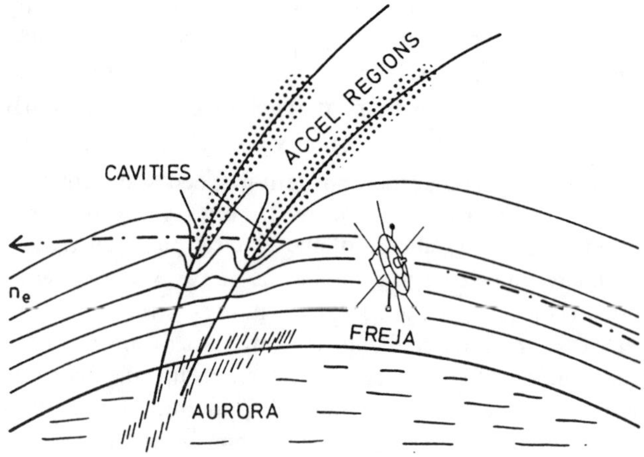

Figure 5 Density depletion regions above auroral arcs as they would be encountered by Freja.

build-up of high wave intensities and non-linear decay instabilities. Thus, Freja scans a region of utmost importance for the production and modification of electrostatic and electromagnetic plasma waves. Particularly interesting is the analysis of reflection, transmission, and damping of Alfvén waves incident from above (e.g. Lysak and Dum, 1983). This is achieved by measuring the Poynting flux and by comparing $\delta E/\delta B$ with the local Alfvén speed. In the sketch contained in Fig 3d, one will often find Freja in what is called the "interference region".

Perhaps the most significant expected advance in auroral physics, which is made possible by the combination of greatly enhanced temporal resolution of the in-situ measurements ($\approx$10 ms for particles) with high spatial resolution in auroral imaging ($\approx$2 km), is the study of the plasmaphysical reason for the omnipresent fine-structure of auroral arcs. For instance, one of the finest structure, namely auroral rays or curls, have typical spacings of 5-10 km at 100 km altitude. Since they move with speeds of several tens of km/s, their passage time through a point fixed in space would be typically 0.2 - 0.3 s. The Freja spacecraft will proceed by less than 2 km during this time interval. Thus, a detailed analysis of the electric field, current, and hot and cold plasma distributions inside and around such structures will become possible, bringing up closer to a complete understanding of the aurora. Similarly, the fine-structure of discrete auroral arcs, frequently with individual arc segments in the hundred meter range, are hitherto only resolved by a relatively limited amount of sounding rocket measurements at altitudes generally much

TABLE 2: AURORAL STRUCTURES, STABILITY AND FREJA TRAVERSAL TIMES

	Diffuse	"Inverted V"	Quiet arc	Active arc	Ray	Vortex	Patch
Aurora							
Width-lat (km)	500	100	30	10	0.5	3	10
Long. ext.(km)	>1000	>1000	1000	1000	0.5	3	10
Aurora stability (s)	>1000	>100	>100	10	5	0.5	0.1
Traversal times							
Latitude (*)	100 s	20 s	6 s	2 s	0.1 s	0.5 s	2 s
Longitude (**)	300 s	100 s	10 s	6 s	0.1 s	0.5 s	2 s

(*) FREJA traversing perpendicularly or at constant longitude over the auroral structures.

(**) FREJA traversing obliquely or at constant latitude over the auroral structures.

below 1000 km. Freja will not only radically improve the data base for such fine-structure measurements, it will also access an altitude range hitherto not attended by instruments utilizing the novel fine-structure measurement technique. Table 2 summarizes typical auroral structures and the corresponding traversal times of Freja near apogee. The table illustrates that the on-board in-situ plasma instruments have sufficient temporal/spatial resolution to resolve the listed structures. Compare for instance the typical auroral ray forms with scale sizes of several hundred meters with e.g. the particle instrument spatial resolution of $\approx$100 meter. The Freja spatial resolution is visualized in Fig. 6. Notice that the smallest auroral structures are marginally resolved by the UV-imager. However, the UV-imager is yet important for identifying boundaries and it is indeed crucial for separating temporal - spatial characteristics.

Observations from sounding rockets (e.g. McFadden et al., 1987, 1990) have stressed the necessity of high-time resolution measurements during structured auroral precipitation. Fig. 7 shows an example of a 6 Hz coherent oscillations of the field-aligned electrons for different energies within an "inverted V". These are the electrons

1.4. SCIENTIFIC CONTRIBUTIONS FROM FREJA

SCIENTIFIC OBJECTIVE

To study the impact of the auroral energization process on the topside ionosphere with greatly enhanced temporal/spatial resolution.

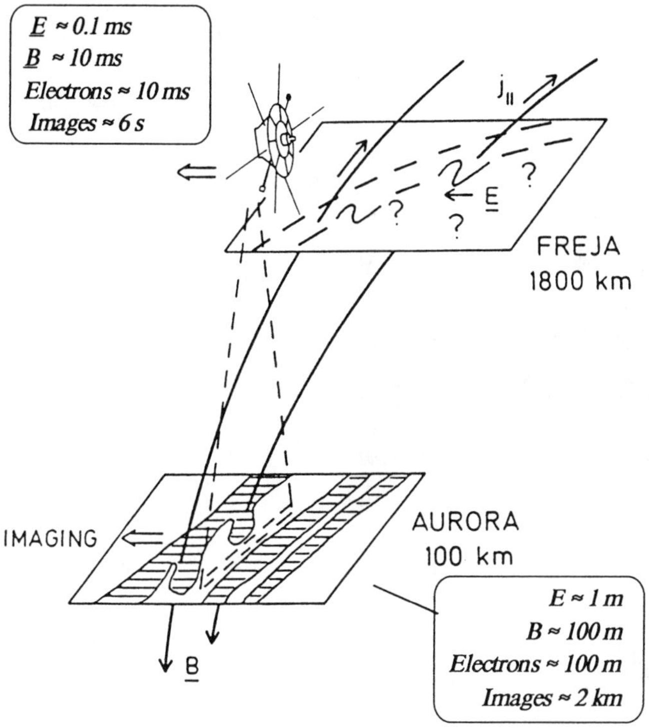

Figure 6 Relationship between the visual auroral fine structure obtained with the Freja imager, and the signatures obtained with the Freja high time resolution in-situ measurements in the acceleration region itself.

which produce flickering aurora, electrons which were in this case probably accelerated by Alfvén waves. The difference in time-of-arrival of electrons at different energies can be translated into a source distance of the order 1 R_E. The oxygen cyclotron frequency is 6 Hz at slightly higher altitude than the source distance, and the energetic electrons can be considered trapped in an Alfvén ion-cyclotron wave at the cyclotron resonance point of the wave and then accelerated with the wave as it moves downward. The Freja wave and particle instruments is capable of making similar measurements as those shown in Fig. 7 on many passes through the auroral oval.

Another important characteristics related with modulations of electron fluxes are the so called "electron conics", first

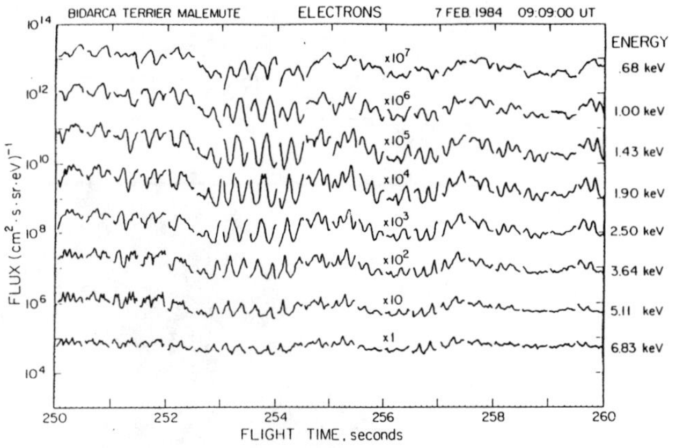

Figure 7 Coherent oscillations observed in the field-aligned electrons at eight different energies below an "inverted-V" (McFadden et al., 1987)

discovered on DE-1 (Menietti and Burch, 1985) and later from Viking interpreted as fluctuations of the field-aligned potential in the few Hz range (André and Eliasson, 1992). "Electron conics" are observed above the field-aligned acceleration region where a strong low-frequency electric field noise (≈1 Hz) is also observed. Thus, one may argue that the effect of the wave resonance observed in Fig. 7 is also related with "electron conics" above the field-aligned acceleration region.

Fig. 8 shows an energy-time spectrogram plot for electrons from a pass over an auroral arc, a small scale "inverted V" with spatial dimension about 30 km. The data is from the same flight as that contained in Fig. 7 (McFadden et al., 1990). Notice here the rich structure within both the electron energy peak (modulatrions near 0.5 - 1 Hz) as well as the presence of strong field-aligned low-energy electron fluxes below the peak energy. Again, this suggests strong dynamics of the acceleration process with cold electrons being injected within the acceleration region in a modular way. Clearly this is another example demonstrating the need for high-time resolution measurements to properly understand the microphysics of the auroral acceleration process. Freja, encountering a similar arc structure as that in Fig. 8, is capable of resolving the auroral arc in quite fine detail. We have illustrated this in Table 3 by giving the number of full particle and wave distributions sampled above an auroral arc corresponding to 30 km in width at 100 km altitude.

4. FREJA DATA TAKING STRATEGY

The key features of the measurements are their high temporal and spatial resolution, made possible by the high data rates which Freja provides. For instance, Freja is able

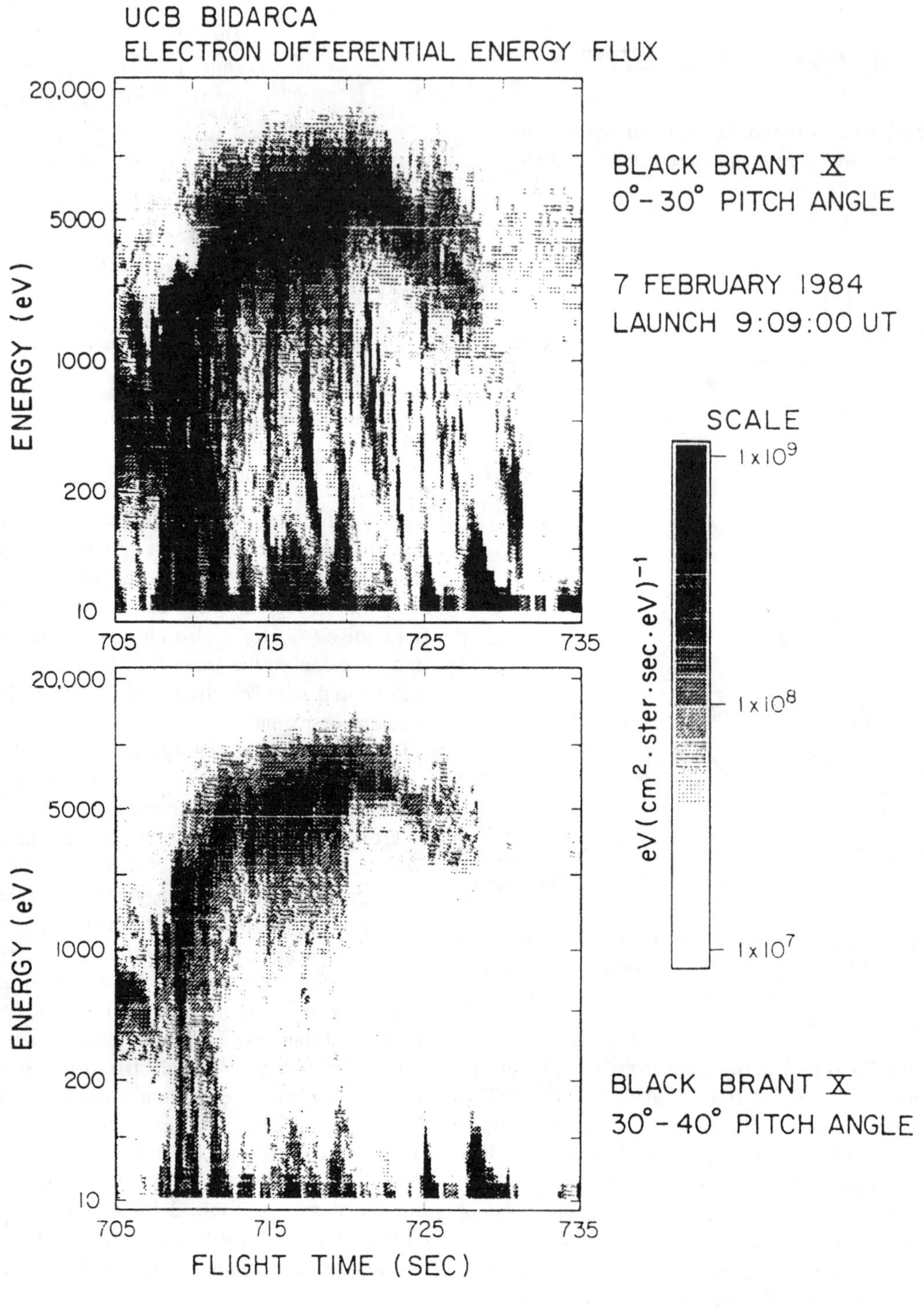

Figure 8 Energy-time spectrogram plot for electrons from a sounding rocket traversal of field lines connected to an "inverted-V" discrete auroral arc (McFadden et al., 1990).

TABLE 3: FREJA SAMPLING RESOLUTION OF A 30 KM WIDE AURORAL ARC (FIG. 8)

		No of samples
Hot/cold plasma:		
2D Ions (4m/q)	0.001 - 10 keV	36
3D Ions (4 m/q)	0.001 - 10 keV	3
2D Electrons (TESP)	0.01 - 30 keV	90
2D Electrons (MATE)	0.1 - 115 keV	600
2D Cold electrons/ions	0.1 - 300 eV	600
Fields:		
2D vector E-field	E-probe	6000
3D vector E-field	electron beam	360
3D vector B-field	fluxgate	750
7 component waves (E, B, n_e)	1 Hz - 4 MHz	4800

to provide a data rate of 524 kilobits per second (kbps) for most 10-15 min. passes over the two ground stations Esrange in Kiruna (Sweden) and Prince Albert (Canada). On board memories (>15 Mbytes) makes it possible to provide burst mode snapshots with >2 Mbps for about 60 seconds. The on board memory can also be used for storing data along orbits out of reach of ground stations (compressed mode) alternatively for providing overview data (mainly F1 and F2 instruments). The Freja data taking is summarized in Fig. 9.

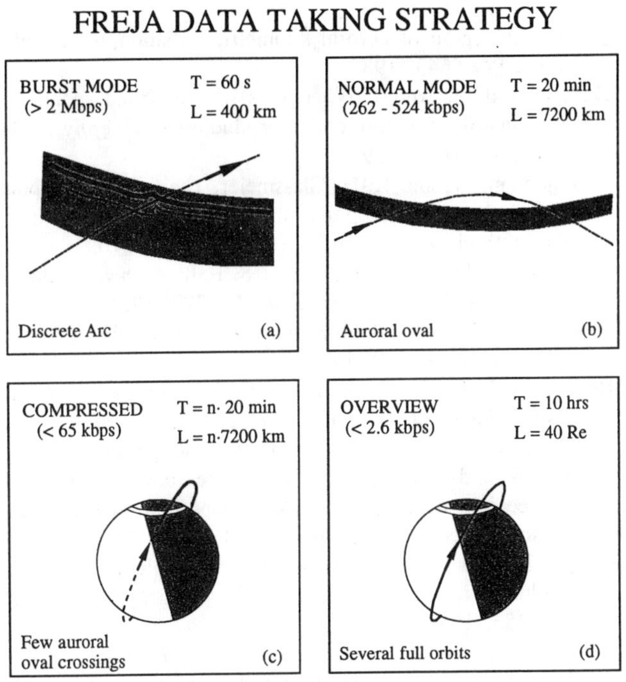

Figure 9 Overview of the Freja data taking strategy showing (a) the burst (memory) mode sequence (≈60 s data taking) and (b) the normal data taking mode when Freja is in contact with the ground TM station. Panels (c) and (d) represent memory mode sequences of either a few oval crossing (c) or several orbits with full data coverage (d) used primarily by F1 (dc E-field) and F2 (vector magnetometer).

CONCLUSIONS

Many problems on the microphysics of the auroral acceleration region require higher time resolution measurements than has been available on satellites before. By a combination of high telemetry rate and on-board memories Freja is able to provide one to several orders of magnitude higher time resolution than previous satellites. Sounding rockets have provided some high resolution in several cases, but the small data set available from them does not provide sufficient statistics to determine which processes are typically important to auroral acceleration, each flight apparently leading to even more questions. Freja will hopefully give a large data base on microphysics phenomena, thus enabling a statistical analysis of such phenomena for the first time. Freja will also provide some higher quality measurements (longer booms, faster distribution function measurements, electron beam instrument) and the ability to reprogram instruments in-flight so as to answer new questions about the microphysics as they come up. The on-board UV-imager may play an important role in this context to distinguish between spatial-temporal phenomena as well as to identify the type of auroral regions within which they occur. This will put micro-scale phenomena in the context of meso-scale structures usually more familiar to us, such as discrete arcs, auroral patches and auroral rays.

REFERENCES

André, M. and L. Eliasson, Electron acceleration by low-frequency electric field fluctuations: Electron conics, *Geophys. Res. Lett., 19*, 1073, 1992.

Albert, R.D., Nearly monoenergetic electron fluxes detected during a visible aurora, *Phys. Rev. Lett.*, 18, 369-372, 1967.

Alfvén, H., On the theory of magnetic storms and aurorae, *Tellus,* 10, 104, 1958.

Alfvén, H., and C.-G. Fälthammar, *Cosmical Electrodynamics, Fundamental Principles*, Oxford University Press, 1963.

André, M. and L. Eliasson, Electron acceleration by low-frequency electric field fluctuations: Electron conics, *Geophys. Res. Lett.*,19, 1073, 1992.

Bahnsen, A., B. M. Pedersen, M. Jespersen, E. Ungstrup, L. Eliasson, J. S. Murphree, R. D. Elphinstone, L. Blomberg, G. Holmgren, and L. J. Zanetti, Viking observations at the source region of auroral kilometric radiation, *J. Geophys. Res.*, 94, 6643, 1989.

Ball, L. and M. André, What parts of broadband spectra are responsible for ion conic production?, *Geophys. Res. Lett.*, 18, 1683, 1991

Baumjohann, W. and K.-H. Glassmeier, The transient response mechanism and Pi2 pulsations at substorm onset: Review and outlook, *Planet. Space Sci.*, 32, 1361-1370, 1984.

Bingham, R., D. A. Bryant, and D. S. Hall, A wave model for the aurora, *Geophys. Res. Lett.*, 11, 327, 1984.

Block, L. P., Potential double layers in the ionosphere, *Cosmic Electrodyn.*, **3**, 349, 1972.

Borovsky, J. E., The production of ion conics by oblique double layers, *J. Geophys. Res.*, *89*, 2251, 1984.

Boström, R., H. Koskinen, and B. Holback, Low frequency waves and solitary structures observed by Viking, in Proceedings of the 21st ESLAB Symposium on Small-Scale Plasma Processes, *ESA SP-291*, 185, 1987.

Chang, T., and B. Coppi, Lower hybrid acceleration and ion evolution in the suprauroral region, *Geophys. Res. Lett.*, **8**, 1253, 1981.

Chang, T., G. B. Crew, N. Hershkowitz, J.R. Jasperse, J.M. Retterer, and J.D. Winningham, Transverse acceleration of oxygen ions by electromagnetic ion cyclotron resonance with broad band left-hand polarized waves, *Geophys. Res. Lett.*, 13, 636, 1986.

Chappell, C. R., T. E. Moore, and J. H. Waite, Jr, The ionosphere as a fully adequate source of plasma for the earth´s magnetosphere, *J. Geophys. Res.*, 92, 5896, 1987.

Chiu, Y. T., and M. Schulz, Self-consistent particle and parallel electrostatic field distribution in the magnetospheric-ionospheric auroral region, *J. Geophys. Res.*, **83**, 629, 1978.

Cole, K. D., Effects of crossed magnetic and (spatially dependent) electric fields on charged particle motion, *Planet. Space Sci.*, 24, 515, 1976.

Evans, D. S., The observations of a near monoenergetic flux of auroral electrons, *J. Geophys. Res.*, 73, 2315, 1968.

Evans, D. S., Precipitating electron fluxes formed by a magnetic field aligned potential difference, *J. Geophys. Res.*, 79, 2853, 1974.

Frank, L. A., and K. L. Ackerson, Observations of charged particle precipitation into the auroral zone, *J. Geophys., Res.*, **76**, 3612, 1971.

Fälthammar, C.-G., Magnetic-field-aligned electric fields, *ESA Journal, 7*, 385, 1983.

Goertz, C. K., and R. W. Boswell, Magnetosphere-ionosphere coupling, *J. Geophys. Res.*, 84, 7239, 1979.

Gurnett, D. A., and L. A. Frank, Observed relationship between electric field and auroral particle precipitation, *J. Geophys. Res.*, 78, 145, 1973.

Harendel G., E. Rieger, A. Valenzuela, H. Föppl, H.C. Stenbaek-Nielsen, E.M. Wescott, First observation of electrostatic acceleration of Barium ions in the magnetosphere, European Space Agency, ESA-SP 115, 203-211, 1976

Haerendel, G., Auroral particle acceleration - An example of a universal plasma process, *ESA Journal,* 4, 197, 1980.

Haerendel, G., An Alfvén wave model of auroral arcs, in *High-Latitude Space Plasma Physics*, Edited by B. Hultqvist and T. Hagfors, Plenum Press, New York, p. 515, 1983.

Haerendel, G., Cosmic linear accelerators, in *Proc. of the international school and workshop on plasma astrophysics*, pp 37-44, ESA SP-285, ESA, Noordwijk, 1989.

Hoffman, R. A., and D. S. Evans, Field-aligned electron bursts at high latitude observed by OGO-4, *J. Geophys. Res.*, 73, 6201, 1968.

Hultqvist, B., On the production of a magnetic-field-aligned electric field by the interaction between the hot magnetospheric plasma and the cold ionosphere, *Planet. Space Sci.*, 19, 749, 1971.

Hultqvist, B., On the acceleration of electrons and positive ions in the same direction along magnetic field lines by parallel electric fields, *J. Geophys. Res.*, 93, 9777, 1988.

Ijima, T., and T. A. Potemra, Large-scale characteristics of field aligned currents associated with substorms, *J. Geophys. Res.*, 83, 559, 1978.

Klumpar, D. M., Transversely accelerated ions: An ionospheric source of hot magnetospheric ions, *J. Geophys. Res.*, 84, 4229, 1979.

Knight, L., Parallel electric fields, *Planet. Space Sci.*, 21, 741, 1973.

Kindel J.M., and C.F. Kennel, Topside current instabilities, *J. Geophys. Res.*, 76, 3055-3077, 1971

Lennartsson, W., On the magnetic mirroring as the basic cause of parallel electric fields, *J. Geophys. Res.*, 81, 5583, 1976.

Lennartsson, W., On the consequences of the interaction between the auroral plasma and the geomagnetic field, *Planet. Space Sci.*, 28, 135, 1980.

Lundin, R., and I. Sandahl, Some characteristics of the parallel electric field acceleration of electrons over discrete auroral arcs as observed from two rocket flights, *ESA SP-135*, 125, 1978.

Lundin, R. and B. Hultqvist, Ionospheric plasma escape by high-altitude electric fields: Magnetic moment pumping, *J. Geophys. Res.*, 94, 6665-6680, 1989.

Lundin, R., G. Gustafsson, A. I. Eriksson, and G. Marklund, On the importance of high-altitude low-frequency electric fluctuations for the escape of ionospheric ions, *J. Geophys. Res.*, 95, 5905, 1990.

Lundin R. and L. Eliasson, Auroral energization processes, *Ann. Geophysicae*, 9, 202-223, 1991.

Lyons, L. R., D. S. Evans, and R. Lundin, An observed relation between magnetic field aligned electric fields and downward electron fluxes in the vicinity of auroral forms, *J. Geophys. Res.*, 84, 457, 1979.

Lysak, R.L., M.K. Hudson, and M. Temerin, Ion heating by strong electrostatic ion cyclotron turbulence, *J. Geophys.*

Res., 85, 678, 1980.

Lysak, R.L. and C.T. Dum, Dynamics of magnetospheric-ionospheric coupling including turbulent transport, *J. Geophys. Res.*, 88, 365, 1983

Mallinckrodt, A.J., and C.W. Carlson, Relations between transverse electric fields and the field aligned currents, *J. Geophys. Res.*, 83, 1426-1432, 1978.

McFadden J. P., C. W. Carlson, M. H. Boehm, and T. J. Hallinan, Field-aligned electron flux oscillations that produce flickering aurora, *J. Geophys. Res.*, 92, 11133, 1987.

McFadden, J.P., C.W. Carlson, and M. Boehm, Structure of an energetic narrow discrete arc, *J. Geophys. Res.*, 95, 6533-6547, 1990.

McIlwain, C. E., Direct measurement of particles producing visible aurora, *J. Geophys. Res.*, 65, 2727, 1960.

Menietti, J. D. and J. L. Burch, "Electron conic" signatures observed in the nightside auroral zone an over the polar cap, *J. Geophys. Res.*, 90, 5345, 1985.

Mizera, P. J., and J. F. Fennell, Signatures of electric fields from high and low altitude particle distributions, *Geophys. Res. Lett.*, 4, 311, 1977.

Mozer, F. S., C. W. Carlson, M. K. Hudson, R. B. Torbert, B. Parady, J. Yatteau, and M. C. Kelley, Observations of paired electrostatic shocks in the polar magnetosphere, *Phys. Rev. Lett*, 38, 292, 1977.

Papadopoulos, K., and T. Coffey, Anomalous resistivity of the auroral plasma, *J. Geophys. Res.*, 79, 1558, 1974.

Papadopoulos, K., A review of anomalous resistivity for the ionosphere, *Rev. Geophys. Space Phys.*, 15, 113, 1977.

Reiff, P. H., H. L. Collin, J. D. Craven, J. L. Burch, J. D. Winningham, E. G. Shelley, L. A. Frank, and M. A. Friedman, Determination of auroral electrostatic potentials using high- and low-altitude particle distributions, *J. Geophys. Res.*, 93, 7441, 1988.

Retterer, J. M., T. Chang, G. B. Crew, J. R. Jasperse, and J. D. Winningham, Monte Carlo modeling of ionospheric oxygen acceleration by cyclotron resonance with broad-band electromagnetic turbulence, *Phys. Rev. Lett.*, 59, 148, 1987.

Shelley, E. G., R. G. Johnson and R. D. Sharp, Satellite observations of an ionospheric acceleration mechanism, *Geophys. Res. Lett.*, **3**, 654, 1976.

Temerin, M., K. Cerny, W. Lotko, and F. S. Mozer, Observations of double layers and solitary waves in auroral plasma, *Phys. Rev. Lett.*, 48, 1175, 1982.

Temerin, M., Evidence for a large bulk ion conic heating region, Geophys. Res. Lett., 13, 1059, 1986.

Temerin, M., J. McFadden, M. Boehm, C. W. Carlson, and W. Lotko, Production of flickering aurora and field aligned electron flux by electromagnetic ion cyclotron waves, *J. Geophys. Res.*, 91, 5769, 1986.

Thelin, B., and R. Lundin, Upflowing ionospheric ions and electrons in the cusp-cleft region, *J. Geomag. Geoelectr.*, 42, 753, 1990.

Thelin, B., B. Aparicio, and R. Lundin, Observations of upflowing ionospheric ions in the mid-altitude cusp/cleft region with the Viking satellite, *J. Geophys. Res.*, 95, 5931, 1990.

Ungstrup, E., D. M. Klumpar, W. J. Heikkila, Low altitude acceleration of ionospheric ions, *J. Geophys. Res.*, 84, 4289, 1979.

Whalen, B.A., W. Bernstein, and P.W. Daley, Low altitude acceleration of ionospheric ions, *Geophys. Res. Lett.*, 5, 55, 1978.

R. Lundin, Swedish Institute of Space Physics, Box 812, S-981 28 Kiruna, Sweden.

G. Haerendel, Max-Planck-Institut für extraterrestrische Physik, D-8046 Garching, b. München, Germany.

Stormtime Ring Current and Radiation Belt Ion Transport: Simulations and Interpretations

Margaret W. Chen, Michael Schulz[1], Larry R. Lyons, and David J. Gorney

Space and Environment Technology Center, The Aerospace Corporation, El Segundo, California

We use a dynamical guiding-center model to investigate the stormtime transport of ring current and radiation-belt ions. We trace the motion of representative ions' guiding centers in response to model substorm-associated impulses in the convection electric field for a range of ion energies. Our simple magnetospheric model allows us to compare our numerical results quantitatively with analytical descriptions of particle transport, (e.g., with the quasilinear theory of radial diffusion). We find that 10–145-keV ions gain access to $L \sim 3$, where they can form the stormtime ring current, mainly from outside the (trapping) region in which particles execute closed drift paths. Conversely, the transport of higher-energy ions ($\gtrsim$ 145 keV at $L \sim 3$) turns out to resemble radial diffusion. The quasilinear diffusion coefficient calculated for our model storm does not vary smoothly with particle energy, since our impulses occur at specific (although randomly determined) times. Despite the spectral irregularity, quasilinear theory provides a surprisingly accurate description of the transport process for $\gtrsim$ 145-keV ions, even for the case of an individual storm. For 4 different realizations of our model storm, the geometric mean discrepancies between diffusion coefficients D_{LL}^{sim} obtained from the simulations and the quasilinear diffusion coefficient D_{LL}^{ql} amount to factors of 2.3, 2.3, 1.5, and 3.0, respectively. We have found that these discrepancies between D_{LL}^{sim} and D_{LL}^{ql} can be reduced slightly by invoking drift-resonance broadening to smooth out the sharp minima and maxima in D_{LL}^{ql}. The mean of the remaining discrepancies between D_{LL}^{sim} and D_{LL}^{ql} for the 4 different storms then amount to factors of 1.9, 2.1, 1.5, and 2.7, respectively. We find even better agreement when we reduce the impulse amplitudes systematically in a given model storm (e.g., reduction of all the impulse amplitudes by half reduces the discrepancy factor by at least its square root) and also when we average our results over an ensemble of 20 model storms (agreement is within a factor of 1.2 without impulse-amplitude reduction). We use our simulation results also to map phase-space densities f in accordance with Liouville's theorem. We find that the stormtime transport of $\gtrsim$ 145-keV ions produces little change in $\bar{f}$ the drift-averaged phase-space density on any drift shell of interest. However, the stormtime transport produces a major enhancement from the pre-storm phase-space density at energies $\sim$ 30–145 keV, which are representative of the stormtime ring current.

1. INTRODUCTION

This work is an outgrowth of our continuing study of energetic charged-particle transport in the magnetosphere. The study began as an effort to understand the development of the stormtime ring current but has expanded to include the radial diffusion of radiation-belt ions. Our study entails a guiding-center simulation of particle motion in the presence of a succession of substorm-associated impulsive enhancements in the magnetospheric convection electric field. We synthesize our model storms by means of a random-number generator and apply them to a simple magnetospheric model in order to make the storm effects realistic but mathematically analyzable.

We have found, in agreement with *Lyons and Williams* [1980], that the access of $\sim$ 10–145 keV ions (having first adiabatic invariants $\mu \sim$ 1–13 MeV/G) to the region ($L \sim 3$) where they can form the stormtime ring current occurs largely as a consequence of the enhanced mean-value of the convection electric field rather than from its impulsive character. Indeed, most of the particles in this energy range that reach $L \sim 3$ in our model storm turn out to have been transported there from outside the (trapping) region in which particles execute closed drift paths. Conversely, the transport of higher-energy particles (having, for example, $\gtrsim$ 145 keV at $L \sim 3$) turns out to resemble radial diffusion across closed drift paths [cf. *Lyons and Schulz*, 1989].

By having formulated the model storm in an eas-

[1]Now at Lockheed Research Laboratory, Palo Alto, California.

Space Plasmas: Coupling Between Small
and Medium Scale Processes
Geophysical Monograph 86

ily analyzed way, we are able to compare the radial diffusion coefficients obtained from our guiding-center simulation with the predictions of quasilinear theory [e.g., *Fälthammar*, 1965; *Cornwall*, 1968] and various refinements thereof. We find that the quasilinear diffusion coefficient calculated for any of our model storms shows a remarkably unsmooth variation with particle energy because the impulses occur at specific (although randomly determined) times. Despite this, quasilinear theory provides a surprisingly accurate description of the transport process for $\gtrsim$ 145-keV ions, even for the case of an individual storm. As expected, the agreement becomes even better when we reduce the impulse amplitudes systematically in a given model storm, and also when we average our results over an ensemble of model storms constructed by the same (random) method.

Of course, a radial diffusion coefficient is not defined for transport from open to closed drift trajectories, such as we have found to occur for ions having 30 keV $\lesssim E \lesssim 145$ keV. For these energies, however, we find a major increase in the drift-averaged phase space density $\bar{f}$ from its pre-storm value upon mapping f in accordance with Liouville's theorem. Particle energies 30–150 keV have been shown by many observational studies [e.g., *Frank*, 1967; *Smith and Hoffman*, 1974; *Williams and Lyons*, 1974; *Lyons and Williams*, 1976; *Hamilton et al.*, 1988] to be representative of the stormtime ring current as a whole. In contrast, we find little change in drift-averaged phase-space density as a consequence of stormtime transport for ions having $\mu \gtrsim 13$ MeV/G ($E \gtrsim 145$ keV at $R = 3$), for which the transport is diffusive.

2. Field Model

The magnetic field model that we use in this study is obtained by adding a uniform southward field $\Delta\mathbf{B}$ to the geomagnetic dipole field. We invoke this simple field configuration because it enables us to make direct comparisons between the simulated transport and previous analytical formulations. An advantage of our model over a purely dipolar field is the presence of a quasi-magnetopause at the boundary between closed and open field lines (see Figure 1). The equation of a field line in this model is

$$[1 + 0.5(r/b)^3]^{-1}(r/R_E)\csc^2\theta = \text{constant} \equiv L \quad (1)$$

where r is the geocentric distance, θ is the magnetic colatitude, R_E is the radius of the Earth, and $b = 1.5L^*R_E = 12.82\ R_E$ is the radius of the equatorial neutral line. This value of b which is obtained by mapping the last closed field line (denoted L^*) to a colatitude of 20° on the Earth, corresponds to $|\Delta B| = 14.474$ nT and $L^* = 8.547$. The limit $b \to \infty$ ($L^* \to \infty$) would correspond to a purely dipolar **B** field. In this study, we

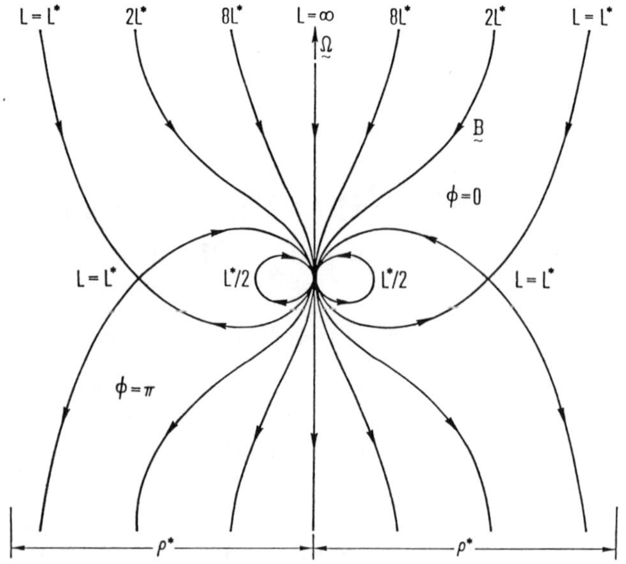

Fig. 1. An illustration of the magnetospheric **B**-field model used in this study. The model is symmetric about the $\sin\theta = 0$ axis and about the equatorial plane, which contains a circular neutral line at $r = b$ on the magnetic shell $L = L^*$, which approaches an asymptotic distance ρ^* from the tail axis at large distances $|z|$ from the equatorial plane.

consider only equatorially mirroring particles in which the equatorial field intensity B_0 is given by

$$B_0 = (\mu_E/r^3) - 14.474\ \text{nT}, \quad (2)$$

where $\mu_E = 3.05\times10^4$ nT-R_E^3 is the geomagnetic dipole moment. Further details of this field model are given by *Schulz* [1991, pp. 98–110].

We assume that the total electric field $\mathbf{E} = -\nabla\Phi_E$ is derivable from the scalar potential

$$\Phi_E = -\frac{V_\Omega}{L} + \frac{V_0}{2}\left(\frac{L}{L^*}\right)^2 \sin\phi + \frac{\Delta V(t)}{2}\left(\frac{L}{L^*}\right)\sin\phi, \quad (3)$$

in which the three separate terms correspond to corotation ($V_\Omega = 90$ kV), the Volland-Stern [*Volland*, 1973; *Stern*, 1975] model of quiescent convection ($V_0 = 50$ kV), and the time-dependent enhancement $\Delta V(t)$ associated with the stormtime convection, respectively. The time-varying term in the potential is assumed to vary as L [cf. *Nishida*, 1966; *Brice*, 1967] rather than as L^2 because electric disturbances are expected to be less well shielded than steady-state convection by the inner magnetosphere.

We model the storm-associated enhancement $\Delta V(t)$ in the cross-tail potential drop,

$$\Delta V(t) = \sum_{i=1}^{N} \Delta V_i \exp[(t_i - t)/\tau]\theta(t - t_i) \quad (4)$$

where $\theta(t)$ is the unit step function ($\equiv 1$ for $t \geq 0$; $\equiv 0$ for $t < 0$), as a superposition of almost randomly occurring impulses that rise sharply and decay exponentially with a "lifetime" $\tau = 20$ min [cf. *Cornwall*, 1968]. The impulses represent the constituent substorms of a storm. The potential drop ΔV_i associated with any individual impulse is chosen randomly from a Gaussian distribution with a 200-kV mean and a 50-kV standard deviation. We have chosen such a large mean value of ΔV_i since our intention is to model a major ($|D_{st}| \sim$ 200 nT) storm, such as those which *Lyons and Williams* [1980] analyzed. Since those storms had a main phase lasting ~ 3 hr, we assume that the N start times t_i in (4) are randomly distributed within a 3-hr time interval corresponding to the main phase of a storm. However, we impose a 10-min "dead time" (after each impulse onset) during which no subsequent impulse can start. This constraint imposes a seemingly realistic delay between consecutive impulse onsets. Without such a dead-time it would be possible for the next impulse to start immediately after the previous one, and this could lead to the build-up of unrealistically large cross-tail potentials. Further details of this model storm are given in [*Chen et al.*, 1992*b*].

We have constructed 100 such random storms so that on average there are 9 impulses per storm or 3 substorms/hr. We have done this by generating 1800 random numbers between 0 and 100 and disqualifying about half of these through the dead-time constraint. We have randomly chosen four model storms for detailed case studies. Figure 2 shows the variation in cross-tail potential for these prototypical storms. The mean enhancement in cross-tail potential drop for these particular storms over the time interval $t_1 < t < t_1 + 3$ hr are $\langle \Delta V(t) \rangle = 180$ kV, 178 kV, 154 kV, and 207 kV, respectively. Since we choose to average over the period t_1 to $t_1 + 3$ hr, we may be excluding a significant portion of the last impulse. Thus, defined in this way, the average cross-tail potential drops are typically somewhat less than 200 kV.

3. Particle Dynamics

Since we simulate the guiding-center motion of nonrelativistic equatorially mirroring particles, we treat the first two adiabatic invariants ($\mu \neq 0$ and $J = 0$, respectively) as conserved quantities. It follows from (2)–(4) that the guiding-center motion of an equatorially mirroring particle subject to $\mathbf{E} \times \mathbf{B}$ and gradient-B drifts is described by

$$\frac{dL}{dt} = \frac{L^2 \cos\phi}{2\mu_E {R_E}^2}\left[V_0\left(\frac{L}{L^*}\right)^2 + \Delta V(t)\left(\frac{L}{L^*}\right)\right] \tag{5}$$

and

$$\frac{d\phi}{dt} = \Omega - 3\frac{\mu\mu_E}{qB_0 r^5} - \frac{R_E}{\mu_E}\left[V_0\left(\frac{L}{L^*}\right)^2 + \frac{\Delta V(t)}{2}\left(\frac{L}{L^*}\right)\right] L\sin\phi, \tag{6}$$

where Ω is the angular velocity of the Earth, q is the charge of the particle, and ϕ is the azimuthal coordinate (local time).

We solve the ordinary differential equations (5) and (6) simultaneously by using the Bulirsh-Stoer extrapolation method with variable time-step [e.g., *Press et al.*, 1986, pp. 563–568] for specified initial conditions. First we obtain steady-state adiabatic drift paths associated with a particular value of the first adiabatic invariant μ by setting $\Delta V(t) = 0$ in (5) and (6). Next, to study the effects of a time-dependent cross-tail potential, we start representative particles at points equally spaced in time on a steady-state drift path. We do this for the purpose of properly calculating a radial diffusion coefficient and for obtaining drift-averaged phase space densities (see below). (We note that this method differs from previous simulation studies [e.g., *Smith et al.*, 1979; *Lee et al.*, 1983; *Takahashi*, 1990] in which particles are injected from a nightside boundary.) We then apply (4), which prescribes a storm-associated variation $\Delta V(t)$ in the cross-tail potential drop and run the simulation to determine the consequent stormtime particle transport. We can run the simulation either backward in time (to determine where any representative particle must have been prior to the storm in order to reach the desired phase on its "final" drift shell) or forward in time (to follow the dispersal of initially co-drifting particles among drift shells during and after the storm.)

4. Simulated Guiding-Center Trajectories

In this section we present results of simulated stormtime transport of singly charged ions having various μ values and vanishing second invariant ($J = 0$). For illustrative purposes, we present only the results that were obtained when we applied the model storm ($\langle \Delta V(t) \rangle = 180$ kV) shown in Figure 2*a*. The dashed outer circle on each particle-trajectory plot shows the location of the neutral line, a circle of radius $R = 12.82$, which marks the boundary between open and closed magnetic field lines in our magnetic-field model (cf. Figure 1). Figure 3*a* illustrates steady-state trajectories of equatorially mirroring ions for $\mu = 3$ MeV/G. This corresponds to an energy of 33 keV at a geocentric radial distance $r = 3\ R_E$. For this particular μ value, an x-type separatrix marks the boundary between open and closed drift trajectories. We label closed drift shells in terms of the dimensionless third adiabatic invariant defined by *Roederer* [1970, p. 1078] as

$$\frac{1}{L} \equiv \left|\frac{\Phi_B R_E}{2\pi\mu_E}\right| = \left[\frac{1}{2\pi}\oint\frac{d\phi}{L(\phi)}\right], \tag{7}$$

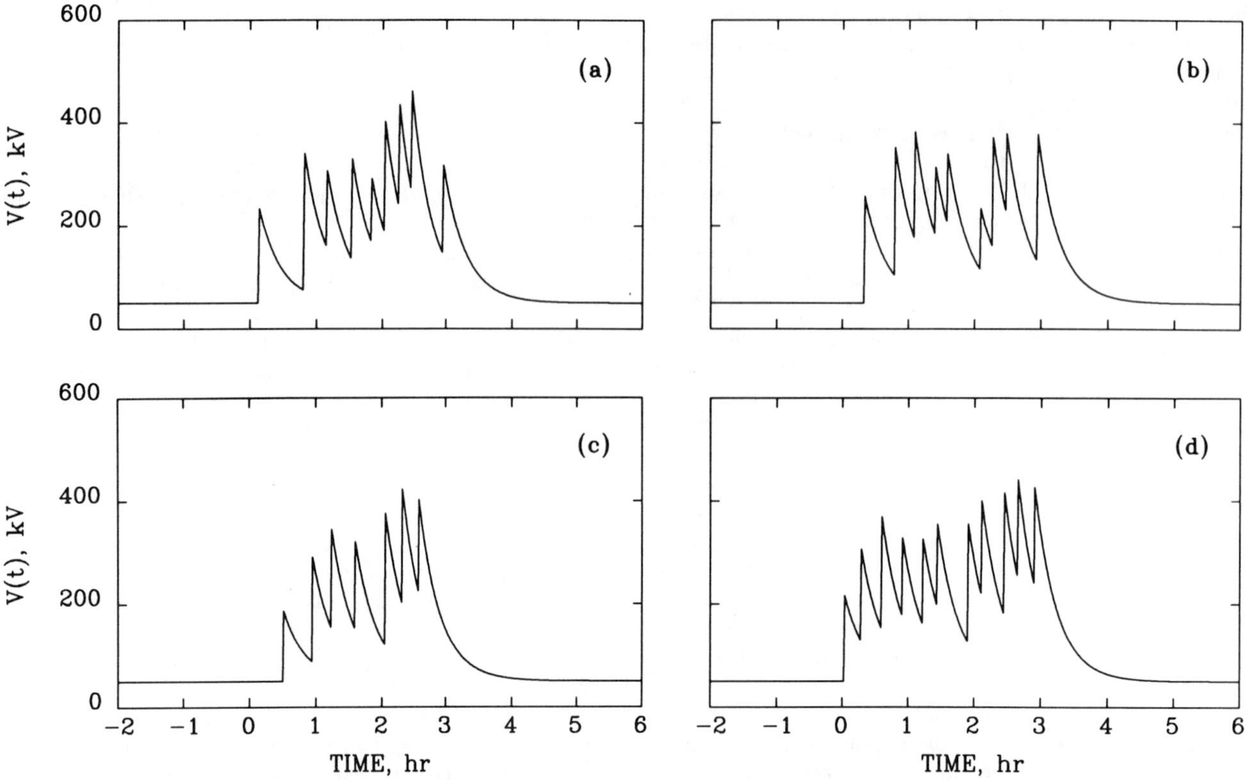

Fig. 2. The cross-tail potential $V(t)$ in our model storm consists of a quiescent value V_0 (= 50 kV) and a superposition of exponentially decaying impulses (decay time $\tau = 20$ min). These impulses represent the constituent substorms of a storm and start at times that are distributed randomly over a 3-hr time interval, except that we impose (after the start of each impulse) a 10-min "dead time" after the start of each impulse during which no subsequent impulse can start. Four realizations of our model storm are shown. The average enhancements $\langle \Delta V(t) \rangle$ in cross-tail potential drop over the respective 3-hr main phases are (a) 180 kV, (b) 178 kV, (c) 154 kV, and (d) 207 kV.

where Φ_B is the magnetic flux enclosed by that drift shell and $L(\phi)$ denotes the field-line label at longitude ϕ on the drift shell. In particular, we denote by L_1 the drift shell that separates open and closed drift paths.

We have examined the access of ions to the quiescent drift shell that intersects the dusk meridian at $R \equiv r/R_E = 3$ for a range of μ values (3 MeV/G $\lesssim \mu \lesssim$ 200 MeV/G). We select this drift shell because it is representative of where particles need to be transported in order to form the stormtime ring current. We find the quiescent ionic drift period τ_3 to be 12 hr, which is longer than the assumed 3-hr main phase. Starting with 12 representative ions (as indicated by the filled circles) equally spaced in time on this drift path of interest, we have run the simulation backward in time.

The time-reversed trajectories shown in Figure 3*b* thus indicate where the particles must have been prior to the storm in order to have reached the final drift shell ($L \approx 3$) of interest. We found that 3 of the 12 representative ions would have been transported outward from closed drift paths having smaller L values. The other 9 representative ions would have been transported inward from the night side along open trajectories to populate the final closed drift path. Eight of these 9 representative ions would have come from beyond the boundary of our model. We have compared the time ($\sim 1-1.3$ hr) required for these 8 representative ions to be transported inward from the neutral line to the final closed drift path of interest with the convection time [cf. *Lyons and Williams*, 1980] obtained by integrating (5) while keeping $\cos\phi$ constant in time. Although *Lyons and Williams* [1980] had envisioned direct-convective access from closed drift trajectories, we have found that the direct convective access occurs mainly along open drift trajectories from the neutral line. However, the access times obtained for the simulated trajectories agree reasonably well in most cases with estimates based on direct-convective access. Agreement is especially good for ions transported from the nightside neutral line to the vicinity of the final drift path of interest in the quadrant centered on midnight. Agreement was not as good for ions that had drifted to other local times be-

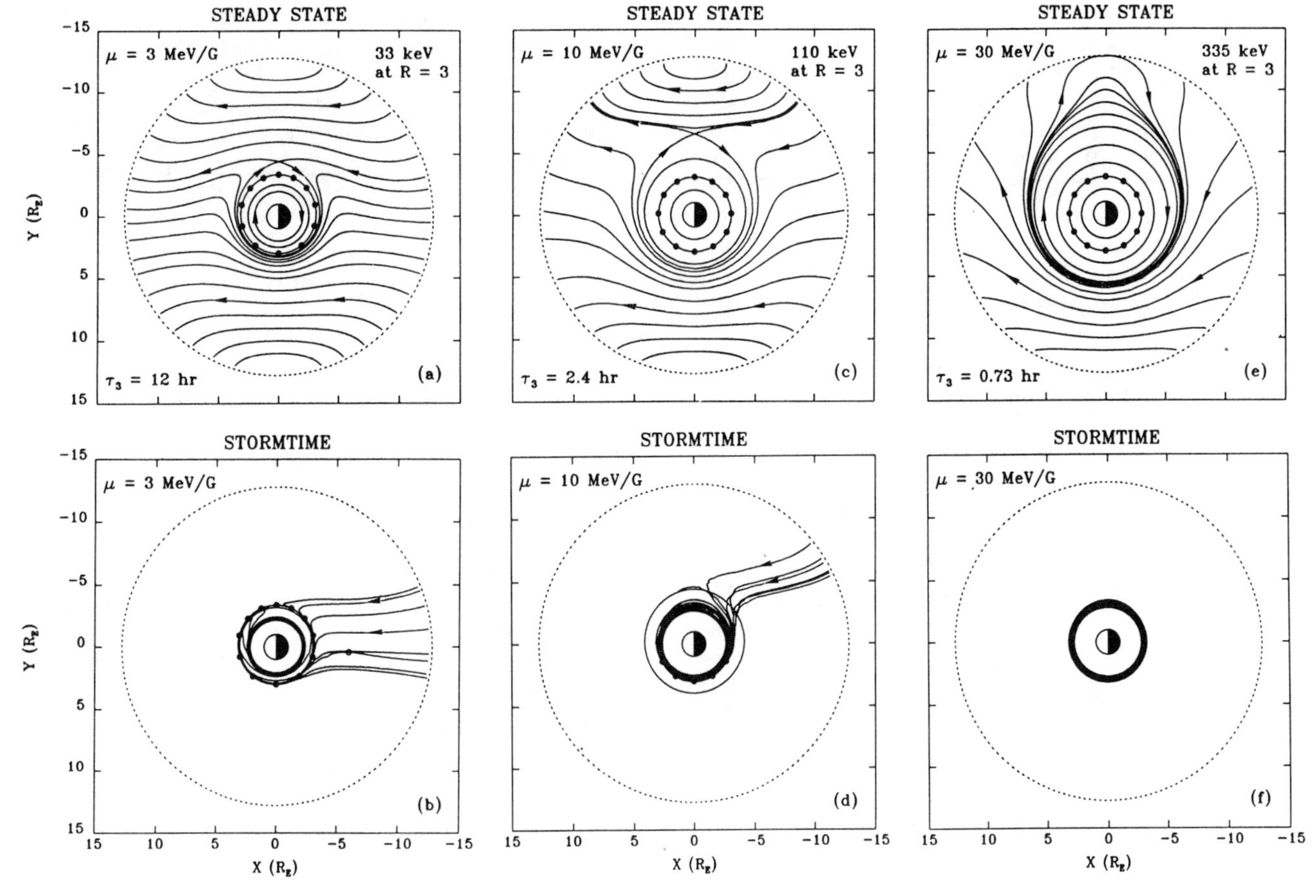

Fig. 3. Quiet-time equatorial trajectories of singly charged ions having $\mu = 3$, 10 and 30 MeV/G are plotted in the upper panels. The outer dashed circle represents the neutral line at $r = b$. Ions whose drift paths cross the dusk meridian at $R = 3$ have drift periods τ_3 as noted. The 12 representative ions' "final" positions on the steady-state drift path of interest are denoted by small "filled" circles. Corresponding stormtime trajectories computed in our time-reversed simulation are shown in the lower panels.

fore reaching the vicinity of the final drift path during the storm. Reasonably good agreement was typical for lower-energy ions ($E \sim 30$–100 keV) [*Chen et al.*, 1992*b*] and thus confirms convective access as their mode of transport.

Equatorial steady-state drift paths for ions having $\mu = 10$ MeV/G or energies of 110 keV at $R = 3$ are shown in Figure 3*c*. Again, in this case, an x-type separatrix marks the boundary between open and closed drift paths. The quiescent drift period ($\tau_3 = 2.4$ hr) for these ions on a drift shell that intersects the dusk meridian at $R = 3$ is comparable to the duration of the storm's main phase.

The time-reversed simulated stormtime trajectories for $\mu = 10$ MeV/G are shown in Figure 3*d*. At this μ value, only half of the representative ions have been transported to the final drift shell of interest by moving inward from the night side along open drift trajectories. The other half have been transported from closed drift paths of either smaller or larger L value. Their stormtime transport begins to resemble radial diffusion. Thus, the mode of access of intermediate-energy ions to the stormtime ring current is transitional between convective and diffusive.

Finally, we consider ions having $\mu = 30$ MeV/G, which would correspond to energies of 335 keV at $R = 3$. Steady-state ion drift paths are illustrated in Figure 3*e*. Here the boundary between open and closed drift paths is a closed drift path tangential to the neutral line [cf. *Brice and Ionannidis*, 1970; *Schulz*, 1976]. Again, we consider the access of ions to the drift path that intersects the dusk meridian at $R = 3$. The quiescent drift period ($\tau_3 = 0.73$ hr) for ions on such a drift shell is less than 1/4 of the main-phase duration.

The simulated storm transports ions to the drift shell of interest from closed drift paths of either smaller or larger L value. Thus, there is a spread among the initial L values of the representative particles (see Figure

3*f*). This transport resembles radial diffusion in this case, except that this was a time-reversed calculation. We have also run a simulation forward in time, so as to follow the dispersal of ions from a common initial drift path. Not surprisingly, the resulting stormtime transport generates a plot qualitatively similar to Figure 3*f* although the time-forward simulation applies to an implicitly different set of particles from those followed in the time-reversed simulations. Transport of the type illustrated in Figure 3*f* was typical for ions having $\mu \gtrsim 13$ MeV/G, which corresponds to energies $\gtrsim 145$ keV at $R = 3$.

Figure 3 thus illustrates a range of modes of particle access to the stormtime ring current. Ions having $\mu \lesssim 5$ MeV/G ($E \lesssim 30$ keV at $R = 3$) undergo mainly direct convection from open (plasmasheet) drift paths to closed drift shells at $R \sim 3$, while ions having $\mu \gtrsim 13$ MeV/G respond to the same enhancement of the convection electric field in a manner which resembles radial diffusion among closed drift shells [cf. *Lyons and Schulz*, 1989]. The transition between these two idealized modes of access occurs at $\mu \sim$ 5–13 MeV/G ($E \sim$ 55–145 keV at $R = 3$) for ring-current particles whose quiescent drift periods are comparable to the duration of the model storm's main phase.

5. Phase-Space Density Mapping

We have performed time-reversed ion simulations at additional μ values (1 MeV/G $\lesssim \mu \lesssim$ 100 MeV/G) for the purpose of mapping phase space distributions in accordance with Liouville's theorem to drift shells that intersect the dusk meridian at $R = 3$. This requires that we specify the distribution at the neutral line before and during the storm, and the distribution on closed trajectories before the storm. At the neutral line, we maintain an exponential spectrum

$$f^* = \exp(-\mu/\mu_0) \tag{8}$$

for the phase-space density at the neutral line and set $\mu_0 = 5$ MeV/G. This leads to a reasonable drop off of the boundary spectrum at high energies [cf. *Williams*, 1981]. We neglect losses for the distribution along open trajectories so that f^* specifies the phase space density everywhere beyond the boundary between open and closed drift trajectories (e.g., see upper panels in Figure 2), which we label $L_1(\mu)$.

We neglect Coulomb drag for simplicity and assume that the pre-storm transport of ions on closed trajectories is governed by an equation of the form

$$\frac{\partial \bar{f}}{\partial t} = L^2\left(\frac{\partial}{\partial L}\right)\left[\frac{D_{LL}}{L^2}\frac{\partial \bar{f}}{\partial L}\right] - \frac{\bar{f}}{\tau_q} \tag{9}$$

where $\bar{f}$ is the drift-averaged phase-space density at fixed μ and J, D_{LL} is the diffusion coefficient for transport in L, and τ_q is the ionic lifetime against charge exchange. The steady-state solution to (9), in which radial diffusion balances charge exchange, can be expressed in closed form in terms of modified Bessel functions of fractional order if D_{LL} and τ_q vary as power laws in L [*Haerendel*, 1968]. Thus, we seek to fit D_{LL} and τ_q accordingly.

We notice, from the plot of H^+ charge exchange lifetime profiles (solid curves) reproduced from selected μ values from *Cornwall* [1972] in our Figure 4*a*, that τ_q tends to vary as a power law in L at the smaller L values. Accordingly, we specify

$$\tau_q \approx L^{-8}10^3(\mu/4)^4 \text{ day}^{-1} \tag{10}$$

as a rough approximation corresponding to the dashed curves in Figure 4*a*. This is a fairly good fit to the charge exchange lifetimes taken from *Cornwall* [1972] for $L \lesssim (2.5\mu)^{(2/5)}$, which covers most of the range of μ and L values of interest. However, we hope to improve upon our fit of the charge exchange lifetimes in future work.

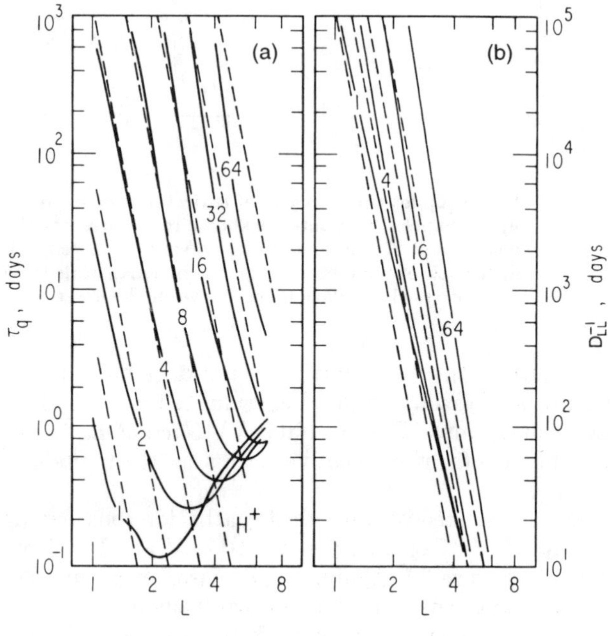

Fig. 4. (a) Profiles of H^+ charge-exchange lifetimes (solid curves taken from *Cornwall* [1968]) for different values (indicated in MeV/G) of the first adiabatic invariant . The formula given by (10) provides a fairly good power-law fit (L^{-8}, dashed lines) to τ_q for $L \lesssim (2.5\mu)^{(2/5)}$. (b) Profiles of D_{LL}^{-1}, reciprocal of the standard diffusion coefficient given by (11), for selected values of μ (indicated in MeV/G). Dashed lines represent power-law (L^{-8}) fits specified by reciprocal of (12).

Similarly, we estimate a power law fit to the diffusion coefficient. The standard model [e.g., *Cornwall*, 1972] leads to a diffusion coefficient of the form

$$D_{LL} \approx \frac{1.4 \times 10^{-5} L^{10}}{\mu^2 + L^4} \text{ day}^{-1}, \qquad (11)$$

where μ is in units of MeV/G. The solid curves in Figure 4*b* are profiles of D_{LL}^{-1} for selected μ values. Because the diffusion coefficient varies as L^6 for $L^2 \gg \mu$ and as L^{10} for $L^2 \ll \mu$, we have compromised on L^8 in order to obtain a single power law. We thus obtain a power-law "fit"

$$D_{LL} \approx 7 \times 10^{-6} \mu^{-1} L^8 \text{ day}^{-1} \qquad (12)$$

to the radial diffusion coefficient by requiring that the dashed curves in Figure 4*b* be tangent to the corresponding solid curves at $L = \mu^{1/2}$. In the future, we plan to refine our power-law model for D_{LL} (as well as for τ_q). However, for the present, our simplistic but reasonable power-law fits to the transport coefficients allow us to express the pre-storm phase-space distribution $f(\mu, L)$ by means of the equation

$$\left(\frac{L}{L_1}\right)^{5/2} \frac{f(\mu, L)}{f^*(\mu)} =$$

$$\left[\frac{I_{-5/2}(\theta)K_{-5/2}(\theta_0) - K_{-5/2}(\theta)I_{-5/2}(\theta_0)}{I_{-5/2}(\theta_1)K_{-5/2}(\theta_0) - K_{-5/2}(\theta_1)I_{-5/2}(\theta_0)}\right], \qquad (13)$$

where $\theta = L(\tau_q D_{LL})^{-1/2}$ and $\tau_q D_{LL}$ depends only on μ. The inner boundary θ_0 in (13) corresponds to the drift shell that grazes the Earth's atmosphere. We obtain L_0 as a very weak function of μ for this purpose by evaluating (7) for the drift shell that intersects the dusk meridian at $R = 1.1$. The outer boundary θ_1 in (13) corresponds to the separatrix $L_1(\mu)$ between closed and open drift paths (cf. Figure 3, upper panels).

Using (8) and (13), we plot (see Figure 5) the pre-storm phase-space density spectrum $f(\mu, L)$ for the drift shell that intersects the dusk meridian at $R = 3$. We distinguish between values of $f(\mu, L)$ on open (dashed curve) and closed (solid curve) drift trajectories ($\mu = 2.7$ MeV/G is the smallest first invariant for which the trajectories that drift through the dusk meridian at $R = 3$ are closed). Our simple model reproduces essential features similar to those found in proton phase-space distributions obtained by *Williams* [1981] from ISEE 1 data. At the higher μ values for which radial diffusion dominates charge exchange, the spectrum drops off like our exponential boundary spectrum. The spectral peak (found at $\mu \sim 22$ MeV/G) had been anticipated by *Spjeldvik* [1977] and occurs mainly because the charge-exchange lifetime decreases with decreasing

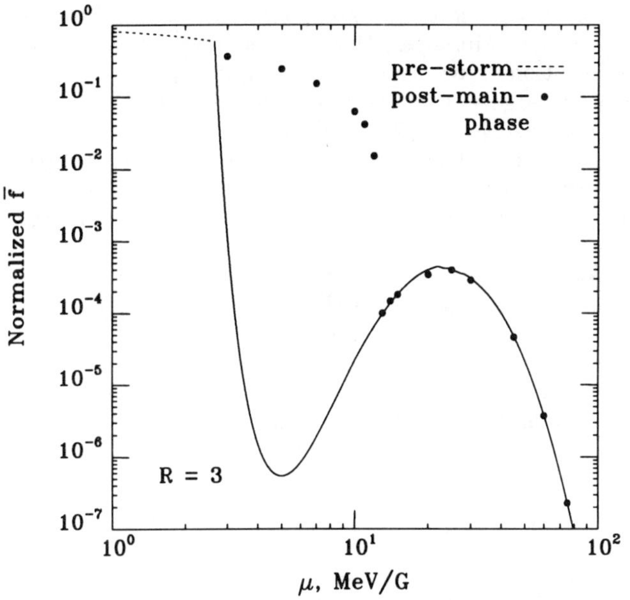

Fig. 5. The pre-storm phase space density $\bar{f}$ spectrum for ions that drift through the dusk meridian at $R = 3$ is represented by the plotted curve. The solid and dashed portions of the curve correspond to closed and open drift trajectories, respectively. The drift-averaged phase-space density distribution (filled circles) denoted post-main-phase is obtained by averaging the mapped values of f for the 24 representative ions.

μ. However, since $L_1(\mu)$ also varies directly with μ for $\mu \gtrsim 1$ MeV/G, ions having small μ values do not have to diffuse as far from their boundary between closed and open drift paths in order to reach $R = 3$ (cf. upper panels in Figure 3). For this reason the solution specified by (13) rises again at low μ to join the exponential boundary spectrum (dashed curve) which corresponds to ions on open drift paths.

We have invoked Liouville's theorem to map the phase-space f for each representative ion from (8) and (13). For this portion of the study, we employ time-reversed tracings of 24 (cf. Figure 3), rather than 12, representative ions from end-point phases equally spaced in time on a drift shell that intersects the dusk meridian at $R = 3$. By averaging the mapped f values for all the 24 representative ions we obtain a good estimate of the drift-averaged phase-space density $\bar{f}$ (filled circles in Figure 5) attained upon completion of the main phase of the model storm. Our approach differs from that of *Kistler et al.* [1989] who made point-to-point mappings of phase space distributions at various local times using pre-storm spectra obtained from AMPTE data.

We find a major enhancement from the pre-storm phase-space for $\mu \sim$ 3–13 MeV/G. This range corresponds to energies $\sim$ 30–150 keV, which are known to be representative of the stormtime ring current [e.g.,

Lyons and Williams, 1976; *Williams*, 1981; *Kistler et al.*, 1989]. Moreover, these are energies for which our simulations have shown that ion transport to $L \sim 3$ occurs largely from open trajectories (e.g., Figure 3*b*). In contrast, for $\mu \gtrsim 13$ MeV/G ($E \gtrsim 145$ keV at $R = 3$) we find little change in $\bar{f}(\mu, L)$ as a consequence of the transport associated with a single storm. This range corresponds to particles whose transport resembles radial diffusion (e.g., Figure 2*f*).

We find our preliminary results of mapping phase-space densities to be particularly satisfying since they are consistent with many observed features of ring-current phase-space distributions. Thus, we are extending this study to other L-shells of interest. In addition, we are refining our model for the pre-storm and boundary phase-space distributions and will report on the results in the near future.

6. Diffusion and Quasi-Diffusion Coefficients

Although the higher-energy ions ($E \gtrsim 145$ keV) do not seem to contribute much to the stormtime ring current, their diffusive transport is nevertheless interesting in the radiation-belt context. Thus, we have made detailed quantitative comparisons of diffusion coefficients obtained from our simulations with predictions from the quasilinear theory of *Fälthammar* [1968]. For this purpose, we consider four realizations (shown in Figure 2) of our random storm model. These were randomly chosen from the 100 storms that we had originally generated.

For each realization of our model storm, we have computed diffusion coefficients for selected values of μ (such that 15 MeV/G $\lesssim \mu \lesssim$ 200 MeV/G) from the distribution of initial L values of the time-reversed trajectories (see Figure 3*f*). We have done this by constructing the quantity

$$D_{LL}^{sim} \equiv \left(\frac{L^4}{24T}\right)\left[\sum_{i=1}^{12}(L_i^{-1}-L_f^{-1})^2-\left[\sum_{i=1}^{12}(L_i^{-1}-L_f^{-1})\right]^2\right], \quad (14)$$

where L_i and L_f denote the drift-shell labels of the initial and final trajectories (respectively) of the 12 representative ions, and where T (= 3 hr) denotes the duration of the main phase of the model storm. The quantity D_{LL}^{sim} is thus a measure of the variance among the initial third adiabatic invariants of particles situated on the final drift shell of interest. We also computed diffusion coefficients D_{LL}^{sim} from time-forward simulations by interchanging the indices i and f in (14). The diffusion coefficients D_{LL}^{sim} obtained from simulations run forward and backward in time are not very different, 80% of them being within a factor of 1.6 of each other, although they pertain implicitly to different sets of particles. The geometric mean discrepancies (among the 12 values of μ) between the values of D_{LL}^{sim} obtained from simulations run forward and backward in time amounted to factors of 1.3, 1.6, 1.5, and 1.8 for the four model storms in Figure 2.

We compare the diffusion coefficients obtained from the simulated trajectories with the resonant-particle formulation [*Fälthammar*, 1965] of radial-diffusion theory in which the diffusion coefficient is of the form

$$D_{LL}^{ql} = \frac{L^6 R_E^4}{4\mu_E^2}\hat{E}\left(\frac{\Omega_3}{2\pi}\right), \quad (15)$$

where $\hat{E}(\omega/2\pi)$ is the spectral-density of the (quasi-uniform) equatorial electric field in the inner magnetosphere and $\Omega_3/2\pi$ is the particles' quiescent drift frequency. When we substitute the spectral-density function for our model storm (see *Chen et al.* [1992*b*] for derivation) into (15), we obtain the quasilinear diffusion coefficient

$$D_{LL}^{ql} = \frac{\tau^2 L^6 R_E^2}{16T\mu_E^2(L^*)^2}\sum_{i=1}^{N}\sum_{j=1}^{N}\frac{\Delta V_i \Delta V_j \cos[\Omega_3(t_j - t_i)]}{1+\Omega_3^2\tau^2} \quad (16)$$

in which correlations between the impulses lead to cross terms ($j \neq i$). By neglecting the cross terms, we could recover essentially the standard diffusion coefficient of *Cornwall* [1968], but here we retain all the cross terms in (16) for comparison with D_{LL}^{sim} as computed for individual storms.

The dashed curves in Figure 6 represent the quasilinear diffusion coefficients at $L \approx 3$ for the corresponding four model storms shown in Figure 2. The respective quasilinear diffusion coefficients are not very smooth functions of μ. This is because the impulse onsets (i) associated with any individual storm modeled by (4) occur at specific (although randomly determined) times t_i, which means that the corresponding spectral density $\hat{E}(\omega/2\pi)$ is not a very smooth function of frequency. We plot as data points in Figure 6 the diffusion coefficients D_{LL}^{sim} obtained from simulations run forward (filled circles) and backward (open circles) in time. For comparison purposes, we chose some of the μ values to correspond with the minima in D_{LL}^{ql} in case (a). Agreement of the diffusion coefficients D_{LL}^{sim} obtained from the simulations with quasilinear theory is surprisingly good despite the strong variability of D_{LL}^{ql} with μ. For cases (a)–(d), we find that the geometric means of the discrepancies amount to factors of 2.3, 2.3, 1.5, and 3.0, respectively. Agreement is best for case (c), in which there were only 7 substorms during the model storm (see Figure 2*c*) and consequently less variability in D_{LL}^{ql} with μ. We find that quasilinear theory even accounts for the μ values (e.g., $\mu = 75$ and 80 MeV/G in Figure

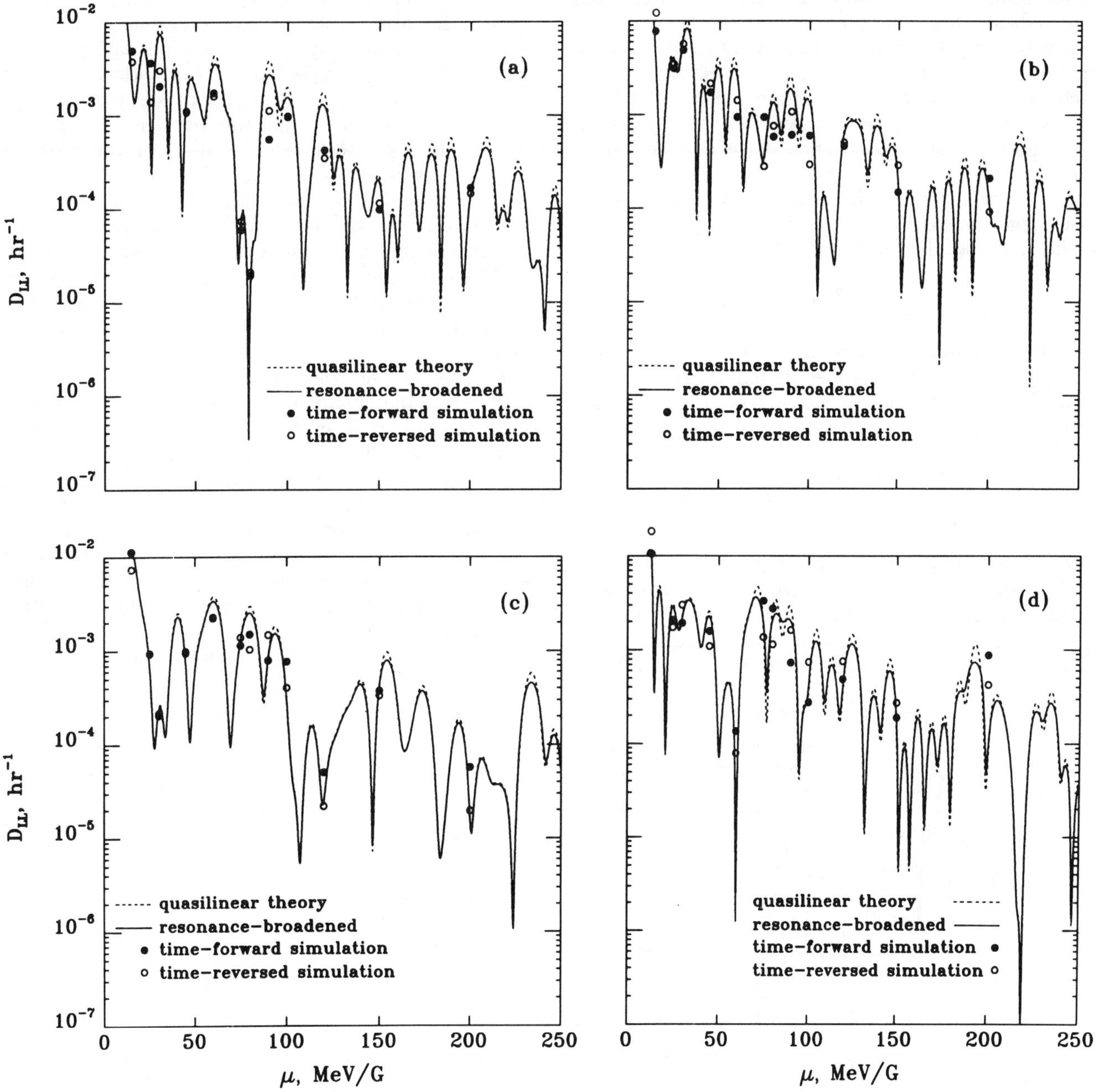

Fig. 6. Plots of diffusion coefficients D_{LL}^{sim} obtained via (14) from time-reversed (open circles) and time-forward (filled circles) simulations, for comparison with the quasilinear diffusion coefficient D_{LL}^{ql} (dashed curve) given by (16) as an implicit function of μ for the four realizations of our model storm shown in Figure 2. The diffusion coefficients D_{LL}^{rb} (corrected for resonance-broadening) are represented by the solid curves.

6a) for which the diffusion coefficients computed from the simulation are especially small.

However, the diffusion coefficients obtained from the simulated trajectories generally do not show quite as much variability with μ as quasilinear theory predicts. This is not surprising, since quasilinear theory postulates a perfectly sharp resonance at the quiescent drift frequency, whereas the simulated transport leads to an eventual spread among the quiescent drift frequencies of the representative ions for each μ. A rough estimate for the anticipated spread in $\Omega_3/2\pi$ is

$$\Delta\omega/2\pi \approx (D_{LL}T/2\pi^2)^{1/2}|(\partial\Omega_3/\partial L)_\mu| \tag{17}$$

since the mean-square spread in L accumulated during the transport is $2D_{LL}T$ [*Chen et al.*, 1992*a*]. An estimate for the diffusion coefficient (corrected for resonance-broadening effects) is thus

$$D_{LL} \approx \frac{L^6 R_E^4}{4\mu_E^2 \Delta\omega} \int_{\Omega_3-(\Delta\omega/2)}^{\Omega_3+(\Delta\omega/2)} \hat{E}\left(\frac{\omega}{2\pi}\right) d\omega. \tag{18}$$

Since the frequency bandwidth given by (17) depends on D_{LL}, we have iterated between (18) and (17) until satisfactory convergence to the desired solution (called D_{LL}^{rb}) is achieved. The results for cases (a)–(d) are plotted as solid curves in Figure 6. We find that inclusion of this nonlinear resonance-broadening effect tends to reduce the discrepancy between quasilinear theory and D_{LL}^{sim} by a smoothing out the sharp relative maxima and minima with respect to μ. Corrections were typically $\sim$ 10–30% at the relative maxima but were as much as 60% at the relative minima (e.g., near $\mu = 184$ MeV/G in Figure 6a). The geometric means of the remaining discrepancies between D_{LL}^{rb} and D_{LL}^{sim} amount to factors of 1.9, 2.1, 1.5, and 2.7 for cases (a)–(d), respectively. To determine whether the remaining discrepancies are attributable to the neglect of nonlinear and/or quasilinear effects, we have made similar comparisons after reducing the impulse amplitudes of the enhanced cross-tail potential drop $\Delta V(t)$ in our model storms.

Figures 7a and 7b show comparisons of D_{LL}^{sim} and D_{LL}^{rb} when the enhanced cross-tail potential drop ΔV for case (a) is reduced by a factors of 2 and 4, respectively. When the average stormtime cross-tail potential drop decreases, the agreement between D_{LL}^{ql} and D_{LL}^{sim} improves considerably. The geometric-mean discrepancy amounts to a factor of 1.4 or 1.1, respectively, when $\Delta V(t)$ is reduced by a factor of 2 or 4. The agreement is quite good despite the fact that the quasilinear diffusion coefficient does not vary smoothly with μ. As the impulse amplitudes in $\Delta V(t)$ are reduced, corrections to D_{LL}^{ql} due to resonance-broadening become

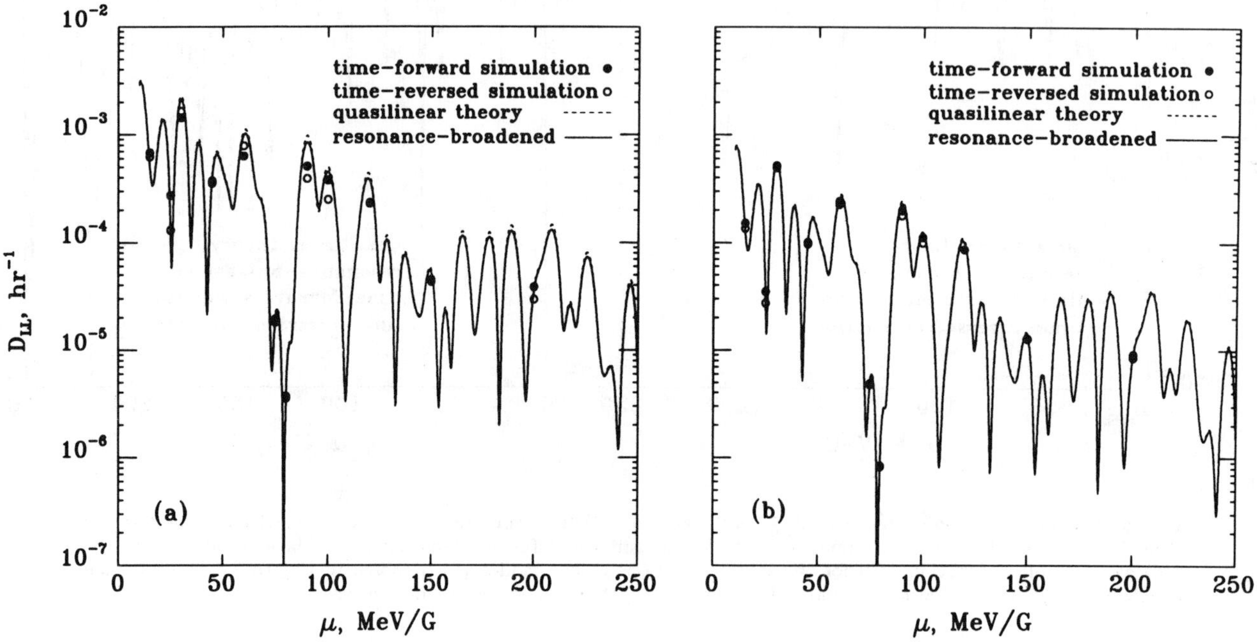

Fig. 7. Plots of diffusion coefficients D_{LL}^{sim} obtained via (14) from time-reversed (open circles) and time-forward (filled circles) simulations, for comparison with the quasilinear diffusion coefficient D_{LL}^{ql} (dashed curve) given by (16) as an implicit function of μ for storms that have the same temporal structure as model storm (a) in Figure 2, but with half (left panel) and a quarter (right panel) of the impulse amplitude in $\Delta V(t)$, respectively. The average enhancements $\langle\Delta V(t)\rangle$ in cross-tail potential drop over the 3-hr main phase are 90 kV and 45 kV, respectively. The diffusion coefficients D_{LL}^{rb} (corrected for resonance-broadening) are represented by the solid curves.

smaller, as the convergence of the solid curve (D_{LL}^{rb}) toward the dashed curve (D_{LL}^{ql}) in Figure 7*b* shows. Moreover, the agreement between diffusion coefficients obtained from simulations run forward (filled circles) and backward (open circles) in time improves. The geometric mean discrepancy between "time-forward" and "time-reversed" diffusion coefficients is only a factor of 1.2 when $\Delta V(t)$ is reduced by a factor of 2 and only a factor of 1.1 when $\Delta V(t)$ is reduced by a factor of 4. This improved agreement is not surprising since time reversal ($t \rightarrow -t$) leaves the quasilinear theory diffusion coefficient invariant under time reversal. As $\max[\Delta V(t)] \rightarrow 0$ quasilinear theory increasingly better approximates the simulated stormtime transport, and so the values of D_{LL}^{sim} estimated from simulations run forward and backward in time become less distinct from each other. We thus conclude that the remaining discrepancies between D_{LL}^{rb} and D_{LL}^{sim} are attributable to unspecified nonlinear effects of which we have not taken account.

In a numerical "experiment" we found that excellent agreement between D_{LL}^{rb} and D_{LL}^{sim} could be achieved by arbitrarily increasing $\Delta\omega/2\pi$ by a factor ~ 3 from the value specified by (17). However, we can think of no physical rationale for actually postulating such a magnification of the bandwidth over which D_{LL}^{ql} should be averaged. We have considered the possibility that the temporal variation of $\Delta V(t)$ in (3) might increase the spread in drift frequencies during transport (i.e., for $0 < t \lesssim T$) beyond that implicit in (17), in which the factor $(\partial\Omega_3/\partial L)_\mu$ pertains to quiescent drift frequencies. The guiding-center simulations presented earlier, however, show little evidence for such an effect. Typically, the mean drift period of a representative ion during the main phase only slightly exceeds the quiescent drift period, as it would if the ion were drifting under the influence of a constant enhancement $\Delta V = \langle\Delta V(t)\rangle$ of the cross-tail potential drop. Thus, the $\Omega_3/2\pi$ which appears in (15)–(18) should perhaps have been interpreted as the mean main-phase drift frequency rather than as the quiescent one. This correction would appear to be small, corresponding to a rightward shift of the plotted curves in Figure 6 by $\Delta\mu \lesssim 1$ MeV/G. However, the stormtime presence of $\langle\Delta V(t)\rangle$ does transform the quiescent drift shell of interest into a stormtime band of drift shells over which $\hat{E}(\Omega_3/2\pi)$ should presumably be averaged. We will explore the ramifications of this refinement elsewhere. In our simulations the transport-induced spread in quiescent drift frequencies has in some cases exceeded (17) by $\sim 40\%$, but this magnification of $\Delta\omega/2\pi$ would be too little to eliminate the remaining discrepancies ~ 1.9 between D_{LL}^{rb} and D_{LL}^{sim} in Figure 6. The very small change in the drift frequency during the storm also seems to eliminate the possibility of trapped-particle effects. A preliminary test suggests that the replacement of L^6 in (15) by its transport-averaged value $\langle L^6\rangle$ would also be a relatively unimportant nonlinear correction.

7. Diffusion Averaged over Storm Ensemble

It could be argued that quasilinear theory is more appropriately applied to an ensemble of model storms than to an individual storm. We have tested this hypothesis by randomly choosing 20 different storms having a 184-kV mean cross-tail potential drop from the 100 storms that we constructed. We averaged the diffusion coefficients obtained from the simulations, standard quasilinear theory, and the resonance-broadened quasilinear theory over the 20 storms. The results are shown in Figure 8 [*Chen et al.*, 1992*b*]. The ensemble-averaged quasilinear diffusion coefficient $\bar{D}_{LL}^{ql}$ (dashed curve) and its resonance-broadened counterpart $\bar{D}_{LL}^{rb}$ (solid curve) are considerably smoother than D_{LL}^{ql} and D_{LL}^{rb}, respectively, for an individual storm. The ensemble-averaged diffusion coefficients $\bar{D}_{LL}^{sim}$ from the time-reversed and time-forward simulations (open circles and filled circles, respectively) typically agree much better with the theoretical diffusion coefficients in Figure 8 than does D_{LL}^{sim} with the theoretical diffusion coefficients for an individual storm in Figure 6. The mean discrepancy between $\bar{D}_{LL}^{sim}$ and $\bar{D}_{LL}^{ql}$ is only a factor of 1.2.

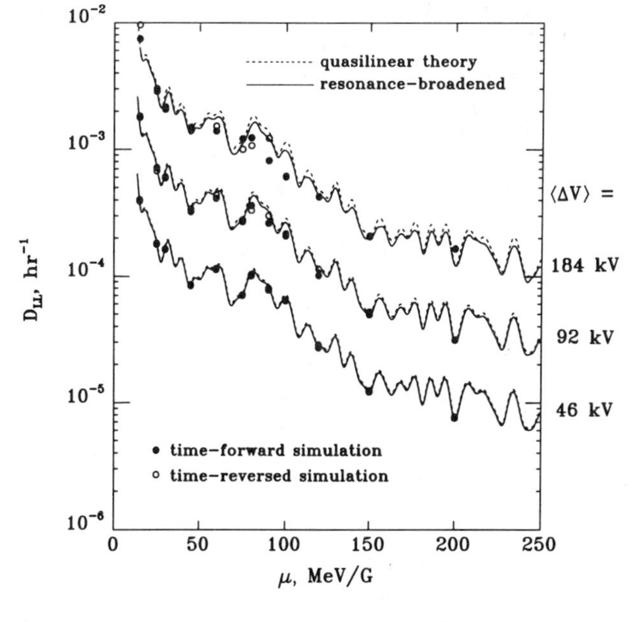

Fig. 8. Ensemble-averaged diffusion coefficients $\bar{D}_{LL}^{sim}$ obtained from the time-reversed (open circles) and time-forward (filled circles) simulations, for comparison with quasilinear theory ($\bar{D}_{LL}^{ql}$ as a function of μ, dashed curve) for equivalent ensembles of 20 storms, but with impulse amplitudes halved and doubled relative to the middle family of curves and data points so as to produce the mean $\Delta V(t)$ values shown. The solid curve represents the ensemble-averaged diffusion coefficients $\bar{D}_{LL}^{rb}$ corrected for resonance broadening effects.

We have also halved the impulse amplitudes of ΔV in the 20 storms of our ensemble and averaged the resulting diffusion coefficients obtained over the 20 storms. The results are shown via the middle family of curves and data points in Figure 8. As expected agreement between $\bar{D}_{LL}^{sim}$ and $\bar{D}_{LL}^{ql}$ is even better (geometric-mean discrepancy is a factor of 1.1 for $\langle \Delta V(t) \rangle = 92$ kV versus 1.2 for $\langle \Delta V(t) \rangle = 184$ kV.) Corrections of D_{LL}^{ql} due to resonance broadening are smaller in this case.

When we reduce all the impulse amplitudes in $\Delta V(t)$ by a factor of 4, we find remarkably good agreement among the ensemble-averaged diffusion coefficients obtained by the various methods. For example, the geometric-mean discrepancy between D_{LL}^{sim} and D_{LL}^{ql} amounts to a factor of 1.03 and the geometric-mean discrepancy between the diffusion coefficients deduced from time-forward and time-reversed simulations amounts to a factor of only 1.02 (see lower family of curves and data points, Figure 8).

8. Summary and Conclusions

We have used a dynamical guiding-center model to investigate energetic charged-particle transport in response to storm-associated impulses in a model of the convection electric field. Our simple magnetospheric model allows us to compare our numerical results with analytical descriptions of particle transport such as the quasilinear theory of radial diffusion [*Fälthammar*, 1965]. Thus, we have tested whether quasilinear theory can appropriately be applied to describe the charged-particle transport caused by electrostatic electric field fluctuations over time intervals as short as an individual storm. Furthermore, we have begun to use our simulation results to map phase-space distributions from the quiet-time to the stormtime ring current by using Liouville's theorem. A summary of our results follows.

Ions having $\mu \lesssim 3$ MeV/G ($E \lesssim 110$ keV at $R = 3$) gain access to closed drift shells at $R \sim 3$ mainly by direct convection from open (plasmasheet) drift paths. At $L \sim 3$ these ions have drift periods that exceed the duration of the main phase of the storm. The mode of access of ions with $\mu \sim$ 5–13 MeV/G ($E \sim$ 55–145 keV at $R = 3$ appears to be transitional between convective and diffusive access. At $L \sim 3$ these ions have drift periods that are comparable to the length of the main phase of the storm. The stormtime transport of ions having $\mu \gtrsim 13$ MeV/G ($E \gtrsim 145$ keV at $R = 3$) resembles radial diffusion across closed drift shells. At $L \sim 3$ these ions have drift periods that are smaller than the duration of the main phase of the storm.

The electric spectral density derived from our model storm is not very smooth, and so the quasilinear diffusion coefficient D_{LL}^{ql} does not vary smoothly with μ. When we compared the diffusion coefficients D_{LL}^{sim} obtained from the simulated trajectories with the quasilinear diffusion coefficient $D_{LL}^{ql}(\mu, L)$, we nevertheless found surprisingly good agreement for 4 distinct individual model storms. The aggregate geometric-mean of discrepancies between D_{LL}^{sim} and D_{LL}^{ql} for the 4 storms amounted to a factor of 2.2 even though D_{LL}^{ql} varied irregularly with μ by 4–5 orders of magnitude for each of the storms studied. When we invoked nonlinear drift-resonance broadening effects, we found that the discrepancies between quasilinear theory and D_{LL}^{sim} were slightly reduced through a smoothing of the sharp relative minima and maxima in D_{LL}^{ql}. The aggregate geometric mean of the remaining discrepancies between D_{LL}^{sim} and D_{LL}^{ql} for the 4 different storms amounted to a factor of 2.0. When we reduced the impulse amplitudes in the enhanced cross-tail potential drop $\Delta V(t)$ of our model storm, the agreement between D_{LL}^{ql} and D_{LL}^{sim} improved considerably. For example, the geometric-mean discrepancy is a factor of 2.3, 1.4, and 1.1, respectively for a particular storm with $\langle \Delta V \rangle = 180$ kV, 90 kV, and 45 kV. This convergence towards 1.0 suggests that the discrepancies between D_{LL}^{sim} and D_{LL}^{ql} are attributable to nonlinear effects. When we averaged D_{LL}^{ql} and D_{LL}^{sim} obtained over an ensemble of 20 random storms, we found even better agreement between $\bar{D}_{LL}^{ql}$ and $\bar{D}_{LL}^{sim}$ (mean discrepancy amounted to a factor of 1.2 for $\langle \Delta V(t) \rangle \approx 184$ kV). When we reduced the impulse amplitudes in $\Delta V(t)$ for all the storms in the ensemble by a factor of 4, we found remarkably good agreement between these ensemble-averaged diffusion coefficients (mean discrepancy factor was 1.02).

We developed a simple model of the pre-storm (steady-state) phase-space density, obtained by balancing radial diffusion against charge exchange, which produces features qualitatively similar to those found in observations. We used our time-reversed simulations to map phase space densities f from pre-storm to stormtime in accordance with Liouville's theorem. The stormtime transport of ions having $\mu \gtrsim 13$ MeV/G ($E \gtrsim 145$ keV at $R = 3$), for which the transport resembles radial diffusion, seems to produce little change in drift-averaged phase space density $\bar{f}$. However, the stormtime transport of f produced a major enhancement in $\bar{f}$ from the pre-storm phase-space density at energies $\sim 30 - 145$ keV, which are representative of the stormtime ring current. These are energies for which our simulations have shown that many of the ions are transported on the nightside from open trajectories to the final drift shell ($L \sim 3$) of interest.

Our preliminary results on the mapping of phase space densities are particularly satisfying since they reproduce many observed features of ring-current phase-space distributions. Thus, we are performing additional mappings to other L-shells of interest. We plan to refine our treatment of the pre-storm and boundary phase-space distributions so as to achieve a more realistic model. We also will include loss processes such as charge exchange in our simulations so that we can map phase space distributions through the recovery phase as well as through the main phase of a model geomagnetic storm.

Acknowledgments. The authors thank Wendall Horton for suggesting to decrease the amplitude of our electric field impulses in order to test for nonlinear effects. Our work is supported by the NASA Space Physics Theory Program under grant NAGW-2126. One of the authors, Margaret W. Chen, was a National Research Council Research (NRC) Associate during this study. Computing resources for this work were provided by the San Diego Supercomputer Center and by the NASA Center for Computational Sciences.

REFERENCES

Brice, N. M., Bulk motion of the magnetosphere, *J. Geophys. Res.*, *72*, 5193–5211, 1967.

Brice, N. M., and G. A. Ioannidis, The magnetospheres of Jupiter and Earth, *Icarus*, *13*, 173, 1970.

Chen, M. W., M. Schulz, L. R. Lyons, and D. J. Gorney, Ion radial diffusion in an electrostatic impulse model for stormtime ring current formation, *Geophys. Res. Lett.*, *19*, 621-624, 1992*a*.

Chen, M. W., M. Schulz, L. R. Lyons, and D. J. Gorney, Stormtime transport of ring-current ions, re-submitted to *J. Geophys. Res.*, August, 1992*b*.

Cornwall, J. M., Diffusion processes influenced by conjugate-point wave phenomena, *Radio Sci.*, *3*, 740–744, 1968.

Cornwall, J. M., Radial diffusion of ionized helium and protons: A probe for magnetospheric dynamics, *J. Geophys. Res.*, *77*, 1756–1770, 1972.

Dungey, J. W., Effects of electromagnetic perturbations on particles trapped in the radiation belts, *Space Sci. Rev.*, *4*, 199–222, 1965.

Fälthammar, C.-G., Effects of time-dependent electric fields on geomagnetically trapped radiation, *J. Geophys. Res.*, *70*, 2503-2516, 1965.

Frank, L. A., On the extraterrestrial ring current during geomagnetic storms, *J. Geophys. Res.*, *72*, 3753–3768, 1967.

Haerendel, G., Diffusion theory of trapped particles and the observed proton distribution, in *Earth's Particles and Fields*, edited by B. M. McCormac, pp. 171–191, Reinhold, New York, 1968.

Hamilton, D. C., G. Gloeckler, F. M. Ipavich, W. Stüdemann, B. Wilken, and G. Kremser, Ring current development during the great geomagnetic storm of February 1986, *J. Geophys. Res.*, *93*, 14,343–14,355, 1988.

Kistler, L. M., F. M. Ipavich, D. C. Hamilton, G. Gloeckler, B. Wilken, G. Kremser, and W. Stüdemann, Energy spectra of the major ion species in the ring current during geomagnetic storms, *J. Geophys. Res.*, *94*, 3579–3599, 1989.

Lee, L. C., G. Corrick, and S.-I.Akasofu, On the ring current energy injection rate, *Planet. Space Sci.*, *31*, 901–911, 1983.

Lyons, L. R., and M. Schulz, Access of energetic particles to storm time ring current through enhanced radial "diffusion", *J. Geophys. Res.*, *94*, 5491–5496, 1989.

Lyons, L. R., and D. J. Williams, Storm associated variations of equatorially mirroring ring current protons, 1–800 keV, at constant first adiabatic invariant, *J. Geophys. Res.*, *80*, 216–220, 1976.

Lyons, L. R., and D. J. Williams, A source for the geomagnetic storm main phase ring current, *J. Geophys. Res.*, *85*, 523–530, 1980.

Nishida, A., Formation of a plasmapause, or magnetospheric plasma knee by combined action of magnetospheric convection and plasma escape from the tail, *J. Geophys. Res.*, *71*, 5669–5679, 1966.

Roederer, J. G., *Dynamics of Geomagnetically Trapped Radiation*, Springer, Heidelberg, 1970.

Schulz, M., Effect of drift-resonance broadening on radial diffusion in the magnetosphere, *Astrophys. Space Sci.*, *36*, 455–458, 1975.

Schulz, M., Plasma boundaries in space in *Physics of Solar Planetary Environments*, edited by D. J. Williams, Vol. 1, pp. 491–504, Am. Geophys. Union, Washington, D. C., 1976.

Schulz, M., The magnetosphere, in *Geomagnetism*, edited by J. A. Jacobs, vol. 4, pp. 87–293, Academic Press, London, 1991.

Smith, P. H., and R. A. Hoffman, Ring current particle distributions during the magnetic storms of December 16–18, 1971, *J. Geophys. Res.*, *78*, 4731–4737, 1973.

Smith, P. H., N. K. Bewtra, and R. A. Hoffman, Motions of charged particles in the magnetosphere under the influence of a time-varying large scale convection electric field, in *Quantitative Modeling of Magnetospheric Processes*, edited by W. P. Olson, pp. 513–535, American Geophysical Union, Washington, D. C., 1979.

Spjeldvik, W. N., Equilibrium structure of equatorially mirroring radiation belt protons, *J. Geophys. Res.*, *82*, 2801–2808, 1977.

Stern, D. P., The motion of a proton in the equatorial magnetosphere, *J. Geophys. Res.*, *80*, 595–599, 1975.

Takahashi, S., T. Iyemori, and M. Takedo, A simulation of the storm-time ring current, *Planet. Space Sci.*, *38*, 1133–1141, 1990.

Volland, H., A semiempirical model of large-scale magnetospheric electric fields, *J. Geophys. Res.*, *78*, 171–180, 1973.

Williams, D. J., Phase space variations of near equatorially mirroring ring current ions, *J. Geophys. Res.*, *86*, 189–194, 1981.

Williams, D. J., and L. R. Lyons, The proton ring current and its interaction with the plasmapause: Storm recovery phase, *J. Geophys. Res.*, *79*, 4195–4207, 1974.

M. W. Chen, L. R. Lyons, and D. J. Gorney, The Aerospace Corporation, P. O. Box 92597, M2-260, Los Angeles, CA 90009.

M. Schulz, Lockheed Research Laboratory, Org 91-20, Bldg. 255, 3241 Hanover Street, Palo Alto, CA 94304.

Investigation of Flow Pattern for Dynamo Effect on Reversed Field Pinch

Shinji Koide, and Jun-Ichi Sakai

Laboratory for Plasma Astrophysics and Fusion Science, Toyama University
Gofuku, Toyama , JAPAN

Reversed field pinch devices (RFPs) are characterized by a reversal of their toroidal magnetic field generated by the poloidal current. Because there is no external poloidal electric field, this current is due to a dynamo effect. In this paper, we study the magnetohydrodynamic (MHD) dynamo effect by means of numerical simulation, investigating the effect of hydrodynamic flow, especially the fluid helicity density, on this calculation. We find that the ordinary fluid helicity density is not appropriate to discuss the dynamo effect, because it is induced only by perpendicular velocity components. We show that the mean magnetic field model supports this point of view. That helicity density explains the dynamo effect in RFP.

1. INTRODUCTION

We want to clarify the relation between the fluid motion and the dynamo effect for sustainment of the reversed field of RFP by means of numerical simulation. First, let us discuss the ordinary flow helicity density. Though that helicity density has a sign opposite to what is expected for the sustainment of the reversed field. Note that the flow parallel to the mean magnetic field does not distort that magnetic field. We therefore discuss the helicity density of the velocity perpendicular to the magnetic field. That fluid helicity is expected to sustain the reversed field.

The validity condition for using that helicity density is given by the quasilinear theory. The mean magnetic field theory provides the expression for the alpha effect in isotropic fluids [Moffatt, 1978]. As the flow in the RFP is not isotropic, we shall generalize the expression to include anisotropic fluids. It is important to distinguish between the flow parallel and perpendicular to the mean magnetic field. The flow along the magnetic field causes no distortion of the magnetic field. This means that, rather than treat the entire flow, we should consider only the flow perpendicular to the mean magnetic field. In the next section, we summarize the MHD simulation and discuss what kind of fluid helicity density profile we expect. We show that the ordinary hydrodynamic helicity density has a sign opposite to what should be expected. The helicity density of the fluid flow perpendicular to the mean magnetic field is important when discussing the relationship between the dynamo effect and the flow.

Space Plasmas: Coupling Between Small
and Medium Scale Processes
Geophysical Monograph 86

2. THE MEAN MAGNETIC FIELD MODEL ON ANISOTROPIC FLUID

We will investigate the creation of the electric field by the uniform, steady, anisotropic random perturbed flow. The velocity and magnetic fields may be separated into mean and fluctuating parts;

$$\mathbf{U}(\mathbf{x},t) = \mathbf{U}_0(\mathbf{x},t) + \mathbf{u}(\mathbf{x},t), \quad (1)$$

$$\mathbf{B}(\mathbf{x},t) = \mathbf{B}_0(\mathbf{x},t) + \mathbf{b}(\mathbf{x},t). \quad (2)$$

The induction equation

$$\frac{\partial \mathbf{B}}{\partial t} = \nabla \times (\mathbf{U} \times \mathbf{B}) + \eta \nabla^2 \mathbf{B}, \quad (3)$$

is separated into its mean and fluctuation part;

$$\frac{\partial \mathbf{B}_0}{\partial t} = \nabla \times (\mathbf{U}_0 \times \mathbf{B}_0) + \nabla \times \mathrm{E} + \eta \nabla^2 \mathbf{B}_0, \quad (4)$$

$$\frac{\partial \mathbf{b}}{\partial t} = \nabla \times (\mathbf{U}_0 \times \mathbf{b}) + \nabla \times (\mathbf{u} \times \mathbf{B}_0) + \nabla \times \mathrm{G} + \eta \nabla^2 \mathbf{b}, \quad (5)$$

where

$$E = \langle \mathbf{u} \times \mathbf{b} \rangle, \tag{6}$$

$$G = \mathbf{u} \times \mathbf{b} - \langle \mathbf{u} \times \mathbf{b} \rangle. \tag{7}$$

E is called the internal electric field. The angular brackets denote averaged values of the local variables.

We separate the fluctuating velocity $\mathbf{u}$ into two parts:

$$\mathbf{u} = \mathbf{u}_\perp + \mathbf{u}_{//}, \tag{8}$$

where $\mathbf{u}_\perp$ is perpendicular to the mean magnetic field and $\mathbf{u}_{//}$ is parallel.

We assume that the mean magnetic field and the mean velocity is uniform. We follow a certain coordinate frame in which the mean flow velocity is zero. Note that the mean magnetic field is not uniform in the RFP plasma, which we neglect in this section.

We define the Fourier transform,

$$\tilde{\mathbf{u}}(\mathbf{k}, \omega) = \frac{1}{(2\pi)^4} \iint \mathbf{u}(\mathbf{x}, t) e^{-i(\mathbf{k}\cdot\mathbf{x} - \omega t)} d^3\mathbf{x}\, dt. \tag{9}$$

We assume that the perturbation velocity is homogeneous and stationary. If $\mathbf{x}'=\mathbf{x}+\mathbf{r}$ and $t'=t+\tau$,

$$\langle u_i^p(\mathbf{x}, t) u_j^q(\mathbf{x}', t') \rangle = R_{ij}^{pq}\delta(\mathbf{r})\delta(\tau), \tag{10}$$

where p and q indicate $//$, $\perp$ or a whole component index of each quantity and i, j, k are coordinate components index 1, 2, 3 corresponding to x, y and z respectively.

We can define the spectrum tensor of the field Φ_{ij}^{pq} [Moffatt, 1978] as,

$$< \tilde{u}_i^p(\mathbf{k}, \omega) \tilde{u}_j^{q*}(\mathbf{k}', \omega') > = \Phi_{ij}^{pq}(\mathbf{k}, \omega)\delta(\mathbf{k}-\mathbf{k}')\delta(\omega - \omega'). \tag{11}$$

The spectrum tensor Φ_{ij}^{pq} satisfies the condition of Hermite symmetry.

We define the helicity spectrum function $F^{pq}(k, \omega)$ by,

$$F^{pq}(k, \omega) = i\int_{S_k} \varepsilon_{ikl} k_k \Phi_{il}^{pq}(\mathbf{k}, \omega)\, dS_k. \tag{12}$$

Here as usual, l_0 and t_0 are characteristic length and time scales of $\mathbf{u}$ and u_0 and b_0, the root mean square values of $\mathbf{u}$ and $\mathbf{b}$, are

$$u_0 = \langle \mathbf{u}^2 \rangle^{1/2}, \quad b_0 = \langle \mathbf{b}^2 \rangle^{1/2}. \tag{13}$$

We employ the random waves approximation;

$$u_0 t_0 / l_0 \ll 1. \tag{14}$$

The induction Eq. (5) reduces to a linear equation;

$$\frac{\partial \mathbf{b}}{\partial t} - \eta\nabla^2\mathbf{b} = (\mathbf{B}_0 \cdot \nabla)\mathbf{u}, \tag{15}$$

where we assume the incompressibility condition $\nabla \cdot \mathbf{u} = 0$. with the Fourier transform, Eq. (15) becomes

$$(-i\omega + \eta k^2)\tilde{\mathbf{b}} = i(\mathbf{B}_0 \cdot \mathbf{k})\tilde{\mathbf{u}}. \tag{16}$$

We find,

$$E = \langle \mathbf{u} \times \mathbf{b} \rangle$$
$$= \iint \frac{\langle \tilde{\mathbf{u}}(\mathbf{k}, \omega) \times \tilde{\mathbf{u}}^*(\mathbf{k}', \omega') \rangle i\mathbf{B}_0 \cdot \mathbf{k}}{-i\omega + \eta k^2}$$
$$\exp\{i(\mathbf{k}-\mathbf{k}')\cdot\mathbf{x} - i(\omega - \omega')t\}\, d\mathbf{k}\, d\mathbf{k}'\, d\omega\, d\omega'. \tag{17}$$

We obtain the expression for the internal electric field,

$$E_i = \alpha_{ij} B_{0j}. \tag{18}$$

where

$$\alpha_{ij} = i\eta\varepsilon_{ikl} \iint \frac{k^2 k_j \Phi_{kl}(\mathbf{k}, \omega)}{\omega^2 + \eta^2 k^4} d\mathbf{k}\, d\omega. \tag{19}$$

This integral is separated into four terms associated with $\Phi_{kl}^{\perp\perp}, \Phi_{kl}^{//\perp}, \Phi_{kl}^{\perp//}$ and $\Phi_{kl}^{////}$. These integrals, except for the first, are zero or pure imaginary (see Appendix A),

$$\alpha_{ij} = i\eta\varepsilon_{ikl} \iint \frac{k^2 k_j \Phi_{kl}^{\perp\perp}(\mathbf{k}, \omega)}{\omega^2 + \eta^2 k^4} d\mathbf{k}\, d\omega$$
$$+ \text{pure imaginary}. \tag{20}$$

The pure imaginary term has no physical meaning. We neglect and omit it from now on.

The direction cosines of the mean magnetic field are $\cos\theta_1, \cos\theta_2, \cos\theta_3$ for x, y and z respectively. From Eq. (18), we write ;

$$E_i = i\eta B_0 \iint \frac{k^2}{\omega^2 + \eta^2 k^4} \varepsilon_{ikl} k_j \Phi_{kl}^{\perp\perp}(\mathbf{k}, \omega) \cos\theta_i\, d\mathbf{k}\, d\omega. \tag{21}$$

After some algebra, we obtain the important equation (see Appendix B),

$$E_i = \alpha B_i, \tag{22}$$

where

$$\alpha = -\eta \iint d\mathbf{k}\, d\omega \frac{k^2}{\omega^2 + \eta^2 k^4} F^{\perp\perp}(k, \omega). \tag{23}$$

Therefore the random wave perturbation flow causes the only parallel internal electric field;

$$\mathbf{E} = \alpha \mathbf{B}_0 \quad . \tag{24}$$

This effect is usually called the alpha effect. This equation holds irrespective of whether $\mathbf{u}$ is isotropic or not.

This result can be understood by following the picture shown in Figure 1[Parker, 1970]. The ordinary explanation of the alpha effect is the case in which mean flow helicity density,

$$h = \langle \mathbf{u} \cdot \nabla \times \mathbf{u} \rangle , \tag{25}$$

is not zero. Such a flow tends to distort a line of force of an initial field $\mathbf{B}_0$ in the manner indicated in Figure 1 (a). The distorted line of force is made by the anti-parallel current. The spatial mean of these elemental currents will have the form $\mathbf{J} = \alpha \mathbf{B}_0 / \eta$,where η is electric resistivity and the coefficient α is negative. Though I is also not zero in the case where the rotational component of the flow is parallel to the mean magnetic field as shown in Figure 1

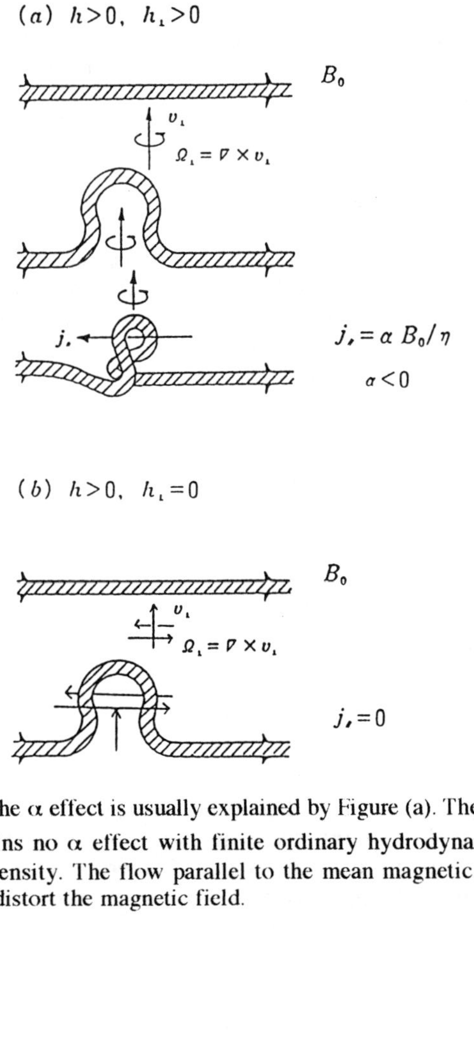

Fig. 1. The α effect is usually explained by Figure (a). The case (b) contains no α effect with finite ordinary hydrodynamical helicity density. The flow parallel to the mean magnetic field does not distort the magnetic field.

(b). The parallel dynamo current is not generated in that case. These cases suggest that hydrodynamic helicity $I_\perp$

$$I_\perp = \int \left(\mathbf{u}_\perp \cdot \nabla \times \mathbf{u}_\perp \right) d^3 \mathbf{x} , \tag{26}$$

is better suited to the investigation of the alpha effect in an anisotropic fluid. It is inferred that this discussion remains valid even if the case were not stationary, uniform and incompressible. In the next section, because it is difficult to solve Eq. (23), we shall discuss hydrodynamic helicity density of the flow perpendicular to the magnetic field, in order to investigate the flow in the RFP simulation. In ideal MHD, Eq. (23) suggests that the alpha effect can easily be caused by low frequency and short wavelength fluid motion. The resistive case is more complicated, and it will be important to compare frequency and magnetic diffusion terms.

3. SIMULATION

The reversed field pinch dynamo is simulated numerically. The code which we use is a three dimensional, time dependent, non-linear, compressible, resistive MHD simulation code in periodic cylindrical geometry. The normalized equations solved here are,

$$\frac{\partial \mathbf{A}}{\partial t} = \mathbf{v} \times \mathbf{B} - \eta \mathbf{J} , \tag{27}$$

$$\rho \frac{D\mathbf{v}}{Dt} = -\nabla p + \mathbf{J} \times \mathbf{B} + \frac{\nu}{S} \nabla^2 \mathbf{v} , \tag{28}$$

$$\frac{Dp}{Dt} = -\frac{5}{3} p \nabla \cdot \mathbf{v} + \eta J^2 + \kappa \nabla^2 p , \tag{29}$$

$$\mathbf{B} = \nabla \times \mathbf{A} , \tag{30}$$

$$\mathbf{J} = \nabla \times \mathbf{B} , \tag{31}$$

$$\frac{D}{Dt} = \frac{\partial}{\partial t} + \mathbf{v} \cdot \nabla ,$$

where $\mathbf{B}$ is the magnetic field measured in units of a characteristic field $B_0 = B_z(r = 0)$, length is measured in units of minor radius of the cylinder a, ρ is the mass density measured in units of a characteristic density $\rho_0 = \rho(r = 0)$. $\mathbf{v}$ is the velocity measured on units of Alfven velocity $v_A = B_0 / \sqrt{2\rho_0}$, t is the time measured in units of Alfven transit time $\tau_A = a / v_A$, p is the thermodynamic pressure measured in units of $p_0 = p(r = 0)$, $\mathbf{A}$ is the vector potential measured in units of aB_0 and $\mathbf{J}$ is the current measured in units of B_0 / a. To keep pressure constant, the dissipation coefficient κ in the equation of pressure is feed-back controlled. The coefficient of viscosity ν is fixed at 2.5. η is a non-dimensional resistivity which is constant in time. We assume that the mass density is uniform in space and time. S is the magnetic Reynolds or Lundquist number which is defined as $S = \tau_R / \tau_A$ where $\tau_R = a^2 / \eta(0)$is the characteristic resistive diffusion time. We simulate the reversed field sustainment for the case in which $S=10^4$.

The solution of Eqs. (27)-(31) are subject to ideal conducting wall boundary conditions, where total axial flux is constant in time. Further, we assume that total toroidal current imply that;

$$A_r = 0, \tag{32}$$

$$A_\theta = \frac{\Phi}{2\pi} = \text{const}, \tag{33}$$

$$\frac{\partial A_z}{\partial r} = -\frac{I_z}{2\pi} = \text{const}, \tag{34}$$

where Φ is the total flux of the magnetic field and I_z is the total current across the poloidal surface.

The boundary condition for the velocity and pressure are ;

$$v_r = 0,\ p = 0 \tag{35}$$

For v_θ and v_z we use the free slip boundary conditions. The boundary condition along z is periodic where the period is $2\pi \times R/a$. R/a is aspect ratio.

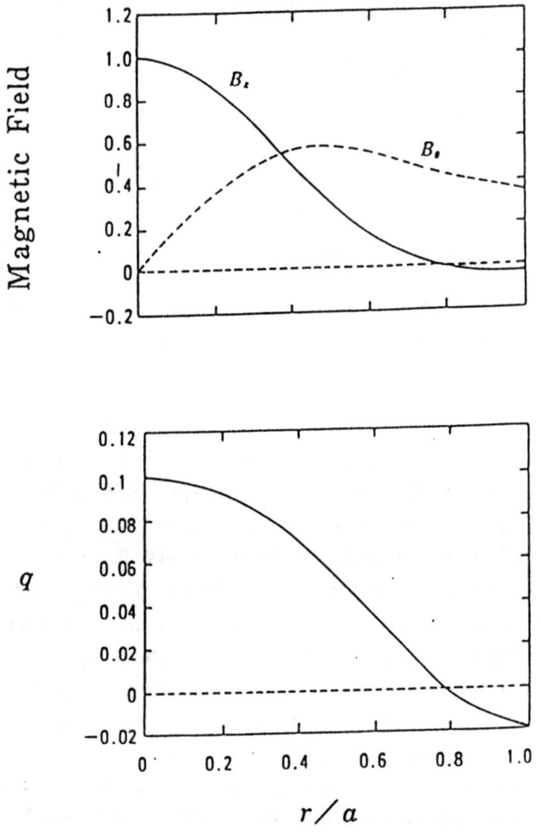

Fig. 2. The initial magnetic field is set according to Robinson. The toroidal magnetic field is already reversed. We simulate the sustainment of the reversed magnetic field.

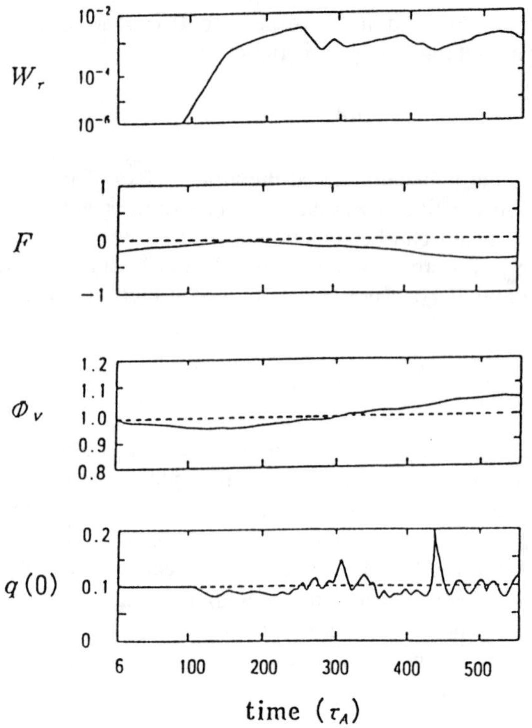

Fig. 3. The time evolution of parameters are plotted. The MHD instability grows. After that, the sustainment of the reversed magnetic field starts.

The initial condition is modeled after Robinson [Robinson,1978] as shown in Figure 2. The initial toroidal magnetic field B_z is already reversed at r_v=0.8 where r_v is called the field reversal radius.

In Figure 3, we show the time evolution of parameters such as reversed magnetic field ratio F, safety factor at the toroidal axis $q(0)$, magnetic flux inside the field reversal surface Φ_V, total energy of radial magnetic field W_r during the simulation which are defined by

$$F \equiv \frac{B_z(a)}{\langle\langle B_z \rangle\rangle}, \tag{36}$$

$$q = \frac{rB_z}{RB_\theta}, \tag{37}$$

$$\Phi_v = \int_0^{r_v} B_z r dr, \tag{38}$$

$$W_r = \int (b_r)^2 d^3\mathbf{r}. \tag{39}$$

Here double brackets denote the average over the whole plasma. F maintains a negative value during the simulation. In the first stage, F increases due to the magnetic diffusion. After $t = 150\tau_A$, F decrease to -0.4 at

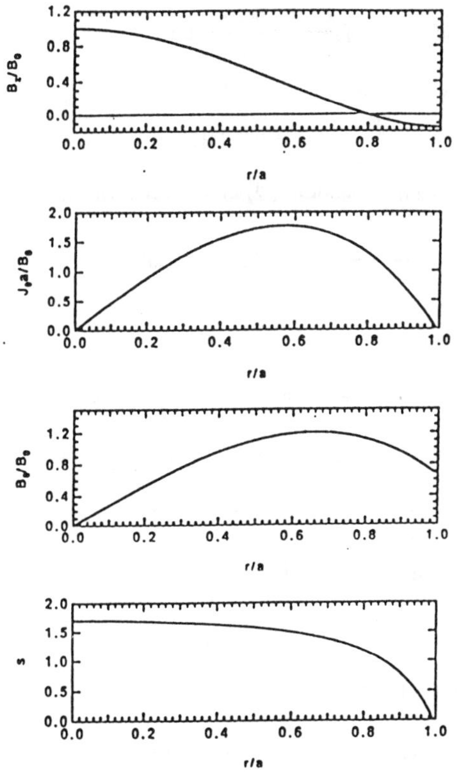

Fig. 4. The expected flow helicity density is represented.

$t = 600\tau_A$. This shows the reversed field is sustained. The field reversal starts after the growth of the instabilities is large enough. The $\theta = B_\theta(a)/ \ll B_z \gg$ value remains almost constant at 1.9. A feed back control of pressure keeps the β value 0.04. We already detected the internal electric field in this calculation to identify the dynamo effect [Koide, 1990].

4. DISCUSSION

Now, let us discuss the relation between the hydrodynamic flow pattern and the internal electric field. What kind of hydrodynamic helicity density of the instability is expected? In reversed field pinch, the external electric field is imposed only in the toroidal direction. If the perturbation flow is uniform, steady and small , we found with Eqs. (23) and (24) in the previous section that;

$$j_\theta = sB_\theta^0 \quad , \qquad (40)$$

where

$$s = -\iint dk\, d\omega \frac{k^2}{\omega^2 + \eta^2 k^4} F^{11}(k,\omega) \quad . \qquad (41)$$

As shown in Figure 4, the toroidal magnetic field is reversed along the radius. $J_\theta^0 = -\frac{dB_z^0}{dr}$ is positive. B_θ^0 is also positive. Therefore s is positive. $h_\perp$ must be negative according to the equation (41).

We discussed the ordinary hydrodynamic helicity density $h = \mathbf{u} \cdot \nabla \times \mathbf{u}$ as shown in Figure 5 (a). The base and top surfaces of the box represents the x-y surface or r-θ surface. The height from bottom to top corresponds to one period. The longitudinal lines are put along the z axis. The side surfaces are tangential to the radial boundary of the cylinder. The circles indicate the value of the hydrodynamic helicity density at each point. The area indicates the absolute value. Black circles are positive values and white circles are negative values. The ordinary helicity density of the flow is positive as not expected. The helicity density perpendicular to the mean magnetic field $h_\perp = \mathbf{u}_\perp \cdot \nabla \times \mathbf{u}_\perp$ is negative in Figure 5 (b). We can employ the model about the relation between flow and magnetic field in the simulation as shown in Figure 1. This example shows the importance of distinguishing total helicity density and perpendicular helicity density.

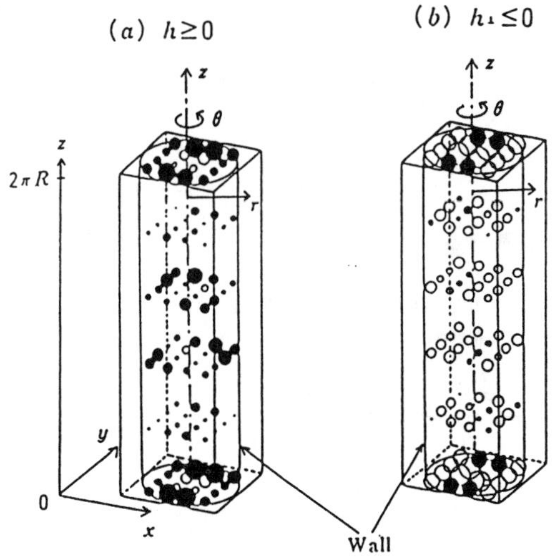

Fig. 5. The helicity density is seen in the numerical simulation. (a) The ordinary flow helicity density is plotted. This helicity density is positive which is opposite to what is expected. (b) The helicity density of flow perpendicular to the mean magnetic field is plotted. This is negative which is expected to sustain the reversal magnetic field.

Acknowledgments. One of authors (S. K.) thanks Dr. T. Amano for his numerical support. He also thanks Dr. M.Okamoto for his kind encouragement. We thank Dr. F. Yasseen of the Center for Space Research at MIT for his editing and correction of this manuscript. We also thank the computer center of National Institute of Fusion Science for providing a good environment in which to perform the numerical calculation.

APPENDIX A

To derive the Eq. (20), we begin with Eq. (19). $\cos\theta_i$ is a directive cosine of the mean magnetic field;

$$\tilde{u}_i^{//}(\mathbf{k},\omega) = \tilde{u}^{//}(\mathbf{k},\omega)\cos\theta_i \tag{42}$$

We can write,

$$\begin{aligned} &< \tilde{u}^{//}(\mathbf{k},\omega)\tilde{u}^{//*}(\mathbf{k}',\omega') > \cos\theta_i\cos\theta_j \\ &\equiv \Phi_{ij}^{////}(\mathbf{k},\omega)\delta(\mathbf{k}-\mathbf{k}')\delta(\omega-\omega') \end{aligned} \tag{43}$$

Defining

$$< \tilde{u}^{//}(\mathbf{k},\omega)\tilde{u}^{//*}(\mathbf{k}',\omega') > \equiv \Phi^{////}(\mathbf{k},\omega)\delta(\mathbf{k}-\mathbf{k}')\delta(\omega-\omega'), \tag{44}$$

we get,

$$\Phi_{ij}^{////}(\mathbf{k},\omega) = \Phi^{////}(\mathbf{k},\omega)\cos\theta_i\cos\theta_j \tag{45}$$

One of the terms corresponding to $\Phi_{kl}^{////}$ of the Eq. (19) becomes

$$\begin{aligned} &i\eta\varepsilon_{ikl}\iint \frac{k^2 k_j}{\omega^2+\eta^2 k^4}\Phi_{kl}^{////}(\mathbf{k},\omega)\,d\mathbf{k}\,d\omega \\ &= i\eta\varepsilon_{ikl}\iint \frac{k^2 k_j}{\omega^2+\eta^2 k^4}\Phi^{////}(\mathbf{k},\omega)\varepsilon_{ikl}\cos\theta_k\cos\theta_l\,d\mathbf{k}\,d\omega \\ &= 0 \end{aligned} \tag{46}$$

using,

$$(\mathbf{n}\times\mathbf{n})_i = \varepsilon_{ijk}\cos\theta_i\cos\theta_j = 0, \tag{47}$$

where $\mathbf{n}$ is a unit vector parallel to the mean magnetic field. The sum of terms corresponding to $\Phi_{kl}^{//\perp}, \Phi_{kl}^{\perp//}$ becomes to

$$\begin{aligned} &i\eta\varepsilon_{ikl}\iint \frac{k^2 k_j}{\omega^2+\eta^2 k^4}\Phi_{kl}^{\perp//}(\mathbf{k},\omega)\,d\mathbf{k}\,d\omega \\ &+ i\eta\varepsilon_{ikl}\iint \frac{k^2 k_j}{\omega^2+\eta^2 k^4}\Phi_{kl}^{//\perp}(\mathbf{k},\omega)\,d\mathbf{k}\,d\omega \\ &= i\eta\varepsilon_{ikl}\iint \frac{k^2 k_j}{\omega^2+\eta^2 k^4}\Phi_{kl}^{\perp//}(\mathbf{k},\omega)\,d\mathbf{k}\,d\omega \\ &+ i\eta\varepsilon_{ikl}\iint \frac{k^2 k_j}{\omega^2+\eta^2 k^4}\Phi_{lk}^{\perp//}(\mathbf{k},\omega)\,d\mathbf{k}\,d\omega \\ &= i\eta\varepsilon_{ikl}\iint \frac{k^2 k_j}{\omega^2+\eta^2 k^4}2\,\mathrm{Re}(\Phi_{kl}^{\perp//}(\mathbf{k},\omega))\,d\mathbf{k}\,d\omega \end{aligned} \tag{48}$$

using

$$\Phi_{kl}^{\perp//}(\mathbf{k},\omega) = \Phi_{lk}^{//\perp*}(\mathbf{k},\omega). \tag{49}$$

This term is pure imaginary. This corresponds to the pure imaginary internal electric field which has no real physical meaning. Therefore we neglect this term. We reduce the Eq. (19) to;

$$\alpha_{ij} = i\eta\varepsilon_{ikl}\iint \frac{k^2 k_j \Phi_{kl}^{\perp\perp}(\mathbf{k},\omega)}{\omega^2+\eta^2 k^4}\,d\mathbf{k}\,d\omega \tag{20}$$

APPENDIX B

We derive the equation (22). From Eq. (20), we find,

$$E_i = i\eta B_0\iint \frac{k^2}{\omega^2+\eta^2 k^4}\varepsilon_{ikl}k_j\Phi_{kl}^{\perp\perp}(\mathbf{k},\omega)\cos\theta_j\,d\mathbf{k}\,d\omega \tag{50}$$

We can write,

$$\{\mathbf{u}^{\perp}(\mathbf{k},\omega)\times\mathbf{u}^{\perp}(\mathbf{k}',\omega')\}\times\mathbf{n} = \mathbf{0}, \tag{51}$$

which, written using coordinate components, becomes

$$\varepsilon_{ijk}\varepsilon_{jpq}\Phi_{pq}^{\perp\perp}(\mathbf{k},\omega)\cos\theta_k = 0 \tag{52}$$

Using this equation, we obtain,

$$\varepsilon_{ilm}\Phi_{lm}^{\perp\perp}(\mathbf{k},\omega)k_j\cos\theta_j = \varepsilon_{jlm}\Phi_{lm}^{\perp\perp}(\mathbf{k},\omega)k_j\cos\theta_i \tag{53}$$

Eq. (50) changes to

$$E_i = i\eta B_0\cos\theta_i\iint \frac{k^2}{\omega^2+\eta^2 k^4}\varepsilon_{jlm}k_j\Phi_{lm}^{\perp\perp}(\mathbf{k},\omega)\,d\mathbf{k}\,d\omega \tag{54}$$

We get the final result;

$$E_i = \alpha B_i^0, \tag{22}$$

where

$$\alpha = -\eta\iint \frac{k^2}{\omega^2+\eta^2 k^4}F^{\perp\perp}(k,\omega)\,dk\,d\omega \tag{23}$$

REFERENCES

Moffatt, H. K. , *Magnetic Field Generation in Electrically Conducting Fluids*, Cambridge University Press,1978.

Robinson, D. C., Tearing-mode-stable diffuse-pinch configuration, *Nucl. Fusion* **18**, 939-953,1978.

Koide, S., 3-dimensional simulation of dynamo effect of reversed field pinch, *Journal of Physical Society of Japan* **59**, 3952-3961,1990.

Parker, E. N. , The generation of magnetic field in astrophysical bodies, I. The dynamo equations, *Astrophys. J.* **162**, 665-673,1970.

S. Koide and J. I. Sakai, Laboratory for Plasma Astrophysics and Fusion Science, Department of Electronics and Information, Faculty of Engineering, Toyama University, Gofuku, Toyama 930 JAPAN.

A Novel Technique for the Numerical Simulation of Collision Free Space Plasma-Vlasov Hybrid Simulation (VHS)

DAVID NUNN

Department of Electronics and Computer Science, University of Southampton, Southampton, Hampshire. S09 5NH. UK

This paper develops a novel method for the numerical simulation of hot collision free space plasma. The method follows the phase trajectories of simulation particles in the phase space simulation box and *interpolates* the value of distribution function from particles onto a fixed phase space grid. The technique is highly efficient and low noise. Fluxes of phase fluid onto and out of the phase box are permitted and this may give great savings. The VHS algorithm is successfully applied to the numerical simulation of triggered VLF emissions.

INTRODUCTION

This paper will describe a novel algorithm for the numerical simulation of hot collision-free plasma. The formalism is a purely general one for the solution of the collision-free Vlasov equation plus Maxwell's equations, and may be applied in principle to any space plasma simulation problem. There will be applications to laboratory plasmas, fusion plasmas and industrial plasmas where collisions may be neglected. The theoretical bases of the algorithm are strictly collision-free but collisional effects and/or velocity space diffusion may be incorporated - albeit inconveniently.

The VHS algorithm is intrinsically more efficient than PIC techniques and is a straightforward, stable and low noise algorithm. It is simpler than other Vlasov algorithms [*Denavit, 1972; Cheng and Knorr, 1976*] and does not need to invoke artificial smoothing in velocity space. The VHS algorithm is particularly characterised by the fact that the population of simulation particles is dynamic and that the algorithm permits a flux of particles at the phase box boundary. This is particularly useful for problems such as nonlinear wave particle interactions in inhomogeneous media.

Space Plasmas: Coupling Between Small
and Medium Scale Processes
Geophysical Monograph 86

BASICS

The relevant equations are Maxwell's equations and the collision-free Vlasov equation

$$\left\{\frac{\partial}{\partial t} + \underline{v}\cdot\frac{\partial}{\partial \underline{x}} + \frac{q_\alpha}{m_\alpha}(\underline{E} + \underline{v}\times\underline{B})\cdot\frac{\partial}{\partial \underline{v}}\right\}F_\alpha = \frac{d}{dt}F_\alpha = 0 \qquad (1)$$

where M species are present. Plasma charge density $\boldsymbol{\rho}$ and current density $\underline{J}$ are of course given by

$$\left\{\frac{J}{\rho}\right. = \sum_{\alpha=1}^{M} q_\alpha \int_{\underline{v}} \left\{\frac{v}{1}\right\} F_\alpha\, d\underline{v} \qquad (2)$$

Not all particle species present need be treated by VHS. Some may be described by fluid equations, or may be cold plasma/cold beams better described analytically or by PIC codes. Henceforth we shall consider only one species treated by VHS. The plasma will be described as a Vlasov or phase fluid filling a phase space of dimensionality n. Granularity of the phase fluid will be ignored. The initial distribution function F_0

$$F_0(\underline{x}, \underline{v}) = F(\underline{x}, \underline{v}, t=0) \qquad (3)$$

is required to be a regular and well behaved function of phase coordinates $\underline{x}$, $\underline{v}$. Discontinuities are acceptable, but delta function beams need to either be given a finite temperature or treated separately either analytically or by a PIC code.

THE PHASE SPACE SIMULATION BOX

The phase box (PSB) is constructed appropriate to the problem at hand. In general the PSB may *vary with time* as the simulation progresses. A phase space grid is defined, with elementary volume $d\Gamma = d\underline{x}.d\underline{v}..$ The grid may be inhomogeneous, non-rectilinear or indeed adaptive.

SIMULATION PARTICLES

The phase space simulation box is evenly filled with simulation particles (SP's) at the start of the simulation (t=0). At each time step each particle is pushed according to the usual equations of motion

$$\dot{\underline{x}} = \underline{v} \; ; \; \dot{v} = \frac{q}{m} (\underline{E} + \underline{v} \times \underline{B}) \tag{4}$$

Each particle trajectory is followed until it leaves the PSB. Trajectories are not restarted at phase space grid points as in the paper by Denavit. [*Denavit, 1972*] New trajectories are continually started at the phase box boundary.

Now each SP is embedded in the phase fluid and moves with it. By Liouville's theorem each SP conserves its value of distribution function F($\underline{x}$, $\underline{v}$, t). The value of F (or δF) is thus defined on the phase trajectories of the simulation particles. During the simulation the value of F is known at a large number of points in the phase box which are the current locations of the SP's. The function of the SP's is solely to *provide information.* At each timestep this information is ued to construct the distribution function on the fixed phase space grid and thus make *estimates* of the zeroth and first moments of F- which are of course plasma charge density and current density.

At each timestep we require to *interpolate* the values of distribution function F_l from the particles onto the fixed phase space grid, giving grid estimates $F_{ijk...}$. This process of interpolation is quite different from that in PIC codes and other Vlasov codes, where charge/current or indeed distribution function are *assigned* or distributed to neighbouring grid points.

Once distribution function $F_{ijk...}$ is defined on the phase space grid estimates of $\underline{J}$ and ρ are readily obtained. For the 6D case

$$\{\frac{\underline{J}}{\rho_{ijk}} = q\, d\Gamma \sum_{lmn} \{\frac{\underline{v}_{lmn}}{1}\} F_{ijklmn} \tag{5}$$

In the evaluation of the above expression it should be noted that values of $F_{ijk...}$ at grid points near the boundary of the phase box will not be accurate and the moments of F are better evaluated using a smaller phase box, eliminating the boundary region.

With charge and current density defined on the spatial grid the EM fields may be pushed using standard field push techniques.

In many plasma simulation problems it might be more convenient to define the quantity δF on each phase trajectory, where

$$\delta F(\underline{x}, \underline{v}, t) = F(\underline{x}, \underline{v}, t) - Fo(\underline{x}, \underline{v}) \tag{6}$$

whence

$$\frac{d}{dt}\partial F = -\underline{v}\frac{\partial Fo}{\partial \underline{x}} - \frac{q}{m}(\underline{E} + \underline{v} \times \underline{B}) \cdot \frac{\partial Fo}{\partial \underline{v}} \tag{7}$$

In the VHS formalism the distinction between defining F or δF on the phase trajectories is fairly trivial - hence VHS may be regarded as an algorithm that pushes δF. Note that the VHS method is *not* the same as a PIC code in which particles are weighted by initial distribution function. The interpolation procedure for F ensures that the available information is treated quite differently.

INTERPOLATION OF DISTRIBUTION FUNCTION FROM PARTICLES TO THE PHASE SPACE GRID

Clearly we need an efficient method for the interpolation of F from the particles onto the fixed phase space grid. Figure 1 shows a section of the grid of a 2D phase space simulation box. Using a derivative of the method of area weighting a suitable expression for the value of distribution function F_{ij} at grid point ij is

$$F_{ij} = \{\sum_{l=1}^{l'} \alpha_l F_l\} / \{\sum_{l} \alpha_l\} \tag{8}$$

where the weight factor α_l for the l'th particle is given by

$$\alpha_l = \{A_1 / \sum_{j=1}^{4} A_j\} \tag{9}$$

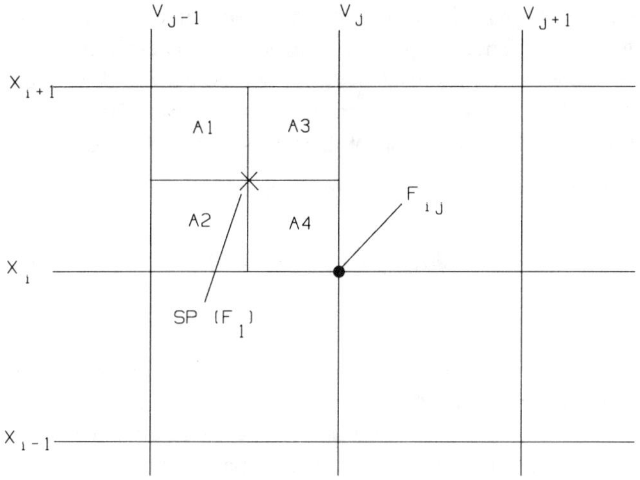

Fig 1. Representation of area weighting scheme for interpolating F from particles to the phase space grid-2D case.

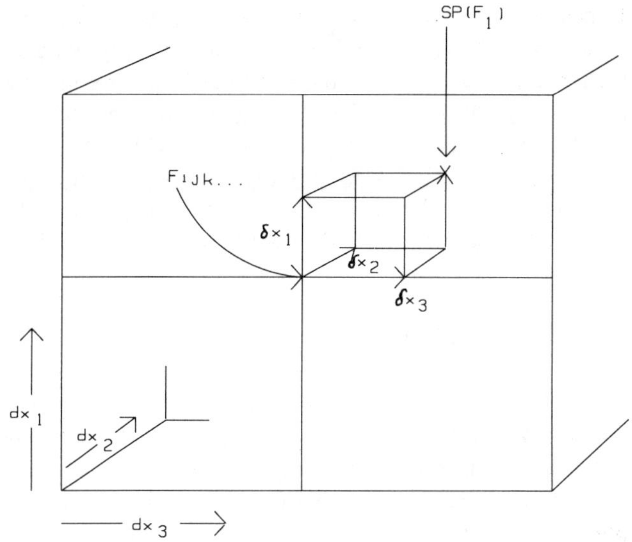

Fig 2. Representation of interpolation scheme for the nD case.

The sum is taken over all l' SP's within the 2^n square area surrounding the grid point in question. The above technique will give a very low noise level in F_{ij} and is simple and easy to encode.

A small number of grid points ij (<1%), particularly near the phase box boundary, will not have any SP's within the surrounding 4-square area. In these cases F_{ij} may be calculated by linear interpolation from surrounding grid points.

The expressions in equations 8 and 9 are readily generalised to the case of an n dimensional phase space. Figure 2 shows a representation of n dimensional phase space with elementary grid volume $d\Gamma = dx.dv..$

A total of 2^n hypercubes will be adjacent to grid point ijk... If a total of l'SP's lie within this volume of phase space a suitable expression for F_{ijk}... is given by

$$F_{ijk\cdots} = \{\sum_{l'=1}^{l} \beta_{l'} F_{l'}\} / \{\sum_{l'=1}^{l} \beta_{l'}\} \tag{10}$$

where the weighting factor for particle l' may be defined by

$$\beta_{l'} = \prod_n (dx_n - \partial x_n^{l'}) / \prod_n dx_n \tag{11}$$

Here $\partial x_n^{l'}$ defines the vector from the grid point ijk... to the l'th particle.

THE REQUIRED DENSITY OF SP'S IN PHASE SPACE

We first note that from Liouville's theorem the density of SP's in the phase fluid is conserved following a particle trajectory, and thus there is no tendency for SP's to bunch and leave grid point `uncovered'. The VHS algorithm requires that 99% of all phase space grid points have at least one SP within the 2^n adjacent hypercubes of volume $d\Gamma = d\underline{x}.d\underline{v}$. Numerical experimentation shows that an average density $\hat{\rho} > \rho_0$ where

$$\hat{\rho}_0 = (4.8) / (2^n \, d\Gamma) \tag{12}$$

will give a 99.5% coverage of grid points. Note that with increasing dimensionality n, fewer SP's are needed per elementary volume $d\Gamma$. For example for a 6D code $\hat{\rho}_0 \, d\Gamma \sim 0.075$ would suffice. To give an idea of the number of particles required by a VHS code, consider a 1 3/2D simulation with 1000 grid points in 1 spatial dimension, and 30 grid points in each of 3 velocity dimensions. The total number of SP's required would be 4.8*30*30*30*1000/16~6M, entirely feasible with current supercomputers.

In some problems such as hot beam excitation there will be regions of phase space where F=0. SP's need not be placed in these regions. It is only necessary to increase SP density somewhat to say 1.3 $\hat{\rho}_0$ in regions where F=0, then all grid points ijk... with no SP's in the surrounding 2^n hypercubes may be regarded as having F_{ijk}...=0.

It must be emphasised that SP density $\hat{\rho}_0$ is a minimum value. Due to the nature of the interpolation procedure higher densities are permitted. Increased densities incur a larger computational workload but give fewer missed grid points, lower noise levels and better averaging over distribution function fine structure. This tolerance of variable density of SP's in the phase fluid makes it relatively easy to incorporate a flux of phase fluid at the phase box boundary.

PARTICLE CONTROL

A VHS code needs to control the particle population such that $\hat{\rho} > \hat{\rho}_0$ everywhere, and the total population of SP's Nt is within certain bounds. Figure 3 shows a representation of a 2D phase box. Over parts of the phase box boundary phase fluid will be flowing out. SP's embedded in the fluid will leave the phase box. These particles are by definition now located in an unimportant region of phase space and are providing information not required. These particles are *discarded* from the simulation. Conversely over parts of the phase box boundary phase fluid will be flowing *in*. New SP's must be inserted into this fluid. At every timestep all grid points on or near the phase box boundary are examined.

New particles are inserted where there are no SP's in adjacent hypercubes. This must be done with care. The exact positions of insertion determine the resulting density in the incoming fluid.

The VHS algorithm is characterised by a *dynamic* particle population. The total number of SP's N_t will fluctuate as the simulation progresses. The algorithm is highly efficient since only particles in important regions of phase space are followed. Thus for example in wave particle interaction simulations non resonant particles will be outside the current phase box and will be discarded from the simulation. The total N_t needs to be kept within reasonable bounds, say

$$d\int N_c\hat{\rho}_0 < N_t < 1.3\, d\int N_c\hat{\rho}_0 \qquad (13)$$

where N_c is the number of hypercubes in the phase box. If N_t falls too low inaccuracy will increase and uncovered grid points will result. The solution is to create new particles in the interior of the phase box at grid points that have 1 or 0 `adjacent' particles. If N_t becomes too large the program will slow up and memory requirements will increase. The remedy here is to remove particles from the phase box interior where SP density if high. In the demonstrator application it was found that these measures were not necessary and that N_t was adequately controlled by careful insertion of new particles at the boundary.

Where the phase space grid is inhomogeneous it will of course be necessary to create or delete particles at internal boundaries between regions of different grid density.

There will exist a wide variety of problems where the flux of phase fluid at the box boundary will be negligible and where F=0 on the boundary. In these cases the simulation may proceed with a fixed population N_t, provided that the initial density is raised to ~$1.2\,\hat{\rho}_0$ so that grid points with no local SP's may safely be assumed to correspond to F=0.

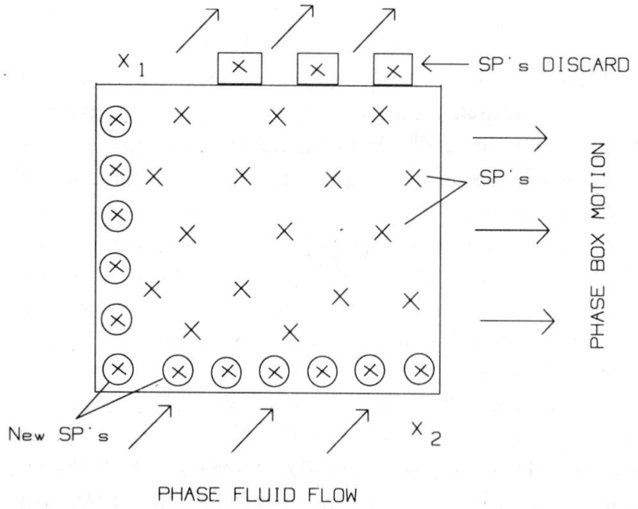

Fig 3. Representation of 2D phase box showing particle discards and insertions of new particles.

VALUES OF DISTRIBUTION FUNCTION FOR SIMULATION PARTICLES

At t=0 simulation particles are given a value for distribution function of $F=F_0$, where F_0 should be self consistent with the presumed initial fields. New SP's inserted into the phase fluid need to be assigned a value for F. For insertions at the phase box boundary $F=F_0$ is often quite sufficient. If the simulation turns out to be unduly sensitive to the choice of F for new particles, this is probably symptomatic of a phase box that is too small. In certain cases a better value for F may be available. In the VLF problem to be described a linear expression for F derived from the EM field history is used. Where a new particle is inserted into the interior of the phase box an initial value for F is readily secured by interpolating from neighbouring grid values $F_{ijk\ldots}$.

SOME COMMENTS ON DISTRIBUTION FUNCTION FINE STRUCTURE

In many plasma simulations the distribution function may develop fine structure in phase space. For example this occurs with nonlinear Landau resonance when resonant particles make many oscillations within the potential trap. Usually fine structure does not have a great deal of physical significance since during the evaluation of $\underline{J}/\rho$ it will be averaged out. However fine structure can be a considerable nuisance for Vlasov codes. Other Vlasov codes such as that of Denavit [*Denavit, 1972*] and Cheng and Knorr [*Cheng and Knorr, 1976*] need to be stabilised by that smooth out fine structure. The VHS method however is intrinsically stable against fine structure, since no attempt is made to evaluate derivatives of F in phase space. The VHS method does not smooth distribution function. This is considered neither desirable nor necessary.

PREVIOUS VLASOV SIMULATIONS

During the 1970's and 1980's a number of papers were written reporting successful simulations using Vlasov type methods. Two methods stand out. The first is that of Cheng and Knorr [*Cheng and Knorr, 1976*] who numerically integrate the Maxwell/Vlasov set of equations. Particles are not used at all in this process. Because of distribution function fine structure the algorithm is complex and only stabilised by a non physical smoothing in phase space. Flux of phase fluid at the box boundary is possible but was not considered.

Denavit [*Denavit, 1972*] uses particles to time advance distribution function defined on a phase space grid. Particles are started off at phase space grid points and assigned appropriate values of F. After M timesteps the values of F defined on particles are *assigned* to nearest grid points. This effects a reconstruction of F on the phase space grid and invokes artificial diffusion. In the limit of M=1 reconstruction occurs at every timestep, and phase fluid flux at the phase box boundary may be accommodated but this was not considered.

A third method by Kotschenreuther [*Kotschenreuther, 1988*] resembles a PIC code except that dF is pushed. [*Kotschenreuther, 1988*]. Kotschenreuther noted that when $\delta F << F_0$ PIC codes are extremely noisy and inefficient and strongly outperformed by Vlasov codes.

SUMMARY OF ADVANTAGES OF VHS

It is worth summarising the advantages of the VHS method at this point. These are

(a) Low noise, intrinsically efficient.
(b) Accommodates phase fluid flux across phase box boundary. The particle population is dynamic. Time is not wasted following particles in unimportant regions of phase space.
(c) Good diagnostics, distribution function immediately available.

SIMULATION OF TRIGGERED VLF EMISSIONS

The demonstrator application is a space plasma simulation in the VLF band at 4-6kHz. The code is a simulation of rising frequency emissions triggered by narrow band VLF pulses transmitted from the VLF facility at Siple, Antarctica. [Helliwell and Katsufrakis, 1974] Triggered emissions result from nonlinear electron cyclotron resonance in the equatorial region of the earth's magnetosphere at L=4.1. The ambient hot electron distribution function is anisotropic and unstable and of a loss cone type. Cyclotron resonant energies are of order keV.

In this paper we shall only make some general remarks and present some sample simulation results. For full details of the VLF problem the reader is referred to [*Nunn,1990*] and [*Nunn,1993*] The triggered VLF emission problem is an excellent one with which to test simulation techniques. The phenomenon is strongly nonlinear and well defined. There is also a vast amount of data available from the active experiment at Siple.

The code has one spatial dimension z, the distance from the equator along the L=4.1 field line. The VLF wavefield is assumed to be parallel propagating (ducted), and restricted to a bandwidth of about 100Hz. The spatial grid is constant with N_j=512 points, and spans the nonlinear trapping zone a few thousand kms either side of the equator.

The ambient magnetic field strength $B_0(z)$ is assumed to have a parabolic dependence upon z. The parabolic inhomogeneity plays a key role in determining the dynamics of resonant particle trapping, and also serves to confine the nonlinear trapping region about the equator.

The code has three velocity dimensions. Perpendicular velocity $V\perp$ is a weak coordinate and may be dealt with in a simplified manner. The code need only use 3 discrete values of $V\perp$. Important velocity components are gyrophase covered by 20 grid points, and parallel velocity $V^*=V_z-V_{res}$, covered by 40-50 grid points.

In order to simulate emissions with a sweeping frequency the centre frequency of the wavefield is allowed to move freely, but at each timestep the wavefield is filtered to a bandwidth of ~50Hz either side of the central frequency. The range of parallel velocity V* covered by the phase box must span the resonant velocities of the current global wavefield, plus several resonant widths either side. Clearly, when simulating a triggered emission frequency changes with time, and so must the V* range. In this problem the phase box is clearly a function of time. The flux of phase fluid across the boundary of the phase box will be highly significant. Due to the magnetic field inhomogeneity and the time varying frequency particles are continually falling out of resonance with the wavefield and are no longer of interest. New particles are constantly being swept into resonance with the wavefield. Thus VHS with its dynamic particle population is highly suited to this problem. Indeed PIC codes and Vlasov codes with fixed phase boxes will require an inordinate number of simulation particles in order to simulate triggered VLF emissions.

The code has a number of specialist features due to the fact that the problem is basically narrowband. The whistler wavefield is defined as a complex quantity R on a separate spatial grid, the field grid. At each time step it is necessary to cross interpolate with the current/particle grid using precomputed coefficients. Since all particles of interest have V_z very close to V_{res} (z=0) particles need only be defined at grid points on the particle spatial grid. Consequently the code applies the VHS interpolation procedure to separate 2D V^*/ϕ planes at each spatial grid point and for each value of $V\perp$. The reader is referred to [*Nunn, 1990*] for further details.

The code employs two discrete values of $V\perp$, referred to as beams. These correspond to pitch angles of 64.4 and 67.3 degrees. Each beam is driven by the corresponding gradient in the unperturbed distribution function F_0.

$$\left[\frac{\partial F_0}{\partial W} + \frac{2}{\omega_0}\frac{\partial F_0}{\partial \mu}\right]_{V\perp,\ V_z = V_{res}} \tag{14}$$

The code actually computes the quantity ΔW, the integrated energy change, along resonant particle trajectories. The code thus explicitly pushes what is effectively δF.

It was found that the 1D code as set up here has no effective saturation mechanism - the field amplitude will go to infinity. Two saturation mechanisms operate in reality. The first of these we will refer to as nonlinear unducting loss. The VLF wavefield is largely confined to a narrow duct of elliptical cross section. When nonlinear trapping sets in at amplitudes >1.4pT the nonlinear resonant particle current radiates power into untrapped VLF modes causing a substantial loss of energy. The second mechanism is as follows. The nonlinear trapping of cyclotron resonant electrons in the inhomogeneous medium results in instability to electrostatic waves, which act to diffuse the phase averaged distribution function in the V_z direction and reduce the whistler growth rate. In the 1D code saturation has to be incorporated phenomenologically with a damping term that switches in above the nonlinear amplitude of 2pT. A rigorous simulation would of course require a full 3D code and inclusion of the electrostatic wavefield.

The code currently runs with about 0.7 million particles. On the IBM3090/600e each timestep takes about 9 secs, and an entire simulation of a VLF emission takes 6 hours of CPU time. All intensive parts of the code vectorise, and in addition the code is highly parallel - having an effective parallelism of 512 corresponding to the spatial particle grid size.

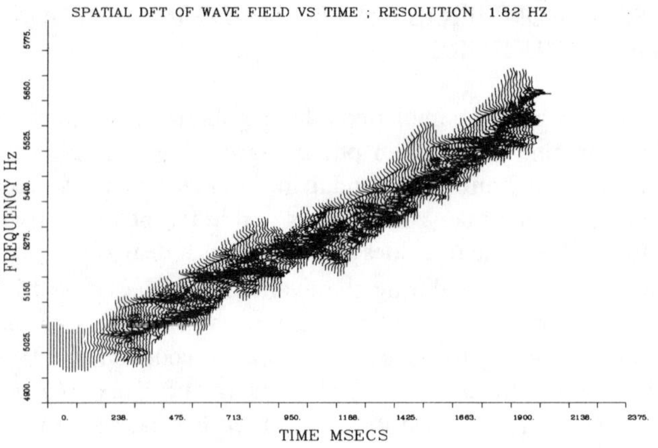

Fig 4. View of simulation zone at t=525 msecs. The curve labelled B is the wave amplitude profile.

The curve labelled γ_{es} is an estimate of electrostatic growth rate. The curve labelled BW is the percentage of available bandwidth that is unstable to electrostatic waves.

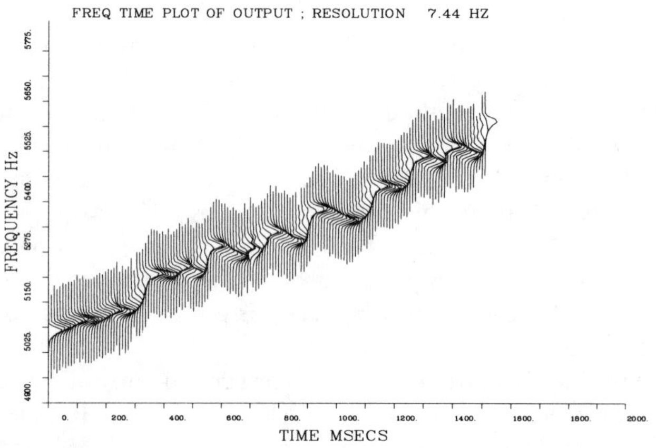

Fig 5. Plot of resonant particle currents Jr (in-phase) and Ji (out-of-phase) as functions of z at time t=525 msecs.

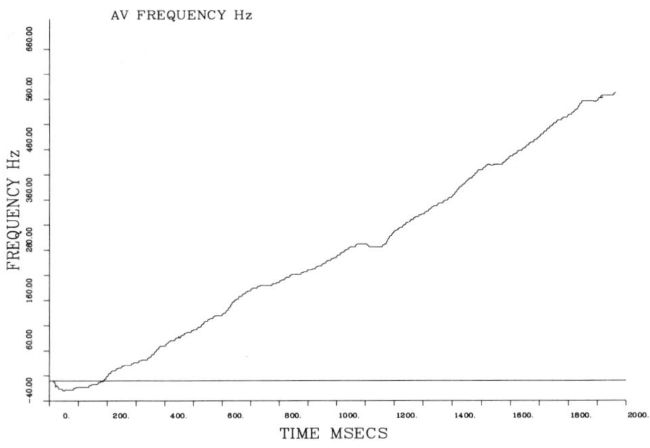

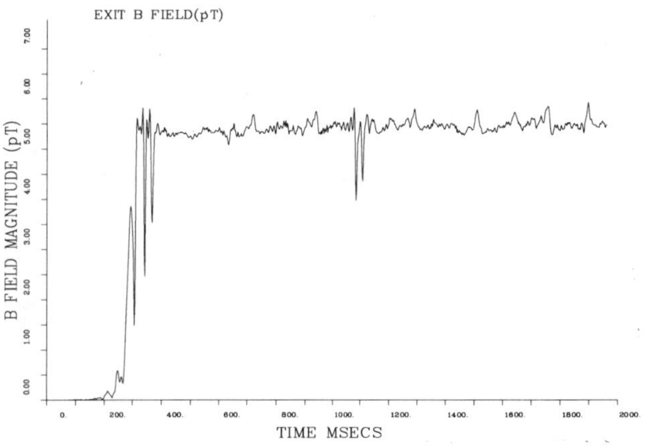

Fig 6. Frequency time plot of wavefield emerging from the right hand boundary of the simulation zone. The resolution of each DFT is 5.64 Hz. The rising frequency emission appears to consist of a sequence of unstable upper sidebands.

Fig 7. Contour plot of integrated energy change ΔW in the +, V* plane after 1800 timesteps, at z=1000kms. The region with large positive ΔW corresponds to a bunch of stably trapped particles.

SIMULATION OF A SIPLE TRIGGERED RISER

In this paper we present the results of the simulation of a rising frequency emission triggered by a ramped Siple VLF pulse with df/dt = -2000 Hz/s. Simulation parameters are as follows:- input pulse length = 89msecs; initial pulse frequency = 5048Hz; input pulse amplitude = 0.3pT; saturation amplitude = 6.0pT; cold plasma density = 400/cc; linear equatorial growth rate = 350dbs/s; N_t = 700k.

Figure 4 is an f/t diagram of the field in the simulation box, presented as a sequence of spatial DFT's of resolution 1.82 Hz and incorporating Hamming weighting. The rate of change of frequency of about 400 Hz/s is in good agreement with observations, though perhaps on the low side.

Figure 5 shows an f/t diagram for the wavefield exiting from the simulation box. The emission is seen to be quite narrow band with a bandwidth of order 50 Hz. The frequency resolution is 7.44 Hz here, and the superior time resolution reveals a pulsating structure in the emission.

Figure 6 is a plot of spatially averaged frequency in the simulation field.

Figure 7 is a plot of wavefield amplitude exiting from the simulation zone. It is seen to rise exponentially and level off below the saturated level at ~5.3 pT.

CONCLUSION

A new algorithm has been devised for the numerical simulation of hot collision-free space plasmas. It is very efficient, low noise, stable and simple to use. Here it has been applied to the triggered VLF emission problem. It is expected that there will be many applications throughout space physics, particularly in the area of wave particle interactions.

REFERENCES

Cheng, C. Z. and G. Knorr, J. Computational Physics, 22, 380, 1976.
Denavit, J., Physics of Fluids, 28 (9), 2773, 1972.
Helliwell, R. A. and J. P. Katsufrakis, J Geophysical Research, 79, 2511, 1974.
Kotschenreuther, M., Physics Abstracts, 33 (9), 2107, 1988.
Nunn, D., Computer Physics Communications, 60, 1, 1990.
Nunn D., J Computational Physics, 108, 1, 180, 1993.

D Nunn, Department of Electronics and Computer Science, University of Southampton, Southampton, SO9 5NH, UK.

Electromagnetic Components of Auroral Hiss and Lower Hybrid Waves in the Polar Magnetosphere

H. K. Wong

Department of Space Science, Southwest Research Institute, San Antonio, Texas

DE-1 has frequently observed waves in the whistler and lower hybrid frequencies range. Besides the electrostatic components, these waves also exhibit electromagnetic components. It is generally believed that these waves are excited by the electron acoustic instability and the electron-beam-driven lower hybrid instability. Because the electron acoustic and the lower hybrid waves are predominately electrostatic waves, they can not account for the observed electromagnetic components. In this work, it is suggested that these electromagnetic components can be explained by waves that are generated near the resonance cone and that propagate away from the source. The role that these electromagnetic waves can play in particle acceleration processes at low altitude is discussed.

INTRODUCTION

The Earth's polar magnetosphere has long been recognized as an active region of wave activities. In the last two decades, numerous spacecraft have sampled this region of space and have identified a large variety of wave modes, most notably, the auroral hiss and the auroral kilometric radiation (for a review, see Shawhan, 1979). These waves are believed to play an important role in various plasma processes in the magnetosphere through wave-particle interactions. Examples of such processes include the diffusion of auroral electrons by electrostatic electron cyclotron waves (Ashour-Abdalla and Kennel, 1978); the generation of the auroral kilometric radiation through relativistic cyclotron resonance (Gurnett, 1974; Wu and Lee, 1979); and the acceleration and heating of ions and electrons by waves in the auroral region, leading to the formation of ion and electron conical distributions (Chang et al., 1986; Lysak, 1986; Wong et al., 1988; Crew et al., 1990; Temerin and Cravens, 1990).

One of the wave modes that has been extensively studied in the past is the whistler mode. This is mainly due to the interest in VLF emissions, auroral hiss, and lighting-related phenomena. In the mid-altitude polar magnetosphere, the electron plasma frequency ω_{pe} is typically less than the electron cyclotron frequency Ω_e. Under this condition, the whistler mode propagates between the lower hybrid frequency, which is approximately the ion plasma frequency ω_{pi}, and the electron plasma frequency. The frequently observed whistler mode auroral hiss is believed to be generated near the resonance cone by either precipitating electrons or upward moving electron beams (Gurnett et al., 1983). Another wave mode that has received considerable attention is the lower hybrid wave. The lower hybrid wave has also been observed in the auroral zone and the polar cusp and can be generated near the resonance cone by either an electron beam or an ion ring distribution (Maggs, 1976; Roth and Hudson, 1985). One distinct feature of the lower hybrid wave is that it can accelerate both electrons and ions and might be the source for auroral precipitating electrons and ion conics (Bingham et al., 1984; Chang and Coppi, 1981).

The whistler and lower hybrid waves excited near the resonance cone are quasi-electrostatic waves with negligible magnetic components. However, the "funnel-shaped" auroral hiss observed by DE-1 signifies that the auroral hiss has considerable magnetic components (Gurnett et al., 1983). Subsequent stability analyses using the observed particle distributions have revealed that the electron acoustic wave,

Space Plasmas: Coupling Between Small
and Medium Scale Processes
Geophysical Monograph 86

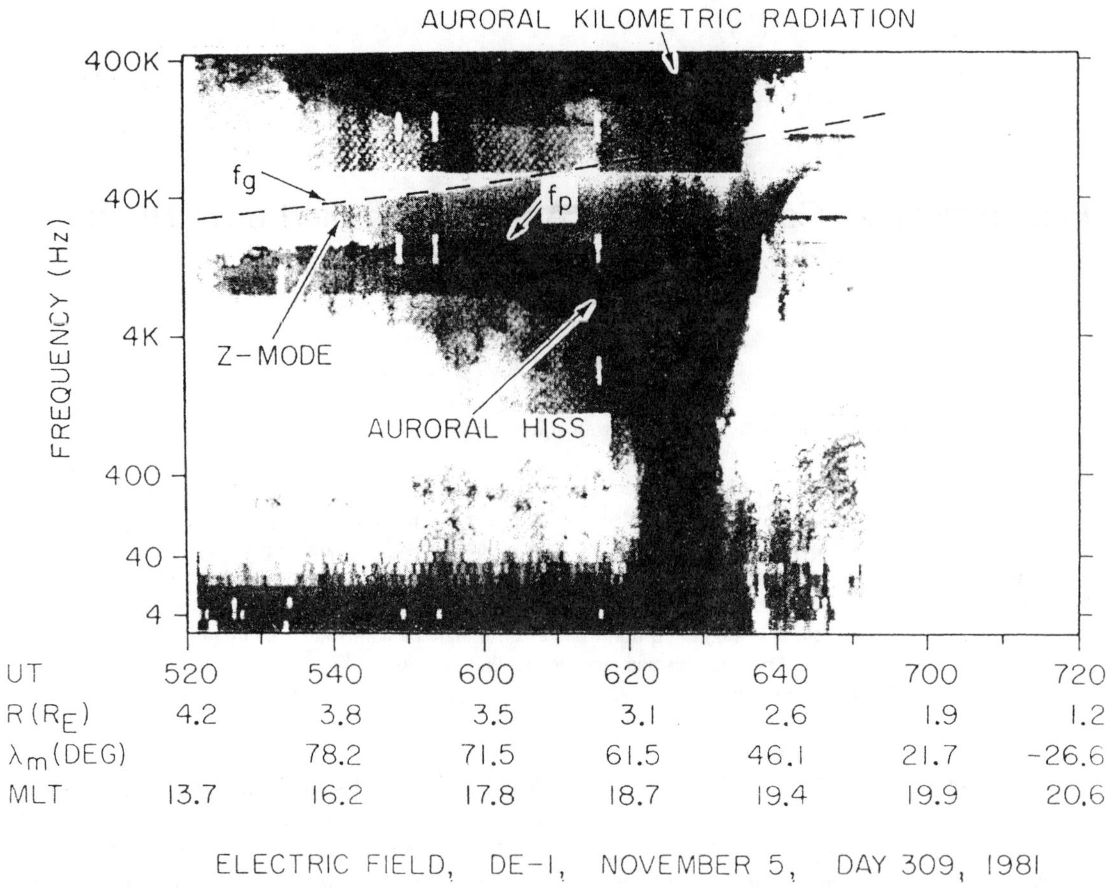

Fig. 1. A representative spectrogram of the electric field intensities for a nightside crossing of the auroral field lines. The plasma density over the polar region is relatively low, with $f_p << f_g$ (from Gurnett et al., 1983).

rather than the whistler wave, is the more unstable mode driven by the electron beam (Lin et al., 1984, 1985; Tokar and Gary, 1984). The electron acoustic wave is electrostatic in nature and thus might account for the electrostatic component of auroral hiss. The origin of the electromagnetic component of the hiss, however, still remains unanswered. In a different context, Benson et al. (1988) have reported the ground-based detection of waves in the frequency range of 150-300 kHz, which indicates the generation of field-aligned waves in the auroral zone. The field-aligned waves observed by Benson et al. are electromagnetic and fall into the frequency range of the whistler mode. Motivated by these observations, Wu et al. (1989) showed that an energetic electron population with a temperature anisotropy or a trapped type distribution is capable of generating field-aligned waves in the observed frequency range.

In this paper, we present data obtained from the Plasma Wave Instrument (PWI) and the High Altitude Plasma Instrument (HAPI) onboard the DE-1 spacecraft. We are mainly interested in electromagnetic waves with frequencies between the ion cyclotron and the electron cyclotron frequencies and the correlation of these waves with the plasmas. As shown below, we have identified broadband, low-frequency electromagnetic waves in the vicinity of the lower hybrid frequency. We have also found examples of electromagnetic waves in the whistler frequency range, which is quite different from the funnel-shaped auroral hiss. We believe that these electromagnetic waves are first generated either by electron beams or ion rings as the lower hybrid or whistler mode near the resonance cone and acquire electromagnetic components when they propagate away from the sources.

OBSERVATIONS

Figure 1 shows a frequency-time spectrogram (from Fig. 4 of Gurnett et al., 1983) during a nightside crossing of the auroral field lines under the condition that the electron plasma frequency, ω_{pe}, is less than the electron cyclotron frequency, Ω_e. The corresponding electric field intensity spectrum is shown in Figure 2 (from Fig. 5 of Gurnett et al., 1983). These figures display some common features of high-frequency electromagnetic waves frequently observed in the low-density auroral region. At or above the electron cyclotron frequency is the auroral kilometric radiation, which is believed to be in the fast extraordinary (R-X) mode. The Z mode radiation is observed at or below the local electron cyclotron frequency, with some possible overlap of ordinary (L-O) mode radiation (the upper cutoff of the Z mode is at the upper hybrid frequency, which is very close to the electron cyclotron frequency for a low-density plasma). The auroral hiss is the most commonly observed and is bounded by the electron plasma frequency (for $\omega_{pe} < \Omega_e$). Auroral hiss falls into the whistler mode range and exhibits a funnel-shape centered on a region of intense electron precipitation (Gurnett et al., 1983).

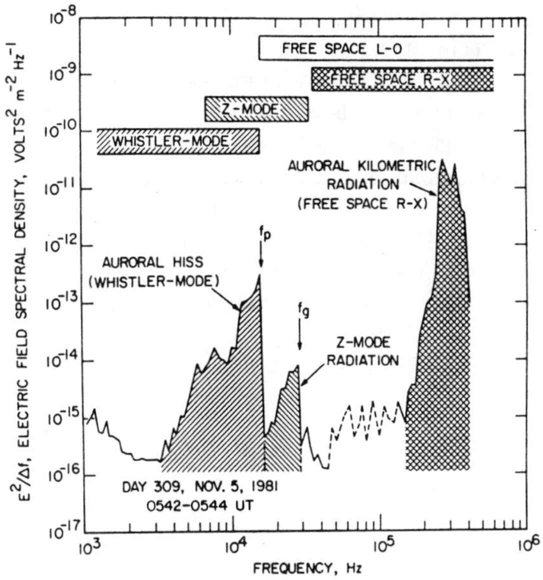

Fig. 2. An electric field intensity spectrum selected from the pass in Figure 1 showing the relative intensities of the auroral kilometric radiation, the Z mode radiation, and the auroral hiss. The sharp upper cutoff of the auroral hiss at f_p and the cutoff of the Z mode radiation at f_g are clearly evident. The dashed line indicates a region where the spectrum is being modified by receiver distortion effects caused by the very intense kilometric radiation (from Gurnett et al., 1983).

Besides the high-frequency electromagnetic waves, low-frequency electromagnetic waves with frequencies at or below the ion cyclotron frequencies have also been observed in the polar region [see Gurnett et al. (1984) for a detailed discussion]. However, not much attention has been paid to electromagnetic waves between the ion cyclotron and the lower hybrid frequencies. One of the main purposes of the present work is to show that electromagnetic waves in this frequency range are a common occurrence in the polar magnetosphere and to explore the roles that these waves can play in particle acceleration processes at low altitude. As a first example, Plate 1 (adapted from Sharber et al., 1988) shows the particle and wave observations in the dayside auroral zone. The upper panel shows the electron data. The middle and lower panels display the electric and magnetic spectra. It is clear from the electric and magnetic field signatures that the most intense electromagnetic waves occurred between 06:50 UT and 07:15 UT. These electromagnetic waves have frequencies between 200 Hz and 700 Hz, whereas the local electron cyclotron frequency (represented by the dotted line) is approximately 25 kHz and the corresponding hydrogen cyclotron frequency is approximately 13.5 Hz. Thus, these electromagnetic waves are above the ion cyclotron frequency and are in the vicinity of the lower hybrid frequency. Two features of these waves are worth mentioning. First, these waves occurred outside the region of intense electron precipitation (~ from 07:45 UT to 08:00 UT) in which broadband electrostatic waves are observed. Second, the electromagnetic waves are observed at the higher magnetic field side of the electron precipitation. These features provide some clue of how these waves are generated, which we will discuss later.

Plate 2 shows an example of particle and wave observations in the nightside auroral zone. The upper panel shows the electron data while the middle and lower panels display the electric and magnetic field data. The electric field data showed a rather weak auroral kilometric radiation, which lies above the electron cyclotron frequency (the upper dotted line), suggesting that the AKR is generated below the satellite. The funnel-shaped auroral hiss emission shows a strong correlation with the auroral precipitating electrons, which is centered at approximately 07:33 UT. There are two distinct bands of emissions just outside the precipitating region, started from 07:43 UT. The first band of emission has a frequency range from 200 Hz to 500 Hz. This band of emission is essentially the same type of emission that we just discussed, with frequencies extending from above the ion cyclotron frequency (the lower dotted line) to the lower hybrid frequency. The second band of emission has a frequency range from 500 Hz to 2 kHz. This band of emission has a frequency above the lower hybrid frequency and falls into the whistler mode range. An examination of the magnetic field data suggested that these two bands of emissions are electromagnetic. These two bands of electromagnetic emissions also have the same features as the case that we discussed above, i.e., they are observed outside the region and at the higher magnetic field side of intense electron precipitation.

In addition to the two examples shown above, our preliminary survey of the DE plasma wave data has found numerous examples of these electromagnetic waves in the auroral zone, polar cusp, and polar cap. The lower band of these electromagnetic waves typically has frequencies between 100 Hz and 1 kHz, falling between the ion cyclotron and lower hybrid frequencies. The upper band emissions have frequencies above the lower hybrid frequency and are in the whistler mode range. However, their characteristics are very different from the auroral hiss whistler mode. In the following, we try to identify these waves and explore the possibilities of how these waves are generated.

DISCUSSION AND SUMMARY

Our initial interpretation of the electromagnetic waves with frequencies between the ion cyclotron and the lower hybrid frequencies are fast magnetosonic waves since fast magnetosonic waves are the only electromagnetic waves in this frequency range that exhibit such characteristics (to our best knowledge). The other band of electromagnetic emissions with frequencies above the lower hybrid frequency (but far below the electron cyclotron frequency) are probably electromagnetic whistler waves. One immediate question to ask is whether these waves are locally generated and by what mechanisms. It seems highly unlikely that these waves are locally generated based on the following considerations. First, the fast magnetosonic wave has a phase speed at or above the Alfvén speed, which is very high for a low density plasma (at least of the order of 0.1 c, with c being the speed of light). Thus Landau resonance can not easily be satisfied between the waves and the particles, except for very energetic electrons (10's of keV or higher). However, electrons with energies 10 keV or higher are seldom observed in the dayside polar region. Second, when wave frequencies are far above the ion cyclotron frequency but far below the electron cyclotron frequency, it is very difficult for cyclotron resonance to occur. This might be the main reason that little attention has been paid to electromagnetic waves in this frequency range since there is no obvious direct means of generating electromagnetic waves in this regime. Third, when these electromagnetic waves are observed, there is little activity among the plasmas.

If one rules out the possibility that these waves are locally generated, one can easily provide an explanation for the occurrence of these waves. We suspect that these waves are generated first as lower hybrid waves near the lower hybrid resonance cone above the satellite in field lines under which the particle precipitation occurs and propagates downward. Lower hybrid waves can be excited either by electron beams in the auroral zone or by ion ring distribution in the polar cusp. As the lower hybrid waves propagate toward higher magnetic field strength, the waves become more electromagnetic and eventually transit into fast magnetosonic waves. As the waves propagate toward lower magnetic field strength, the waves become more electrostatic and are absorbed by the background plasmas. The same mechanism can also apply to the electromagnetic whistler waves except whistler waves can only be generated by electron beams since the frequency of the waves is too high for ions to respond. This scenario explains naturally why these waves are mainly observed outside the region and at the higher magnetic field side of intense electron precipitation.

We now divert our attention to what role the fast magnetosonic waves can play in particle acceleration processes at low altitude. The acceleration of ionospheric ions and the formation of ion conics have been extensively studied in the past. Despite the fact that electrostatic ion cyclotron and lower hybrid heating are regarded as the promising ion acceleration mechanisms and these waves have frequently been observed in the polar regions, there is no conclusive evidence that these waves are correlated with ion conics (Kitner and Gorney, 1984). Also, there is considerable doubt whether these processes are operative at low altitudes, probably as low as 400 km, where ion conics have been observed (Yau et al., 1983). The electrostatic ion cyclotron waves are difficult to excite at low altitudes since it requires a very large critical current. The lower hybrid wave has too high a phase velocity to resonate with the cold ionospheric ions. One way to get around this difficulty is to assume that the lower hybrid waves undergo a parametric decay process. The resulting daughter lower hybrid waves have lower phase velocity and can resonate with the thermal ions (Koskinen, 1985; Retterer et al., 1986). However, this process requires intense lower hybrid pump waves, but there is no strong observational support of intense lower hybrid waves at very low altitudes. As was just mentioned, fast magnetosonic waves have a very high phase velocity when the wave frequency is not too close to lower hybrid frequency. Thus, the wave can propagate a long distance downward practically undamped, until the wave frequency is very close to the local ion cyclotron frequency in which ion cyclotron damping occurs. (The fast magnetosonic wave consists of both right and left hand components. The left hand component contributes to cyclotron damping). This process leads to the heating or acceleration of ions at low altitudes by fast magnetosonic waves originating at high altitudes and also provides additional support that the fast

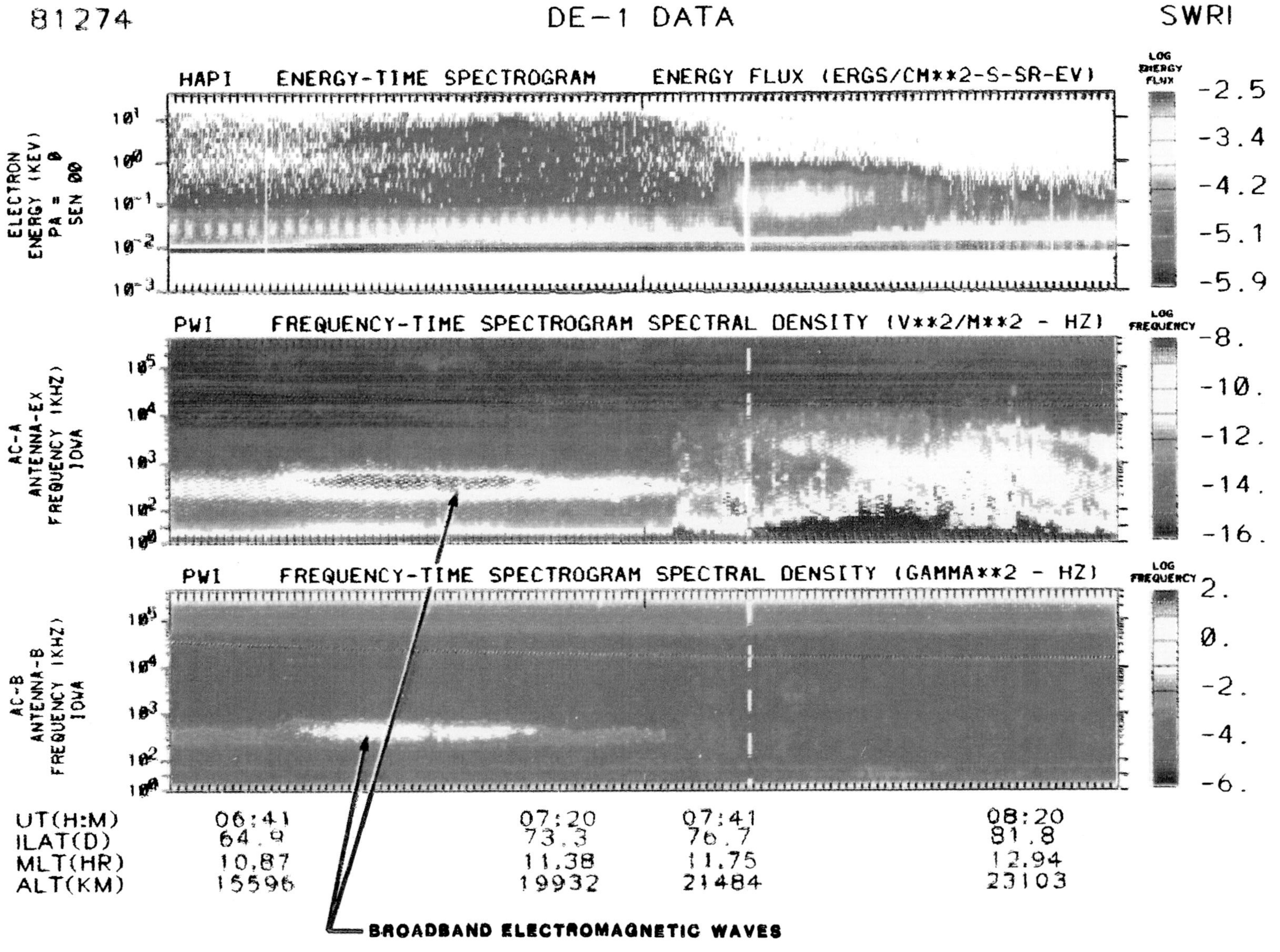

Plate 1. Dayside example of simultaneous particle and wave measurements from DE-1. The upper panel shows the electron data. The middle and lower panels display the electric and magnetic spectra (adapted from Sharber et al., 1988).

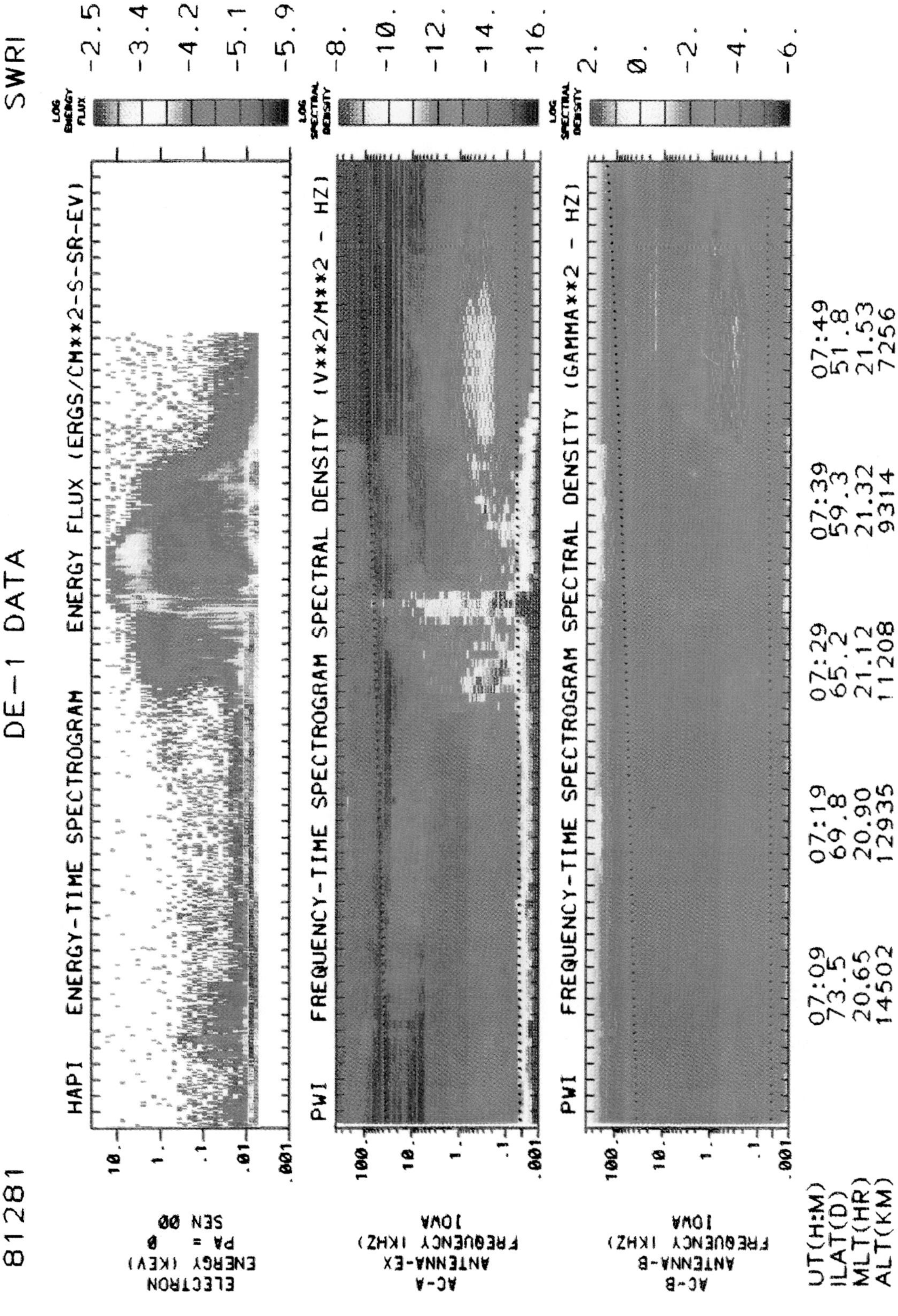

Plate 2. Nightside example of particle and wave measurements. The two distinct bands of electromagnetic emissions centered at approximately 07:50 UT.

magnetosonic waves are often observed at above, not below, the ion cyclotron frequency.

In summary, it has been shown through examples from DE-1 wave and plasma data that electromagnetic waves with frequencies between the ion cyclotron and the electron cyclotron frequencies are common occurrences in the Earth's polar magnetosphere. It is suggested that these electromagnetic waves are first generated either by electron beams or ion rings as the lower hybrid or whistler mode nears the resonance cone and then they acquire electromagnetic components when propagated away from the sources toward higher magnetic field strengths. The fast magnetosonic waves might play an important role in the acceleration of ions at low altitudes as they propagate toward the ionosphere. Further studies which include stability analysis using the observed particle distributions, ray tracing, and quasilinear and nonlinear analysis are desirable to examine quantitatively the origin of these waves, how they propagate down to the ionosphere, and their subsequent heating of cold ionosphere ions through cyclotron damping.

Acknowledgements. The author would like to thank Doug Menietti and Chin Lin for valuable discussions. This work was supported by NASA grants NAGW-1620 and NAG5-1552.

REFERENCES

Ashour-Abdalla, M., and C. F. Kennel, Diffuse auroral precipitation, *J. Geomag. Geoelectr.*, *30,* 239, 1978.

Benson, R. F., M. D. Desch, R. D. Hunsucker, and J. G. Romick, Ground-level detection of low and medium-frequency auroral emissions, *J. Geophys. Res.*, *93,* 277, 1988.

Bingham, R., D. A. Bryant, and D. S. Hall, A wave model for the aurora, *Geophys. Res. Lett.*, *11,* 327, 1984.

Chang, T., and B. Coppi, Lower hybrid acceleration and ion evolution in the suprauroral regions, *Geophys. Res. Lett.*, *8,* 1253, 1981.

Chang, T., G. B. Crew, N. Hershkowitz, J. R. Jasperse, J. M. Retterer, and J. D. Winningham, Transverse acceleration of oxygen ions by electromagnetic ion cyclotron resonance with broadband left-hand polarized waves, *Geophys. Res. Lett.*, *13,* 636, 1986.

Crew, G. B., T. Chang, J. M. Retterer, W. K. Peterson, D. A. Gurnett, and R. L. Huff, Ion cyclotron resonance heated conics: Theory and observations, *J. Geophys. Res.*, *95,* 3959, 1990.

Gurnett, D. A., The Earth as a radio source: Terrestrial kilometric radiation, *J. Geophys. Res.*, *79,* 4227, 1974.

Gurnett, D. A., S. D. Shawhan, and R. R. Shaw, Auroral hiss, Z mode radiation, and auroral kilometric radiation in the polar magnetosphere: DE-1 observations, *J. Geophys. Res.*, *88,* 329, 1983.

Gurnett, D. A., R. L. Huff, J. D. Menietti, J. L. Burch, J. D. Winningham, and S. D. Shawhan, Correlated low-frequency electric and magnetic noise along the auroral field lines, *J. Geophys. Res.*, *89,* 8971, 1984.

Kitner, P. M., and D. J. Gorney, A search for the plasma processes associated with perpendicular ion heating, *J. Geophys. Res.*, *89,* 937, 1984.

Koskinen, H. E. J., Lower hybrid parametric processes on auroral field lines in the topside ionosphere, *J. Geophys. Res.*, *90,* 8361, 1985.

Lin., C. S., J. L. Burch, S. D. Shawhan, and D. A. Gurnett, Correlation of auroral hiss and upward electron beams near the polar cusp, *J. Geophys. Res.*, *89,* 925, 1984.

Lin, C. S., D. Winske, and R. L. Tokar, Simulation of the electron acoustic instability in the polar cusp, *J. Geophys. Res.*, *90,* 8269, 1985.

Lysak, R. L., Ion acceleration by wave-particle interaction, in *Ion Acceleration in the Magnetosphere and Ionosphere*, edited by T. Chang, p. 261, A. G. U., Washington, D. C., 1986.

Maggs, J. E., Coherent generation of VLF hiss, *J. Geophys. Res.*, *81,* 1707, 1976.

Retterer, J. M., T. Chang, and J. R. Jasperse, Ion acceleration by lower hybrid waves in the suprauroral region, *J. Geophys. Res.*, *91,* 1609, 1986.

Roth, I., and M. K. Hudson, Lower hybrid heating of ionospheric ions due to ion ring distribution in the cusp, *J. Geophys. Res.*, *90,* 4191, 1985.

Sharber, J. R., J. D. Menietti, H. K. Wong, J. L. Burch, D. A. Gurnett, J. D. Winningham, and P. J. Tanskanen, Plasma waves associated with diffuse auroral electrons at mid-altitudes, *Ad. Space Res.*, *8,* no. 9, 447, 1988.

Shawhan, S. D., Magnetospheric plasma waves, in *Solar System Plasma Physics*, vol. III, edited by L. J. Lanzerotti, C. F. Kennel, and E. N. Parker, North-Holland, Amsterdam, 1979.

Temerin, M. A., and D. Cravens, Production of electron conics by stochastic acceleration parallel to the magnetic field, *J. Geophys. Res.*, *95,* 4285, 1990.

Tokar, R. L., and S. P. Gary, Electrostatic hiss and the beam driven electron acoustic instability in the dayside polar cusp, *Geophys. Res. Lett.*, *11,* 1180, 1984.

Wong, H. K., J. D. Menietti, C. S. Lin, and J. L. Burch, Generation of electron conical distributions by upper hybrid waves in the Earth's polar region, *J. Geophys. Res.*, *93,* 10025, 1988.

Wu, C. S., and L. C. Lee, A theory of the terrestrial kilometric radiation, *Astrophys. J.*, *230,* 621, 1979.

Wu, C. S., P. H. Yoon, and H. P. Freund, A theory of electron cyclotron waves generated along auroral field lines observed by ground facilities, *Geophys. Res. Lett.*, *16,* 1461, 1989.

Yau, A. W., B. A. Whalen, A. G. McNamara, P. J. Kellogg, and W. Bernstein, Particle and wave observations of low altitude ionospheric ion acceleration events, *J. Geophys. Res.*, *88,* 341, 1983.

H. K. Wong, Department of Space Science, Southwest Research Institute, PO Drawer 28510, San Antonio, TX 78228-0510.

Solar Wind-Magnetosphere Interaction as Simulated by A 3D, EM Particle Code

Oscar Buneman[1]

STAR Laboratory, Department of Electrical Engineering, Stanford University, Stanford, California

Ken-Ichi Nishikawa

Department of Physics and Astronomy, The University of Iowa, Iowa City, Iowa

Torsten Neubert

Space Physics Research Laboratory, The University of Michigan, Ann Arbor, Michigan

We present here our results of simulating the solar wind-magnetosphere interaction with a 3-dimensional, electromagnetic particle code. Hitherto such global simulations were done with MHD codes while lower-dimensional particle or hybrid codes served to account for microscopic processes and such transport parameters as have to be introduced ad hoc in MHD. Our kinetic model attempts to combine the macroscopic and microscopic tasks. It relies only on the Maxwell curl equations and the Lorentz equation for particles, which are ideally suited for computers. The preliminary results shown here are for an unmagnetized solar wind plasma streaming past a dipolar magnetic field. The results show the formation of a bow shock and a magnetotail, the penetration of energetic particles into cusp and radiation belt regions, and dawn-dusk asymmetries.

1. INTRODUCTION

The solar wind interaction with the Earth's magnetic field gives rise to a number of important and intriguing phenomena many of which are only partially understood. These include reconnection between the solar wind magnetic field at the dayside magnetopause and flux transfer events, the drag of the solar wind exerted on the magnetotail and associated instabilities at the magnetopause, plasma convection in the magnetosphere/ionosphere, and the generation of field-aligned current systems. Reviews of these and other phenomena can be found in *Haerendel and Paschmann* [1982], *Elphic* [1987], *Huang* [1987], and *Mauk and Zanetti* [1987]. The range of physical processes involved in the solar wind-magnetosphere system is quite overwhelming, and consequently a wide array of methods have been used to study these, ranging from detailed studies of select phenomena with the assumption of a specific field geometry or set of boundary conditions, to fully three dimensional simulations with less spatial resolution but self-consistent particle or field geometries and with no (little) influence of the boundary conditions.

The first 3-dimensional (3D), global magnetohydrodynamic (MHD) simulations of the solar wind-magnetosphere system was reported by *Brecht et al.* [1981], *Leboeuf et al.* [1981], and *Wu et al.* [1981]. Since then, global 3D MHD simulations have been used with some success to study a range of processes including reconnection in the tail [*Brecht et al.*, 1982; *Walker et al.*, 1987], reconnection at the day-side magnetopause [*Sato et al.*, 1986], the dependence on the interplanetary magnetic field (IMF) orientation of magnetospheric convec-

[1]Deceased January 24, 1993

Space Plasmas: Coupling Between Small
and Medium Scale Processes
Geophysical Monograph 86

tion and field-aligned currents [*Ogino et al.*, 1985, 1986; *Ogino*, 1986], the solar wind-magnetosphere-ionosphere current-voltage relationship [*Fedder and Lyon*, 1987], the self-excitation of auroral arcs [*Watanabe and Sato*, 1988], and the reconnection voltage between the closed geomagnetic field and the IMF as a function of the IMF clock angle [*Fedder et al.*, 1991].

In MHD-codes the micro-processes are represented by statistical (macroscopic) constants such as diffusion coefficients, anomalous resistivity, viscosity, temperature, and the adiabatic constant. Recent studies have pointed out that results from MHD codes are sensitive to the choice of these parameters, as well as to the accuracy of the iteration scheme and have cast doubt on some results obtained in the past. Thus it has been found that in ideal MHD simulations using a high-precision code, no Kelvin-Helmholtz instability is generated along the magnetopause and there is no entry of mass, energy and momentum through the magnetopause contrary to conventional knowledge [*Watanabe and Sato*, 1990].

To include particle dynamics explicitly, a number of simulations have been performed using hybrid codes (fluid electrons, particle ions) and full particle codes in one and two dimensions. With such codes local problems have been investigated such as shocks formed in the solar wind as it encounters the magnetosheath [*Quest*, 1988; *Burgess*, 1989; *Omidi et al.*, 1990; *Thomas et al.*, 1990; *Thomas and Winske*, 1990; *Winske et al.*, 1990] and the current layer at the dayside magnetopause [*Gary and Sgro*, 1990; *Berchem and Okuda*, 1990; *Okuda*, 1991; *Okuda*, 1992a,b]. Recently also global problems have been investigated in three dimensions with hybrid codes, namely the solar wind interaction with the dayside of Venus [*Moore et al.*, 1991] and the solar wind interaction with Mars and Venus [*Brecht and Ferrante*, 1991].

With the model presented here, we intend to take the full step, namely the global simulation of the solar wind interaction with a planetary magnetic field using a particle code which contains the complete particle dynamics. As will be shown below, the advantage is that the model contains the complete physics. The price to be paid is that some scaling of plasma parameters must be done.

2. A NEW 3D EM PARTICLE SIMULATION MODEL

Our code is a successor to the TRISTAN code [*Buneman et al.*, 1980; *Peratt et al.*, 1980; *Peratt*, 1986]. Its new features are: (1) Poisson's equation and Fourier transforms have been eliminated by updating the fields locally from the curl equations and depositing the particle currents according to charge-conserving formulas [*Villasenor and Buneman*, 1992], (2) radiating boundary conditions are applied to the fields using a first order Lindman approximation [*Lindman*, 1975], (3) filtering is done locally, (4) localisation makes the code ideally suited to modern parallel machines which call for minimising data paths, (5) the code is in Fortran and fully transportable: modest versions run on PCs and on workstations. In the past, the TRISTAN code has successfully simulated large scale space plasma phenomena such as the formation of systems of galaxies [*Peratt*, 1986]. The new version of the code has been applied to the study of the dynamics of low-β plasma clouds [*Neubert et al.*, 1992] and the whistler wave driven by Spacelab 2 electron beam [*Nishikawa et al.*, 1992], and a tutorial on the code will appear in the proceedings of the ISSS-4 (International School of Space Simulation) held in Kyoto, Japan, 1991 [*Buneman*, 1991].

For the simulation of solar wind-magnetosphere interaction the following boundary conditions were used for the particles: (1) Fresh particles representing the incoming solar wind (in our test run unmagnetized) are continuously injected across the yz-plane at $x = xmin$ with a thermal velocity plus a bulk velocity in the $+x$ direction, (2) thermal solar particle flux is also injected across the sides of our rectangular computation domain, (3) escaping particles are arrested in a buffer zone and redistributed there more uniformly by making the zone conducting in order to simulate their escape to infinity. They are then written off.

For the fields, boundary conditions were imposed just outside these zones: radiation is prevented from being reflected back inward, following Lindman's ideas [Lindman,1975]. The lowest order Lindman approximation was found adequate: radiation at glancing angles was no problem. However, special attention was given to conditions on the edges of the computational box.

In order to bring naturally disparate time- and space-scales closer together in this simulation of phenomena dominated by ion inertia and magnetic field interaction, the natural electron mass was raised to 1/16 of the ion mass and the velocity of light was lowered to twice the incoming solar wind velocity. This means that charge separation and anomalous resistivity phenomena are accounted for qualitatively but perhaps not with quantitative certainty. Likewise, radiation related phenomena (e.g. whistler modes) are covered qualitatively only.

3. SIMULATION RESULTS

While on a CRAY-2, with 100Mword core memory, a 200^3 grid simulation with some 10 million particles would be possible in principle, a first test exploring the solar wind-magnetosphere interaction was run on the CRAY-YMP at NCAR using a modest 105 by 55 by 55 grid and only 200,000 electron-ion pairs [*Buneman et al.*, 1992]. Here, we report on our second test run on the CRAY-2 at NCSA using a larger 215 by 95 by 95 grid and about 1,000,000 electron-ion pairs. Initially, these fill the entire box uniformly and drift with a velocity $v_D = 0.5c$ in the $+x$ direction, representing the solar wind. The electron thermal velocity is $v_{th} = 0.02c$ while the magnetic field is initially zero. A circular current generating the dipole magnetic field is increased smoothly from 0 to a maximum value reached at time step 65 and kept constant at that value for the rest of the simulation. The center of the current loop is located at (70.5, 47.5, 48) with the current in the xy-plane and the axis in the z-direction. The initial expansion of the magnetic field cavity is found to expel a large fraction of the initial plasma. The solar wind density is about 0.7 electron-ion pairs per cell, the mass ratio is $m_i/m_e = 16$, and $\omega_{pe}\Delta t = 0.84$.

In Figure 1 is shown the ion density in the center xz-layer containing the dipole center at time step 768. (Data were recorded at 64 step intervals and while no fully steady state was approached, no major change was seen between steps 768 and 1024 [*Buneman et al.*, 1992].) The ion density is color coded and the magnetic field component in the plane is shown with arrows at every third grid point. The magnitude of the field has been scaled in order to see the field direction for weak fields. Thus the length of the vectors is not a true representation of the field magnitude. The plasma is flowing through the simulation domain from left to right (low to high x-values). In the process the dipolar field is compressed at the side facing the plasma wind and is extended to a long tail on the down-wind side, just as the Earth's magnetic field in the solar wind. Note that in these simulations the conventional direction of the Earth's magnetic field is used (The opposite direction was used in the previous report [*Buneman et al.*, 1992]).

Some particles have entered the cusp regions and, on the down-wind side (night side), particles have gathered on closed field lines in what could be equivalent to the radiation belt. The particles at the up-wind magnetopause are forming a bow shock. The temperature of these particles is much elevated from the background temperature. Also seen up-wind from the magnetopause is what could be a second shock or a foreshock. The magnitude of both shocks has a minimum at the sub-solar point and is increasing to a maximum at some distance from the sun-Earth line.

The current density associated with the electron and ion fluxes and the magnetic field are shown in Figure 2. The current density component shown is perpendicular to the xz-plane (J_y), with current into the plane (the positive y-direction) colored red and current out of the plane colored blue. Again, the values are for the center layer containing the center of the dipole. The current densities have been determined from the y-component of the particle fluxes at each cell in the layer.

The current system shown in Figure 2 is equivalent to the Chapman-Ferraro current system of the Earth. At the up-wind, low latitude magnetopause it runs in a direction which cancels the original magnetic field on the up-wind side, and doubles the field at the down-wind side. At the cusp regions the current changes polarity with the magnetic field as expected. The current with a positive J_y in the neutral sheet is carried by the very few particles located here.

The ion density and magnetic field in the yz-plane is shown in Figure 3. The values are those of the layer which contains the center of the dipole and the dipole axis. The direction of the solar wind is out of the plane towards the viewer. Note that the shape of the magnetosphere is asymmetric, with more ions located at the right (dawn) side. This is caused by ions **grad**-B drifting towards this side as they move around the magnetic obstacle. The current density J_x in the yz-plane containing the dipole axis is shown in Figure 4. The current is running in a belt around the dipole with a polarity that cancels fields on the outside, thereby creating a magnetopause.

Some key parameters are shown in Figure 5 as functions of x, for center y and center z (sun-Earth line). These are the ion density (top panel), the ion thermal pressure (second panel), the ion dynamic pressure (third panel), the magnetic pressure (fourth panel), and the ion Larmor radius (bottom panel). The magnetic field is small in the wind and in the tail, and rises to a peak at the location of the dipole center. Up-wind, the increase is quite sudden and corresponds to the magnetopause. Up-wind of the magnetopause, the density is high and some little oscillations in the density are present here. The improvements with using a larger system are found in the ion number density and the ion dynamic pressure which have much less fluctuations in the front of the magnetopause. At the magnetopause the density drops, but recovers and reaches a local maximum at at

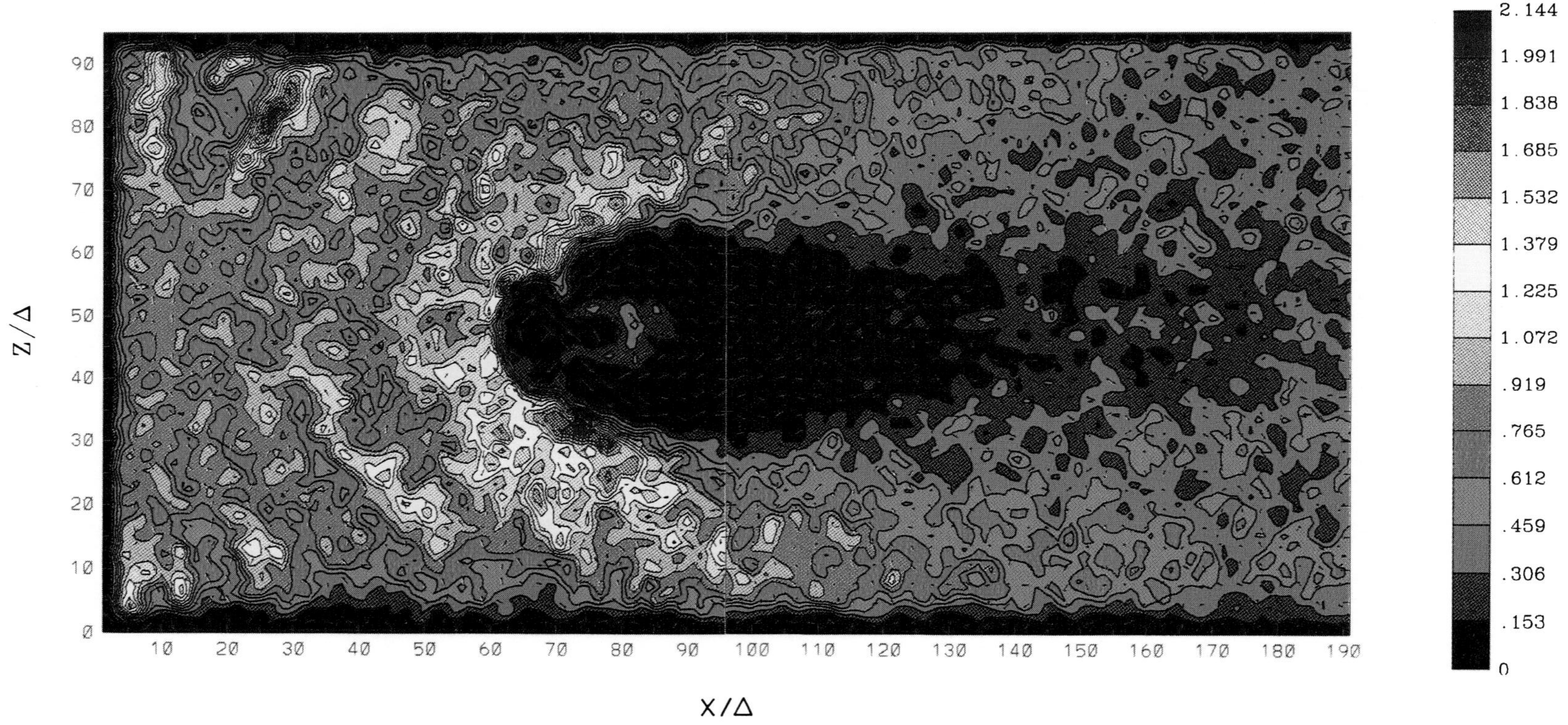

Fig. 1. Ion density in the center xz-layer containing the dipole center. "Relative amplitude" on color bar signifies simulation units.

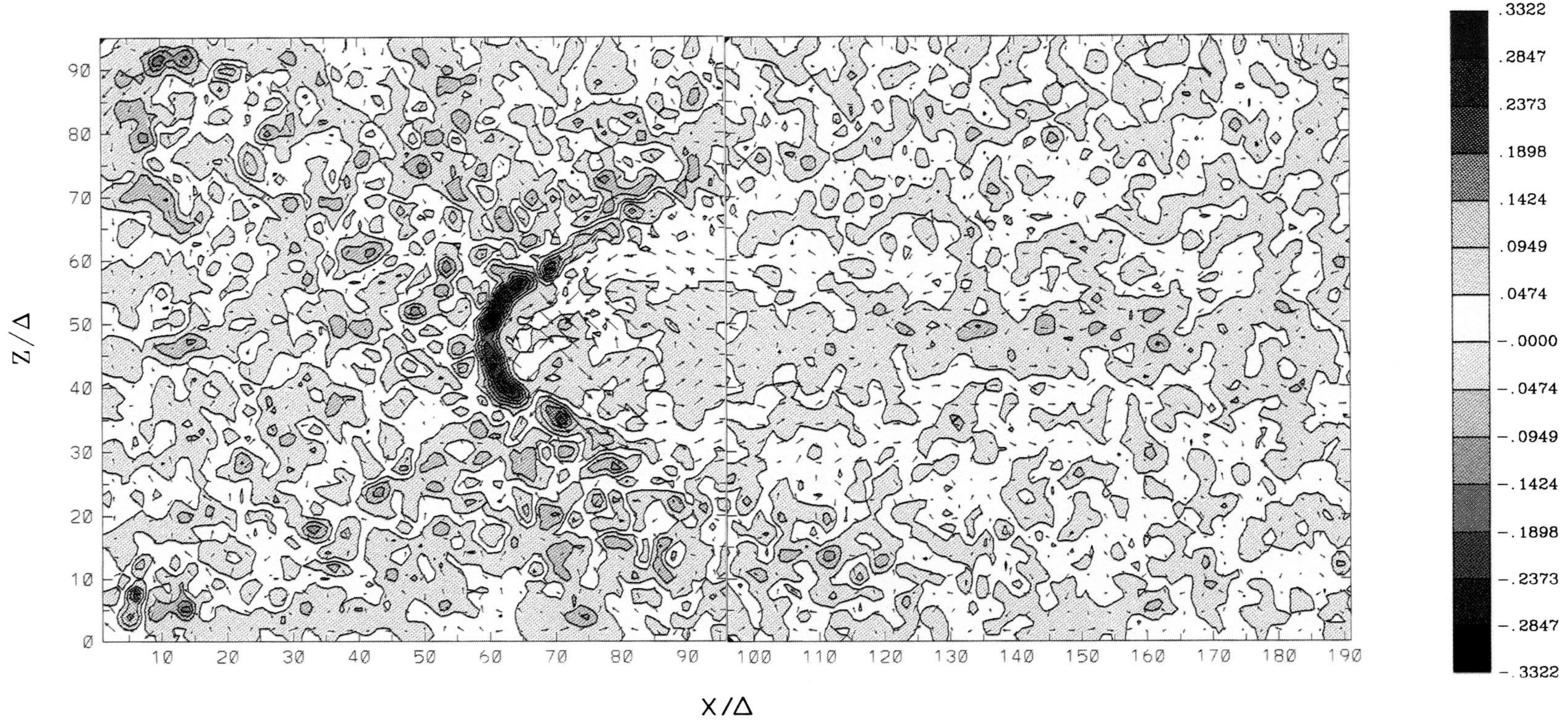

Fig. 2. Current density J_y in the center xz-layer containing the dipole center. Positive currents flow into the plane and negative out of the plane.

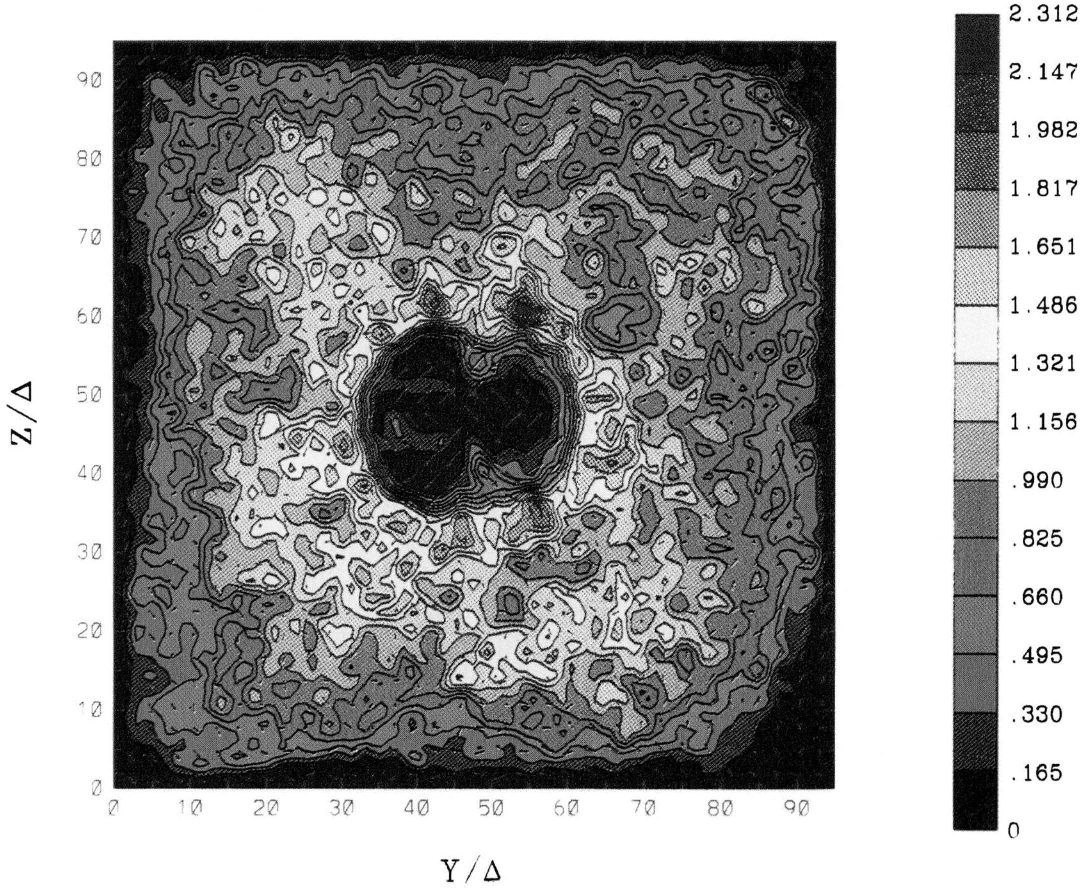

Fig. 3. Ion density in the center yz-layer containing the dipole center.

$x = 70$. This maximum is associated with cusp particles that have penetrated to the center of the dipole (there is no planet to stop them). At $x = 82$ a second maximum is reached which correspond to the radiation belt-type particles mentioned earlier. Finally, the density increases slowly with distance in the tail.

The thermal pressure increases dramatically in a region up-wind of the magnetopause. Thus the structures at the up-wind side of the magnetopause bear much resemblance to the structures of the Earth's bow shock. The first peak from the left in the thermal pressure corresponds to the foreshock. The fluctuation in the magnetosheath is seen the second peak. The increase of the magnetic field due to the Chapman-Ferraro current shown in Figure 2 is seen in the magnetic field around $x = 60$. The ion Larmor radius has large variations with location as does the magnetic field. In the wind and in the tail (we are in the neutral sheet) the Larmor radius is of the order of 10-90, with peaks in places where the magnetic field becomes small. The magnetic fluctuations in the tail are caused by the tail waving like a flag. (In regions of rapidly varying low field strength "Larmor radius" looses its meaning.)

4. DISCUSSION

The results presented in Figures 1 through 5 show that even with the modest grid-size of 215 by 95 by 95 cells, our 3D fully kinetic model is able to generate the complete magnetosphere with its basic characteristics. This encourages us to continue our work using finer grids (larger array sizes) and including such features as an IMF, a tilt of the magnetic axis, and an inner conducting surface simulating the ionosphere. Larger dimensions will allow finer resolution of fields. The number of particles should also be increased. Longer runs would allow us to enhance realism by using smaller m_e/m_i and larger c/v_D. On presently available sin-

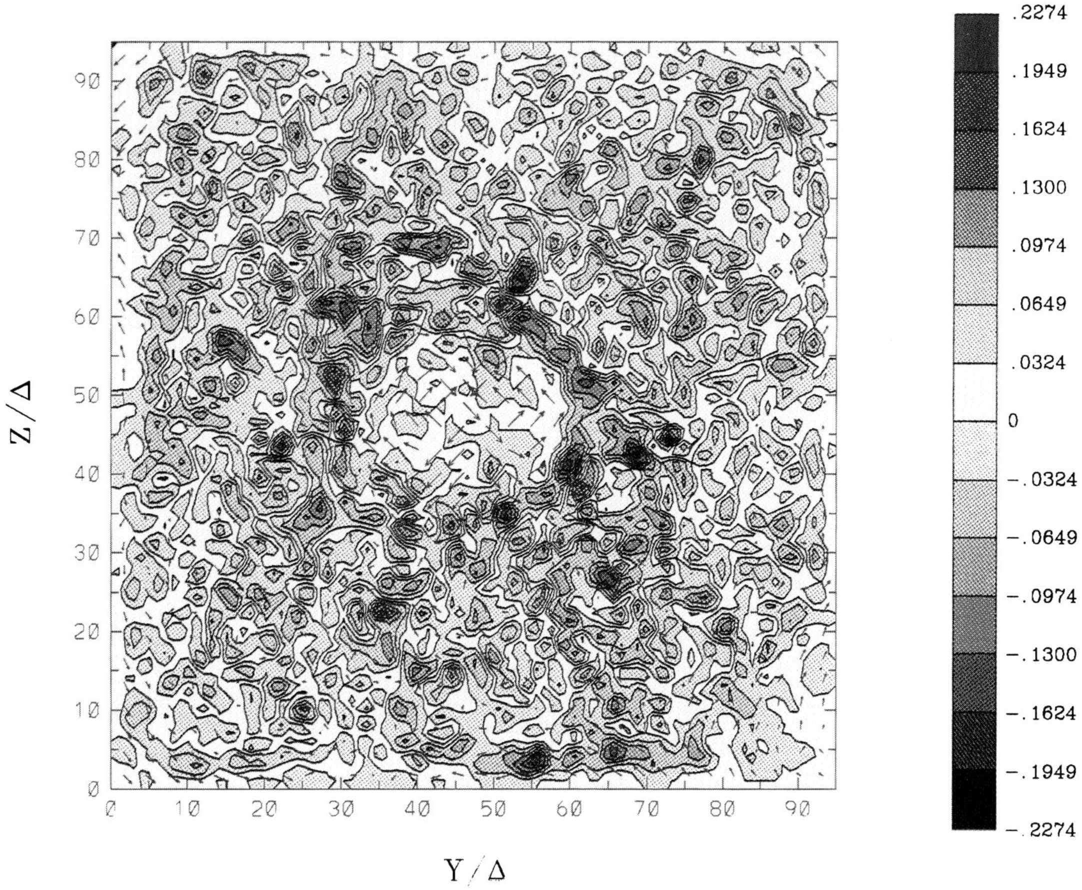

Fig. 4. Current density J_x in the center yz-layer containing the dipole center. Positive currents flow out of the plane and negative into the plane.

gle processor computers the grid resolution could be increased by a factor of 3 in all dimensions. Multi-processor machines in the future will allow even greater improvement, since the code allows simultaneous updating of many cells.

Some important issues that with advantage can be addressed with a kinetic code are:

1) In MHD simulations reconnection phenomena are sometimes studied using artificial numerical dissipation to drive the process [*Fedder et al.*, 1991]. Furthermore, artificially high diffusion coefficients are often used in order to achieve a stable solution [*Ogino et al.*, 1985]. With a full kinetic code such problems may be addressed directly.

2) The motions of individual particles are tracked. In many regions of the magnetosphere, such as the magnetopause and the tail, gradients in the magnetic field are of such magnitude that the magnetic field changes significantly during a particle gyration. This gives rise to non-isotropic, non-Maxwellian distribution functions which drives the physics of those regions. The recent research into chaotic particle orbits in the magnetotail [*Chen and Palmadesso*, 1986; *Burkhart and Chen*, 1991] is an example. The complex particle motion gives rise to such phenomena as collisionless electron viscosity [*Dungey*, 1988; *Sonnerup*, 1988; *Lyons and Pridmore-Brown*, 1990] and collisionless conductivity [*Horton and Tajima*, 1991]. Furthermore, a particle code calculates the particle dynamics in self-consistent rather than model magnetic fields.

3) The complete wave dynamics from low frequency magnetohydrodynamic modes (Kelvin-Helmholtz instabilities at the magnetopause and ULF pulsations) to electron waves (Langmuir oscillations) is included. An example where electron kinetics is important is illustrated by recent studies of the dynamics of the magnetopause current layer [*Berchem and Okuda*, 1990; *Okuda*, 1991; *Okuda*, 1992a,b]. It is thought that a

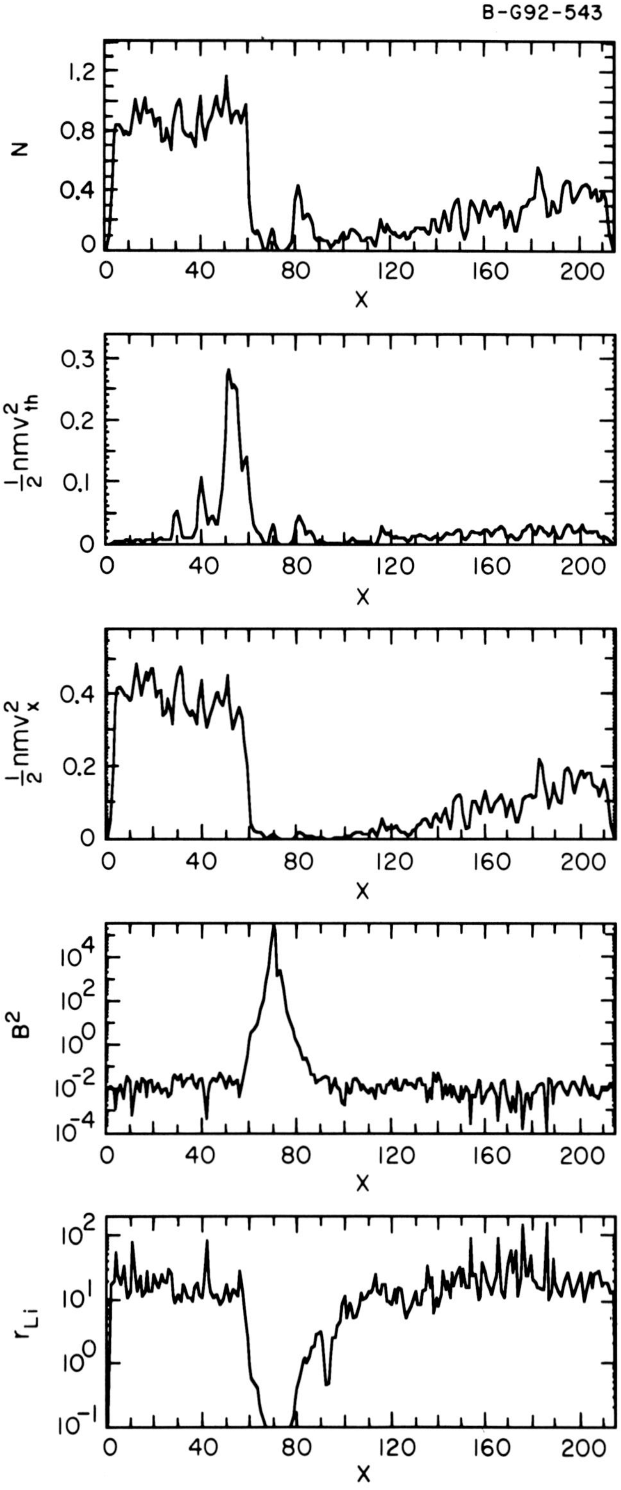

Fig. 5. Key parameters as a function of x at center y and center z (along the sun-Earth line). From top to bottom are: density, the thermal pressure, the dynamic pressure, the magnetic pressure, and the ion Larmor radius.

generalized lower hybrid drift instability is generated, which in the case of the magnetopause has the effect of broadening the current layer and causing filamentation of the current.

4) Any complicated magnetic field configuration can be simulated by a simple modification of the source-current. This makes it straight-forward to simulate not only other planetary magnetospheres such as that of Uranus, but also the effect of the diurnal rotation of the Earth's dipole axis with respect to the direction to the sun.

Questions that need to be addressed in the future are those of scaling from the micro-world of the simulation to the real macro-world. In particular, the scaling of the ion to electron mass ratio, the Larmor radii and the Debye length to the Earth's radius or other characteristic dimensions. An example is that in our simulation the Debye length is a significant fraction of the Earth's radius or the thickness of the bow shock. The Earth's radius itself can be determined by the distance from the dipole center to the up-wind magnetopause. This distance is typically equal to 10 Earth radii. In the test simulation described earlier an Earth radius is of the order of one grid spacing.

However, some form of scaling is usually needed in particle simulations, even in one and two dimensions, and still such simulations are able to reveal much of the physics behind natural phenomena. Similar problems are also encountered, and to some degree overcome, in laboratory plasma simulations of macro-scale space plasma phenomena. Examples include the study of magnetic field line reconnection [*Gekelman and Stenzel*, 1984; *Gekelman and Pfister*, 1988], and the interaction of a magnetized plasma flow with a dipole magnetic field [*Rahman et al.*, 1991]. We propose, therefore, that a kinetic particle model will be a useful tool for the study of the Earth's and other planetary magnetospheres in the solar wind, by supplementing present models that do not contain the complete particle dynamics.

Acknowledgments. Support for this work was provided by NASA contract NAS8-38772, NASA grant NAGW-2350, and NSF grants ATM-9106639 and ATM-9121116. The development of the simulation code was performed at the National Center for Supercomputing Applications, University of Illinois at Urbana-Champaign and test runs were performed at National Center for Atmospheric Research. Both centers are supported by the National Science Foundation.

REFERENCES

Berchem, J., and H. Okuda, A two-dimensional particle simulation of the magnetopause current layer, *J. Geophys. Res., 95*, 8133, 1990.

Brecht, S. H., and J. R. Ferrante, Global hybrid simulation of unmagnetized planets: Comparison of Venus and Mars *J. Geophys Res., 96*, 11,209, 1991.

Brecht, S. H., J. Lyon, J. A. Fedder, and K. Hain, A simulation study of East-West IMF effects on the magnetosphere, *Geophys. Res. Lett., 8*, 397, 1981.

Brecht, S. H., J. G. Lyon, J. A. Fedder, and K. Hain, A time-dependent three-dimensional simulation of the Earth's magnetosphere: Reconnection events, *J. Geophys. Res., 87*, 6098, 1982.

Buneman, O., C. W. Barnes, J. C. Green, and D. E. Nielsen, Review: Principles and capabilities of 3-d, E-M particle simulations, *J. Comput. Phys., 38*, 1, 1980.

Buneman, O., TRISTAN: The 3-D, E-M Particle Code, *Proceedings of ISSS-4*, Kyoto, Japan, 1991.

Buneman, O., T. Neubert, and K.-I. Nishikawa, Solar wind-magnetosphere interaction as simulated by a 3D, EM particle code, *IEEE Trans. Plasma Sci., 20*, 810, 1992.

Burgess, D., Cyclic behavior of quasi-parallel collisionless shocks, *Geophys. Res. Lett, 16*, 345, 1989.

Burkhart, G. R., and J. Chen, Differential memory in the earth's magnetotail, *J. Geophys. Res., 96*, 14,033, 1991.

Chen, J., and P. Palmadesso, Chaos and non-linear dynamics of single particle orbits in a magnetotail-like magnetic field, *J. Geophys. Res., 91*, 1499, 1986.

Dungey, J. W., Noise-free neutral sheets. Proceedings of an International Workshop in Space Plasma, held in Potsdam, GDR, European Space Agency Special Report,*ESA SP-285 (II)*, 15, 1988.

Elphic, R. C., The bowshock and magnetopause, *Rev. Geophys., 25*, 510, 1987.

Fedder, J. A., and J. G. Lyon, The solar wind-magnetosphere-ionosphere current-voltage relationship, *Geophys. Res. Lett, 14*, 880, 1987.

Fedder, J. A., C. M. Mobarry, and J. G. Lyon, Reconnection voltage as a function of IMF clock angle,*Geophys. Res. Lett., 18*, 1047, 1991.

Gary, S., and A. G. Sgro, The lower hybrid drift instability at the magnetopause, *Geophys. Res. Lett., 17*, 909, 1990.

Gekelman, W., and H. Pfister, Experimental observations of the tearing of an electron current sheet, *Phys. Fluids, 31*, 2017, 1988.

Gekelman, W., and R. L. Stenzel, Magnetic field line reconnection experiments 6. Magnetic turbulence, *J. Geophys. Res., 89*, 2715, 1984.

Haerendel, G., and G. Paschmann, Interaction of the solar wind with the dayside magnetosphere, in *Magnetospheric Plasma Physics*, edited by A. Nishida, Reidel Publishing Company, Dordrecht, Boston, London, 1982.

Horton, W., and T. Tajima, Collisionless conductivity and stochastic heating of the plasma sheet in the geomagnetic tail, *J. Geophys. Res., 96*, 15,811, 1991.

Huang, C. Y., Quadrennial review of the magnetotail, *Rev. Geophys., 25*, 529, 1987. Huang, C. Y., L. A. Frank, W. K. Peterson, D. J. Williams, W. Lennartsson, D. G. Mitchell, R. C. Elphic, and C. T. Russell, Filamentary structure in the magnetotail lobes, *J. Geophys. Res., 92*, 2349, 1987.

Leboeuf, J. N., T. Tajima, C. F. Kennel, and J. M. Dawson, Global simulation of the three-dimensional magnetosphere, *Geophys. Res. Lett, 8*, 257, 1981.

Lindman, E. L., Free-space boundary conditions for the time dependent wave equation, *J. Comp. Phys., 18*, 66, 1975.

Lyons, L. R., and D. C. Pridmore-Brown, Force balance near an X-line in a collisionless plasma, *J. Geophys. Res., 95*, 20903, 1990.

Mauk, B. H., and L. J. Zanetti, Magnetospheric electric fields and currents, *Rev. Geophys., 25*, 541, 1987.

Moore, K. R., V. A. Thomas, and D. J. McComas, Global hybrid simulation of the solar wind interaction with the dayside of Venus, *J. Geophys. Res., 96*, 7779, 1991.

Nishikawa, K.-I., O. Buneman, and T. Neubert, New aspects of whistler waves driven by an electron beam studied by a 3-D electromagnetic code, submitted to *Geophys. Res. Lett.*, 1992.

Neubert, T., R. H. Miller, O. Buneman, and K.-I. Nishikawa, The dynamics of low-β plasma clouds as simulated by a 3-dimensional, electromagnetic particle code, *J. Geophys. Res., 97*, 12,057, 1992.

Ogino, T., R. J. Walker, M. Ashour-Abdalla, and J. M. Dawson, An MHD simulation of the By-dependant magnetospheric convection and field-aligned current during northward IMF, *J. Geophys. Res., 90*, 10,853, 1985.

Ogino, T., A three-dimensional simulation of the interaction of the solar wind with the Earth's magnetosphere: the generation of field-aligned currents, *J. Geophys. Res., 91*, 6791, 1986.

Ogino, T., R. J. Walker, M. Ashour-Abdalla, and J. M. Dawson, An MHD simulation of the effects of the interplanetary magnetic field By component on the interaction of the solar wind with the Earth's magnetosphere during southward interplanetary magnetic field, *J. Geophys. Res., 91*, 10,029, 1986.

Okuda, H., Numerical simulations on the magnetopause current layer, *Modeling Magnetospheric Plasma Processes*, American Geophysical Union, Geophysical Monograph 62, p. 9, 1991.

Okuda, H., Structure of the magnetopause current layer at the subsolar point, *J. Geophys. Res., 97*, 1389, 1992.

Okuda, H., Numerical simulation of the subsolar magnetopause current layer in the Sun-Earth meridian plane, *J. Geophys. Res.*, (in press), 1992.

Omidi, N., K. B. Quest, and D. Winske, Low Mach number parallel and quasi-parallel shocks, *J. Geophys. Res., 95*, 20717, 1990.

Quest, K. B., Theory and simulation of collisionless parallel shocks, *J. Geophys. Res., 93*, 9649, 1988.

Peratt, A. L., J. Green, and D. Nielsen, Evolution of colliding plasmas, *Phys. Rev. Lett., 44*, 1767, 1980.

Peratt, A. L., Evolution of the plasma universe: II. The formation of systems of galaxies, *IEEE Trans. Plasma*

Sci., PS-14, 763, 1986.

Rahman, H. U., G. Yur, R. S. White, J. Birn, and F. J. Wessel, On the influence of the magnetization of a model solar wind on a laboratory magnetosphere, *J. Geophys. Res., 96*, 7823, 1991.

Sato, T., T. Shimada, M. Tanaka, T. Hayashi, and K. Watanabe, Formation of twisting flux tubes on the magnetopause and solar wind particle entry into the magnetosphere, *Geophys. Res. Lett., 13*, 801, 1986.

Sonnerup, B. U. O, On the theory of steady-state reconnection, *Comput. Phys. Commun., 49*, 143, 1988.

Thomas, V. A., D. Winske, and N. Omidi, Re-forming supercritical quasi-parallel shocks 1. One- and two-dimensional simulations, *J. Geophys. Res., 95*, 18,809, 1990.

Thomas, V. A., and D. Winske, Two dimensional hybrid simulations of a curved bow shock, *Geophys. Res. Lett., 17*, 1247, 1990.

Villasenor, J., and O. Buneman, Rigorous charge conservation for local electromagnetic field solvers, *Comp. Phys. Comm., 69*, 306, 1992.

Walker, R. J., T. Ogino, and M. Ashour-Abdalla, A magnetohydrodynamic simulation of reconnection in the magnetotail during intervals of with southward interplanetary magnetic field, in *Magnetotail Physics*, Edited by A. T. Y Lui, Johns Hopkins University Press, Baltimore, p. 183, 1987.

Watanabe, K., and T. Sato, Self-excitation of auroral arcs in a three-dimensionally coupled magnetosphere-ionosphere system, *Geophys. Res. Lett., 15*, 717, 1988.

Watanabe, K., and T. Sato, Global simulation of the solar wind-magnetosphere interaction: The importance of its numerical validity, *J. Geophys. Res., 95*, 75, 1990.

Winske, D., N. Omidi, K. B. Quest, and V. A. Thomas, Re-forming supercritical quasi-parallel shocks 2. Mechanism for wave generation and front re-formation, *J. Geophys. Res., 95*, 18,821, 1990.

Wu, C. C., R. J. Walker, and J. M. Dawson, A three-dimensional MHD model of the Earth's magnetosphere, *Geophys. Res. Lett, 8*, 523, 1981.

Ken-Ichi Nishikawa, Department of Physics and Astronomy, The University of Iowa, Iowa City, IA 52242-1479

Torsten Neubert, Space Physics Research Laboratory, The University of Michigan, Ann Arbor, MI 48109-2143

AMPTE/IRM Observations of the MHD Structure of the Plasmasheet Boundary: Evidence for a Normal Component of the Magnetic Field

C. A. Cattell and C. W. Carlson

Space Sciences Laboratory, University of California, Berkeley

W. Baumjohann

Max Planck Institut für Extraterrestriche Physik, Garching, Germany

H. Lühr

Institut für Geophysik und Metrologie Technische, Universität Braunschweig, Braunschweig, Germany

We present preliminary results of a study of the MHD structure of the plasmasheet boundary in the near tail, utilizing plasma and magnetic field data from the AMPTE/IRM spacecraft. To understand the MHD structure of the boundary, it is first necessary to determine whether there is a normal component of the magnetic field. Three methods were utilized herein: (1) the tangential momentum balance relationship for a discontinuity with a nonzero B_n; (2) examination of the 3d electron distributions; and (3) direct determination from the measured magnetic field for various calculated normals. The first two do not depend on the calculated normal to the boundary, while the third does. Statistical results on tangential momentum relation for ~80 crossings are presented. The statistical study shows that the measured change in the tangential velocity is almost always less than that predicted by the Rankine-Hugonoit relations for a boundary with non-zero B_n. This suggests that either: (1) There is not usually a normal component of the magnetic field across the boundary; combined with previous work on pressure balance, this implies that the boundary is a tangential discontinuity; or (2) The boundary is not usually well-modeled as a planar, time-stationary MHD discontinuity. For selected events, B_n has been directly calculated and the 3d electron distribution functions have been examined for evidence of a normal magnetic field component as discussed by Feldman et al. [1984]. The expected signature of tailward field aligned electron beams was found both during events for which the measured change in the tangential velocity was close to that predicted and during cases for which the measured change was much smaller than the predicted value. The explanation may be that there are other mechanisms which produce tailward, field-aligned electron beams in the plasmasheet boundary. There is a small subset of the crossings which are consistent with the identification of part of the plasmasheet boundary as a slow mode shock.

1. INTRODUCTION

An understanding of the MHD structure of the plasma sheet boundary at radial distances from ~12-~20 R_e is important in determining whether and how reconnection occurs in the near-tail, where particle acceleration is occurring, and how the plasma sheet boundary maps to low altitudes (i.e., the connection between the plasma sheet boundary and the aurora). It is also important to understand whether there are times when the boundary cannot be classified as an MHD discontinuity or combination thereof due to rapid time variations, short spatial scales, or other causes (for example, lack of dissipation mechanisms) which invalidate the assumptions of MHD. In fact, the

Space Plasmas: Coupling Between Small
and Medium Scale Processes
Geophysical Monograph 86

structure of the plasma sheet boundary layer (and the plasma sheet) has generally been explained by single particle effects [see, for example, *Cowley*, 1980, 1984; *Speiser*, 1987; *Zelenyi et al.*, 1990].

The study described herein was motivated by the results of MHD simulations of the magnetotail and observations of the plasmasheet boundary in both the near (~20 R_e) and far (~100-200 R_e) tail, all of which provided evidence for the occurrence of slow mode shocks. Data from the ISEE-3 satellite [*Feldman et al.*, 1984, 1985] shows that the deep tail (~200 R_e) plasma sheet boundary is often a slow-mode shock tailward of the location of the "steady-state" neutral line. This result is not unexpected since reconnection in the deep tail should be quasi-steady and, therefore, Petschek-type [*Petschek*, 1964] reconnection might occur. MHD simulations of the magnetotail [see, for example, *Sato*, 1979; *Birn and Hones*, 1981; *Scholer and Roth*, 1987; *Walker and Sato*, 1984] have shown the development of slow-mode shocks at the boundary of the plasma sheet, sometimes in conjunction with other MHD structures. In the near tail (~20 R_e), however, only one candidate slow-mode event has been identified [*Feldman* et al., 1987] and, in this case, only a small part of the entire boundary layer was consistent with a slow-mode shock. MHD simulations with boundary conditions designed to mimic the ionosphere do not have developed slow mode shocks earthward of the neutral line [*Hesse and Birn*, 1991; *Birn*, private communication, 1992]. These studies, indicating that there is a clear difference between the structure of the boundary in the near and the far magnetotail, justify a statistical study of the MHD structure of the plasmasheet boundary in the near tail. In this paper, we present preliminary results of such a study utilizing data from the AMPTE/IRM spacecraft.

It is well known that, as one moves from the lobe into the plasmasheet boundary, the density and velocity increase, and the magnetic field decreases. These changes can occur in a slow mode shock or a tangential discontinuity. A key difference between the two types of discontinuities is the existence of a normal component of the magnetic field, B_n, across a slow mode shock but not across a tangential discontinuity. Because the determination of the normal to the boundary is often not very accurate when there is not a large change in the magnetic field, the direct calculation of the normal component of the magnetic field can also be inaccurate. In this study, therefore, we utilize three methods to examine the question of the existence of a magnetic field component normal to the boundary. The first is to determine whether the change in the velocity component tangential to the boundary is consistent with that predicted by the Rankine-Hugoniot relations for a discontinuity with a normal component of the magnetic field. Although this relation is not strongly dependent on the accuracy of the normal, it provides a good indication of whether there is a normal component of the magnetic field. For several crossings, the 3d electron distributions and directly determined B_n will be examined in addition to the tangential momentum change to determine whether the individual events are consistent with the hypothesis that a portion of the boundary is a slow mode shock. In Section II, the data base and methodology are briefly described; in Section III discusses several specific events; statistical results are shown in Section IV; and conclusions and discussions of future work are given in Section V.2.

2. DATA AND METHODOLOGY

This study utilizes data from the 3d plasma instrument [*Paschmann et al.*, 1985] and the triaxial fluxgate magnetometer [*Lühr et al.*, 1985] on the AMPTE/IRM satellite. Crossings of the plasmasheet boundary during both the 1985 and 1986 tail season were examined. The statistical data base consists of approximately 110 events for which plasma moment and magnetic field data were available. An event consists of an approximately 30 minute to 4 hour interval and contains from 1 to more than 4 crossings of the plasma sheet boundary. The data base contains the following quantities: (1) the magnetic field, B_x, B_y, B_z, B_t and δ B in GSE (4s spin period averages are used so that the magnetic field data will have the same time resolution as the plasma moment data); and, for both ions and electrons, (2) the density in three energy ranges and the total density; (3) the velocity in GSE, V_x, V_y, V_z and V_t; (4) the average temperature, T_{ave}, and the three components of the temperature tensor in the principle axis system, T_1, T_2, and T_3; and (5) the three components of the heat flux in GSE. For most cases, the moments were calculated on-board and are 4s averages obtained every 4 s. In some events, however, specific moments had to be calculated on the ground from the transmitted 3d distributions so that the moments are not available for every spin period.

For each plasmasheet boundary crossing, upstream, downstream and crossing intervals were chosen by examining the moment data. Using these intervals, the normal to the boundary is calculated in six different ways (as described in Table 1. References to discussions of the methods are also listed). In general, the six normals do not agree. The best agreement is usually found between the magnetic field minimum variance and the electric field $\mathbf{v} \times \mathbf{B}$ maximum variance directions. The electric field maximum variance direction was usually the best deter-

TABLE 1. March 3, 1985 0220 UT

	v x B normal		B normal	
$\hat{n}$	(-0.15, -0.27, -0.95)		(-0.33, -0.33, -0.89)	
$(V_{tu} - V_{td})$, measured, km/s		839		837
$(V_{tu} - V_{td})$, predicted km/s		1082		1077
ratio		.78		.78
V_{shock}, km/s		24		-226
shock angle		89.6°		79.8°
	Up	Down	Up	Down
$\|B\|, \gamma$	35.47	28.97	35.47	28.97
n_i, cm^{-3}	0.07	0.21	0.07	0.21
B_n, γ	0.23	0.98	6.31	5.79
V_n, km/s	-35	4	-32	-161
$C_{Alfvèn}$, km/s	2957	1401	2957	1401
C_{slow}, km/s	1	1.5	27	42

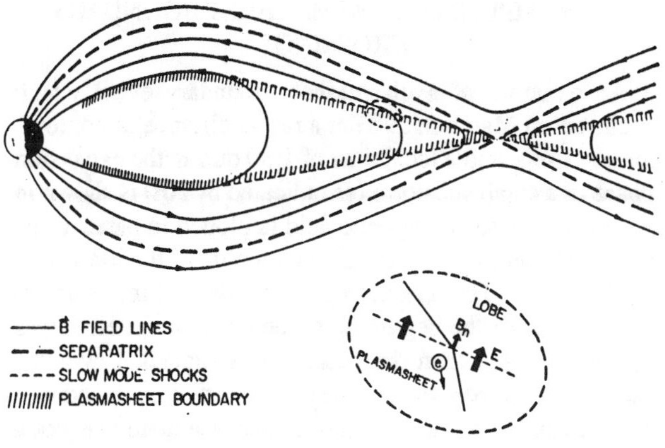

Fig. 1. A schematic drawing of the plasma sheet with slow mode shocks, indicating the mechanism which produces field aligned electron beams.

mined as indicated by the ratios of the eigenvalues. These comparisons will be discussed in detail in another paper. To determine the type of MHD discontinuity which best describes the boundary, it is first necessary to determine whether or not there is a normal component of the magnetic field. Although it is straightforward to calculate the normal component once the normal direction has been determined, in practice, the results are often inaccurate particularly when the normal component is small.

Another method is to examine the 3d distributions for cases when the electron distribution functions provide evidence for the existence of a normal component of the magnetic field [see *Feldman et al.*, 1984, 1987; *Schwartz et al.*, 1987]. Feldman et al. [1984; 1987] used the existence of an electron beam accelerated tailward into the plasma sheet as evidence for a non-zero B_n. A schematic drawing of the magnetotail (Figure 1) illustrates this method. The difference in the density and temperatures across the boundary causes a thermoelectric field normal to the boundary. If there is also a normal component of the magnetic field, then there will be a component of the electric field parallel to the magnetic field which can accelerate electrons into the plasmasheet. We have observed and presented [*Cattell et al.*, 1987] several examples of this type in which there was a very intense field-aligned tailward beam which may have been accelerated across the shock layer into the plasma sheet (the hot quasi-isotropic background) by the thermoelectric field normal to the boundary.

A second indirect way to test for the existence of a normal magnetic field is to examine whether the measured change in the tangential momentum matches that expected from the jump conditions across a discontinuity with a normal component of the magnetic field and plasma velocity [see, e.g., *Hudson*, 1970]. This method has been applied by *Sonnerup et al.*, [1981], *Paschmann et al.*, [1986] and others at the magnetopause and by *Feldman et al.*, [1987] in the tail. The change in tangential velocity is given by:

$$V_{td} - V_{tu} = \left[\frac{(B_{tu} - B_{td})\,(B_{tu} - B_{td}(n_u/n_d))}{4\pi m_p n_u} \right]^{1/2} \qquad (1)$$

where n is the plasma density, V_t (B_t) is the tangential components of the velocity (magnetic field), and the subscripts u and d refer to averages over the intervals upstream and downstream of the discontinuity.

This test is not strongly dependent on the normal determination since both B_n and v_n are expected to be small. The crossings where the tangential momentum balance is definitely not met may be examples either of tangential discontinuities, or of periods when the boundary is not locally planar or time-stationary, or when MHD does not provide a reasonable description of the boundary. In addition to tangential momentum balance, there are a number of other relations that are checked by the data base program. Because of the number of physical variables measured by the satellite instrumentation, there are many different ways that consistency with the Rankine-Hugoniot relations can be checked. For any chosen normal, we chose to begin by calculating the shock velocity from the continuity of normal momentum. A number of other quantities required to understand the boundary structure are then calculated including the shock angle, the normal component of the magnetic field, the predicted normal velocity in the upstream region, pressure balance, and the Alfvèn, sound, slow mode and fast mode speeds in the upstream and downstream regions.

3. EXAMPLES OF PLASMASHEET BOUNDARY CROSSINGS

An example of a plasmasheet boundary event which occurred on March 3, 1985 at a radial distance of approximately 15 R_E and a local time of 1:30 during the expansion phase of a small substorm (as indicated by AE) is shown in Figure 2. The total magnetic field is plotted in panel a, the ion density in panel b, and the magnitude of the ion velocity in panel c. The upstream and downstream intervals are also shown. At the beginning of the event the satellite was in the lobe, and then the plasmasheet expanded over the satellite. It can be seen that the only a portion of the boundary is included in the calculations because there was not a well-defined "downstream" across the entire boundary. This is frequently true for the cases in the data base and was also true for the event in Feldman *et al.*, (1987). The $\mathbf{v} \times \mathbf{B}$ maximum variance normal, (-.15,-.27,-.95), was well determined with a maximum to intermediate eigenvalue ratio of ~13. The **B** minimum variance normal was also well determined with an intermediate to minimum eigenvalue ratio of ~15. The angle between the two normals was 11°. Table 2 lists the values of various measured and calculated quantities for the two variance normals. Direct calculation of the normal component of the magnetic field yields different values depending on the method used to determine the normal direction. For the $\mathbf{v} \times \mathbf{B}$ maximum variance direction, the upstream (downstream) value of B_n is 0.23 nT (0.98 nT); for the B_n minimum variance direction, the value is 6.3 nT (5.8 nT). It can be seen that the difference between the two normals has almost no effect on the measured and calculated values of the change in the tangential velocity. The ratio of the measured to the predicted change is approximately 0.8. For this event, therefore, the change in the tangential velocity agrees well with the prediction for a discontinuity with a normal magnetic field. Cattell et al. [1992] present an example which does not agree well with the prediction.

The electron distribution functions were examined for three minutes beginning just prior to the start of the "upstream" interval. Selected distributions show the features described by [*Feldman et al.*, 1984; 1987] in that there were no beam features in upstream region and narrow tailward beams superimposed on the warmer background plasma in some of the downstream samples. However, the actual picture is much more complex. The electron distributions in the crossing and downstream regions had large variations, probably reflecting both spatial and temporal changes. For example, the energy at which the beam flux peaked varied from ~ 0.6 to ~11 keV but did not consistently increase as regions deeper into the

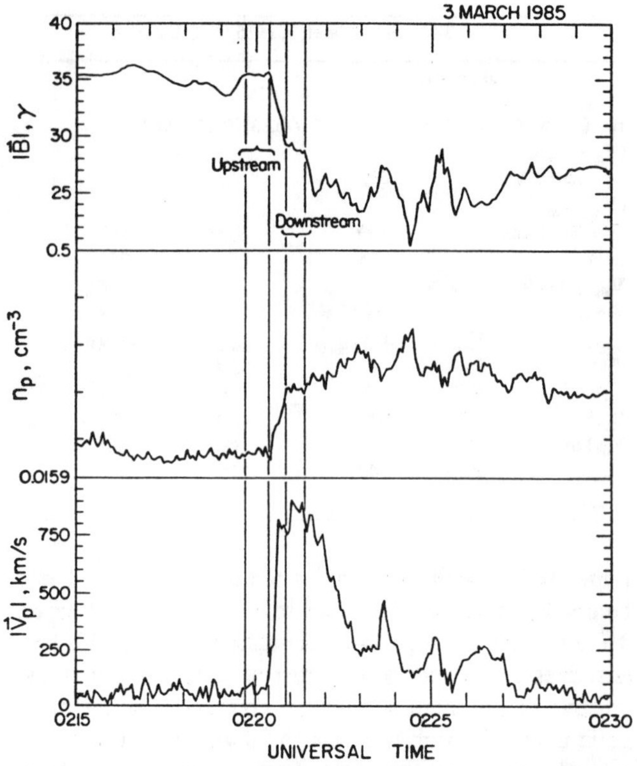

Fig. 2. An example of a crossing of the plasma sheet boundary by AMPTE/IRM showing the magnitude of the magnetic field, |B|, the ion density and the ion flow velocity. The upstream and downstream intervals are indicated.

TABLE 2. Normal Calculations

1.	Magnetic coplanarity: Hudson (1970); Lepping and Argentiero (1971); Viñas and Scudder (1986) $\hat{n} = (\mathbf{B}_u \times \mathbf{B}_d) \times (\mathbf{B}_u - \mathbf{B}_d) / \vert(\text{same})\vert$
2.	Velocity coplanarity: Hudson (1970); Viñas and Scudder (1986) $\hat{n} = (\vec{v}_d - \vec{v}_u) / \vert(\text{same})\vert$
3.	Abraham-Shrauner: Abraham-Schrauner (1972); Viñas and Scudder (1986) $\hat{n} = \dfrac{[(\mathbf{B}_d - \mathbf{B}_u) \times (\mathbf{v}_d - \mathbf{v}_u) \times (\mathbf{B}_u - \mathbf{B}_d)]}{\vert(\text{same})\vert}$
4.	Tangential: Hudson (1970) $\hat{n} = (\mathbf{B}_u \times \mathbf{B}_d) / \vert(\text{same})\vert$
5.	Magnetic minimum variance: Sonnerup and Cahill (1968)
6.	$(-\mathbf{v} \times \mathbf{B})$ maximum variance: Wygant, et al. (1987); Sonnerup, et al. (1987)

downstream region were sampled. In addition, counter-streaming electron beams were frequently observed and the earthward beams were sometimes more intense than the tailward beams. For this event, therefore, the electron distributions do not provide clear evidence for the existence of a normal component of the magnetic field. The observed features may be due to other mechanisms such as those discussed by *Cattell et al.*, [1987], *Parks et al.*, [1984; 1992], or *Onsager et al.*, [1990].

The fact that two of the three methods support the existence of a normal component of the magnetic field suggests that, during this crossing, a portion of the plasma sheet boundary was a slow mode shock. Further work must be done to determine whether the variability in electron distributions invalidate this conclusion.

In other cases, the electron distributions were more consistent with the Rankine-Hugoniot analysis. For example, in a crossing which occurred on March 28, 1985 at ~18:30 UT, there were no beam features in the electron distributions during the event and the measured change in the tangential velocity was <0.15 times the predicted value. None of the three methods indicated that there was a normal component of the magnetic field. During a crossing on March 7, 1986 at ~03:05 UT where the upstream distributions were almost isotropic and most of the downstream distributions had tailward field-aligned beams, the tangential velocity ratio was ~0.64. In this case the MHD analysis and the distributions both provided evidence for a normal magnetic field. Because there is not a one-to one correspondence between events identified as having a normal component using the Rankine-Hugoniot analysis and cases with tailward field-aligned beams in the downstream region, we plan to further study the characteristics of the electron distributions.

4. STATISTICAL RESULTS

When all the events in the data base were examined, 98 crossings were found for which reasonably well defined upstream (downstream) regions could be found in the lobe (plasmasheet boundary or plasmasheet). Cases with either no flow or low flow speeds in the plasmasheet boundary were not included. The results of the statistical study of tangential momentum balance are shown in Figure 3. The ratio of the measured to the predicted change in the tangential velocity is plotted for each crossing meeting certain limits on the angles between the tangential velocity change and the tangential magnetic field and limits on the ratio of the eigenvalues in the variance calculations of the normal direction. For an event to be included, the ratio of the maximum eigenvalue to the intermediate eigenvalue in the

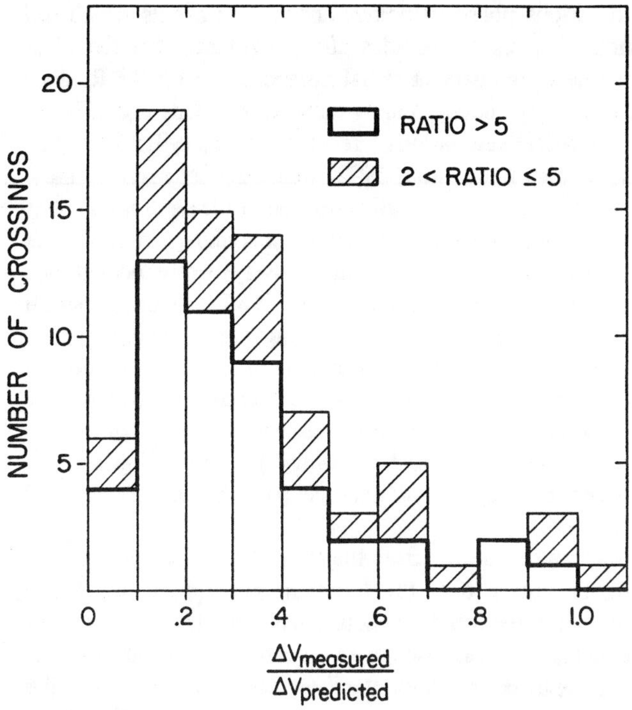

Fig. 3. The number of plasma sheet boundary crossings for which a given ratio of the measured value of the change in the tangential velocity to that predicted was observed. The open bars are cases for which the eigenvalue ratio in the normal calculation was >5.

v × **B** maximum variance calculation (the ratio of the intermediate to the minimum eigenvalue in the **B** minimum variance calculation) had to be greater than 5 (or greater than 2 and less than or equal to 5 for the hatched cases). Seventy-six cases met these criteria. Figure 3 shows that the measured change in the tangential velocity was much less than the change predicted for a boundary with a normal component of the magnetic field for most of the crossings. In fewer than 20% of the cases was it within 50% of the prediction. Comparison of the values of the ratio for cases where both normals met the criteria indicates that the results were not dependent on the choice of normal as was also shown in the previous section.

5. CONCLUSIONS

We have presented the results of a statistical study of the MHD structure of the plasmasheet boundary which examined the agreement between the measured change in the tangential velocity across the boundary to that predicted for discontinuities with a normal component of the magnetic field. The measured change is usually much less than that predicted, both for events which met the slow mode shock criteria for the size of the normal velocity compared to various phase velocities and for the events which did not. This result provides strong evidence that the plasmasheet boundary at radial distances of 10 to 18 R_e does not usually contain a slow mode shock. This is consistent with MHD simulations [*Hesse and Birn*, 1991; *Birn*, personal communication, 1992] which do not result in slow mode shocks on the earthward side of the neutral line. It is very different from the plasmasheet boundary in the distant tail ($\gtrsim 100R_e$) which is almost always well modeled by a slow mode shock [*Feldman et al.*, 1984]. During a small percentage of the events, the tangential momentum test, combined with other tests on the normal velocity, showed that a portion of the boundary was consistent with a slow mode shock. These may be cases when near-earth reconnection began in close proximity to the satellite as happened in the near-earth example published by [*Feldman et al.*, 1987].

It can be inferred from the above that some combination of the presence of the line-tying ionosphere, the dense plasma in the inner magnetosphere, and the rapid communication with the ionosphere via Alfvèn waves and mirroring particles changes the boundary conditions on the magnetotail system (compared to the open boundary conditions tailward of the distant neutral line) in such a way that slow mode shocks either do not have sufficient time to develop or are very restricted in both time and space to being in close proximity to the substorm neutral line. If the latter is true, satellites such as ISEE-1 and -2 and AMPTE/IRM would rarely observe slow mode shocks since the substorm neutral line usually forms tailward of ~20Re [*Cattell and Mozer*, 1984; *Baumjohann et al.*, 1988]. It is also interesting to note that the ISEE-3 distant tail observations of slow mode shocks all occurred tailward of the neutral line [*Feldman et al.*, 1985] and, therefore, in regions with no magnetic connection to the earth. If, in fact, slow mode shocks are rare earthward of the neutral line in the deep tail as well as the near tail, then some factor operating earthward of the neutral line (for example, magnetic connection to the ionosphere) must prevent the formation of slow mode shocks.

For a selected number of crossings, the electron distribution functions were examined for tailward field-aligned beams in the downstream region which have been interpreted as signatures of a normal component of the magnetic field [*Feldman et al.*, 1984; *Schwartz et al.*, 1987]. There was not a one-to-one correspondence between the crossings identified as slow mode shocks using tangential momentum balance and cases identified as having a normal magnetic field using the electron distributions. This may be due to the fact that there are other mechanisms which could produce tailward field-aligned beams. The electron distributions are often quite complicated in the neighborhood of the plasmasheet boundary. For example, there are often cases of tailward electron beams occurring in the upstream region. In order to clarify these observations, we plan to compare the energy of the observed beams to that predicted for a slow mode shock [*Schwartz et al.*, 1987] to determine if the cases identified as slow mode shocks by the tangential momentum relation also have electron beams with the predicted energy.

Future papers will add the following to the study: (1) the $\mathbf{v}\times\mathbf{B}$ maximum variance analysis will be iterated to include the velocity of the discontinuity; (2) errors in the normal vector and B_n will be calculated for the different normals, and the consistency between the different normals will be checked; (3) magnetic activity, radial distance and local time will be added to determine whether crossings consistent with slow mode shocks occurred at particular locations or during particular phases of substorms; and (4) selected events will be run through the self-consistent Rankine-Hugoniot program of [*Viñas and Scudder* 1986].

Acknowledgments: This work was supported by NASA grant NAGW-1565 at University of California, Berkeley. G. Paschman is the PI of the IRM 3d plasma instrument. We thank J. Caron, J. Bonnell and N. Danner for programming assistance.

REFERENCES

Abraham-Schrauner, B., Determination of magnetohydrodynamic shock normals, *J. Geophys. Res.*, **77**, 736, 1972.

Acuña, M. H., and R. P. Lepping, Modification to shock fitting program, *J. Geophys. Res.*, **89**, 11,004, 1984.

Baumjohann, W., G. Pashmann, N. Sckopke, C. A. Cattell, and C. W. Carlson, Average ion moments in the plasma sheet boundary layer, *J. Geophys. Res.*, **93**, 11,507, 1988.

Birn, J., and E. W. Hones, Jr., Three-dimensional computer modeling of dynamic reconnection in the geomagnetic tail, *J. Geophys. Res.*, **86**, 6802, 1981.

Cattell, C. A., C. W. Carlson, W. Baumjohann, G. Paschmann and H. Lühr, Observations of the structure of the plasma sheet boundary, *EOS Trans. AGU*, **68**, 1440, 1987.

Cattell, C. A., C. W. Carlson, W. Baumjohann and H. Lühr, The MHD structure of the plasma sheet boundary: (1) Tangential momentum balance, *Geophys. Res. Lett.*, **19**, 2083, 1992.

Cowley, S. W. H., Plasma populations in a simple open model magnetosphere, *Space Sci. Rev.*, **28**, 217, 1980.

Cowley, S. W. H., The distant magnetotail in theory and observation, in Magnetic Reconnection in Pace and Laboratory Plasma, *Geophys. Monograph*, **30**, ed. by E. W. Hones, Jr., p. 288, American Geophysical Union, Washington, DC 1984.

Feldman, W. C., S. J. Schwartz, S. J. Bame, D. N. Baker, J. Birn, J. T. Gosling, E. W. Hones, Jr., D. J. McComas, J. A. Slavin, E. J. Smith, and R. D. Zwickl, Evidence for slow-mode shocks in the deep geomagnetic tail, *Geophys. Res. Lett.*, **11**,

599, 1984.

Feldman, W., D. Baker, S. Bame, J. Birn, J. Gosling, E. Hones and S. Schwartz, Slow mode shocks: A semipermanent feature of the distant geomagnetic tail, *J. Geophys. Res.*, **90**, 233, 1985.

Feldman, W. C., R. L. Tokas, J. Birn, E. W. Hones, Jr., S. J. Bame, and C. T. Russell, Structure of a slow mode shock observed in the plasma sheet boundary layer, *J. Geophys. Res.*, **92**, 83, 1987.

Hesse, M. and J. Birn, Plasmoid evolution in an extended magnetotail, *J. Geophys. Res.*, **96**, 5683, 1991.

Hudson, P. D., Discontinuties in an anisotropic plasma and their identification in the solar wind, *Planet. Space Sci.*, **18**, 1611, 1970.

Lepping, R.P. and P.D. Argentiero, Single spacecraft method of estimating shock normals, *J. Geophys. Res*, **77**, 2957, 1971.

Lühr, H., N. Klöckler, W. Oelschlägel, B. Häugles and M. Acuña, The IRM fluxgate magnetometer, *IEEE Trans-Geosi. Elec.*, **GE-23**, 259, 1985.

Onsager, T.G., M.F. Thomsen, J.T. Gosling and S.J. Barne, Electron distributions in the plasma sheet boundary layer: Time of Flight Effects, *Geophys. Res. Lett.*, **17**, 1837, 1990.

Parks, G., R. Fitzenreiter, K. Ogilvie, C. Huang, K. Anderson, J. Dandouras, L. Frank, R. Lin, M. McCarthy, H. Rème, J. Sauvaud and S. Werden, Low energy particle layer outside of the plasma sheet boundary, *J. Geophys. Res.*, **97**, 2943, 1992.

Parks, G., M. McCarthy, R. Fitzenreiter, J. Etcheto, K. Anderson, T. Eastman, L. Frank, D. Gurnett, C. Huang, R. Lin, L. Lui, K. Ogilvie, A. Pedersen, H. Rème and D. Williams,, Particle and field characteristics of the high-latitude plasma sheet boundary, *J. Geophys. Res.*, **89**, 8885, 1984.

Paschmann, G., H. Loidl, P. Obermeyer, M. Ertl, R. Laborenz, N. Sckopke, W. Baumjohann, C. W. Carlson, and D. W. Curtis, The plasma instrument for AMPTE/IRM, *IEEE Trans. Geosci. Elec.*, **GE-23**, 262, 1985.

Paschmann, T., I. Papamastorakis, W. Baumjohann, N. Sckopke, C. W. Carlson, B. U. O. Sonnerup, and H. Lühr, The magnetopause for large magnetic shear: AMPTE/IRM observations, *J. Geophys. Res.*, **91**, 11099, 1986.

Petschek, H. E., Magnetic field annihilation, in *Physics of Solar Flares*, **NASA SP-50**, 425, 1964.

Sato, T., Strong plasma acceleration by slow shocks resulting from magnetic reconnection, J. *Geophys. Res.*, **84**, 7177, 1979.

Scholer, M., and D. Roth, Simulation study on reconnection and small-scale plasmoid formation, *J. Geophys. Res.*, **92**, 3223, 1987.

Schwartz, S., M. Thomsen, W. Feldman, and F. Douglas, Electron dynamics and potential jump across slow mode shocks, *J. Geophys. Res*, **92**, 3165, 1987.

Sonnerup, B. U. Ö., and L.J. Cahill, Explorer 12 observations of the magnetopause current layer, *J. Geophys. Res.*, **73**, 1757, 1968.

Sonnerup, B. U. Ö., G. Pashmann, I. Papamastorakis, N. Sckopke, G. Haerendel, S. J. Bame, J. R. Asbridge, J. T. Gosling, and C. T. Russell, Evidence for magnetic field reconnection at the earth's magnetopause, *J. Geophys. Res.*, **86**, 10049, 1981.

Sonnerup, B. U. Ö., I. Papamastorakis, G. Pashmann, and H. Lühr, Magnetopause properties from AMPTE/IRM observations of the convection electric field: method development, *J. Geophys. Res.*, **92**, 12137, 1987.

Speiser, T. W., Kinetic aspects of tail dynamics: Theory and simulation, in *Magnetotail Physics*, ed. by A. T. Y. Lui, Johns Hopkins Univ. Press, Baltimore, 1987.

Viñas, A., and J. D. Scudder, Fast and optimal solution to the "Rankine-Hugoniot Problem," *J. Geophys. Res.*, **91**, 39, 1986.

Walker, J. R., and T. Sato, Externally driven magnetic reconnection, in magnetic reconnection in space and laboratory plasmas, *Geophys. Monograph* **30**, ed. by E. W. Hones, Jr., 272, American Geophysical Union, Washington, D.C., 1984.

Wygant, J.R., M. Bensadoun, and F.S. Mozer, Electric field measurements at subcritical, oblique bow shock crossings, *J. Geophys. Res.*, **92**, 11109, 1987.

Zelenyi, L., A. Galeev, and C. Kennel, Ion precipitation from the inner plasma sheet due to stochastic diffusion, *J. Geophys. Res.*, **95**, 3871, 1990.

C.A. Cattell, C.W. Carlson, Space Sciences Laboratory, University of California, Berkeley, CA 94720

W. Baumjohann, Max Planck Institut fur Extraterrestriche Physik, 8046 Garching bei Munchen, Germany

H. Luhr, Institut fur Geophysik und Metrologie Technische, Universitat Braunschweig, Braunschweig, Germany

Observation of High Speed Flows ($V > V_{SW}$) in the Magnetosheath During an Interval of Strongly Northward IMF

SHENG-HSIEN CHEN AND MARGARET G. KIVELSON

Institute of Geophysics and Planetary Physics and Department of Earth and Space Sciences, University of California, Los Angeles

JACK T. GOSLING

Los Alamos National Laboratory, Los Alamos, New Mexico

ALLAN J. LAZARUS

Center for Space Research, Department of Physics, Massachusetts Institute of Technology, Cambridge, Massachusetts

On February 15, 1978, the orientation of the interplanetary magnetic field (IMF) remained steadily northward for several hours. The ISEE 1 and 2 spacecraft were located near apogee on the dawnside flank of the magnetotail. IMP 8 was almost symmetrically located in the magnetosheath on the dusk flank and IMP 7 was upstream in the solar wind. Using plasma and magnetic field data, we show that the magnetosheath flow speed on the flanks of the magnetotail steadily exceeded the solar wind speed by 20%. We propose that the acceleration of the magnetosheath flow is achieved by magnetic tension in the draped field configuration for northward IMF.

INTRODUCTION

The magnetosheath separates the bow shock from the magnetopause. The solar wind plasma, which is super-fast-magnetosonic, slows to a speed below the local fast-magnetosonic speed across the bow shock. The slowed flow in the magnetosheath is diverted around the magnetopause, continues along the flanks, and then speeds up to the solar wind velocity in the downstream magnetotail. Considerable attention has been directed to the consequences of the coupling of the magnetosheath to the magnetosphere. In particular, it is well known that the magnetosheath flow drives plasma convection in the magnetosphere both through a viscous interaction between the magnetosheath and magnetospheric plasmas at the magnetopause boundary layers [*Axford and Hines*, 1961; *Axford*, 1964] and through magnetic reconnection between the magnetosheath magnetic field and the magnetic field of the magnetosphere [*Dungey*, 1961]. However, less attention has been directed at fully characterizing the magnetosheath itself [but see *Saunders*, 1990; *Song et al.*, 1990; *Southwood and Kivelson*, 1992] or in examining details of the coupling between the magnetosheath and the magnetosphere along the flanks of the tail [but see *Gosling et al.*, 1986; *Saunders*, 1990]. In this paper, we use data from an interval of persistent northward interplanetary magnetic field (IMF) to study aspects of magnetosheath flow.

Space Plasmas: Coupling Between Small and Medium Scale Processes
Geophysical Monograph 86

A detailed description of the observations is given in the next section. We show that the flow velocity on the magnetosheath side of the magnetopause exceeded the solar wind speed. This observation can be linked to earlier observations of the magnetosheath on the flanks reported by *Howe and Binsack* [1972]. They showed that the flow velocity in the tail magnetosheath ($-20 \gtrsim X_{GSM} \gtrsim -60\ R_E$) correlated weakly with distance from the magnetopause. Some of the flow velocities at distances $X_{GSM} \gtrsim -40\ R_E$ in the magnetosheath were $\sim$10% higher than in the upstream solar wind, but the study did not investigate the effects of IMF orientation.

Flows at speeds greater than the solar wind speed are not found in gasdynamic models of the magnetosheath. For example, the gasdynamic model of *Spreiter et al.* [1966] shows that the magnetosheath flow accelerates gradually downstream of the subsolar point and its speed varies with distance from the magnetopause but the flow velocity never becomes greater than that of the upstream solar wind. *Gosling et al.* [1986] showed evidence of accelerated plasma flows at the near-tail magnetopause ($0 > X_{GSM} > -12.5\ R_E$). They found that the accelerated flows were likely to occur when the terrestrial

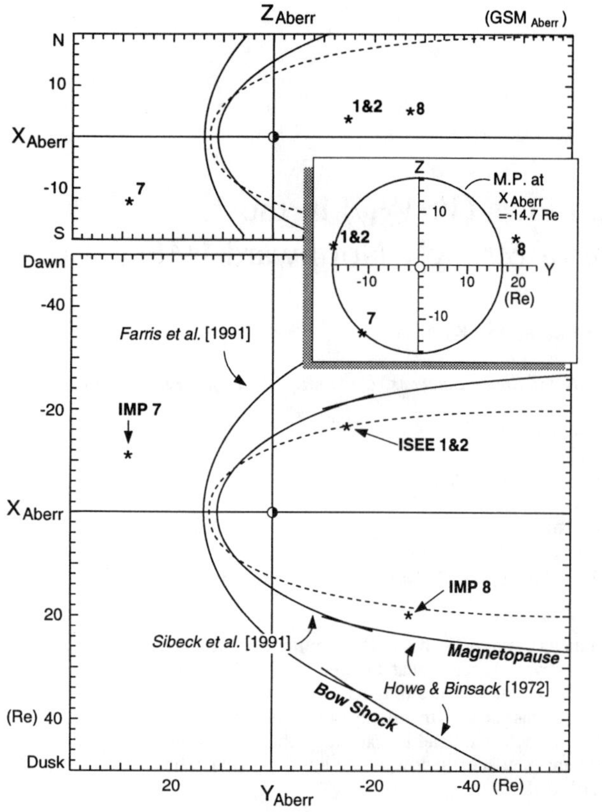

Fig. 1. Positions of the spacecraft ISEE 1&2, IMP 7 and IMP 8 at 1600 UT, Feb. 15, 1978. Models of the magnetopause [*Sibeck et al.*, 1991; *Howe and Binsack*, 1972] and bow shock [*Farris et al.*, 1991; *Howe and Binsack*, 1972] are plotted. The dynamic pressure in the upstream solar wind for the models is ~2 nPa without considering the IMF B_z contribution. The dotted curve is a possible magnetopause for our event, obtained by extrapolating Sibeck's B_z dependent model subsolar and dawn-dusk distances to IMF $B_z \simeq 10$ nT and combining this with the observed magnetopause position from ISEE 2 by using a rescaled Howe and Binsack nightside magnetopause model. We have assumed symmetry about the aberrated X_{GSM} axis.

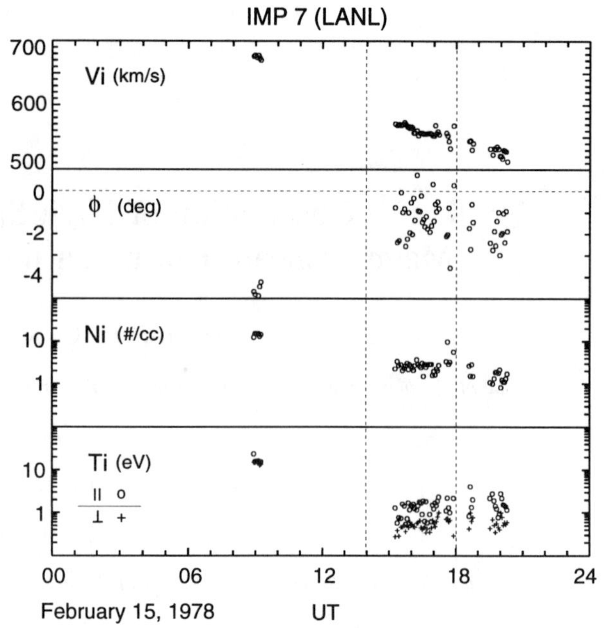

Fig. 2. Data from the IMP 7 Los Alamos National Laboratory plasma instrument in the upstream solar wind. From top to bottom, the components are the ion bulk flow velocity, the azimuthal flow angle, the ion density, and the parallel and perpendicular ion temperatures. The azimuthal flow angle is zero antisunward and positive from dawn to dusk. Note that the azimuthal flow angle is the actual flow angle. The flow aberration due to the motion of the Earth's inertial frame with respect to Sun has been already corrected.

and magnetosheath magnetic fields were antiparallel and to be found on the magnetospheric side of the magnetopause. They attributed the acceleration to antiparallel merging. However, in our case the accelerated flows were observed in the near-tail magnetosheath and occurred for nearly parallel magnetotail and magnetosheath fields.

Observations

The positions of the four spacecraft used in our analysis relative to a nominal magnetosphere are shown in Figure 1. Empirical bow shock [*Howe and Binsack*, 1972; *Farris et al.*, 1991] and magnetopause [*Howe and Binsack*, 1972; *Sibeck et al.*, 1991 as labelled] models are indicated (the solid curves). The dotted curve shows a possible position of the magnetopause in our event. It is obtained by fitting a curve through two points: the position of a model subsolar, dawn-dusk magnetopause of *Sibeck et al.* [1991] for an IMF $B_Z \simeq 10$ nT, and the position of ISEE 2 during boundary layer crossings and assuming that the magnetopause is symmetric about an aberrated X_{GSM}. The Earth's dipole tilt at the time of our event is small and therefore the hinging of the magnetotail is negligible. The position of the neutral sheet then is close to $Z_{GSM}= 0$.

Figure 2 shows the upstream solar wind conditions observed by IMP 7. The upstream solar wind speed was between 540 and 580 km/s, the ion temperature was a few eV and the density was ~3 cm^{-3} during the interval 1400–1800 UT. Although there was no magnetic field measurement in the upstream solar wind, we use the auroral electrojet (*AE*) index [*Akasofu et al.*, 1973, 1983] and the characteristics of magnetic field in the magnetosheath measured by the IMP 8 and ISEE spacecraft to infer the orientation of IMF. The value of *AE* was below 100 nT and the B_Z in the magnetosheath was dominantly positive after 1140 UT. The phenomena persisted for at least 12 hours [*Chen et al.*, 1993]. Accordingly we assume that the orientation of the IMF was mainly northward during the interval.

To measure the plasma flow in the magnetosheath, we have ISEE 1 and 2 at dawnside and IMP 8 at the other side (Figure 1). According to the ion and electron distribution functions of Fast Plasma Experiment (FPE), ISEE was moving between the magnetosheath and the low-latitude boundary layer (LLBL) during the time interval of 1400 to 1800 UT, February 15, 1978 (see Figure 6 in the paper of *Chen et al.* [1993] for the detail). To study the abnormal high speed flow ($V > V_{SW}$) in the magnetosheath, we examine Cross-Fan plasma data from ISEE 1. Figure 3 shows a representative interval of the plasma data from 1500 to 1600 UT. The vertical dashed lines are placed at times just before the ion density starts to increase significantly. The high (low) density regions are the magnetosheath (LLBL) identified by using plasma measurement from both FPE and Cross-Fan plasma instruments. The ion bulk velocity in the higher density plasma of the magnetosheath exceeds the upstream solar wind velocity V_{SW} (560 km/s) with peak values of ~780 km/s. There were also flows in the LLBL with bulk velocities exceed V_{SW}. These flow speeds are comparable with those measured at IMP 8 in the dusk magnetosheath (Figure 4). The IMP 8 observations shows that the flow velocity was ~700 km/s, the ion number density was 4 cm^{-3} in average and, as mentioned, the magnetic field was mainly in Z_{GSM} direction.

Table 1 summarizes the plasma properties observed in the solar wind and on the two flanks of the magnetosheath during the period from 1400 to 1800 UT. The solar wind velocity was 560 km/s, making a small angle with respect to the X-axis, the ion temperature was of the order of 1 eV and the number density was 3 cm^{-3}. Note that the other plasma instrument of IMP 7 was the Faraday cup plasma analyzer of MIT. In the magnetosheath, the flow velocity was around 670 km/s and the flow angle was small but with large fluctuations (±5°). The small flow angle in the magnetosheath indicates a very small flaring of the magnetopause. In summary, the magnetosheath flow speeds on both flanks of the magnetosphere exceeded the solar wind velocity (560 km/s) by ~20%.

EXPLANATIONS

Magnetosheath flows are often analyzed by use of gas dynamic models. In particular, the steady state gasdynamic model of *Spreiter et al.*, [1966] provides useful profiles of plasma parameters throughout much of the magnetosheath. In the gas dynamic model, the fluid equations are solved without a magnetic field. After the flow field is determined, the magnetic fields are included by adding the solar wind field and mapping it into the magnetosheath by using the frozen-in field condition $\nabla \times (\mathbf{V} \times \mathbf{B}) = 0$, where $\mathbf{V}$ is the flow velocity vector and $\mathbf{B}$ is the magnetic field vector. The results of *Spreiter et al.* [1966; 1969; 1980] predict that the

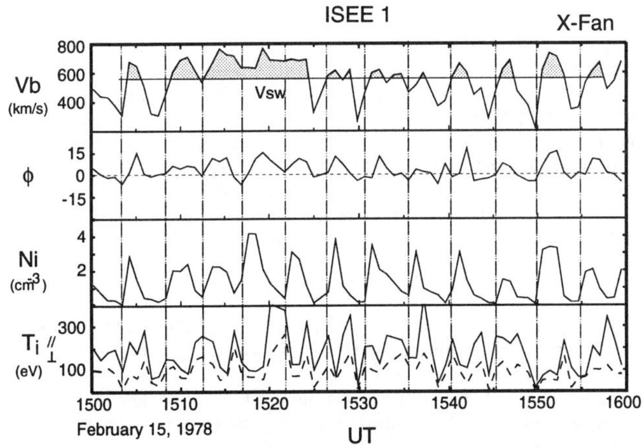

Fig. 3. Data from the Cross-Fan plasma instrument of ISEE 1. From top to bottom, the components are ion bulk velocity, azimuthal flow angles, ion number density and ion temperature.

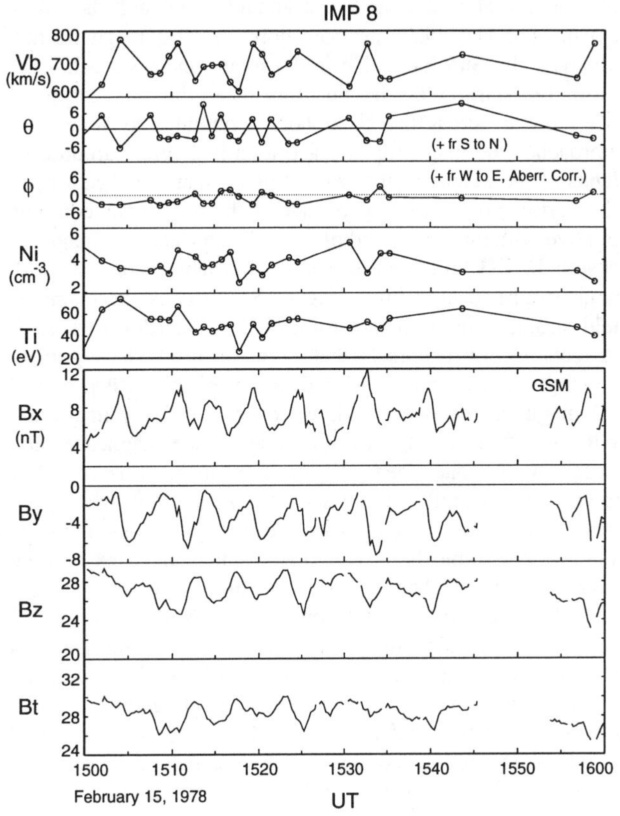

Fig. 4. A 1-hour interval of IMP 8 plasma and magnetic field components in the duskside magnetosheath.

Table 1. Summary of the Observation

	Solar Wind		Magnetosheath	
	IMP 7*	IMP 7**	IMP 8†	ISEE 1‡
Vb (km/s)	560 ±8	550	683 ±49	667 ±35
ϕ (deg)	-1.2 ±0.7	2.5	-1.7 ±5.2	3.0 ±5.3
Ti (eV)	1.0 ±0.2	5	46 ±10	158 ±59
Ni (#/cc)	3.1 ±0.8	3	3.5 ±0.8	2.7 ±0.5

Time interval: 1400-1800 UT, Feb 15, 1978.
ϕ is positive from W to E.
*LANL (37 pts); **MIT (estimate); †MIT (100 pts); ‡X-Fan (Vb>600, 120 pts).

flow velocity in the magnetosheath never exceeds that of the upstream solar wind.

Phenomenologically, the solar wind plasma near the magnetopause of the Earth or the other planets acquires considerable inhomogeneity because it is confined to flux tubes that extend through the bow shock, linking the unperturbed solar wind at the ends to plasma that has been diverted around the magnetopause. The consequences are most extreme for flux tubes that pass near the stagnation point and then flow antisunward very near the magnetopause. If a magnetic flux tube near the Sun-Earth line in the solar wind approaches the Earth, the portion of the flux tube near the subsolar point slows down across the bow shock, continues to slow as it approaches the magnetopause and then is diverted around the flanks of the magnetopause as part of the magnetosheath flow. The portions of a flux tube outside of the bow shock continue to move with the solar wind plasma. This means that magnetic tension $\mathbf{B}\cdot\nabla\mathbf{B}$ builds up unless the flux tube reconnects with magnetospheric field lines. Figure 5 illustrates the magnetic field structure in the magnetosheath for a northward IMF if no reconnection occurs at the magnetopause. The argument is equally valid for any other direction of the IMF provided that magnetic reconnection is not overwhelmingly important. If the portions of magnetic flux tubes in the magnetosheath, especially the ones near the subsolar magnetopause, slow down in the vicinity of the subsolar point, they must speed up elsewhere to catch up with the ends that remain in the solar wind. Somewhere their flow velocity must exceed the solar wind speed. The question remaining is where do these flux tubes start to accelerate or, in other words, where does the magnetic tension in the flux tubes begin to release? As an analogy, the release of magnetic tension is like launching a stone from a slingshot. If there is no drag on the band, the stone will be launched and the tension of the band will be released. Around the dayside, the magnetopause provides the drag on the magnetic field in the magnetosheath. On the nightside flanks, the drag decreases and the tension is released. The shape of the magnetopause is important in controlling the drag, and pressure balance at the magnetopause is crucial.

SUMMARY

We have found that magnetic effects are important in controlling the plasma flow velocity in the magnetosheath near to the magnetopause boundary. Observationally, one often sees high speed flows in the magnetosphere near the magnetopause which can be attributed to magnetic reconnection but such events do not show accelerated flows in the magnetosheath [for example, *Gosling et al.,* 1986]. Our study brings up a possibility that the plasma near, but external to the magnetopause can be accelerated by magnetic tension on one side of the boundary or the other; this means that neither reconnection nor merging processes are needed to account for the flow acceleration.

We summarize the principal points of our study:

1. We have presented evidence of high speed flows ($V > V_{SW}$) in the low-latitude magnetosheath during a strongly northward IMF.
2. Magnetic tension in the magnetosheath and the geometry of the magnetopause both may play important roles in accelerating plasmas in the magnetosheath.

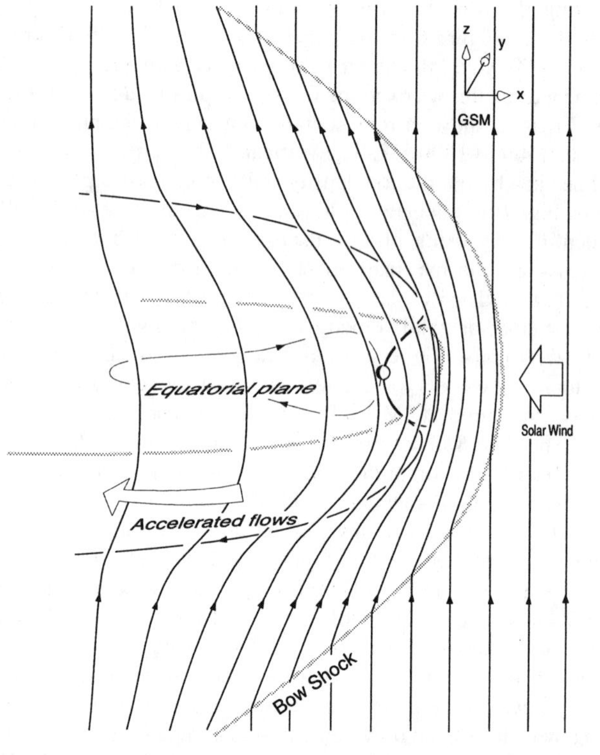

Fig. 5. A schematic diagram to illustrate a draping configuration of the IMF near the magnetopause. The diagram shows magnetic field lines in the magnetosheath near the magnetopause for a northward IMF. Field line bending develops as the flux tubes flow around the dayside magnetopause. Along the flanks, the release of magnetic tension of draped magnetic fields is proposed as an important mechanism for accelerating plasmas in the antisolar magnetosheath.

Acknowledgments. We thank C.T. Russell for supplying the ISEE magnetometer data. Computing resources were supplied both by the Institute of Geophysics and Planetary Physics and the Department of Earth and Space Sciences of UCLA. The work was supported by the National Aeronautics and Space Administration under grant NAG 5–1529 and by the Division of Atmospheric Sciences of the National Science Foundation under grant ATM 91–15557. Work at Los Alamos was performed under the auspices of the U.S. Department of Energy with support from NASA under 5–04039–D.

REFERENCES

Akasofu, S.-I., P. D. Perreault, F. Yasuhara, and C.-I. Meng, Auroral substorm and the interplanetary magnetic field, *J. Geophys. Res., 78,* 7490, 1973.

Akasofu, S.-I., B.-H. Ahn, Y. Kamide, and J. H. Allen, A note on the accuracy of the auroral electrojet indices, *J. Geophys. Res., 88,* 5769, 1983.

Axford, W. I., and C. O. Hines, A unifying theory of high-latitude geophysical phenomena and geomagnetic storms, *Can. J. Phys., 39,* 1433, 1961.

Axford, W. I., Viscous interaction between the solar wind and the

Earth's magnetosphere, *Planet. Space Sci., 12*, 45, 1964.
Chen, S.-H., M. G. Kivelson, J. T. Gosling, R. J. Walker, and A. J. Lazarus, Anomalous aspects of magnetosheath flow and of the shape and oscillations of the magnetopause during an interval of strongly northward interplanetary magnetic field, *J. Geophys. Res., 98,* 5727, 1993.
Dungey, J. W., Interplanetary magnetic field and the auroral zones, *Phys. Rev. Lett., 6,* 47, 1961.
Farris, M. H., S. M. Petrinec, and C. T. Russell, The thickness of the magnetosheath: Constraints on the polytropic index, *Geophys. Res. Lett., 18*, 1821, 1991.
Gosling, J. T., M. F. Thomsen, and S. J. Bame, Accelerated plasma flows at the near-tail magnetopause, *J. Geophys. Res., 91*, 3029, 1986.
Howe, H. C., Jr., and J. H. Binsack, Explorer 33 and 35 plasma observations of magnetosheath flow, *J. Geophys. Res., 77*, 3334, 1972.
Saunders, M. A., Magnetosheath, magnetopause and low latitude boundary layer research, 1987–1989, *J. Atmos. Terr. Phys., 52*, 1107, 1990.
Sibeck, D. G., R. E. Lopez, and E. C. Roelof, Solar wind control of the magnetopause shape, location, and motion, *J. Geophys. Res., 96*, 5489, 1991.
Song, P., C. T. Russell, J. T. Gosling, M. Thomsen, and R. C. Elphic, Observations of the density profile in the magnetosheath near the stagnation streamline, *Geophys. Res. Lett., 17*, 2035, 1990
Southwood, D. J., and M. G. Kivelson, On the form of the flow in the magnetosheath, *J. Geophys. Res., 97,* 2873, 1992.
Spreiter, J. R., A. L. Summers, and A. Y. Alksne, Hydromagnetic flow around the magnetosphere, *Planet. Space Sci., 14*, 223, 1966.
Spreiter, J. R., and A. Y. Alksne, Plasma flow around the magnetosphere, *Rev. Geophys. Space Phys., 7,* 11, 1969.
Spreiter, J. R., and S. S. Stahara, Solar wind flow past Venus: Theory and comparisons, *J. Geophys. Res., 85* 7715, 1980.

S.-H. Chen and M. G. Kivelson, Institute of Geophysics and Planetary Physics and Deparment of Earth and Space Sciences, University of California, Los Angeles, CA 90024.

J. T. Gosling, Los Alamos National Laboratory, Los Alamos, New Mexico 87545.

A. J. Lazarus, Center for Space Research, Department of Physics, Massachusetts Institute of Technology, Cambridge, Massachusetts 02139.

The Dynamical Plasma Sheet Boundary Layer: A New Perspective

G. Ganguli[1] and H. Romero[2]

Space Plasma Branch, Plasma Physics Division
Naval Research Laboratory, Washington, DC 20375

P. Dusenbery

Department of Astrophysical, Planetary and Atmospheric Sciences
University of Colorado, Boulder, CO 80309

We investigate the dynamical and steady state features caused by the inherent inhomogeneities of the Plasma Sheet Boundary Layer (PSBL). Evidence for these inhomogeneities has recently been presented [G. Parks, JGR, 97, 2943, 1992, and Takahashi and Hones, JGR, 93, 8558, 1988], and existing models of the PSBL do not incorporate these effects satisfactorily. We have developed a Vlasov equilibrium picture of the PSBL and have analyzed the linear stability of this equilibrium. We have also developed a 2 1/2 D particle code to study the nonlinear evolution of the PSBL. Interesting new features and their observational evidence are provided. Using the Vlasov description we find that a transverse localized d.c. electric field exists at the interface of the plasma sheet with the lobe in the north-south direction, and that there are highly inhomogeneous flows and currents both parallel and perpendicular to the magnetic field in this region. During active periods the d.c electric field will be intensified and localized over smaller scale-sizes, leading to a stressed boundary layer. Here we describe the dynamical evolution and the relaxation mechanism of a stressed PSBL. It is found that a new class of plasma modes appear which can cascade from high frequency short wavelength to low frequency long wavelength, provide large viscosity and a broadband frequency spectrum, and are capable of causing much larger transport than that available in existing models.

1. INTRODUCTION

The plasma sheet boundary layer (PSBL) is one of the primary regions of transport in the magnetosphere [Eastman et al., 1984; Huang et al., 1987]. It separates the hot (thermal energy > 1 keV), dense (density ~ 1 cm^{-3}) central plasma sheet (CPS) from the cool (thermal energy ~ 10's of eV), tenuous (density ~ 0.01 cm^{-3}) lobe. It extends from the tail current sheet where magnetospheric boundary particles are accelerated to keV energies [Lyons and Speiser, 1982], to the high-latitude ionosphere which is the source of cold ions streaming tailward [Sharp et al., 1981]. While most recent studies of the PSBL have focused on the region where field-aligned ion beams are present [Williams, 1981; Eastman et al., 1984], Parks et al. [1984; 1992] note that the PSBL exhibits a more complex structure which includes thin layers of energetic electrons confined to its outermost region and whose appearance is generally correlated with active periods [Parks, private communication, 1991]. In the PSBL, plasmas of different origins (e.g., solar, ionosphere, plasma sheet), with different characteristics, are mixed and energized due to local dissipation processes. Other signatures of the PSBL include the occurrence of intense, localized electric fields in the north-south direction [Cattel et al., 1982; Orsini et al.,

Space Plasmas: Coupling Between Small
and Medium Scale Processes
Geophysical Monograph 86

1984] and the existence of significant gradients in the background plasma parameters [Lui, 1987; Takahashi and Hones, 1988]. Finally, electrostatic and electromagnetic wave activity is detected at the PSBL-Lobe interface and extends throughout the PSBL [Scarf et al., 1974; Gurnett et al., 1976; Grabbe and Eastman, 1984; Parks et al., 1984; Cattel et al., 1986]. Less intense waves are at times observed in the lobe and CPS regions adjacent to the PSBL.

The dynamical state of the PSBL is closely related to that of the magnetosphere and its degree of coupling with the solar wind. Therefore, the level of PSBL activity can be used as an indicator of 'magnetospheric weather'. For instance, when the magnetosphere is highly stressed, the plasma sheet and its boundary layer become narrow and the micro-processes discussed here would be most active. During these times particles originating from different regions of the magnetosphere are vigorously mixed and energized by the dissipative processes taking place within the PSBL. This in turn gives rise to unique distributions which can be accelerated at lower altitudes in the auroral zone and provide the spectacular auroral displays. As the magnetosphere relaxes leading to thicker boundaries, the PSBL will likely become less active. The boundary plasma distributions and the plasma wave activity can therefore be considered as a diagnostic of the global magnetospheric conditions. Thus, it is important to understand the dynamics of the PSBL and how it is affected by global changes in the magnetosphere. Specifically, we have to quantify such observables as composition, structure, width of the PSBL, and its activity level in relation to its dynamical state. In doing so the subject of wave generation and the nonlinear evolution of strongly inhomogeneous boundary layers will have to be addressed. With the availability of high-time-resolution detectors in the upcoming Cluster and Geotail missions, small scale-sizes will be resolved more accurately. The new data can then be used to validate the predictions of waves and their influence on the dynamics of the PSBL.

Sufficient observational evidence has now been gathered to warrant a comprehensive and systematic modelling effort of the PSBL. To date, most previous studies of this region have focussed on the influence of counterstreaming uniform beams on the generation of broadband electrostatic noise (BEN). Virtually no attention has been given to the study of the origin and dynamical evolution of the strong inhomogeneities inherent to this region. Also, while effects due to parallel beams have been extensively studied, no attention has been given to the role played by the observed shear in the beam velocities which occur throughout the PSBL [see Fig. 1, taken from Takahashi and Hones, 1988 and Lui 1987]. We find that

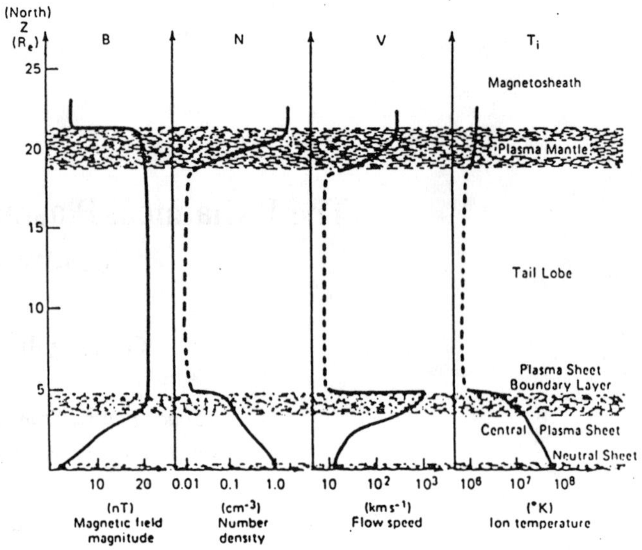

Fig. 1a. A schematic of spatial variation of ion beams at the PSBL (from Takahashi and Hones, 1988).

inhomogeneities can profoundly affect the nonlinear evolution of the PSBL. In particular, it appears that the PSBL width is determined from a balance between the stress generated by the solar wind-magnetosphere coupling and that dissipated in the PSBL. During active periods the rate of solar wind energy input can exceed the rate of dissipation in the PSBL. This leads to narrower PSBL widths with their own unique properties and signatures, while the quiet time broad PSBL follows a different evolution scheme and is typified by a different signature.

Boundary layers are generally dynamic regions which, during active periods when the magnetosphere is stressed, may contain structures of small scale sizes of a few hundred kms (approximately, 1 to 2 ρ_i, the ion gyroradius) but during moderate to low activity periods, the scale sizes may be a few thousand kms (around 6 to 10 ρ_i) [Dandouras et al., 1986, Parks et al., 1992]. The highly active nature of boundary layers offers considerable challenge as far as modelling this region is concerned. Major outstanding issues concerning the PSBL are: (1) the origin and nature of the d.c. electric fields and inhomogeneous flows [Cattell et al., 1982; Orsini et al., 1984; Takahashi and Hones, 1988], (2) the origin, nature and evolution of stressed ($L \leq 2\rho_i$) boundary layers [Parks, 1992, Parks, 1990, private communication], (3) the nature of dissipation and transport processes in the PSBL and their influence on its structure (i.e., thickness), and (4) the nature and origin of the broadband electrostatic noise and its relationship to the dynamical evolution of the PSBL. In the following we address some of these issues.

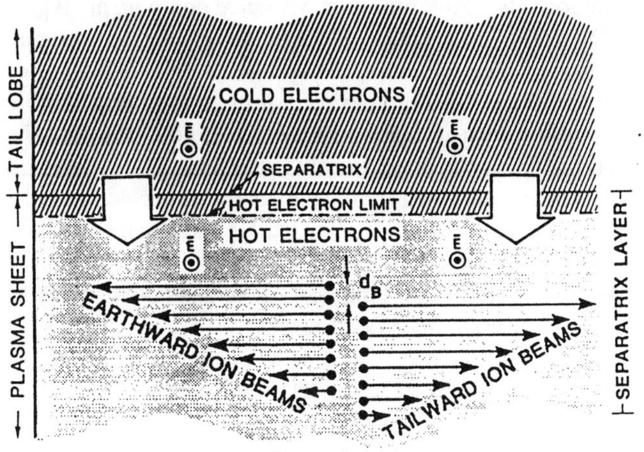

Fig. 1b. North-south profiles of plasma parameters in the midnight region of the magnetotail (from Lui, 1987).

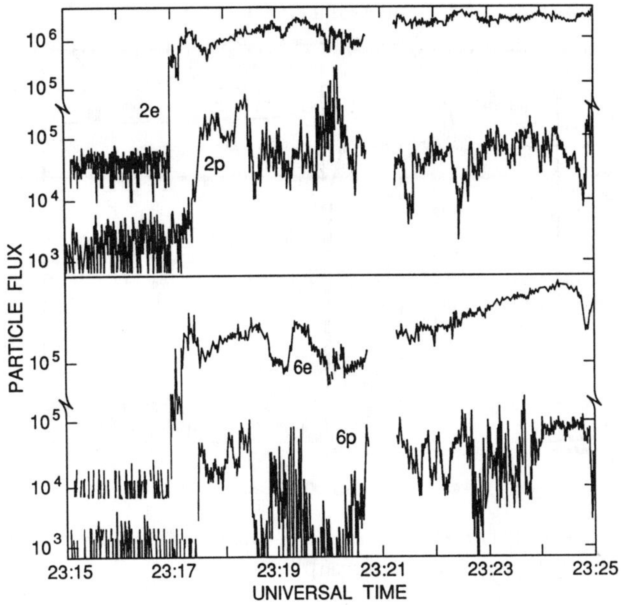

Fig. 2. Ion (p) and electron (e) flux data from ISEE 1 (March 31, 1979) versus UT for two energy channels (2 keV and 6 keV) [courtesy, G. Parks].

2. EXISTING MODELS OF BOUNDARY LAYERS

While some attention has been devoted to developing models of the magnetopause [Okuda, 1991, Cargill and Eastman, 1991, Winske et al. [1990], Lee [1990] and references therein], comparatively little work is available concerning models of the PSBL. Nonetheless, a great number of studies concerning wave generation in the PSBL region are available [e.g., Huba et al., 1978; Grabbe and Eastman, 1984; Grabbe, 1985; Dusenbery and Lyons, 1985, 1988, 1989; Dusenbery, 1986; Omidi, 1985; Akimoto and Omidi, 1986; Ashour-Abdalla and Okuda, 1986; Schriver and Ashour-Abdalla, 1987, Schriver et al., 1990, Ashour-Abdalla and Schriver, 1989, and references therein]. Recent works of Romero et al. [1990] and Onsager et al., [1990] are the first attempts to systematically model the inhomogeneous equilibrium properties of the PSBL. The Onsager et al., model uses time of flight concepts to obtain the plasma density variation in the PSBL. It treats the PSBL as a time stationary region while observations indicate that it is a very dynamic structure. Also, Ashour-Abdalla et el., [1990], have developed a model based on non-self-consistent single particle orbits which exhibits the existence of field-aligned beams in the PSBL. Test particle models preclude plasma effects and hence their validity is restricted primarily to non-dynamic cases where collective effects are non-existent. Such models cannot therefore be applied to a stressed PSBL. We have taken the initial steps to understand the structure and stability of the PSBL by developing a Vlasov model for its inhomogeneous structure [Romero et al., 1990], a nonlocal theory to study its linear stability properties [Romero et al., 1992a], and a 2-1/2 dimension Particle In Cell (PIC) code to study its nonlinear evolution, dissipation and transport properties [Romero et al., 1992b]. In the following we provide a description of our PSBL model.

3. THE PSBL MODEL

3.1 *Evidence of Strong Inhomogeneities*

To illustrate the nature of the nonuniformities present in the PSBL, we note here the existence of thin electron layers on its outer edge. These layers were first identified in the ISEE particle data by Parks et al. [1979], and were later shown to be a persistent feature of the PSBL by Parks et al. [1984; 1992] and by Takahashi and Hones [1988]. Observations have shown that the width of these layers is typically of the order of 100 km and at times may even be smaller [G. Parks, private communication, 1990]. Fig. 2 (courtesy of G. Parks) illustrates these features by showing 1/4 second averages of the electron and ion fluxes from the two lowest energy channels (2 keV and 6 keV, respectively) of the fixed energy detector on ISEE 1 [Anderson et al., 1978]. The particles measured by this instrument have pitch angles in the range 70-90 degrees. It is seen that electron fluxes in both energy channels increase substantially at 23:17 UT. This is followed about

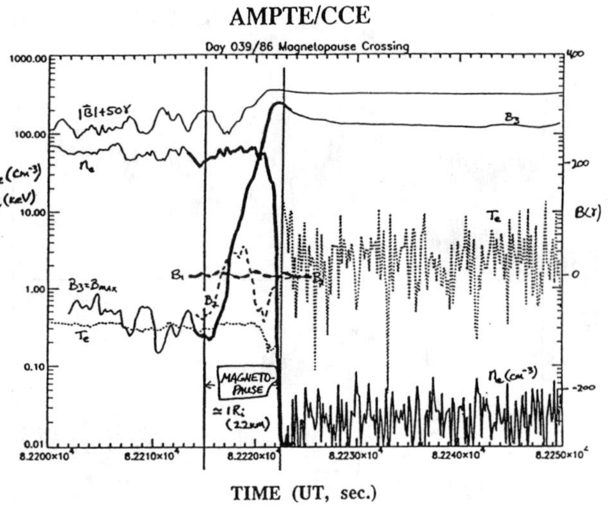

Fig. 3. AMPTE/CCE magnetopause crossing of February 8, 1986 (day 039) [courtesy, T. Eastman].

30 seconds later by an ion flux increase. The width of the electron layer can be estimated to be ~ 100 km but smaller distances cannot be ruled out because the motion of the PSBL is difficult to determine accurately. Since the ion gyroradius (ρ_i) for keV range protons in a 40 nT magnetic field is around 125 km, Fig. 2 reveals that significant variations in plasma parameters may occur at the PSBL-lobe interface on a scale length smaller than ρ_i. At least 30 more cases with similar features have been recently identified and are currently under investigation [G. Parks, private communication, 1992]. Although the observation of the energetic electron layers are infrequent, their frequency of occurrence is not insignificant. Therefore, they deserve attention. The origin of these layers, their contribution to PSBL dynamics, and their relaxation mechanism(s) are not established. These aspects and the tendency of these electron layers to be more prominent after strong geomagnetic activity are important issues to be addressed in this article.

The occurrence of strong density gradients on the outer edge of the PSBL is not a unique feature of this region. Strong density gradient is also evident in the recent AMPTE/CCE magnetopause data. This data supports the notion that strong density (or pressure) gradients are common to all boundary layers in the magnetosphere [Eastman et al., 1991]. Fig. 3 (courtesy of T. Eastman) illustrates an AMPTE/CCE magnetopause crossing on April 14, 1987 (day 104). This magnetopause crossing occurs at 12.6 hours local time and 8.8R_E radial distance. The current layer itself is clearly identified by the changes in the intermediate (B_2) and maximum variance (B_3) field components. The minimum variance component (B_1) is close to zero on average. Electron density and temperature plots are based on non-spin-averaged moments integrated above 50 eV. Using energetic ions to remotely sample boundary motion leads to an estimate of 1.2ρ_i for the magnetopause thickness in this crossing. Note the sharp drop in the electron density near the inner edge of the magnetopause. The scale length for the primary density gradient is less than 0.1 of the magnetopause thickness. This observation shows that extremely thin boundary layers may be rather common occurrences in space plasmas.

In the following sections of this paper (i) we develop a Vlasov model for the PSBL which can explain the existence of the observed energetic electron layers, (ii) investigate its stability properties, and (iii) follow its nonlinear evolution.

3.2 *Vlasov Equilibrium Model*

We assume an ideal PSBL (electron-hydrogen plasma, one temperature, no beams) but retain its highly inhomogeneous nature. Fig. 4 gives the equilibrium configuration under consideration. We construct an equilibrium distribution function using the constants of motion [Romero et al., 1990], which are: (1) $E_j = m_j v^2/2 + q_j \phi(x)$, the total energy, (2) $X_g = x + v_y/\Omega_j$, the guiding center position where Ω_j is the gyrofrequency and (3) v_z, the velocity along the magnetic field. The magnetic field is assumed to be nearly uniform and in the z (earthward, or x in a GSE co-ordinate system) direction and the strong inhomogeneities are assumed to be in the x (north-south, or z in a GSE co-ordinate system) direction. This leads to,

$$f_0(E_j, X_g, v_z) = \frac{N_j Q_j(\zeta_j)}{(\pi v_j^2)^{3/2}} \exp\left(-\frac{E_j}{T_j}\right), \quad (1)$$

where $\zeta_j = C_j P_j - D_j v_z$, $P_j = \Omega_j X_g$, v_j is the thermal velocity, C_j and D_j are constants, N_j is a normalization constant, and 'j' denotes the species. The function Q_j is the distribution of guiding centers which determines the density profile and is defined by,

$$Q_j(\zeta) = \begin{cases} R_j, & \zeta < \zeta_1 \\ R_j + (S_j - R_j)\left(\dfrac{\zeta - \zeta_1}{\zeta_2 - \zeta_1}\right), & \zeta_1 < \zeta < \zeta_2 \\ S_j, & \zeta > \zeta_2 \end{cases}$$

(2)

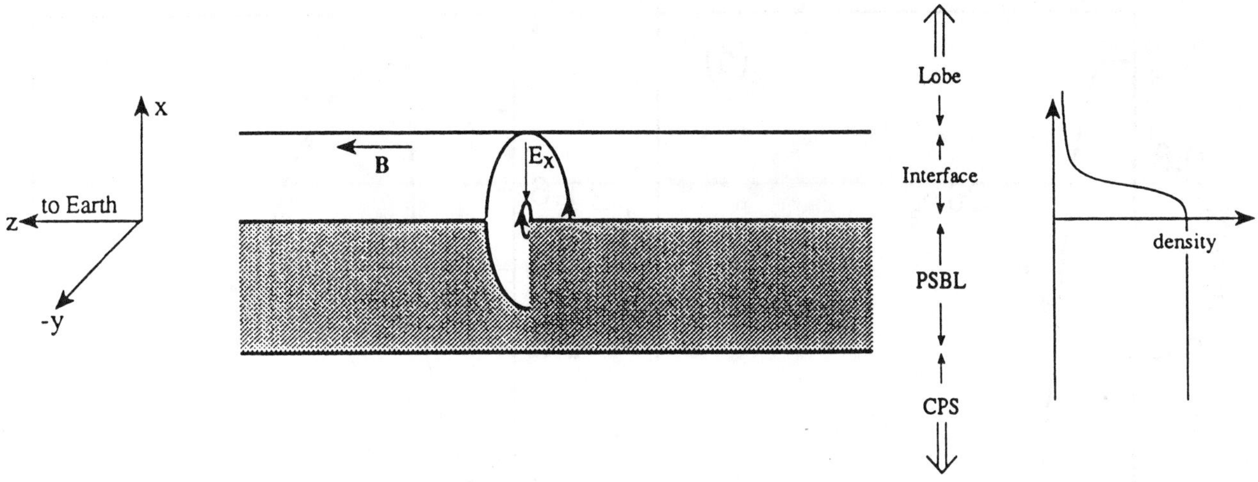

Fig. 4. A sketch of the equilibrium configuration at the PSBL.

We define $N_jR_j = n_{sj}$ the density in the PSBL, while $N_jS_j = n_{Lj}$ is the density in the lobe. During active periods, especially during a substorm growth phase, there is enhanced compression on the magnetotail due to increased solar wind coupling. This tends to increase the density gradient at the intersection of the PSBL with the lobe. On the other hand, during less active periods the spatial scale size is considerably larger and the gradient in the density is correspondingly smaller. In Eq. (2), ζ_1 and ζ_2 are input parameters determining the width of the structure and can be obtained from observations. Other inputs are the magnetic field strength B_0 (assumed constant since the plasma β is low) and the ion and electron temperatures. In terms of these input quantities f_0 can be integrated over velocity space to obtain the density profiles across the boundary layer. Imposing quasi-neutrality, we solve for the electrostatic potential $\phi(x)$ self-consistently. With $\phi(x)$ determined, f_0 is fully specified and the flows (therefore, currents) and pressures can be readily calculated by taking higher moments of f_0.

Our Vlasov model is similar in spirit to the magnetopause models of Lee and Kan [1979], and Whipple et al., [1984] (also, Sestero [1964; 1966]). Analyzing magnetopause models, Whipple et al., [1984] and Rogers and Whipple [1988] have pointed out the nonuniqueness of models of this nature. Nonuniqueness arises because the choice of the function Q_j is arbitrary, which is chosen primarily for convenience. We note however, that many of the qualitative features of Vlasov equilibria can also be derived from fluid theory where this ambiguity is not a problem.

Using numbers pertinent to Fig. 2 (described in more detail by Romero et al. [1990]) the model predicts the existence of a d.c. electric field in the north-south direction. A rough average value of this d.c. electric field is estimated to be around 5 - 10 mV/m which compares favorably with observed north-south electric fields [Cattel et al., 1982 and Orsini et al., 1984]. Fig. 5a is the density profile and Fig. 5b is the corresponding self-consistent potential at the PSBL-lobe interface.

We also find highly localized flows and currents both perpendicular and parallel to the magnetic field. In Fig. 6 we show the equilibrium transverse ion and electron flows normalized by their respective thermal velocities. In Fig. 7 we plot the gradient in these flows normalized by the species gyrofrequency (i.e., $(dV/dx)/\Omega$). The peak value of this dimensionless number is defined as the shear parameter $\alpha_j = \omega_{sj}/\Omega_j$, where $\omega_{sj} = V^0_j/L_j$ is the shear frequency, V^0_j is the peak flow, L_j is the scale-size of this flow, and j is the species. For electrons the shear parameter is less than unity implying that the electron gyrofrequency dominates its shear frequency and hence the electrons are magnetized. On the other hand the ion shear frequency is larger than its gyrofrequency, implying that the ions are essentially unmagnetized. The magnitude of the flows and their spatial gradients are found to be closely related to the gradient in the density at the PSBL-Lobe interface which in turn is related to the magnetotail compression by the solar wind. A smaller density gradient scale-size leads to a stronger flow with larger shear in it. Thus, the thin energetic electron current layer as observed by Parks et al. [1984; 1992] and Takahashi and Hones [1988] at the outer edge of the PSBL may be a consequence of the density (or pressure) gradient at the PSBL-Lobe interface and are intensified during active periods when the magnetotail is highly compressed.

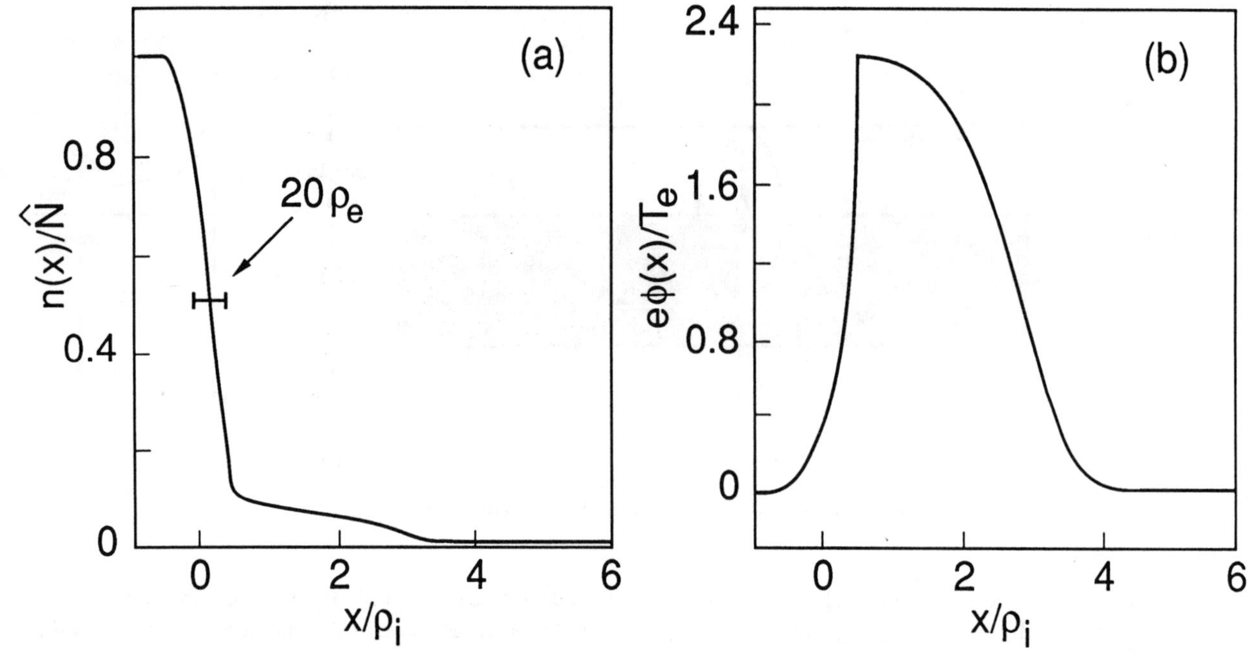

Fig. 5. Self-consistent density (a) and the electrostatic potential (b) at the boundary layer. Note x < 0 represents the PSBL and x > 0 represents the lobe.

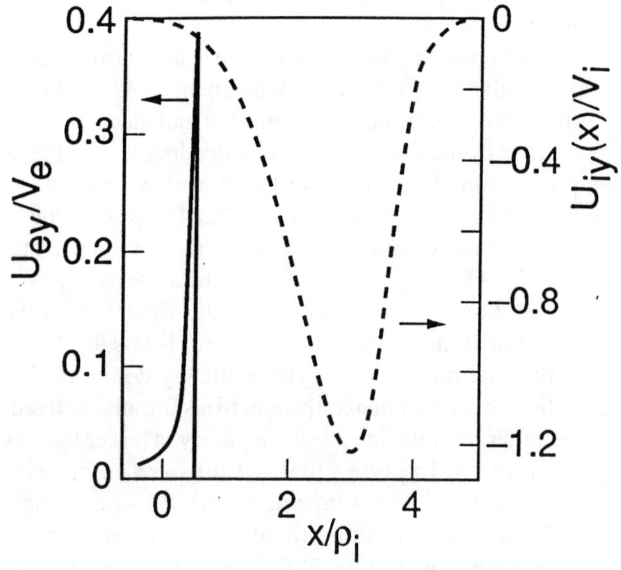

Fig. 6. Self-consistent ion (dotted) and electron (solid) flows normalized by their respective thermal velocities as a function of position.

3.3 *Linear Stability*

An essential next step is to examine the stability of the equilibrium described in the previous section. One of the most prominent results of our Vlasov model is the existence of sheared electron flows both along and across the magnetic field. Moreover, ion flows also exhibit shears with large values of the shear parameter, α. From the stability analysis we find that the shear parameter α_j determines the character of the waves. For example, the condition $\alpha_j > 1$ implies that the shear frequency dominates over the species gyrofrequency. Consequently, the species is effectively unmagnetized. These features directly affect the ensuing turbulence in the system which strongly influences the nonlinear evolution and anomalous transport.

The stability of sheared flow layers, perpendicular to a magnetic field, has been extensively discussed in the published literature and need not be repeated here. In general, we find that for $\alpha_i < 1$ and $\alpha_e < 1$, both electrons and ions are magnetized. This is the likely scenario in a less stressed PSBL, whose width could be $(6 - 10)\rho_i$. The resulting shear driven waves are of low frequency ($\omega \sim \Omega_i$ or less) [Ganguli et al., 1985a,b; 1988a; 1989a; 1991; Nishikawa et al., 1988]. As discussed in Ganguli et al. 1988a and 1991, these modes, along with the well known Kelvin-Helmholtz (KH) modes, form the two branches of a magnetized plasma with sheared flow perpendicular to it.

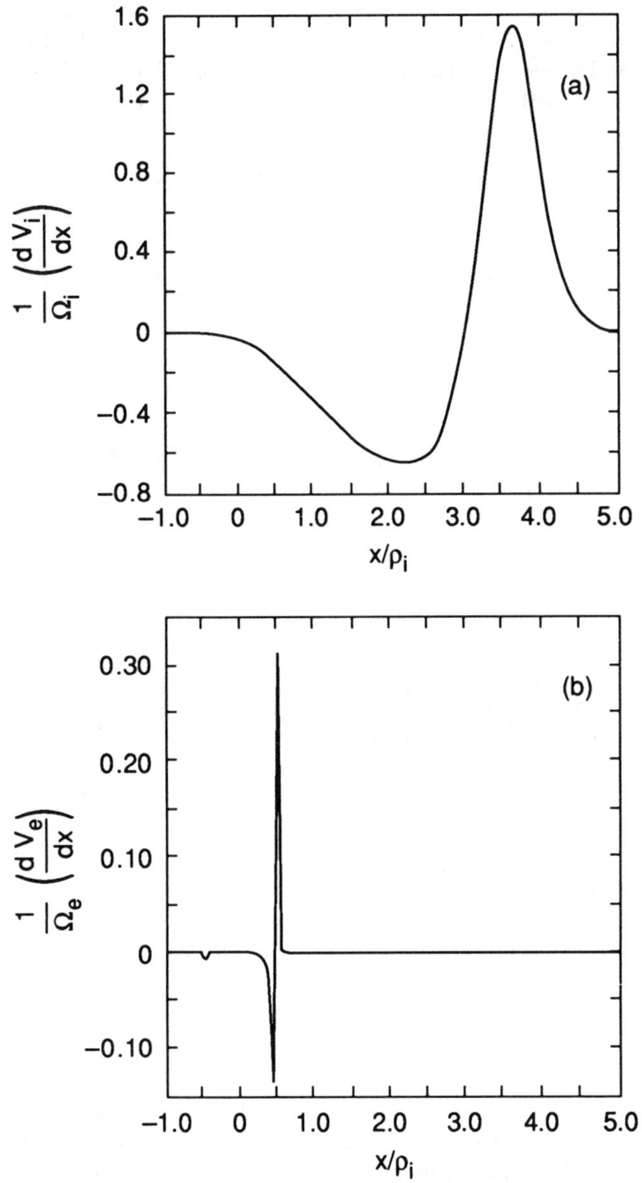

Fig. 7. The spatial gradients of the ion (a) and electron (b) flows as a function of position. Note the peak value of the ion flow gradient exceeds its gyrofrequency while the corresponding parameter for the electrons does not.

Since these modes are generated by an inhomogeneity in the energy density introduced by the localized nature of the perpendicular flow, they have been referred to as the Inhomogeneous Energy Density Driven Instability (IEDDI). While the KH branch is characterized by low frequency ($\omega << \Omega_i$) and long wavelengths ($k_\perp \rho_i << 1$), the IEDDI branch is characterized by higher frequency ($\omega \sim \Omega_i$) and shorter wavelengths ($k_\perp \rho_i \sim 1$). The strong density gradients, such as those found in the PSBL-lobe interface and at the magnetopause, have a damping effect on the KH branch [Satyanarayana et al., 1987], while they can enhance the IEDDI branch. PIC simulations of Nishikawa et al. [1988; 1990] have shown that the IEDDI is capable of contributing significantly to particle diffusion and viscosity.

For $\alpha_i > 1$ and $\alpha_e < 1$, electrons are magnetized but the ions become unmagnetized. This is the likely case in a stressed PSBL during highly active periods, whose widths could be $(1 - 2)\rho_i$. The resulting shear driven waves are in the intermediate frequency range ($\omega \sim \omega_{LH}$, the lower hybrid frequency to Ω_e) [Ganguli et al., 1988b; 1989c; Romero et al., 1992a,b]. These modes have been referred to as the Electron-Ion-Hybrid (EIH) modes. In Section 3.4 we describe the nonlinear evolution of this mode and its relevance to stressed boundary layers. For $\alpha_i > 1$ and $\alpha_e > 1$, both ions and electrons are unmagnetized and the resulting waves are of high frequency ($\omega \sim \omega_{pe}$, the plasma frequency) [Romero et al., 1992a; Mikhailovskii, 1974].

A similar hierarchy of parallel flow shear driven modes also exist and are relevant to the PSBL since field-aligned beams [Eastman et al., 1984; Takahashi and Hones, 1988] and currents [Frank et al., 1981, Frank 1985] are important constituents of the PSBL. Accordingly, there are a great number of theoretical studies in the published literature concerning the generation of BEN by PSBL beams [for example see Ashour-Abdalla and Schriver, 1989, and references therein]. These studies neglected the effects of transverse shear in the beam and current velocities which is unrealistic for the PSBL (see Fig. 1). Velocity shear can significantly affect the properties of parallel flow and current driven waves. The threshold for some modes, for example the low frequency D'Angelo waves [D'Angelo, 1965] which are driven by shear in the parallel ion flow, is significantly increased by transverse velocity shears [Ganguli et al., 1989b]; while the threshold for the current driven ion-cyclotron instability (CDICI) is lowered [Ganguli et al., 1989a; 1991; Nishikawa et al., 1990]. This prediction has recently been verified in a series of well diagnosed laboratory experiments [Koepke et al., 1994; Amatucci et al., 1994]. In view of this, it is important to ask whether and how the observed shear in beam velocities in the PSBL will alter the conclusions of previous investigations and affect the PSBL's nonlinear state.

Our preliminary investigation of the effect of velocity shear on parallel current driven modes is encouraging. For example, significant changes take place in the well known dispersion relation of the Current Driven Ion -Cyclotron Instability [Drummond and Rosenbluth, 1962; Kindel and Kennel, 1971] when shear in the field-aligned current

(FAC) is included. By including shear in V_d, the field-aligned electron drift, the dispersion relation for the CDICI in the local limit becomes [Ganguli et al, 1991],

$$1 + \sum_n \Gamma_n(b)\left(\frac{\omega}{\sqrt{2}|k_z|v_i}\right) Z\left(\frac{\omega - n\Omega}{\sqrt{2}|k_z|v_i}\right)$$

$$+ \tau'\left[1 + \left(\frac{\omega - k_z V_d}{\sqrt{2}|k_z|v_e}\right) Z\left(\frac{\omega - k_z V_d}{\sqrt{2}|k_z|v_e}\right)\right] = 0, \quad (3)$$

where v_i and v_e are ion and electron thermal velocities, $\tau' = \tau(1 - V_d'/u\Omega_e)$, $\tau = T_i/T_e$, $u = k_z/k_\perp$, $b = k_\perp^2 \rho_i^2$ and $\Gamma_n(b) = \exp(-b)I_n(b)$ where $I_n(b)$ are the modified Bessel function. Thus, if shear in the parallel flow $V_d' = dV_d/dx = 0$, then $\tau' = \tau$, and Eq. (3) reduces to the dispersion relation for the CDICI, otherwise shear affects the system by reducing the temperature ratio and thereby reducing the threshold for this instability. The effective temperature ratio τ', can even become negative. This is an interesting domain never explored before. For $\tau = 10$, Kindel and Kennel predict a threshold velocity at $V_d \sim 50v_i$, where v_i is the ion thermal velocity. However, for the same τ we find that there is a strongly growing ion cyclotron root ($\omega = (1.03+0.60i)\Omega_i$) for $V_d = 15v_i$, where $k_y\rho_i = 44.7$, $u = 0.0015$, and $\mu = 1837$, even if we assume a small $V_d' = 0.0019\Omega_e$. Thus, shear in the parallel drift velocity can substantially reduce the threshold drift for the CDICI.

When ions are unmagnetized but electrons are magnetized, FAC will give rise to higher frequency ($\omega_r \sim (\omega_{LH} - \Omega_e)$) instabilities. The nonlocal differential dispersion relation for these modes including density and velocity gradients is given by,

$$\frac{d}{dx}\left[A(x)\frac{d\tilde{\phi}}{dx}\right] - \left\{k_y^2 A(x) + k_\parallel^2 C(x) - E(x)\right\}\tilde{\phi} = 0, \quad (4)$$

where,

$$A = 1 + \frac{\omega_{pe}^2(x)}{2\Omega_e^2}\xi_0\left\{Z(\xi_1) - 2Z(\xi_0) + Z(\xi_{-1})\right\} - \frac{\omega_{pi}^2(x)}{\omega^2},$$

$$C = 1 + \frac{\omega_{pe}^2(x)}{k_\parallel^2 v_e^2}\left\{1 + \xi_0 Z(\xi_0)\right\} - \frac{\omega_{pi}^2(x)}{\omega^2}$$

$$- \frac{k_y}{k_\parallel^2 \Omega_e}\frac{d}{dx}\left\{\frac{\omega_{pe}^2(x)}{k_\parallel v_e}\left[Z(\xi_0) + \xi_0 Z'(\xi_0)\right]\right\},$$

$$E = -\frac{d}{dx}\frac{k_y \omega_{pe}^2}{2\Omega_e^2}\xi_0\left\{\frac{2\Omega_e}{k_\parallel v_e}Z'(\xi_0) + Z(\xi_1) - Z(\xi_{-1})\right\},$$

and

$$\xi_n = \left[\omega - n\Omega_e - k_\parallel V_d(x)\right] / |k_\parallel v_e| .$$

Preliminary results indicate that the threshold for the electron acoustic and two stream branches is lowered due to velocity shear. For a homogeneous plasma the electron acoustic branch is stabilized when the thermal velocity approaches the drift velocity, i.e., $v_e \sim V_d$, making it less relevant to PSBL conditions where the flow speed can be of the order of the thermal velocity. However, by including velocity shear effects, the thermal velocity can be raised to 5 times the drift velocity. Even for such strong thermal effects there is growth around $\omega_r \sim 0.2\Omega_e$ and $\gamma \sim \omega_{LH}$. Thus, this mode may contribute substantially to the dynamics of the PSBL and deserves further attention. A detailed analysis of these modes is deferred to a future article.

Mikhailovskii, [1974] has shown that inhomogeneous electron flows lead to an instability around the electron plasma frequency, ω_{pe}, in an unmagnetized plasma. He considered the ions to be stationary with a relative drift between ions and electrons. We have extended this to the case where both ions and electrons are drifting with no relative velocity between them. We find that an instability near ω_{pe} can still be sustained by shear in the flow velocity [Romero et al., 1992a]. From our preliminary work we see that shear-driven instabilities around ω_{pe} need very high values of velocity shear. Further investigation of the properties of this branch is needed to examine whether the expected velocity shear in the PSBL can sustain this instability and, if so, can it explain the observed power around the plasma frequency?

3.4 *Nonlinear Evolution*

In this section, we consider the investigation of the dynamical evolution of a stressed PSBL by developing an electrostatic, 2-1/2D PIC code [Romero et al., 1992b]. In its current version, 800,000 particles are used in the simulation with, typically, 40 particles per cell. The electric field is decomposed into two constituents: the first is doubly periodic in the two spatial dimensions (hence, it is represented in a double Fourier series and its solution is facilitated by the use of fast Fourier transform techniques using a grid employing 64 X 128 nodes), and the second one is a time-independent, externally imposed field. This latter component models the effects of a constant driver in the system. It represents the effects of compression of the magnetotail owing to coupling with the solar wind. Increased coupling, especially during a substorm growth phase, will tend to steepen the PSBL as well as cause the well known plasma sheet thinning prior to substorm onset. This effect is modelled by intensifying the driving electric field. To begin the simulation, the particle loading is accomplished in a way such that both the electrons and ions are in force balance with the externally imposed electric field. This is achieved by a density gradient in the system similar in form to that which occurs at the PSBL-Lobe interface (or at the magnetopause). As predicted by our Vlasov model we find a cross-field electron flow with respect to the ions. The grid spacing in both directions is chosen to be 1.4 to 2 times smaller than the Debye length. The time evolution of the system is then observed through the use of various diagnostic tools.

For magnetospheric boundary layer plasmas in which density gradients are important, the modes of interest are the Lower Hybrid Drift (LHD) Instability [Krall and Liewer, 1971] and the Electron Ion Hybrid (EIH) Instability [Ganguli et al., 1988b]. Also of interest, but not dependent on density gradients, is the Modified Two Stream (MTS) Instability [McBride et al., 1972]. The value of the shear parameter is obtained from the Vlasov analysis of the equilibrium. From a number of simulation runs with different α_e (from 0.25 to 0.01) we find that if the shear frequency is larger than the wave frequency (in this case the lower hybrid frequency), shear effects are substantial and the EIH instability dominates the onset and character of the ensuing wave turbulence as well as the corresponding transport properties of the system. Fig. 8 shows the time evolution of a typical stressed boundary layer simulation. The plasma parameters corresponding to this simulation are as follows: $L_x/\lambda_D = 32$, $L_y/\lambda_D = 76.8$, $M_i/m_e = 400$, $T_i/T_e = 1$, and $\omega_{pe}/\Omega_e = 0.5$. Here λ_D is the Debye length and ω_{pe} is the electron plasma frequency. In more recent simulations we have increased the value of ω_{pe}/Ω_e to larger than unity without any appreciable modification to these results. In this example the electric field is chosen such that $\alpha_e = 0.25$, and the width of the region over which the electric field is localized is $L_E/\lambda_D = 1.25$. It is found that 6 vortices are formed after 5 lower hybrid times. This number is in close accordance with the fastest growing mode in the system as derived from linear theory for the EIH mode [Romero et al., 1992a]. The wavelength of this mode is much larger than that of the LHD instability.

Figure 9 shows the time evolution of the average velocity in the y direction, U_y, normalized by the ion thermal velocity. Considerable viscosity is observed as a result of the turbulence. Preliminary results indicate that the viscosity scales as $0.33(\rho_i^2\Omega_i)(\alpha_e - .08)$ for $\alpha_e > 0.08$; although more simulations are needed to arrive at an accurate scaling law.

Fig. 10 is the initial and final density profiles for this run. It is seen that no appreciable particle diffusion occurs. At the end of the simulation (after 20 lower hybrid times) particle diffusion is found to be restricted to within $(1 - 2)\rho_i$.

The preliminary results also indicate that in the nonlinear stage there is strong transverse ion acceleration. Fig. 11 shows the initial and the final ion distribution functions after only 20 lower hybrid times.

The power spectrum for this simulation is given in Fig. 12. The spectrum is broadband in frequency, extending from below ω_{LH} to around ω_{pe} with more power in the lower frequency portion. This is consistent with observations [Cattell and Mozer, 1988]. Inclusion of electron beams in the model is expected to extend the spectrum toward higher frequency and possibly provide an enhancement near ω_{pe}, while the inclusion of ion beams and lower frequency ($\omega \sim \Omega_i$) IEDDI contributions is expected to reinforce the lower frequency portion of the spectrum. However, even in this preliminary simulation

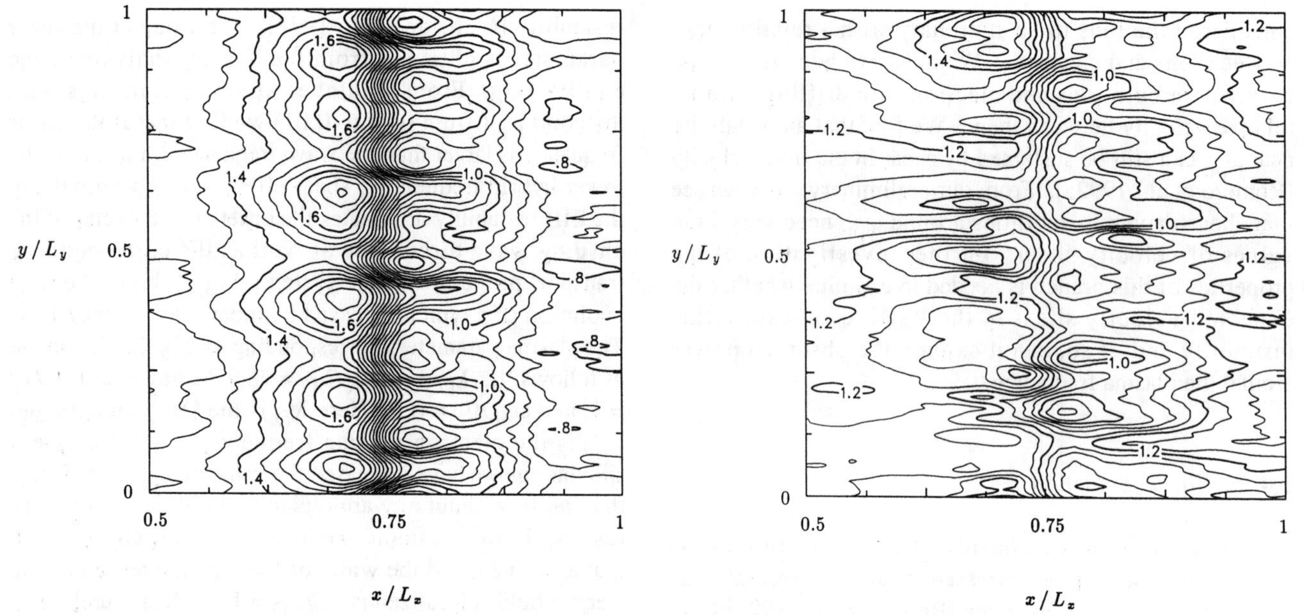

Fig. 8. Time sequence of the nonlinear evolution of a stressed boundary layer. The electrostatic potential, electron cross-field flow, and electron density are shown at times (a) $\omega_{LH}t = 7.2$, and (b) 19.8, respectively.

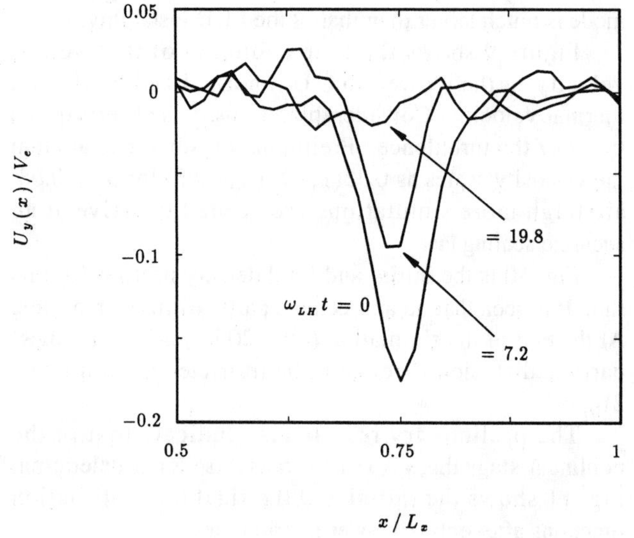

Fig. 9. The transverse flow layer at $\omega_{LH}t = 0$, 7.2, and 19.8. Substantial anomalous viscosity leads to the depletion of cross-field electron flow within 20 lower hybrid times.

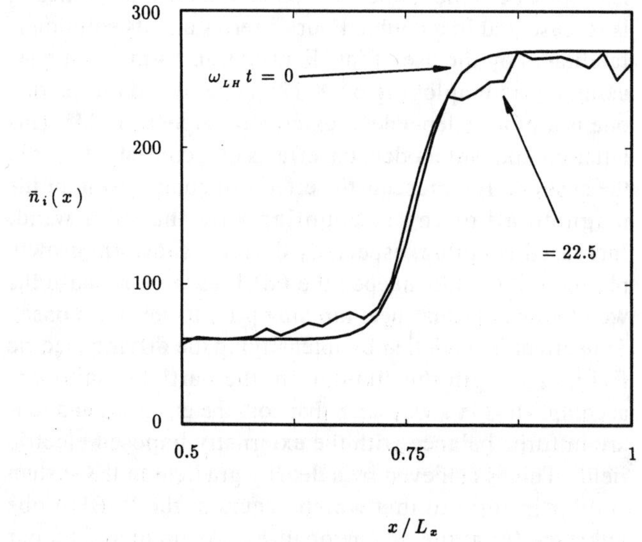

Fig. 10. The initial and final density profiles in the simulation run. Within 22 lower hybrid times not much particle diffusion results.

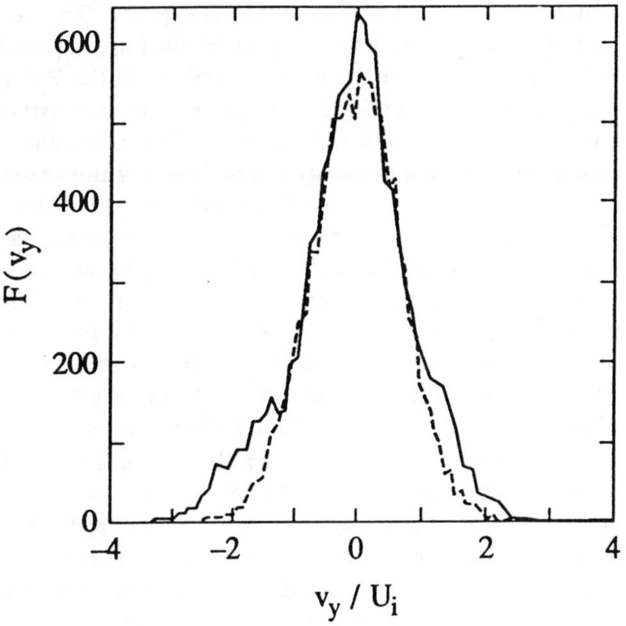

Fig. 11. Ion acceleration due to the EIH modes. Initial (dotted) and final (solid) ion distribution functions after only 20 lower hybrid times.

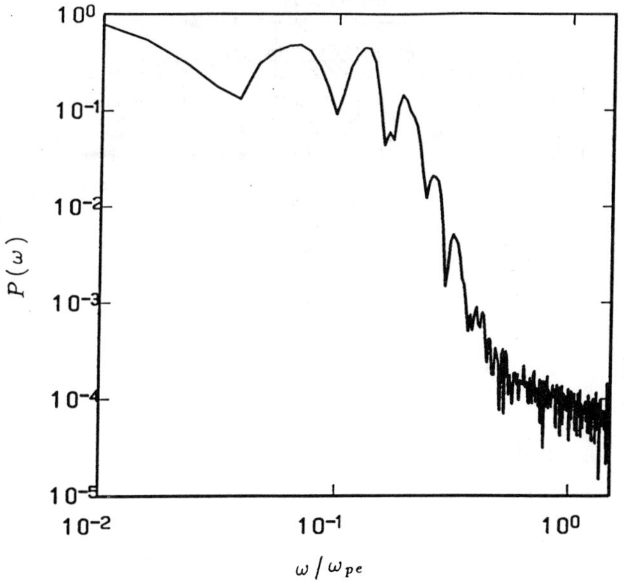

Fig. 12. Wave power spectrum from the PIC simulation of a stressed boundary layer. Note thet the broadband character is due to the inherent inhomogeneities of the PSBL.

the spectrum is similar in character to that seen by satellites. A remarkable feature is that the broadband nature is a direct consequence of the inherent inhomogeneity of the PSBL and is not critically dependent on the existence of a large amount of cold plasma. Nevertheless, it may be interesting to investigate the effects of a cold plasma component on these wave modes.

Another effect, seen in the simulation results, is a decrease in the electron cross field flow by over a factor of 3 in 17 lower hybrid times while its the flow width broadens. Concomitant with this decrease there is a reduction in the source of free energy for the EIH mode and the intensity of this instability is reduced. As the scale-size of the flow broadens to become larger than ρ_i, the ions become magnetized, and conditions are met for cascading toward lower frequencies and longer wavelengths. As seen in the simulations of Nishikawa et al. [1988; 1990] the shift towards lower frequency is observed during the nonlinear evolution. Since this run was limited to 20 lower hybrid times, the full development of the downward cascade remains to be investigated. Cascade from high ($> \omega_{LH}$) to low ($< \Omega_i$) frequency and the attendant consequences and nonlinear signatures are extremely relevant to the boundary layer dynamics, especially with regard to the relaxation of highly stressed layers, which will be the subject of our ongoing research.

We expect that during active periods enhanced stress can lead to strong gradients in the background plasma parameters in the PSBL. From the simulations discussed here, we find that there is a strong tendency to relax the stress via generation of plasma waves. In the evolution of a typical stressed PSBL discussed here (Fig. 9) it took only 20 lower hybrid times to relax a flow layer of $\rho_i/2$ to $2\rho_i$. In order to maintain the layer width at its initial value the solar wind stress will have to be sufficient to overcome the strong dissipative agents acting on the system. This situation may be attained only during highly active periods, which is in keeping with observed characteristics that (1) the energetic electron layers are infrequently observed and (2) they are generally observed during active periods.

4. A POSSIBLE SCENARIO OF PSBL DYNAMICS

From this study, a global picture of PSBL dynamics is emerging as summarized in Fig. 13. The dynamics is initiated as the solar wind interacts with the magnetosphere via tail reconnection processes and creates the boundary layer along with its dominant features, such as the density gradient and the associated sheared flows. During active periods, when large amounts of solar wind energy are being deposited, smaller scale-size structures with large values of

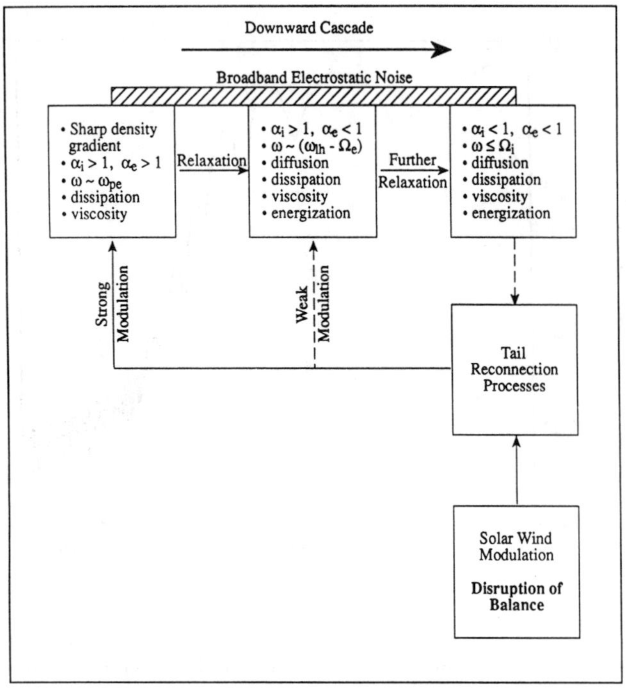

Fig. 13. A schematic description of the PSBL dynamics.

the shear parameter, α, are formed. A large value of α leads to higher frequency waves and implies a more unstable situation. Hence, the system will rapidly relax to lower values of α by exciting waves and thereby broadening the scale-size of the structure. With lower α, lower frequency waves will be excited. In the absence of a continuous driving mechanism, the structure will continue to broaden until α is too weak to support further wave growth. Thus, the observed α (and hence the width of the PSBL) represents a balance between the rates at which energy is being supplied by the solar wind and dissipated by the system. Modulation in the energy input disturb this balance and the cycle can start again, leading to a broadband frequency response. As the micro-turbulence sets in there is rapid dissipation, mixing, and particle energization. The cascade toward lower frequency longer wavelength waves can provide increasing cross-field transport and viscosity which can substantially alter the initial state.

5. CONCLUSIONS

We have presented a new model for the dynamics and the transverse structure of the PSBL which takes into account the inherent inhomogeneities of this region. Although the model reveals interesting features of PSBL dynamics, it is incomplete in many respects. The key purpose of this study is to demonstrate the importance of velocity shear and other nonuniformities in the PSBL (ignored thus far) and their unique effects on particle energization and the nonlinear evolution of boundary layers. We find that boundary layer inhomogeneities relax by giving rise to instabilities generated by velocity shear. These instabilities influence the nonlinear behavior of the PSBL significantly. We find that the ensuing turbulence is broadband in frequency and is not dependent on a cold plasma component. The dissipation due to micro-processes balances the steepenening effects due to solar wind compression and thereby determines the thickness of the PSBL. More recently, we [Romero and Ganguli, 1994] have examined the effects of the magnetic field inhomogeneity and find that it intensifies the processes discussed here. The model, however, needs to be further improved in a number of ways in order to make it more realistic. For example, multi-temperature plasma components and counter streaming beams are now being added to the model. This will allow us to investigate any additional role played by cold electron and ion components on the waves generated in a spatially inhomogeneous boundary layer.

Acknowledgments. Numerous discussions with G. Parks, T. Eastman, and J. Fedder are gratefully acknowledged. This work is supported by ONR. P.B. Dusenbery is supported by NSF grant ATM-9002447.

REFERENCES

Akimoto, K and N. Omidi, The generation of broadband electrostatic noise by an ion beam in the magnetotail, *Geophys. Res. Lett.*, 13, 97, 1986.

Amatucci, W.E., M.E. Koepke, J.J. Carroll, and T.E. Sheridan, Observation of ion-cyclotron turbulence at small values of magnetic-field-aligned current, *Geophys. Res. Lett.*, 21, 1595, 1994.

Anderson, K.A., R.P. Lin, R.J. Paoli, G.K. Parks, C.S. Lin, H. Reme, J.S. Bosqued, F. Martel, F. Cotin, *IEEE Trans. Geosci*, GE-16, 213, 1978.

Ashour-Abdalla, M. and H. Okuda, Theory and simulations of broadband electrostatic noise in the geomagnetic tail, *J. Geophys. Res.*, 91, 6833, 1986.

Ashour-Abdalla, M. and D. Scriver, Acceleration and transport in the plasma sheet boundary layer, *Geophysical Monograph*, # 54, 305, 1989.

Ashour-Abdalla, M., J. Berchem, J. Buchner, and L.M. Zelenyi, Large and small scale structures in the plasma sheet: A signature of chaotic motion and resonance effects, *Geophys Res Lett.*, 18, 1603, 1991.

Baumjohann, W., G. Paschmann, N. Sckopke, C.A. Cattell and C.W. Carlson, Average ion moments in the plasma sheet boundary layer, *J. Geophys. Res.*, 93, 11507, 1988.

Cattell, C.A., M. Kim, R.P. Lin, and F.S. Mozer, Observations of large electric fields near the plasma sheet boundary by ISEE 1, *Geophys. Res. Lett.*, 9, 539, 1982.

Cattell, C.A. and F.S. Mozer, Experimental determination of the dominant wave mode in the active near-earth magnetotail, *Geophys. Res. Lett.*, 13, 221, 1986.

Cattell, C.A., F.S. Mozer, R.R. Anderson, E.W. Hones, Jr., and R.D. Sharp, ISEE observations of the plasma sheet boundary, plasma sheet, and neutral sheet 2. Waves, *J. Geophys. Res.*, 91, 5681, 1986.

Cargill, P.J. and T.E. Eastman, The structure of tangential discontinuities 1. Results of hybrid simulations, *J. Geophys Res.*, 13763, 1991.

D'Angelo, N., Kelvin-Helmholtz instability in a fully ionized plasma in a magnetic field, *Phys. Fluids.*, 8, 1748, 1965.

Dandours, J., A. Saint-Marc, H. Reme, J. Sanvaud and G. Parks, *Adv Sp Res.*, 6, 159, 1986.

Drummond, W.E. and M.N. Rosenbluth, Anomalous diffusion arising from microinstabilities in a plasma, *Phys. Fluids.*, 5, 1507 (1962).

Dusenbery, P.B. and L.R. Lyons, Generation of electrostatic noise in the plasma sheet boundary layer, *J. Geophys. Res.*, 90, 10935, 1985.

Dusenbery, P.B., Generation of broadband noise in the magnetotail by the beam acoustic instability, *J. Geophys. Res.*, 91, 12005, 1986.

Dusenbery, P.B. and L.R. Lyons, Unmagnetized diffusion for azimuthally symmetric wave and particle distributions, *J. Plasma Physics*, 40, 179, 1988.

Dusenbery, P.B. and L.R. Lyons, Ion diffusion coefficients from resonant interactions with broadband turbulence in the magnetotail, *J. Geophys. Res.*, 94, 2484, 1989.

Eastman, T.E., L.A. Frank, W.K. Peterson and W. Lennartsson, The plasma sheet boundary layer, *J. Geophys. Res.*, 89, 1553, 1984.

Eastman, T.E., B.J. Anderson, P.J. Cargill, S.A. Fuselier, and J.T. Gosling, Printine Magnetopause crossings, *EOS Transc. AGU*, 72, 492, 1991.

Frank, L.A., R.L. McPherron, R.J. DeCoster, B.G. Burek, K.L. Ackerson, and C.T. Russel, Field-aligned currents in the Earth's magnetotail, *J. Geophys. Res.*, 86, 687, 1981.

Frank, L.A., Plasmas in the Earth's magnetotail, in *Space Plasma Simulations*, edited by M. Ashour-Abdalla and D.A. Dutton, p. 211, D. Reidel Publ. Co., Dordrecht, Holland, 1985.

Ganguli, G., Y.C. Lee and P. Palmadesso, Electrostatic ion cyclotron instability due to a nonuniform electric field perpendicular to the external magnetic field, *Phys. Fluids.*, 28, 761, 1985a.

Ganguli, G., P. Palmadesso and Y.C. Lee, Electrostatic ion cyclotron instability due to a nonuniform electric field perpendicular to the external magnetic field, *Geophys. Res. Lett.*, 12, 643, 1985b.

Ganguli G., Y.C. Lee, and P.J. Palmadesso, A new mechanism for excitation of waves in a magnetoplasma I. Linear theory, *Geophysical Momograph* # 38, Proc. Chapman Meeting on Ion Acceleration, Boston, p 297, 1985c.

Ganguli, G. and P.J. Palmadesso, Electrostatic ion instabilities in the presence of parallel currents and transverse electric fields, *Geophys. Res. Lett.*, 15, 103, 1988.

Ganguli, G., Y.C. Lee and P.J. Palmadesso, Kinetic theory for electrostatic waves due to transverse velocity shears, *Phys. Fluids*, 31, 823, 1988a.

Ganguli, G., Y.C. Lee and P.J. Palmadesso, Electron-Ion hybrid mode due to transverse velocity shear, *Phys. Fluids.*, 31, 2753, 1988b.

Ganguli, G., Y.C. Lee, P.J. Palmadesso and S.L. Ossakow, Oscillations in a plasma with parallel currents and transverse velocity shears, in Physics of Space Plasmas (1988), SPI Conference Proceedings and Reprint Series, edited by T. Chang, G.B. Crew and J.R. Jasperse, 8, pp. 231, Scientific Publishers, Inc., Cambridge, MA, 1989a.

Ganguli, G., Y.C. Lee, P.J. Palmadesso and S.L. Ossakow, D.C. electric field stabilization of plasma fluctuations due to a velocity shear in the parallel ion flow, *Geophys. Res. Lett.*, 16, 735, 1989b.

Ganguli, G., P.J. Palmadesso, Y.C. Lee and J.D. Huba, High Frequency Waves due to Velocity Shear in the Electron Flow, in *Proceedings of the 1989 International Conference on Plasma Physics*, New Delhi, India, p 197, 1989c.

Ganguli G., Y.C. Lee, and P.J. Palmadesso, Role of small scale processes in global plasma modelling, *AGU Monograph* # 62, The second Huntsville Workshop on Magnetosphere/Ionosphere Plasma Models, Huntsville, Alabama, 17, 1991.

Ganguli, S.B. and P.J. Palmadesso, Plasma transport in the auroral return current region, *J. Geophys. Res.*, 92, 8673, 1987.

Grabbe, C.L. and T. E. Eastman, Generation of broadband electrostatic noise by ion beam instabilities in the magnetotail, *J. Geophys. Res.*, 89, 3865, 1984.

Grabbe, C.L., New results on the generation of broadband electrostatic waves in the magnetotail, *Geophys. Res. lett.*, 12, 483, 1985.

Gurnett, D.A., L.A. Frank and R.P. Lepping, Plasma waves in distant magnetotail, *J. Geophys. Res.*, 81, 6059, 1976.

Huang, C.Y., L.A. Frank, and T.E. Eastman, Observations of plasma distributions during the coordinated data analysis workshop substorms of March 31 to April 1, 1979, *J. Geophys. Res.*, 92, 2377, 1987.

Huba, J.D., N.T. Gladd and K. Papadopolous, Lower hybrid drift wave turbulence in the distant magnetotail, *J. Geophys. Res.*, 83, 5217, 1978.

Koepke, M.E., W.E. Amatucci, J.J. Carroll III, and T.E. Sheriden, Experimental verification of the inhomogeneous energy density driven instability, *Phys. Rev. Lett.*, 72, 3355, 1994.

Kindel, J.M. and C.F. Kennel, Topside current instabilities, *J. Geophys. Res.*, 76, 3055, 1971.

Lee, L.C., The Magnetopause: A tutorial review, in *Physics of Space Plasmas (1990)*, SPI Conference Reprint Series, # 10, T. Chang, G.B. Crew, and J. Jasperes, eds., (Scientific Publishers Inc., Cambridge, MA, 1991), p.33.

Lee, L.C. and J.R. Kan, A unified kinetic model of the tangential magnetopause structure, *J. Geophys Res.*, 84, 6417, 1979.

Lui, A.T.Y., Road map to magnetotail domains, in *Magnetotail Physics*, edited by A.T.Y. Lui, John Hopkins Press, Baltimore and London, p 3, 1987.

Lyons, L.R. and T.W. Speiser, Evidence for current sheet acceleration in the geomagnetic tail, *J. Geophys. Res.*, 87, 2276, 1982.

Mcbride J.B., E. Ott, J.P. Boris and J.H. Orens, *Phys. Fluids.*, Theory and simulation of turbulent heating by the modified two stream instability, 15, 2367, 1972.

Mikhailovskii, A.B., *Theory of Plasma Instabilities*, Vol. I, p. 18, (Consultants Bureau, New York, 1974).

Nishikawa, K.-I., G. Ganguli, Y.C. Lee and P.J. Palmadesso, Simulation of ion-cyclotron-like modes in a magnetoplasma with transverse inhomogeneous electric field, *Phys. Fluids.*, 31, 1568, 1988.

Nishikawa, K.-I., G. Ganguli, Y.C. Lee and P.J. Palmadesso, Simulation of electrostatic turbulence due to sheared flows parallel and transverse to the magnetic field, *J. Geophys. Res.*, 95, 1029, 1990.

Okuda, H., Numerical simulations on the magnetopause current layer, *AGU Monograph # 62*, The second Huntsville Workshop on Magnetosphere/Ionosphere Plasma Models, Huntsville, Alabama, 9, 1991.

Omidi, N., Broadband electrostatic noise produced by ion beams in the Earth's magnetotail, *J. Geophys. Res.*, 90, 12330, 1985.

Onsager, T., M.F. Thomsen, J.T. Gosling, and S.J. Bame, Electron distributions in the plasma sheet boundary layer: Time-of-flight effects, *Geophys. Res. Lett.*, 17, 1837, 1990.

Orsini, S., M. Candidi, V. Formisano, H. Balsiger, A. Ghielmetti and K.W. Ogilvie, The structure of plasma sheet-lobe boundary in the Earth's magnetotail, *J. Geophys. Res.*, 89, 1573, 1984.

Parks, G.K., C.S. Lin, K.A. Anderson, R.P. Lin, and H. Reme, ISEE-1 and -2 observations of the outer plasma sheet boundary, *J. Geophys. Res.*, 84, 6471, 1979.

Parks, G.K., M. McCarthy, R.J. Fitzenreiter, J. Etcheto, K.A. Anderson, R.R. Anderson, T.E. Eastman, L.A. Frank, D.A. Gurnett, C. Huang, R.P. Lin, A.T.Y. Lui, K.W. Ogilvie, A. Pedersen, H. Reme and D.J. Williams, Particle and field characteristics of the high-latitude plasma sheet boundary layer, *J. Geophys. Res.*, 89, 8885, 1984.

Parks, G.K., R. Fitzenreiter, K.W. Ogilvie, C. Huang, K.A. Anderson, J. Dandouras, L. Frank, R.P. Lin, M. McCarthy, H. Reme, J.A. Sauvaud, and S. Werden, Low-energy particle layer outside of the plasma sheet boundary, *J. Geophys Res.*, 97, 2943, 1992.

Romero, H., G. Ganguli, P.B. Dusenbery and P.J. Palmadesso, Equilibrium structure of the plasma sheet boundary layer-lobe interface, *Geophys. Res. Lett.*, 17, 2313, 1990.

Romero, H., G. Ganguli, Y.C. Lee and P.J. Palmadesso, Electron-ion hybrid instabilities driven by velocity shear in a magnetized plasma, *Phys. Fluids. B*, 4, 1708, 1992a.

Romero, H., G. Ganguli, and Y.C., Lee, Ion acceleration and coherent structures generated by lower hybrid shear-driven instabilities, *Phys. Rev. Lett.*, 69, 3505, 1992b.

Romero, H. and G. Ganguli, Relaxation of the stressed plasma sheet boundary layer, *Geophys. Res. Lett.*, 21, 645, 1994.

Rogers, S.H. and E.C. Whipple, *Astrophys and Space Sciences*, 144, 231, 1988.

Satyanarayana, P., Y.C. Lee, and J.D. Huba, Stability of a stratified shear layer, *Phys Fluids.*, 30, 81, 1987.

Scarf, F., L.A. Frank, K.L. Ackerman and R.P. Lepping, Plasma wave turbulence at distant crossings of the plasma sheet boundaries and neutral sheet, *Geophys. Res. Lett.*, 1, 189, 1974.

Schriver, D. and M. Ashour-Abdalla, Generation of high-frequency broadband electrostatic noise: The role of cold electrons, *J. Geophys. Res.*, 92, 5807, 1987.

Schriver, D., M. Ashour-Abdalla, R. Treumann, M. Nakamura, and L.M. Kistler, The lobe to plasma sheet boundary layer transition: Theory and observations, *Geophys. Res. Lett.*, 17, 2027, 1990.

Sestero, A., Structure of plasma sheaths, *Phys. Fluids.*, 7, 44, 1964.

Sestero, A., Vlasov equation study of plasma motion across magnetic fields, *Phys. Fluids.*, 9, 2006, 1966.

Sharp, R.D., D.L., Carr, W.K. Peterson and E.G. Shelley, Ion streams in the magnetotail, *J. Geophys. Res.*, 86, 4639, 1981.

Takahashi, K. and E.W. Hones, Jr., ISEE 1 and 2 observations of ion distributions at the plasma sheet-tail lobe boundary, *J. Geophys. Res.*, 93, 8558, 1988.

Whipple, E. C., J.R. Hill and J.D. Nichols, Magnetopause structures and the question of particle accessibility, *J. Geophys. Res.*, 89, 1508, 1984.

Williams, D.J., Energetic ion beams at the edge of the plasma sheet: ISEE 1 observations plus a simple explanatory model, *J. Geophys. Res.*, 86, 5507, 1981.

Winske, D., S.P. Gary, and D.S. Lemmons, Diffusive transport at the magnetopause, in *Physics of Space Plasmas (1990)*, SPI Conference Reprint Series, # 10, T. Chang, G.B. Crew, and J. Jasperes, eds., (Scientific Publishers Inc., Cambridge, MA, 1991), p. 397.

---

[1] Beam Physics Branch, Plasma Physics division, Naval Research Laboratory, Washington DC 20375

[2] Applied Optics Branch, Optical Sciences Division, Naval Research Laboratory, Washington DC 20375

Comments and Questions on the Plasma Sheet Boundary and Boundary Layer

George K. Parks and Michael P. McCarthy

Geophysics Program, University of Washington, Seattle, Washington

This article will present a short discussion of boundaries and boundary layers, with emphasis on the plasma sheet boundary layer. A key question regarding the plasma sheet boundary (PSB) and the plasma sheet boundary layer (PSBL) concerns the relationships among the structures that are observed in the ionosphere and those observed in the distant plasma sheet regions. There are now several suggestions about these relationships which were derived using data obtained principally during substorms. Because of the constraints under which the data were obtained, the resulting picture is not one of steady state PSB and PSBL. Are there steady state PSB and PSBL structures in the geomagnetic tail? Do the PSB and PSBL even exist in the absence of substorm activity? To what locations do ionospheric auroral structures map to in the geomagnetic tail? There are many questions about the PSB and PSBL, but we have no clear answers. The main purpose of this paper is to ask questions on results that have been published but not yet answered. We hope the questions raised will keep the dialogue alive and stimulate discussions.

INTRODUCTION

Shortly after the discovery of the Earth's Van Allen radiation belt in 1958, *Gringauz et al.* [1961] observed a population of low energy electrons (≈100 eV) outside the trapping boundary. Later observations [*Freeman*, 1964; *Bame et al.*, 1966] showed that these low energy particles are a part of the plasma sheet in the geomagnetic tail [see also *Vasyliunas*, 1968]. These observations produced a picture of the nightside magnetosphere [Figure 1, from *Ness*, 1969] which has been used as a basis for discussions of solar wind interaction with the geomagnetic field and particle and auroral dynamics in the magnetosphere. In 1972, a boundary layer was discovered at the flanks of the plasma sheet [*Hones et al.*, 1972]; in 1976, another boundary layer was found near the magnetopause [*Eastman et al.*, 1976]. The presence of these layers modified our understanding of the magnetosphere. A diagram that includes these boundary layers is shown in Figure 2 [from *Eastman et al.*, 1984].

A boundary is an interface that separates two regions of space. A boundary layer is a region adjacent to a boundary that contains plasmas from both sides. For example, a magnetopause boundary layer contains both magnetosheath and magnetospheric plasmas. The frontside magnetopause boundary layer has been extensively studied by spacecraft such as ISEE which had an apogee of ≈22.5 R_e (R_e=earth radius), and an inclination of ≈23° permitting detailed observations of the low latitude boundary layer on the sunlit hemisphere. The ISEE spacecraft, however, only occasionally encountered the boundaries on the tailward flanks of the magnetosphere. These boundaries are subject to strong solar wind shears and data show that the Kelvin-Helmholtz instability may be active in these regions [*Couzens et al.*, 1985; *Larson and Parks*, 1992]. ISEE also could not reach the high altitude polar boundary and observations of this boundary mainly come from Heos 2 which had an apogee over the pole at ≈38.7 R_e. Heos 2 observations identified a region just inside the magnetopause that contains tailward flowing magnetosheath-like plasma [*Haerendel et al.*, 1978; *Rosenbauer et al.*, 1975]. Magnetosheath-like plasma was first detected inside the magnetopause at lower polar altitudes by *Heikkila and Winningham* [1971] and *Frank and Ackerson* [1971].

There are many complex features in the frontside boundary layer, some of which have been recently discussed by *Sonnerup et al.* [1992]. It is also important to note that the frontside magnetopause boundary layer contains "cold" electrons of ionospheric origin [*Ogilvie et al.*, 1984]. The magnetopause boundary layer can be formed

Space Plasmas: Coupling Between Small
and Medium Scale Processes
Geophysical Monograph 86

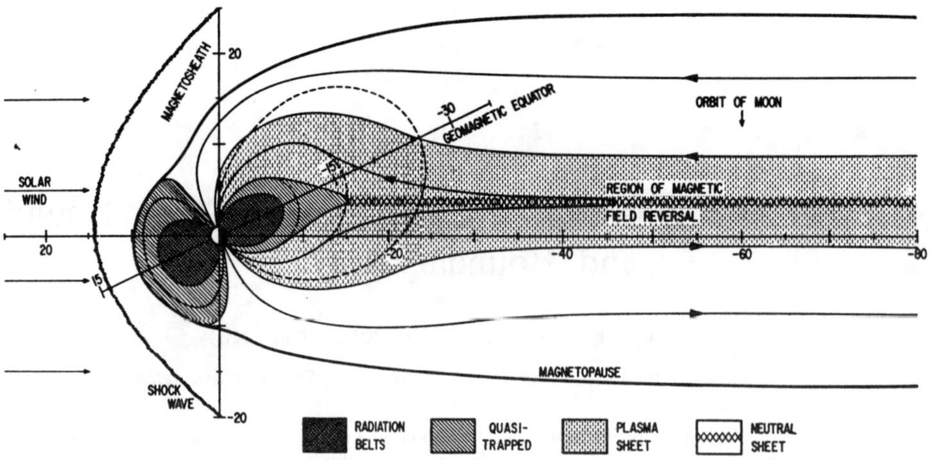

Fig. 1. A schematic diagram of the magnetosphere proposed in 1969 [from *Ness*, 1969].

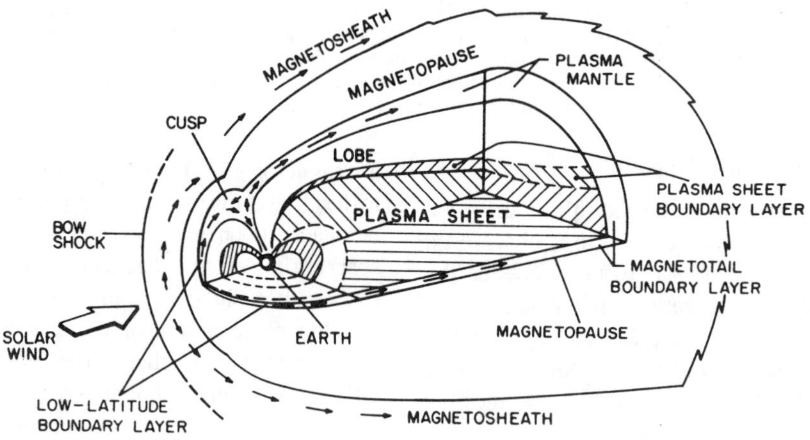

Fig. 2. A schematic diagram of the magnetosphere including boundary layers [from *Eastman et al.*, 1988].

in both open and closed models of the magnetosphere but the current models cannot totally explain all of the observed features. We will not discuss any further the problems of frontside boundaries other than to point out that a considerable amount of research has been and is being done to understand these boundary features. The remainder of this paper will emphasize the problems of the distant plasma sheet boundary and the boundary layer on the nightside.

PLASMA SHEET BOUNDARY AND BOUNDARY LAYER

A three-dimensional statistical analysis of the plasma sheet boundary layer (PSBL) crossings using three years of data from ISEE spacecraft shows that the PSBL can be encountered everywhere in the tail for $X_{GSM} \approx 8\text{–}22R_e$, independent of A_E values [*Dandouras et al.*, 1986b]. This result indicates that even weak substorms ($A_E < 100$ nT) can produce large perturbations of the plasma sheet. Flux decreases are detected in the east-west extent with equal probability from $Y_{GSM} = -20R_e$ to $Y_{GSM} = +20R_e$ and in the north-south direction from -10 to +10 R_e (distance from the predicted neutral sheet). This statistical study also shows that thinning can be observed by one spacecraft, and not by another, even when the two spacecraft are separated by only a few hundred km. Also, a significant number of substorm events were not accompanied by thinning of particle fluxes.

Understanding the structure and dynamics of the plasma

sheet boundary (PSB) and the PSBL is fundamentally important to substorm physics. Plasma distributions in the PSBL include beams of electrons and ions of both plasma sheet and ionospheric origins [*Eastman et al.*, 1984, 1988; *Parks et al.*, 1984]. These distributions are unstable to a variety of instabilities, including the generation of broadband electrostatic waves in the frequency range from ≈10 Hz to the local electron plasma frequency, ≈10 kHz [*Dusenbery and Lyons*, 1985; *Ashour-Abdalla and Okuda*, 1986]. Electron flow velocity shear at the PSBL provides another free energy source, which can excite the electron ion hybrid instability [*Romero et al.*, 1990]. The dimensions of the PSB and PSBL vary from a few to several tens of ion Larmor radii.

A question that has been asked for many years concerns the relationships between features observed in the ionosphere and in the distant plasma sheet regions. Where do the various regions of the plasma sheet map to in the ionosphere, and vice versa? *Eastman et al.* [1988] suggested that the PSBL maps to discrete auroral arcs, and the plasma sheet into the diffuse precipitation region. On the other hand, *Zelenyi et al.* [1990] proposed that discrete auroral structures map to the central plasma sheet, and the PSBL to the subvisual precipitation region poleward of the discrete auroral structures identified in the ground-based data by *Feldstein and Galperin* [1985] and *Galperin and Feldstein* [1989].

These models were recently modified after finding that the plasma sheet boundary also supports a layer of low energy particles just outside the PSBL [*Parks et al.*, 1992]. This low energy layer (LEL) is distinct from the PSBL and is characterized by <100 eV electrons streaming in the tailward direction and <100 eV ions streaming in the earthward direction. The presence of a layer outside the PSBL was also deduced by *Nishida et al.* [1993] using data obtained from a low altitude polar orbiting satellite. A picture of the plasma sheet that includes the LEL and possible projections of these features in the ionosphere is shown in Figure 3 [from *Parks et al.*, 1992].

The main uncertainty of this model is the relationship between the PSBL and discrete arc structures. Observations show that the average thickness of the PSBL at ≈20 R_e maps to ionospheric regions <0.5° [*Dandouras et al.*, 1986a]. However, the inverted V structure associated with discrete auroras is several degrees wide in the ionosphere [*Lin and Hoffman*, 1982]. Using Tsyganenko's magnetic field model, *Elphinstone et al.* [1991] further predicted that the discrete auroral structures project to the central plasma sheet (CPS). The final concern is that high speed flows are not confined to the PSBL alone, but are also observed in the CPS [*Baumjohann et al.*, 1990].

DISCUSSION

Boundaries and boundary layers are usually viewed as steady-state spatial features. This is the implicit message conveyed by Figures 1, 2, and 3. This view is roughly correct for the frontside magnetopause boundary, even though this boundary undergoes large variations from quiet to disturbed geomagnetic conditions. However, the picture of the plasma sheet boundary needs to be clarified, since the conditions under which data are obtained in the geomagnetic tail are quite different. Unlike the magnetopause boundary and layer, the PSB and PSBL are observed principally during substorm activity. This is because the PSB and PSBL at distances >12 R_e are observed only when a spacecraft detects decreases and increases of particle fluxes as a result of plasma sheet thinning and recovery during substorm activities [*Hones et al.*, 1972]. The PSB and PSBL are observed as the spacecraft exits from the plasma sheet into the lobe region and when the spacecraft returns to the plasma sheet as the substorm activity subsides. Upon recovery, the spacecraft usually detects higher fluxes than before the spacecraft exited the plasma sheet due to substorms generating "freshly accelerated" particles. Note also that flux decreases in the tail region coincide with injections of particles closer to Earth in the outer radiation belt [Figure 4, from *Sauvaud et al.*, 1984], and precipitation of energetic electrons in the auroral zone [*Parks and Winckler*, 1968].

This evolutionary sequence describes a simplified ideal case, but it conveys the information when the PSB and PSBL are observed by a spacecraft. The PSB and PSBL have not received much attention under quiet geomagnetic conditions, and very little is known about these boundaries at these times. Due to the dynamic conditions under which the PSB and PSBL structures are observed, many of the observed structures may well be temporal features produced by substorms. This raises an important question about the spatial and temporal characteristics of boundaries and boundary layers. Which features of the PSBL are temporal? Are beam distributions and high speed flows temporal consequences of substorm dynamics? What features characterize a geomagnetically quiet PSBL? Do boundary layers even exist in the absence of substorm activity? What is the relationship between the PSB and PSBL and the various responses of the magnetosphere to a southward turning of the interplanetary magnetic field [*Sauvaud et al.*, 1987]?

These questions are not easily answered. We note, however, that a considerable amount of information exists on auroral behavior, including how auroras vary and evolve in space and time [*Akasofu*, 1964]. The relation-

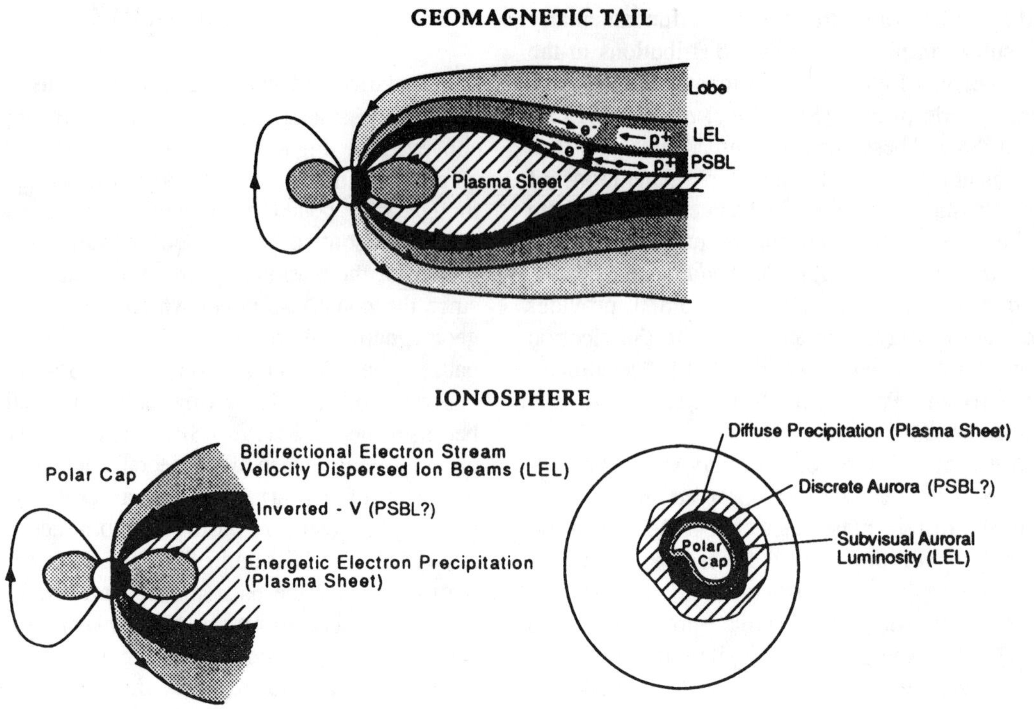

Fig. 3. A schematic diagram of the geomagnetic tail that includes the low energy particle layer outside of the PSBL [from *Parks et al.*, 1992].

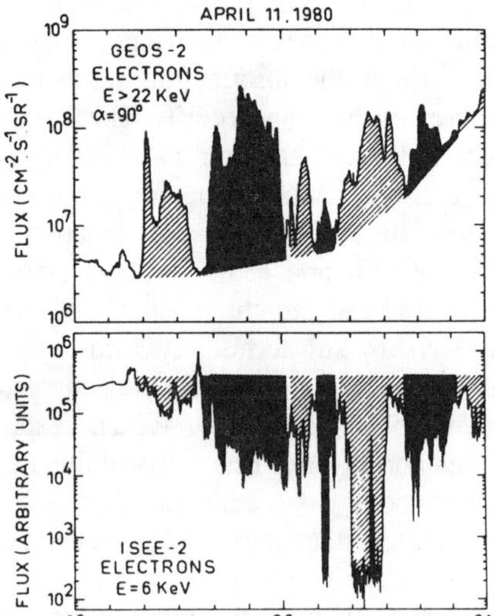

Fig. 4. The left hand panel shows an example of particle decreases observed in the geomagnetic tail, particle injection at synchronous altitudes that occurred in coincidence with the flux decrease, and a magnetogram that shows the onset of a substorm. The right hand panel shows repeated observations of flux decreases and injection events [from *Sauvaud et al.*, 1984].

ships of the various plasma sheet and auroral zone structures shown in Figures 2 and 3 assume that auroras are footprints of particle structures in the distant tail regions, and vice versa. Thus, it seems reasonable to attempt to characterize the distant tail regions on the basis of auroral observations. For example, ground-based data show that auroral intensities are very low during weak geomagnetic disturbances. This would imply that the fluxes in the tail will also be low during these disturbances. What will the flux level be and will the fluxes be below the threshold for spacecraft particle detectors? When there are no auroral displays in the auroral zone, will this mean that the gradient of particle fluxes in the neighborhood of boundaries will be small, and that the PSB and PSBL will appear diffuse and not clearly discernible in the satellite data? The visual aurora includes many complex and dynamic forms. How will these features be reflected in the plasma sheet regions?

If these auroral features are tied to plasma sheet dynamics, the PSB and PSBL clearly cannot be viewed as steady state structures of the geomagnetic tail. The physics is further complicated by the fact that the ionosphere and magnetosphere can both be sources of auroral particles. Most of the precipitated electrons with energies <10 keV are accelerated by a field-aligned potential close to the ionosphere [*Mozer et al.*, 1977; *Evans*, 1974], but those with energies >10 keV are accelerated by processes in the more distant regions of the magnetosphere. Both of these sources contribute to auroral luminosity, and thus auroras are more complex than just being footprints. Another complication comes from observations of field-aligned currents [*Zmuda et al.*, 1966] which may extend to the magnetosphere. These observations indicate that plasmas in the distant regions cannot be treated as ideal. Magnetic lines of force are very likely twisted, and the magnetic topology will be complex. Magnetic surfaces cannot be considered to be equipotential surfaces, and the simple mapping procedure we invoked to construct the above pictures will not be valid.

The coordinated studies of substorms during the ISTP era using multiple spacecraft, global imaging, and worldwide ground-based observations should help to answer some of these questions.

Acknowledgments. The research work at the University of California at Berkeley, the University of Iowa, the University of Washington, and the Goddard Space Flight Center is supported in part by grants from the National Aeronautics and Space Administration. The helpful efforts of the ISEE Project Office are acknowledged. The work at CESR is in part supported by CNES. G.K. Parks appreciates helpful discussions with J. Sauvaud.

REFERENCES

Akasofu, S., The development of auroral substorms, *Planet. Space Sci., 12,* 273, 1964.

Ashour-Abdalla, M., and H. Okuda, Theory and simulations of broadband electrostatic noise in the geomagnetic tail, *J. Geophys. Res., 91,* 6833-6844, 1986.

Bame, S., J. Asbridge, H. Felthauser, R. Olson, and I. Strong, Electrons in the plasma sheet of the earth's magnetic tail, *Phys. Rev. Lett., 16,* 138-142, 1966.

Baumjohann, W., G. Paschmann, and H. Lühr, Characteristics of high-speed ion flows in the plasma sheet, *J. Geophys. Res., 95,* 3801-3809, 1990.

Couzens, D., G. Parks, K. Anderson, R. Lin, and H. Rème, ISEE particle observations of surface waves at the magnetopause boundary layer, *J. Geophys. Res., 90,* 6343-6353, 1985.

Dandouras, J., H. Rème, A. Saint-Marc, J. Sauvaud, G. Parks, K. Anderson, and R. Lin, A statistical study of plasma sheet dynamics using ISEE 1 and 2 energetic particle flux data, *J. Geophys. Res., 91,* 6861-6870, 1986a.

Dandouras, J., A. Saint-Marc, H. Rème, J. Sauvaud, and G. Parks, Particle dynamics of the plasma sheet boundary layer, *Adv. Space Res., 6,* 159, 1986b.

Dusenbery, P., and L. Lyons, The generation of electrostatic noise in the plasma sheet boundary layer, *J. Geophys. Res., 90,* 10935-10943, 1985.

Eastman, T., E. Hones, Jr., S. Bame, and R. Asbridge, The magnetospheric boundary layer: Site of plasma, momentum and energy transfer from the magnetosheath into the magnetosphere, *Geophys. Res. Lett., 3,* 685, 1976.

Eastman, T., L. Frank, W. Peterson, and W. Lennartsson, The plasma sheet boundary layer, *J. Geophys. Res., 89,* 1553-1572, 1984.

Eastman, T., G. Rostoker, L. Frank, C. Huang, and D. Mitchell, Boundary layer dynamics in the description of magnetospheric substorms, *J. Geophys. Res., 93,* 14411-14432, 1988.

Elphinstone, R., D. Hearn, J. Murphree and L. Cogger, Mapping using the Tsyganenko long magnetospheric model and its relationship to Viking auroral images, *J. Geophys. Res., 96,* 1467-1480, 1991.

Evans, D., Precipitating electron fluxes formed by a magnetic field aligned potential difference, *J. Geophys. Res., 79,* 2853-2858, 1974.

Feldstein, Y., and Y. Galperin, The auroral luminosity structure in the high-latitude upper atmosphere: Its dynamics and relationship to the large-scale structure of the Earth's magnetosphere, *Rev. Geophys., 23,* 217-275, 1985.

Frank, L., and K. Ackerson, Observations of charged particle precipitation into the auroral zone, *J. Geophys. Res., 76,* 3612-3643, 1971.

Freeman, J., The morphology of the electron distribution in the outer radiation zone and near the magnetospheric boundary as observed by Explorer 12, *J. Geophys. Res., 69,* 1691-1723, 1964.

Galperin, Y., and Y. Feldstein, Auroral luminosity and its relationship to magnetospheric plasma domains, in *Auroral Physics,* edited by C. Meng, pp. 207-222, Cambridge University Press, New York, NY, 1989.

Gringauz, K., V. Bezrukikh, V. Ozerov, and R. Rybchinskii, A study of the interplanetary ionized gas, high energy electrons, and corpuscular radiation from the sun by means of the three-electrode trap for charged particles on the second Soviet cosmic rocket (English translation), *Soviet Phys. Dokl., 5,* 361-364, 1961 (originally *Dokl. Akad. Nauk SSSR, 131,* 1301-1304, 1960).

Haerendel, G., G. Paschmann, N. Sckopke, H. Rosenbauer, and P. Hedgecock, The front side boundary layer of the magnetosphere and the problem of reconnection, *J. Geophys. Res., 83,* 3195, 1978.

Heikkila, W., and J. Winningham, Penetration of magnetosheath plasma to low altitudes through the dayside magnetospheric cusps, *J. Geophys. Res., 76,* 883-891, 1971.

Hones, E. Jr., J. Asbridge, S. Bame, M. Montgomery, S. Singer, and S. Akasofu, Measurements of magnetotail plasma flow made with Vela 4B, *J. Geophys. Res., 77,* 5503-5522, 1972.

Larson, N., and G. Parks, Motions of particle microstructures in the magnetopause boundary layer, *J. Geophys. Res., 97,* 10733-10749, 1992.

Lin, C., and R. Hoffman, Observations of inverted-V electron precipitation, *Space Sci. Rev., 33,* 415, 1982.

Mozer, F., C. Carlson, M. Hudson, R. Torbert, B. Parady, J. Yatteau, and M. Kelley, Observations of paired electrostatic shocks in the polar magnetosphere, *Phys. Rev. Lett., 38,* 292, 1977.

Ness, N., The geomagnetic tail, *Rev. Geophys., 7,* 97-127, 1969.

Nishida, A., T. Mukai, H. Hayakawa, A. Matsuoka, K. Tsuruda, N. Kaya, and H. Fukunishi, Unexpected features of the ion precipitation in the so-called cleft/low-latitude boundary layer region; Association with sunward convection and occurrence on open field lines, *J. Geophys. Res., 98,* 11161-11176, 1993.

Ogilvie, K., R. Fitzenreiter, and J. Scudder, Observations of electron beams in the low-latitude boundary layer, *J. Geophys. Res., 89,* 10723-10732, 1984.

Parks, G., and J. Winckler, Acceleration of energetic electrons observed at the synchronous altitude during magnetospheric substorms, *J. Geophys. Res., 73,* 5786-5791, 1968.

Parks, G., M. McCarthy, R. Fitzenreiter, J. Etcheto, K. Anderson, R. Anderson, T. Eastman, L. Frank, D. Gurnett, C. Huang, R. Lin, A. Lui, K. Ogilvie, A. Pedersen, H. Rème, and D. Williams, Particle and field characteristics of the high-latitude plasma sheet boundary layer, *J. Geophys. Res., 89,* 8885-8906, 1984.

Parks, G., R. Fitzenreiter, K. Ogilvie, C. Huang, K. Anderson, J. Dandouras, L. Frank, R. Lin, M. McCarthy, H. Rème, J. Sauvaud, and S. Werden, Low-energy particle layer outside of the plasma sheet boundary, *J. Geophys. Res., 97,* 2943-2954, 1992.

Romero, H., G. Ganguli, P. Palmedesso, and P. Dusenbery, Equilibrium structure of the plasma sheet boundary layer-lobe interface, *Geophys. Res. Lett., 17,* 2313-2316, 1990.

Rosenbauer, H., H. Grünwaldt, M. Montgomery, G. Paschmann, and N. Sckopke, Heos 2 plasma observations in the distant polar magnetosphere: The plasma mantle, *J. Geophys. Res., 80,* 2723-2737, 1975.

Sauvaud, J., A. Saint-Marc, J. Dandouras, H. Rème, A. Korth, G. Kremser, and G. Parks, A multisatellite study of the plasma sheet dynamics at substorm onset, *Geophys. Res. Lett., 11,* 500-503, 1984.

Sauvaud, J., J. Treilhou, A. Saint-Marc, J. Dandouras, H. Rème, A. Korth, G. Kremser, G. Parks, A. Zaitzev, V. Petrov, L. Lazutine, and R. Pellinen, Large scale response of the magnetosphere to a southward turning of the interplanetary magnetic field, *J. Geophys. Res., 92,* 2365-2376, 1987.

Sonnerup, B., G. Paschmann, T. Phan, and H. Lühr, Magnetic field maxima in the low latitude boundary layer, *Geophys. Res. Lett., 19,* 1727-1730, 1992.

Vasyliunas, V., A survey of low-energy electrons in the evening sector of the magnetosphere with OGO 1 and OGO 3, *J. Geophys. Res., 73,* 2839-2884, 1968.

Zelenyi, L., R. Kovrazkhin, and J. Bosqued, Velocity-dispersed ion beams in the nightside auroral zone: AUREOL 3 observations, *J. Geophys. Res., 95,* 12119-12139, 1990.

Zmuda, A., J. Martin, and F. Heuring, Transverse magnetic disturbances at 1100 kilometers in the auroral region, *J. Geophys. Res., 71,* 5033-5045, 1966.

G. Parks and M. McCarthy, Geophysics Program AK-50, University of Washington, Seattle, WA 98195.